D0643985

ADDISON-WESLEY ONTARIO

Advanced Functions 12 and Introductory Calculus

Addison-Wesley Secondary Mathematics Authors

Elizabeth Ainslie
Paul Atkinson
Maurice Barry
Cam Bennet
Barbara J. Canton
Ron Coleborn
Fred Crouse
Garry Davis
Jane Forbes
George Gadanidis
Liliane Gauthier
Florence Glanfield
Katie Pallos-Haden
Carol Besteck Hope
Terry Kaminski
Brendan Kelly
Stephen Khan
Ron Lancaster
Duncan LeBlanc
Kevin Maguire
Rob McLeish
Jim Nakamoto
Nick Nielsen
Paul Pogue
Brent Richards
David Sufrin
Paul Williams
Elizabeth Wood
Rick Wunderlich
Paul Zolis
Leanne Zorn

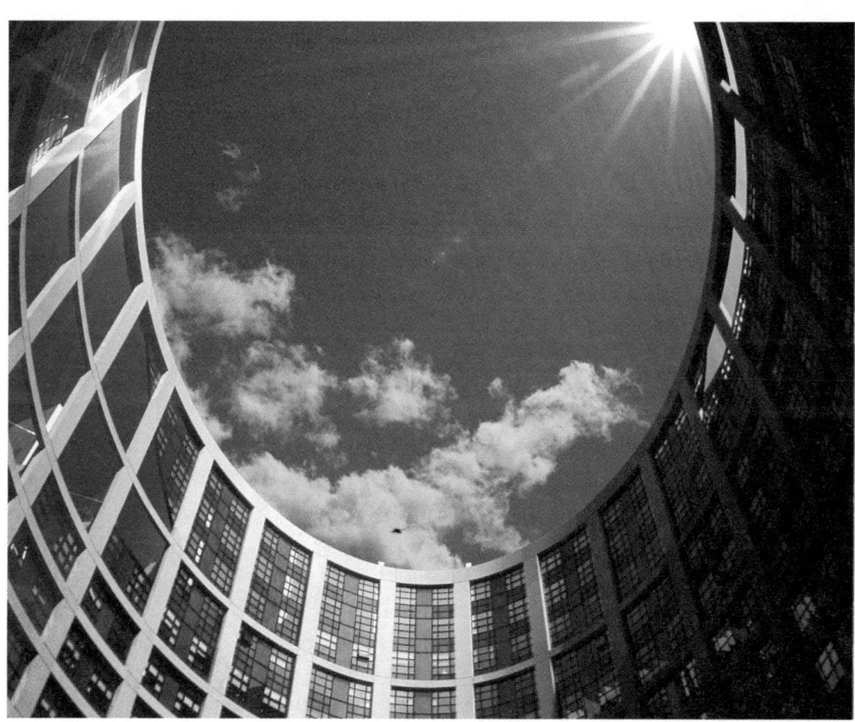

Robert Alexander
Bonnie Edwards
Peter J. Harrison
Antonietta Lenjosek
Linda Rajotte
Peter Taylor

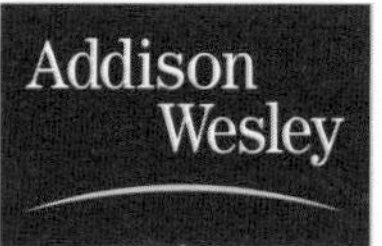

Toronto

Senior Consulting Mathematics Editor
Lesley Haynes

Coordinating Editor
Mei Lin Cheung

Production Coordinator
Stephanie Cox

Editorial Contributors
Kelly Davis
Melissa Lee
Gay McKellar
Dan Pignat

Product Manager
Susan Cox

Managing Editor
Enid Haley

Publisher
Claire Burnett

Design/Production
Pronk&Associates

Art Direction
Pronk&Associates

Electronic Assembly/Technical Art
Pronk&Associates

The publisher has taken every care to meet or exceed industry specifications for the manufacturing of textbooks. The spine and the endpapers of this sewn book have been reinforced with special fabric for extra binding strength. The cover is a premium, polymer-reinforced material designed to provide long life and withstand rugged use. Mylar gloss lamination has been applied for further durability.

ISBN: 0-201-77104-7

This book contains recycled product and is acid free.

Printed and bound in Canada

1 2 3 4 5 GG 06 05 04 03 02

Program Consultants and Reviewers

Ron Bender
Faculty of Engineering
University of Ottawa

Bill Chambers
Formerly Head of Mathematics
John Fraser Secondary School
Mississauga

Joseph Cunsolo
Associate Professor
University of Guelph
Guelph

Rod Forcier
Head of Mathematics and Computer Science
Christ the King Secondary School
Georgetown
Halton Catholic District School Board

John Kitney
Head of Mathematics
Bayridge Secondary School
Limestone District School Board
Kingston

Kevin Maguire
Mathematics Consultant
Toronto District School Board

Sandra G. McCarthy
Head of Mathematics
South Carleton High School
Ottawa Carleton District School Board
Richmond

John McGrath
Formerly Head of Mathematics
Adam Scott Secondary School
Peterborough

Jamie Pyper
ESSO Centre for Mathematics Education
University of Western Ontario

Wendy Solheim
Head of Mathematics
Thornhill Secondary School
Thornhill

Dr. Eric Wood
Faculty of Education
University of Western Ontario

Assessment Consultant

Lynda E.C. Colgan
Professor
Department of Education
Queen's University
Kingston

Overview

Chapter 1 makes a strong connection to previous work with slope as rate of change. Students use a graphical approach to develop a visual understanding of the derivative, using examples of graphs generated from authentic data and from equations of functions they have studied before.

Chapter 1 lays the groundwork for the formal definition of the derivative developed in Chapter 2. It establishes calculus as a new way of thinking about functions.

Chapter 2 provides a formal definition of the derivative as a limit. Students learn to differentiate some simple functions, including power functions. This chapter establishes all the calculus that students need for their work with polynomial functions, in **Chapter 3**.

In **Chapter 4**, students complete the required set of differentiation rules.

In **Chapters 5, 6, 7,** and **8**, students apply these rules to all the function types that the curriculum specifies.

By working with the derivative early, students have time to assimilate concepts, and to see the power of calculus for analysing functions, and modelling applications.

Contents

Contents

5 *Rational Functions*

6 *Exponential and Logarithmic Functions*

Contents

About *Addison-Wesley Advanced Functions and Introductory Calculus 12...*

Welcome to the grade 12 calculus course!

In your grade 9, 10, and 11 courses you examined a variety of functions, such as linear and quadratic functions. You've seen how the graph of a basic function, such as $y = x^2$, is transformed when the equation takes a new form, such as $y = -2x^2 + 3x - 7$. Calculus gives you new tools that help you predict the behaviour of these functions, and other functions that you will encounter in this course.

In earlier courses, you examined how two related quantities change. You represented relationships using numbers (tables of values), symbols (equations of functions), and graphs. Calculus provides efficient new methods for analysing how two related quantities change.

This book offers a visual approach to one of the basic tools of calculus, the derivative. It shows you how to apply the derivative in a variety of contexts, and with some new types of functions. We hope your progress through this book will highlight the powerful uses of calculus for mathematicians, scientists, and engineers.

Mathematical Modelling: An Organizing Influence for the Book

Calculus enables us to model authentic applications that involve related quantities, by giving us tools to analyse how the relationship is affected by changes in our assumptions. **Mathematical Modelling** sections in this book feature authentic applications where functions, and calculus, help us to solve a range of applied problems. The pages lead you through the mathematical thinking behind an initial problem, then present related problems for you to solve.

You'll encounter modelling problems related to:

- ***Optimal Cooking Time***
- ***Optimal Profits***
- ***Predicting Populations***
- ***Quality Control***
- ***Optimal Consumption Time***

The **Mathematical Modelling** sections build from your core chapter experiences, and so they appear regularly after most chapters. Your teacher may have you work on **Mathematical Modelling** when he or she wants to emphasize the Thinking/Inquiry/Problem Solving and Application categories of the provincial Achievement Chart.

Chapter Features

Necessary Skills

In mathematics, new concepts build on what has gone before. **Necessary Skills** give a quick refresher in the prerequisite skills you need for the chapter. Your teacher may ask you to prepare for a chapter by completing these exercises as homework, or use them as a diagnostic before starting a chapter.

Occasionally, a "New" skill comes up in Necessary Skills. You didn't learn this skill in previous grades, usually because it wasn't included in the curriculum. We teach it in **Necessary Skills** because you have all the related concepts you need to develop an understanding quickly.

Numbered Sections

These develop the new content of the course. **Investigations** are included regularly, to help you build a strong conceptual understanding. As your conceptual framework develops, other sections consolidate the results through guided Examples.

Watch for **Take Note** boxes: they highlight important results or definitions, and should be part of your study notes.

The Big Picture appears in each chapter. It describes how the content of that chapter relates to the whole field of mathematics, or where it's heading as you proceed through this course.

Exercises are organized into A, B, and C categories according to their level of difficulty.

 You'll see that some exercises have a check mark beside them; try these exercises to be sure you have covered all core curriculum requirements.

Each exercise set identifies one exercise for each of the four categories of the provincial **Achievement Chart.**

- Knowledge/Understanding
- Thinking/Inquiry/Problem Solving
- Communication
- Application

These show you what to expect when you are assessed on any of the four categories. Our selection provides examples only. Each exercise set has many exercises that relate to each of the categories of achievement. Exercises that are labelled are not limited to one category only, but the focus helps to simplify assessment.

Ongoing Review

A **Self-Check** is a one-page review that occurs two or three times in a chapter. It lets you check your knowledge and understanding of the preceding sections, and provides a sample **Performance Assessment** suggestion.

The **Mathematics Toolkit** in each Chapter Review summarizes important chapter results. Use the toolkit and the **Review Exercises** to study for a chapter test.

Self-Test at the end of each chapter helps you prepare for a class test. Each Self-Test concludes with one or more **Performance Assessment** tasks: these exercises consolidate chapter concepts, and allow for a full range of responses, up to a level-4 performance.

Cumulative Reviews after Chapters 3, 6, and 8 provide additional review for all your work up to that point in the course.

Communication

Communication is a key part of all learning. This book provides many ways for you to improve your mathematical communication.

Presentations of **Solutions** to Examples model clear, concise mathematical communication. Following an Example solution will help you develop clear communication skills.

Discuss questions prompt you to reflect on solutions or the implications of new concepts, and share your thinking.

Selected **Exercises** ask you to explain your reasoning, or describe your findings. Each numbered section also contains an exercise highlighted with a "Communication" emphasis.

Technology

Graphing technologies give us a powerful tool for exploring the behaviour of functions. Investigations that require technology provide explicit instructions for its use. Exercises that appear with a technology icon indicate when a technology tool is recommended.

Assessment

Several features of this book relate to a balanced assessment approach:

- **Achievement Chart Categories** are highlighted in each exercise set.
- **Communication** opportunities appear in Examples, **Discuss** suggestions, and Exercises.
- **Self-Check** supports your knowledge and understanding.
- **Self-Test** provides a sample of the type of chapter test your teacher might use.
- **Performance Assessment** exercises appear in every Self-Test and in every Self-Check, providing opportunities for you to demonstrate a level-4 performance.
- **Mathematical Modelling** sections with rich contextual problems address all four categories of the Achievement chart, with special emphasis on Thinking/Inquiry/Problem Solving, and Application.

Analysing Change 1

Curriculum Expectations

By the end of this chapter, you will:

- Calculate and interpret average rates of change from various models of functions drawn from the natural and social sciences.
- Estimate and interpret instantaneous rates of change from various models of functions drawn from the natural and social sciences.
- Explain the difference between average and instantaneous rates of change within applications and in general.
- Demonstrate an understanding that the slope of a secant on a curve represents the average rate of change of the function over an interval, and that the slope of the tangent to a curve at a point represents the instantaneous rate of change of the function at that point.
- Demonstrate an understanding that the slope of the tangent to a curve at a point is the limiting value of the slopes of a sequence of secants.
- Demonstrate an understanding that the instantaneous rate of change of a function at a point is the limiting value of a sequence of average rates of change.
- Demonstrate an understanding that the derivative of a function at a point is the instantaneous rate of change or the slope of the tangent to the graph of the function at that point.
- Sketch, by hand, the graph of the derivative of a given graph.
- Identify examples of functions that are not differentiable.
- Determine the equation of the tangent to the graph of a function.
- Sketch the graph of a function, given the graph of its derivative function.

Necessary Skills

1. Review: Slope of a Line Segment

Line segment PQ has endpoints $P(x_1, y_1)$ and $Q(x_2, y_2)$.
To calculate the slope of PQ, when $x_1 \neq x_2$, divide the rise by the run:

$$\begin{aligned}\text{slope of PQ} &= \frac{\text{rise}}{\text{run}}\\ &= \frac{\text{change in } y}{\text{change in } x}\\ &= \frac{y_2 - y_1}{x_2 - x_1}\\ &= \frac{\Delta y}{\Delta x}\end{aligned}$$

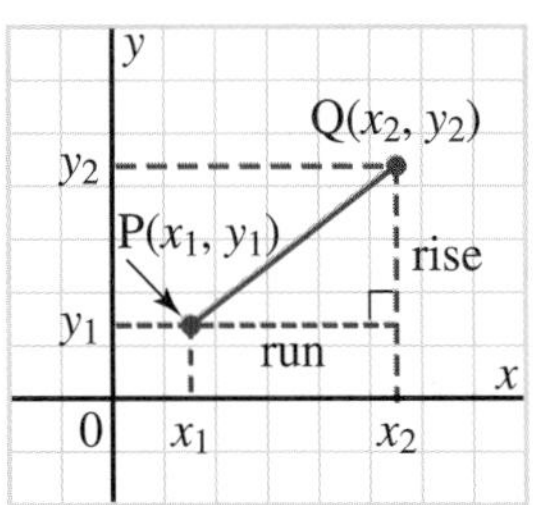

The symbol Δ above is the Greek letter delta and means "change in."

Example

The endpoints are given for 3 line segments. For each line segment, calculate the rise, the run, and the slope. Describe the orientation of the segment.

a) P(–3, 2) and Q(5, 18)

b) A(5, 2) and B(1, 14)

c) C(3, 4) and D(5, 4)

Solution

a) P(–3, 2) and Q(5, 18)

$$\begin{aligned}\text{slope of PQ} &= \frac{\Delta y}{\Delta x}\\ &= \frac{18 - 2}{5 - (-3)}\\ &= \frac{16}{8}\\ &= 2\end{aligned}$$

Since the slope is positive, line segment PQ goes up to the right.

b) A(5, 2) and B(1, 14)

$$\begin{aligned}\text{slope of AB} &= \frac{\Delta y}{\Delta x}\\ &= \frac{14 - 2}{1 - 5}\\ &= \frac{12}{-4}\\ &= -3\end{aligned}$$

Since the slope is negative, line segment AB goes down to the right.

c) C(3, 4) and D(5, 4)

$$\text{slope of CD} = \frac{\Delta y}{\Delta x} = \frac{4-4}{5-3} = \frac{0}{2} = 0$$

Since the slope is 0, line segment CD is horizontal.

Discuss
Why is the slope of a vertical line segment undefined?

Exercises

1. For each pair of endpoints, calculate the slope of the line segment. Describe its orientation.
   a) N(–14, 3) and M(6, 13)
   b) R(7, 12) and S(7, –3)
   c) C(124, 0) and D(0, –31)

2. For points A(–35, 7) and B(5, k), calculate k so that:
   a) Slope of AB is 2.
   b) Slope of AB is –2.
   c) AB is horizontal.

2. Review: Equations of a Line

The equation of a line can be written in different forms.

Equation of a Line with Slope m and y-intercept b

P(x, y) is any point on the line.
Slope of AP = m

$$\frac{y-b}{x-0} = m$$

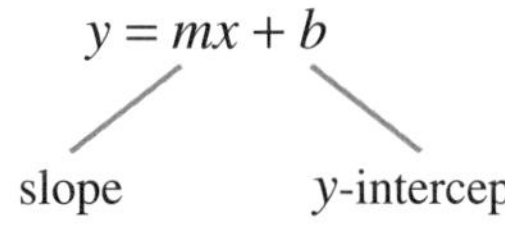

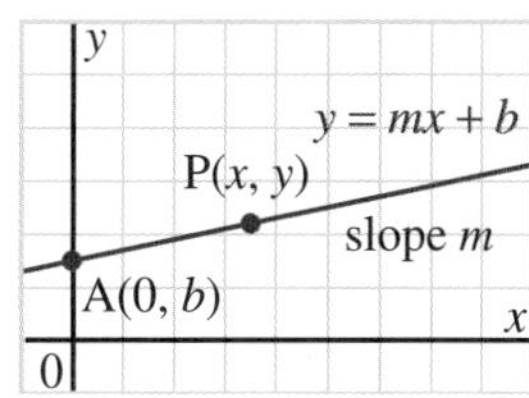

Necessary Skills

Equation of a Line with Slope *m* passing through the Point (*p*, *q*)

P(x, y) is any point on the line.

Slope of AP = m

$$\frac{y-q}{x-p} = m$$

$y = m(x - p) + q$

slope — coordinates of the point on the line

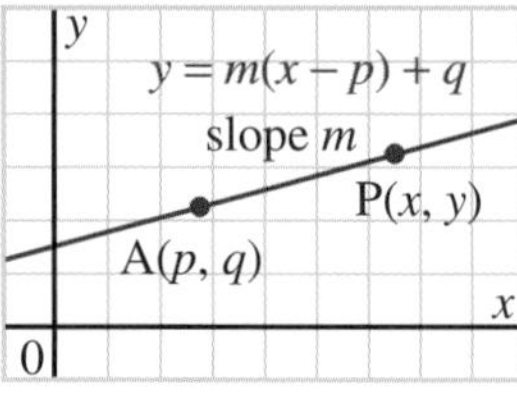

Equation of a Line in Standard Form

$Ax + By + C = 0$ where A, B, and C are integers and $A > 0$

Special Cases of the Equation of a Line

Horizontal line: $y = b$

Vertical line: $x = a$

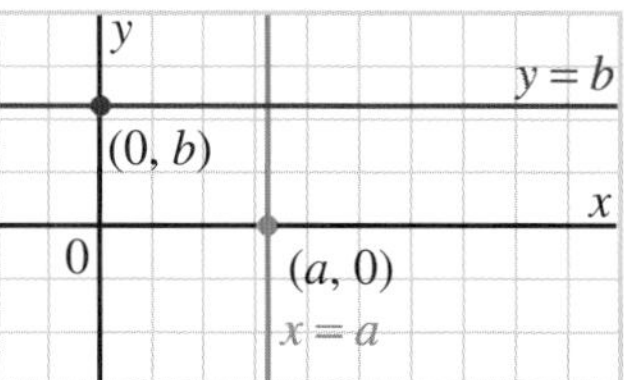

Example 1

Write the slope-intercept form of the equation of the line that passes through the points (–1, 10) and (3, 20). Rewrite the equation of the line in standard form.

Solution

Calculate the slope of the line.

$$m = \frac{20-10}{3-(-1)}$$
$$= \frac{5}{2}$$

Choose one point on the line; for example, (–1, 10).

Substitute $m = \frac{5}{2}$, $p = -1$, and $q = 10$ in $y = m(x - p) + q$.

$$y = \frac{5}{2}(x+1) + 10$$

Expand and simplify.

$$y = \frac{5}{2}x + \frac{5}{2} + 10$$
$$y = \frac{5}{2}x + \frac{25}{2}$$

The equation of the line in slope-intercept form is $y = \frac{5}{2}x + \frac{25}{2}$.

To change to standard form, multiply each term by the common denominator 2, then collect all the terms on one side of the equation.

$$2y = 5x + 25$$

$-5x + 2y - 25 = 0$

Multiply by -1 so the coefficient of x is positive.

$5x - 2y + 25 = 0$

The equation of the line in standard form is $5x - 2y + 25 = 0$.

Example 2

The equation of a line in standard form is $4x - 7y + 21 = 0$.

a) Rewrite the equation in slope-intercept form.

b) State the slope of the line.

Solution

a) $4x - 7y + 21 = 0$

Isolate the y-term. $-7y = -4x - 21$

Divide each term by -7. $y = \frac{4}{7}x + 3$

The equation of the line in slope-intercept form is $y = \frac{4}{7}x + 3$.

b) The slope is $\frac{4}{7}$.

Exercises

1. For the line through each pair of points below:

i) Write the slope-intercept form of its equation.

ii) Write the standard form of its equation.

a) A(−1, −1), B(1, 3) **b)** C(1, 3), D(0, 7) **c)** E(−6, −4), F(2, 2)

2. Rewrite each equation in slope-intercept form, then state the slope of the line it represents.

a) $3x + y - 11 = 0$ **b)** $2x - 5y + 15 = 0$ **c)** $x - 6y + 2 = 0$

3. Determine the standard form of the equation of the line with x-intercept 5 and y-intercept 15.

3. Review: Function Concepts

A *function* is a rule that gives a single output number for every valid input number. A *relation* is a rule that gives one or more output numbers for every valid input number.

The input is the *independent* variable. We often use x as the independent variable. The output is the *dependent* variable. We often use y as the dependent variable. To stress the dependence of the function f on the independent variable x, we may use $f(x)$ instead of y.

The *domain* of a function is the set of permissible values of the independent variable. The domain is the set of x-values for which the function is defined.

We can use *set notation* to describe the domain; for example, when the domain is the set of all real numbers, except, say, $x = 1$, we write $D = \{x | x \in \Re, x \neq 1\}$. We say, "$x$ is such that x belongs to the set of all real numbers except $x = 1$."

The *range* of a function is the set of resulting values of the dependent variable. The range is the set of y-values of the function. In set notation, when the range is the set of all real numbers, we write $R = \Re$.

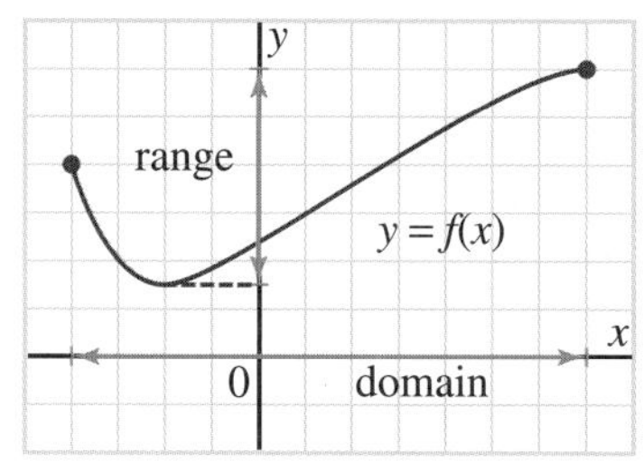

We use the *vertical line test* on the graph of a relation to check whether the relation is a function. Imagine drawing a vertical line through each point on the x-axis. If any vertical line intersects the graph in two or more points, then the relation is not a function.

Example

For each graph below:

i) Does the graph represent a function? Explain.
ii) If the graph does represent a function, state its domain and range.

a)
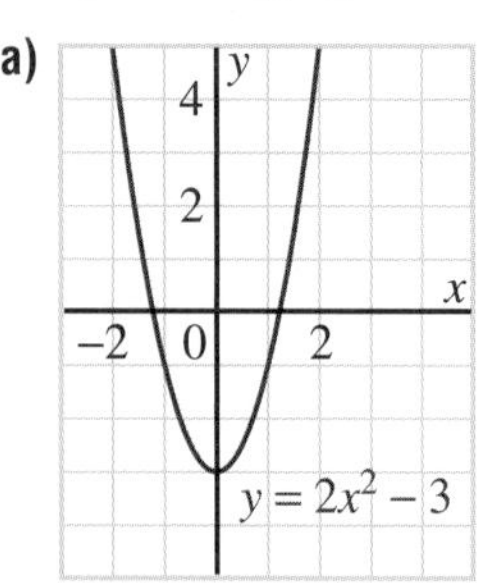

b)
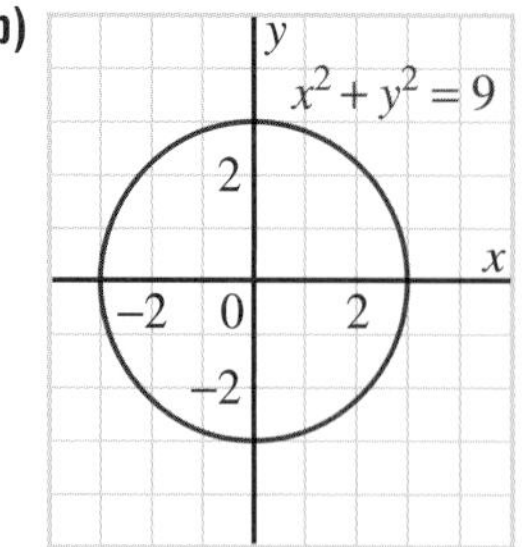

c)
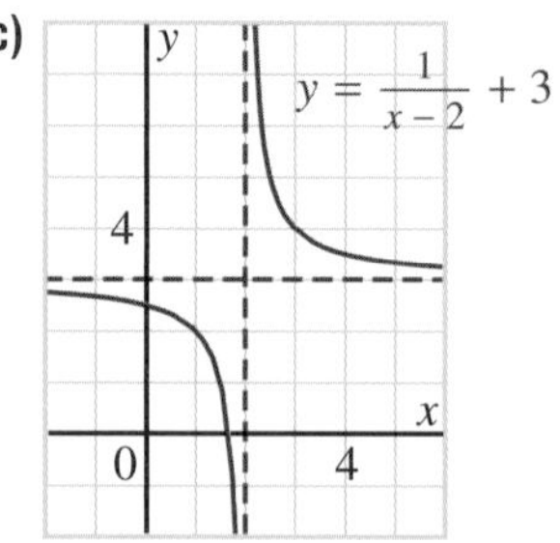

Solution

a) The quadratic relation $y = 2x^2 - 3$ is a function. Any vertical line intersects the graph at exactly 1 point. From the graph, the domain is the set of all real numbers. The range is the set of all real numbers greater than or equal to -3.
In set notation, $D = \Re$
$R = \{y | y \in \Re, y \geq -3\}$

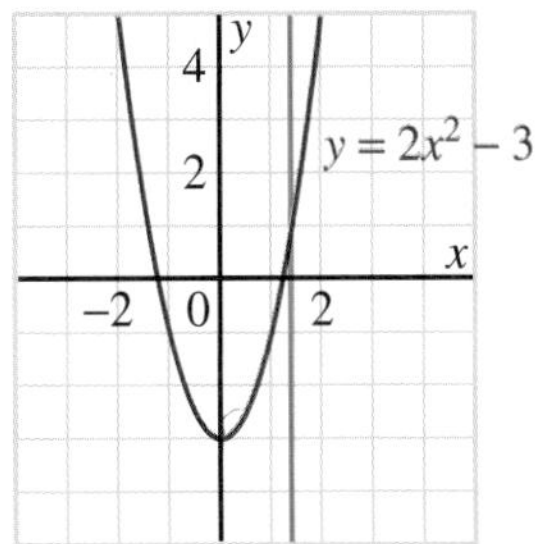

b) The circle $x^2 + y^2 = 9$ is not a function.
A vertical line intersects the graph at 2 points.

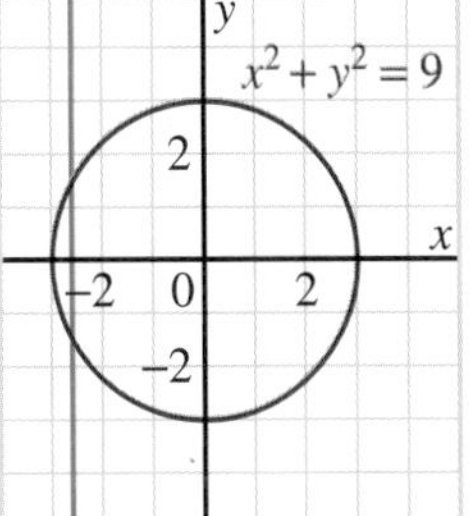

c) The relation $y = \frac{1}{x - 2} + 3$ is a function.
Any vertical line intersects the graph at exactly 1 point.
From the graph, the domain is the set of all real numbers except 2. The range is the set of all real numbers except 3. In set notation,
$D = \{x | x \in \Re, x \neq 2\}$
$R = \{y | y \in \Re, y \neq 3\}$

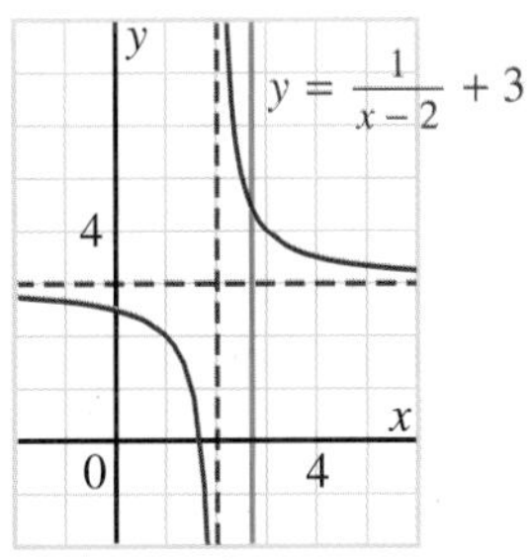

Exercises

1. Graph each relation for values of x between -5 and 5. Identify which relations are functions.

a) $y = 3(x - 2)^2$ **b)** $4x + y^2 = 0$ **c)** $y = x^3 - 16x$

2. For each function in exercise 1, state its domain and range.

1.1 Determining Rates of Change from Data

Much of our understanding of the world depends on describing how things change. Here are examples of things that change:

- the motion of a gymnast doing floor exercises
- the path of a satellite
- the temperatures of the oceans
- the fluctuations of the stock market
- the propagation of radio waves
- the power produced by a nuclear reaction

Algebra and geometry are useful tools for describing relationships among static quantities, but they do not include concepts appropriate for describing change. For this we need a new mathematical operation that measures the way quantities change. This operation will be defined in Chapter 2. The background for understanding this new operation is given in this chapter.

Consider the following examples of change:

- A child with a mass of 20 kg has gained 12 kg in a year.
- The temperature in a town at the base of a mountain is 20°C. On a ridge 1000 m above the town, the temperature is 5°C lower.
- A family on vacation travelled 240 km in 3 h by car.

In each case, we are comparing two quantities that changed — the age and mass of the child; the altitude and temperature on a mountain; the time and location of the family on vacation.

For the family on vacation, the average speed is $\frac{\text{distance travelled (km)}}{\text{time taken (h)}} = \frac{240\text{ km}}{3\text{ h}} = 80\text{ km/h}$. The *average speed* of the family's car is 80 km/h. During the trip, the speedometer shows the *instantaneous speed*, which would be greater than 80 km/h at some times and less than 80 km/h at other times.

The following situation involves both average rate of change and instantaneous rate of change.

A Bunsen burner is used to heat water in a beaker. The table on page 9 shows the temperature every 5 s. The graph shows how the temperature, T degrees Celsius, of the water increases with time, t seconds. At $t = 80$ s, the water reaches 100°C and begins to boil.

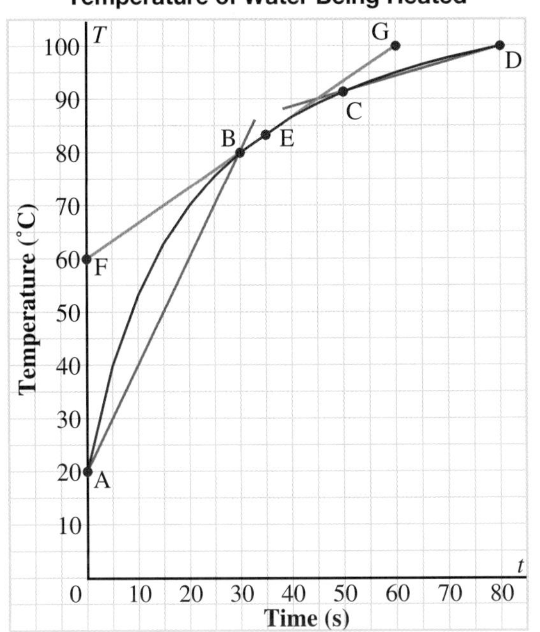

Time, t (s)	Temperature, T (°C)
0	20.0
5	40.0
10	53.3
15	62.9
20	70.0
25	75.6
30	80.0
35	83.6
40	86.7
45	89.2
50	91.4
55	93.3
60	95.0
65	96.5
70	97.8
75	98.9
80	100.0

The temperature of the water increased by 80°C in 80 s. That is an average rate of increase of 1°C/s. However, we can tell from the graph that the temperature increased at a greater rate near the beginning and at a lower rate near the end.

Calculating Average Rate of Change

To calculate the average rate of temperature increase in any time interval, we determine the change in temperature, ΔT, and the change in time, Δt. Suppose we use the time interval from $t = 0$ s to $t = 30$ s. From the table or the graph, we find:

$$\begin{aligned}\Delta T &= 80.0^\circ\text{C} - 20.0^\circ\text{C}\\ &= 60.0^\circ\text{C}\\ \Delta t &= 30\text{ s} - 0\text{ s}\\ &= 30\text{ s}\end{aligned}$$

The average rate of temperature increase during this time interval is

$$\begin{aligned}\frac{\Delta T}{\Delta t} &= \frac{60.0^\circ\text{C}}{30\text{ s}}\\ &= 2^\circ\text{C/s}\end{aligned}$$

This is the slope of the line through A and B.

Similarly, the slope of the line through C and D represents the average rate of temperature increase during the last 30 s:

$$\frac{\Delta T}{\Delta t} = \frac{100.0°\text{C} - 91.4°\text{C}}{80\text{ s} - 50\text{ s}}$$
$$= \frac{8.6°\text{C}}{30\text{ s}}$$
$$\doteq 0.3°\text{C/s}$$

A line through two points on a graph is a *secant*. The slope of any secant on the temperature-time graph represents the average rate of temperature increase from the first point to the second.

The graph shows temperature as a function of time. It illustrates one way to consider average rate of change for a function.

Take Note

Average Rate of Change

For any function $y = f(x)$, the *average rate of change* of y with respect to x between $x = x_1$ and $x = x_2$ is:

$$\frac{\text{change in } y}{\text{change in } x} = \frac{\Delta y}{\Delta x} = \frac{y_2 - y_1}{x_2 - x_1}$$

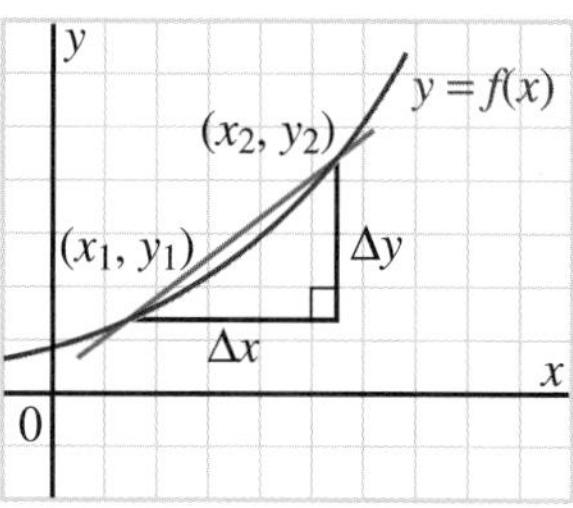

On a graph of the function, this is the slope of the secant through (x_1, y_1) and (x_2, y_2).

Estimating Instantaneous Rate of Change

To determine the instantaneous rate of temperature increase at point E where $t = 35$ s, we use a tangent. This is a line that passes through a point on the graph and has the same direction as the graph at that point. The tangent at E has been drawn on the graph, using visual estimation. Its slope is an estimate of the instantaneous rate of change of temperature at $t = 35$ s.

We will define a tangent later in this chapter.

To determine the slope of the tangent, we use any two points on this line, such as F(0, 60) and G(60, 100). The slope of FG is:

$$\frac{\Delta T}{\Delta t} = \frac{100°\text{C} - 60°\text{C}}{60\text{ s} - 0\text{ s}}$$
$$= \frac{40°\text{C}}{60\text{ s}}$$
$$\doteq 0.67°\text{C/s}$$

The instantaneous rate of change of temperature at 35 s is about 0.67°C/s. This is only an estimate because the tangent at E was drawn by visual estimation, and the coordinates of F and G were estimated.

In practice, estimates of instantaneous rates of change obtained from a graph may be inaccurate because errors are introduced in each step of the process.

We cannot calculate the instantaneous rate of temperature change at 35 s exactly from the data in the table on page 9, but we can estimate it using a short interval. For example, suppose we use the interval from t = 35 s to t = 40 s. Over this interval, the average rate of change of temperature is:

$$\begin{aligned}\frac{\Delta T}{\Delta t} &= \frac{86.7^\circ\text{C} - 83.6^\circ\text{C}}{40\text{ s} - 35\text{ s}} \\ &= \frac{3.1^\circ\text{C}}{5\text{ s}} \\ &= 0.62^\circ\text{C/s}\end{aligned}$$

This estimate of the instantaneous rate of change of temperature at t = 35 s is slightly lower than the previous estimate, as we would expect from the graph. If we had used the interval from t = 30 s to t = 35 s, we would have obtained a slightly higher estimate. We could also use the interval from t = 30 s to t = 40 s (see exercise 8).

Take Note

Instantaneous Rate of Change

For any function $y = f(x)$, the *instantaneous rate of change* of y with respect to x at x_1 is the slope of the tangent to the graph of the function at the point where $x = x_1$.

To estimate the instantaneous rate of change from a graph, draw a tangent at the point where $x = x_1$ and determine its slope.

To estimate instantaneous rate of change from a table, calculate the average rate of change over a short interval containing the point where $x = x_1$. Different intervals can be used.

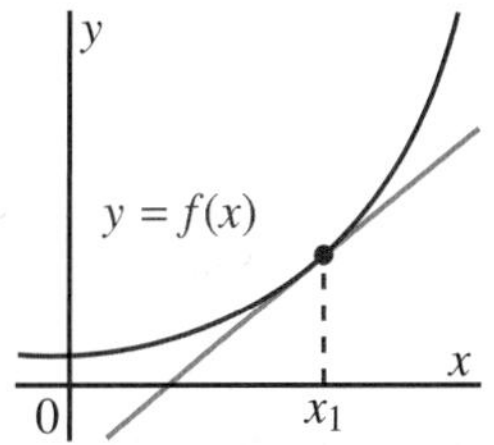

Following interval
From the point where $x = x_1$ to the next data point:

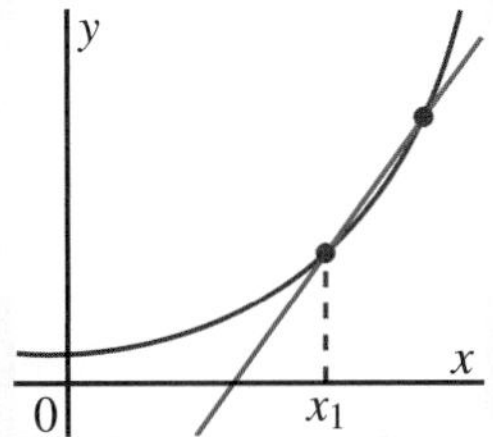

Preceding interval
From the previous data point to the point where $x = x_1$:

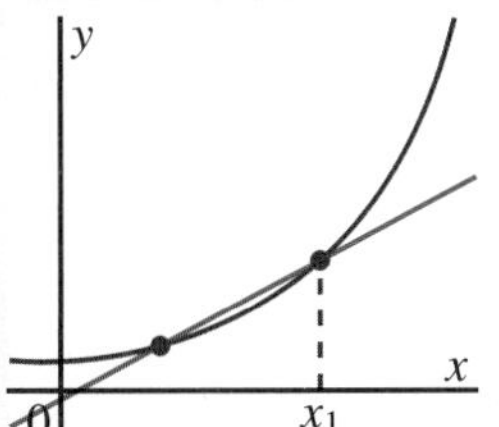

Centred interval
From the previous data point to the next data point:

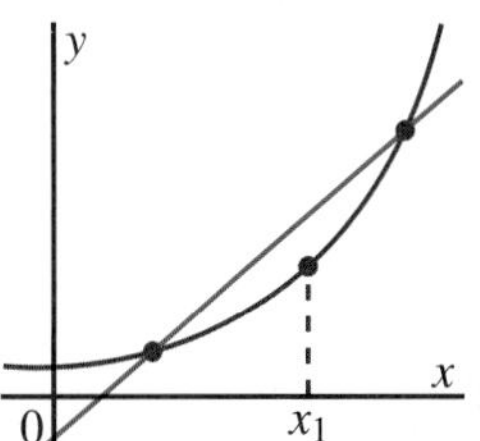

Estimates of instantaneous rates of change may be inaccurate.

Example

An antiseptic spray is applied to a culture of bacteria. The percent of bacteria, b, that are alive after t minutes is shown on the graph. A secant has been drawn through the points for $t = 5$ min and $t = 35$ min. A tangent has been drawn at the point where $t = 20$ min.

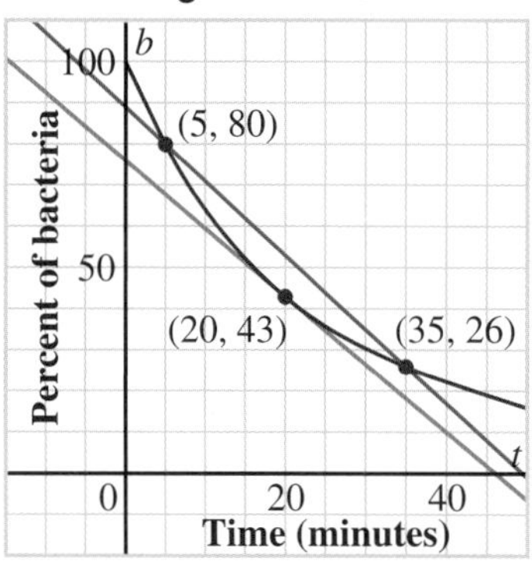

a) Calculate the slope of the secant. Describe what the slope represents. Explain the sign of the slope.

b) Estimate the instantaneous rate of change of b with respect to t at $t = 20$ min.

Solution

a) The slope of the secant is

$$\frac{\Delta b}{\Delta t} = \frac{26\% - 80\%}{35 \text{ min} - 5 \text{ min}}$$
$$= \frac{-54\%}{30 \text{ min}}$$
$$= -1.8\%/\text{min}$$

This slope is the average rate of change of the percent of live bacteria between 5 min and 35 min. The slope is negative, which indicates that the percent of live bacteria is decreasing.

b) The slope of the tangent at $t = 20$ min is the instantaneous rate of change of b with respect to t.

Two points on the tangent are D(0, 75) and E(40, 10).

The slope of the tangent is approximately

$$\frac{(10 - 75)\%}{(40 - 0) \text{ min}} = \frac{-65\%}{40 \text{ min}}$$
$$= -1.625\%/\text{min}$$

The instantaneous rate of change in the bacteria population is approximately $-1.6\%/\text{min}$ at $t = 20$ min.

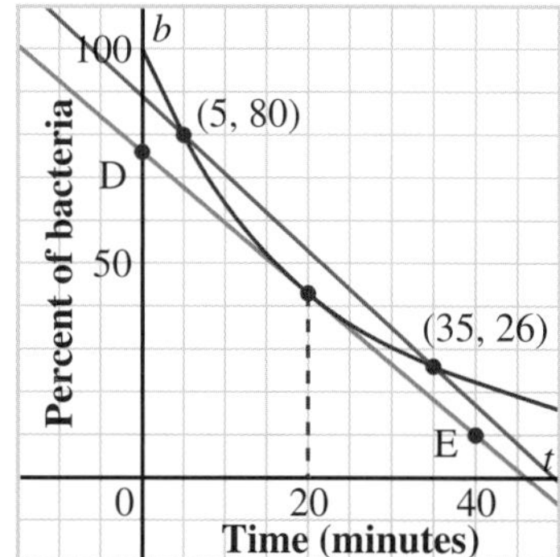

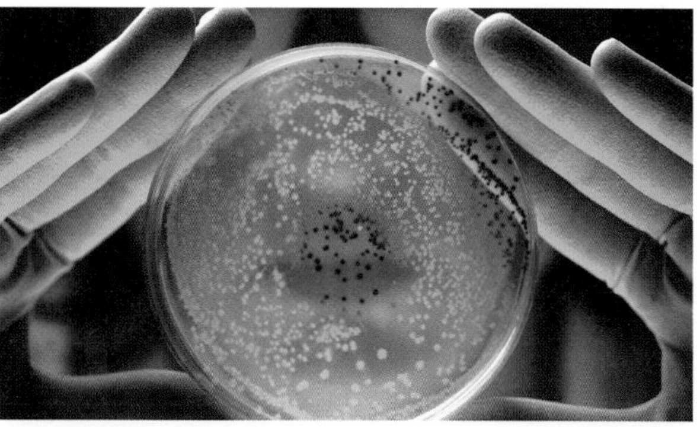

1.1 Exercises

1. The average maximum daily temperature in Sault Ste. Marie in March is –4°C. The average maximum daily temperature increases about 7°C per month during March, April, May, and June. Estimate the average maximum daily temperature in Sault Ste. Marie in each month.

   a) April b) May c) June

2. a) Graph these data.

x	0	2	4	6	8	10
y	5	7	13	23	37	55

   b) Calculate the average rate of change of y over each interval.

   i) $x = 0$ to $x = 2$ ii) $x = 0$ to $x = 4$

   iii) $x = 0$ to $x = 10$ iv) $x = 6$ to $x = 10$

3. For each data set, calculate the average rate of change of y between each consecutive pair of values of x.

a)

x	0	1	2	3	4	5	6
y	1	3	9	27	81	243	729

b)

x	50	100	150	200	250	300	350
y	7.0	6.0	4.6	2.9	0.9	–1.3	–3.8

4. On each graph, a tangent has been drawn at the point where $x = 5$. Determine the instantaneous rate of change of y with respect to x, at $x = 5$.

a)

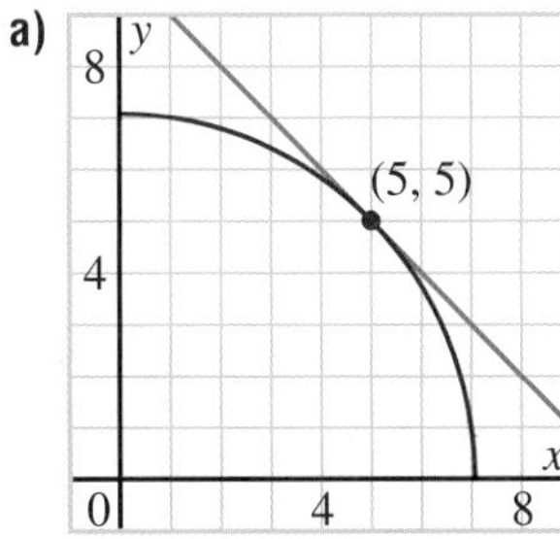

b)

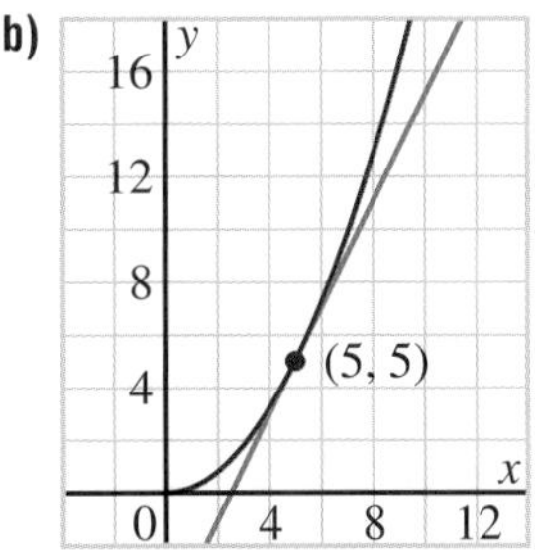

c)

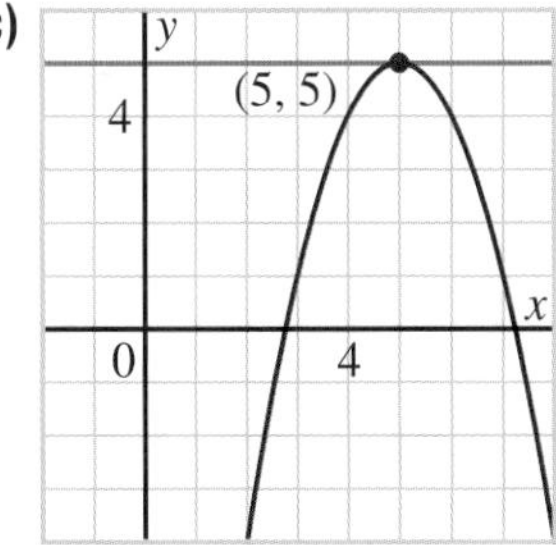

5. Explain the use of the word "average" in *average* rate of change. Illustrate your explanation with examples from daily life.

✓ **6.** Determine the average rate of change of temperature for each time interval in the table on page 9. Use the graph on page 9 to check that your results are reasonable.

a) 0 s to 10 s **b)** 10 s to 25 s

c) 25 s to 45 s **d)** 45 s to 80 s

✓ **7.** Earlier, we estimated the instantaneous rate of temperature increase at $t \doteq 35$ s using the interval from $t = 35$ s to $t = 40$ s. Other intervals can be used. Use the table on page 9.

a) Estimate the instantaneous rate of temperature increase for each interval.

i) the interval from $t = 30$ s to $t = 35$ s

ii) the interval from $t = 30$ s to $t = 40$ s

b) Which interval do you think gives the best estimate of the instantaneous rate of temperature increase at $t = 35$ s? Explain.

Exercise 7 is significant because it involves different ways of estimating the instantaneous rate of change at a point, P, from data. There are at least four ways to do this. Each method provides an estimate only.

✓ **8.** The following three estimates of the instantaneous rate of temperature increase at $t = 35$ s were obtained for the graph on page 9. See exercise 7 above.

Using the following interval from $t = 35$ s to $t = 40$ s: 0.62°C/s
Using the preceding interval from $t = 30$ s to $t = 35$ s: 0.72°C/s
Using the centred interval from $t = 30$ s to $t = 40$ s: 0.67°C/s

The estimate for the centred interval is the mean of the estimates for the following and preceding intervals.

Choose a different value of t. For the value you chose, determine the three estimates of the instantaneous rate of temperature increase using the following interval, the preceding interval, and the centred interval. Is the estimate from the centred interval the mean of the estimates from the other two intervals? Explain.

9. Suppose water in a beaker is heated with a Bunsen burner. Use the graph on page 9. Sketch an example of the temperature-time graph you would expect for each situation below. Explain.

a) There is more water in the beaker.

b) There is less water in the beaker.

c) Warmer water from the hot water tap at 40°C is used.

d) Colder water from the refrigerator at 5°C is used.

10. Choose one part of exercise 9. In this situation, how would the various rates of change you determined previously for the graph on page 9 be affected? Explain.

✓ **11.** A chinook is a warm winter wind that blows across western Canada. It can cause temperatures to rise dramatically. An example is shown in the graph below. Tangents corresponding to 10 A.M. and 12 noon are drawn on the graph.

a) Estimate the instantaneous rate of temperature increase at these two times.

b) At what time is the temperature increasing most rapidly? Explain how you can tell this from the graph.

The Effect of a Chinook on Temperature

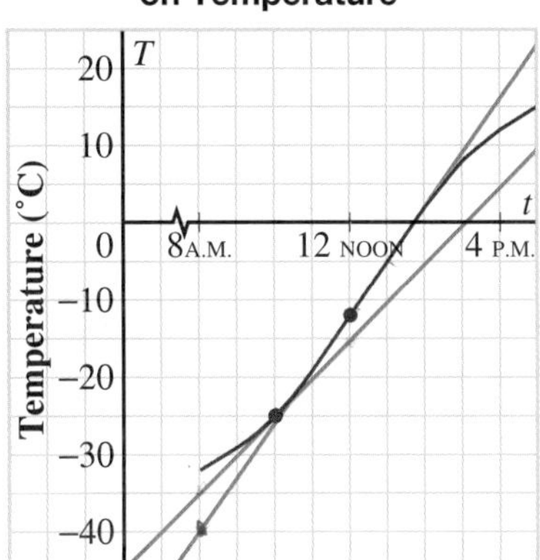

✓ **12. Application** A 2-L plastic pop bottle is filled with water. A hole is made in the base of the bottle. As the water drains out, the height of the water level in the bottle drops. The table shows the height of the water at 20-s intervals.

a) Graph water height as a function of time.

b) Draw a secant on the graph to correspond to each time interval below. On each secant, show Δt and Δh, then calculate the average rate of change.

i) 0 s to 100 s **ii)** 100 s to 200 s

iii) 0 s to 200 s

c) Make as good an estimate as possible of the instantaneous rate of change of water height at $t = 60$ s.

Time, t(s)	Height, h(mm)
0	130
20	105
40	83
60	64
80	48
100	34
120	22
140	13
160	6
180	2
200	0

✓ **13. Communication** Refer to exercise 12c. Describe as many different reasons as you can why your estimate of the instantaneous rate of change of water height at $t = 60$ s may be inaccurate.

✓ **14. Thinking/Inquiry/Problem Solving** Suppose the experiment in exercise 12 is repeated with each change described below. Sketch the graph you would expect. Justify your thinking.

a) The hole in the bottle is larger.

b) The hole is higher up.

c) A 750-mL bottle is used, which has a smaller diameter.

✓ **15. Knowledge/Understanding** A chemistry class performed an experiment in which some helium gas was compressed in a sealed insulated cylinder. The gas pressure, measured in kilopascals, was recorded for different volumes, and used to make the graph below. A secant has been drawn through the points for $V = 200$ mL and $V = 600$ mL. A tangent has been drawn at the point for $V = 400$ mL.

a) Calculate the slope of the secant. Describe what the slope represents. Explain the sign of the slope.

b) Estimate the instantaneous rate of change of P with respect to V at $V = 400$ mL.

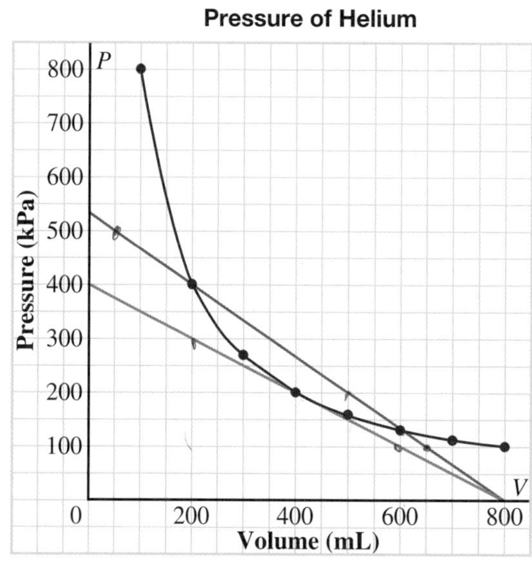

16. This growth chart shows average data for North American girls from birth to age 3 years.

Age, t (months)	0	3	6	9	12	15	18	21	24	27	30	33	36
Mass, m (kg)	3.4	5.5	7.2	8.5	9.5	10.3	11.0	11.5	12.0	12.5	12.9	13.5	13.8

a) Graph mass against age. Draw a smooth curve through the points. This graph is a growth curve.

b) Describe the shape of the graph.

c) Describe how mass changes as a girl ages.

d) Explain how you can recognize this trend on the graph.

e) Explain why this trend makes sense.

f) Estimate the instantaneous rate of growth at several points on the graph. Explain how these results justify your answer to part c.

17. When the data from a certain investigation are graphed, the relation is found to be linear. A student then calculates the average rate of change of the function over several intervals, some long and some short. The student also calculates the slope of the graph. Describe how these rates and the slope are related.

18. Collect and analyse your own data for the experiment in exercise 12.

a) Use a 2-L plastic pop bottle with the label removed. The bottle is cylindrical for most of its height. Punch a small hole in the side of the bottle near the bottom.

b) Cover the hole with your finger. Fill the bottle with water to the top of the cylindrical region. Set the bottle over a large pan.

c) Remove your finger and measure the time for the water to drain out.

d) Use this time to select a convenient time interval for your measurement of heights. You should use at least 10 time intervals. Make a table similar to that in exercise 12.

e) Refill and reposition the bottle. Place a ruler beside it.

f) Start timing when you remove your finger. Record the height of the water after every time interval until the water has stopped draining.

g) Graph the data. Compare your graph with the graph you drew in exercise 12.

h) Repeat exercise 12b and c for your graph.

C

19. In exercise 8, you should have found that the estimated instantaneous rate of temperature change calculated using the centred interval is always the mean of the estimates calculated using the following and preceding intervals. Prove that this is true for any three consecutive data points in any experiment, as long as they are equally spaced.

1.2 Determining Rates of Change from an Equation

In Section 1.1, we determined average and instantaneous rates of change for functions determined by data in a table or on a graph. We can also determine average and instantaneous rates of change for functions defined by an equation.

Recall the graph of temperature against time when water is heated. The graph is shown at the right. The equation that expresses the temperature, T degrees Celsius, as a function of time, t seconds, is:

$$T = \frac{120t + 400}{t + 20},\ 0 \le t \le 80$$

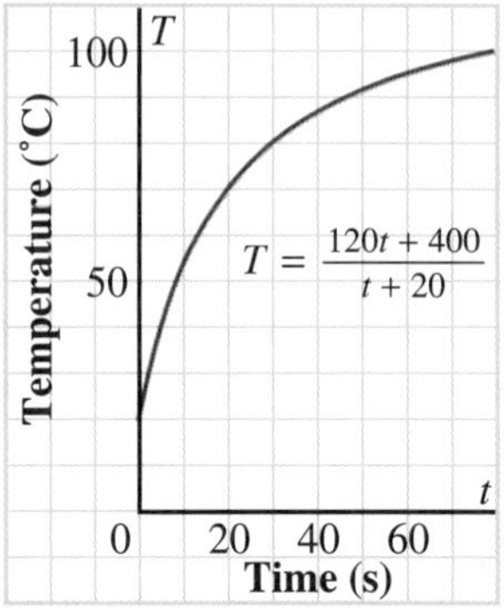

We can use this equation to determine average and instantaneous rates of change.

Example 1

Use the equation $T = \frac{120t + 400}{t + 20},\ 0 \le t \le 80$.

a) Calculate the average rate of increase in temperature from $t = 0$ s to $t = 20$ s.

b) Estimate the instantaneous rate of increase in temperature at $t = 35$ s.

Solution

$$T = \frac{120t + 400}{t + 20}$$

a) Calculate the temperatures at $t = 0$ s and at $t = 20$ s.

When $t = 0$, $T = \frac{0 + 400}{0 + 20}$
$= 20$

The temperature is 20°C.

When $t = 20$, $T = \frac{120(20) + 400}{20 + 20}$
$= 70$

The temperature is 70°C.

The average rate of increase in temperature is $\frac{\Delta T}{\Delta t} = \frac{70°\text{C} - 20°\text{C}}{20\,\text{s} - 0\,\text{s}}$
$= 2.5°\text{C/s}$

The average rate of increase in temperature is 2.5°C/s.

b) Use a 1-s following interval starting at $t = 35$ s. Calculate the temperatures at $t = 35$ s and at $t = 36$ s.

When $t = 35$, $T = \frac{120(35) + 400}{35 + 20}$
$\doteq 83.636\,363$

When $t = 36$, $T = \dfrac{120(36) + 400}{36 + 20}$

$\doteq 84.285\ 714$

The average rate of increase in temperature from 35 s to 36 s is

$$\frac{\Delta T}{\Delta t} \doteq \frac{84.285\ 714°\text{C} - 83.636\ 363°\text{C}}{36\ \text{s} - 35\ \text{s}}$$

$\doteq 0.649\ 351°\text{C/s}$

The average rate of increase in temperature is 0.6494°C/s.

So, the instantaneous rate of increase in temperature at $t = 35$ s is approximately 0.65°C/s.

The results of *Example 1* agree with earlier calculations from the temperature-time graph on pages 9 to 11 in Section 1.1. In part b, we can get a better estimate of the instantaneous rate of temperature increase at $t = 35$ s by calculating the average rate of temperature increase over a shorter interval. We cannot do this with a table or a graph.

Suppose we use a 0.1-s following interval starting at $t = 35$ s. Calculate the temperature at $t = 35.1$ s.

When $t = 35.1$, $T = \dfrac{120(35.1) + 400}{35.1 + 20}$

$\doteq 83.702\ 359$

The average rate of increase in temperature from 35 s to 35.1 s is:

$$\frac{\Delta T}{\Delta t} = \frac{83.702\ 359°\text{C} - 83.636\ 363°\text{C}}{35.1\ \text{s} - 35\ \text{s}}$$

$\doteq 0.659\ 96°\text{C/s}$

The average rate of increase in temperature is 0.6600°C/s to 4 decimal places. If we had used a 0.01-s interval from $t = 35$ s to $t = 35.01$ s, we would have found that the average rate of increase in temperature is 0.6611°C/s to 4 decimal places.

If we had used preceding intervals or centred intervals, the results would be slightly different. As the intervals become shorter, these differences become smaller. For a very small interval, it does not matter which kind of interval is used.

These calculations suggest that the instantaneous rate of temperature increase at $t = 35$ s is slightly greater than 0.66°C/s.

We can use the above methods to determine average and approximate instantaneous rates of change for any function from its equation.

For example, consider the function $y = 2x^3$.

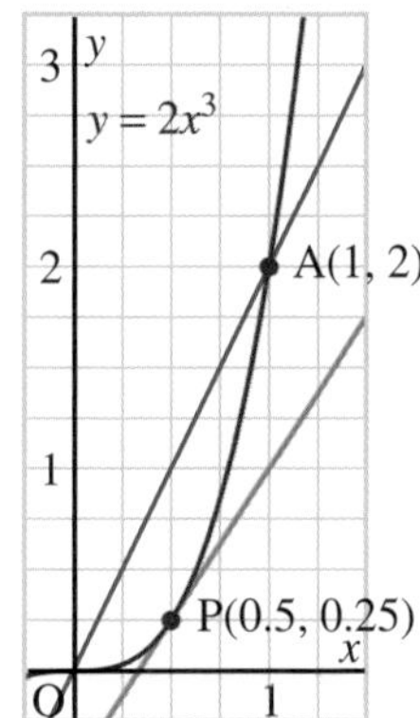

Two points on the graph of this function are O(0, 0) and A(1, 2).

The average rate of change of y with respect to x from O to A is the slope of the secant OA:

$$\frac{2-0}{1-0} = 2$$

Another point on the graph of this function is P(0.5, 0.25).

The instantaneous rate of change of y with respect to x at $x = 0.5$ is the slope of the tangent at P, which is slightly less than the slope of the secant OA.

Hence, the instantaneous rate of change of y with respect to x at P is less than 2.

To estimate this rate of change, we calculate the slope of a secant through P and another point close to P, such as $Q(0.6,\ 2(0.6)^3)$. The graph below shows this secant. This graph is an enlarged version of part of the graph above.

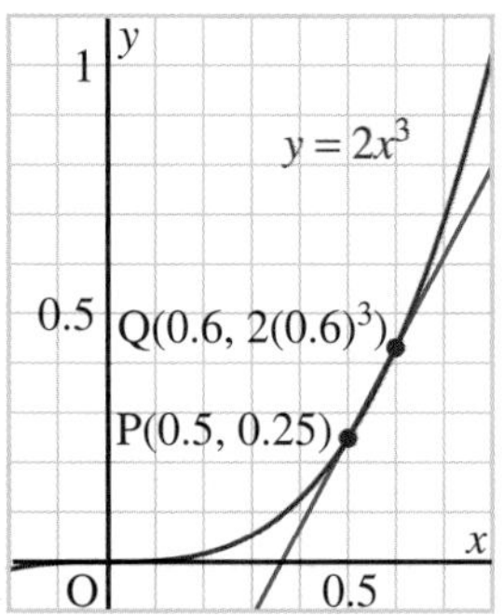

The average rate of change of y with respect to x from P to Q is:

$$\begin{aligned}\frac{\Delta y}{\Delta x} &= \frac{2(0.6)^3 - 2(0.5)^3}{0.6 - 0.5}\\ &= 1.82\end{aligned}$$

Hence, 1.82 is an estimate of the instantaneous rate of change of y with respect to x at P. We can improve the estimate by calculating the average rate of change using points that are closer and closer to P. The table below summarizes the results for points whose x-coordinates are 0.01, 0.001, and 0.0001 greater than the x-coordinate of P. We use the letter h to represent Δx, the horizontal distance from P to each point.

h	P	Second point	Average rate of change
0.01	$(0.5, 2(0.5)^3)$	$(0.51, 2(0.51)^3)$	$\frac{2(0.51)^3 - 2(0.5)^3}{0.51 - 0.5} = 1.5302$
0.001	$(0.5, 2(0.5)^3)$	$(0.501, 2(0.501)^3)$	$\frac{2(0.501)^3 - 2(0.5)^3}{0.501 - 0.5} = 1.503\ 002$
0.0001	$(0.5, 2(0.5)^3)$	$(0.5001, 2(0.5001)^3)$	$\frac{2(0.5001)^3 - 2(0.5)^3}{0.5001 - 0.5} = 1.500\ 300\ 02$
	⋮	⋮	⋮

These results suggest that the instantaneous rate of change of y with respect to x at $x = 0.5$ is 1.5. We cannot be certain that it is 1.5, but this is a good estimate of its value.

We can use the zoom feature of a graphing calculator to illustrate this situation graphically. The first screen below shows the graphs of the function $y = 2x^3$ and the line $y = 1.5x - 0.5$. This line was chosen because it has slope 1.5 and passes through P(0.5, 0.25). If we zoom in on this point a few times, the graph of $y = 2x^3$ and the line become indistinguishable. This line is the tangent to the graph of $y = 2x^3$ at P(0.5, 0.25).

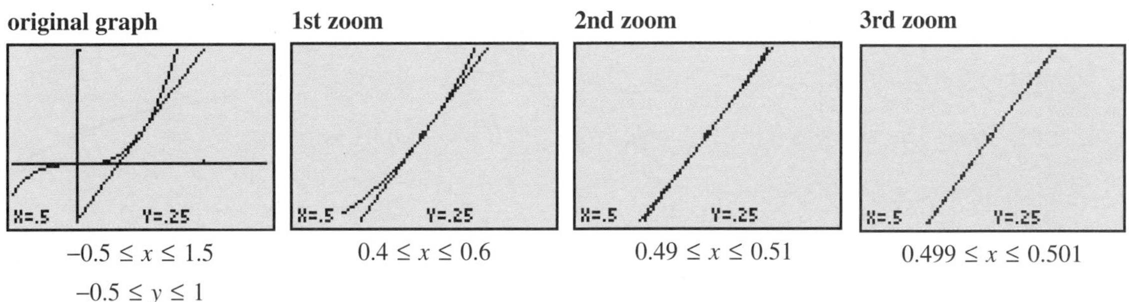

Although this is only one example, it provides strong evidence of the following statement. In general, for a line to be a tangent to a curve at a point on the curve, the line and the curve become indistinguishable over smaller and smaller intervals.

Take Note

Estimating Instantaneous Rate of Change from an Equation

Let $P(x_1, y_1)$ be a point on the graph of the function $y = f(x)$. To estimate the instantaneous rate of change of y with respect to x at P, calculate the slope of a secant over a very short interval with its left end at P:

$$\frac{f(x_1 + h) - f(x_1)}{h}$$

Use a very small number for h, such as 0.01 or 0.001. The smaller the number, the more accurate the estimate.

The expression $\frac{f(x_1 + h) - f(x_1)}{h}$ is called a *difference quotient.*

You can use this method to estimate instantaneous rates of change, when they are meaningful. For some functions, the instantaneous rate of change may not be meaningful at certain points. We will encounter examples in Section 1.5.

Example 2

The function $y = \sqrt{x}$ is given. Estimate the instantaneous rate of change of y with respect to x at $x = 6$.

Solution

$y = \sqrt{x}$

Calculate the average rate of change over a very short interval with its left end at $x = 6$.

Use $h = 0.01$. The average rate of change from $x = 6$ to $x = 6.01$ is:

$$\frac{\sqrt{6.01} - \sqrt{6}}{0.01} \doteq 0.204\ 039\ 164\ 3$$

The instantaneous rate of change of y with respect to x at $x = 6$ is approximately 0.20.

In *Example 2*, we cannot be certain from this single calculation that the instantaneous rate of change is 0.20 to 2 decimal places, but this is a good estimate of its value. To confirm this, we repeat the calculation with $h = 0.001$.

$$\frac{\sqrt{6.001} - \sqrt{6}}{0.001} \doteq 0.204\ 115\ 640\ 7$$

This result provides further evidence that the instantaneous rate of change of y with respect to x at $x = 6$ is 0.20 to 2 decimal places.

1.2 Exercises

A

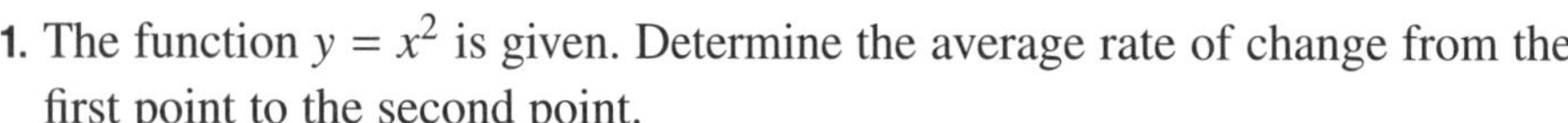

1. The function $y = x^2$ is given. Determine the average rate of change from the first point to the second point.
   a) (0, 0), (3, 9) **b)** (1, 1), (3, 9) **c)** (2, 4), (3, 9)
   d) (–2, 4), (0, 0) **e)** (–2, 4), (1, 1) **f)** (–2, 4), (2, 4)

2. The function $y = 0.5x + 3$ is given.
   a) Determine the average rate of change from the first point to the second point.
   i) (0, 3), (2, 4) **ii)** (0, 3), (4, 5) **iii)** (0, 3), (6, 6)
   b) Explain the pattern in the results in part a.
   c) What do the results of part b tell you about the instantaneous rate of change of y with respect to x for this function? Explain.

B

3. The expression $\frac{f(x_1 + h) - f(x_1)}{h}$ is called a difference quotient. Explain why this is a reasonable description of this expression.

4. Repeat *Example 1b* for each time. Use 1-s following intervals.

   a) 15 s   **b)** 40 s   **c)** 60 s   **d)** 0 s

5. For water draining from a certain pop bottle, the height, h millimetres, of the water as a function of time, t seconds, is $h = 0.003\,25(200 - t)^2$.

   a) Calculate the average rate of change in height from $t = 50$ s to $t = 100$ s.

   b) Estimate the instantaneous rate of decrease in height for each time. Use 1-s following intervals.

   i) 0 s   **ii)** 60 s   **iii)** 120 s   **iv)** 180 s

6. For a certain volume of compressed helium gas, the pressure, P kilopascals, as a function of volume, V millilitres, is $P = \frac{79\,500}{V}$.

   a) Calculate the average rate of change in pressure from $V = 450$ mL to $V = 500$ mL.

   b) Estimate the instantaneous rate of decrease in pressure for each volume. Use 1-mL following intervals.

   i) 100 mL   **ii)** 300 mL   **iii)** 500 mL   **iv)** 700 mL

7. **Application** An art student used a light meter to measure the light intensity at different distances from a bright light bulb. She used the results to make this graph. The graph expresses the intensity, I watts per square metre, as a function of the distance, d metres. The equation of the function is $I = \frac{5}{d^2}$. Tangents to the graph are drawn at $d = 1.5$ m and at $d = 3.0$ m.

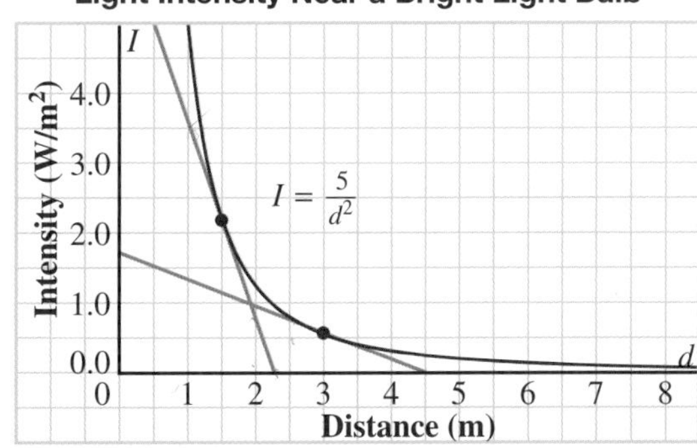

   a) Use the graph to determine the slope of each tangent. Describe what the slope represents.

   b) Use 0.1-m following intervals to estimate the instantaneous rate of change of intensity at $d = 1.5$ m and at $d = 3.0$ m.

   c) Compare your results in parts a and b. Explain any similarities and differences.

8. A rock falls from a cliff. The distance it falls, d metres, is a function of the time, t seconds, with equation $d = 4.9t^2$. Tangents are drawn at $t = 1$ and at $t = 2$ on the graph below.

a) Use the graph to determine the slope of each tangent. Describe what the slope represents.

b) Use 0.01-s following intervals to estimate the instantaneous speed of the rock at $t = 1.0$ s and at $t = 2.0$ s.

c) Compare your results in parts a and b. Explain any similarities and differences.

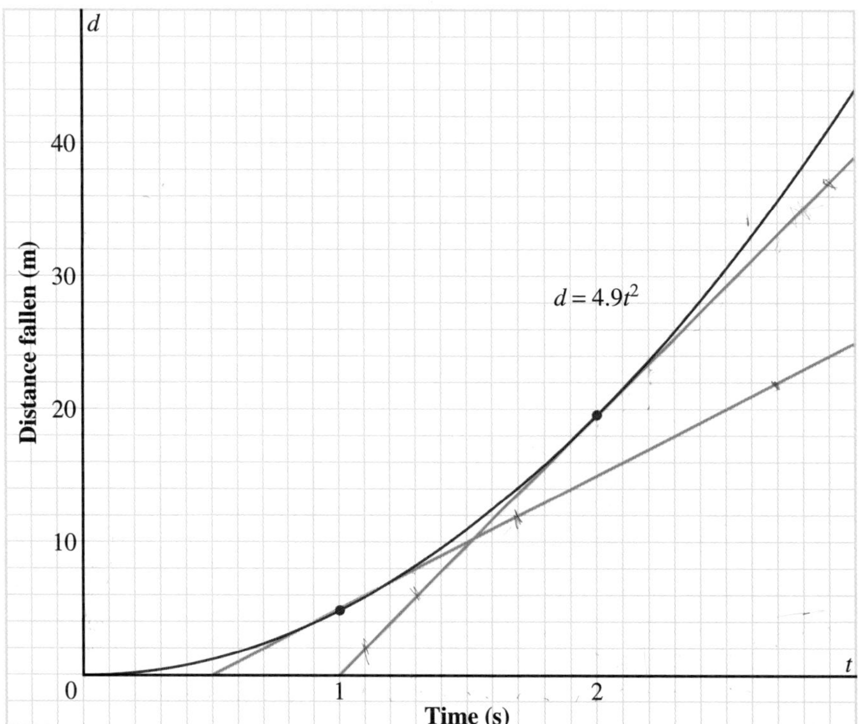

9. **Knowledge/Understanding** The equations of three functions are given below. For each function, estimate the instantaneous rate of change of y with respect to x at each value of x.

i) $x = 4$ ii) $x = 1$ iii) $x = 0.25$

a) $y = 5x^2$ b) $y = \sqrt{x}$ c) $y = \frac{1}{x}$

10. **Communication** Answer the following question for each function in exercise 9. What appears to be happening to the instantaneous rates of change as x decreases? Explain.

11. The functions $y = \sqrt{x}$, $y = 2\sqrt{x}$, and $y = 3\sqrt{x}$ are given.

a) For each function, estimate the instantaneous rate of change of y with respect to x at $x = 2.5$.

b) Compare the results in part a. How are the results related?

c) Explain the relationship you discovered in part b.

12. The functions $y = x^3$, $y = x^3 + 2$, and $y = x^3 - 2$ are given.

a) For each function, estimate the instantaneous rate of change of y with respect to x at $x = 1.5$.

b) Compare the results in part a. How are the results related?

c) Explain the relationship you discovered in part b.

13. Estimate each instantaneous rate of change of y with respect to x.

a) for the function $y = 2^x$ at $x = 0$ and $x = 1$

b) for the function $y = 3^x$ at $x = 0$ and $x = 1$

14. Estimate each instantaneous rate of change of y with respect to x.

a) for the function $y = \sin x$ at $x = 0$ and $x = \frac{\pi}{6}$

b) for the function $y = \cos x$ at $x = 0$ and $x = \frac{\pi}{6}$

✓ 15. We used following intervals to estimate the instantaneous rate of change of $y = 2x^3$ at P(0.5, 0.25), on page 20. The intervals do not need to be following intervals. Draw a diagram similar to the second graph on page 20.

a) Using preceding intervals
Plot point P and point $C(0.4, 2(0.4)^3)$. Calculate the average rate of change from C to P. Then complete a table similar to that on page 20 for intervals to the left of P.

b) Using centred intervals
Calculate the average rate of change from C to Q. Then complete a table similar to that on page 20 for intervals centred at P.

Exercise 15 is significant because it shows that there is more than one way to estimate the instantaneous rate of change of y with respect to x at a point P on the graph of a function. It does not matter which method you use. If the intervals are very small, the estimates will be very close to the slope of the tangent at P.

16. Consider the function $y = x^2$.

a) Estimate the instantaneous rate of change of y with respect to x at $x = 3$ for each centred interval.

i) 2.99 to 3.01 ii) 2.9 to 3.1 iii) 2 to 4 iv) 1 to 5

b) What do the results of part a suggest about the instantaneous rate of change of y with respect to x at $x = 3$?

17. Repeat exercise 16, but use a different value of x in part a. Use centred intervals of different lengths. As in exercise 16, you should find that all the average rates of change are equal.

18. **Thinking/Inquiry/Problem Solving** Let P be any point on the graph of the parabola $y = x^2$. Let the x-coordinate of P be x_1. Points A and B are on the graph and have x-coordinates $x_1 - h$ and $x_1 + h$, respectively.

   a) Determine the average rate of change from A to B.

   b) What is the instantaneous rate of change at P? Explain.

19. The TI-83 or TI-83 Plus calculator can estimate the instantaneous rate of change at a point on the graph of a function. It calculates an average rate of change using a centred interval. The first calculation on the screen at the right shows the result for *Example 2,* using an interval from 5.99 to 6.01. The second calculation shows the result using an interval from 5.999 to 6.001. Use a calculator to verify the results shown on the screen. (For information about nDeriv, consult the TI-83 or TI-83 Plus manual.)

C

20. On grid paper, draw an accurate graph of the parabola $y = x^2$ that is large enough to include the points (±5, 25). Draw the chord joining each pair of points: (0, 0) and (2, 4); (–1, 1) and (3, 9); (–2, 4) and (4, 16); (–3, 9) and (5, 25).

   a) Show that all the average rates of change represented by the chords are equal.

   b) Determine the coordinates of the midpoints of the chords. What do you notice about these midpoints?

   c) Show that the midpoint of every chord with slope 2 has x-coordinate 1.

   d) What is the slope of the tangent at (1, 1)? Explain.

21. a) Consider the function $f(x) = 3x^2 + x - 2$. Show that the average rate of change of f between $x = 5$ and $x = 9$ is the mean of the average rate of change of f between $x = 5$ and $x = 7$ and the average rate of change of f between $x = 7$ and $x = 9$.

   b) Prove the following generalization of the result of part a. For any function f, the average rate of change of f over an interval is the mean of the average rate of change of f over the first half of the interval and the average rate of change of f over the second half of the interval.

1.3 The Derivative Function

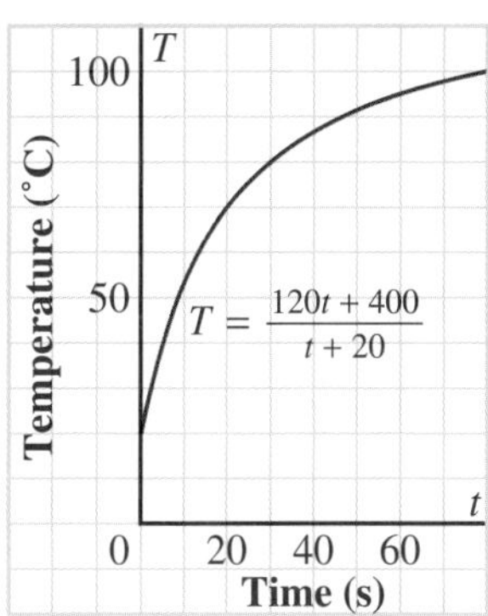

In Sections 1.1 and 1.2, we estimated instantaneous rates of change for different functions. Suppose we do this at many different points on the graph of a function. For example, consider once again the graph of temperature against time when water in a beaker is brought to a boil.

Earlier, we used a 1-s following interval to estimate the instantaneous rate of temperature increase at 35 s. The result was approximately 0.65°C/s (highlighted in the table below). Then, you used 1-s following intervals to estimate the instantaneous rate of temperature increase at 15 s, 40 s, 60 s, and 0 s (exercise 4, page 23). The table includes these results, and corresponding results for other times.

Time (s)	Estimated instantaneous rate of temperature change (°C/s)
0	4.76
5	3.08
10	2.15
15	1.59
20	1.22
25	0.97
30	0.78
35	0.65
40	0.55
45	0.47
50	0.40
55	0.35
60	0.31
65	0.27
70	0.24
75	0.22
80	

We can graph these instantaneous rates of temperature change against time. The result is shown below.

Rate of Change of Water Temperature

This graph was drawn using estimated instantaneous rates of change calculated from 1-s following intervals. It is an approximation to a similar graph that shows the actual instantaneous rates of change. Such a graph illustrates a new function. This function is called the *derivative* of the temperature-time function.

The fact that we can take a given function and create another function, called the derivative, is very significant. On page 8, we stated that we need a new mathematical operation to describe the way quantities change. This operation is finding the derivative of a function. We will be determining and using derivative functions many times in this book.

Take Note

The Derivative of a Function

The *derivative* of a function f is a new function denoted f', defined as follows:

For each x, $f'(x)$ is the instantaneous rate of change of f at x when f' exists.

The two graphs on page 27 are shown below. The second graph has a smaller horizontal scale to facilitate comparing the two functions (the second graph shows the actual instantaneous rates of change). Both graphs show how the temperature is changing, but in a different way.

Temperature-Time Function

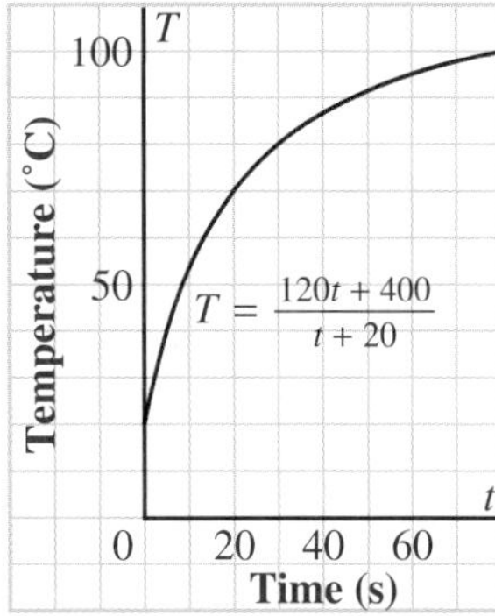

Derivative of Temperature-Time Function

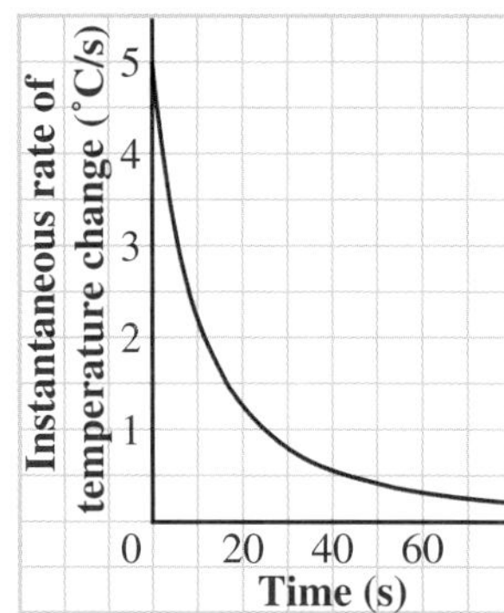

The temperature is increasing.

We can tell this from the temperature graph because the curve goes up to the right.

The temperature is increasing.

All the points lie above the horizontal axis. This means that all the instantaneous rates of change are positive. The temperature is increasing.

The temperature rises more rapidly at first, and less rapidly later on.

We can tell this from the shape of the curve on the temperature graph.

The temperature rises more rapidly at first, and less rapidly later on.

The curve goes down to the right. This means that the instantaneous rates of temperature change are decreasing. The temperature rises less rapidly as time passes.

Example

Digital introduced the VAX 11/750 computer in Europe during the 1980s. The VAX was ideal for the type of data handling physicists and engineers needed. As a result, Digital's sales were good. The table shows the total number of computers sold since this computer was introduced.

Time, t years	Total sales, S (number of computers)
0	0
0.5	160
1.0	320
1.5	560
2.0	1040
2.5	1680
3.0	2720
3.5	4000
4.0	5440
4.5	6400
5.0	7200
5.5	7680
6.0	8000
6.5	8240
7.0	8400

a) Graph total sales as a function of time.

b) For each time, t, estimate the instantaneous rate of increase in total sales.

c) Use the results of part b. Draw a graph that approximates the derivative of total sales with respect to time.

d) Explain how both the total sales graph and the derivative of the total sales graph show how the total sales are changing.

Solution

a) **Total Sales**

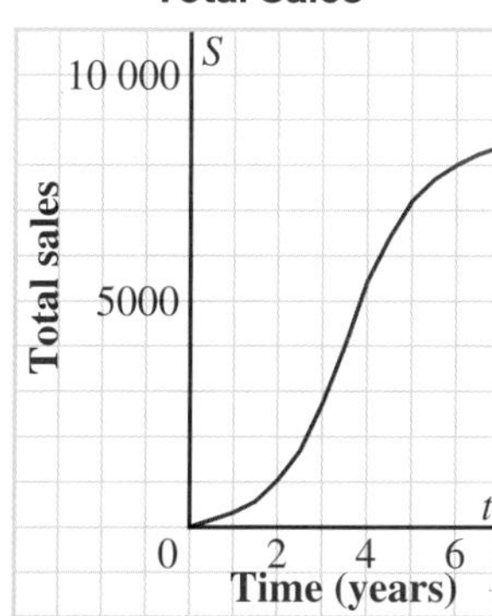

b) To estimate the instantaneous rates of increase in total sales, use 0.5-year following intervals.

For $t = 0$, calculate the average rate of increase from $t = 0$ to $t = 0.5$.

$$\frac{\Delta S}{\Delta t} = \frac{160 - 0}{0.5 - 0}$$
$$= 320$$

At $t = 0$, the rate of increase is about 320 computers/year.

For $t = 2.5$, calculate the average rate of increase from $t = 2.5$ to $t = 3.0$.

$$\frac{\Delta S}{\Delta t} = \frac{2720 - 1680}{3.0 - 2.5}$$
$$= 2080$$

At $t = 2.5$, the rate of increase is about 2080 computers/year.

These results are highlighted in the table on page 30. Repeat this method to determine the instantaneous rate for each half-year. The table shows these results.

c)

Time, *t* years	Estimated rate of increase in sales (computers/year)
0	320
0.5	320
1.0	480
1.5	960
2.0	1280
2.5	2080
3.0	2560
3.5	2880
4.0	1920
4.5	1600
5.0	960
5.5	640
6.0	480
6.5	320
7.0	

Derivative of Total Sales

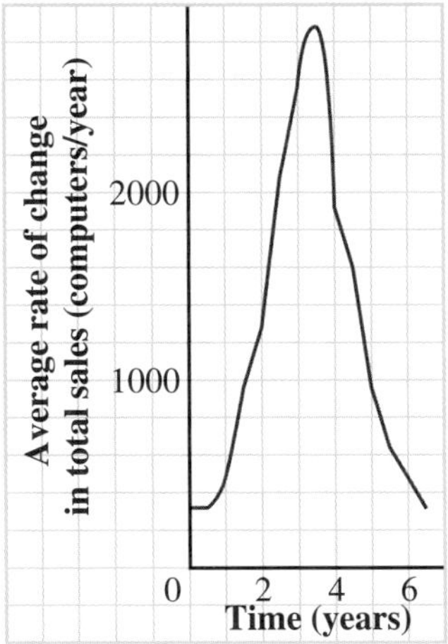

d) The total sales are increasing. The total sales graph shows this because the curve goes up to the right. The derivative of the total sales graph shows this because all points lie above the horizontal axis.

From the total sales graph, the total sales rise rapidly at first, then less rapidly later on. The derivative of the total sales shows this because the curve goes up to the right, then down to the right.

1.3 Exercises

B

1. **Knowledge/Understanding** The graph of mass against age for girls during the first 3 years of life is shown below left. The instantaneous growth rates at several points on the graph were estimated, and these results were used to draw the graph below right.

Growth of Girls

Mass (kg)

14
12
10
8
6
4
2
0

6 12 18 24 30 36

Age (months)

Derivative of Growth of Girls

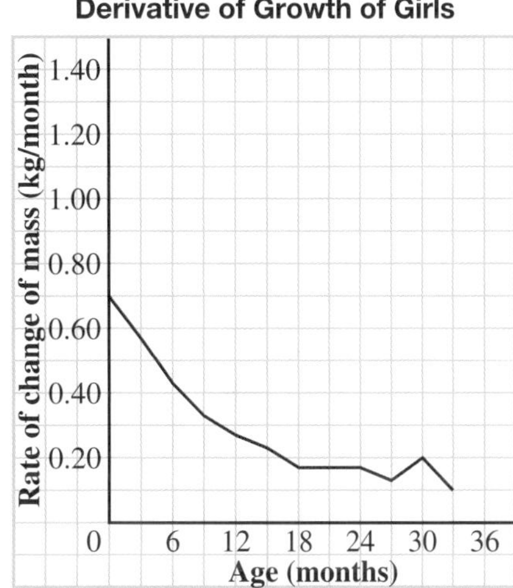

a) Describe how both graphs show how the growth rate changes.

b) Explain why the growth graph is increasing but the growth rate graph is decreasing.

2. **Application** The equation, $h = 0.003\,25(200 - t)^2$, represents the height, h millimetres, of water in a pop bottle as a function of time, t seconds. This equation was used to estimate the instantaneous rate of change of the height of the water at four different times. The results are shown in this table.

Time (s)	Estimated instantaneous rate of change in height (mm/s)
0	−1.30
60	−0.91
120	−0.52
180	−0.13

a) Graph the data in the table.

b) The graph in part a approximates the derivative of the height function. What special feature does the derivative function appear to have?

c) To confirm your answer in part b, choose a different time. Use the equation to estimate the instantaneous rate of change of water height at that time. Plot the point for this time on the derivative graph.

d) Compare the graph of the derivative function with the graph of the height function shown below. Describe how each graph shows how the height of the water is decreasing.

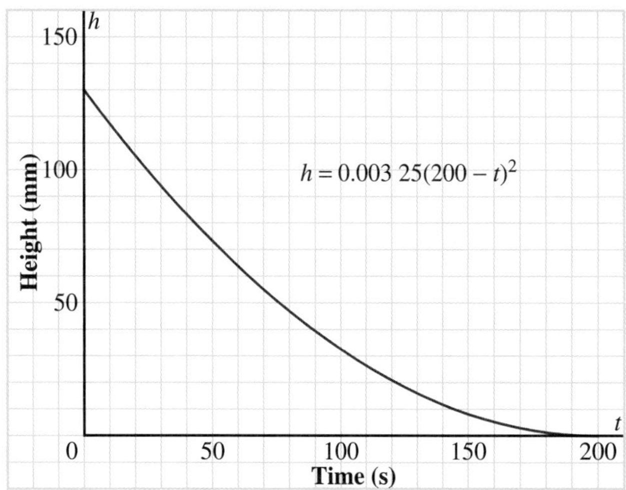

In exercises 2 and 3, the functions $h = 0.003\,25(200 - t)^2$ and $h = -4.89t^2 + 8.09t$ are quadratic functions. You should find that the derivatives of these functions are linear functions. This is true for all quadratic functions. We will show this in Chapter 2.

3. In a physics experiment, a stroboscopic picture of a ball thrown almost straight upward is made. By taking measurements from the picture and using the known time interval between the flashes of the stroboscope, the data in the table were obtained.

a) Graph the height of the ball against time.

b) The equation $h = -4.89t^2 + 8.09t$ expresses the height of the ball as a function of time. Use this equation to estimate the instantaneous rate of change of height for each time. Use 0.01-s following intervals.
   i) 0 s ii) 0.5 s iii) 1.0 s iv) 1.5 s

c) Graph the data from part b to make a graph that approximates the derivative of the height function. What special feature does the derivative function appear to have?

d) To confirm your answer in part c, choose a different time. Use the equation to estimate the instantaneous rate of change of height at that time. Plot the point for this time on the derivative graph.

Time (s)	Height of ball (m)
0.0	0.00
0.1	0.76
0.2	1.43
0.3	1.99
0.4	2.46
0.5	2.83
0.6	3.10
0.7	3.27
0.8	3.35
0.9	3.31
1.0	3.20
1.1	2.99
1.2	2.66
1.3	2.26
1.4	1.74
1.5	1.13
1.6	0.43

e) Compare the graph of the derivative function with the graph of the height function. Describe how each graph shows how the height of the ball is changing.

4. In exercise 2, you should have found that the graph of the derivative function is a straight line.

   a) Determine the equation of this line, and write it in the form $y = mx + b$.

   b) What do m and b represent? Explain.

5. In exercise 3, you should have found that the graph of the derivative function is a straight line.

   a) Determine the equation of this line, and write it in the form $y = mx + b$.

   b) What do m and b represent? Explain.

6. Under certain conditions, a colony of bacteria can double in number about every 20 min. Suppose there are 1000 bacteria in a colony. The population, P, after n hours is given by the formula $P = 1000(2)^{3n}$. The graph of this function is shown. Tangents to the graph are drawn at the times corresponding to 0 h, 1 h, and 2 h.

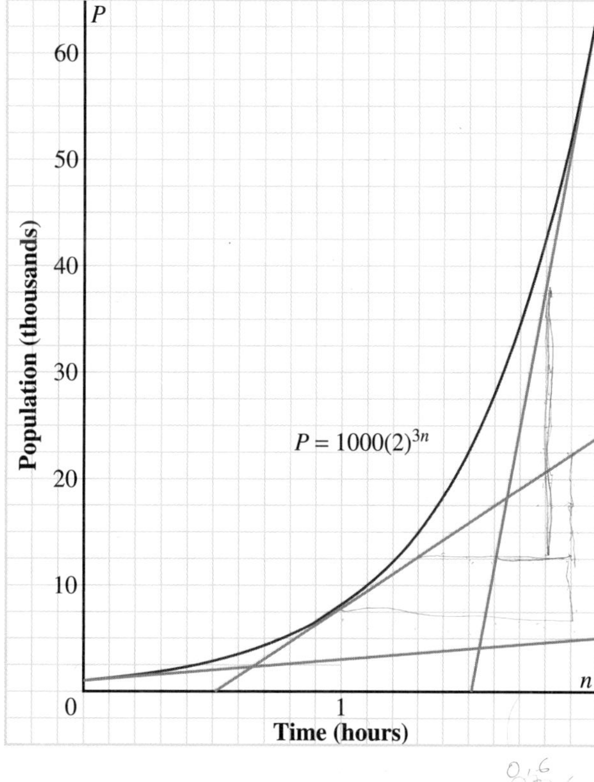

   a) Use the graph. Estimate the instantaneous rate of change in population at each time.
      i) 0 h
      ii) 1 h
      iii) 2 h

   b) Use the equation. Estimate the instantaneous rate of change in population at each time.
      i) 0.5 h   ii) 1.5 h

   c) Use your results from parts a and b to graph the instantaneous rate of change of population as a function of time.

   d) Compare the graph in part c with the graph above. Describe how each graph shows how the population is growing.

7. Coffee, tea, cola, and chocolate contain caffeine. When you consume caffeine, the percent, P, left in your body is expressed as a function of the elapsed time, n hours, by the equation $P = 100(0.87)^n$. The graph of the function is shown. Tangents to the graph are drawn at the times corresponding to 0 h, 12 h, and 24 h after consumption.

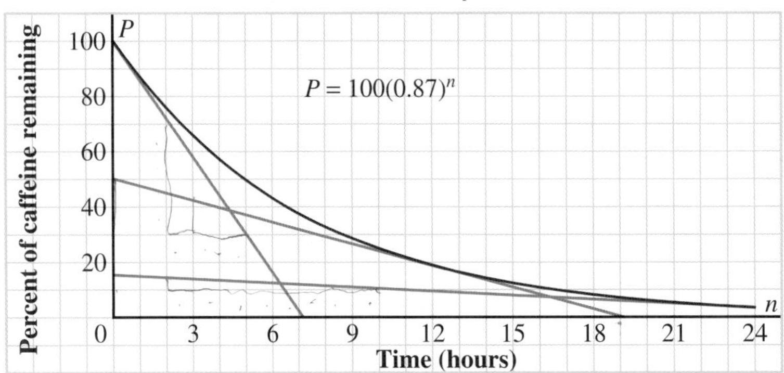

a) Use the graph. Estimate the instantaneous rate of change in percent of caffeine in the body at each time.

i) 0 h ii) 12 h iii) 24 h

b) Use the equation. Estimate the instantaneous rate of change in percent of caffeine in the body at each time.

i) 6 h ii) 18 h

c) Use the results from parts a and b to graph the instantaneous rate of change in percent of caffeine as a function of time.

d) Compare the graph of the derivative function with the graph of the percent function. Describe how each graph shows how the percent of caffeine in the body is changing.

8. In 1.2 Exercises, you used the equation $P = \frac{79\,500}{V}$ to estimate the instantaneous rate of change in pressure for four different volumes of helium. The table at the right includes these results, and corresponding results for other times.

Volume (mL)	Estimated instantaneous rate of change in pressure (kPa/mL)
100	−7.87
200	−1.98
300	−0.88
400	−0.50
500	−0.32
600	−0.22
700	−0.16
800	−0.12

a) Graph these data.

b) The graph in part a approximates the derivative of the pressure function. What special feature does the derivative function appear to have?

c) To confirm your answer in part b, choose a different volume. Use the equation to estimate the instantaneous rate of change of pressure at that volume. Plot the point for volume on the derivative graph.

d) Compare the graph of the derivative function with the graph of the pressure function. Describe how each graph shows how the pressure of the helium is changing.

✓ **9. Communication** A jogger's heart rate increases during the first few minutes of running. It then levels off for the duration of the run. After the run, the heart rate returns to its normal resting rate within half an hour. Sketch a graph of heart rate against time and a graph of the derivative of heart rate against time. Describe how each graph shows the changes in heart rate.

✓ **10.** The height of a tomato plant is recorded every 2 weeks from germination until the tomatoes ripen.

Time, *t* (weeks)	0	2	4	6	8	10	12	14	16	18
Height, *h* (cm)	0	5	10	20	40	60	75	85	90	90

a) Graph height as a function of time.

b) Estimate the instantaneous growth rate at 2-week intervals, starting at $t = 1$ week. Use 2-week centred intervals.

c) Graph the growth rate as a function of time. Draw a smooth curve to represent the trend in growth rate.

d) During which weeks is the tomato plant growing fastest? Describe how this can be seen on the graphs.

e) Explain why the tomato plant grows slowly, then quickly, then slowly again.

✓ **11.** The type of tomato plant in exercise 10 produces ripe tomatoes in 18 weeks. Use the graphs in exercise 10 as a guide. Sketch a graph of height against time and a graph of growth rate against time for each type of tomato plant described below. Justify the graphs you drew.

a) maximum height of 60 cm, 18 weeks to ripen

b) maximum height of 90 cm, 16 weeks to ripen

c) maximum height of 120 cm, 20 weeks to ripen

12. Thinking/Inquiry/Problem Solving A bungee jump is an experience of change: change of distance fallen, change of speed, change of heart rate. An engineer recorded data for the speed of a jumper as a function of distance fallen.

Distance, *d* (m)	0	5	10	15	20	25	30	35	40	45	50
Speed, *v* (m/s)	0.0	10.0	14.1	16.9	18.4	19.0	18.7	17.6	15.4	11.6	0.0

a) Graph speed as a function of distance.

b) Describe the various stages of the motion.

c) Estimate the distance travelled when the bungee cord begins to tighten. Explain your thinking.

d) Where is the jumper moving fastest? Explain.

e) Estimate the instantaneous rate of change of speed with distance at 5-m intervals, starting at 2.5 m.

f) Graph the rate of change of speed with distance against distance.

g) Over which distances is the speed increasing? Explain.

h) Where is the rate of change of speed with distance zero? Explain how this appears on the speed against distance graph. Explain how this relates to the jumper's motion.

13. Gerhard Hertzberg was the first Canadian to be awarded the Nobel Prize for chemistry. His research into atoms and molecules spanned almost 70 years, and he published 255 scientific articles. These data are given below.

Years	1925–1934	1935–1944	1945–1954	1955–1964	1965–1974	1975–1984	1985–1994
Number of papers in this decade, *D*	48	35	40	47	43	30	12

a) Graph *D*, the number of papers each decade, as a function of time.

b) Calculate *T*, the total number of papers published to date, starting in 1925 and increasing in 10-year increments. Graph the total number of papers published as a function of time.

c) You now have the graphs of two functions, *D* and *T*. These two functions are closely related.

i) You obtained *T* from *D* by summing entries. How can you obtain *D* from *T*?

ii) Which function is almost the rate of change of the other? Explain.

d) Hertzberg's productivity was approximately constant throughout much of his career. Explain how you can recognize this in each case.

i) from the table

ii) from the graph of the function *D*

iii) from the graph of the function *T*

C

14. Refer to the graph of rate of change of water temperature on page 27. This graph was drawn using estimates of instantaneous rates of change of temperature calculated from 1-s following intervals. It is an approximation of the graph of the actual derivative function. How does the graph of the actual derivative function compare with this graph? Explain.

Self-Check 1.1 – 1.3

1. Here is the average airfare for a one-way domestic trip from Ottawa for the years 1988 to 1997.

Year	1988	1989	1990	1991	1992
Air fare ($)	132.50	152.90	164.90	172.60	173.90

Year	1993	1994	1995	1996	1997
Air fare ($)	184.60	202.40	199.00	186.50	179.50

a) Graph the average airfare as a function of time.

b) Calculate the average rate of change of price for each 1-year interval from 1988 to 1997.

c) Over what years are the fares increasing? Is the increase constant? Explain.

2. The Van de Graaff generator at Science North can be charged to several thousand volts to make your hair stand on end! The electric field, E newtons per coulomb, is a function of distance, d metres, from the generator. The equation is $E = \frac{6000}{d^2}$.

a) Graph the function $E = \frac{6000}{d^2}$.

b) Calculate the average rate of change of electric field as a function of distance for each distance interval.

i) 0.5 m to 4.5 m **ii)** 1.0 m to 5.0 m

iii) 1.5 m to 5.5 m **iv)** 2.0 m to 7.0 m

c) Estimate the instantaneous rate of change of electric field at 2.5 m from the generator.

3. The function $y = x^2 + 3$ is given. Estimate the instantaneous rate of change of y with respect to x at each value of x.

a) 1.0 **b)** –2.0 **c)** 2.0

PERFORMANCE ASSESSMENT

4. Use the equation in exercise 2.

a) Estimate the instantaneous rate of change of electric field at 0.75 m. Repeat this calculation at 0.50-m increments up to 4.25 m.

b) Graph the instantaneous rate of change of electric field against distance.

c) Describe the derivative of the electric field function.

d) Compare the graph in part b with the graph in exercise 2a. Describe how each graph shows how the electric field is changing.

1.4 The Derivatives of Some Basic Functions

In Section 1.3, we defined the derivative of a function f as a new function f', which represents the instantaneous rate of change of f at each point on its graph where f' exists. The derivative function also represents the slope of the tangent at each point on the graph of the function.

The simplest function is $y = x$. The rate of change of y with respect to x is the slope of the line, which is 1.

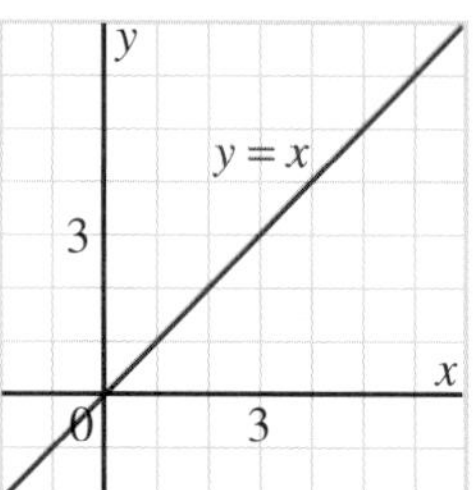

The derivative of the function $y = x$ is $y' = 1$. Note that the derivative of y is denoted y'.

In the following investigation, you will determine the derivatives of the functions $y = x^2$ and $y = x^3$.

Investigation

The Derivatives of $y = x^2$ and $y = x^3$

Work with 3 other people. Two people use the function $y = x^2$. The other two people use the function $y = x^3$.

1. Make a table of values with these headings. Use integer values of x from –3 to 3 in the first column. Use the equation of the function to complete the second column.

x	*y*	slope of the tangent at *x*

2. A graph of the function is shown on the next page. Tangents are drawn to the graph at A, B, C, and D, where $x = 0$, 1, 2, and 3, respectively. Another point is marked on each tangent, A_1, B_1, C_1, and D_1, respectively.

a) Calculate the slope of each tangent. Record the slopes in the third column of the table.

b) Use the symmetry of the graph to determine the slopes of the tangents at $x = -1$, -2, and -3. Record the slopes in the table.

3. a) The slopes of the tangents form a pattern. Describe the pattern.

b) Predict the slopes of the tangents at the points where $x = \pm 5$ and $x = \pm 10$.

c) Suppose you know the x-coordinate of a point on the graph of the function. How can you calculate the slope of the tangent at that point?

4. a) On the same grid, graph the function and its derivative.

   b) Compare the two graphs. Explain how the graph of the derivative provides information about how the values of y are changing on the graph of the function as x increases.

   c) Write the equation of the derivative function.

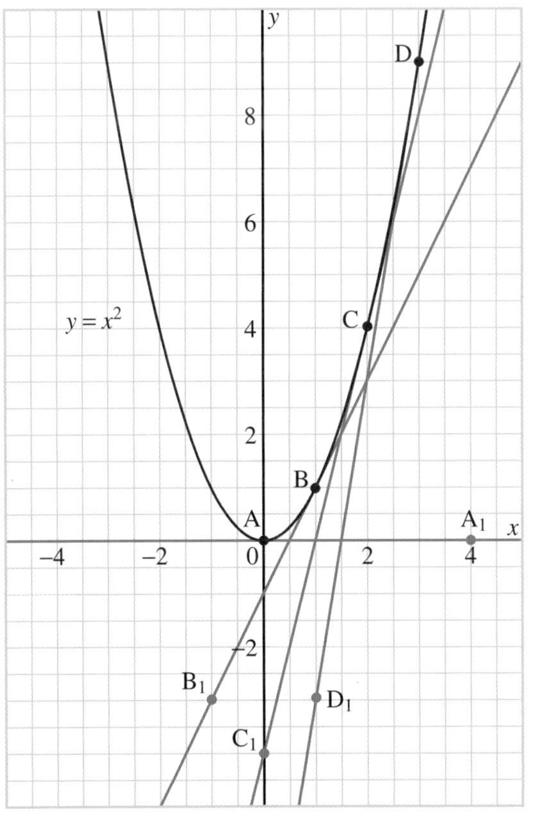

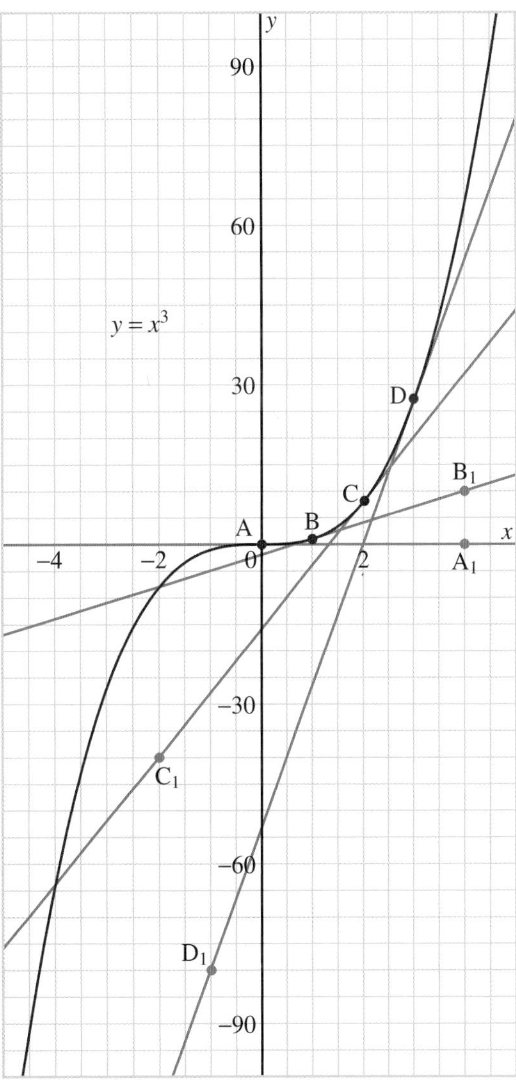

In the *Investigation*, you should have conjectured that:

- The derivative of the function $y = x^2$ is $y' = 2x$.
- The derivative of the function $y = x^3$ is $y' = 3x^2$.

Take Note

The Derivatives of $y = x^2$ and $y = x^3$

The derivative of $y = x^2$ is $y' = 2x$.

This means that the instantaneous rate of change of y with respect to x is $2x$.

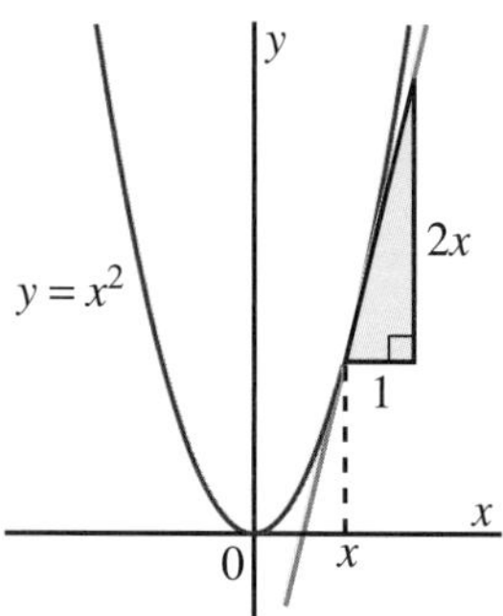

Visualize a point moving from left to right along the graph of $y = x^2$. The slope of the tangent is always double the x-coordinate of the point.

The derivative of $y = x^3$ is $y' = 3x^2$.

This means that the instantaneous rate of change of y with respect to x is $3x^2$.

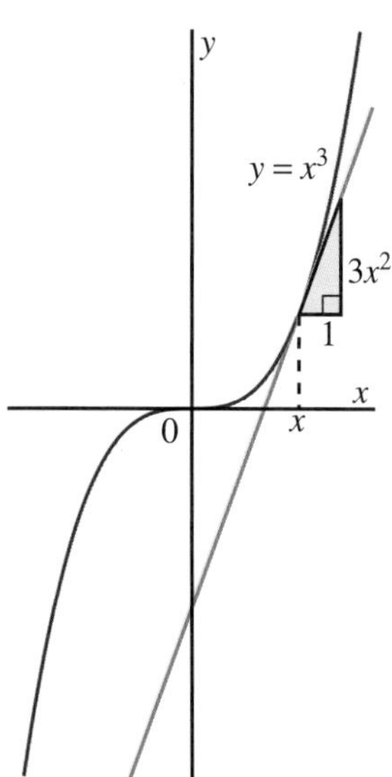

Visualize a point moving from left to right along the graph of $y = x^3$. The slope of the tangent is always 3 times the square of the x-coordinate of the point.

We can use the derivative of a function to determine the instantaneous rate of change of y with respect to x at any point on the graph of the function.

Example 1

Consider the function $y = x^2$.

a) At each value of x below, what is the instantaneous rate of change of y with respect to x?

i) $x = 3$ **ii)** $x = -2$

b) Interpret each result in part a graphically.

Solution

a) The derivative of $y = x^2$ is $y' = 2x$.

i) Substitute $x = 3$ in $y' = 2x$ to obtain $y' = 2(3) = 6$.
When $x = 3$, y is increasing 6 times as fast as x is increasing.

ii) Substitute $x = -2$ in $y' = 2x$ to obtain $y' = 2(-2) = -4$.
When $x = -2$, y is decreasing 4 times as fast as x is increasing.

b) On the graph of $y = x^2$:

i) when $x = 3$, the slope of the tangent is 6

ii) when $x = -2$, the slope of the tangent is –4

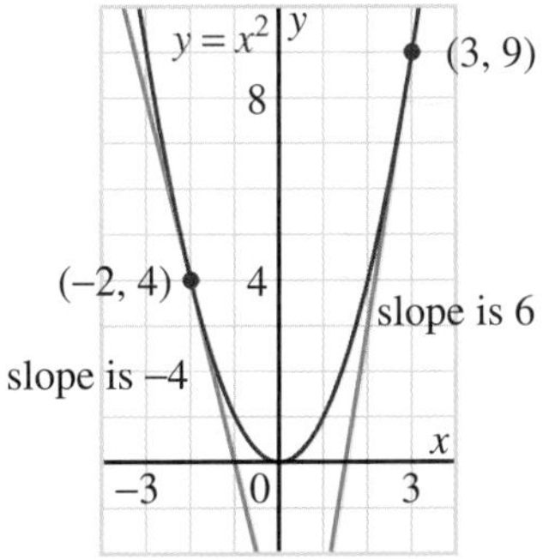

We can use the derivative of a function to determine the equation of the tangent at a point on the graph of the function.

Example 2

a) Determine the equation of the tangent to the graph of $y = x^3$ at the point (2, 8) on the graph.

b) Illustrate the result of part a on a graph.

Solution

a) The derivative of $y = x^3$ is $y' = 3x^2$. This represents the slope of the tangent at a point on the graph.

Substitute $x = 2$ in $y' = 3x^2$ to obtain $y' = 3(4) = 12$.
The slope of the tangent at (2, 8) is 12.

The equation of the tangent has the form $y = m(x - p) + q$.
Substitute $m = 12$, $p = 2$, and $q = 8$.
The equation of the tangent is $y = 12(x - 2) + 8$, or $y = 12x - 16$.

b)

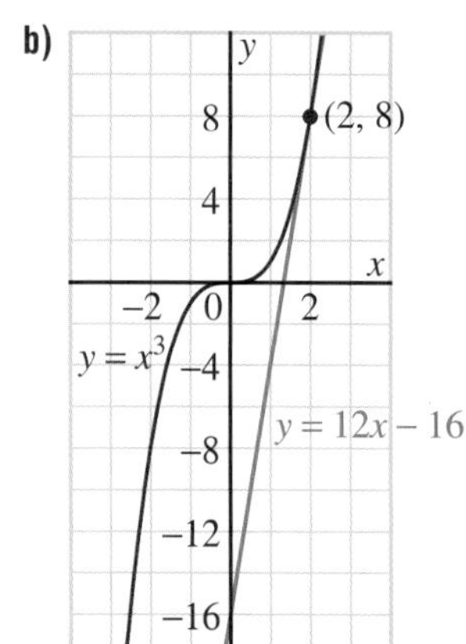

The Derivative of a Linear Function

The graph of a linear function, such as $y = -0.5x + 3$, is a straight line.
The rate of change of y with respect to x is the slope of the line, which is –0.5.
The derivative of the function $y = -0.5x + 3$ is $y' = -0.5$.

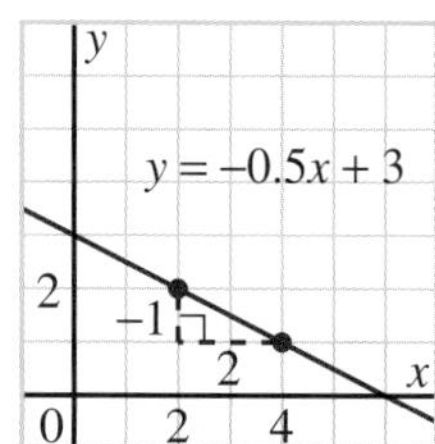

Take Note

The Derivative of y = mx + b

The derivative of $y = mx + b$ is $y' = m$.

Visualize a point moving from left to right along the graph of $y = mx + b$. The instantaneous rate of change of y with respect to x is always m.

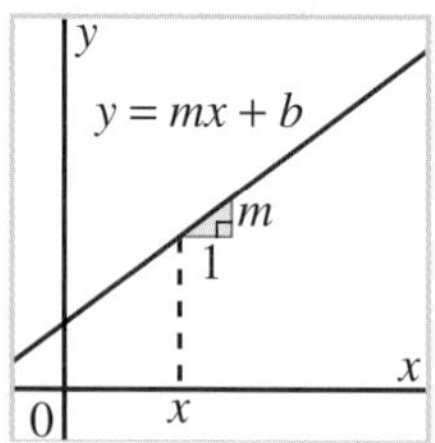

There are two important rules that will be used many times in this book. You will verify these rules in exercises 13 to 17.

- ***Vertical Translation Rule***
  When the graph of a function is translated vertically, its derivative is not affected.
  The derivative of $y = f(x) + c$ is $y = f'(x)$, where c is any constant.

- ***Vertical Stretch Rule***
  When the graph of a function is expanded or compressed vertically, the graph of the derivative is also expanded or compressed vertically.
  The derivative of $y = cf(x)$ is $y = cf'(x)$, where c is any constant.

Another example of a simple relationship between a function and its derivative occurs with trigonometric functions. In exercises 20 and 21, you will verify the following results. We shall use them in this book, but we shall not prove them algebraically.

The derivative of $y = \sin x$ is $y' = \cos x$.

The derivative of $y = \cos x$ is $y' = -\sin x$.

THE BIG PICTURE

What Is Calculus?

This chapter is an introduction to calculus. Calculus is a tool that mathematicians, scientists, and engineers use to describe how related quantities change. To use calculus, we need to know the derivatives of the functions we are working with. Hence, an important part of calculus deals with developing methods for finding the equation of the derivative when the equation of the function is given. These methods must be more efficient than the graphical method used in the *Investigation*, page 38. In the rest of this book, we will determine and use the derivatives of many functions.

There is often a simple relationship between the equation of a function and the equation of its derivative. In this section, you conjectured that the derivative of $y = x^2$ is $y' = 2x$, and the derivative of $y = x^3$ is $y' = 3x^2$. These results can be proved algebraically, and they are part of a general pattern. In Chapter 2, we will show that the derivative of $y = x^n$ is $y' = nx^{n-1}$. This result applies for all values of n.

1.4 Exercises

A

1. What is the derivative of each function?

a) $y = 3x + 1$ **b)** $y = -0.5x + 2$ **c)** $y = 4 - x$

d) $y = x$ **e)** $y = x^2$ **f)** $y = x^3$

2. Suppose the graph of a function f and its derivative f' are given. For each description of the graph of f', describe what this tells you about the graph of f. Explain.

a) The graph of f' is a horizontal line above the x-axis.

b) The graph of f' is a horizontal line below the x-axis.

B

3. The function $y = -2x + 3$ is given.

a) Determine the derivative of the function.

b) Graph the function and its derivative on the same grid.

c) Write the equation of another function that has the same derivative.

4. Repeat exercise 3 for each function.

a) $y = 0.5x - 1$ b) $y = 4 - x$ c) $y = 5$

5. **Knowledge/Understanding** Consider the function $y = x^2$.

a) At each value of x below, what is the instantaneous rate of change of y with respect to x?

i) $x = 4$ ii) $x = -5$ iii) $x = 0.5$ iv) $x = 0$

b) Interpret each result in part a graphically.

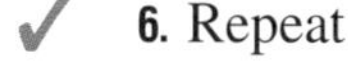

6. Repeat exercise 5 for each function.

a) $y = x^3$ b) $y = 2x + 1$

7. In exercise 5, you determined the derivative of $y = x^2$ at $x = 4$. When you know the derivative at $x = 4$, at what other point do you also know the derivative? Explain.

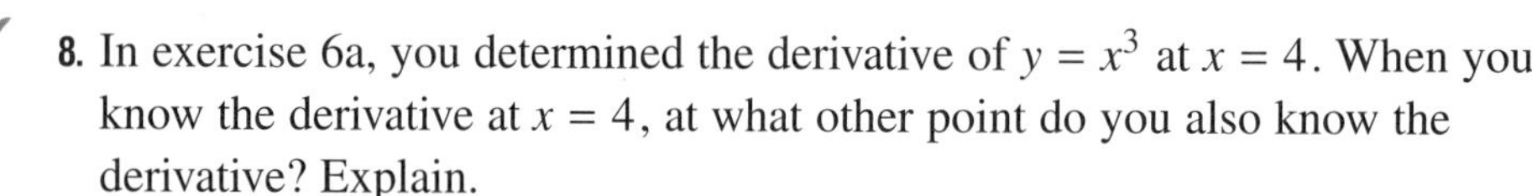

8. In exercise 6a, you determined the derivative of $y = x^3$ at $x = 4$. When you know the derivative at $x = 4$, at what other point do you also know the derivative? Explain.

9. a) Determine the equation of the tangent to the graph of $y = x^2$ at each point on the graph.

i) (1, 1) ii) (2, 4) iii) (3, 9)

b) Illustrate the results of part a on a graph.

10. a) Determine the equation of the tangent to the graph of $y = x^3$ at each point on the graph.

i) (1, 1) ii) (–2, –8) iii) (0, 0)

b) Illustrate the results of part a on a graph.

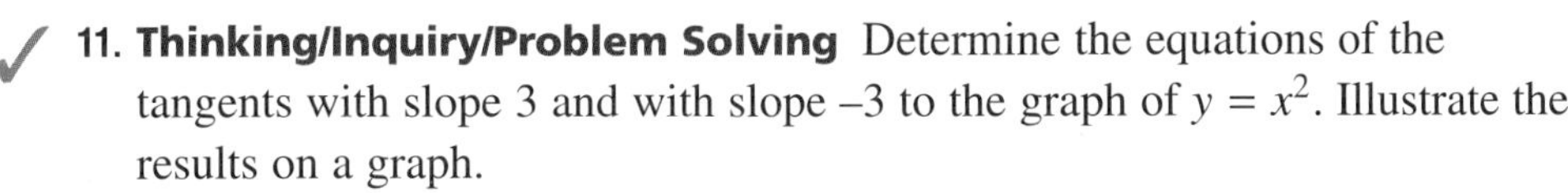

11. Thinking/Inquiry/Problem Solving Determine the equations of the tangents with slope 3 and with slope –3 to the graph of $y = x^2$. Illustrate the results on a graph.

12. There are two tangents with slope $\frac{3}{4}$ to the graph of $y = x^3$. Determine their equations and illustrate the results on a graph.

13. The graphs of three related functions and their derivatives are shown. Write the equations of the three derivatives. Explain.

a) the graph of $y = x^2$ and its derivative

b) the graph of $y = x^2$ is translated 2 units up

c) the graph of $y = x^2$ is translated 2 units down

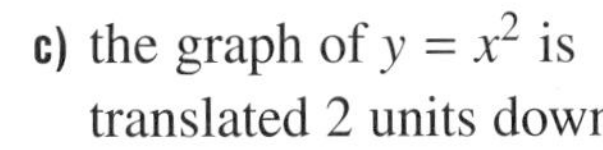

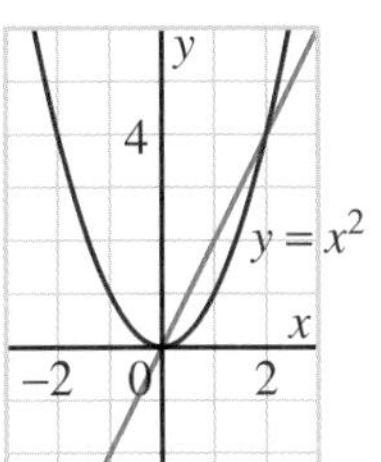

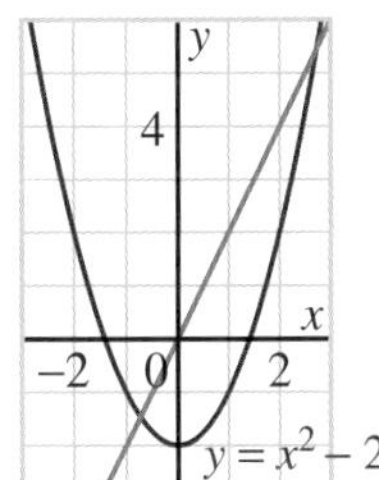

14. Use the results in exercise 13. Write the equation of the derivative of each function.

a) $y = x^2 + 1$ **b)** $y = x^2 + 5$ **c)** $y = x^2 - 3$

d) $y = x^3$ **e)** $y = x^3 - 2$ **f)** $y = x^3 + 4$

15. Communication Explain why translating the graph of a function vertically does not affect its derivative.

16. Application The graphs of three related functions and their derivatives are shown. Write the equations of the three derivatives. Explain.

a) the graph of $y = x^2$ and its derivative

b) the graph of $y = x^2$ is expanded vertically by a factor of 2

c) the graph of $y = x^2$ is compressed vertically by a factor of $\frac{1}{2}$

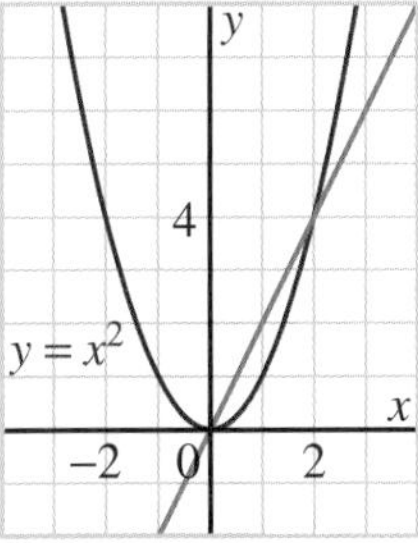

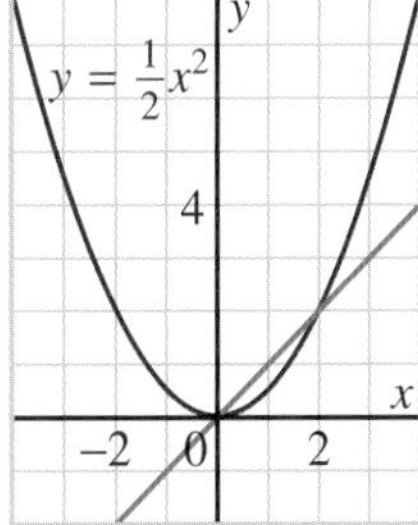

17. Use the relationships you discovered in exercise 16. Write the derivative of each function. Explain.

a) $y = 4x^2$ **b)** $y = 0.25x^2$ **c)** $y = -2x^2$ **d)** $y = 2x^3$

18. The graph shows the tangent at the point $B\left(2, \frac{1}{2}\right)$ on the graph of $y = \frac{1}{x}$. The slope of this tangent is $-\frac{1}{4}$.

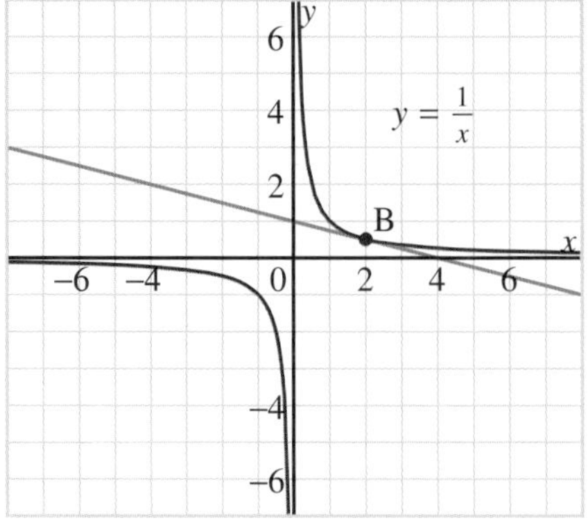

a) Use the symmetries of the graph to determine the slopes of the tangents at three other points on the graph.

b) In Chapter 2, we will prove that the derivative of $y = \frac{1}{x}$ is $y' = -\frac{1}{x^2}$. Use this result to verify your answers in part a.

19. In Chapter 2, we shall prove that the derivative of

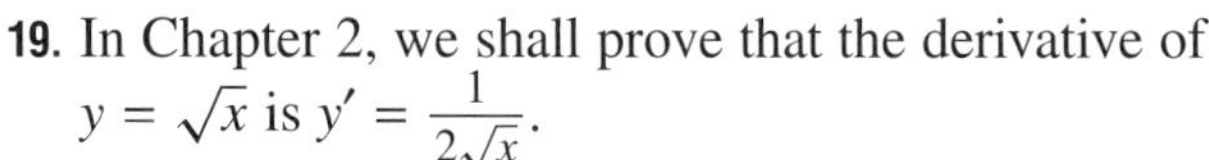

$y = \sqrt{x}$ is $y' = \frac{1}{2\sqrt{x}}$.

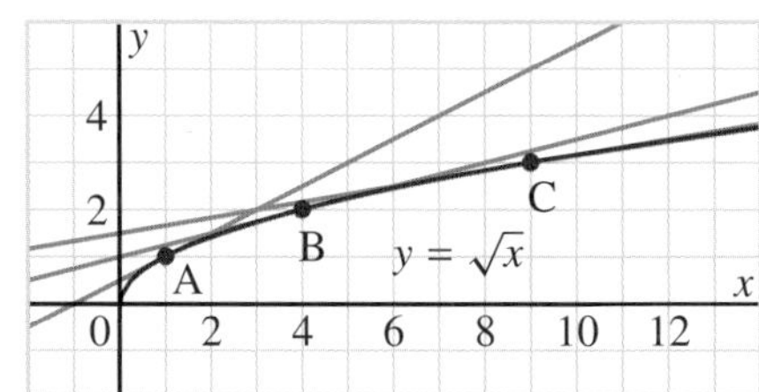

a) Use this result to calculate the slopes of the tangents to the graph of $y = \sqrt{x}$ at the points A(1, 1), B(4, 2), and C(9, 3).

b) These tangents are shown on the graph. Verify your results in part a by using the graph to determine the slopes of the tangents.

20. Although we shall not do so in this book, it can be

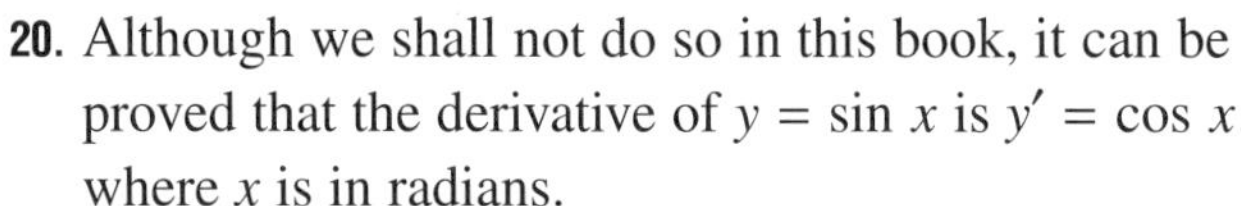

proved that the derivative of $y = \sin x$ is $y' = \cos x$, where x is in radians.

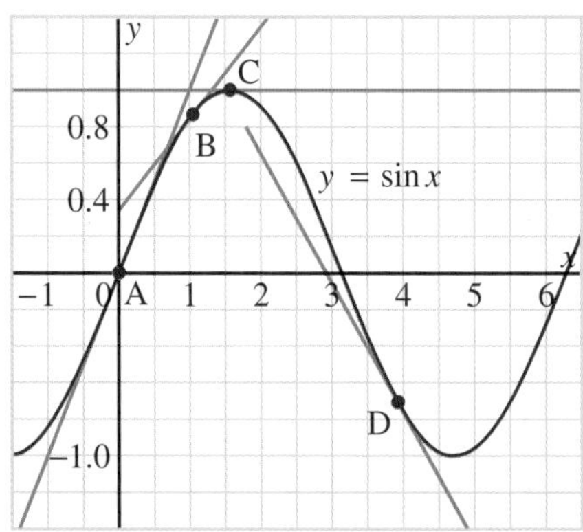

a) Use this result to calculate the slopes of the tangents to the graph of $y = \sin x$ at the points $A(0, 0)$, $B\left(\frac{\pi}{3}, \frac{\sqrt{3}}{2}\right)$, $C\left(\frac{\pi}{2}, 1\right)$, and $D\left(\frac{5\pi}{4}, -\frac{1}{\sqrt{2}}\right)$.

b) These tangents are shown on the graph. Verify your results in part a by using the graph to determine the slopes of the tangents.

21. It can be proved that the derivative of $y = \cos x$

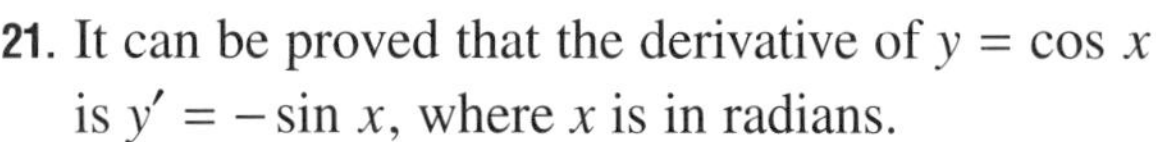

is $y' = -\sin x$, where x is in radians.

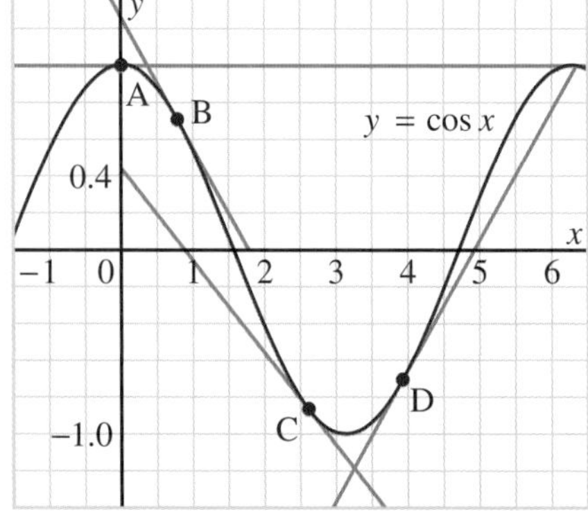

a) Use this result to calculate the slopes of the tangents to the graph of $y = \cos x$ at the points $A(0, 1)$, $B\left(\frac{\pi}{4}, \frac{1}{\sqrt{2}}\right)$, $C\left(\frac{5\pi}{6}, -\frac{\sqrt{3}}{2}\right)$, and $D\left(\frac{5\pi}{4}, -\frac{1}{\sqrt{2}}\right)$.

b) These tangents are shown on the graph. Verify your results in part a by using the graph to determine the slopes of the tangents.

22. On the TI-83 or TI-83 Plus calculator, follow these steps to determine the equation of the tangent at a point on the graph of a function. The calculator uses the formula in the screen in exercise 19, page 26, to estimate the slope of the tangent.

- Graph any function in an appropriate window.
- With the graph on the screen, press 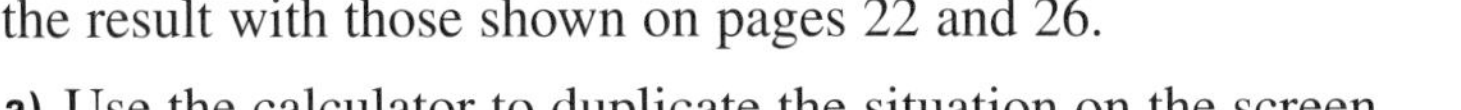 2nd PRGM for DRAW. Then press **5** to select Tangent(.
- The equation of the function appears, with a flashing cursor at a point on the graph. Press the x-coordinate of the point where you want the tangent.
- Press ENTER. The calculator draws the tangent and displays its equation.

The screen shows the result for *Example 2,* page 22. Compare the result with those shown on pages 22 and 26.

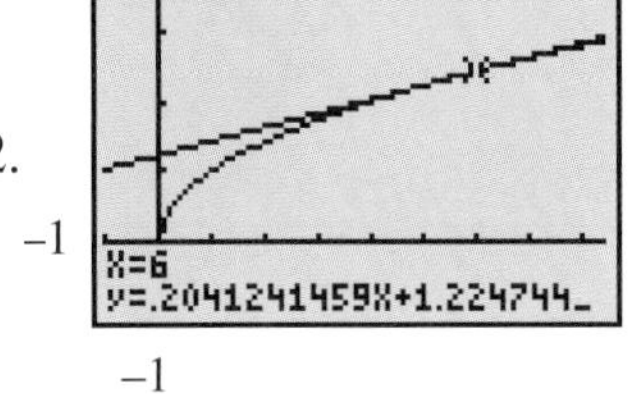

a) Use the calculator to duplicate the situation on the screen.

b) Use the calculator to verify the result in *Example 2,* page 42.

c) Use the calculator to determine equations of tangents to points on the graphs of other functions.

For more information about the Tangent command, consult the TI-83 or TI-83 Plus manual.

23. a) The function $y = x^4$ is given. Use the Tangent command to determine the slopes of the tangents at $x = -3, -2, -1, 0, 1, 2,$ and 3.

b) Look for a pattern in the results in part a. What is the derivative of the function $y = x^4$?

c) Compare the derivative of $y = x^4$ with the derivatives of $y = x^2$ and $y = x^3$. Predict the derivative of $y = x^5$. Use the Tangent command to check your prediction.

24. The graphs of three functions that involve absolute value are given. Sketch the graph of the derivative of each function.

a)

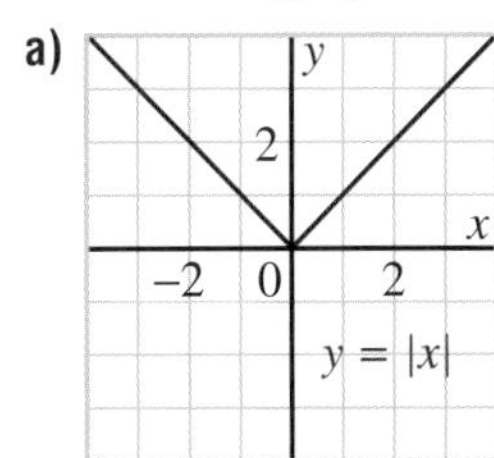

b)

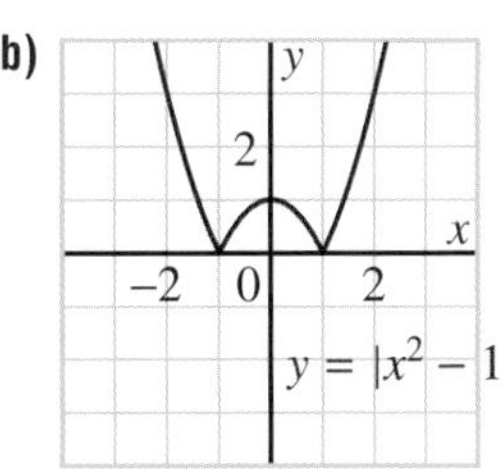

c)

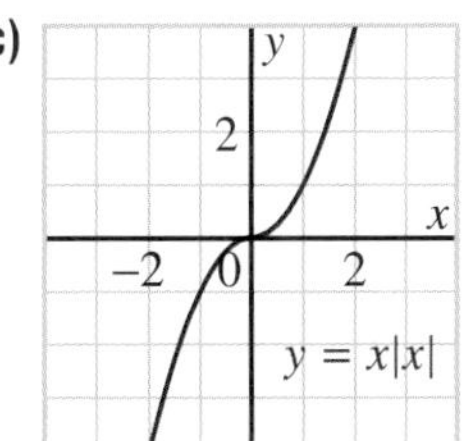

25. a) Determine a formula for the equation of the tangent with slope m to the graph of $y = x^2$.

b) Determine a formula for the equation of the tangent at the point (x_1, y_1) on the graph of $y = x^2$.

26. Repeat exercise 25 for the graph of $y = x^3$.

1.5 Some Features of the Derivative

Derivatives are a new way of thinking about functions. When we determine the derivative of a function, we *differentiate* the function. The derivative of a function represents the slopes of the tangents at the points on its graph. Frequently, the slopes of the tangents change gradually as the point moves along the graph, as the following examples show.

The derivative of $y = x^2$ is $y' = 2x$. The slopes of the tangents increase steadily, and their values are given by the equation $y' = 2x$. The function $y = x^2$ is *differentiable* for all values of x.

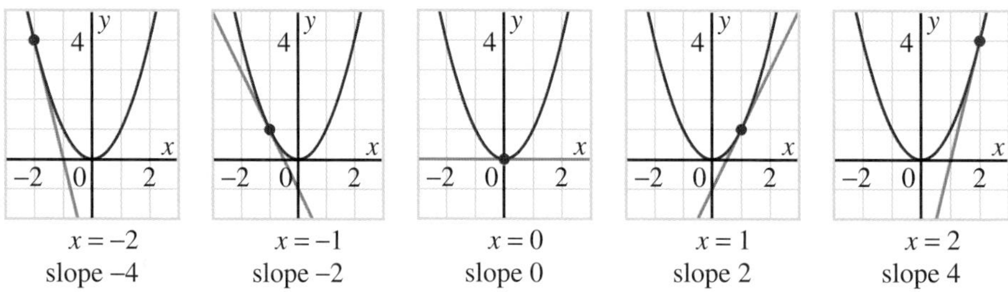

The derivative of $y = x^3$ is $y' = 3x^2$. The slopes of the tangents decrease to 0, then increase, and their values are given by the equation $y' = 3x^2$. The function $y = x^3$ is differentiable for all values of x.

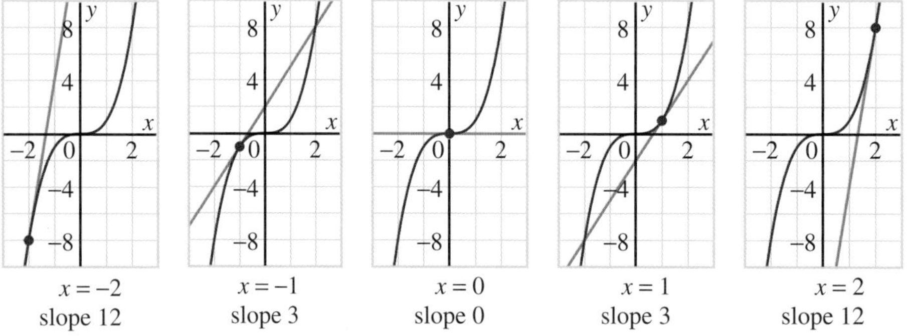

For some functions, there may be points on the graph where the tangent does not exist. Before we consider an example of this kind of function, we need to define what we mean by a tangent to a curve. The word *tangent* comes from the Latin verb “tangere,” which means to touch. We think of a tangent as a line that touches a curve, as in the examples above.

However, not all lines that touch a curve are tangents. All three lines touch this curve at A, but none of them is a tangent.

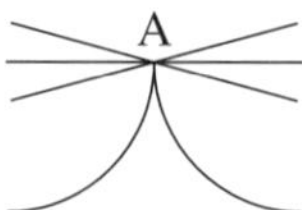

A tangent at a point on a curve is defined as follows. Let P be a point on the curve. Let Q be another point on the curve, on either side of P. Construct the secants PQ. If the secants approach the same line, as Q approaches P from either side, this line is called the tangent at P.

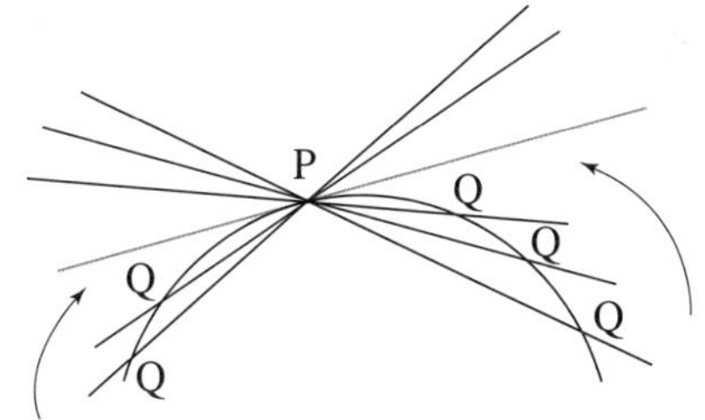

The diagram at the right shows the graph of a function, f, where the tangent does not exist at P(3, 4). Suppose Q approaches P from the left. Secant PQ has a positive slope, and gets closer and closer to a line through P with slope 4. Suppose Q approaches P from the right. Secant PQ has a negative slope, and gets closer and closer to a line with slope –4. Since these lines are different, the graph of f has no tangent at P. Hence, f' is not defined at P.

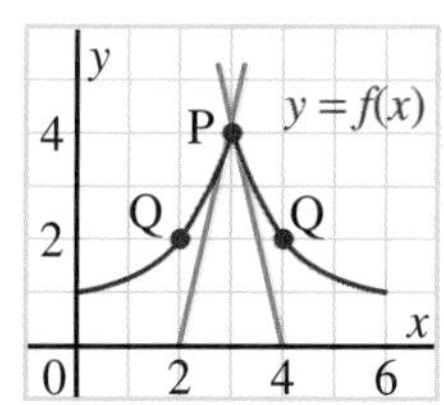

The diagrams below show a few tangents to the graph of f. As x increases from 0 to 3, the slope of the tangent increases from 0 to 4. When $x = 3$, the slope changes abruptly from 4 to –4. This means that the derivative is not defined when $x = 3$. As x increases from 3 to 6, the slope of the tangent increases from –4 to 0.

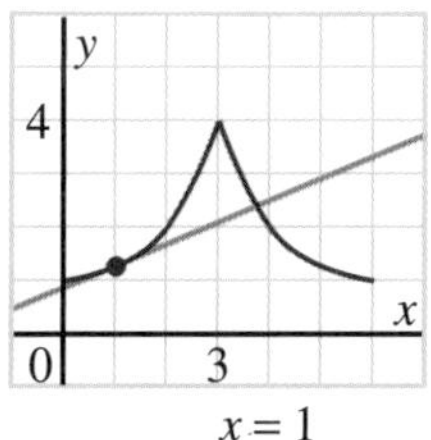
$x = 1$
slope 0.48

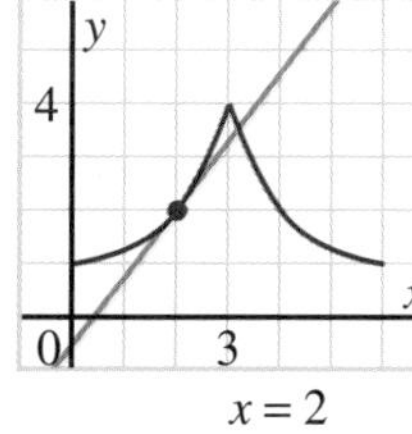
$x = 2$
slope 1.15

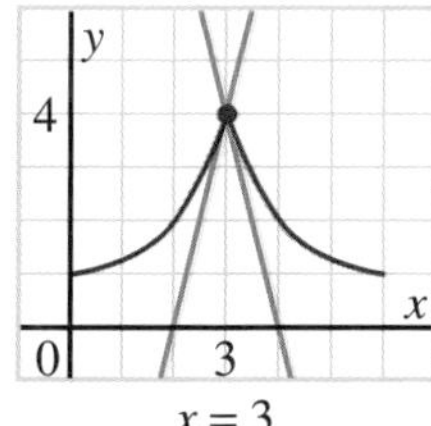
$x = 3$
derivative not defined

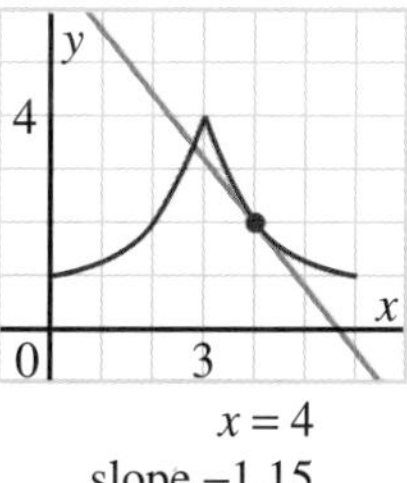
$x = 4$
slope –1.15

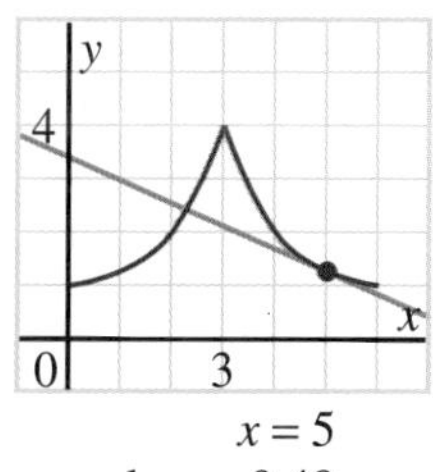
$x = 5$
slope –0.48

The graph of f' is shown at the right. Although the function is defined at $x = 3$, its derivative is not defined at $x = 3$. We say that f is *not differentiable* at $x = 3$.

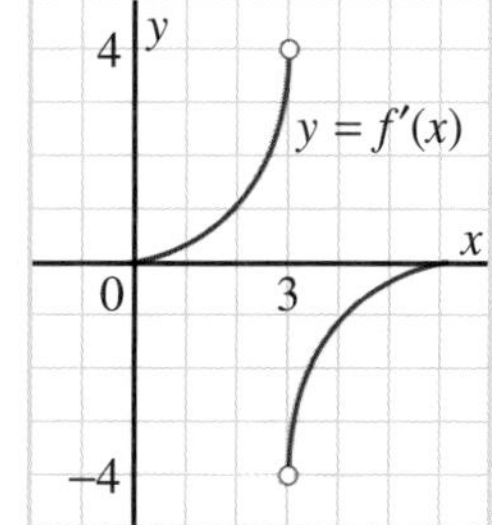

Example

The graph of the function $y = -|x - 2| + 3$ is shown at the right.

a) Describe the graph.

b) Describe the derivative function.

c) Sketch the graph of the derivative function.

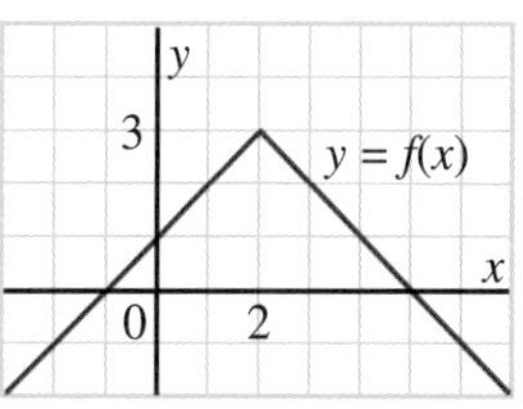

Solution

a) The graph consists of parts of two straight lines that meet at (2, 3).

b) When $x < 2$, the slope of the graph is 1, so the values of the derivative are 1.
When $x > 2$, the slope of the graph is –1, so the values of the derivative are –1.
When $x = 2$, the graph of the function has no tangent. Hence, the derivative function is not defined when $x = 2$.

c) The graph of the derivative has two parts. When $x < 2$, it consists of the horizontal line $y = 1$. When $x > 2$, it consists of the horizontal line $y = -1$. When $x = 2$, the slope changes abruptly from 1 to –1. So, the derivative is not defined, and there is no point on the derivative graph at $x = 2$. We indicate this with open circles at the points where $x = 2$.

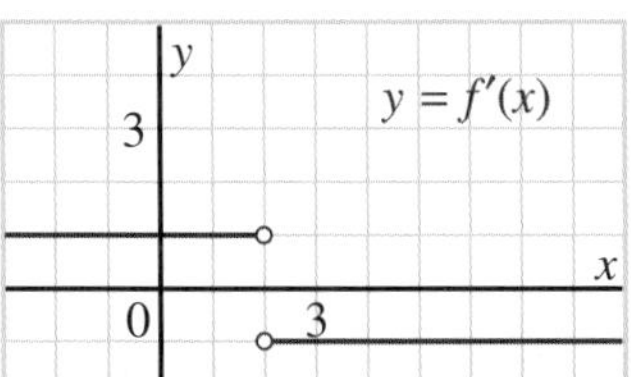

Here is another way to see why the function in the *Example* is not differentiable at $x = 2$. Its graph is pointed at (2, 3). Suppose we graph this function on a graphing calculator. If we zoom in on the point (2, 3) several times, there is no change in the graph. It is still pointed, no matter how many times we zoom in.

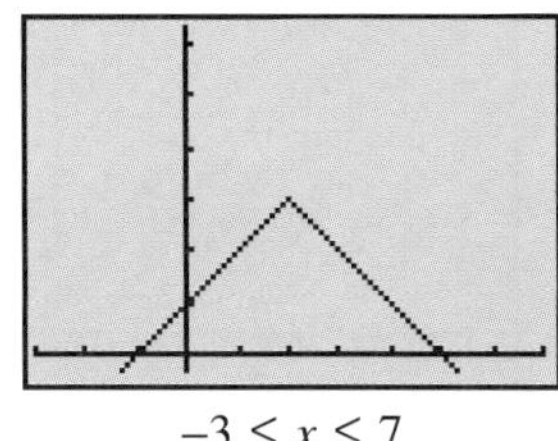

$-3 \le x \le 7$

1st zoom

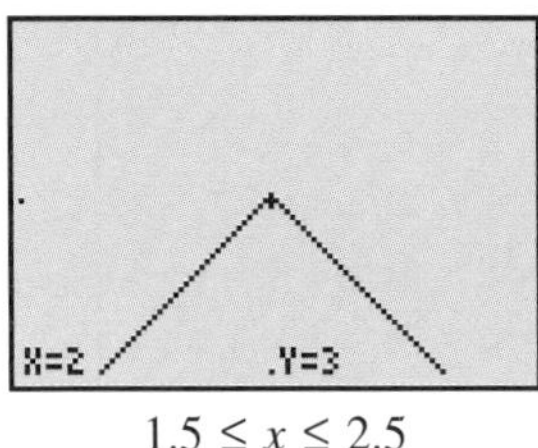

$1.5 \le x \le 2.5$

2nd zoom

X=2 Y=3

$1.95 \le x \le 2.05$

3rd zoom

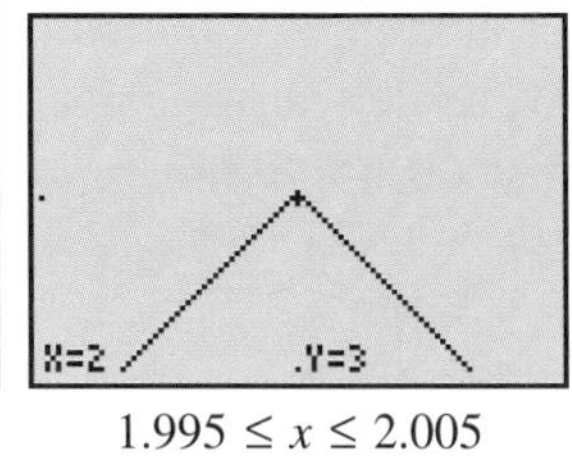

$1.995 \le x \le 2.005$

4th zoom

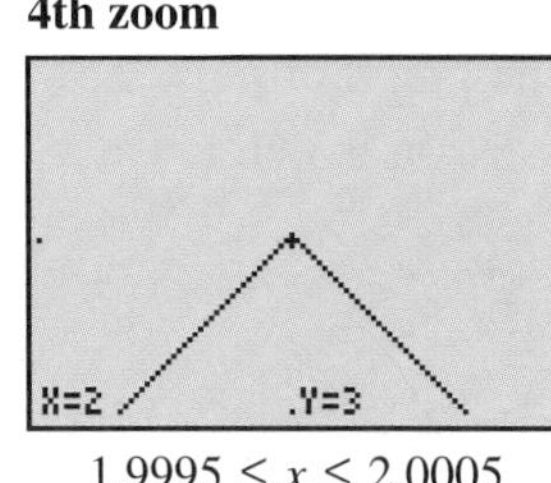

$1.9995 \le x \le 2.0005$

The situation is different for a differentiable function. This screen shows the graph of $y = -\sqrt{(x-2)^2 + 0.0001} + 3.01$, which looks identical to the graph of the function in the *Example*. If we zoom in on the point (2, 3) several times, the graph appears to change and it eventually becomes indistinguishable from a line with slope 0 (see the screens on the next page). This function is differentiable at $x = 2$, and the derivative is 0.

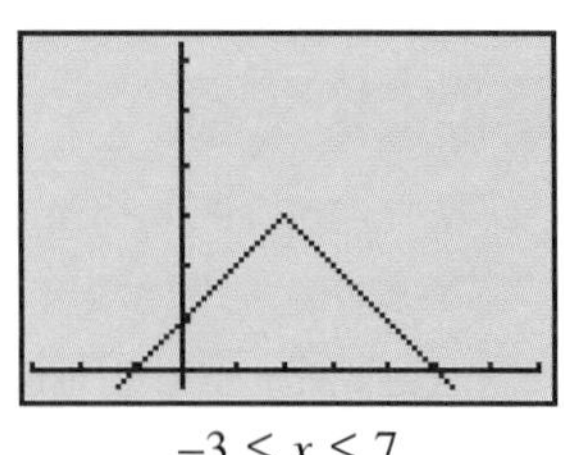

$-3 \le x \le 7$

1st zoom

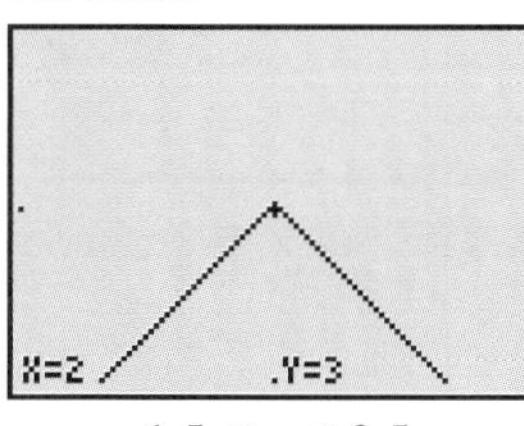

$1.5 \le x \le 2.5$

2nd zoom

X=2 .Y=3

$1.95 \le x \le 2.05$

3rd zoom

X=2 .Y=3

$1.995 \le x \le 2.005$

4th zoom

X=2 .Y=3

$1.9995 \le x \le 2.0005$

1.5 Exercises

B

1. **Knowledge/Understanding** The graphs of two functions are given. Sketch the graph of each derivative.

a)

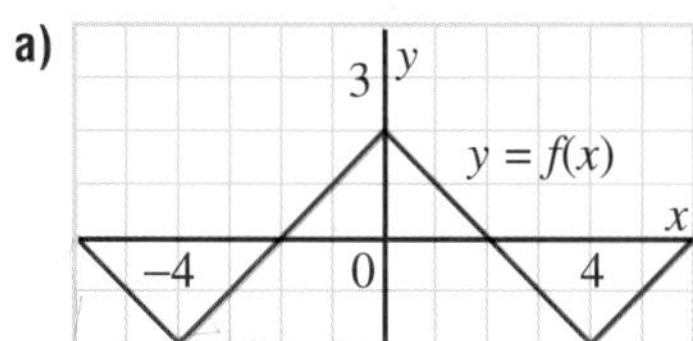

b)

3
y
y = f(x)
x
−4
0
4

2. The graphs of three functions are given. Sketch the graph of each derivative.

a)

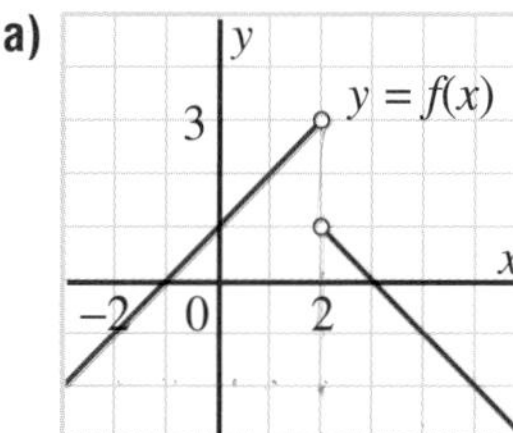

b)

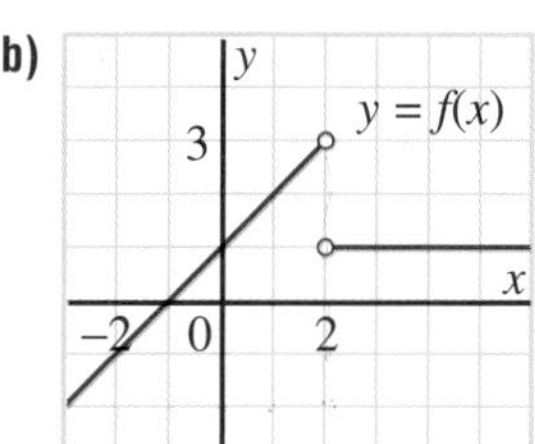

c)

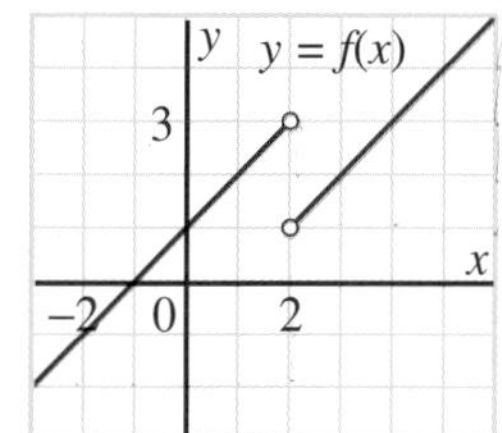

3. Sketch a diagram to illustrate your answer to the following question. In the definition of a tangent, does it matter how far Q is from P at the beginning?

4. Use a diagram to explain your answer to each question.
   a) Could a line be a tangent to a curve at two different points on the curve?
   b) Could a line be a tangent to a curve at a point and cross the curve at that point?
   c) Could a line be a tangent to a curve at a point and intersect the curve at a different point?

5. The graph of the absolute-value function $y = |x|$ is given.

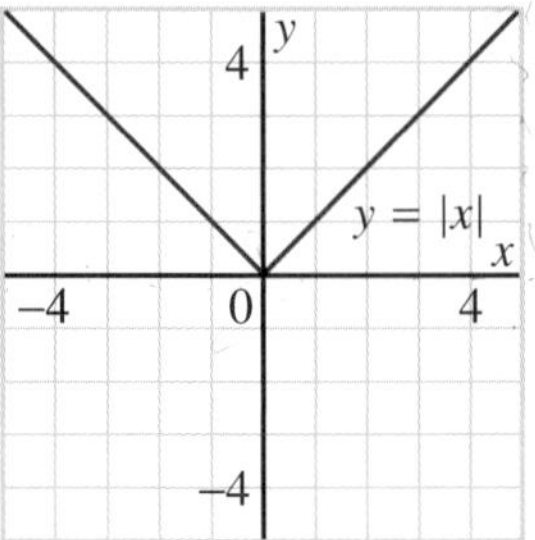

   a) Sketch the graph of the derivative.

   b) The two parts of the graph of the absolute-value function are called *rays*. Suppose the ray on the right rotates about the origin (without crossing the y-axis). Describe the effect on the graph of the derivative.

   c) Suppose the ray on the left rotates about the origin (without crossing the y-axis). Describe the effect on the graph of the derivative.

   d) In parts b and c, why is it necessary that the rays not cross the y-axis?

6. **Application** The function $y = \text{ipart}(x)$ is defined so that $\text{ipart}(x)$ represents the greatest integer less than or equal to x. We say that $\text{ipart}(x)$ is the *integral part* of x. For example, $\text{ipart}(2.75) = 2$, and $\text{ipart}(-2.75) = -3$

   a) Determine each value.

      i) $\text{ipart}(3.15)$ ii) $\text{ipart}(5)$ iii) $\text{ipart}(-4)$ iv) $\text{ipart}(-2.15)$

   b) Explain why the graph below left represents $y = \text{ipart}(x)$.

   c) Sketch the graph of the derivative of $y = \text{ipart}(x)$.

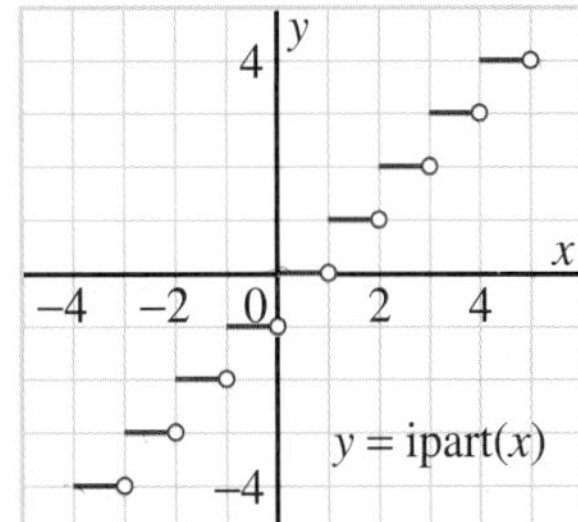

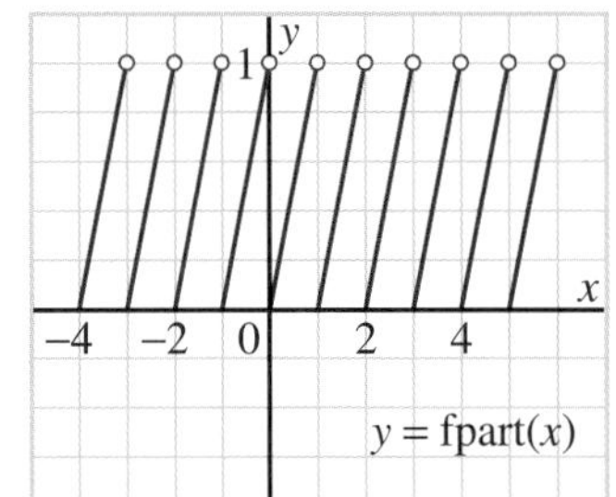

7. The function $y = \text{fpart}(x)$ is defined as follows: $\text{fpart}(x) = x - \text{ipart}(x)$
   We say that $\text{fpart}(x)$ is the *fractional part* of x.
   For example, $\text{fpart}(2.75) = 2.75 - \text{ipart}(2.75) = 2.75 - 2 = 0.75$

   a) Determine each value.

      i) $\text{fpart}(3.75)$ ii) $\text{fpart}(5)$ iii) $\text{fpart}(-4)$ iv) $\text{fpart}(-2.15)$

   b) Explain why the graph above right represents $y = \text{fpart}(x)$.

   c) Sketch the graph of the derivative of $y = \text{fpart}(x)$.

8. **Thinking/Inquiry/Problem Solving** Graphs of the function $y = -\sqrt{(x-2)^2 + 0.0001} + 3.01$ are shown at the bottom of page 50 and on page 51.

   a) Describe how the graph of this function differs from the graph of the function in the *Example*.

   b) Use the graph in the solution of the *Example* as a guide. Sketch a graph of the derivative of the function $y = -\sqrt{(x-2)^2 + 0.0001} + 3.01$.

✓ **9.** The graphs of three functions are shown. The graphs of their derivatives are also shown, but not in the same order as the functions. For each function, identify the graph of its derivative. Explain.

Three Functions

a)

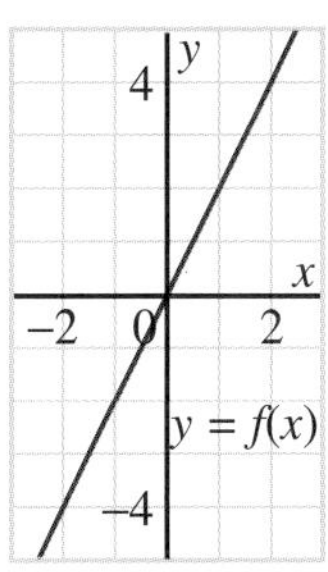

b)

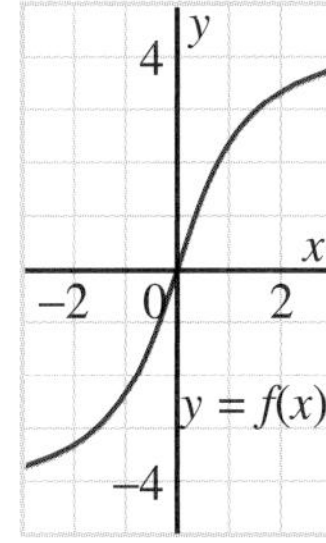

c)

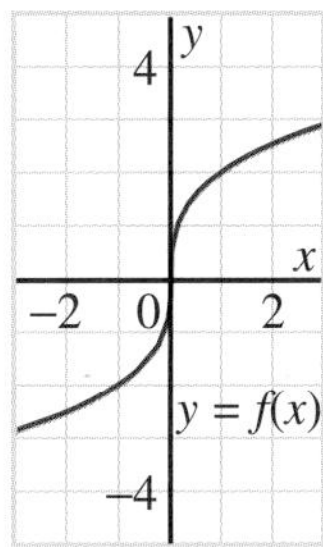

Three Derivatives

i)

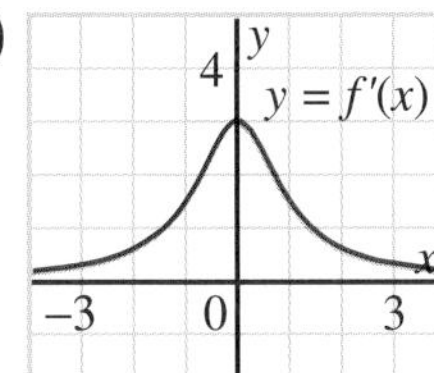

ii)

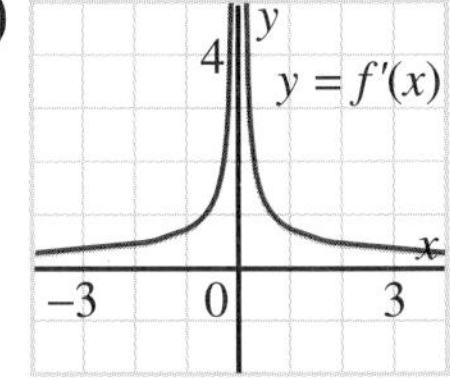

iii)

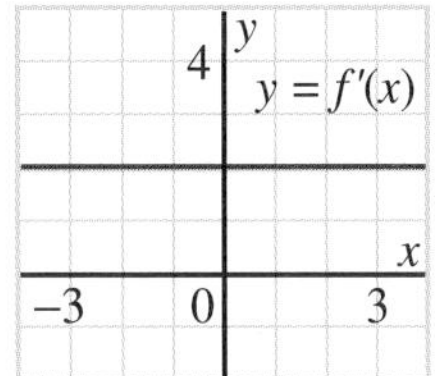

✓ **10. Communication** Refer to the three functions in exercise 9. One of these functions is not differentiable at the origin. Which function is it? Explain why this function is not differentiable at the origin.

In previous examples and exercises, functions that were not differentiable at particular points had either sharp corners or gaps at these points. Exercise 10 is significant because it provides an example of a function with a smooth graph that has a point where it is not differentiable.

11. The graph of the function $y = g(x)$, below left, consists of two quarter circles with radius 3 and centres (0, 4) and (6, 4).

a) Compare this graph with the graph of the function $y = f(x)$ at the top of page 49. In what ways are they similar? In what ways are they different?

b) Sketch the graph of the derivative function, $y = g'(x)$.

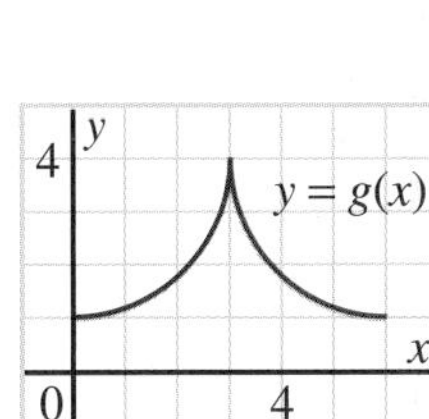

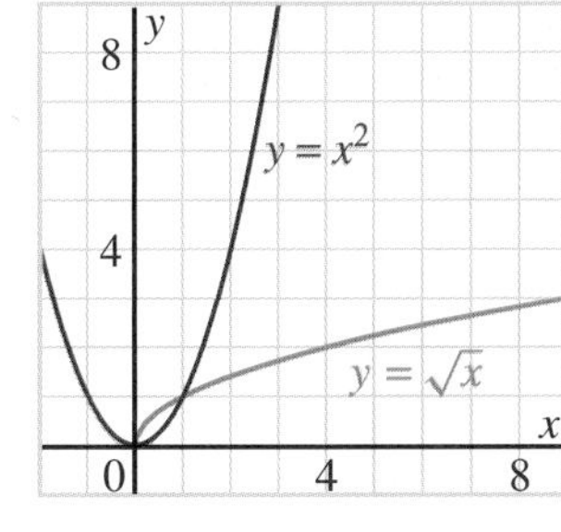

12. The graph of $y = x^2$, above right, is a parabola that opens up. The graph of $y = \sqrt{x}$ is part of a congruent parabola that opens to the right. The derivative of $y = x^2$ is $y' = 2x$. Use this information to determine the derivative of $y = \sqrt{x}$.

1.6 Relating Graphs of Functions and Their Derivatives

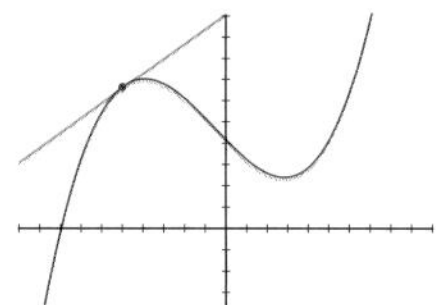

You can use a shareware program called *Winplot* to graph a function, mark a point on the graph, and construct a tangent at that point. Then you can move the point along the graph. As it moves, the tangent moves with it. The diagrams at the right show the screen for the function $y = f(x)$, when $x = -7, -5, -3, -1, 1, 3$, and 5.

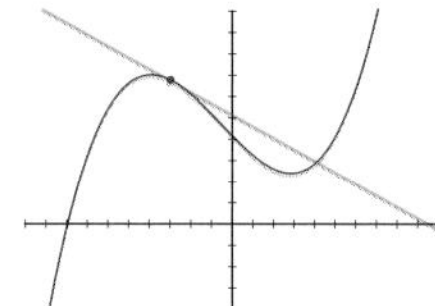

The program can graph the derivative, either with or without the original function. The graph below shows both functions.

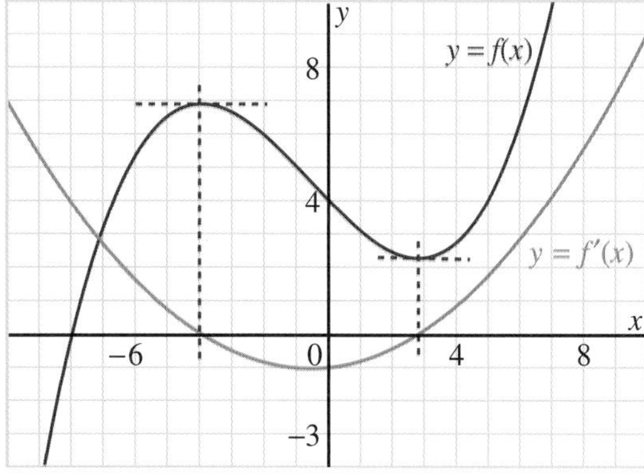

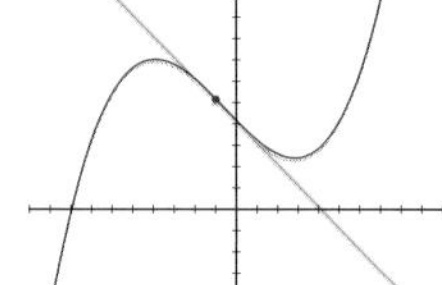

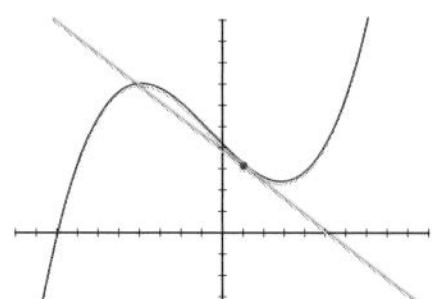

Notice how the graphs are related.

At the maximum and minimum points of $y = f(x)$, the tangents are horizontal, so the value of $f'(x)$ is 0.

When $y = f(x)$ is increasing, the slopes of the tangents are positive, so the values of $f'(x)$ are positive.

When $y = f(x)$ is decreasing, the slopes of the tangents are negative, so the values of $f'(x)$ are negative.

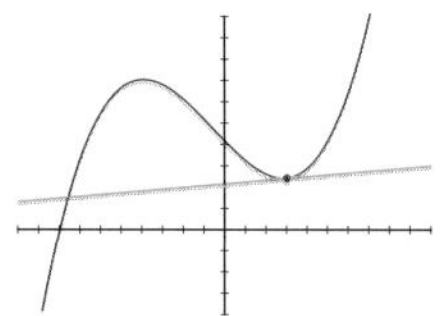

The derivative of a function $y = f(x)$ provides useful information about how the values of y are changing with respect to the values of x. Much of the rest of this book involves the relationship between the graph of a function $y = f(x)$ and the graph of its derivative $y = f'(x)$.

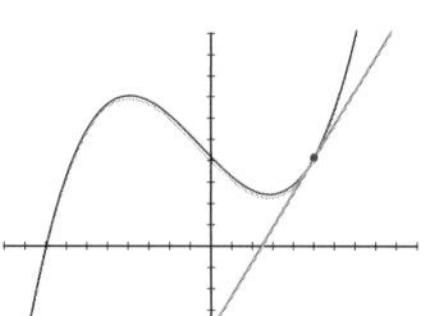

Example 1

A function f and its derivative f' are graphed on the same grid. Which graph is the function and which graph is its derivative? Explain.

a)

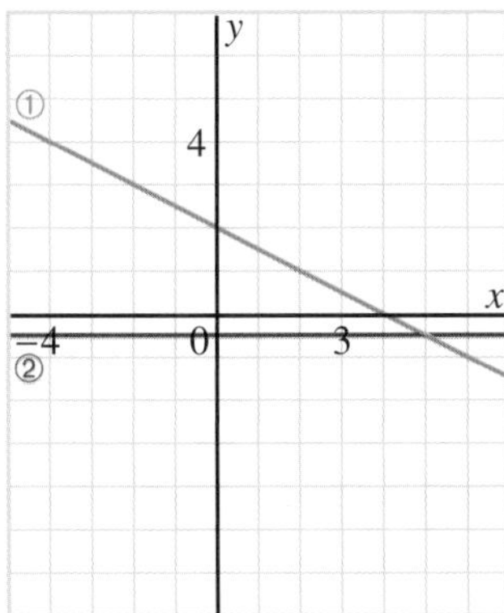

b)

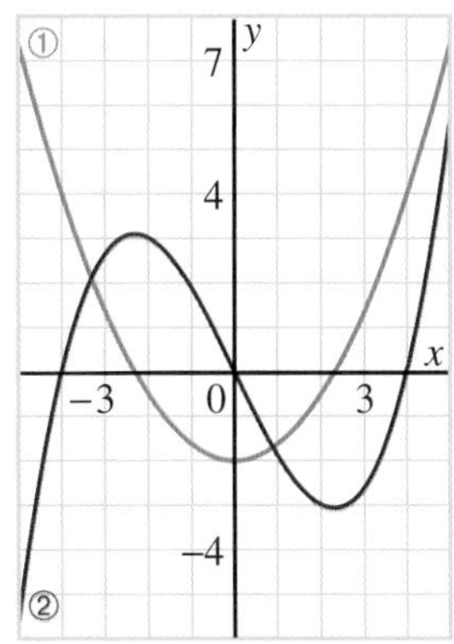

Solution

a) Graph ① is an oblique line with slope –0.5. It has a constant rate of change, –0.5, that is represented by graph ②.
Therefore, graph ① is the function and graph ② is its derivative.

b) Look at the points where each graph intersects the x-axis. Graph ① has two such points, A and B, and graph ② has horizontal tangents at the corresponding values of x.

This suggests that graph ② is the function and graph ① is the derivative.

Graph ② intersects the x-axis at three points, C, D, and E. If this graph were the derivative, then graph ① would have horizontal tangents at the corresponding values of x. This occurs at D, but not at C or E. Therefore, graph ② cannot be the derivative.

Graph ② is the function and graph ① is its derivative.

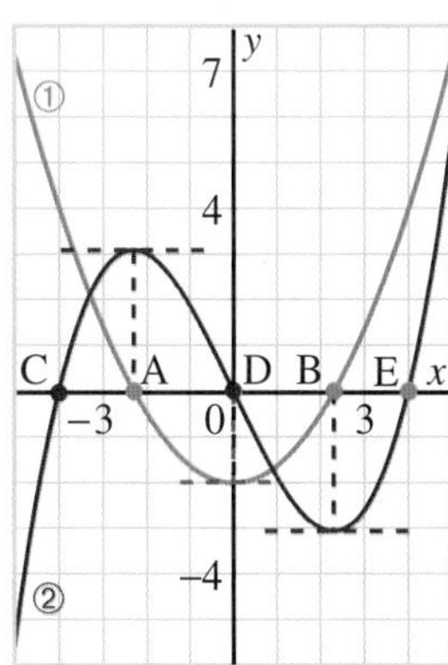

Example 2

A graph of a function f is given. Sketch the graph of its derivative f'.

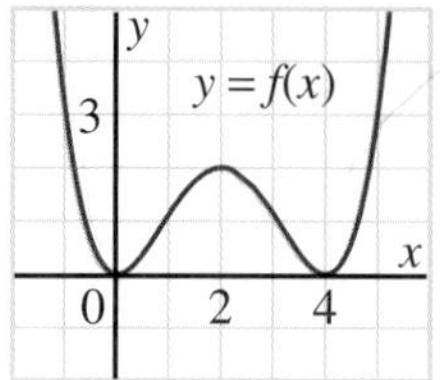

Solution

Draw x- and y-axes. The graph of f has horizontal tangents when $x = 0$, $x = 2$, and $x = 4$. Mark points on the x-axis at these values of x. These are the points A, B, and C, respectively, on the graph of f'.

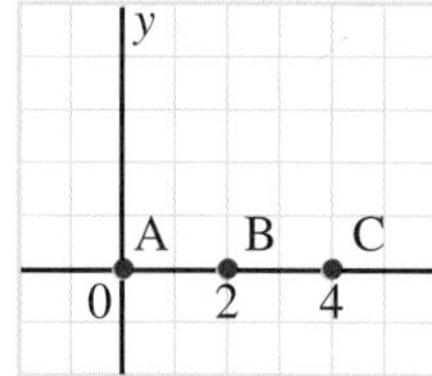

Discuss

Why do we not insert numbers on the *y*-axis?

Visualize x increasing in each of the following cases, and refer to the given graph.

Case 1. $x < 0$

The y-values are decreasing. The tangents have negative slopes, but the slopes are increasing until they reach 0. So, the values of $f'(x)$ are negative but increasing to 0. On the graph of f', draw a curve below the x-axis, up to A.

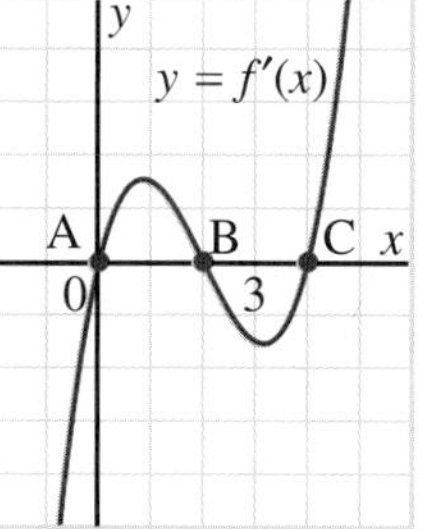

Case 2. $0 < x < 2$

The y-values are increasing. The tangents have positive slopes, increasing at first, then decreasing to 0. So, the values of $f'(x)$ are positive; increasing at first, then decreasing to 0. On the graph of f', draw a curve up from A, then down to B.

Case 3. $2 < x < 4$

The y-values are decreasing. The tangents have negative slopes, decreasing at first, then increasing to 0. So, the values of $f'(x)$ are negative; decreasing at first, then increasing to 0. On the graph of f', continue the curve down from B, then up to C.

Case 4. $x > 4$

The y-values are increasing. The tangents have positive slopes, which are increasing. So, the values of $f'(x)$ are positive and increasing. On the graph of f', draw a curve above the x-axis, from C.

1.6 Exercises

A

1. A function f and its derivative f' are graphed on the same grid. Which graph is the function f and which graph is the derivative f'? Explain how you can tell.

a) 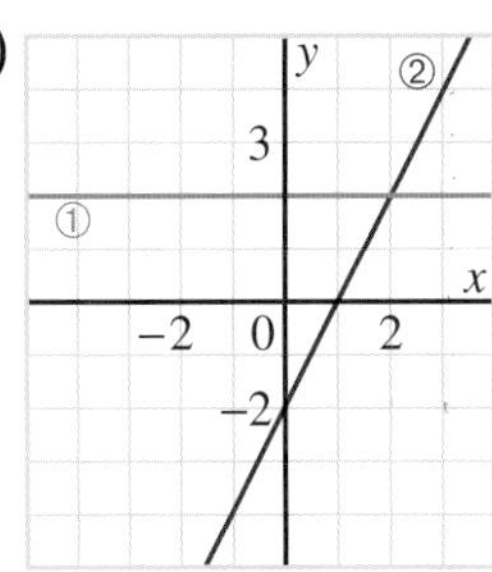

b) 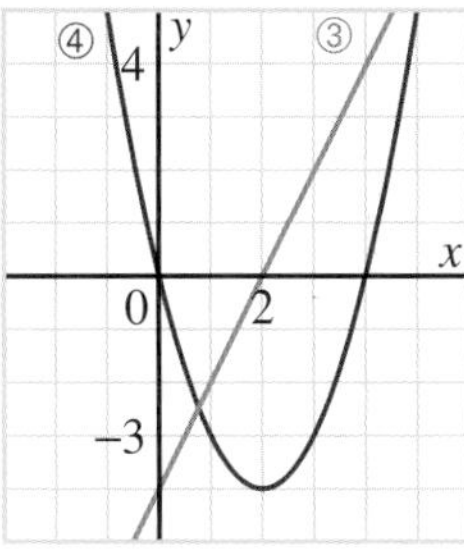

c) 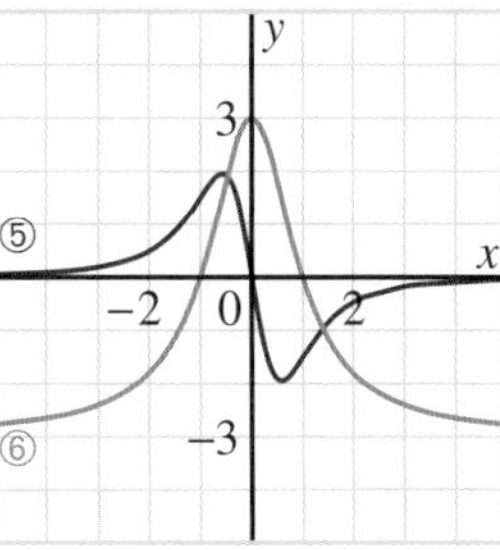

B

2. In exercise 1, the graph of each function and its derivative intersect at one or more points. Suppose the graph of a function $y = f(x)$ and its derivative $y = f'(x)$ intersect at a point with coordinates (x_1, y_1). What special property do these coordinates have? Explain.

3. A function f and its derivative f' are graphed on the same grid. Which graph is the function f and which graph is the derivative f'? Explain how you can tell.

a) 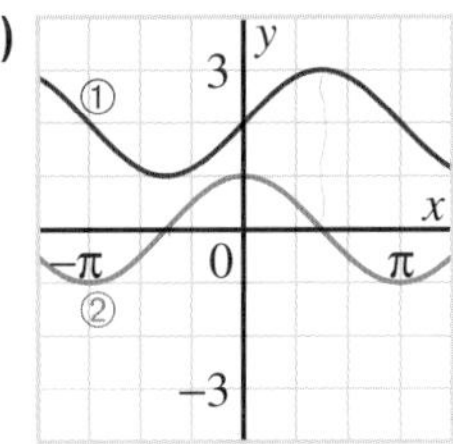

b) 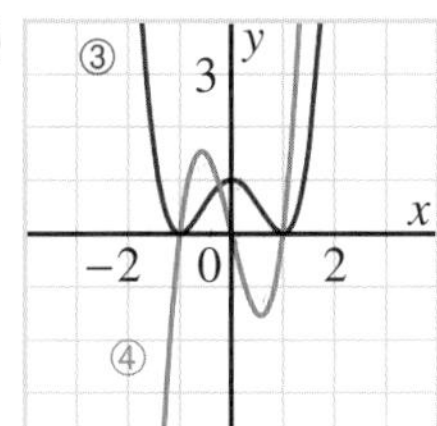

c) 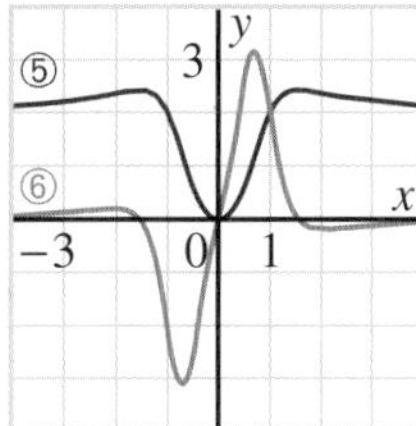

4. **Knowledge/Understanding** Sketch the graph of the derivative of each function.

a) 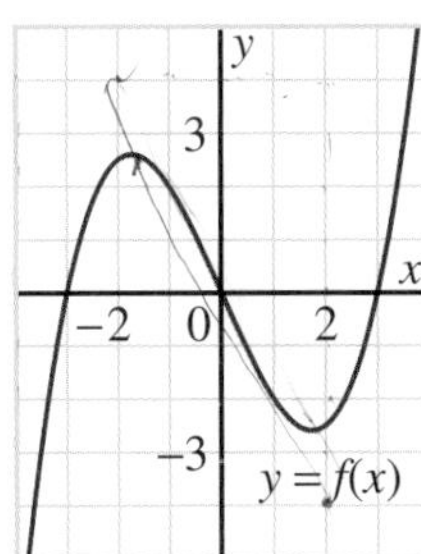

b) 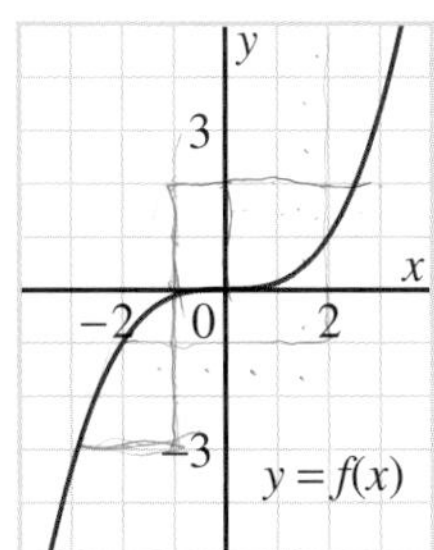

c) 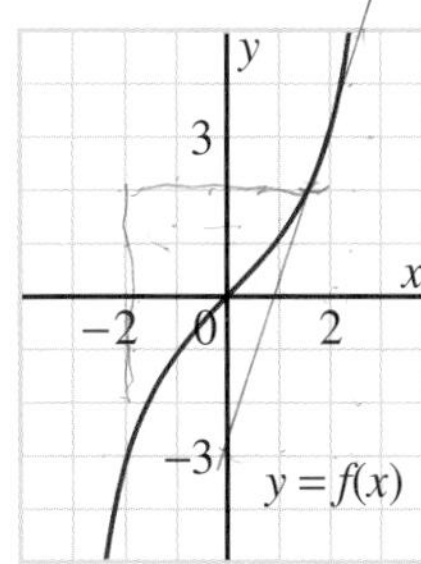

d) 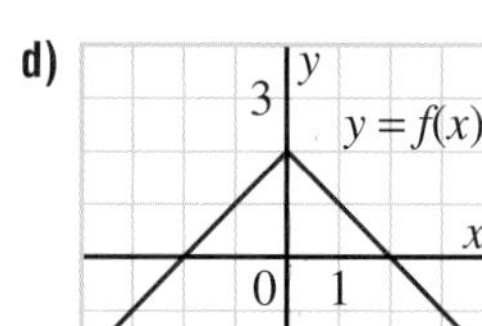

e) 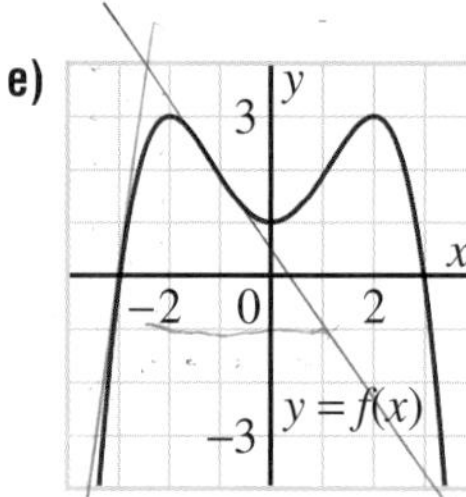

f) 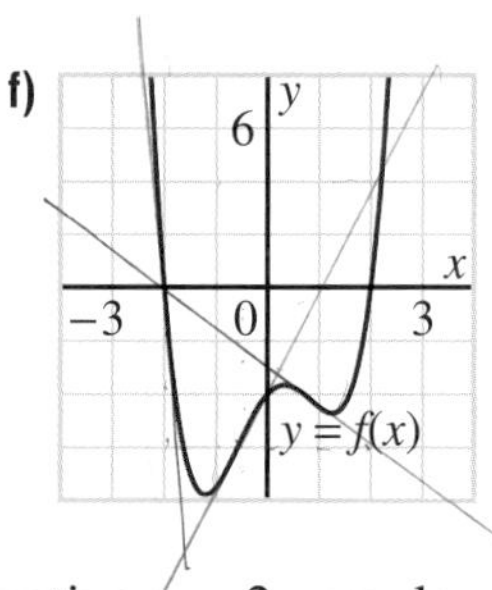

5. **Communication** The graphs of $y = x^2$ and its derivative $y = 2x$ are drawn on the same grid (below left). Explain your answers to each question.

a) For what values of x is y increasing? How can you tell this from:
   i) the graph of the function?
   ii) the graph of the derivative?

b) For what values of x is y decreasing? How can you tell this from:
   i) the graph of the function?
   ii) the graph of the derivative?

c) At what value of x is y neither increasing nor decreasing? How can you tell this from:
   i) the graph of the function?
   ii) the graph of the derivative?

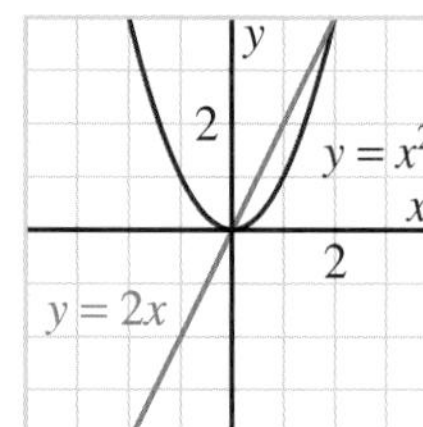

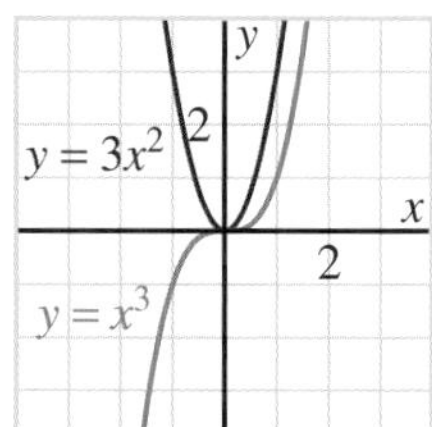

6. Repeat exercise 5 for the graphs of $y = x^3$ and its derivative $y = 3x^2$ (above right).

7. A function f and its derivative f' are graphed on the same grid. Which graph is the function f and which graph is the derivative f'? Explain how you can tell.

a) 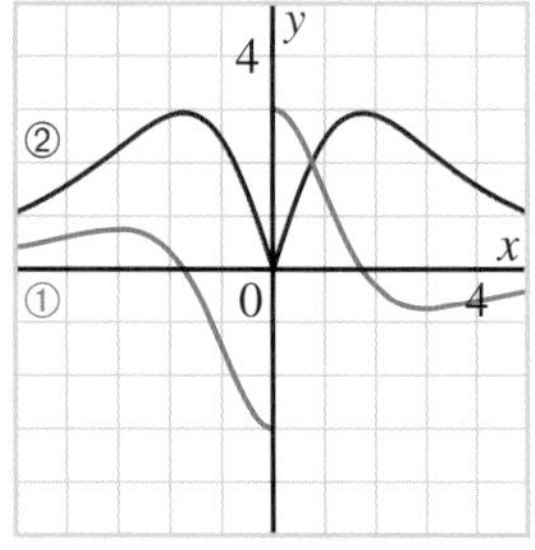

b) 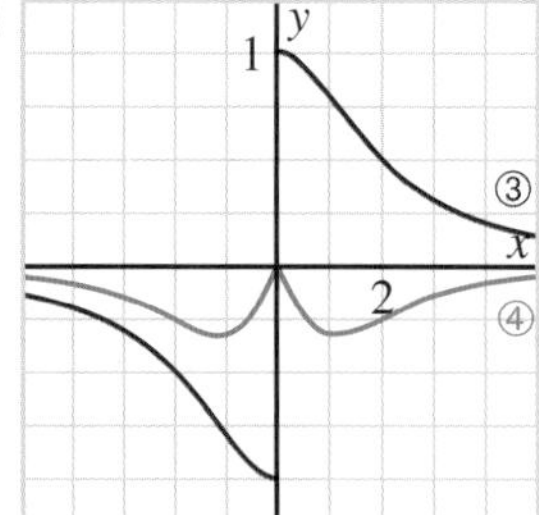

c) 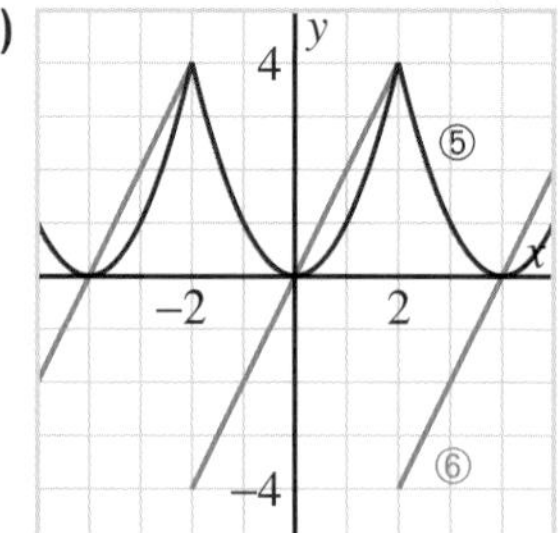

8. The graphs of three functions and their derivatives are given. For each function, identify the graph of its derivative. Explain.

Three Functions

a)
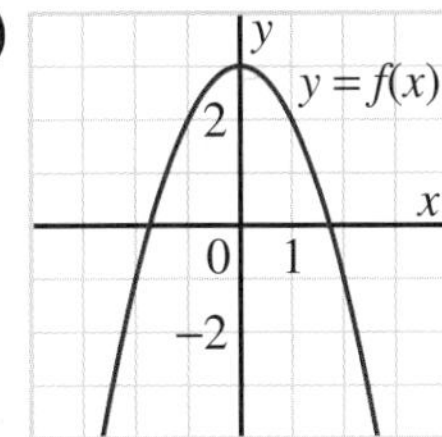

b)
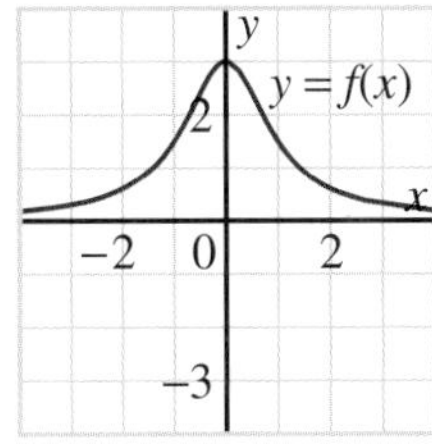

c)
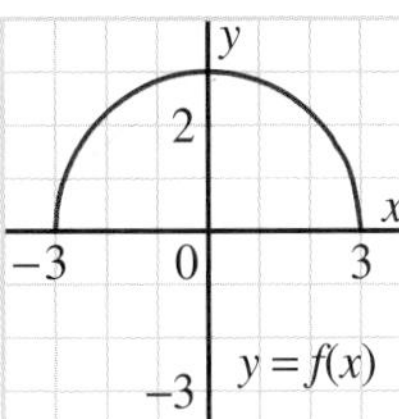

Three Derivatives

Graph 1

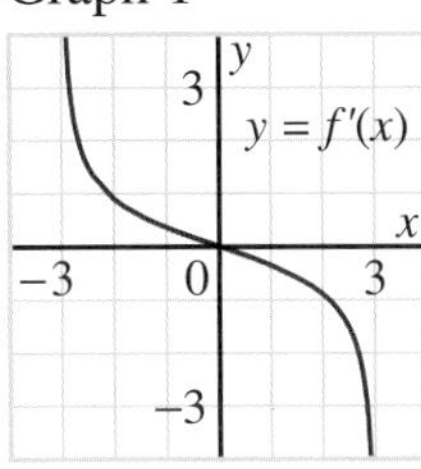

Graph 2

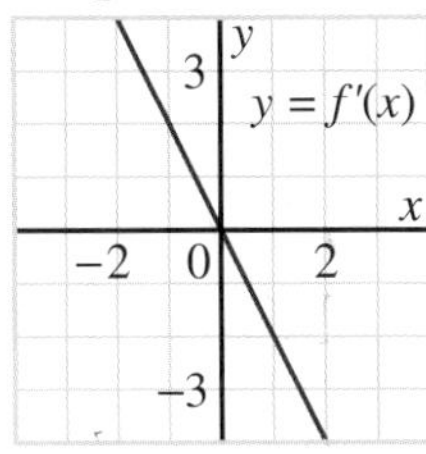

Graph 3

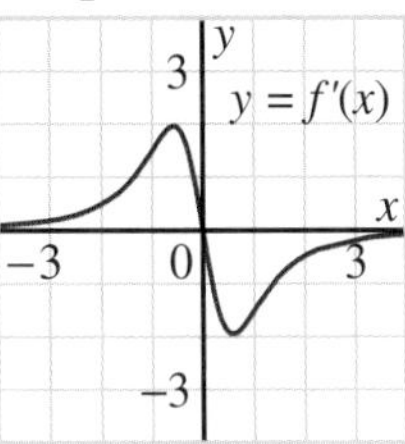

9. **Thinking/Inquiry/Problem Solving** In exercise 8, the graphs of the derivatives have certain common properties. For each property below, what information (if any) does this tell you about the graphs of the given functions?

a) The graphs of the derivatives pass through the origin.

b) The graphs of the derivatives are not changed if they are rotated through an angle of 180° about the origin.

10. **Application** Compare the three graphs below with the graphs of the three functions in exercise 8. Use the graphs of the derivatives in exercise 8 to sketch the graph of the derivative of each function.

a)
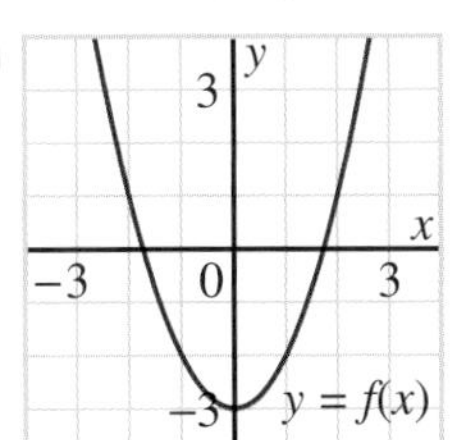

b)
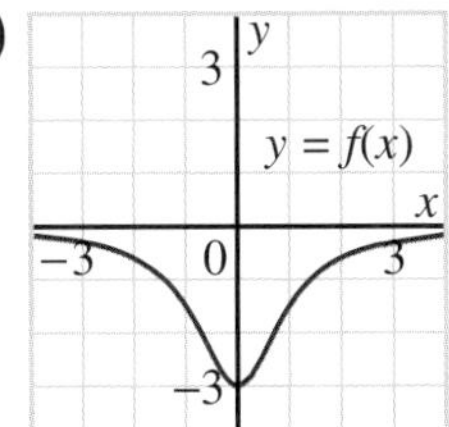

c)
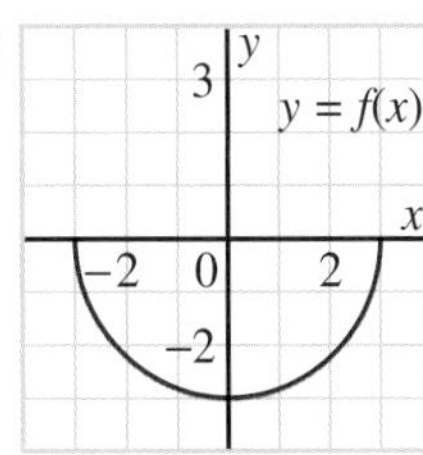

✓ **11.** Each graph below is a graph of the derivative of a function. Sketch a possible graph of the function.

a) 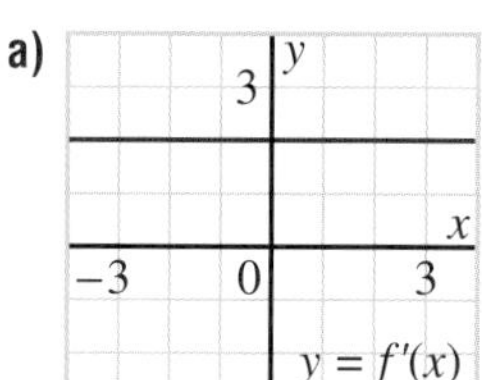

b) 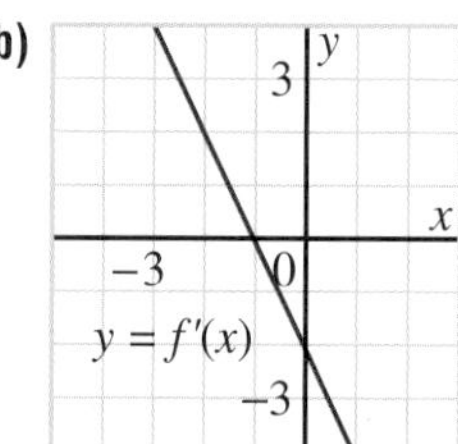

c) 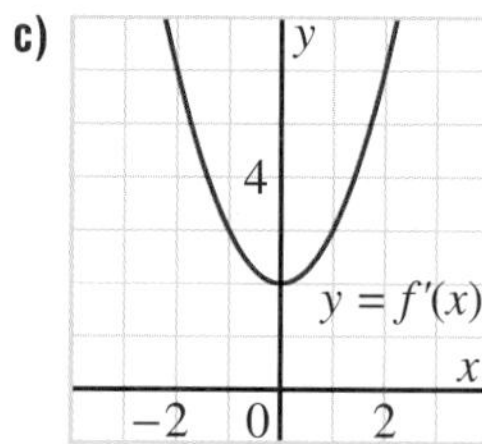

Exercise 11 is significant because more than one function can have the same derivative. The graphs of these functions are related by vertical translations, and their equations differ only by a constant. It is not possible to determine the original function uniquely without additional information. See *The Vertical Translation Rule* on page 42.

Refer to exercise 9, page 53, where the graphs of three functions and their derivatives are shown. Use these graphs to complete exercises 12 to 15.

12. The graphs of the derivatives have certain common properties. For each property below, what information (if any) does this tell you about the graphs of the original functions?

a) The graphs of the derivatives lie above the x-axis.

b) The graphs of the derivatives are symmetrical about the y-axis.

13. Suppose the graphs of the three functions are reflected in the y-axis.

a) Sketch each image function and its derivative.

b) Explain how the graph of each derivative is related to the graph of the corresponding derivative in exercise 9, page 53.

14. Suppose the graphs of the three functions are rotated counterclockwise about the origin through 90°.

a) Sketch each image function and its derivative.

b) Explain how the graph of each derivative is related to the graph of the corresponding derivative in exercise 9, page 53.

15. Suppose the graphs of the three functions are reflected in the line $y = x$.

a) Sketch each image function and its derivative.

b) Explain how the graph of each derivative is related to the graph of the corresponding derivative in exercise 9, page 53.

C

16. It is possible for a function to have a graph that is the same as the graph of its derivative. Using only this information, sketch a graph of this function.

Self-Check 1.4 – 1.6

1. What is the derivative of each function?

   a) $y = x + 1$ b) $y = 2x$ c) $y = 0.5x - 3$
   d) $y = x^2 + 4$ e) $y = -2x^3$ f) $y = 5x^3$

2. For each derivative in exercise 1, write a different function that has the same derivative.

3. a) The derivative of $y = x^2$ is $2x$. Use this information to write the derivative of each function.
   i) $y = 3x^2$ ii) $y = \frac{1}{2}x^2$ iii) $y = -4x^2$ iv) $y = -3x^2$
   b) What rule did you use in part a?

4. a) Sketch the graph of the function $y = |x + 3| - 4$.
   b) Sketch the graph of the derivative function.
   c) Where is the derivative function in part b not defined? Explain.

5. A function f and its derivative f' are graphed on the same grid (below left). Which graph is the function and which graph is its derivative? Explain.

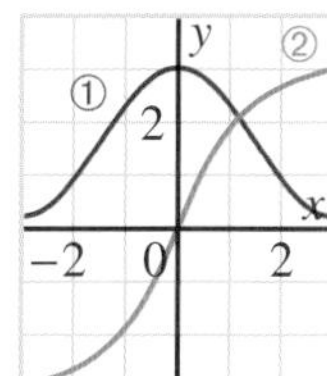

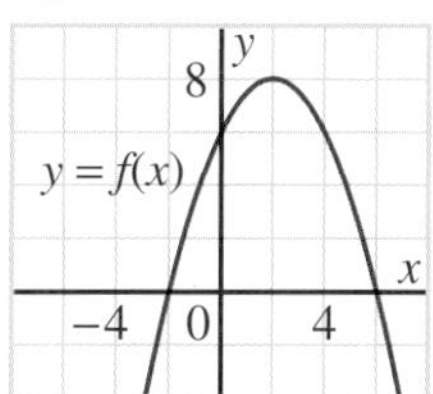

6. The graph of $y = f(x)$ is given (above right). Sketch the graph of its derivative, $y = f'(x)$.

PERFORMANCE ASSESSMENT

7. a) Calculate the instantaneous rate of change of $y = -x^3$ with respect to x at $x = 2.5$.
   b) Determine the equation of the tangent to $y = -x^3$ at $x = 2.5$.
   c) There is another point on the graph of $y = -x^3$ at which the instantaneous rate of change of y with respect to x is equal to its value at $x = 2.5$. Determine the equation of the tangent to the curve at this point.

Review Exercises

Mathematics Toolkit

Derivative Tools

- For any function $y = f(x)$, the *average rate of change* of y with respect to x between $x = x_1$ and $x = x_2$ is $\frac{\Delta y}{\Delta x} = \frac{y_2 - y_1}{x_2 - x_1}$. On a graph of the function, this is the slope of the secant through (x_1, y_1) and (x_2, y_2).
- For any function $y = f(x)$, the instantaneous rate of change of y with respect to x at x_1 is the slope of the tangent to the graph of the function at the point where $x = x_1$.
- From the equation of the function $y = f(x)$, an estimate of the *instantaneous rate of change* at $x = x_1$ is the average rate of change over a very short interval that contains the point $x = x_1$. When x_1 is to the left of the interval, the instantaneous rate of change is $\frac{f(x_1 + h) - f(x_1)}{h}$. This expression is the difference quotient.
- The *derivative* of a function f is a new function denoted f', such that for each x, $f'(x)$ is the instantaneous rate of change of f at x when f' exists.
- Some derivatives you have learned:

Function	Derivative
$y = x$	$y' = 1$
$y = x^2$	$y' = 2x$
$y = x^3$	$y' = 3x^2$
$y = \frac{1}{x}$	$y' = -\frac{1}{x^2}$
$y = \sqrt{x}$	$y' = \frac{1}{2\sqrt{x}}$
$y = \sin x$	$y' = \cos x$
$y = \cos x$	$y' = -\sin x$

- When the graph of a function is translated vertically, its derivative is not affected; that is, the derivative of $y = f(x) + c$ is $y = f'(x)$.
- When the graph of a function is expanded or compressed vertically, the graph of the derivative is also expanded or compressed vertically; that is, the derivative of $y = cf(x)$ is $y = cf'(x)$.
- A function is *not differentiable* at points on the graph where the tangent does not exist.
- The graphs of $y = f(x)$ and its derivative $y = f'(x)$ are related:
  – At the maximum and minimum points of $y = f(x)$, $f'(x) = 0$
  – When $y = f(x)$ increases, $f'(x) > 0$
  – When $y = f(x)$ decreases, $f'(x) < 0$

1.1 1. For each data set, sketch the graph. Then calculate the average rates of change of y between each consecutive pair of values of x.

a)

x	0	1	2	3	4
y	3	4	7	12	19

b)

x	0	5	10	15	20
y	0	16	24	28	30

c)

x	−20	−10	0	10	20
y	10	9	7	4	0

2. The salt content of seawater is its *salinity*. This is measured in parts per thousand (ppt). These salinity measurements were recorded in the Bay of Fundy.

Depth, *d* (m)	0.0	5.0	10.0	15.0
Salinity, *S* (ppt)	21.0	28.0	30.0	31.0

Calculate the average rate of change of salinity with depth over each interval.

a) 0.0 m to 15.0 m   b) 0.0 m to 5.0 m   c) 10.0 m to 15.0 m

1.2 3. The drag, D kilonewtons, experienced by a racing car is a function of the speed, v metres per second, of the car. The equation is $D = 0.08v^2$.

a) Graph the function.

b) Calculate the average rate of change of drag with respect to speed from $v = 20$ m/s to $v = 50$ m/s.

c) Estimate the instantaneous rate of change of drag with respect to speed when $v = 35$ m/s.

4. Calculate the average rate of change of each function between $x = 1$ and $x = 5$.

a) $y = x^2 + x$ b) $y = \sqrt{x + 3}$ c) $y = \frac{6}{x}$ d) $y = 7$

5. For each function, estimate the instantaneous rate of change of y with respect to x at $x = -2$ and at $x = 3$.

a) $y = 4x$ b) $y = x^2 + 1$ c) $y = -x^3$ d) $y = 3x + 20$

6. Estimate the instantaneous rate of change of $y = 5x^2$ at $x = 0.5$.

1.3 7. A car travelling at 50 km/h can stop in about 23 m. A car travelling at 100 km/h needs about 85 m to stop. Here are the data for stopping distances at different speeds.

Speed, v (km/h)	Average stopping distance, d (m)
20	9
30	12
40	16
50	23
60	31
70	42
80	54
90	69
100	85
110	104
120	124
130	147

a) Graph stopping distance against speed.

b) The equation $d = 0.01v^2 - 0.25v + 10$ expresses the stopping distance as a function of speed. Use the equation to estimate the instantaneous rate of change of stopping distance for each speed. Use 1-km/h following intervals.

i) 20 km/h
ii) 40 km/h
iii) 60 km/h
iv) 80 km/h
v) 100 km/h

c) Graph the data from part b to make a graph that approximates the derivative function. What special feature does the derivative appear to have?

d) Compare the graphs in parts a and c. Describe how each graph shows how the stopping distance changes.

8. When a person stands at the end of a diving board, the board bends and feels bouncy. This situation can be modelled with a metre stick and some weights. The metre stick is clamped to a table with a length, L centimetres, extended over the edge. The weights are suspended from a paper clip taped to the end of the stick. The downward deflection, d millimetres, of the stick is recorded.

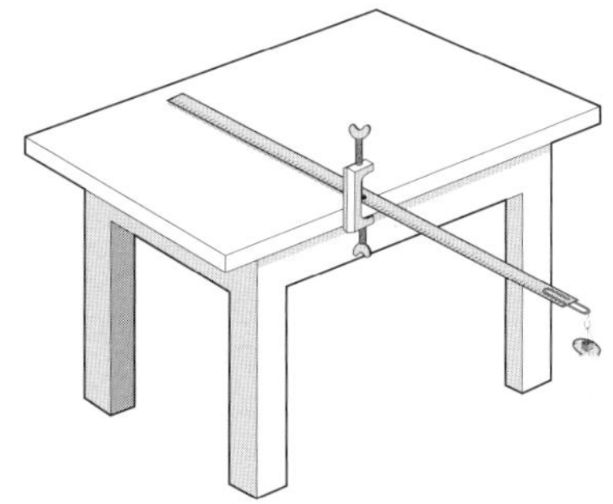

L (cm)	5	10	15	20	25	30	35
d (mm)	0	1	2	3	7	13	21

L (cm)	40	45	50	55	60	65
d (mm)	34	49	68	93	122	157

a) Graph downward deflection against length.

b) Estimate the instantaneous rate of change of deflection every 5 cm, starting at 7.5 cm. Use 5-cm centred intervals.

c) Graph the rate of change of deflection as a function of length.

d) Where is the rate of change of deflection greatest? Use the results of parts b and c to justify your answer.

1.4 9. Determine the derivative of each function.

a) $y = -2$ b) $y = 4x^3$ c) $y = 3 - x^2$ d) $y = 7x + 5$

10. For each function, calculate the value of the derivative at $x = 0.5$.

a) $y = x^2 + 5$ b) $y = -3x + 2$ c) $y = x^4$ d) $y = -2x^2 + 1$

11. Determine the equation of the tangent to the graph of the function $y = -4x^2 + 3$ at $x = 1$.

12. Determine the derivative of each function.

a) $y = \sqrt{x} + 3$ b) $y = -\frac{1}{x}$ c) $y = 2\sin x$ d) $y = \cos x - 4$

13. For each function, determine the equation of the tangent to the graph of the function at $x = 4$.

a) $y = 3\sqrt{x}$ b) $y = \frac{2}{x}$ c) $y = \sqrt{x} - 4$

14. For each function y, determine the value(s) of x where $y' = 2$, if it exists.

a) $y = x^2 + 1$ b) $y = x^3$ c) $y = 4x - 1$

d) $y = 2\sqrt{x}$ e) $y = -\frac{4}{x}$ f) $y = 2x$

1.5 15. The graph of the function $y = 2|x + 1| - 3$ is shown below left.

a) Describe the graph.

b) Describe the derivative function.

c) Sketch the graph of the derivative function.

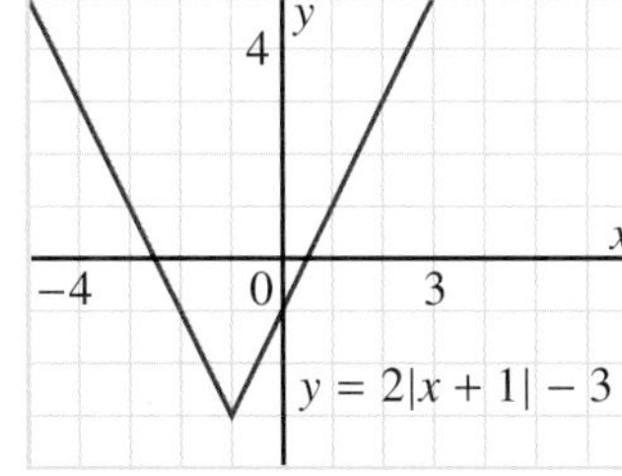

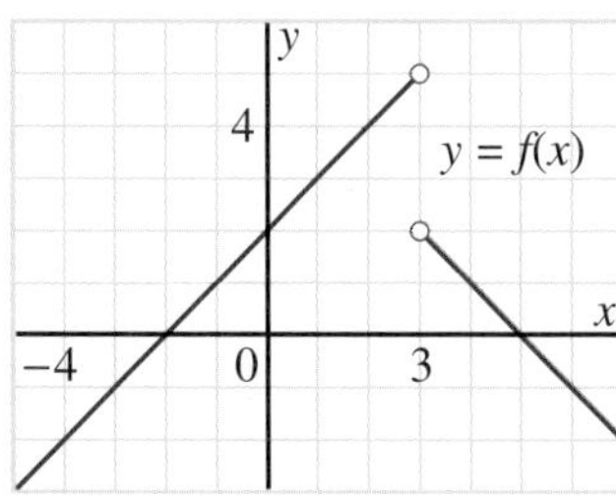

16. Sketch the graph of the derivative of the function above right.

17. The graphs of three functions (parts a to c) and their derivatives (parts i to iii) are shown on the next page. For each function, identify the graph of its derivative. Explain.

a) 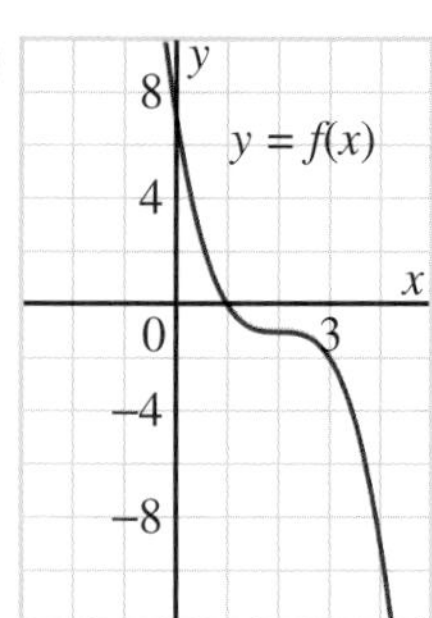

b) 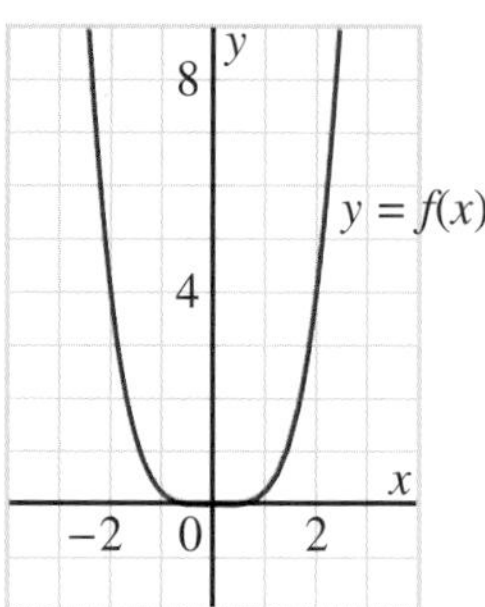

c) 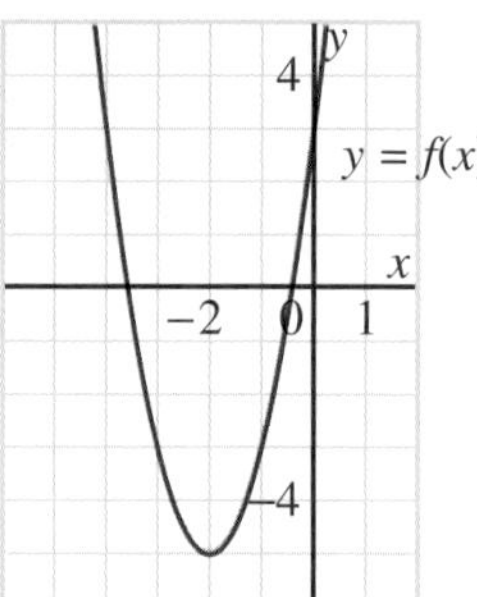

i) 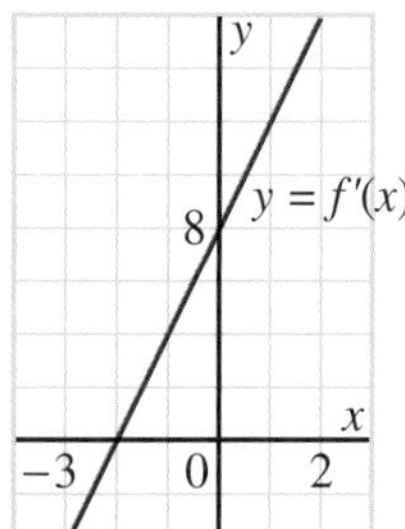

ii) 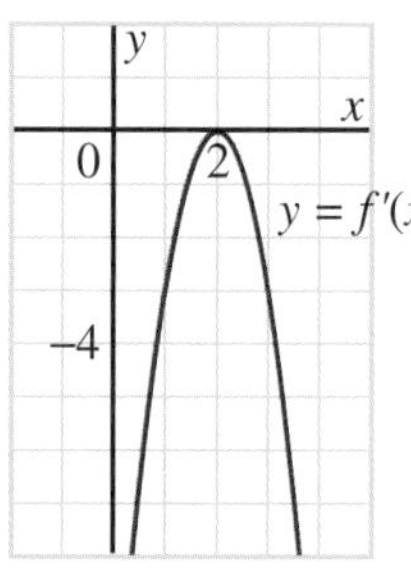

iii) 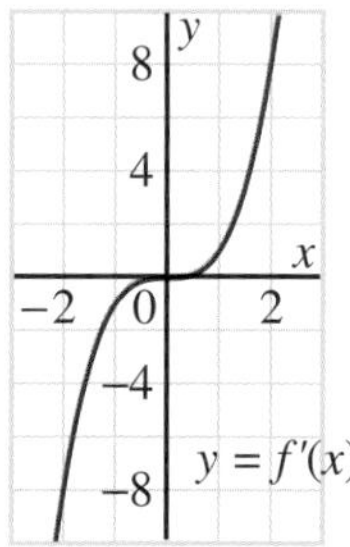

1.6 **18.** List 4 different functions $y = f(x)$ for which $y' = 3$.

19. List 4 different functions $y = f(x)$ for which $y' = 2x$.

20. For each function $y = f(x)$, sketch the derivative of y as a function of x.

a) 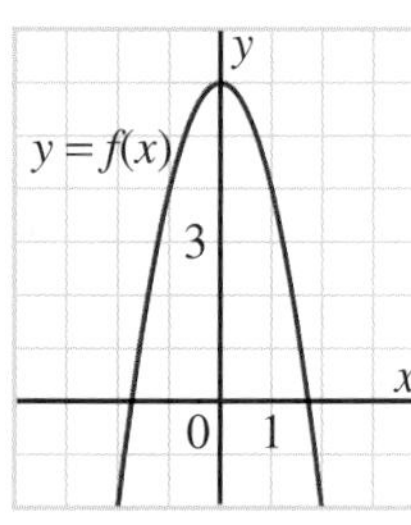

b) 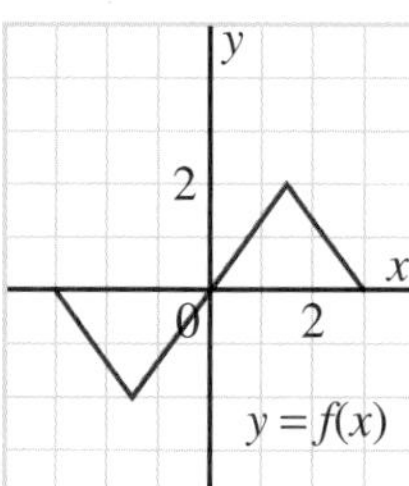

c) 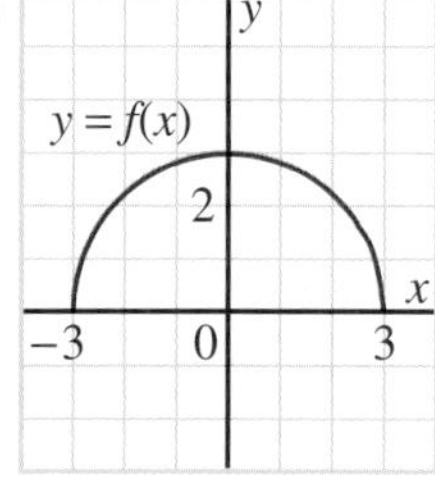

21. For each derivative function $y = f'(x)$, sketch a possible graph of the function $y = f(x)$.

a) 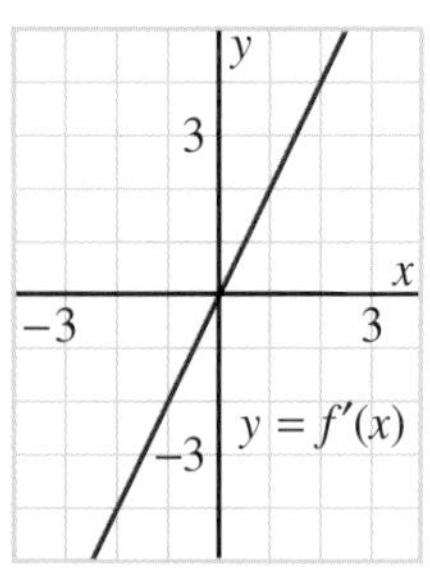

b) 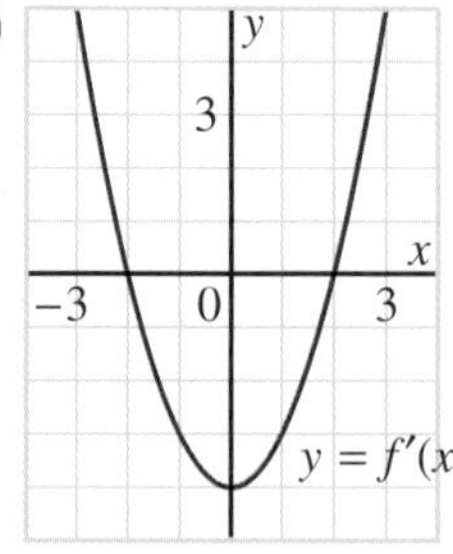

c) 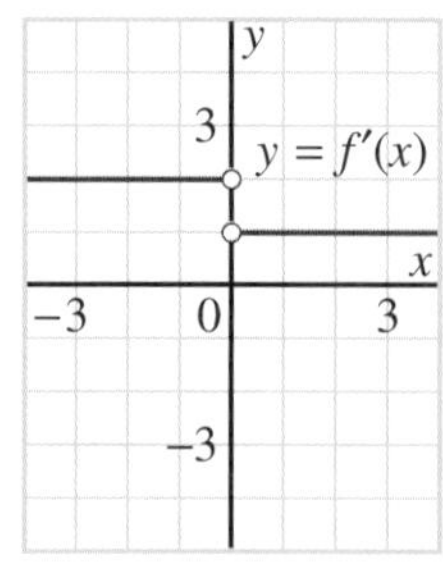

Self-Test

1. **Knowledge/Understanding** Statistics Canada has documented the Canadian infant mortality rate for many years. The infant mortality rate is the number of deaths of children up to 1 year of age per 1000 live births.

Year	1960	1965	1970	1975	1980	1985	1990	1995
Infant mortality rate	27.3	23.5	18.7	14.3	10.1	7.8	6.7	6.1

   a) Calculate the average rate of change of the mortality rate for each 5-year period.

   b) Describe how the average rates of change show that the mortality rate is falling.

   c) Describe how the average rates of change show that the mortality rate is levelling off.

   d) Explain why the infant mortality rate in Canada has changed substantially since 1960.

2. Consider the function $y = x^2 + 4$.

   a) Estimate the instantaneous rate of change of y with respect to x at $x = 1$.

   b) Determine the derivative of y with respect to x.

   c) Calculate the value of y' at $x = 1$.

   d) Explain why your answers for parts a and c should be the same.

3. **Communication** Explain how to sketch the graph of the derivative of the function below left; then sketch the graph.

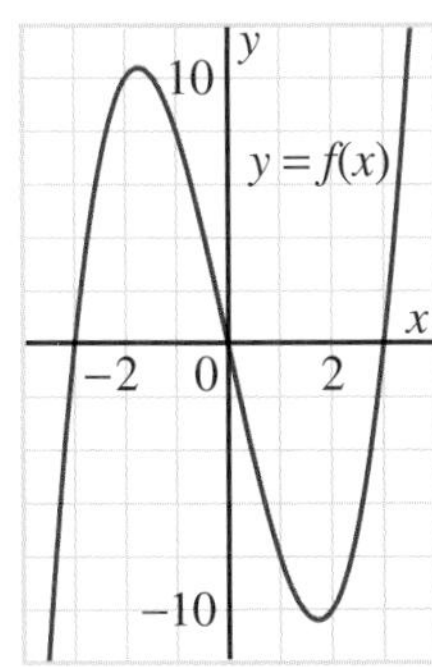

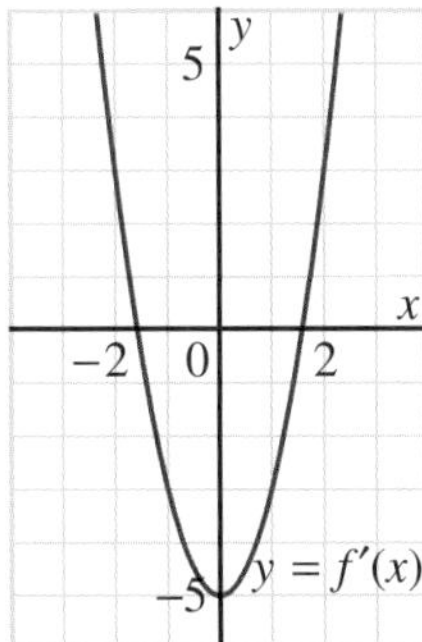

4. The graph of the derivative of a function is shown above right. Sketch a possible graph of the function.

Self-Test

5. Consider the function $y = \sqrt{x}$.

   a) Calculate the average rate of change of y with respect to x over each interval. Give the answers to 2 decimal places where necessary.

   i) $x = 0$ to $x = 10$ ii) $x = 0$ to $x = 1$
   iii) $x = 0$ to $x = 0.1$ iv) $x = 0$ to $x = 0.01$

   b) Explain why the instantaneous rate of change does not exist at $x = 0$.

6. **Application** A wood stove operates efficiently at flue temperatures between 150°C and 200°C (423K and 473K). Its energy output is given by the function $E = \frac{T^4}{5\,000\,000} - 1600$, where E is the energy output in watts and T is the Kelvin temperature.

   a) Determine the average rate of change of energy output with temperature between 423K and 473K.

   b) Estimate the instantaneous rate of change of energy output at 450K.

7. Determine the equation of the tangent to the graph of $y = -2x^2 + 4$ at the point (1, 2).

8. **Thinking/Inquiry/Problem Solving** Consider the function $y = \frac{1}{x} + 2$.

   a) Graph the function for x between –5 and 5.

   b) Describe the graph of the function for values of x greater than 5 and less than –5.

   c) Sketch the graph of the derivative y'.

   d) Use the behaviour of y' for large values of x to explain your answer in part b.

PERFORMANCE ASSESSMENT

9. The cross section of a hill is parabolic with equation $y = -0.1x^2 + 4$.

   a) Describe what x and y represent.

   b) Explain why the steepness of the hill at a point on the hill is given by the derivative of y with respect to x.

   c) A ramp is placed on the hill so that one end of the ramp is a tangent to the hill at the point (–4, 2.4). Determine the equation of the ramp.

   d) Determine the coordinates of the point where the other end of the ramp meets the level ground.

Mathematical Modelling

An Introduction

Each of 5 chapters in this book is followed by a Mathematical Modelling section. It features a significant problem from the physical, natural, or social sciences. These problems are described briefly on the next page. Each section is organized under these headings:

Developing a Mathematical Model suggests a graph or a function that can represent the situation.

Formulating a Hypothesis describes a likely explanation for some aspect of the situation.

In ***Analysing the Model*** and ***Key Features of the Model***, you examine the model in detail.

In ***Making Inferences from the Model***, you use the model to form conclusions about the situation.

Related Problems provides you with an opportunity to solve other problems similar to the one that was developed.

Curriculum Expectations

In these featured sections, you will:

- Pose problems and formulate hypotheses regarding rates of change within applications drawn from the natural and social sciences.
- Make inferences from models of applications and compare the inferences with the original hypotheses regarding rates of change.
- Determine the key features of a mathematical model of an application drawn from the natural or social sciences, using the techniques of differential calculus.
- Compare the key features of a mathematical model with the features of the application it represents.
- Predict future behaviour within an application by extrapolating from a mathematical model of a function.
- Pose questions related to an application and answer them by analysing mathematical models, using the techniques of differential calculus.
- Communicate findings clearly and concisely, using an effective integration of essay and mathematical forms.

Mathematical Modelling

Optimal Cooking Time (page 113) – an application from physics

A yam can be cooked in 20 min in boiling water or a microwave oven. A combination of the two methods can reduce the cooking time by about 25%. This is possible because the rates of temperature increase are different for the two cooking methods.

Optimal Profits (page 289) – an application from economics

A company hires workers to manufacture a product. Too many workers will increase costs and too few workers will not produce enough of the product to make a profit. Hence, there must be an optimum number of workers that maximizes the profit. You will determine this number in a typical situation.

Predicting Populations (page 359) – an application from demographics

The world population has been increasing for centuries, and it is continuing to increase. However, since about 1965 the rate of growth has been decreasing. You will analyse this rate of growth and apply the results to predict the annual world population up to the year 2025.

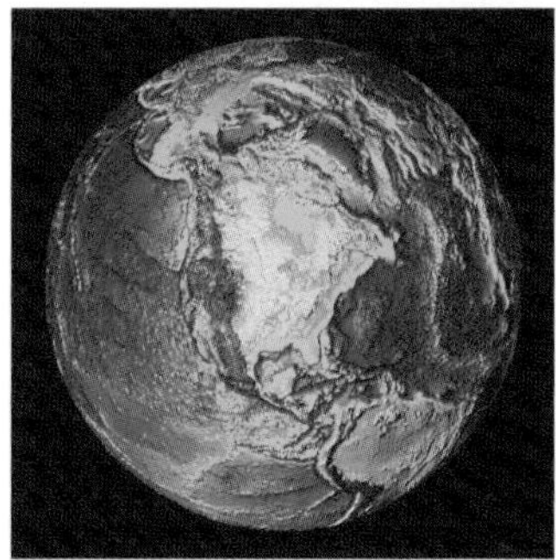

Quality Control (page 477) – an application from economics

To reduce costly warranty claims, a manufacturer can spend extra money to develop a more reliable product. If too much is spent, the average cost of the product will increase. You will determine the amount to be spent to minimize the average cost.

Optimal Consumption Time (page 531) – an application from biology

A bird foraging for food must maximize its net energy gain. You will determine the optimal time a bird should spend in a berry patch before moving on to search for a new food source.

Calculating Derivatives from First Principles

2

Curriculum Expectations

By the end of this chapter, you will:

- Identify the nature of the rate of change of a given function as it relates to the key features of the graph of that function.
- Sketch, by hand, the graph of the derivative of a given graph.
- Determine the limit of a polynomial function.
- Determine the derivatives of polynomial and simple rational functions from first principles, using the definitions of the derivative function,
  $f'(x) = \lim_{h \to 0} \frac{f(x+h) - f(x)}{h}$ and
  $f'(a) = \lim_{x \to a} \frac{f(x) - f(a)}{x - a}$.
- Justify the constant, power, and sum and difference rules for determining derivatives.
- Determine the derivatives of polynomial and rational functions using the constant, power, and sum and difference rules for determining derivatives.
- Determine the equation of a tangent to the graph of a polynomial and a rational function.

Necessary Skills

1. New: Powers of Binomials

Recall how to expand the square of a binomial.

$$(a + b)^2 = a^2 + 2ab + b^2$$

We can use this result to expand higher powers of $(a + b)$.

$$\begin{aligned}(a + b)^3 &= (a + b)(a + b)^2 \\ &= (a + b)(a^2 + 2ab + b^2) \\ &= a^3 + 2a^2b + ab^2 + a^2b + 2ab^2 + b^3 \\ &= a^3 + 3a^2b + 3ab^2 + b^3\end{aligned}$$

Similarly, $(a + b)^4 = a^4 + 4a^3b + 6a^2b^2 + 4ab^3 + b^4$

Take Note

Powers of Binomials

$$(a + b)^2 = a^2 + 2ab + b^2$$
$$(a + b)^3 = a^3 + 3a^2b + 3ab^2 + b^3$$
$$(a + b)^4 = a^4 + 4a^3b + 6a^2b^2 + 4ab^3 + b^4$$

We can use these patterns to expand powers of other binomials.

Example

Expand $(1 - 2x)^3$.

Solution

In the equation $(a + b)^3 = a^3 + 3a^2b + 3ab^2 + b^3$, replace a with 1 and b with $-2x$.

$$\begin{aligned}(1 - 2x)^3 &= 1^3 + 3(1)^2(-2x) + 3(1)(-2x)^2 + (-2x)^3 \\ &= 1 - 6x + 12x^2 - 8x^3\end{aligned}$$

Exercises

1. Expand.

a) $(a - b)^3$ **b)** $(a - b)^4$ **c)** $(2x + 1)^3$ **d)** $(1 + 3x)^3$

e) $(1 - 2x)^4$ **f)** $(3a + 2b)^2$ **g)** $(3a + 2b)^3$ **h)** $(3a + 2b)^4$

2. a) Verify that $(a + b)^4 = a^4 + 4a^3b + 6a^2b^2 + 4ab^3 + b^4$.

b) Determine $(a + b)^5$.

2. New: Factoring Differences of Powers

Recall how to factor a difference of squares. This can be verified by multiplication.

$$a^2 - b^2 = (a - b)(a + b)$$

Differences of higher powers can also be factored. The results below can be verified by multiplication.

Take Note

Factoring Differences of Powers

$$a^2 - b^2 = (a - b)(a + b)$$
$$a^3 - b^3 = (a - b)(a^2 + ab + b^2)$$
$$a^4 - b^4 = (a - b)(a^3 + a^2b + ab^2 + b^3)$$
$$\vdots \qquad \vdots$$
$$a^n - b^n = (a - b)(a^{n-1} + a^{n-2}b + a^{n-3}b^2 + \ldots + a^2b^{n-3} + ab^{n-2} + b^{n-1})$$

We can use these patterns to factor any expression that is a difference of the same two powers.

Example

Factor $x^3 - 125$.

Solution

Use $a^3 - b^3 = (a - b)(a^2 + ab + b^2)$.

Since $125 = 5^3$, replace a with x and b with 5:

$$x^3 - 5^3 = (x - 5)(x^2 + 5x + 5^2)$$

Hence, $x^3 - 125 = (x - 5)(x^2 + 5x + 25)$

Exercises

1. Factor.

a) $x^3 - 1$ **b)** $x^3 - 8$ **c)** $x^3 - 27$ **d)** $x^3 - 64$

e) $x^4 - 1$ **f)** $x^4 - 16$ **g)** $x^5 - 1$ **h)** $x^5 - 32$

2. Factor and simplify.

a) $(a + b)^2 - a^2$ **b)** $(a + b)^3 - a^3$ **c)** $(a + b)^4 - a^4$

2.1 The Derivative as a Limit

In Chapter 1, we defined the derivative of a function f as a new function, f'. This represents the instantaneous rate of change of f at x when f' exists. Geometrically, $f'(x)$ is the slope of the tangent to the graph of f at $(x, f(x))$. We used graphs on which a few tangents were drawn to conjecture the derivatives of $y = x^2$ and $y = x^3$. Although the graphs provided strong evidence to support these conjectures, we cannot be certain that they are true. To establish them convincingly, and to determine the derivatives of other functions, we need more precise definitions and techniques.

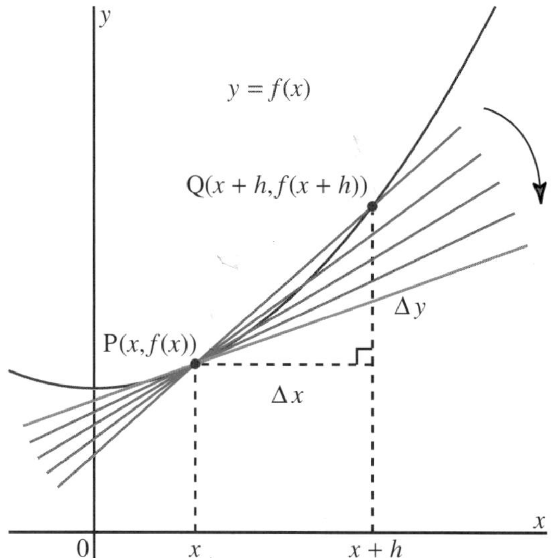

Let P be any point on the graph of a function $y = f(x)$.

Then the coordinates of P are $P(x, f(x))$.

Let Q be a point on the graph near P.

The coordinates of Q are $Q(x + h, f(x + h))$, for some small number h, positive or negative.

The slope of secant PQ is:

$$\frac{\Delta y}{\Delta x} = \frac{f(x+h) - f(x)}{(x+h) - x}$$

or $$\frac{\Delta y}{\Delta x} = \frac{f(x+h) - f(x)}{h} \quad \ldots \text{①}$$

As h gets smaller and smaller, Q moves closer and closer to P, secant PQ comes closer and closer to the tangent at P, and the slope of PQ comes closer and closer to the slope of the tangent. We cannot substitute $h = 0$ in equation ① to determine the slope of the tangent because the result is $\frac{0}{0}$, which is not defined. However, when we know the equation of the function, we can use equation ① to determine the slope of the tangent at P, when h is very small.

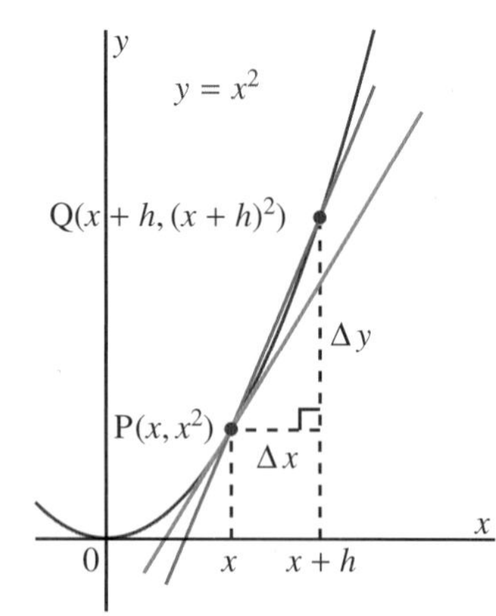

For example, in Chapter 1, we used a graph to show that the derivative of $f(x) = x^2$ is $f'(x) = 2x$. We can show this algebraically, as follows.

Let $P(x, x^2)$ be any point on the graph of f.

Let $Q(x + h, (x + h)^2)$ be a point on the graph near P.

The slope of PQ is:

$$\frac{\Delta y}{\Delta x} = \frac{(x+h)^2 - x^2}{h}$$

$$\frac{\Delta y}{\Delta x} = \frac{x^2 + 2hx + h^2 - x^2}{h}$$

$$\frac{\Delta y}{\Delta x} = \frac{h(2x+h)}{h}$$

We could substitute very small values of h into any of the above expressions, and the result is the slope of PQ. We cannot determine the slope of the tangent at P by substituting $h = 0$ because the result is $\frac{0}{0}$. But we can simplify further to obtain:

$$\frac{\Delta y}{\Delta x} = 2x + h, h \neq 0$$

The slope of secant PQ is $2x + h$. As h becomes smaller and smaller, PQ becomes closer and closer to the tangent at P, and the slope of PQ becomes closer and closer to $2x$. When h approaches 0, the secant approaches the tangent, and the slope of the secant approaches $2x$, which is the slope of the tangent. Therefore, the derivative of $f(x) = x^2$ is $f'(x) = 2x$.

We have shown that as h gets closer and closer to 0, $\frac{(x+h)^2 - x^2}{h}$ gets closer and closer to $2x$. We write:

$$\lim_{h \to 0} \frac{(x+h)^2 - x^2}{h} = 2x$$

We say: "The *limit* of $\frac{(x+h)^2 - x^2}{h}$ as h approaches 0 is $2x$."

We can repeat the above steps for other functions $y = f(x)$. As h gets closer and closer to 0, $\frac{f(x+h) - f(x)}{h}$ gets closer and closer to the derivative, $f'(x)$. We write:

$$f'(x) = \lim_{h \to 0} \frac{f(x+h) - f(x)}{h}$$

This equation is called the *first-principles definition* of the derivative of a function.

Take Note

First-Principles Definition of the Derivative

The *derivative* of a function f is a new function f', defined as follows:

$$f'(x) = \lim_{h \to 0} \frac{f(x+h) - f(x)}{h}$$

We will use the first-principles definition to determine derivatives of functions.

Example 1

Use the first-principles definition to determine the derivative of the function $f(x) = 4x^2 + x$.

Solution

Use $f'(x) = \lim_{h \to 0} \frac{f(x + h) - f(x)}{h}$.

Replace $f(x)$ with $4x^2 + x$.

$$
\begin{aligned}
f'(x) &= \lim_{h \to 0} \frac{(4(x + h)^2 + (x + h)) - (4x^2 + x)}{h} \\
&= \lim_{h \to 0} \frac{4x^2 + 8xh + 4h^2 + x + h - 4x^2 - x}{h} \\
&= \lim_{h \to 0} \frac{8xh + 4h^2 + h}{h} \\
&= \lim_{h \to 0} \frac{h(8x + 4h + 1)}{h} \\
f'(x) &= \lim_{h \to 0}(8x + 4h + 1)
\end{aligned}
$$

As h approaches 0, $8x + 4h + 1$ approaches $8x + 1$.

So, $\lim_{h \to 0}(8x + 4h + 1) = 8x + 1$

That is, $f'(x) = 8x + 1$

Therefore, the derivative of $f(x) = 4x^2 + x$ is $f'(x) = 8x + 1$.

Recall that determining the derivative of a function is called differentiating the function.

Example 2

a) Use the first-principles definition to differentiate the function $y = x(6 - x)$.

b) Sketch the function in part a, and its derivative.

c) Determine the equation of the tangent at the point (1, 5) on the graph. Draw the tangent on the graph of the function.

Solution

a) Let $f(x) = x(6 - x)$.

Then, $f(x) = 6x - x^2$

Use $f'(x) = \lim_{h \to 0} \frac{f(x + h) - f(x)}{h}$.

Replace $f(x)$ with $6x - x^2$.

$$f'(x) = \lim_{h \to 0} \frac{(6(x+h) - (x+h)^2) - (6x - x^2)}{h}$$

$$= \lim_{h \to 0} \frac{(6x + 6h - x^2 - 2xh - h^2) - (6x - x^2)}{h}$$

$$= \lim_{h \to 0} \frac{6h - 2xh - h^2}{h}$$

$$= \lim_{h \to 0} \frac{h(6 - 2x - h)}{h}$$

$$f'(x) = \lim_{h \to 0}(6 - 2x - h)$$

As h approaches 0, $6 - 2x - h$ approaches $6 - 2x$.

So, $\lim_{h \to 0}(6 - 2x - h) = 6 - 2x$

That is, $f'(x) = 6 - 2x$

Therefore, the derivative of $y = x(6 - x)$ is $y' = 6 - 2x$.

b) The graph of $y = x(6 - x)$ is a parabola, opening down, with x-intercepts 0 and 6.

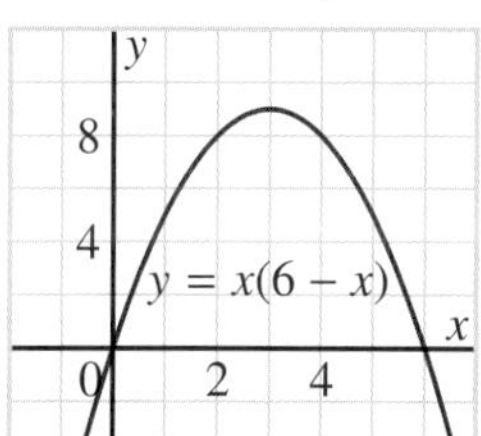

The graph of $y = 6 - 2x$ is a straight line with x-intercept 3 and y-intercept 6.

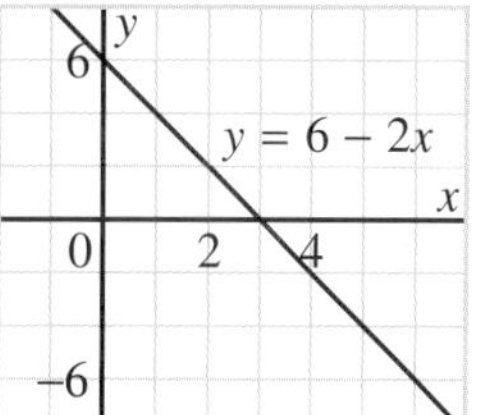

c) The derivative of $y = x(6 - x)$ is $y' = 6 - 2x$.

This equation represents the slope of the tangent at all points on the graph.

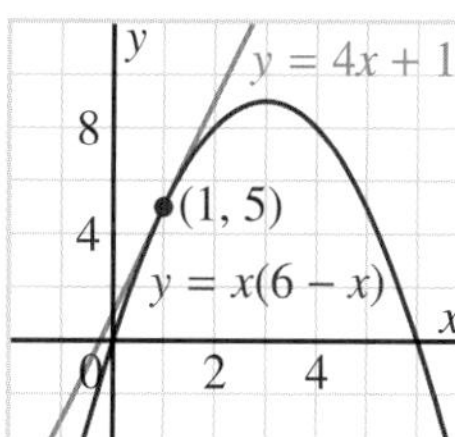

Substitute $x = 1$.

$$y' = 6 - 2(1)$$
$$= 4$$

The slope of the tangent at (1, 5) is 4.

Use the point-slope form of the equation of a line.

The equation of this tangent is:

$$y = m(x - p) + q$$
$$y = 4(x - 1) + 5$$
$$y = 4x + 1$$

THE BIG PICTURE

Defining the Derivative

Calculus is a powerful tool for solving problems in which quantities are changing. It should not be surprising that there are different ways to express the definition of the derivative. In Chapter 1, we expressed it graphically as an instantaneous rate of change and as the slope of a tangent. On page 75, the first-principles definition expresses the derivative as the limiting value of a certain difference quotient.

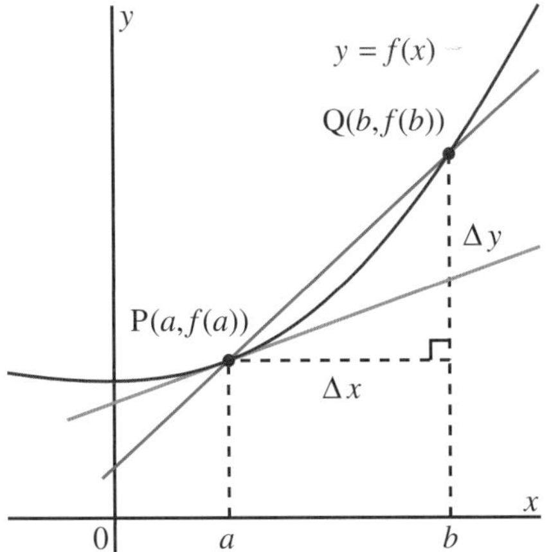

The first-principles definition can be expressed in a slightly different way.

Let P$(a, f(a))$ represent a fixed point on the graph of a function $y = f(x)$.

Let Q$(b, f(b))$ represent another point on the graph close to P.

Then, $\Delta y = f(b) - f(a)$ and $\Delta x = b - a$

And, the derivative of f at $x = a$ can be defined as:

$$f'(a) = \lim_{b \to a} \frac{f(b) - f(a)}{b - a}$$

This definition gives a formula for the slope of the tangent at the point P. To apply this formula to all points in the domain of the function, we replace a with x. The result is the derivative function, $f'(x)$.

2.1 Exercises

A

1. Use the first-principles definition to determine the derivative of each function.

a) $y = x$ **b)** $y = 3x$ **c)** $y = 3x + 4$ **d)** $y = mx + b$

2. a) Use the first-principles definition to determine the derivative of the constant function $y = 2$.

b) Use the first-principles definition to determine the derivative of any constant function $y = c$.

3. Refer to *Example 2*. The first step in the solution was to expand $x(6 - x)$ so that the first-principles definition was applied to $6x - x^2$. Repeat the solution of this example, without expanding, so that the first-principles definition is applied directly to $x(6 - x)$.

4. Why do you think the definition of the derivative on page 75 is called the first-principles definition?

5. a) Use the first-principles definition to differentiate the function $y = x^2 + 3$. Compare this derivative with the derivative of $y = x^2$ on page 75.

   b) Recall the geometric meaning of the derivative from Chapter 1. Explain why the derivative of the function in part a is the same as the derivative of $y = x^2$.

6. **Knowledge/Understanding** Use the first-principles definition to determine the derivative of the function $f(x) = 2x^2 - 3x$.

7. Use the first-principles definition to differentiate each function.

   a) $y = x^2 - 5x$ b) $y = x^2 + x + 1$ c) $f(x) = 4x^2 + 2x - 3$

   d) $f(x) = x(x - 1)$ e) $g(x) = x(2x + 1)$ f) $g(x) = x(3 + 2x)$

8. a) Use the first-principles definition to determine the derivative of $y = 5x^2$. Compare the result with the derivative of $y = x^2$, on page 75.

   b) Use the first-principles definition to determine the derivative of $y = -x^2$.

   c) Based on your results in parts a and b, predict how the derivative of $y = cf(x)$ compares with the derivative of $y = f(x)$.

Exercise 8 is significant because it involves a differentiation rule called the *constant multiple rule*, which will be justified in Section 2.3. We will use this rule frequently in the rest of this book.

9. Given $f(x) = (x + 1)^2$, determine $f'(x)$ in two ways:

   a) by applying the first-principles definition of the derivative

   b) by expanding the binomial square, then applying the first-principles definition

10. **Communication** Recall the geometric meaning of the derivative of a function as the slope of the tangent at x on the graph of the function. You know that the derivative of $f(x) = x^2$ is $f'(x) = 2x$. Explain why the derivative of $f(x) = (x + 1)^2$ is $f'(x) = 2(x + 1)$.

11. **Application** Determine the derivative of $y = x^3$ from first principles. Use the formula from *Necessary Skills* for expanding a binomial cube: $(a + b)^3 = a^3 + 3a^2b + 3ab^2 + b^3$

12. Determine the derivative of $y = x^4$ from first principles. Use the formula from *Necessary Skills* for expanding the fourth power of a binomial: $(a + b)^4 = a^4 + 4a^3b + 6a^2b^2 + 4ab^3 + b^4$

13. Use the results of exercises 11 and 12. Predict the derivatives of $y = x^5$ and $y = x^6$.

14. The graphs of $f(x) = x^2$, $f(x) = x^3$, and $f(x) = x^4$ and their derivatives are shown. Complete parts a and b for each graph.

a) Suppose $x < 0$. Visualize tangents to the graph of f as x increases. Describe what happens to the slopes of the tangents as x increases. Explain how the graph of the derivative function shows what is happening to the slopes of the tangents.

b) Suppose $x > 0$. Repeat part a.

i)

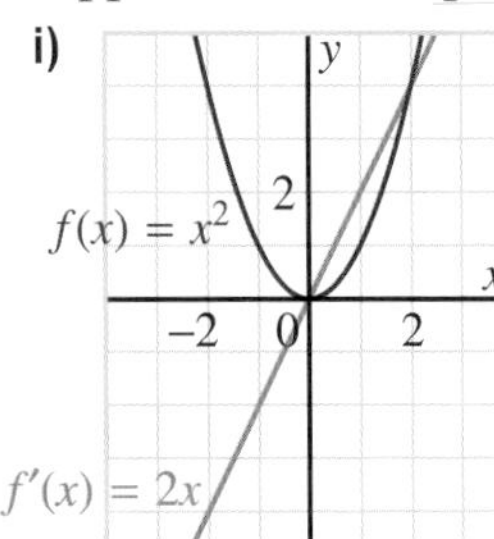

ii)

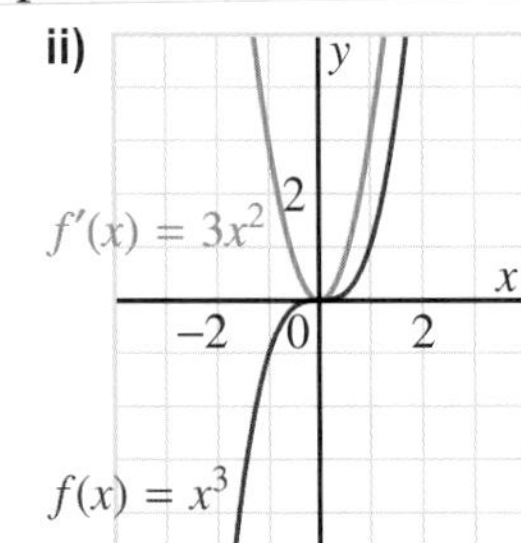

iii)

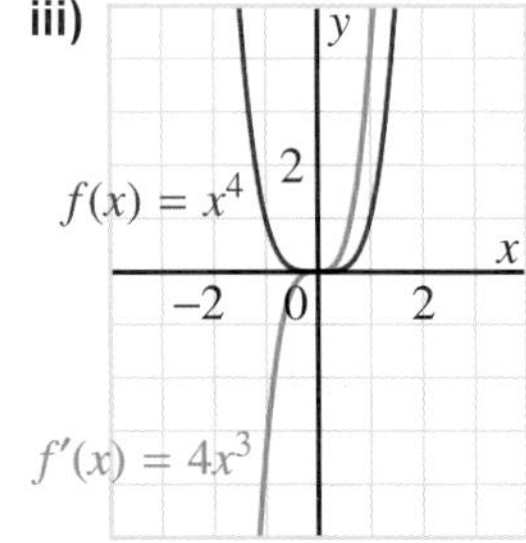

15. a) Use the first-principles definition to differentiate the function $y = x^2 - x$.

b) Sketch the function in part a, and its derivative.

c) Determine the equation of the tangent at the point (2, 2) on the graph. Draw the tangent on the graph of the function.

16. a) Use the first-principles definition to differentiate the function $y = 4 - x^2$.

b) Sketch the function in part a, and its derivative.

c) Determine the instantaneous rate of change of y when $x = 3$. Explain the result in terms of the graph.

17. **Thinking/Inquiry/Problem Solving**

Recall that a *tangent* to a curve is defined as follows:

Let P be a point on the curve. Let Q be another point on the curve, on either side of P. Construct the secant PQ. Now let Q approach P. If the secant PQ approaches the same line as Q approaches P from either side, this line is the tangent at P.

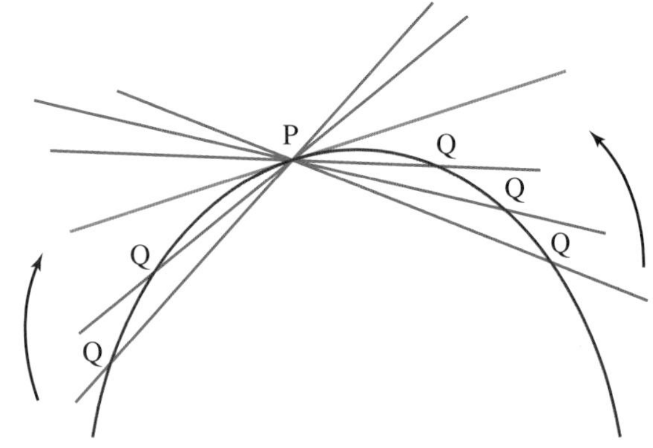

Determine whether each statement is true or false. Use a diagram to explain each answer.

a) A tangent is a line that intersects a curve at only one point.

b) A line cannot be a tangent to a curve at two different points.

c) A tangent to a curve at a point never crosses the curve at that point.

2.2 Differentiating Power Functions

A function that has an equation of the form $y = x^n$ is called a *power function*. The exponent n in the equation of a power function can be any real number. In this section, we will assume that n is a natural number (unless stated otherwise). The graphs of some power functions are shown at the right.

Power Functions
Even exponents

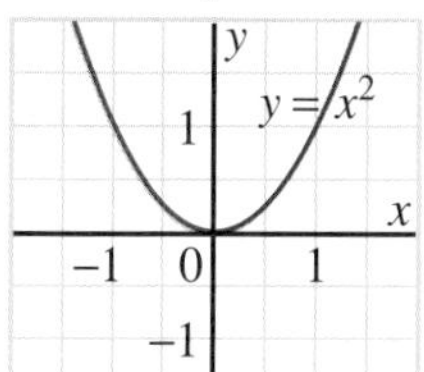

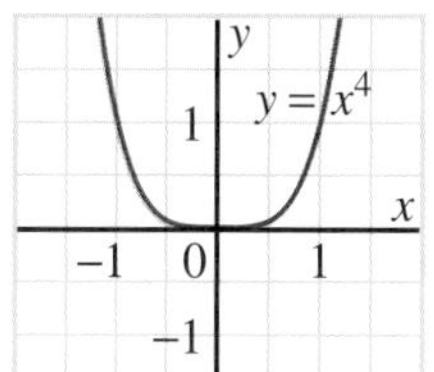

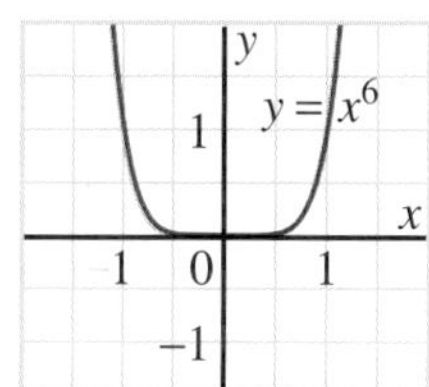

Odd exponents

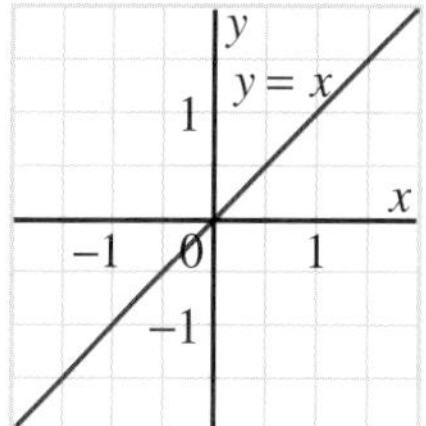

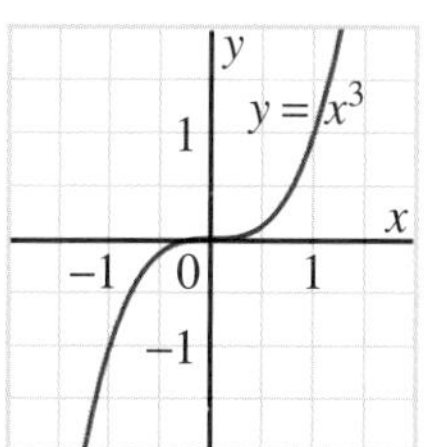

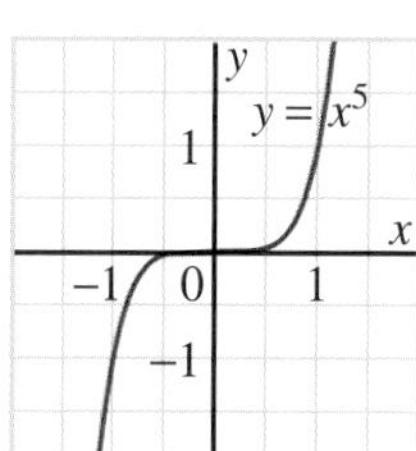

In Section 2.1, you used the first-principles definition to determine the following derivatives:

Function	**Derivative**
$y = x$	$y' = 1$
$y = x^2$	$y' = 2x$
$y = x^3$	$y' = 3x^2$
$y = x^4$	$y' = 4x^3$

If the pattern in the derivatives continues, the derivative of $y = x^n$ should be $y' = nx^{n-1}$, where n is a natural number. However, we cannot be certain that the pattern continues. We use the first-principles definition to prove that the pattern does continue.

When you determined the derivatives of $y = x^2$, $y = x^3$, and $y = x^4$, you expanded $(x + h)^2$, $(x + h)^3$, and $(x + h)^4$. To determine the derivative of $y = x^n$ in the same way, we need to expand $(x + h)^n$. This involves a topic called the binomial theorem, which is beyond the scope of this course. So, we use a different method. Instead of expanding, we factor. Before we apply this method to $y = x^n$, we will apply it to $y = x^2$ and $y = x^3$.

Alternative development of the derivative of $y = x^2$

Use the first-principles definition.

$$f'(x) = \lim_{h\to 0} \frac{f(x + h) - f(x)}{h}$$

Replace $f(x)$ with x^2.

$$f'(x) = \lim_{h\to 0} \frac{(x + h)^2 - x^2}{h}$$

Factor the numerator as a difference of squares; use $a^2 - b^2 = (a - b)(a + b)$.

$$f'(x) = \lim_{h\to 0} \frac{((x + h) - x)((x + h) + x)}{h}$$

$$f'(x) = \lim_{h\to 0} \frac{h(2x + h)}{h}$$

$$f'(x) = \lim_{h\to 0}(2x + h)$$

As h approaches 0, $2x + h$ approaches $2x$.
$f'(x) = 2x$
The derivative of $y = x^2$ is $y' = 2x$.

Alternative development of the derivative of $y = x^3$

Use the first-principles definition.
$$f'(x) = \lim_{h \to 0} \frac{f(x+h) - f(x)}{h}$$
Replace $f(x)$ with x^3.
$$f'(x) = \lim_{h \to 0} \frac{(x+h)^3 - x^3}{h}$$
Factor the numerator as a difference of cubes; use $a^3 - b^3 = (a - b)(a^2 + ab + b^2)$.
$$f'(x) = \lim_{h \to 0} \frac{((x+h) - x)((x+h)^2 + (x+h)x + x^2)}{h}$$
$$f'(x) = \lim_{h \to 0} \frac{h((x+h)^2 + (x+h)x + x^2)}{h}$$
$$f'(x) = \lim_{h \to 0} ((x+h)^2 + (x+h)x + x^2)$$
As h approaches 0, $(x + h)^2 + (x + h)x + x^2$ approaches $x^2 + x^2 + x^2$.
$f'(x) = x^2 + x^2 + x^2$
$f'(x) = 3x^2$
We could continue in this way to determine the derivatives of $y = x^4$, $y = x^5$, and so on. However, the algebraic steps become more awkward. It is better to determine the derivative of $y = x^n$ directly. The algebraic steps need only be completed once, and the result will apply for all natural number exponents.

The derivative of $y = x^n$

To determine the derivative of $y = x^n$, where n is a natural number, we use the first-principles definition.
$$f'(x) = \lim_{h \to 0} \frac{f(x+h) - f(x)}{h}$$
Replace $f(x)$ with x^n.
$$f'(x) = \lim_{h \to 0} \frac{(x+h)^n - x^n}{h} \quad \ldots ①$$
To factor the numerator, we use the formula in *Take Note*, page 73:
$a^n - b^n = (a - b)(a^{n-1} + a^{n-2}b + \ldots + a^2b^{n-3} + ab^{n-2} + b^{n-1})$
Replace a with $(x + h)$ and b with x.
$(x + h)^n - x^n$
$= ((x + h) - x)((x + h)^{n-1} + (x + h)^{n-2}x + \ldots + (x + h)^2x^{n-3} + (x + h)x^{n-2} + x^{n-1})$
$= h((x + h)^{n-1} + (x + h)^{n-2}x + \ldots + (x + h)^2x^{n-3} + (x + h)x^{n-2} + x^{n-1})$
Therefore, equation ① becomes:

$$f'(x) = \lim_{h \to 0} \frac{h((x + h)^{n-1} + (x + h)^{n-2}x + \ldots + (x + h)^2x^{n-3} + (x + h)x^{n-2} + x^{n-1})}{h}$$

$$f'(x) = \lim_{h \to 0}[(x + h)^{n-1} + (x + h)^{n-2}x + \ldots + (x + h)^2x^{n-3} + (x + h)x^{n-2} + x^{n-1}]$$

As h approaches 0, $(x + h)^{n-1} + (x + h)^{n-2}x + \ldots + (x + h)^2x^{n-3} + (x + h)x^{n-2} + x^{n-1}$ approaches $x^{n-1} + x^{n-2}x + \ldots + x^2x^{n-3} + xx^{n-2} + x^{n-1}$.

$f'(x) = x^{n-1} + x^{n-2}x + \ldots + x^2x^{n-3} + xx^{n-2} + x^{n-1}$

To multiply the powers of x in each term, add the exponents.

$f'(x) = x^{n-1} + x^{n-1} + \ldots + x^{n-1} + x^{n-1} + x^{n-1}$

There are n terms on the right side, and each one is x^{n-1}.

Therefore, $f'(x) = nx^{n-1}$.

Take Note

Power Rule

Let n be any natural number.

The derivative of $y = x^n$ is $y' = nx^{n-1}$.

Notation for derivatives

Just as there are different notations for writing functions, there are different notations for writing their derivatives. We have used the prime notation, y', and $f'(x)$. Two other notations are introduced below.

Leibniz notation

According to the power rule, the derivative of $y = x^5$ is $y' = 5x^4$. We may write the derivative this way:

$$\frac{dy}{dx} = 5x^4$$

We say "dee y by dee x equals ...."

The symbol $\frac{dy}{dx}$ is not a fraction. It is a single symbol that represents the derivative of y with respect to x. This notation was introduced by Gottfried Leibniz about 300 years ago. It is suggested by the definition of the derivative. The slope of secant PQ is $\frac{\Delta y}{\Delta x}$, and the derivative is the limiting value of this slope as the secant approaches the tangent at P: $\frac{dy}{dx} = \lim_{\Delta x \to 0} \frac{\Delta y}{\Delta x}$.

Operator notation

This notation is a variation of Leibniz notation:

$$\frac{d}{dx}(x^5) = 5x^4$$

In this notation, the symbol $\frac{d}{dx}$ is used as an operator, which means "the derivative with respect to x of ...."

Earlier, we stated that we need a new mathematical operation to measure the way quantities change. This is the operation of differentiating a function $f(x)$, indicated by $\frac{d}{dx}$.

Example 1

a) Given $y = x^8$, determine $\frac{dy}{dx}$.

b) Determine $\frac{d}{dx}(x^{11})$.

Solution

Use the power rule.

a) $y = x^8$

$\frac{dy}{dx} = 8x^7$

b) $\frac{d}{dx}(x^{11}) = 11x^{10}$

Example 2

Determine the equation of the tangent to the graph of $y = x^4$ with slope 4. Illustrate the result on a graph.

Solution

$y = x^4$

Differentiate each side.

$y' = 4x^3$

The derivative represents the slope of the tangent at all points on the graph. We need to know the point where the slope of the tangent is 4. Substitute $y' = 4$.

$4 = 4x^3$

$x^3 = 1$

Solve for x.

$x = 1$

The x-coordinate of the point of contact is 1.

Substitute $x = 1$ in the equation $y = x^4$ to obtain $y = 1$.

The coordinates of the point of contact are (1, 1).

The equation of the tangent is:

$y = m(x - p) + q$

$y = 4(x - 1) + 1$

$y = 4x - 3$

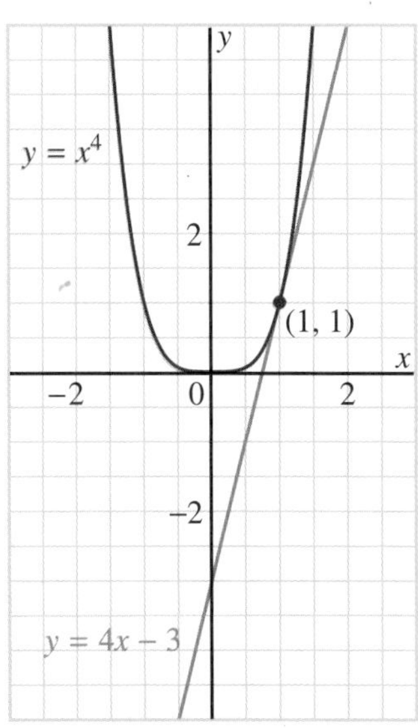

2.2 Exercises

1. **Knowledge/Understanding** Differentiate each function.

   a) $y = x^5$ b) $y = x^{10}$ c) $f(x) = x^{17}$

2. Determine each derivative.

   a) $\frac{d}{dx}(x^9)$ b) $\frac{d}{dx}(x^{20})$ c) $\frac{d}{dx}(x^1)$

B

3. **Communication** Look at the graphs of power functions on page 81. Explain your answer to each question.

   a) Assume n is even.
      i) Why are there no points on the graph of $y = x^n$ below the x-axis?
      ii) Why is the graph of $y = x^n$ symmetrical about the y-axis?
      iii) Suppose the graphs were drawn on the same grid. Would the graphs have any common points of intersection? Explain.

   b) Assume n is odd.
      i) Why are there no points on the graph of $y = x^n$ in the second or fourth quadrants?
      ii) Why is the graph of $y = x^n$ not symmetrical about the y-axis?
      iii) Suppose the graphs were drawn on the same grid. Would the graphs have any common points of intersection? Explain.

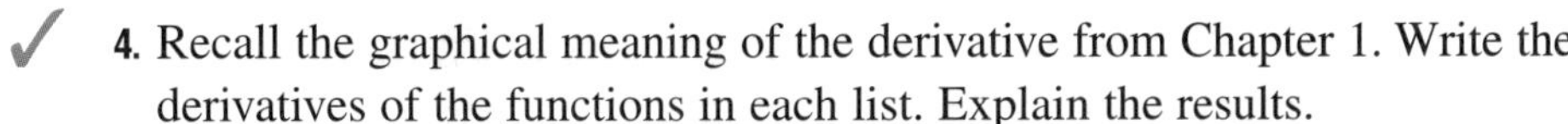

4. Recall the graphical meaning of the derivative from Chapter 1. Write the derivatives of the functions in each list. Explain the results.

a)	b)	c)
$y = x^3$	$y = x^4$	$y = x^3$
$y = x^3 + 1$	$y = x^4 - 2$	$y = (x - 1)^3$
$y = x^3 + 2$	$y = x^4 - 4$	$y = (x - 2)^3$
$y = x^3 + 3$	$y = x^4 - 6$	$y = (x - 3)^3$

5. a) Determine the equation of the tangent to the graph of the function $y = x^3$ at the point $\left(\frac{1}{2}, \frac{1}{8}\right)$.

   b) The tangent in part a intersects the graph at another point. What are the coordinates of this point?

6. **Thinking/Inquiry/Problem Solving** In *Example 2*, we determined the equation of the tangent with slope 4 to the graph of $y = x^4$. The slope is the same as the exponent in the power. Make up some other examples like this, using different powers. Describe a pattern in the equations of the tangents. Explain the pattern.

7. **Application** The graphs of two power functions $f(x) = x^n$ and their derivatives are shown. For each diagram:

i) Explain how to tell which is the function and which is the derivative. Try to find two different ways to do this.

ii) What can you infer about the value of n? Explain.

a)

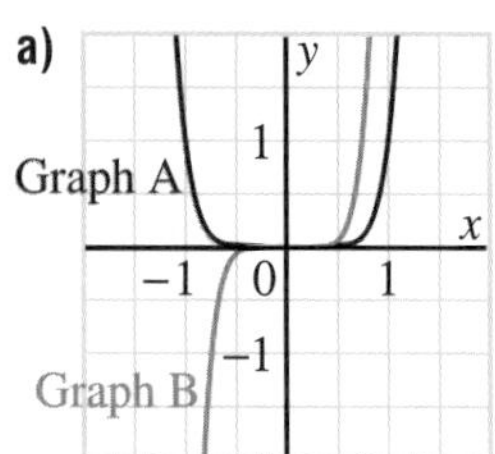

b)

y
1
Graph A
x
−1
0
1
−1
Graph B

8. The function $y = \frac{1}{x}$ can be written in power form as $y = x^{-1}$. In Section 2.4, we shall use the first-principles definition to show that $\frac{d}{dx}(x^{-1}) = -x^{-2}$.

a) Does the power rule justify that $\frac{d}{dx}(x^{-1}) = -x^{-2}$? Explain.

b) The graphs of the functions $f(x) = x^{-1}$ and $g(x) = -x^{-2}$ are shown below left. Does it appear from the graphs that $\frac{d}{dx}(x^{-1}) = -x^{-2}$? Explain.

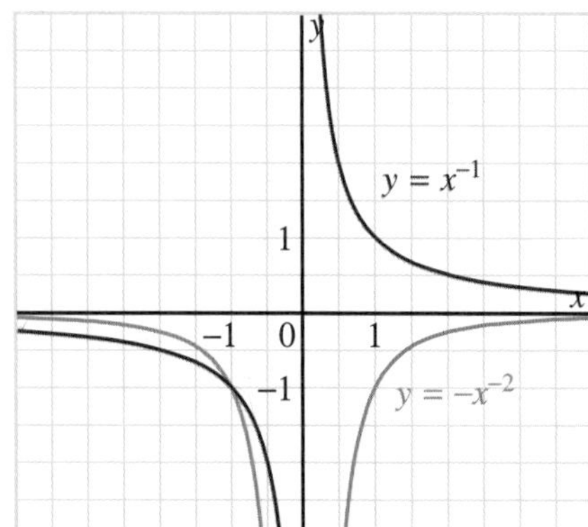

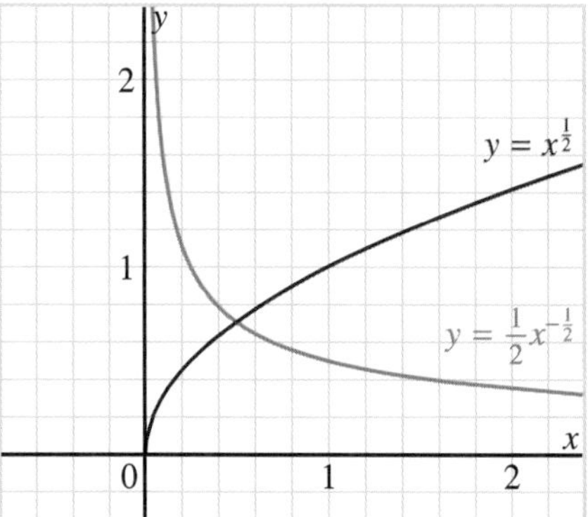

9. The function $y = \sqrt{x}$ can be written in power form as $y = x^{\frac{1}{2}}$. In Section 2.5, we shall use the first-principles definition to show that $\frac{d}{dx}(x^{\frac{1}{2}}) = \frac{1}{2}x^{-\frac{1}{2}}$.

a) Does the power rule justify that $\frac{d}{dx}(x^{\frac{1}{2}}) = \frac{1}{2}x^{-\frac{1}{2}}$? Explain.

b) The graphs of the functions $f(x) = x^{\frac{1}{2}}$ and $g(x) = \frac{1}{2}x^{-\frac{1}{2}}$ are shown above right. Does it appear from the graphs that $\frac{d}{dx}(x^{\frac{1}{2}}) = \frac{1}{2}x^{-\frac{1}{2}}$? Explain.

Exercises 8 and 9 are significant because they involve special cases of the power rule that have not yet been justified using the first-principles definition. It can be proved that the power rule is true for all real numbers *n*, but the proof is beyond the scope of this course.

C

10. a) On the same grid, sketch the graphs of the function $y = |x^n|$ for some positive and negative values of n.

b) What is the derivative of $y = |x^n|$? Explain.

2.3 Some Differentiation Rules

When we differentiate functions using the first-principles definition of the derivative, patterns emerge that we can use to formulate some rules for differentiation. Then we can differentiate functions without using the first-principles definition.

The Derivative of a Constant Multiple

Suppose we use the first-principles definition to differentiate $y = 6x^2$ and $y = x^2$. Compare the results:

$$\frac{dy}{dx} = \lim_{h\to 0}\frac{6(x+h)^2 - 6x^2}{h}$$
$$\frac{dy}{dx} = \lim_{h\to 0}\frac{6(x^2 + 2xh + h^2 - x^2)}{h}$$
$$\frac{dy}{dx} = \lim_{h\to 0}\frac{6h(2x+h)}{h}$$
$$\frac{dy}{dx} = \lim_{h\to 0}(6(2x+h))$$

As h gets closer and closer to 0, $6(2x + h)$ gets closer and closer to $6(2x)$, or $12x$. Therefore, $\frac{dy}{dx} = 12x$

$$\frac{dy}{dx} = \lim_{h\to 0}\frac{(x+h)^2 - x^2}{h}$$
$$\frac{dy}{dx} = \lim_{h\to 0}\frac{x^2 + 2xh + h^2 - x^2}{h}$$
$$\frac{dy}{dx} = \lim_{h\to 0}\frac{h(2x+h)}{h}$$
$$\frac{dy}{dx} = \lim_{h\to 0}(2x+h)$$

As h gets closer and closer to 0, $2x + h$ gets closer and closer to $2x$. Therefore, $\frac{dy}{dx} = 2x$

The derivative of $y = 6x^2$ is 6 times the derivative of $y = x^2$. This occurs because the constant 6 is carried through the calculation as a common factor.

We can use the first-principles definition to prove that the derivative of $y = cf(x)$ is $y = cf'(x)$, where c is a constant.

$$\frac{d}{dx}(cf(x)) = \lim_{h\to 0}\frac{cf(x+h) - cf(x)}{h}$$

Remove c as a common factor.

$$\frac{d}{dx}(cf(x)) = \lim_{h\to 0}\left(c \times \frac{f(x+h) - f(x)}{h}\right)$$

As h gets closer and closer to 0, $\frac{f(x+h) - f(x)}{h}$ gets closer and closer to $f'(x)$, so $c \times \frac{f(x+h) - f(x)}{h}$ gets closer and closer to $cf'(x)$.

Therefore, $\frac{d}{dx}(cf(x)) = c\frac{d}{dx}f(x)$

Take Note

Constant Multiple Rule

The derivative of a constant multiple of a function is the same constant multiple of the derivative of the function.

$$\frac{d}{dx}(cf(x)) = c\frac{d}{dx}f(x)$$

Example 1

a) Given $y = 5x^3$, determine $\frac{dy}{dx}$.

b) Determine $\frac{d}{dx}\left(-\frac{1}{2}x^4\right)$.

Solution

a)
$$\begin{aligned} y &= 5x^3 \\ \frac{dy}{dx} &= 5\frac{d}{dx}(x^3) \\ &= 5(3x^2) \\ &= 15x^2 \end{aligned}$$

b)
$$\begin{aligned} \frac{d}{dx}\left(-\frac{1}{2}x^4\right) &= -\frac{1}{2}\frac{d}{dx}(x^4) \\ &= -\frac{1}{2}(4x^3) \\ &= -2x^3 \end{aligned}$$

The Derivative of a Sum or a Difference

Suppose we use the first-principles definition to differentiate $y = x^2 + x$. We may arrange the calculations as follows.

$$\frac{dy}{dx} = \lim_{h\to 0}\frac{((x+h)^2 + (x+h)) - (x^2 + x)}{h}$$

$$\frac{dy}{dx} = \lim_{h\to 0}\frac{((x+h)^2 - x^2) + ((x+h) - x)}{h}$$

$$\frac{dy}{dx} = \lim_{h\to 0}\left[\frac{x^2 + 2xh + h^2 - x^2}{h} + \frac{x + h - x}{h}\right]$$

$$\frac{dy}{dx} = \lim_{h\to 0}\left[\frac{h(2x+h)}{h} + \frac{h}{h}\right]$$

$$\frac{dy}{dx} = \lim_{h\to 0}[(2x+h) + 1]$$

As h gets closer and closer to 0, $(2x + h) + 1$ gets closer and closer to $2x + 1$.

Therefore, $\frac{dy}{dx} = 2x + 1$

The derivative of $y = x^2 + x$ is the derivative of $y = x^2$ plus the derivative of $y = x$. This occurs because the calculations in the difference quotient can be arranged so those for each function occur separately.

We can use the first-principles definition to prove that the derivative of $y = f(x) + g(x)$ is $y = f'(x) + g'(x)$.

$$\frac{d}{dx}(f(x) + g(x)) = \lim_{h\to 0}\frac{(f(x+h) + g(x+h)) - (f(x) + g(x))}{h}$$

$$\frac{d}{dx}(f(x) + g(x)) = \lim_{h\to 0}\left[\frac{f(x+h) - f(x)}{h} + \frac{g(x+h) - g(x)}{h}\right]$$

As h gets closer and closer to 0, $\frac{f(x+h) - f(x)}{h}$ gets closer and closer to $\frac{d}{dx}f(x)$ and $\frac{g(x+h) - g(x)}{h}$ gets closer and closer to $\frac{d}{dx}g(x)$.

Therefore, $\frac{d}{dx}(f(x) + g(x)) = \frac{d}{dx}f(x) + \frac{d}{dx}g(x)$

A corresponding result is also true for $\frac{d}{dx}\big(f(x) - g(x)\big)$.

Take Note

Sum and Difference Rules

The derivative of a sum or a difference of two functions is the sum or difference of the derivatives.

$$\frac{d}{dx}\big(f(x) + g(x)\big) = \frac{d}{dx}f(x) + \frac{d}{dx}g(x)$$
$$\frac{d}{dx}\big(f(x) - g(x)\big) = \frac{d}{dx}f(x) - \frac{d}{dx}g(x)$$

A *polynomial function* has an equation of the form $y = f(x)$, where the expression on the right side of the equation is a polynomial. To differentiate a polynomial function, we use the constant multiple rule and the sum and difference rules together, as shown in *Example 2*.

Example 2

Differentiate each function.

a) $f(x) = 7x^3 - 5x^2 + x + 6$ **b)** $y = 3 - x + 4x^4$ **c)** $y = (2x + 3)(x - 1)$

Solution

a) $f(x) = 7x^3 - 5x^2 + x + 6$
$f'(x) = 21x^2 - 10x + 1$

b) $y = 3 - x + 4x^4$
$\frac{dy}{dx} = -1 + 16x^3$

c) $y = (2x + 3)(x - 1)$
Expand the binomial product.
$y = 2x^2 + x - 3$
$\frac{dy}{dx} = 4x + 1$

In *Example 2c*, the function is quadratic. The derivative is a linear function. We can use the method of this example to show that the derivative of every quadratic function is a linear function.

Let $y = ax^2 + bx + c$, where $a \neq 0$.
Then $y' = 2ax + b$
Since this is the equation of a linear function, we know that the derivative of every quadratic function is a linear function.

Examples of quadratic functions occurred in Chapter 1. One example involves the height of the water in a pop bottle as it drains out a hole in the bottom. In an earlier exercise (Section 1.1, exercise 12) you drew the graph at the right, and used the tangent to estimate the instantaneous rate of change of water height at $t = 60$ s. You should have found that the result is approximately –0.9 mm/s. When we know the equation of the function, we can use the derivative to determine the exact value of the instantaneous rate of change of water height.

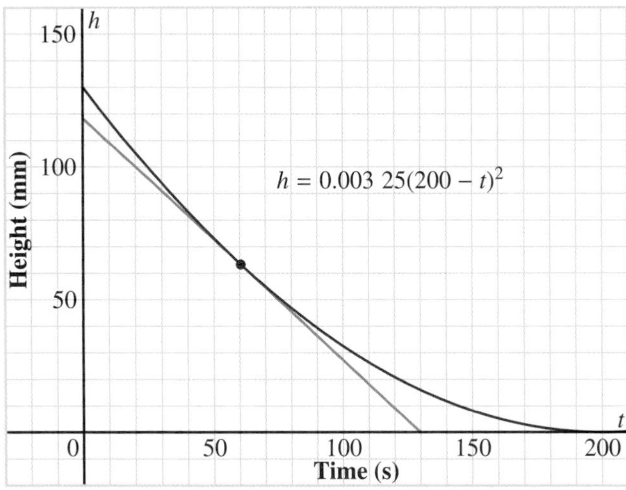

Example 3

Water is draining out a hole in the bottom of a pop bottle. The height of the water, h millimetres, is expressed as a function of time t seconds by the equation $h = 0.003\ 25(200 - t)^2$.

a) Determine the instantaneous rate of change of water height at $t = 60$ s.

b) Draw a graph to show the instantaneous rate of change of water height as a function of time.

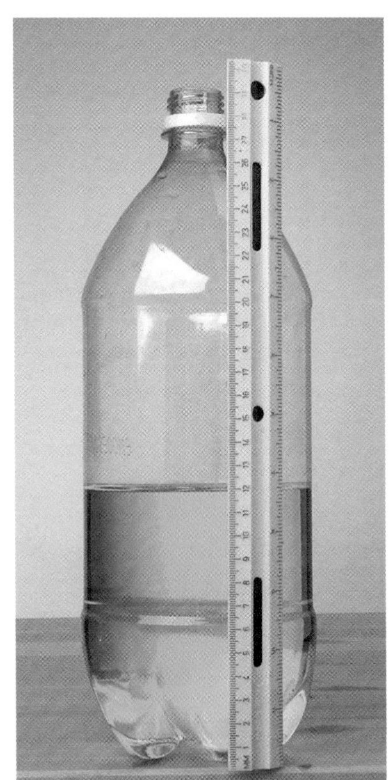

Solution

a) Differentiate the function $h = 0.003\ 25(200 - t)^2$.
Expand the binomial square.

$$h = 0.003\ 25(40\ 000 - 400t + t^2)$$

$$h = 130 - 1.3t + 0.003\ 25t^2$$

$$\frac{dh}{dt} = -1.3 + 0.0065t$$

This expression gives the instantaneous rate of change of water height at any time.
Substitute $t = 60$ to obtain

$$\frac{dh}{dt} = -1.3 + 0.0065(60)$$

$$= -0.91$$

The instantaneous rate of change of water height at $t = 60$ s is –0.91 mm/s.

b) Graph the function $\frac{dh}{dt} = -1.3 + 0.0065t$.

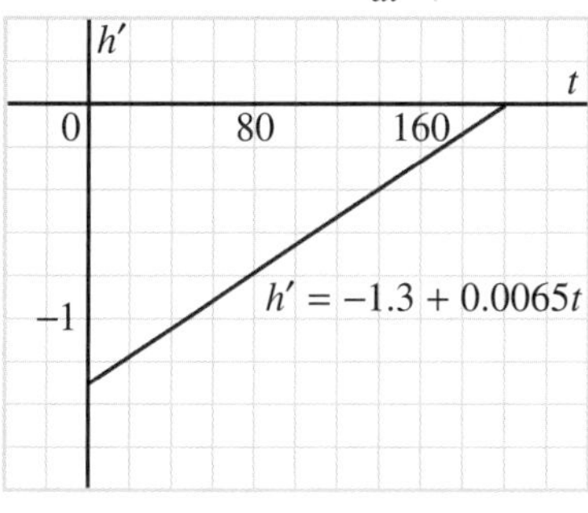

The differentiation rules apply to all functions that have derivatives. Some derivatives from Chapter 1 are shown at the right. Although we have not used the first-principles definition to prove these, the differentiation rules apply to them.

$\frac{d}{dx}(\sin x) = \cos x$

$\frac{d}{dx}(\cos x) = -\sin x$

Example 4

Differentiate each function.

a) $y = 3x + 2\sin x$ **b)** $f(x) = 2x^3 + 4\cos x$

Solution

a) $y = 3x + 2\sin x$

$\frac{dy}{dx} = 3 + 2\cos x$

b) $f(x) = 2x^3 + 4\cos x$

$f'(x) = 6x^2 - 4\sin x$

2.3 Exercises

A

1. Differentiate each function.

a) $y = 4x^2$ **b)** $y = -2x^5$

c) $y = x^3 + x$ **d)** $y = 1 + x^4$

2. Knowledge/Understanding Differentiate each polynomial function.

a) $y = 4x^5 - 2x^3 + 1$ **b)** $y = 2x^7 + 3x^4 - 5x$

c) $f(x) = 2x^4 - 7x^2 - x$ **d)** $f(x) = 2 - 3x + 6x^5$

3. Determine the derivative of each polynomial function.

a) $y = 3x^2 - 2x$ **b)** $y = 6x^4 - 2x^3 + x$

c) $f(x) = 5x^8 + 3x^2 - 1$ **d)** $f(x) = 2 - 4x + 9x^3$

B

4. Recall the formulas for the area, A, and the circumference, C, of a circle: $A = \pi r^2$ and $C = 2\pi r$. Show that $\frac{dA}{dr} = C$. Write this derivative in words.

5. The formulas for the volume, V, and the surface area, A, of a sphere are $V = \frac{4}{3}\pi r^3$ and $A = 4\pi r^2$. Show that $\frac{dV}{dr} = A$. Write this derivative in words.

6. Differentiate each function.

   a) $y = x(x - 5)$ b) $y = 2x(5 - 3x)$

   c) $y = (3x + 4)(2x - 1)$ d) $y = (2x + 3)^2$

7. Determine the derivative of each function.

   a) $y = x + \sin x$ b) $y = 2x^3 + 3\cos x$

   c) $f(x) = \sin x + \cos x$ d) $f(x) = 5\sin x - 2\cos x$

8. **Communication** Recall the graphical meaning of the derivative. Explain your answer to each question. Use graphs in your explanations.

   a) Why is it reasonable that the derivative of $y = 6x^2$ is 6 times the derivative of $y = x^2$?

   b) Why is it reasonable that the derivative of $y = x^2 + x$ is the derivative of $y = x^2$ plus the derivative of $y = x$?

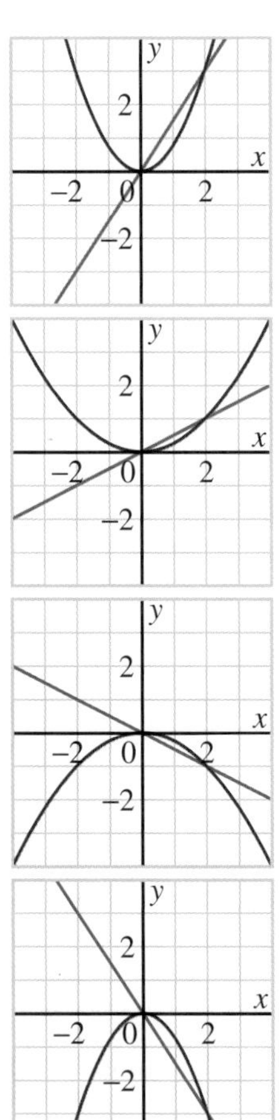

9. **Thinking/Inquiry/Problem Solving** The program *Graphs! Graphs! Graphs!* graphs the function $y = ax^2$ and its derivative on the same grid. You can use a slider to vary a, and observe the effect on the graphs. A few examples are shown at the right.

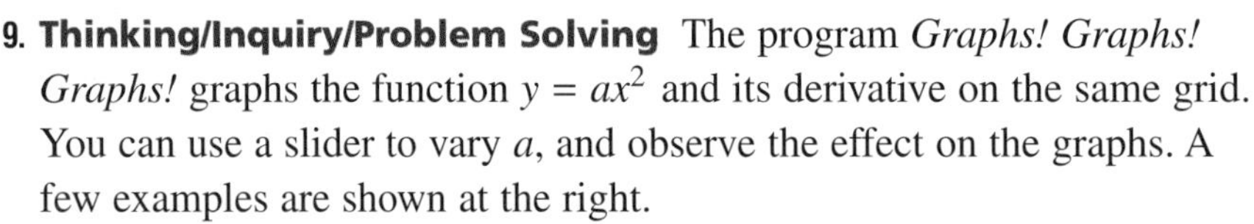

   a) Visualize how the graphs change as a varies. The graph of each function and its derivative intersect at the origin and at one other point. All those other points have a common property. What is this common property?

   b) Differentiate $y = ax^2$ and use the result to prove this property.

   c) What is the significance of the points of intersection of the graphs of a function and its derivative?

10. Exercise 9 suggests investigating a similar problem for other power functions.

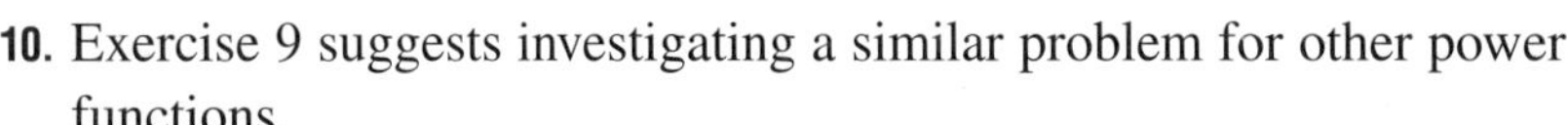

   a) Do the functions $y = ax^3$ and $y = ax^4$ have a similar property? Explain.

   b) Does the function $y = ax^n$ have a similar property? Explain.

11. Refer to *Example 3*, page 90. Calculate the instantaneous rate of change of water height at $t = 0$ s, $t = 120$ s, and $t = 180$ s. Compare your results with those in the table on page 31.

12. The equation $d = 4.9t^2$ expresses the distance, d metres, that a rock falls as a function of time, t seconds. In an earlier exercise (Section 1.2, exercise 8) a graph was given on which tangents are drawn at $t = 1.0$ s and 2.0 s. The graph is repeated here. You determined the slopes of the tangents from the graph and also from the equation using 0.01-s following intervals.

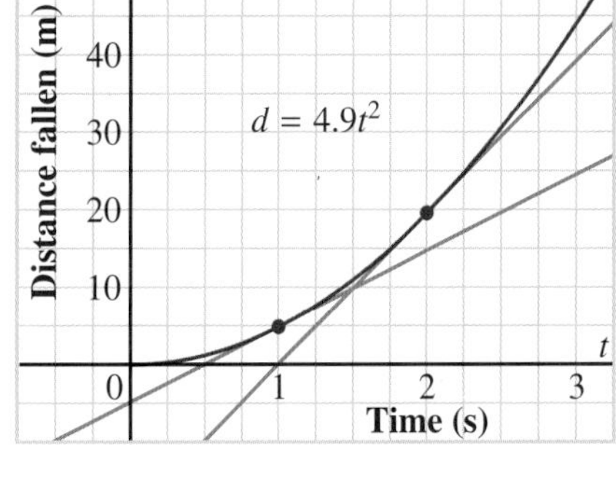

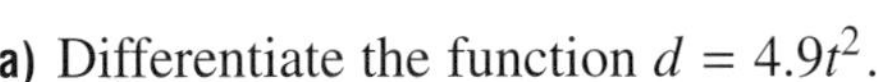

a) Differentiate the function $d = 4.9t^2$.

b) Use your result in part a to determine the slopes of the tangents to the graph of $d = 4.9t^2$ at $t = 1.0$ s and $t = 2.0$ s.

c) Graph the derivative function. Indicate the results of part b on your graph.

13. **Application** The equation $h = -4.89t^2 + 8.09t$ expresses the height of a ball thrown upward, h metres, as a function of time, t seconds. The graphs of h and its derivative h' are shown below.

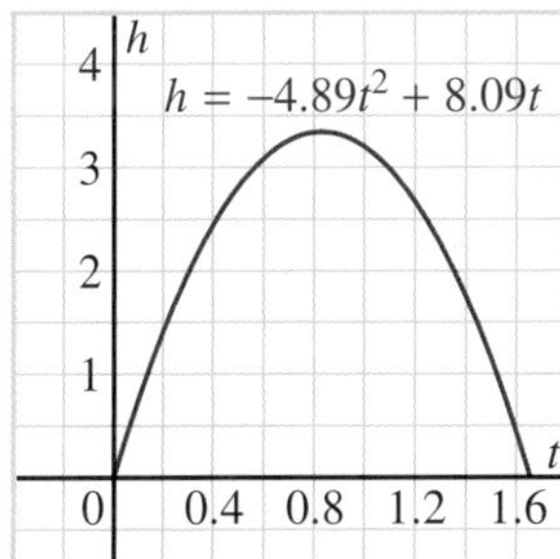

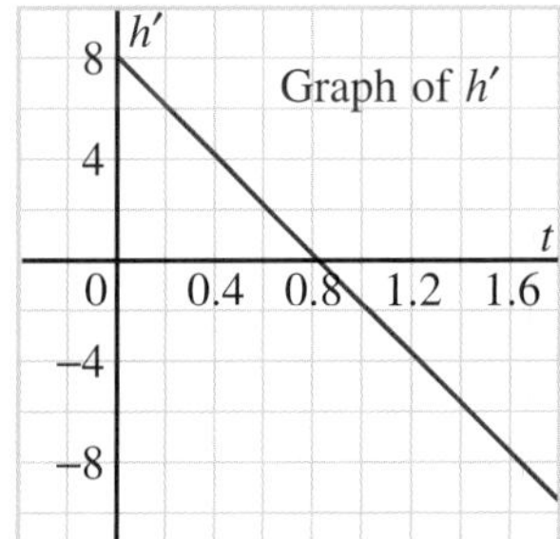

a) Differentiate the function $h = -4.89t^2 + 8.09t$ to determine the equation of the derivative function.

b) Explain what the derivative function represents.

c) Explain how you can tell from both graphs how long it takes the ball to reach its maximum height.

d) Determine the speed of the ball as it hits the ground.

14. Recall the graphical meaning of the derivative.

a) You know that the derivative of $y = x^2$ is $y' = 2x$. Explain how you can determine the derivative of $y = (x - 3)^2$ without expanding the right side.

b) Check your answer to part a by expanding $(x - 3)^2$ then differentiating.

15. Could you use the method of exercise 14a to differentiate these functions? Explain.

a) $y = (x + 5)^3$ **b)** $y = (2x - 1)^2$ **c)** $y = (x^3 - 1)^2$

✓ **16.** The graph of the function $f(x) = x^3 - 4x$ is shown below left.

a) Use this graph to sketch the graph of the derivative function f'.

b) What is the equation of the derivative function?

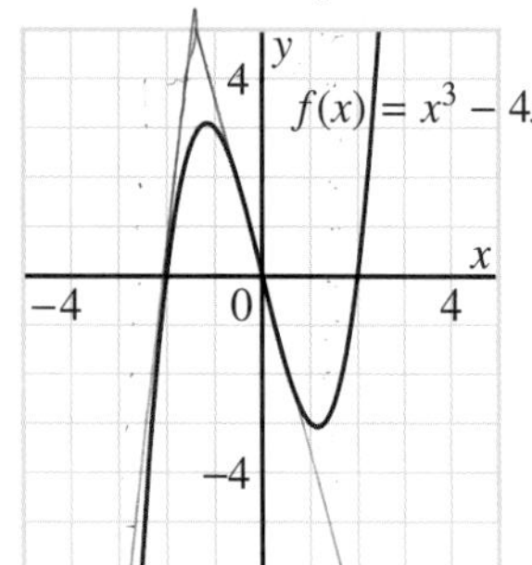

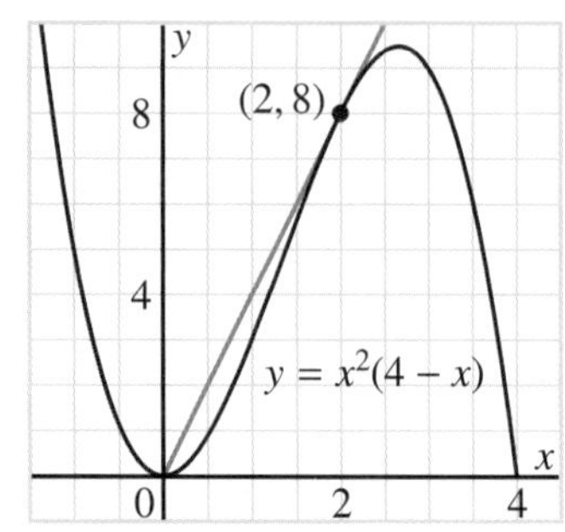

✓ **17.** The diagram above right shows the graph of the function $y = x^2(4 - x)$ and the tangent at the point (2, 8).

a) Determine the equation of the tangent.

b) There is a second tangent parallel to the tangent shown on the graph. Determine the equation of this tangent.

C

18. Consider the functions $f(x) = (x^2 + 1)^2$ and $g(x) = (x^2 + 1)^3$. We might expect that their derivatives are $f'(x) = 2(x^2 + 1)$ and $g'(x) = 3(x^2 + 1)^2$, respectively.

a) Show that this is not true by expanding $(x^2 + 1)^2$ and $(x^2 + 1)^3$, then differentiating the results.

b) Determine functions $h(x)$ and $k(x)$ such that:
$f'(x) = 2(x^2 + 1) \times h(x)$ and $g'(x) = 3(x^2 + 1)^2 \times k(x)$

c) How are the functions h and k in part b related to the given functions f and g?

d) The functions f and g are examples of functions of the form $y = (f(x))^n$. Make a prediction about the derivatives of functions in this form. Make up three examples to check your prediction.

Exercise 18 is significant because the functions being differentiated are powers of a function. In Chapter 4, we will develop a rule, called *the power of a function rule*, that can be used to differentiate functions of the form $y = (f(x))^n$. We will use this rule frequently in this book.

19. The diagram shows a circle with radius r. The radius is increased by a small amount, h, to form a slightly larger circle. Use this diagram and the results of exercise 4. Explain why the instantaneous rate of increase, with respect to r, of the area of the circle equals its circumference.

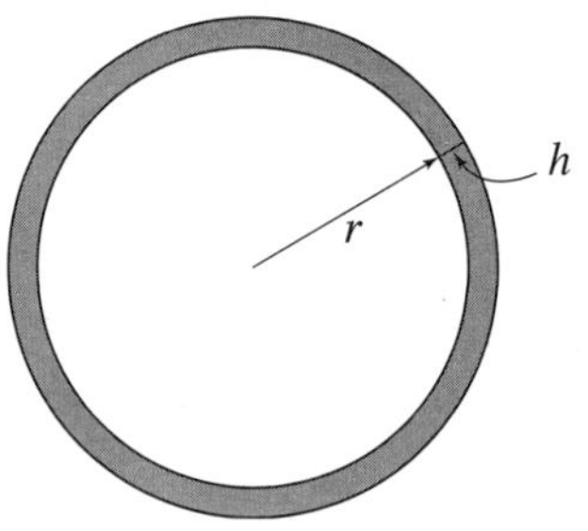

20. Explain why the instantaneous rate of increase, with respect to r, of the volume of a sphere equals the surface area. Use the results of exercise 5 in your explanation.

Self-Check 2.1 – 2.3

1. Use the first-principles definition to determine the derivative of each function.

   a) $y = x + 1$
   b) $y = 2x$
   c) $y = 3x - 1$
   d) $y = -3x + 4$

2. Use the first-principles definition to differentiate each function.

   a) $y = x^2 + 2x$
   b) $y = x^2 - x - 2$
   c) $f(x) = 2x^2 - 3x$
   d) $g(x) = -x(x + 1)$

3. Differentiate each function.

   a) $y = x^3$
   b) $y = x^7$
   c) $f(x) = x^5$
   d) $f(x) = (x - 2)^4$

4. Determine each derivative.

   a) $\frac{d}{dx}(x^8)$
   b) $\frac{d}{dx}(x^{12})$
   c) $\frac{d}{dx}(x^{10})$
   d) $\frac{d}{dx}(x - 3)^7$

5. Write the derivatives of the functions in each list.

   a) $y = x^2$
      $y = x^2 - 1$
      $y = x^2 - 2$
      $y = x^2 - 3$
   b) $y = x^4$
      $y = x^4 + 2$
      $y = x^4 + 3$
      $y = x^4 + 4$
   c) $y = x^5$
      $y = (x - 1)^5$
      $y = (x - 2)^5$
      $y = (x - 3)^5$

6. Determine the equation of the tangent to the graph of $y = x^4$ at the point $(-2, 16)$.

7. Determine the coordinates of the point on the graph of $y = 2x^4 - 1$ where the slope of the tangent is 8.

8. Differentiate each function.

   a) $y = 3x^3 - 2x + 1$
   b) $y = -2x^4 + 3x^2$
   c) $y = (2x - 3)(x + 5)$
   d) $y = (-3x - 5)^2$

9. a) Sketch a graph of $f(x) = x(x - 3)$.
   b) Sketch a graph of the derivative function f', without finding its equation first.

PERFORMANCE ASSESSMENT

10. a) Use the first-principles definition to differentiate the function $y = -x^2 + 2x$.
    b) Sketch the function in part a, and its derivative.
    c) Determine the equation of the tangent at the point $(3, -3)$ on the graph. Draw the tangent on the graph of the function.

2.4 Differentiating Simple Rational Functions

A *rational function* has an equation of the form $y = f(x)$, where the expression on the right side of the equation is a rational expression. A simple example is the *reciprocal function*, $f(x) = \frac{1}{x}$, $x \neq 0$. We can use the first-principles definition to differentiate this function.

$$f'(x) = \lim_{h\to 0} \frac{f(x+h) - f(x)}{h}$$

Replace $f(x)$ with $\frac{1}{x}$.

$$f'(x) = \lim_{h\to 0} \frac{\frac{1}{x+h} - \frac{1}{x}}{h}$$

$$= \lim_{h\to 0} \left(\frac{1}{x+h} - \frac{1}{x}\right) \times \frac{1}{h} \quad \text{To divide by } h \text{, multiply by } \frac{1}{h}.$$

$$= \lim_{h\to 0} \frac{x - (x+h)}{x(x+h)} \times \frac{1}{h}$$

$$= \lim_{h\to 0} \frac{-h}{x(x+h)} \times \frac{1}{h}$$

$$f'(x) = \lim_{h\to 0} \frac{-1}{x(x+h)}$$

As h approaches 0, $\frac{-1}{x(x+h)}$ approaches $\frac{-1}{x(x)}$.

$$f'(x) = -\frac{1}{x^2},\ x \neq 0$$

The derivative of $f(x) = \frac{1}{x}$ is $f'(x) = -\frac{1}{x^2}$, $x \neq 0$.

In previous sections, we simplified $f(x + h) - f(x)$ by expanding then collecting like terms, or by factoring. We cannot do this here, but we can simplify to express $\frac{1}{x+h} - \frac{1}{x}$ as a single rational expression. We multiply numerator and denominator by the common denominator $x(x + h)$ to get $\frac{-h}{x(x+h)}$.

Take Note

The Derivative of the Reciprocal Function

$$\frac{d}{dx}\left(\frac{1}{x}\right) = -\frac{1}{x^2},\ x \neq 0$$

or $\quad \frac{d}{dx}(x^{-1}) = -x^{-2},\ x \neq 0$

Visualize a point moving from left to right along the graph of $y = \frac{1}{x}$. The slope of the tangent is $-\frac{1}{x^2}$, except when $x = 0$. There is no point on the graph when $x = 0$.

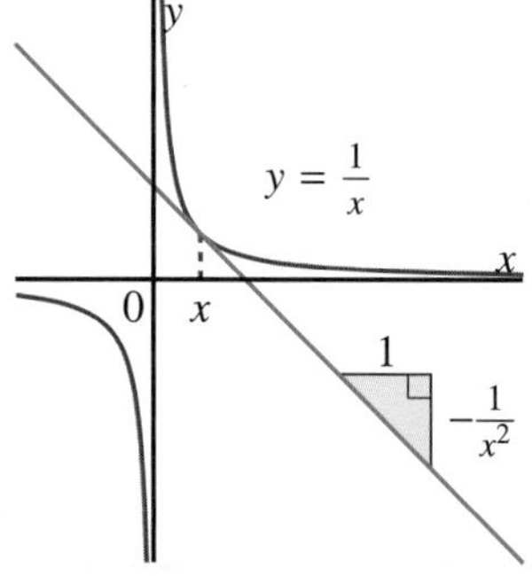

We have now proved that the power rule $\frac{d}{dx}(x^n) = nx^{n-1}$ is true when $n = -1$.

Example 1

Determine the equations of the tangents to the graph of $y = \frac{1}{x}$, $x \neq 0$, with slope $-\frac{1}{4}$. Illustrate the results on a graph.

Solution

The derivative of $y = \frac{1}{x}$ is $y' = -\frac{1}{x^2}$, $x \neq 0$.

Since the slopes of the tangents are $-\frac{1}{4}$, substitute $y' = -\frac{1}{4}$:

$$-\frac{1}{4} = -\frac{1}{x^2}$$

$$x^2 = 4$$

$$x = \pm 2$$

The x-coordinates of the points of contact of the tangents with the graph are ± 2.

Substitute $x = \pm 2$ in the equation $y = \frac{1}{x}$ to obtain $y = \pm\frac{1}{2}$.

The coordinates of the points of contact are $(2, \frac{1}{2})$ and $(-2, -\frac{1}{2})$.

The equation of each tangent has the form $y = m(x - p) + q$.

The equation of the tangent through $(2, \frac{1}{2})$ is:

$$y = -\frac{1}{4}(x - 2) + \frac{1}{2}$$

$$y = -\frac{1}{4}x + 1$$

The equation of the tangent through $(-2, -\frac{1}{2})$ is:

$$y = -\frac{1}{4}(x + 2) - \frac{1}{2}$$

$$y = -\frac{1}{4}x - 1$$

The equations of the tangents are $y = -\frac{1}{4}x \pm 1$.

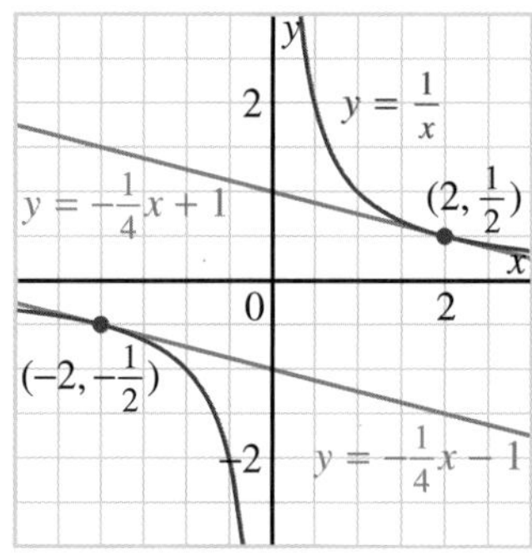

We can use the first-principles definition to determine the derivatives of other simple rational functions. The strategy is always the same. We simplify the difference quotient $\frac{f(x + h) - f(x)}{h}$ that appears in the definition of the derivative, then consider its value as h approaches 0.

In Chapter 4 we will develop a rule, called the quotient rule, which we will use to differentiate rational functions. Rational functions will be studied in detail in Chapter 5.

Example 2

Use the first-principles definition to differentiate the function $f(x) = \frac{1}{1 - 2x}, x \neq 0.5$.

Solution

Use $f'(x) = \lim_{h \to 0} \frac{f(x + h) - f(x)}{h}$.

Replace $f(x)$ with $\frac{1}{1 - 2x}$.

$$f'(x) = \lim_{h \to 0} \frac{\frac{1}{1 - 2(x + h)} - \frac{1}{1 - 2x}}{h}$$

$$= \lim_{h \to 0} \left(\frac{1}{1 - 2(x + h)} - \frac{1}{1 - 2x}\right) \times \frac{1}{h}$$

To divide by h, multiply by $\frac{1}{h}$.

$$= \lim_{h \to 0} \frac{(1 - 2x) - (1 - 2(x + h))}{(1 - 2x)(1 - 2(x + h))} \times \frac{1}{h}$$

Multiply numerator and denominator by $(1 - 2x)(1 - 2(x + h))$.

$$= \lim_{h \to 0} \frac{1 - 2x - 1 + 2x + 2h}{h(1 - 2x)(1 - 2(x + h))}$$

$$= \lim_{h \to 0} \frac{2h}{h(1 - 2x)(1 - 2(x + h))}$$

$$f'(x) = \lim_{h \to 0} \frac{2}{(1 - 2x)(1 - 2(x + h))}$$

As h approaches 0, $\frac{2}{(1 - 2x)(1 - 2(x + h))}$ approaches $\frac{2}{(1 - 2x)(1 - 2x)}$.

$$f'(x) = \frac{2}{(1 - 2x)^2}$$

The derivative of $f(x) = \frac{1}{1 - 2x}$ is $f'(x) = \frac{2}{(1 - 2x)^2}, x \neq 0.5$.

Discuss

Why do we expand and simplify the numerator, but not the denominator?

Example 3

Use the first-principles definition to determine $\frac{d}{dx}\left(\frac{x}{1 - 2x}\right), x \neq 0.5$.

Solution

Let $f(x) = \frac{x}{1 - 2x}$.

Use $f'(x) = \lim_{h \to 0} \frac{f(x + h) - f(x)}{h}$.

Replace $f(x)$ with $\frac{x}{1 - 2x}$.

$$f'(x) = \lim_{h \to 0} \frac{\frac{x+h}{1-2(x+h)} - \frac{x}{1-2x}}{h}$$

$$= \lim_{h \to 0} \frac{(x+h)(1-2x) - x(1-2(x+h))}{(1-2x)(1-2(x+h))} \times \frac{1}{h}$$

$$= \lim_{h \to 0} \frac{x - 2x^2 + h - 2hx - x + 2x^2 + 2xh}{h(1-2x)(1-2(x+h))}$$

$$= \lim_{h \to 0} \frac{h}{h(1-2x)(1-2(x+h))}$$

$$f'(x) = \lim_{h \to 0} \frac{1}{(1-2x)(1-2(x+h))}$$

As h approaches 0, $\frac{1}{(1-2x)(1-2(x+h))}$ approaches $\frac{1}{(1-2x)(1-2x)}$.

$$f'(x) = \frac{1}{(1-2x)^2}$$

Therefore, $\frac{d}{dx}\left(\frac{x}{1-2x}\right) = \frac{1}{(1-2x)^2}, x \neq 0.5$

Example 4

Differentiate the function $f(x) = 4x^2 + \frac{3}{x}, x \neq 0$.

Solution

$f(x) = 4x^2 + \frac{3}{x}, x \neq 0$

Use the constant multiple and sum rules.

$f'(x) = 8x + 3\left(-\frac{1}{x^2}\right)$

$f'(x) = 8x - \frac{3}{x^2}, x \neq 0$

In *Example 4*, we could have added to express the function in a different form:

$$f(x) = 4x^2 + \frac{3}{x}$$

$$f(x) = \frac{4x^3 + 3}{x}$$

We did not do this because we do not have a rule, at this time, for differentiating the quotient of two functions. This rule will be developed in Chapter 4.

In 2.4 Exercises, it is assumed that each rational function is defined for all permissible values of the variable. So, we do not include the restrictions on the variable when we write a rational function.

2.4 Exercises

1. **Application** Use the results of *Examples 2* and *3*. Write the derivative of each rational function.

   a) $f(x) = \frac{2}{1 - 2x}$ b) $f(x) = \frac{3}{1 - 2x}$

   c) $f(x) = \frac{-x}{1 - 2x}$ d) $f(x) = \frac{5x}{1 - 2x}$

2. The function $y = \frac{1}{x}$ is given.

   a) At each value of x, what is the instantaneous rate of change of y with respect to x?

      i) $x = 0.25$ ii) $x = 1$ iii) $x = 2$ iv) $x = 4$

   b) Interpret each result in part a graphically.

B

3. **Knowledge/Understanding** Use the first-principles definition to differentiate each rational function.

   a) $f(x) = \frac{1}{x - 1}$ b) $f(x) = \frac{x}{x - 1}$

4. **Communication**

   a) When two different functions have the same derivative, how are their graphs related?

   b) Explain why the graphs of the functions in exercise 3 are related in this way.

5. In *Example 2*, we found that the derivative of $f(x) = \frac{1}{1 - 2x}$ is $f'(x) = \frac{2}{(1 - 2x)^2}$. The graph of f is shown at the right. Use this graph as a guide to sketch the graph of the derivative function f'.

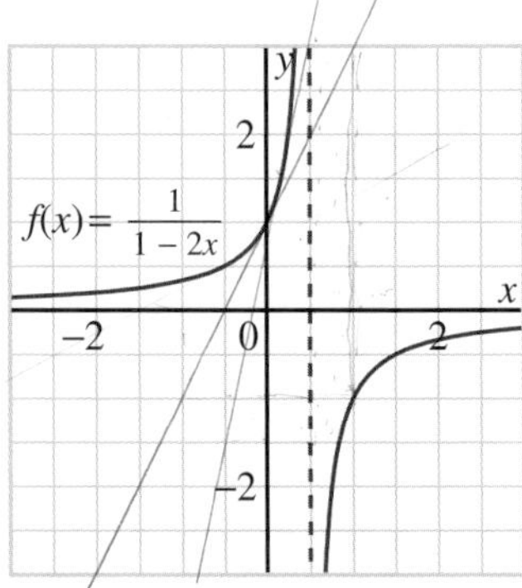

6. Differentiate each function. Use the differentiation rules.

   a) $y = 3x^4 + \frac{1}{x}$ b) $y = 2x^2 - \frac{3}{x}$

   c) $f(x) = 2x^2 + 3x + \frac{4}{x}$ d) $f(x) = \frac{2}{x} - 6x + 3x^3$

7. Use the first-principles definition to differentiate each rational function.

   a) $y = \frac{x}{x + 3}$ b) $y = \frac{-3}{x + 3}$ c) $y = \frac{x - 3}{x + 3}$

8. Explain how the functions and their derivatives in exercise 7 are related.

9. Determine the equation of the tangent to the graph of $y = \frac{1}{1 - 2x}$ at the point $(1, -1)$.

10. The relationship between the pressure and the volume of a sample of helium in a chemistry experiment is illustrated in Section 1.1, exercise 15. The graph from that exercise is repeated at the right. You used the tangent on the graph to estimate the instantaneous rate of change of pressure at $V = 400$ mL.

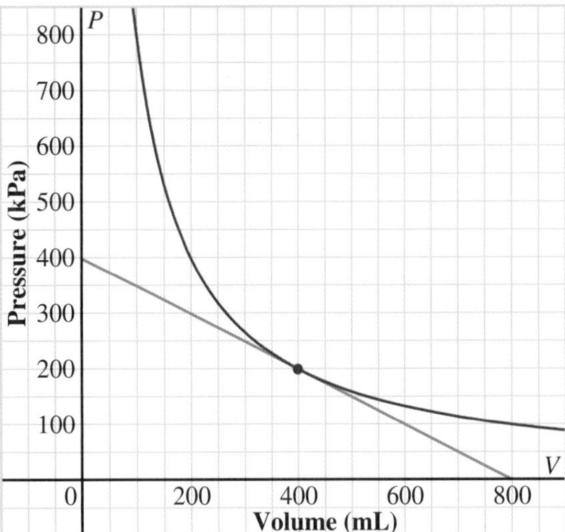

a) The equation of the function is $P = \frac{79\,500}{V}$. Differentiate this function. Use the result to calculate the instantaneous rate of change of pressure at $V = 400$ mL.

b) The table on page 34 shows the estimated instantaneous rates of change of pressure for other volumes. Choose two of these volumes. Use the derivative you calculated in part a to calculate the instantaneous rate of change of pressure at each volume.

11. Determine the equation of the tangent to the graph of $y = \frac{x}{1 - 2x}$ at the point $(2, -\frac{2}{3})$.

12. Use the first-principles definition to determine each derivative.

a) $\frac{d}{dx}\left(\frac{2}{x - 1}\right)$ b) $\frac{d}{dx}\left(\frac{x}{2x - 1}\right)$ c) $\frac{d}{dx}\left(\frac{2}{x^2 - 1}\right)$

13. **Thinking/Inquiry/Problem Solving** You know that the derivative of the sum of two functions is the sum of their derivatives. Is the derivative of the quotient of two functions equal to the quotient of their derivatives? Use an example from this section to justify your answer.

14. In *Examples* 2 and 3, each function f is not defined when $x = 0.5$. Each derivative f' is also not defined when $x = 0.5$. Suppose a function is not defined for some value of x. Will its derivative also not be defined for the same value of x? Explain.

15. Determine the equation of the tangent to the graph of $y = \frac{1}{x}$ at each point on the graph.

a) $(1, 1)$ b) $(4, \frac{1}{4})$ c) $(\frac{1}{4}, 4)$

16. a) Determine the equation(s) of the tangent(s) to the graph of $y = \frac{1}{x}$ with each given slope, if possible.

i) $-\frac{1}{9}$ ii) $\frac{1}{9}$ iii) 0

b) Explain why there are two tangents in part a) i but none in parts a) ii and iii.

17. Use the first-principles definition to differentiate each rational function.

a) $y = \frac{x}{x^2 + 1}$ b) $y = \frac{x^2}{x + 1}$ c) $y = \frac{1}{x^2 + x}$

18. The graphs of the functions in exercise 17 are shown below. Sketch the graphs of their derivatives.

a)

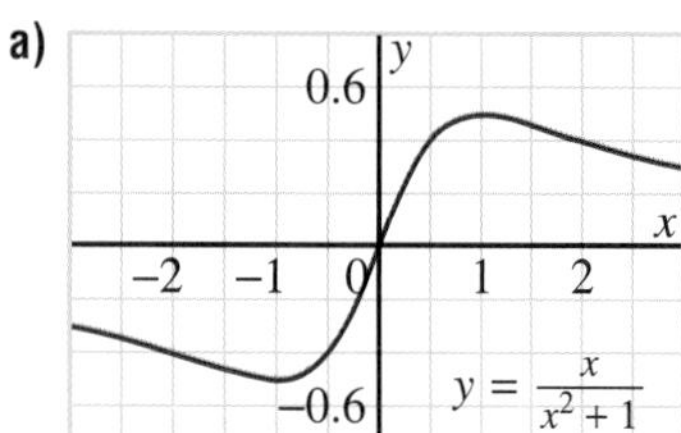

b)

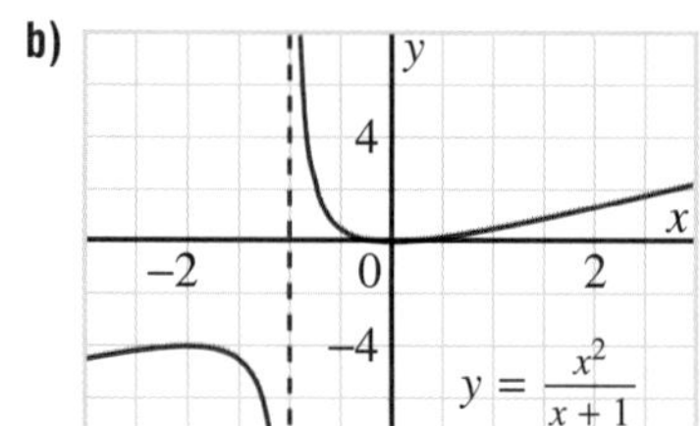

c)

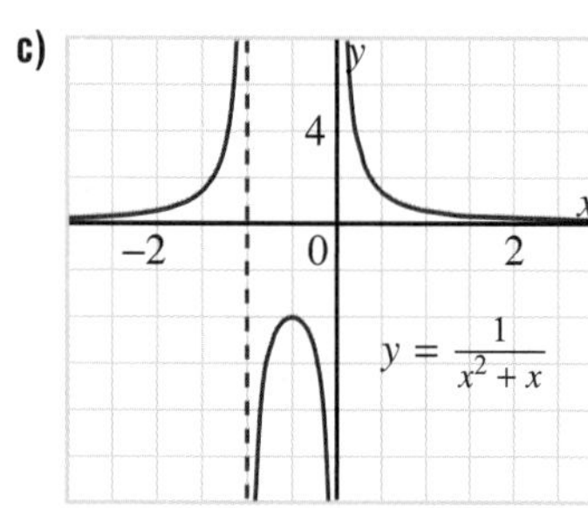

19. A tangent to the graph of $y = \frac{x}{x - 1}$ has slope –4.

a) Determine the equation of the tangent.

b) Is there more than one tangent with slope –4? Explain.

20. Differentiate the function $f(x) = \frac{x - 1}{x}$ in two different ways.

C

21. a) A tangent to the graph of $y = \frac{1}{x}$ has slope m. Determine the equation of the tangent.

b) Determine a formula for the equation of the tangent at the point (x_1, y_1) on the graph of $y = \frac{1}{x}$.

22. Use the first-principles definition to differentiate each function. In part c, n is a natural number.

a) $y = \frac{1}{x^2}$ b) $y = \frac{1}{x^3}$ c) $y = \frac{1}{x^n}$

Exercise 22 is significant because the expressions can be written in power form as x^{-2}, x^{-3}, and x^{-n}. Part c shows that the power rule applies when n is any integer.

23. Differentiate the function $f(x) = \frac{1}{x^2 + 2x + 1}$ in three different ways.

2.5 Differentiating Square-Root Functions

We can use the first-principles definition to determine the derivative of the square-root function $f(x) = \sqrt{x}$, $x \geq 0$.

Use $f'(x) = \lim\limits_{h \to 0} \frac{f(x+h) - f(x)}{h}$.

Replace $f(x)$ with $\sqrt{x}$.

$$f'(x) = \lim_{h \to 0} \frac{\sqrt{x+h} - \sqrt{x}}{h} \qquad \ldots ①$$

We cannot expand $\sqrt{x+h}$, so there is no obvious way to simplify the numerator. We write the expression above in an equivalent form by multiplying it by an expression that equals 1. If we use $\frac{\sqrt{x+h} + \sqrt{x}}{\sqrt{x+h} + \sqrt{x}}$, the numerator is simplified. Hence, we write equation ① as follows:

$$f'(x) = \lim_{h \to 0} \left(\frac{\sqrt{x+h} - \sqrt{x}}{h} \times \frac{\sqrt{x+h} + \sqrt{x}}{\sqrt{x+h} + \sqrt{x}} \right)$$ Use the distributive law in the numerator.

$$f'(x) = \lim_{h \to 0} \left(\frac{\left(\sqrt{x+h}\right)^2 + (\sqrt{x})(\sqrt{x+h}) - (\sqrt{x})(\sqrt{x+h}) - \left(\sqrt{x}\right)^2}{h(\sqrt{x+h} + \sqrt{x})} \right)$$

Recall that $\left(\sqrt{x+h}\right)^2 = x + h$ and $\left(\sqrt{x}\right)^2 = x$.

$$f'(x) = \lim_{h \to 0} \frac{(x+h) - x}{h(\sqrt{x+h} + \sqrt{x})}$$

$$f'(x) = \lim_{h \to 0} \frac{h}{h(\sqrt{x+h} + \sqrt{x})}$$

$$f'(x) = \lim_{h \to 0} \frac{1}{\sqrt{x+h} + \sqrt{x}}$$

As h approaches 0, $\frac{1}{\sqrt{x+h} + \sqrt{x}}$ approaches $\frac{1}{\sqrt{x} + \sqrt{x}}$.

$$f'(x) = \frac{1}{\sqrt{x} + \sqrt{x}}$$

$$f'(x) = \frac{1}{2\sqrt{x}}$$

The derivative of $f(x) = \sqrt{x}$ is $f'(x) = \frac{1}{2\sqrt{x}}$, $x > 0$.

Take Note

The Derivative of the Square-Root Function

$$\frac{d}{dx}\left(\sqrt{x}\right) = \frac{1}{2\sqrt{x}},\ x > 0$$

or $$\frac{d}{dx}\left(x^{\frac{1}{2}}\right) = \frac{1}{2}x^{-\frac{1}{2}},\ x > 0$$

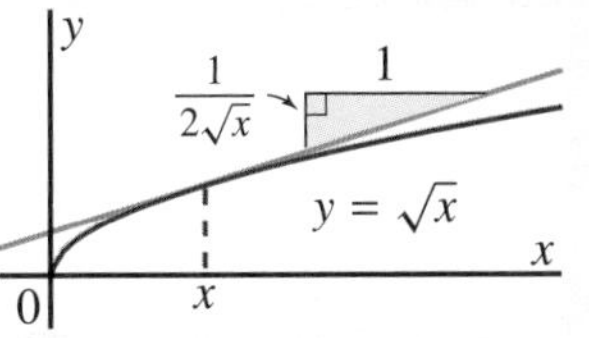

Visualize a point moving from left to right along the graph of $y = \sqrt{x}$. The slope of the tangent is $\frac{1}{2\sqrt{x}}$.

We have now proved that the power rule $\frac{d}{dx}(x^n) = nx^{n-1}$ is true when $n = \frac{1}{2}$.

Example 1

Differentiate each function.

a) $f(x) = 4\sqrt{x},\ x \geq 0$

b) $y = 7x^2 - 3\sqrt{x},\ x \geq 0$

Solution

Use the constant multiple and difference rules.

a)
$$f(x) = 4\sqrt{x}$$
$$f'(x) = 4\left(\frac{1}{2\sqrt{x}}\right)$$
$$= \frac{2}{\sqrt{x}},\ x > 0$$

b)
$$y = 7x^2 - 3\sqrt{x}$$
$$\frac{dy}{dx} = 14x - 3\left(\frac{1}{2\sqrt{x}}\right)$$
$$= 14x - \frac{3}{2\sqrt{x}},\ x > 0$$

Discuss

Why is the restriction on *x* different for *f* and *f'*?

We can use the first-principles definition to determine the derivatives of other square-root functions.

Example 2

Use the first-principles definition to differentiate the function $f(x) = \sqrt{2x + 1},\ x \geq -0.5$.

Solution

Use $f'(x) = \lim_{h \to 0} \frac{f(x + h) - f(x)}{h}$.

Replace $f(x)$ with $\sqrt{2x + 1}$.

$$f'(x) = \lim_{h \to 0} \frac{\sqrt{2(x + h) + 1} - \sqrt{2x + 1}}{h}$$

Multiply the numerator and the denominator of the difference quotient by $\sqrt{2(x + h) + 1} + \sqrt{2x + 1}$.

$$f'(x) = \lim_{h \to 0}\left(\frac{\sqrt{2(x + h) + 1} - \sqrt{2x + 1}}{h} \times \frac{\sqrt{2(x + h) + 1} + \sqrt{2x + 1}}{\sqrt{2(x + h) + 1} + \sqrt{2x + 1}}\right)$$
$$= \lim_{h \to 0} \frac{(2(x + h) + 1) - (2x + 1)}{h(\sqrt{2(x + h) + 1} + \sqrt{2x + 1})}$$
$$= \lim_{h \to 0} \frac{2h}{h(\sqrt{2(x + h) + 1} + \sqrt{2x + 1})}$$
$$f'(x) = \lim_{h \to 0} \frac{2}{\sqrt{2(x + h) + 1} + \sqrt{2x + 1}}$$

As h approaches 0, $\frac{2}{\sqrt{2(x + h) + 1} + \sqrt{2x + 1}}$ approaches $\frac{2}{\sqrt{2x + 1} + \sqrt{2x + 1}}$.

$$f'(x) = \frac{2}{\sqrt{2x + 1} + \sqrt{2x + 1}}$$
$$= \frac{2}{2\sqrt{2x + 1}}$$
$$f'(x) = \frac{1}{\sqrt{2x + 1}},\ x > -0.5$$

In 2.5 Exercises, it is assumed that each square-root function is defined for all permissible values of the variable. So, we do not include the restrictions on the variable when we write a square-root function.

2.5 Exercises

A

1. **Application** Write the derivative of each function.

   a) $f(x) = 2\sqrt{x}$ b) $f(x) = 3\sqrt{x}$

   c) $f(x) = x + \sqrt{x}$ d) $f(x) = x^2 - \sqrt{x}$

2. Use the result of *Example 2*. Write the derivative of each function.

   a) $f(x) = 4\sqrt{2x + 1}$ b) $f(x) = x - \sqrt{2x + 1}$

B

3. **Knowledge/Understanding** Differentiate each function. Use the differentiation rules.

   a) $y = 5\sqrt{x}$ b) $y = x^2 + 2\sqrt{x}$ c) $f(x) = \sqrt{x} + \frac{1}{x}$

4. Use the first-principles definition to differentiate each function.

   a) $f(x) = \sqrt{x + 1}$ b) $f(x) = \sqrt{1 - 2x}$

5. **Communication**

   The derivative of the function $f(x) = \sqrt{x}$ is $f'(x) = \frac{1}{2\sqrt{x}}$.

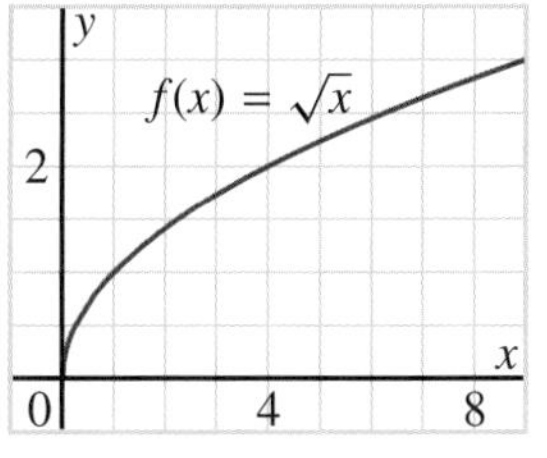

   a) Use these equations to explain why f is defined when $x = 0$, but f' is not defined when $x = 0$.

   b) The graph of f is shown. Use this graph as a guide to sketch the graph of f'.

   c) Use the graphs to explain why f is defined when $x = 0$, but f' is not defined when $x = 0$.

6. **Thinking/Inquiry/Problem Solving** Is the derivative of the square root of a function equal to the square root of the derivative? Use an example to explain your answer.

Many properties of derivatives can be explained both algebraically and geometrically. Exercise 5 is significant because it illustrates one of these properties. The domain of the derivative of a function is not necessarily the same as the domain of the function.

7. Determine the equation of the tangent to the graph of $y = \sqrt{x}$ at each point.

   a) (1, 1) b) (4, 2) c) (9, 3)

Self-Check 2.4, 2.5

1. The derivative of $f(x) = \frac{1}{1+2x}$ is $f'(x) = -\frac{2}{(1+2x)^2}$.
Write the derivative of each function.
a) $f(x) = \frac{2}{1+2x}$ **b)** $f(x) = \frac{1}{2(1+2x)}$ **c)** $f(x) = -\frac{2}{1+2x}$

2. The derivative of $f(x) = \frac{x}{1+2x}$ is $f'(x) = \frac{1}{(1+2x)^2}$.
Write the derivative of each function.
a) $f(x) = \frac{2x}{1+2x}$ **b)** $f(x) = -\frac{2x}{1+2x}$ **c)** $f(x) = -\frac{x}{3(1+2x)}$

3. The graph of $f(x) = \frac{1}{1+2x}$ is shown below left. Use this graph as a guide to sketch the graph of the derivative function f'.

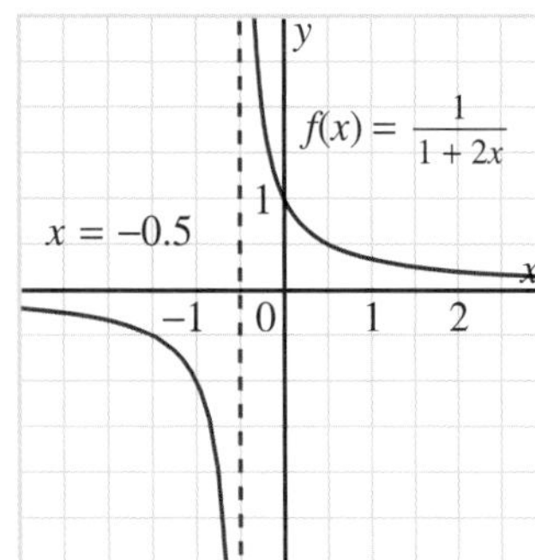

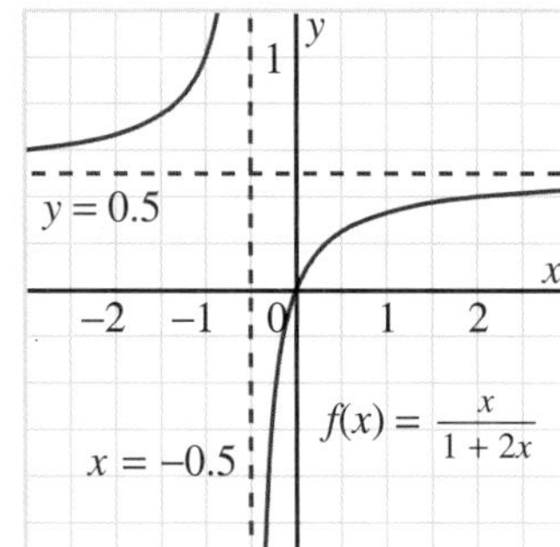

4. The graph of $f(x) = \frac{x}{1+2x}$ is shown above right. Use this graph as a guide to sketch the graph of the derivative function f'.

5. Differentiate each function.
a) $y = 2x^2 - \frac{1}{x}$ **b)** $y = x^3 - x^2 + \frac{1}{x}$ **c)** $y = -4x^2 - x - \frac{3}{x}$

6. Use the first-principles definition to differentiate each function.
a) $y = \frac{1}{x+1}$ **b)** $y = -\frac{1}{x+2}$ **c)** $y = \frac{-x}{x+3}$

7. Differentiate each function.
a) $y = -2\sqrt{x}$ **b)** $y = \sqrt{x} - 3x^3$ **c)** $y = 4\sqrt{x} - x^4 + 2x^2$

PERFORMANCE ASSESSMENT

8. **a)** Determine the equation of the tangent to the graph of $y = -\frac{x}{1+2x}$ at the point $(-1, -1)$.
b) Sketch the graph of $y = -\frac{x}{1+2x}$ and its derivative.

Review Exercises

Mathematics Toolkit

Differentiation Tools

- First-principles derivative definitions

  $f'(x) = \lim_{h \to 0} \frac{f(x+h) - f(x)}{h}$

  $f'(a) = \lim_{b \to a} \frac{f(b) - f(a)}{b - a}$

- Derivative notation

  $y', f'(x), \frac{d}{dx}, \frac{dy}{dx}$

- Derivative of a constant k

  $\frac{d}{dx}(k) = 0$

- Power rule

  $\frac{d}{dx}(x^n) = nx^{n-1}$, where n is a real number

- Constant multiple rule

  $\frac{d}{dx}(cf(x)) = c\frac{d}{dx}f(x)$

- Sum and difference rules

  $\frac{d}{dx}(f(x) + g(x)) = \frac{d}{dx}f(x) + \frac{d}{dx}g(x)$

  $\frac{d}{dx}(f(x) - g(x)) = \frac{d}{dx}f(x) - \frac{d}{dx}g(x)$

- Derivative of the reciprocal function

  $\frac{d}{dx}\left(\frac{1}{x}\right) = -\frac{1}{x^2}, x \neq 0$

  or $\frac{d}{dx}\left(x^{-1}\right) = -x^{-2}, x \neq 0$

- Derivative of the square-root function

  $\frac{d}{dx}\left(\sqrt{x}\right) = \frac{1}{2\sqrt{x}}, x > 0$

  or $\frac{d}{dx}\left(x^{\frac{1}{2}}\right) = \frac{1}{2}x^{-\frac{1}{2}}, x > 0$

Review Exercises

2.1 1. Use the first-principles definition to determine the derivative of each function.

a) $y = 4x + 3$ b) $y = 2x^2 + 5x - 10$

c) $y = 7x(4x - 1)$ d) $y = 9$

2. a) Use the first-principles definition to differentiate the function $y = x^2 + x - 3$.

b) Sketch a graph of the function in part a, and its derivative.

c) Determine the equation of the tangent at the point $(-2, -1)$ on the graph. Draw the tangent on the graph of the function.

3. Determine the equation of the tangent at the point $(1, -3)$ on the graph of $y = 2x^2 - 5$.

2.2 4. Differentiate each function.

a) $y = x^6$ b) $y = x^{15}$ c) $y = x^{30}$

5. Determine the equation of the tangent to the graph of the function $y = x^4$ at the point $(1, 1)$.

6. A tangent to the graph of $y = x^6$ has slope -6.

a) Determine the equation of the tangent.

b) Is there more than one tangent with slope -6? Explain.

2.3 7. Differentiate each polynomial function.

a) $y = 7x^2 + 3x - 8$ b) $y = -x^3 - 5x - 2$

c) $y = (11x + 4)(4x + 9)$ d) $y = 4x^9 + 6x^7 - 5x^6 + 3x$

8. The formulas for the distance fallen s and the velocity v of a falling object are $s = ut + \frac{1}{2}at^2$ and $v = u + at$, where u and a are constants. Show that $\frac{ds}{dt} = v$.

9. Differentiate each function.

a) $y = (2x - 3)^3$ b) $y = \frac{1}{10}(x^4 - 2x)$

c) $y = 7 \sin x + 2 \cos x$ d) $y = -\cos x + 2 \sin x$

10. a) You know that the derivative of $y = x^4$ is $y' = 4x^3$. Explain how you can determine the derivative of $y = (x + 2)^4$ without expanding the right side.

b) Check your answer to part a by expanding $(x + 2)^4$ and differentiating the result.

11. The graph of the function $f(x) = 9x - x^3$ is shown below left.

a) Use this graph to sketch the graph of the derivative function f'.

b) What is the equation of the derivative function?

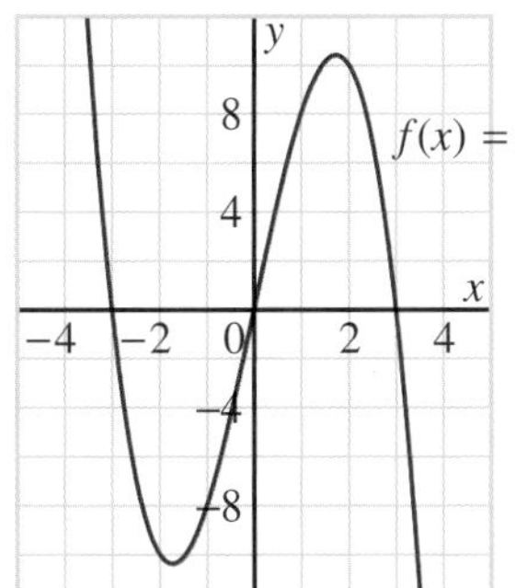

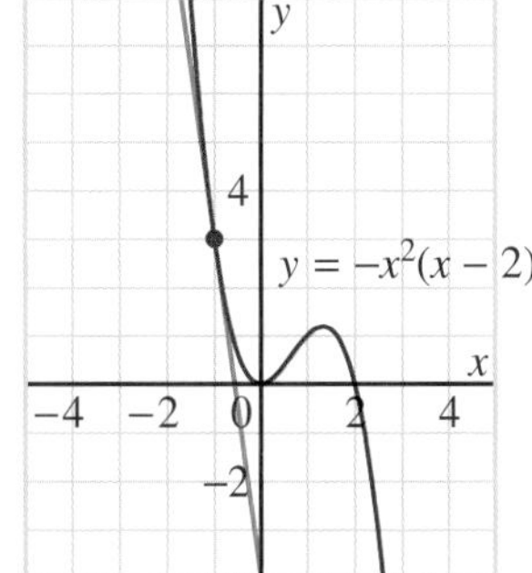

12. The graph of the function $y = -x^2(x - 2)$ and the tangent at the point $(-1, 3)$ are shown above right.

a) Determine the equation of the tangent.

b) There is a second tangent parallel to the tangent shown on the graph. Determine the equation of this tangent.

2.4 13. Use the first-principles definition to determine the derivative of each function.

a) $y = \dfrac{3}{x + 2}$ b) $y = \dfrac{2x^2}{2x - 1}$

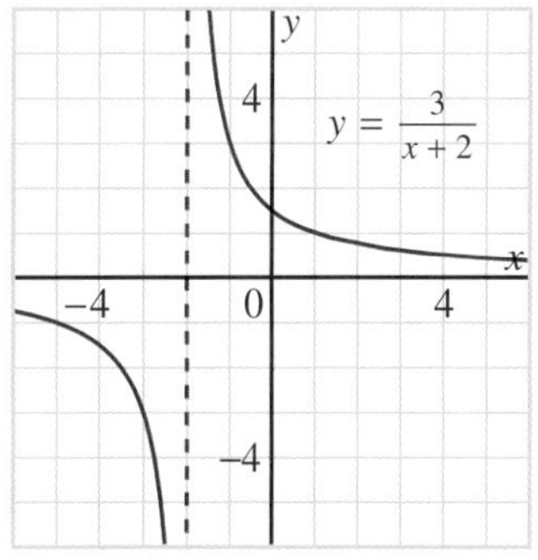

14. The graph of $y = \dfrac{3}{x + 2}$ is shown at the right. In exercise 13a, you determined the derivative of $y = \dfrac{3}{x + 2}$. Use the graph of $y = \dfrac{3}{x + 2}$ as a guide to sketch the graph of the derivative function.

15. Differentiate each function.

a) $y = 4x - \dfrac{5}{x}$ b) $y = 2x^2 + \dfrac{1}{4x}$

c) $y = (x + 5)\left(4 - \dfrac{3}{x}\right)$ d) $y = \dfrac{3}{2x} - \dfrac{7}{6x}$

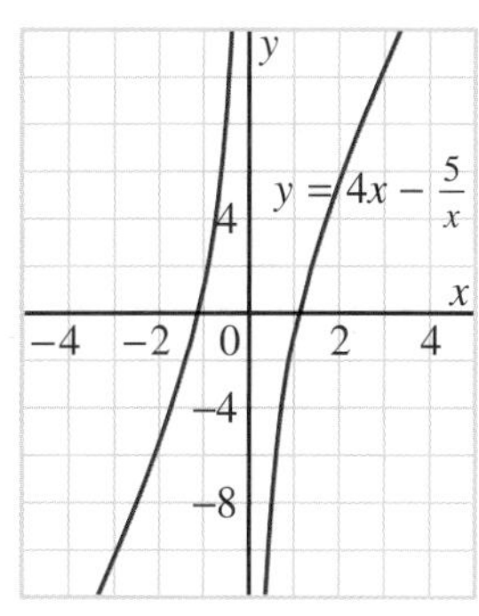

16. The graph of $y = 4x - \dfrac{5}{x}$ is shown at the right. In exercise 15a, you determined the derivative of $y = 4x - \dfrac{5}{x}$. Use the graph of $y = 4x - \dfrac{5}{x}$ as a guide to sketch the graph of the derivative function.

Review Exercises

17. The graph of $y = \frac{3}{x+3}$ is shown below left. Determine the equation of the tangent to the graph at the point $(-4, -3)$.

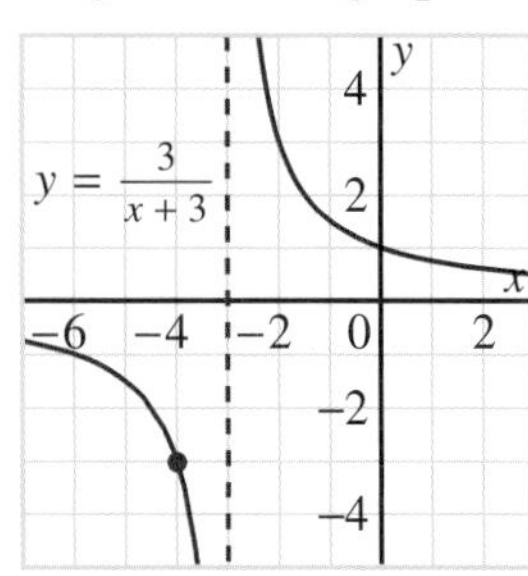

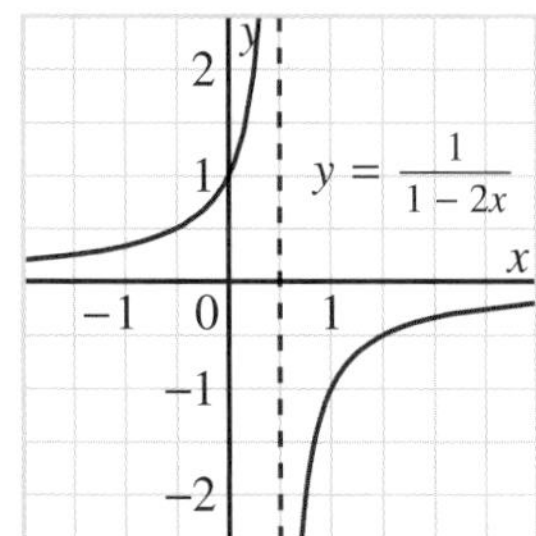

18. The graph of $y = \frac{1}{1-2x}$ is shown above right. A tangent to the graph has slope 2.

a) Determine the equation of the tangent.

b) Is there more than one tangent with slope 2? Explain.

19. Determine the equation of the tangent(s) to the graph of $y = \frac{2}{x}$ with each given slope, if possible.

a) 4

b) $-\frac{1}{4}$

2.5 **20.** Write the derivative of each function.

a) $y = -4\sqrt{x}$

b) $y = 3x - \sqrt{x}$

c) $y = 2x^2 - 5\sqrt{x}$

d) $y = -\sqrt{x} - \frac{3}{x}$

21. Use the first-principles definition to differentiate each function.

a) $y = 4\sqrt{3-x}$

b) $y = \sqrt{8x+2}$

22. Determine the equation of the tangent to the graph of $y = 2\sqrt{x}$ at the point $(1, 2)$.

23. A tangent to the graph of $y = \sqrt{3x+1}$ has slope $\frac{3}{2}$.

a) Determine the coordinates of the point of contact of the tangent.

b) Is there more than one tangent with slope $\frac{3}{2}$? Explain.

24. A tangent to the graph of $y = \sqrt{2x-1}$ has slope 2.

a) Determine the equation of the tangent.

b) Is there more than one tangent with slope 2? Explain.

Self-Test

1. Determine the derivative of each function from first principles.

   a) $y = 2x^2 - x + 3$ b) $y = \frac{-x}{1 - x}$

2. **Knowledge/Understanding** Differentiate each function.

   a) $y = -5x^2 + 6x + 10$ b) $y = \frac{1}{x} - 5x + x^3$

   c) $y = x^4 - 3\sqrt{x} + 2x^3$ d) $y = \frac{2}{x} - 3 - \frac{1}{2}x^4$

3. Determine the equation of the tangent to the graph of each function, at the point indicated.

   a) $y = x(x + 3)$ at the point where $x = 2$

   b) $y = \frac{-x}{1 - x}$ at the point where $x = -3$

4. Use the graph of each function to sketch the graph of the derivative function. Explain how you sketched each graph.

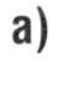

a)

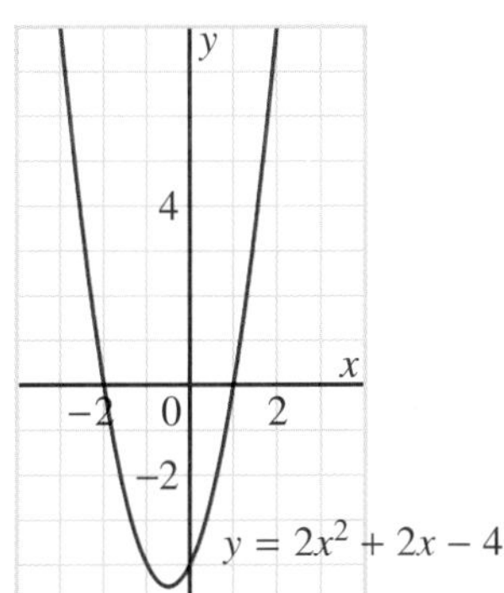

b)

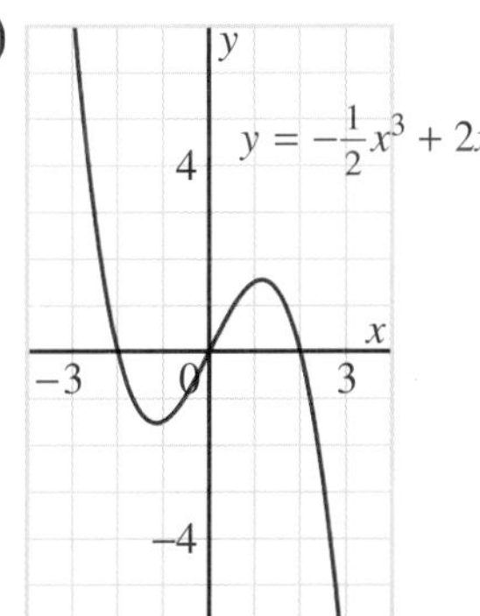

5. **Application** A soccer ball is kicked vertically, with an initial speed of 30 m/s. Its height, h metres, at any time, t seconds, is given by $h = -4.9t^2 + 30t$.

   a) Determine $\frac{dh}{dt}$. Explain what it represents.

   b) Determine the speed of the soccer ball after 1 s. Describe its position.

6. **Communication** Use the graphical meaning of the derivative to explain how to determine the derivative of $y = 2x^2 - 5$ from the derivative of $y = x^2$.

7. **Thinking/Inquiry/Problem Solving** Determine the equations of the tangents to the graph of $y = 2x^2 - 4x - 7$ at the point where $y = -1$.

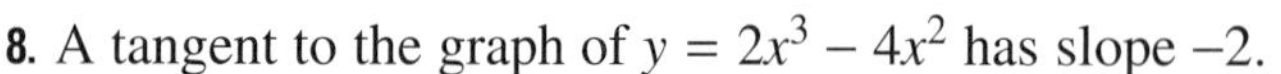

Self-Test

8. A tangent to the graph of $y = 2x^3 - 4x^2$ has slope -2.

   a) Determine the coordinates of the point of contact.

   b) Is there more than one point of contact? Explain.

9. The function $y = \sqrt{x}$ is given.

   a) At each value of x, what is the instantaneous rate of change of y with respect to x?

   i) $x = 1$ ii) $x = 2$ iii) $x = 4$ iv) $x = 0.25$

   b) Interpret each result in part a graphically.

10. Determine the equation of the tangent to the graph of $y = \sqrt{x}$ with each given slope, if possible.

    a) 1 b) 2 c) 3

PERFORMANCE ASSESSMENT

11. a) Write the first-principles definition of the derivative of the function $y = f(x)$.

    b) Use the definition in part a to determine the derivative of each function.

    i) $f(x) = 2x^2$ ii) $f(x) = 2x^2 - 3x$ iii) $f(x) = \frac{1}{2x + 1}$, $x \neq -0.5$

12. a) Graph the function $y = x(4 - x)$.

    b) Determine the equations of the tangents to the graph at the points where the graph intersects the x-axis.

    c) The tangent at (4, 0) forms a right triangle with the coordinate axes. Determine the area of this triangle.

    d) Describe how the tangent at (0, 0) is related to the triangle in part c.

13. Determine the derivative of each function from first principles.

    a) $y = x^5$ b) $y = x^6$

14. You have learned two definitions of the derivative: on pages 75 and 78.

    a) Compare the two definitions. In what ways are they similar? In what ways are they different?

    b) Use the definition on page 78 to determine the derivative of each function. Compare the results with those on pages 75–77.

    i) $f(x) = x^2$ ii) $f(x) = 4x^2 + x$ iii) $y = x(6 - x)$

Mathematical Modelling

Optimal Cooking Time

Suppose we wish to cook a yam. We take it from the fridge, then place it on the counter until its temperature is 10°C. A yam is cooked when the temperature at its centre reaches 90°C. There are two ways to cook the yam: immerse it in boiling water (at a constant temperature of 100°C), or microwave it at medium setting. The cooking time is the same for both methods—exactly 20 minutes.

We shall consider whether a combination of the two methods might reduce the cooking time.

Developing a Mathematical Model

Consider the temperature-time graph for each cooking method. The two graphs are drawn at the right.

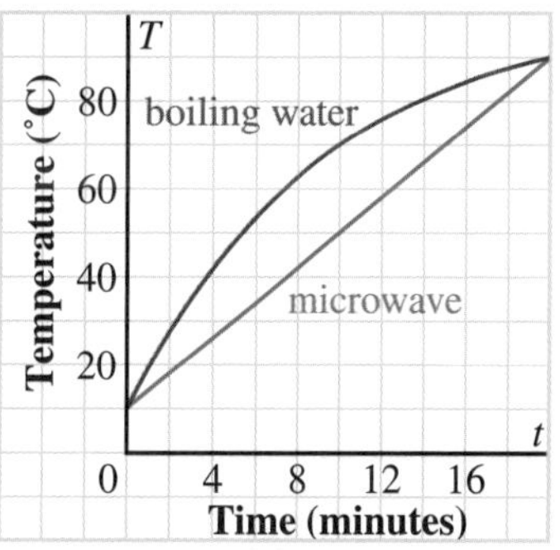

The microwave oven draws power from the electrical outlet at a constant rate, and all this electrical energy is converted to heat energy. Thus, the yam absorbs heat energy at a constant rate and this makes its temperature rise at a constant rate. The microwave graph is a straight line.

For the boiling water, the flow of heat from the water to the yam depends on the temperature difference between them. This flow of heat is greater when the temperature difference is greater. Thus, the temperature of the yam rises more quickly at the beginning than at the end, and the graph has a decreasing slope.

Formulating a Hypothesis

We can see from the slopes of the graphs that the boiling water method produces a large rate of increase in the temperature of the yam at the beginning and a small rate at the end, while the microwave gives a constant intermediate rate. We make the following hypothesis:

If we boil the yam for a time at the beginning then at some point move the yam to the microwave oven, we should obtain a shorter cooking time.

Analysing the Model

A local argument

We want the rate of increase of temperature to be as large as possible; that is, we want the slope of the new temperature-time graph to be as large as we can make it. The slope of the microwave graph is always the same, but the boiling water graph starts with a greater slope and ends with a lesser slope. We place the yam in the boiling water until the rate of increase in temperature is equal to that of the microwave oven, then move the yam to the oven.

This argument is called "local" because it only uses information about the functions at one time. In this case, we only compare the two slopes at the current temperature.

To determine the time we move from the water to the microwave oven, we identify the time at which the boiling water graph has the same slope as the microwave graph. The microwave graph has slope 4, so we want the point on the boiling water graph at which the slope is 4. We place a ruler along the microwave line, then slide the ruler up, keeping it parallel to the microwave graph, until the ruler is a tangent to the boiling water curve. This appears to be when t is just greater than 8, $t \doteq 8.2$.

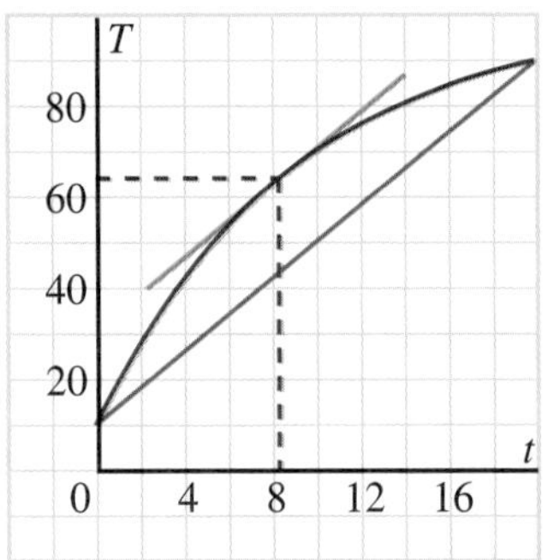

To find the point on the curve where the tangent has slope 4, draw a tangent parallel to the microwave graph.

So, the optimal solution is to put the yam in boiling water for 8.2 min, then move the yam to the microwave oven to finish cooking.

From the graph, the yam is at approximately 64°C when it is placed in the microwave oven.

Its temperature increase is 90°C – 64°C = 26°C.

The rate of increase is 4°C/min.

The time taken is $\frac{26°\text{C}}{4°\text{C/min}} = 6.5$ min

The total cooking time is 8.2 min + 6.5 min = 14.7 min.

This combination method is approximately 5 min less than the conventional time of 20 min.

Key Features of the Model

The two methods of heating the yam have graphs with different shapes. The microwave graph is linear because heat is delivered to the yam at a constant rate. The boiling water graph is concave down because heat is delivered at a decreasing rate as the temperature of the yam gets closer to the water temperature. We use this difference to combine the two methods to get the least cooking time. We begin with the method with the highest rate of temperature increase. That means we use the boiling water for the first part and the microwave oven for the second.

1. a) How do we know the slope of the microwave graph is 4?

   b) What is the equation of the microwave graph?

Making Inferences from the Model

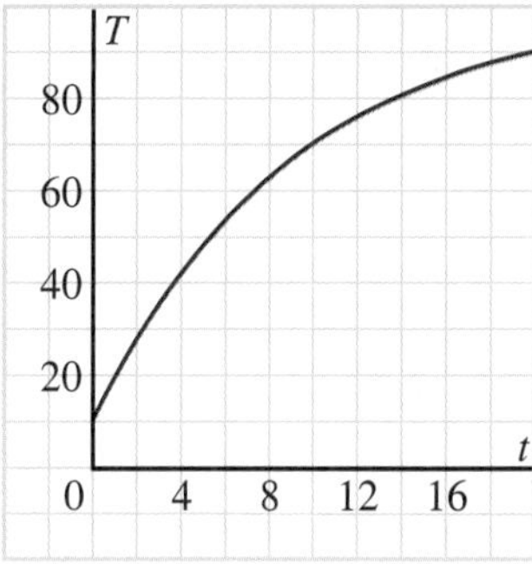

2. A yam can safely be heated in a microwave oven at a slightly higher setting, which produces a constant temperature increase of 5°C/min. Suppose that the boiling water method is still available (and is illustrated in the graph at the right).

   a) Suppose the yam is cooked in the microwave oven only. What is the cooking time?

   b) Copy or trace this graph. Alternatively, your teacher may give you a copy of the graph. On the same grid, draw the graph for the microwave cooking.

   c) Find the strategy that minimizes the total cooking time. Illustrate your solution on your graph. Estimate the total cooking time. Communicate your findings clearly and precisely.

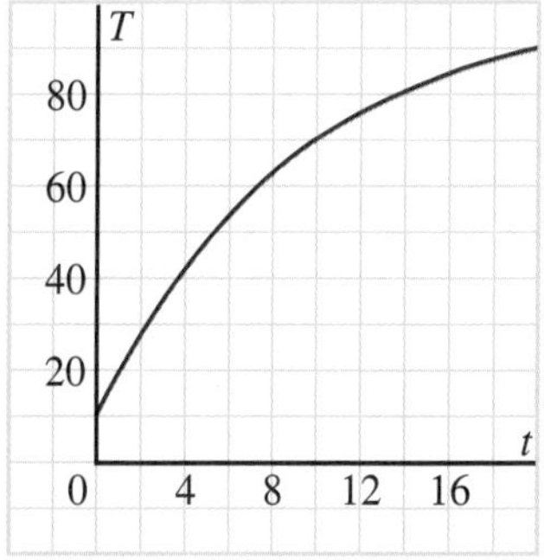

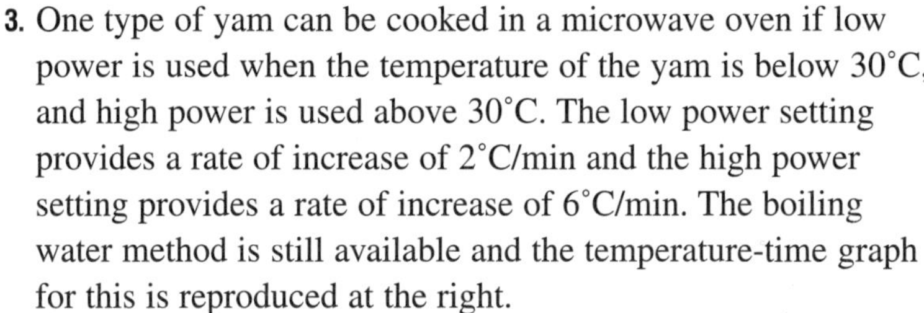

3. One type of yam can be cooked in a microwave oven if low power is used when the temperature of the yam is below 30°C, and high power is used above 30°C. The low power setting provides a rate of increase of 2°C/min and the high power setting provides a rate of increase of 6°C/min. The boiling water method is still available and the temperature-time graph for this is reproduced at the right.

   a) Copy or trace this graph. On the grid, draw the microwave graph for this type of yam.

   b) What strategy will minimize the cooking time? Use a construction on the graph to obtain and justify your answer.

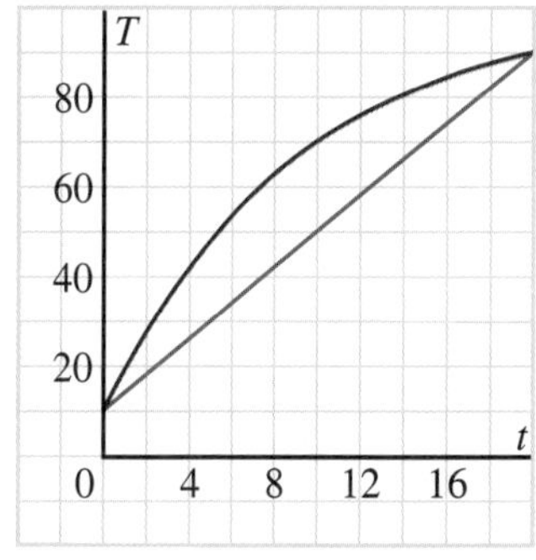

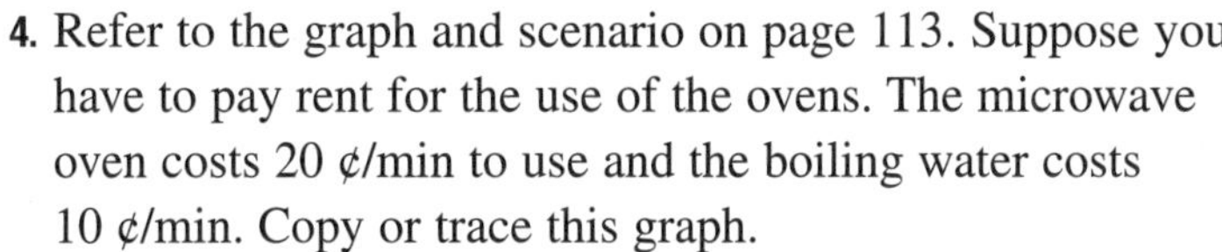

4. Refer to the graph and scenario on page 113. Suppose you have to pay rent for the use of the ovens. The microwave oven costs 20 ¢/min to use and the boiling water costs 10 ¢/min. Copy or trace this graph.

   a) Suppose you start the yam in the boiling water and move it to the microwave oven at $t = 4$ min. What is the total cost?

   b) Suppose you want to minimize the total cost rather than the total time. Use a construction on the graph to determine the optimal time to move the yam from the boiling water to the microwave oven.

Related Problem

5. A large tank that contains 800 L of heavy water must be emptied as quickly as possible. There is an outlet in the bottom that can be opened, allowing the water to run out. With this method, the water flows out more quickly at the beginning than at the end, because the pressure at the outlet is greater when there is more water in the tank. The emptying time for this method is 100 min. The graph of the volume, V litres, in the tank against time is drawn at the right. Another method is to pump out the water at a constant rate of 8 L/min. For technical reasons, these two methods cannot be used simultaneously, but it is possible to use one method, then switch (instantaneously) to the other. Copy or trace this graph.

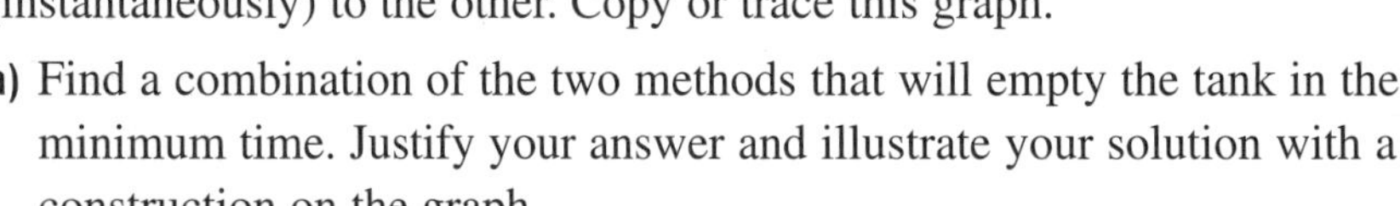

   a) Find a combination of the two methods that will empty the tank in the minimum time. Justify your answer and illustrate your solution with a construction on the graph.

   b) Calculate the minimum emptying time.

   c) The curve is a parabola with vertex at (100, 0). Use this fact to determine the equation of $V(t)$.

   d) Determine $\frac{dV}{dt}$, then use it to check your answer in part b.

6. Write a report to describe what you have learned about the optimal time for cooking a yam or emptying a tank. Include graphs and equations in your report.

Polynomial Functions 3

Curriculum Expectations

By the end of this chapter, you will:

- Determine, through investigation, using graphing technology, various properties of the graphs of polynomial functions.
- Describe the nature of change in polynomial functions of degree greater than two, using finite differences.
- Compare the nature of change in polynomial functions of higher degree with that in linear and quadratic functions.
- Sketch the graph of a polynomial function whose equation is given in factored form.
- Determine an equation to represent a given graph of a polynomial function.
- Demonstrate an understanding of the Remainder Theorem and the Factor Theorem.
- Factor polynomial expressions of degree greater than two, using the Factor Theorem.
- Determine, by factoring, the real or complex roots of polynomial equations of degree greater than two.
- Determine the real roots of non-factorable polynomial equations.
- Write the equation of a family of polynomial functions, given the real or complex zeros.
- Describe intervals and distances, using absolute-value notation.
- Solve factorable polynomial inequalities.
- Solve non-factorable polynomial inequalities.
- Solve problems involving the abstract extension of algorithms.
- Describe the key features of a given graph of a function.
- Identify the nature of the rate of change of a given function, and the rate of change of the rate of change, as they relate to the key features of the graph of that function.
- Determine second derivatives.
- Solve problems of rates of change involving polynomial functions.
- Determine, from the equation of a polynomial function, the key features of the graph of the function using differential calculus, and sketch the graph by hand.
- Sketch the graphs of the first and second derivative functions, given the graph of the original function.
- Sketch the graph of a function, given the graph of its derivative function.

Necessary Skills

1. Review: Quadratic Functions

You studied quadratic functions in grade 10. These functions are examples of polynomial functions.

Recall that the equation of a quadratic function, $y = ax^2 + bx + c$, can be written in the form $y = a(x - p)^2 + q$. The graph is a parabola that is congruent to the parabola $y = ax^2$ and has vertex at (p, q). When $a > 0$, the parabola opens up and is said to be concave up (below left). The vertex is the minimum point on the graph. It is where the least y-value occurs, at $y = q$.

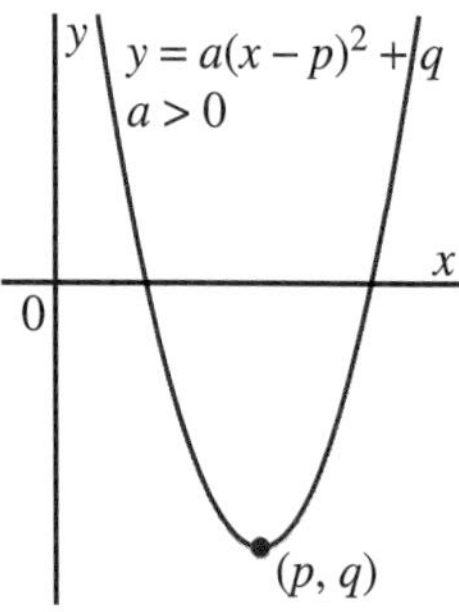

The parabola is concave up.

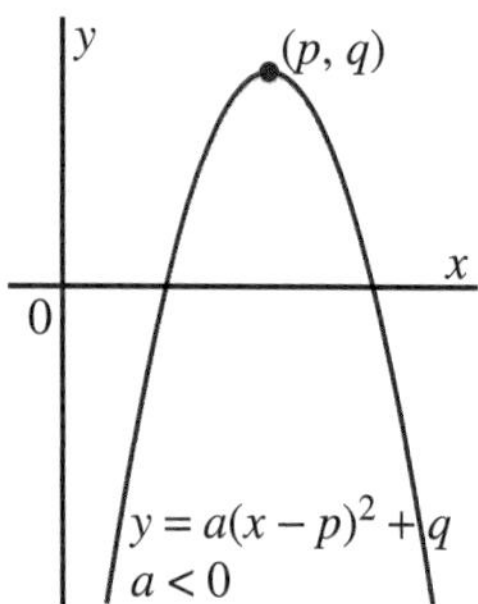

The parabola is concave down.

When $a < 0$, the parabola opens down and is said to be concave down (above right). The vertex is the maximum point on the graph. It is where the greatest y-value occurs, at $y = q$.

Example 1

For the quadratic function $y = 2(x - 1)^2 - 3$:

a) State the coordinates of the vertex.

b) State the equation of the axis of symmetry.

c) State the minimum or maximum value of the function.

d) Sketch a graph of the function.

Solution

Compare $y = 2(x - 1)^2 - 3$ with $y = a(x - p)^2 + q$.

a) The vertex is (p, q); that is, $(1, -3)$.

b) The equation of the axis of symmetry is $x - p = 0$; that is, $x - 1 = 0$, or $x = 1$.

c) Since $a > 0$; that is, $2 > 0$, the parabola is concave up. The vertex is the minimum point on the graph. The minimum value is $y = q$ when $x = p$; that is, $y = -3$ when $x = 1$.

d) The vertex is $(1, -3)$. Choose another x-value, such as $x = -1$.
Substitute $x = -1$ in $y = 2(x - 1)^2 - 3$.

$$y = 2(-1 - 1)^2 - 3$$
$$= 5$$

Another point on the graph is $(-1, 5)$.
Reflect this point in the axis of symmetry, $x = 1$.
Another point is $(3, 5)$.
Join the points to form a smooth curve.

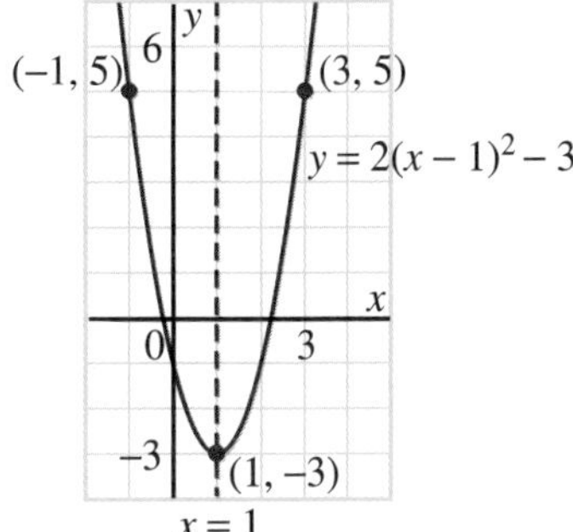

From the graph in *Example 1*, we see that as x increases, y decreases to the vertex, then increases. We say that the function is *decreasing* for $x < 1$, and the function is *increasing* for $x > 1$.

Example 2

For the quadratic function $y = -3(x + 2)^2 + 5$:

a) State the coordinates of the vertex. Is the vertex a maximum or a minimum point on the graph?

b) Sketch a graph of the function.

c) State the domain and range of the function.

d) State the values of x where the function is increasing and decreasing.

Solution

$$y = -3(x + 2)^2 + 5$$

a) The parabola is concave down. Therefore, the vertex $(-2, 5)$ is the maximum point.

b) The vertex is $(-2, 5)$. Choose another point, such as the y-intercept.
Substitute $x = 0$ in $y = -3(x + 2)^2 + 5$.

$$y = -3(0 + 2)^2 + 5$$
$$= -7$$

Another point on the graph is $(0, -7)$.
Reflect this point in the axis of symmetry, $x = -2$.

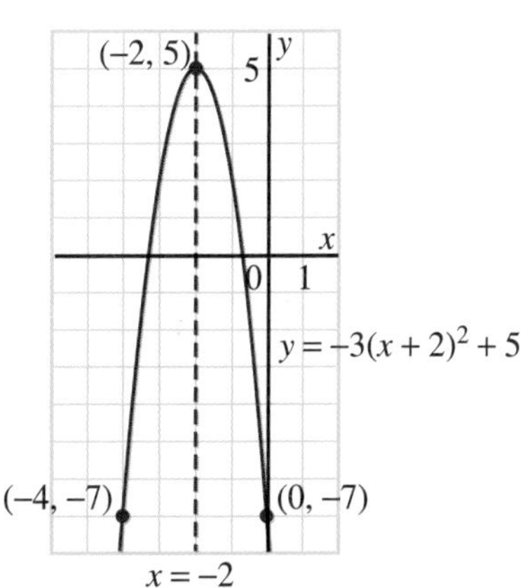

Another point is (–4, –7).
Join the points to form a smooth curve.

c) From the graph,
$D = \Re$
$R = \{y \mid y \in \Re, y \leq 5\}$

d) As x increases, the function is increasing for $x < -2$ and the function is decreasing for $x > -2$.

Exercises

1. For each quadratic function:

i) State the coordinates of the vertex.
ii) State the maximum or minimum value of the function.
iii) State the equation of the axis of symmetry.
iv) Sketch a graph of the function.
v) State the domain and range.

a) $y = 2(x - 1)^2 + 3$ **b)** $y = -3(x + 4)^2 - 5$ **c)** $y = 4(x + 5)(x - 3)$

2. For each quadratic function in exercise 1:

i) State the values of x for which the function is increasing and decreasing.
ii) Describe a relationship between the axis of symmetry and the values of x for which the function is increasing and decreasing.

3. Determine the equation for each graph. State the domain and range of each function.

a)

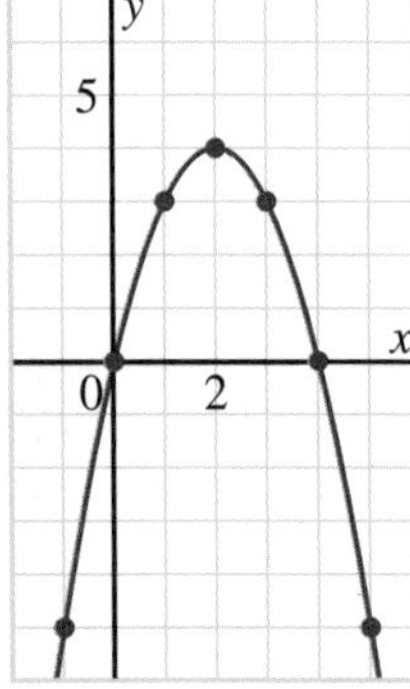

b)

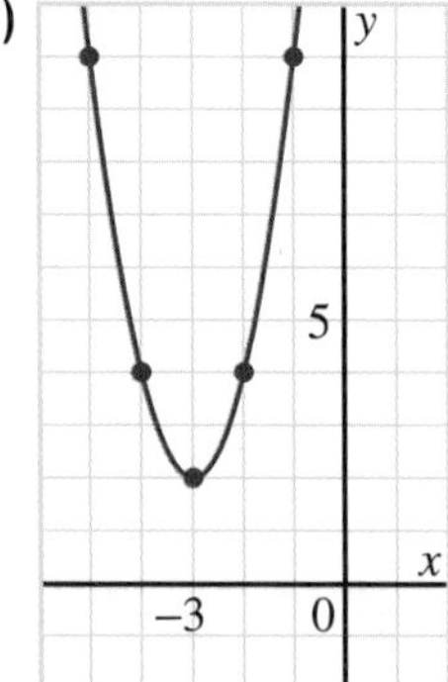

2. New: Solving Quadratic Inequalities

In an inequality, the "greater than," the "greater than or equal to", the "less than," or the "less than or equal to" sign is used to compare two expressions. For example, these are quadratic inequalities:

$2x^2 > 0$; $-3x^2 + 2x \geq 0$; $x^2 + 2x - 3 < 0$; $-x^2 - 3x + 7 \leq 0$

An inequality may be solved algebraically or graphically.

Recall that the quadratic equation $x^2 - 5x + 4 = 0$ can be solved by factoring to obtain $(x - 1)(x - 4) = 0$. The roots are $x = 1$ and $x = 4$. This equation can also be solved by graphing the function $f(x) = x^2 - 5x + 4$, and noting that the graph intersects the x-axis at $x = 1$ and $x = 4$. The x-intercepts of the graph of $f(x) = x^2 - 5x + 4$ are the zeros of f and the roots of $x^2 - 5x + 4 = 0$.

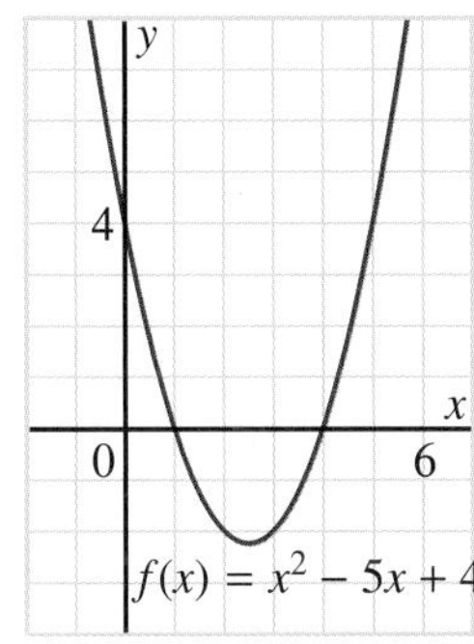

The graph divides the x-axis into three different sets of values for x:

- The values of x that satisfy the equation $x^2 - 5x + 4 = 0$
  These are the x-intercepts of the graph. The roots of the equation are 1 and 4.
- The values of x that satisfy the inequality $x^2 - 5x + 4 > 0$
  These are the values of x where the graph is above the x-axis. The solution of the inequality is $x < 1$ or $x > 4$.
- The values of x that satisfy the inequality $x^2 - 5x + 4 < 0$
  These are the values of x where the graph is below the x-axis. The solution of the inequality is $1 < x < 4$.

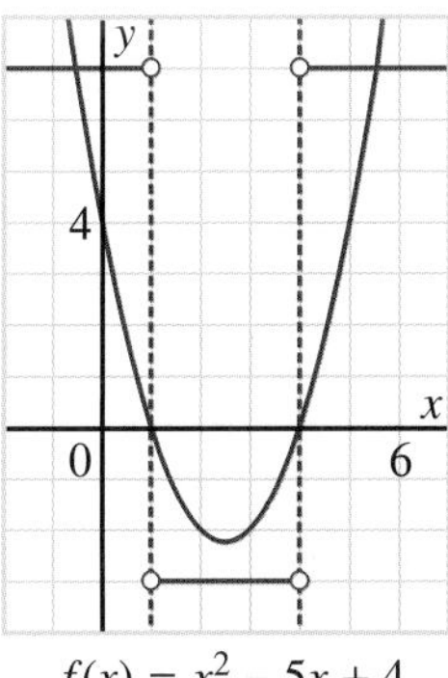

$f(x) = x^2 - 5x + 4$

In general, to solve an inequality, such as $f(x) > 0$, graphically means to determine all the values of x where the graph of the function lies above the x-axis. Similarly, to solve an inequality, such as $f(x) < 0$, graphically means to determine all the values of x where the graph of the function lies below the x-axis.

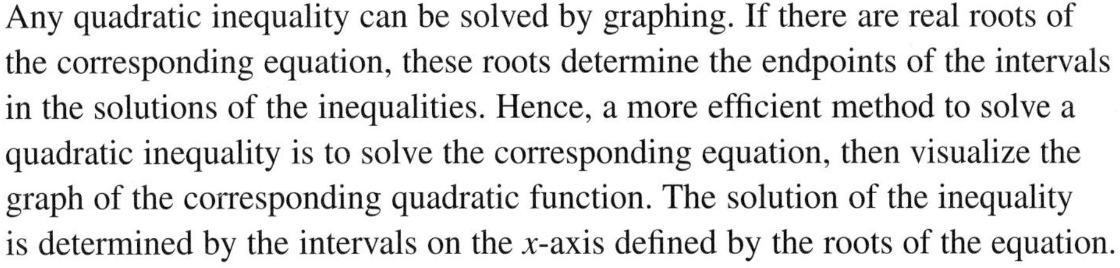

Any quadratic inequality can be solved by graphing. If there are real roots of the corresponding equation, these roots determine the endpoints of the intervals in the solutions of the inequalities. Hence, a more efficient method to solve a quadratic inequality is to solve the corresponding equation, then visualize the graph of the corresponding quadratic function. The solution of the inequality is determined by the intervals on the x-axis defined by the roots of the equation.

Necessary Skills

Example 1

Solve the inequality $5x^2 + 13x - 6 > 0$.

Solution

$5x^2 + 13x - 6 > 0$

Solve the corresponding equation.

$5x^2 + 13x - 6 = 0$

$(5x - 2)(x + 3) = 0$

Either $\quad 5x - 2 = 0 \quad$ or $\quad x + 3 = 0$

$\qquad\quad x = 0.4 \qquad\qquad\quad x = -3$

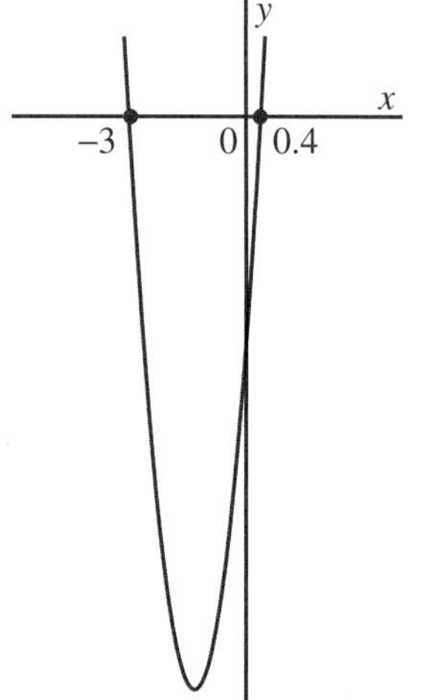

Visualize the graph of the corresponding quadratic function, $f(x) = 5x^2 + 13x - 6$. Since the coefficient of x^2 is positive, the graph is concave up. It intersects the x-axis at $x = -3$ and $x = 0.4$.

The solution of the inequality $5x^2 + 13x - 6 > 0$ is the set of values of x for which the graph of $f(x) = 5x^2 + 13x - 6$ lies above the x-axis. Hence, the solution is $x < -3$ or $x > 0.4$.

In *Example 1*, we did not need to know the coordinates of the vertex; so, a sketch of the graph showing the x-intercepts was sufficient.

The solution of the inequality in *Example 1* can be represented on a number line:

0.4

−6 −5 −4 −3 −2 −1 0 1 2 3 4

In *Example 1*, the numbers −3 and 0.4 are not part of the solution; therefore, they are represented by open circles on the number line. If the inequality had been $5x^2 + 13x - 6 \geq 0$, then these numbers would satisfy the inequality, and the solution would be $x \leq -3$ or $x \geq 0.4$. These numbers would be represented by solid circles on the number line.

Not all inequalities can be solved by factoring. In such cases, we use a graphing calculator to estimate the roots of the quadratic equation and the zeros of the corresponding function.

Example 2

Solve $2x^2 + 3x - 1 \leq 0$.

Solution

The expression $2x^2 + 3x - 1$ cannot be factored. The roots of the equation $2x^2 + 3x - 1 = 0$ may be determined from the graph of the corresponding function $f(x) = 2x^2 + 3x - 1$.

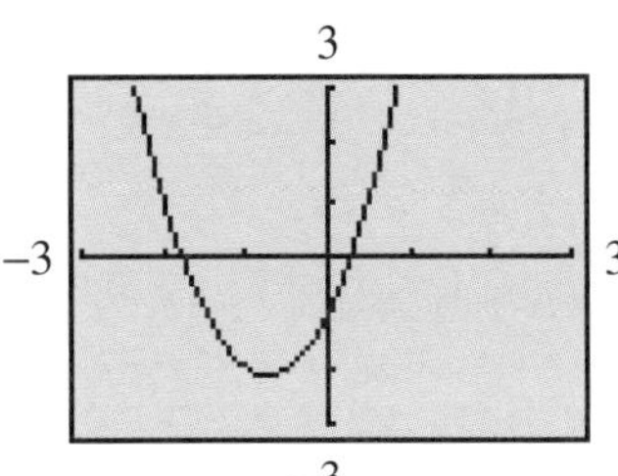

Input $y = 2x^2 + 3x - 1$.
Set the window as shown at the right.
Press GRAPH.
Press 2nd TRACE for CALC.
Then press **2** for zero.

Use ◄ to move the cursor to the left of one zero.
Press ENTER.
Move the cursor to the right of the zero.
Press ENTER ENTER.
$x = -1.78077641$ is displayed.
Repeat the process for the second zero.
$x = 0.28077641$ is displayed.

These two values of x are the roots of the equation $2x^2 + 3x - 1 = 0$ and the zeros of the function $f(x) = 2x^2 + 3x - 1$. Between these two values, the curve is below the x-axis. Therefore, the solution of $2x^2 + 3x - 1 \leq 0$ is approximately $-1.8 \leq x \leq 0.3$.

The roots of the equation in *Example 2* may be found algebraically using the quadratic formula:

$$x = \frac{-b \pm \sqrt{b^2 - 4ac}}{2a}$$

For the equation $2x^2 + 3x - 1 = 0$, the roots are:

$$x = \frac{-3 \pm \sqrt{3^2 - 4(2)(-1)}}{2(2)}$$

$$= \frac{-3 \pm \sqrt{17}}{4}$$

$$x = \frac{-3 + \sqrt{17}}{4} \quad \text{or} \quad x = \frac{-3 - \sqrt{17}}{4}$$

$$\doteq 0.280\,776\,41 \qquad \doteq -1.780\,776\,41$$

The two roots are approximately 0.3 and −1.8.

Necessary Skills

Exercises

1. Use each graph to write the solutions of the inequalities below it.

a) 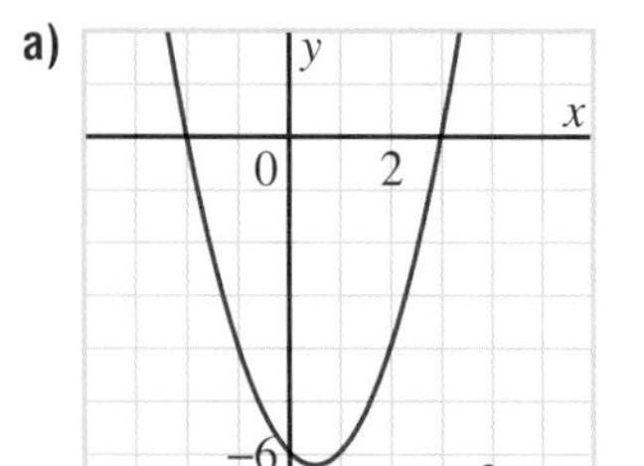

i) $x^2 - x - 6 < 0$

ii) $x^2 - x - 6 \geq 0$

b) 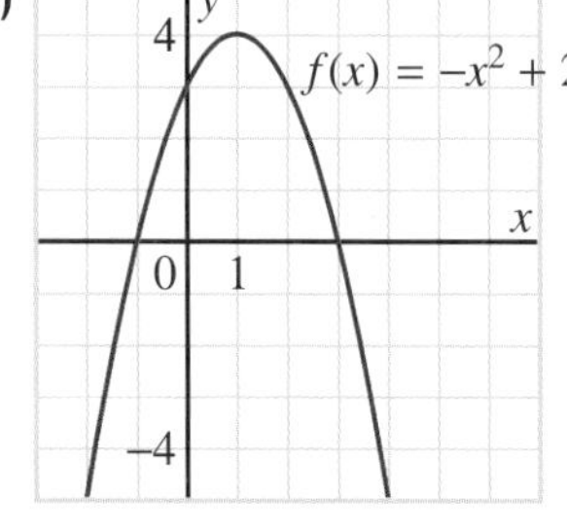

i) $-x^2 + 2x + 3 \leq 0$

ii) $-x^2 + 2x + 3 > 0$

2. Solve each inequality algebraically.

a) $(x - 2)(x + 5) > 0$

b) $2x(x - 5) < 0$

c) $(n - 1)(n + 4) \geq 0$

d) $3(x + 1)(x - 2) \leq 0$

3. Solve each inequality. Show each solution on a number line.

a) $x^2 + 5x + 6 > 0$

b) $m^2 - 2m - 8 > 0$

c) $18 - 3y - y^2 < 0$

d) $3 - x^2 + 2x > 0$

e) $x^2 - 10x + 9 \leq 0$

f) $2x^2 - x - 15 < 0$

4. Solve each inequality.

a) $x^2 - 4x + 4 > 0$

b) $1 - 4x^2 \leq 0$

c) $3x^2 - 4x - 5 \geq 0$

d) $6x^2 + 7x - 20 < 0$

e) $2x^2 + x + 6 > 0$

f) $2x - 2 - x^2 \geq 0$

g) $12x - 4 - 9x^2 \geq 0$

h) $4x^2 + 10x - 7 \leq 0$

5. For each polynomial function, determine the values of x for which $f(x) > 0$.

a) $f(x) = (x + 3)^2 - 2$

b) $f(x) = -2(x - 1)^2 + 5$

c) $f(x) = 4(x + 6)^2 - 3$

3. New: Dividing a Polynomial by a Binomial

In this chapter, we need to be able to divide a polynomial by a binomial. The method of long division in arithmetic may be applied here. We shall divide 783 by 35, then compare the steps as we divide the polynomial $6x^2 + 5x - 5$ by the binomial $3x + 1$.

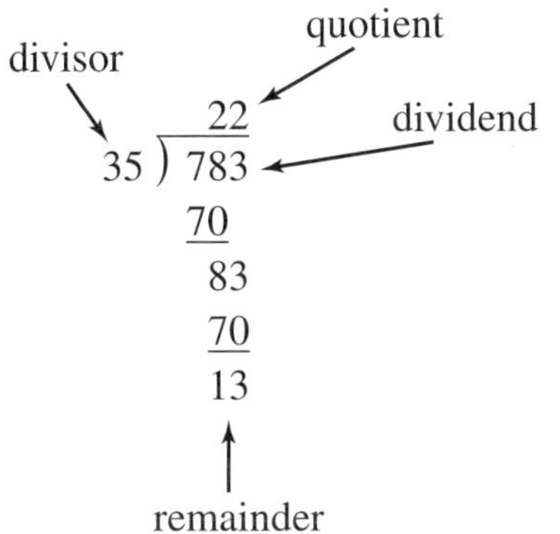

Divide 78 by 35 to get 2.
Multiply 2 by 35 to get 70.
Subtract 70 from 78 to get 8.
Bring down 3.
Divide 83 by 35 to get 2.
Multiply 2 by 35 to get 70.
Subtract 70 from 83 to get 13.
$783 \div 35 = 22$, remainder 13

divisor, quotient, dividend

$$
\begin{array}{r}
2x + 1 \\
3x + 1 \overline{) \; 6x^2 + 5x - 5} \\
\underline{6x^2 + 2x} \\
3x - 5 \\
\underline{3x + 1} \\
-6
\end{array}
$$

remainder

Divide $6x^2$ by $3x$ to get $2x$.
Multiply $2x$ by $3x + 1$ to get $6x^2 + 2x$.
Subtract $6x^2 + 2x$ from $6x^2 + 5x$ to get $3x$.
Bring down -5.
Divide $3x$ by $3x$ to get 1.
Multiply 1 by $3x + 1$ to get $3x + 1$.
Subtract $3x + 1$ from $3x - 5$ to get -6.
$(6x^2 + 5x - 5) \div (3x + 1) = (2x + 1)$, remainder -6

In any division problem, the dividend = (divisor) × (quotient) + remainder; this is the *division statement*.

That is, $783 = 35 \times 22 + 13$ Similarly, $6x^2 + 5x - 5 = (3x + 1)(2x + 1) - 6$

To check, multiply the divisor by the quotient, then add the remainder. The result should be the dividend.

$35 \times 22 + 13 = 783$

$$
\begin{aligned}
(3x + 1)(2x + 1) - 6 &= 6x^2 + 3x + 2x + 1 - 6 \\
&= 6x^2 + 5x - 5
\end{aligned}
$$

The divisor cannot be zero because division by zero is undefined.
So, there is a restriction on the divisor $3x + 1$; that is, $3x + 1 \neq 0$, so $x \neq -\frac{1}{3}$.
When dividing a polynomial by a binomial, it is easier when both expressions are written in descending powers of x.

Necessary Skills

Example 1

a) Divide $-x + 5x^3 + 3 - 7x^2$ by $x - 1$.

b) Express the result as a division statement.

c) State the restriction on the divisor.

Solution

a) Rearrange the dividend so it is written in descending powers of x: $5x^3 - 7x^2 - x + 3$.

$$
\begin{array}{r l}
 & 5x^2 - 2x - 3 \\
x - 1 & \overline{)\, 5x^3 - 7x^2 - x + 3} \\
 & \underline{5x^3 - 5x^2} \\
 & -2x^2 - x \\
 & \underline{-2x^2 + 2x} \\
 & -3x + 3 \\
 & \underline{-3x + 3} \\
 & 0
\end{array}
$$

b) The division statement is $5x^3 - 7x^2 - x + 3 = (x - 1)(5x^2 - 2x - 3)$.

c) The divisor cannot be zero; that is, $x - 1 \neq 0$, so $x \neq 1$.

In *Example 1*, since the remainder is 0, $x - 1$ is a factor of the given polynomial. The quotient can be factored; thus,

$$
\begin{aligned}
5x^3 - 7x^2 - x + 3 &= (x - 1)(5x^2 - 2x - 3) \\
&= (x - 1)(5x + 3)(x - 1)
\end{aligned}
$$

Example 1 illustrates that if one factor of a polynomial is known, the other factors may be found by dividing, then, if possible, factoring the quotient.

If a power of x is missing in the dividend, it may be included using zero as the coefficient. This will facilitate long division.

Example 2

a) Determine the quotient of $(x^4 - 25x^2 + 62x - 36) \div (x + 6)$.

b) Write the division statement.

Solution

a) Write the dividend in descending powers of x. Since there is no x^3 term in the dividend, it is inserted with 0 as its coefficient.

$$
\begin{array}{r}
x^3 - 6x^2 + 11x - 4 \\
x+6 \overline{) x^4 + 0x^3 - 25x^2 + 62x - 36} \\
\underline{x^4 + 6x^3} \\
-6x^3 - 25x^2 \\
\underline{-6x^3 - 36x^2} \\
11x^2 + 62x \\
\underline{11x^2 + 66x} \\
-4x - 36 \\
\underline{-4x - 24} \\
-12
\end{array}
$$

b) The division statement is $x^4 - 25x^2 + 62x - 36 = (x+6)(x^3 - 6x^2 + 11x - 4) - 12$.

Exercises

1. Determine each quotient and remainder. Write the division statement. State the restrictions on the divisor.

a) $(2x^2 - x + 5) \div (x + 3)$

b) $(x^3 - x - 10) \div (x + 4)$

c) $(-10x^3 + x - 8 + 21x^2) \div (x - 2)$

d) $(4x^3 - 10x^2 + 6x - 18) \div (2x - 5)$

2. Divide each polynomial by $2x - 1$, then factor the quotient. Write the division statement.

a) $6x^3 + 31x^2 + 3x - 10$

b) $2x^3 + 3x^2 - 8x + 3$

c) $2x^3 - x^2 - 8x + 4$

d) $4x^3 + 4x^2 - x - 1$

3. When a certain polynomial is divided by $5x + 2$, the quotient is $x^2 - 2x + 3$ and the remainder is -1. What is the polynomial?

4. When a certain polynomial is divided by $3x - 4$, the quotient is $2x^3 - x + 1$ and the remainder is 8. What is the polynomial?

5. Determine each quotient. Write the division statement. Factor the quotient completely where possible.

a) $(2m^3 - 3m^2 - 8m - 3) \div (2m + 1)$

b) $(-a^4 + 3a^2 - a + 1) \div (a + 2)$

c) $(c^3 - 39c + 70) \div (c - 2)$

d) $(2x^3 - x^2 - 13x - 6) \div (2x + 1)$

e) $(10x^3 - 21x^2 + 6 - x) \div (5x - 3)$

f) $(5x^4 + x^2 - 3 + 4x) \div (x - 1)$

6. Complete the division in *Example 2* without the $0x^3$ term in the dividend. Compare your solution with that in *Example 2*. Explain why the $0x^3$ term is included.

4. New: Factoring by Grouping

To factor a polynomial means to express it as a product. A factor may be common only to some terms of a polynomial. Sometimes, these terms may be grouped so the polynomial can be factored.

Example

Factor.

a) $x^3 + 3x^2 + 4x + 12$

b) $2x^3 - 2x^2 - 3x + 3$

Solution

a) $x^3 + 3x^2 + 4x + 12$

The first 2 terms have a common factor x^2, and the last 2 terms have a common factor 4.

$$\begin{aligned} x^3 + 3x^2 + 4x + 12 &= x^2(x + 3) + 4(x + 3) \\ &= (x + 3)(x^2 + 4) \end{aligned}$$

b) $2x^3 - 2x^2 - 3x + 3$

The first 2 terms have a common factor $2x^2$, and the last 2 terms have a common factor -3.

$$\begin{aligned} 2x^3 - 2x^2 - 3x + 3 &= 2x^2(x - 1) - 3(x - 1) \\ &= (x - 1)(2x^2 - 3) \end{aligned}$$

Exercises

1. Factor.

a) $x^3 - 3x^2 + 2x - 6$

b) $x^3 + 2x^2 - 5x - 10$

c) $x^3 + x^2 - 6x - 6$

d) $-x^3 + x^2 - 2x + 2$

e) $x^3 + 4x^2 + x + 4$

f) $x^3 - 8x^2 - 3x + 24$

2. Factor.

a) $2x^3 + x^2 - 4x - 2$

b) $2x^3 - x^2 + 6x - 3$

c) $-2x^3 + x^2 - 8x + 4$

d) $2x^3 - 3x^2 + 4x - 6$

e) $3x^3 + 2x^2 - 9x - 6$

f) $4x^3 + x^2 - 16x - 4$

5. Review: Absolute Value

On the number line, each of the numbers –6 and 6 is located 6 units from 0.
Each number is said to have an *absolute value* of 6.
We write $|-6| = 6$ and $|6| = 6$.
The distance from –6 to 0 is $|-6|$.
The distance from 6 to 0 is $|6|$.

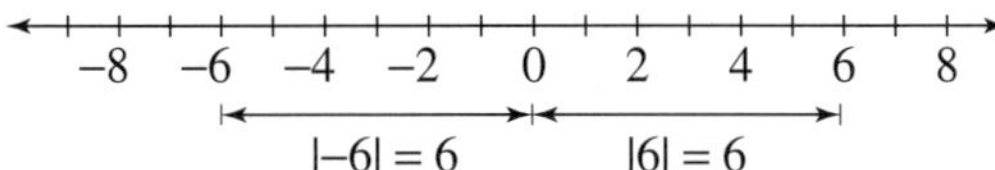

The definition of the absolute value of a number depends on whether the number is positive or negative:

- When a number is positive or zero, its absolute value is the number itself.
  That is, when $x \geq 0$, then $|x| = x$
- When a number is negative, its absolute value is the opposite number.
  That is, when $x < 0$, then $|x| = -x$

Example 1

Simplify.

a) $|16|$ **b)** $|-16|$

Solution

a) The absolute value of a positive number is the number itself.
$|16| = 16$

b) The absolute value of a negative number is the opposite number.
$|-16| = 16$

An equation that contains the variable inside an absolute-value sign is called an *absolute-value equation*.
An inequality that contains the variable inside an absolute-value sign is called an *absolute-value inequality*.

Example 2

Mark on a number line all the values of x that satisfy each equation or inequality.

a) $|x| = 3$ **b)** $|x| \leq 2$ **c)** $|x| > 5$

Solution

a) $|x| = 3$

x is located 3 units from 0; that is, x is 3 or -3.

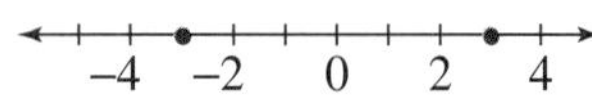

b) $|x| \leq 2$

x is located less than or equal to 2 units from 0; that is, x is between -2 and 2 inclusive. This is written as $-2 \leq x \leq 2$.

–4 –2 0 2 4

The solid dots at -2 and 2 indicate that they are part of the solution.

c) $|x| > 5$

x is located greater than 5 units from 0; that is, x is greater than 5 or less than -5. This is written as $x > 5$ or $x < -5$.

–8 –6 –4 –2 0 2 4 6 8

The open dots at -5 and 5 indicate that they are not part of the solution.

Exercises

1. Simplify.

a) $|-8|$ **b)** $|3.9|$ **c)** $\left|-\frac{4}{3}\right|$ **d)** $|23|$

2. Mark on a number line all the values of x that satisfy each equation.

a) $|x| = 6$ **b)** $|x| = 7$ **c)** $|x| = 4$ **d)** $|5x| = 10$

3. Mark on a number line all the values of x that satisfy each inequality.

a) $|x| \leq 5$ **b)** $|x| > 1.5$ **c)** $|x| \geq 4$ **d)** $|x| < 1$

3.1 Introduction to Polynomial Functions

In Chapter 1, we introduced the derivative as a tool for studying how a function changes. In Chapter 2, we developed some rules for differentiating functions to determine their derivatives. You will use these new tools in this chapter to study certain kinds of functions called *polynomial functions*. Linear and quadratic functions are simple examples of polynomial functions.

Linear functions

$y = 3x + 4$ (thin line)

$y = -x - 3$ (thick line)

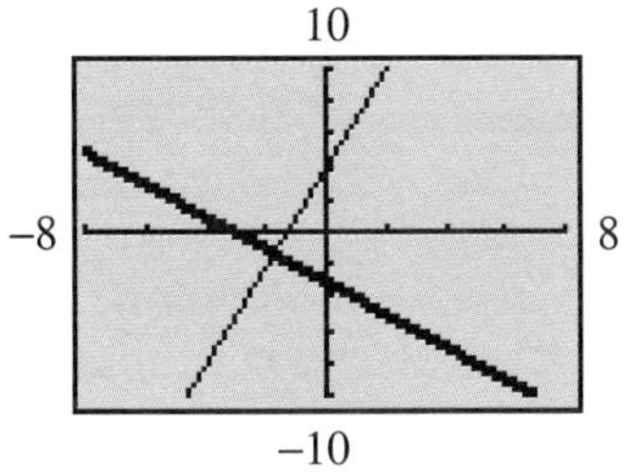

Quadratic functions

$y = x^2 + 3$ (thin curve)

$y = -0.25x^2 + x + 2$ (thick curve)

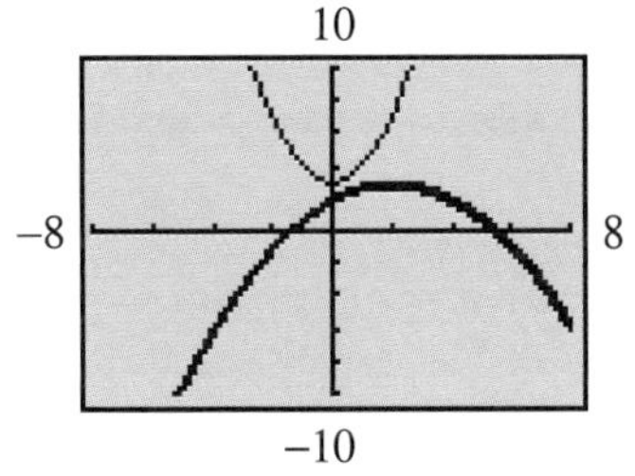

The coordinate axes divide the plane into 4 quadrants. We use these quadrants to describe the graphs.

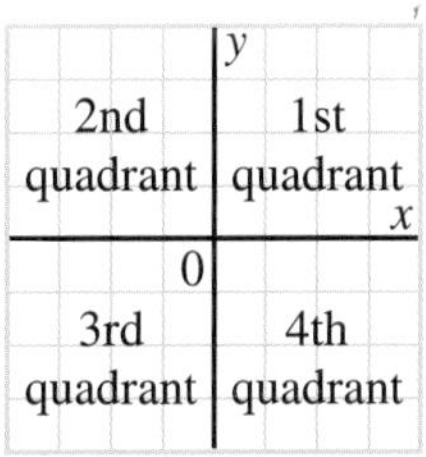

$y = 3x + 4$
The line extends from the 3rd quadrant to the 1st quadrant. The line has one x-intercept.

$y = -x - 3$
The line extends from the 2nd quadrant to the 4th quadrant. The line has one x-intercept.

$y = x^2 + 3$
The parabola is concave up. It extends from the 2nd quadrant to the 1st quadrant, forming a valley. There is a line of symmetry through the vertex. The curve has no x-intercepts.

$y = -0.25x^2 + x + 2$
The parabola is concave down. It extends from the 3rd quadrant to the 4th quadrant, forming a hill. There is a line of symmetry through the vertex. The curve has two x-intercepts.

You will use descriptions similar to those above in the *Investigation*, page 133.

Here are further examples of polynomial functions:

Linear function (first degree)	$f(x) = -3x + 7$
Quadratic function (second degree)	$f(x) = 5x^2 - 4x + 1$
Cubic function (third degree)	$f(x) = 2x^3 + 6x^2 - 8x + 3$
Quartic function (fourth degree)	$f(x) = -x^4 - 7x^3 + 4$
Quintic function (fifth degree)	$f(x) = 4x^5 + 2x^3 - 8x^2 - 2x + 5$

The degree of a polynomial function is the exponent of the highest power of x in the equation. The leading coefficient is the coefficient of this power of x. For example, the cubic function $f(x) = 2x^3 + 6x^2 - 8x + 3$ has degree 3 and leading coefficient 2.

Take Note

Polynomial Functions

A *polynomial function* of *degree n* is a function whose equation can be written in the form

$$f(x) = a_nx^n + a_{n-1}x^{n-1} + a_{n-2}x^{n-2} + \ldots + a_1x + a_0$$

where n is a non-negative integer and $a_n, a_{n-1}, a_{n-2}, \ldots, a_1, a_0$ are real numbers.

The coefficient, a_n, of the highest power of x is the *leading coefficient.*

The *degree* of the function is n, the exponent of the highest power of x.

In the *Investigation* that follows, you will use a graphing calculator to explore the graphs of some polynomial functions. Polynomial functions involve powers of x, which become large even for relatively small values of x. Hence, you may need to adjust the window settings of the calculator to show the main features of a graph on the screen. You may need a narrower interval along the x-axis or a wider interval along the y-axis.

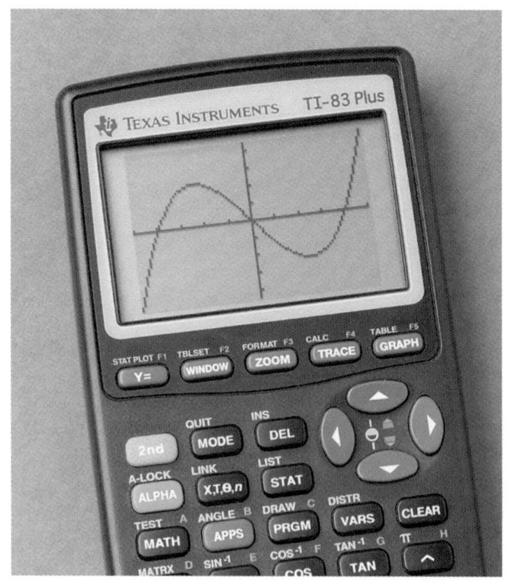

Graphing Polynomial Functions

When forming conclusions about the graphs, use the descriptions of the linear and quadratic functions on page 131 as a guide.

1. Graph each cubic function. Sketch the results. Use the results to form some conclusions about the graphs of cubic functions.
   a) $y = x^3$
   b) $y = -2x^3 + 15x - 7$
   c) $y = x^3 - 2x^2 + 4x - 8$
   d) $y = -5(x + 2)(x + 1)(x - 1)$

2. Repeat exercise 1 for each quartic function.
   a) $y = x^4$
   b) $y = x^4 + 4x^3 - 16x - 25$
   c) $y = -\frac{1}{2}(x - 4)(x + 4)(x - 2)(x + 2)$
   d) $y = -x^4 + 16x^2 - 3x - 40$

3. Repeat exercise 1 for each quintic function.
   a) $y = x^5$
   b) $y = -x^5 - 5x^4 - 2x^3 + 14x^2 + 19x + 17$
   c) $y = -x^5 - 2x^3 - 9x^2 + 19x - 14$
   d) $y = x(x + 3)(x + 1)(x - 2)(x - 3)$

4. Use the results of exercises 1 and 3. Make a statement about:
   a) the graph of a function with an odd degree and a positive leading coefficient
   b) the graph of a function with an odd degree and a negative leading coefficient
   c) the number of x-intercepts of the graph of a function with an odd degree

5. Use the results of exercise 2 and the description of quadratic functions on page 131. Make a statement about:
   a) the graph of a function with an even degree and a positive leading coefficient
   b) the graph of a function with an even degree and a negative leading coefficient
   c) the number of x-intercepts of the graph of a function with an even degree

A summary of the results from the *Investigation* is on page 134.

Graphs of Polynomial Functions

Functions with odd degree

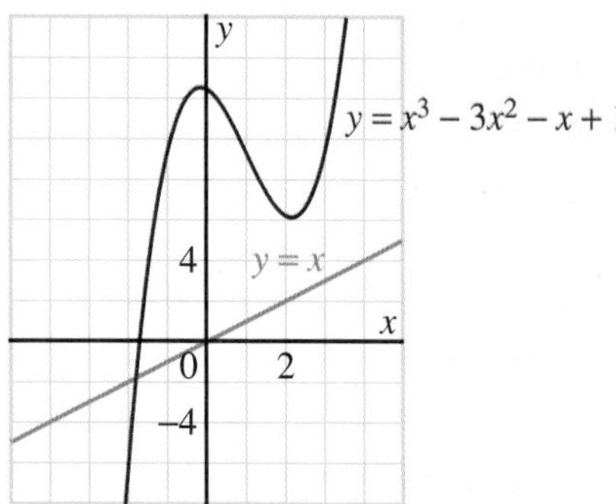

When the leading coefficient is positive, the graph extends from the 3rd quadrant to the 1st quadrant, as the graph of $y = x$ does.

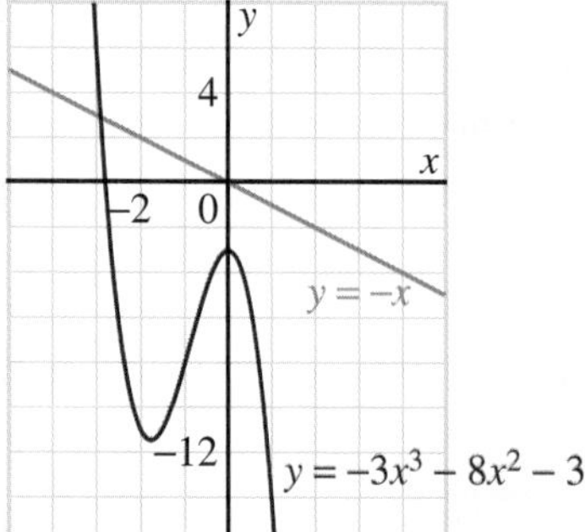

When the leading coefficient is negative, the graph extends from the 2nd quadrant to the 4th quadrant, as the graph of $y = -x$ does.

Functions with even degree

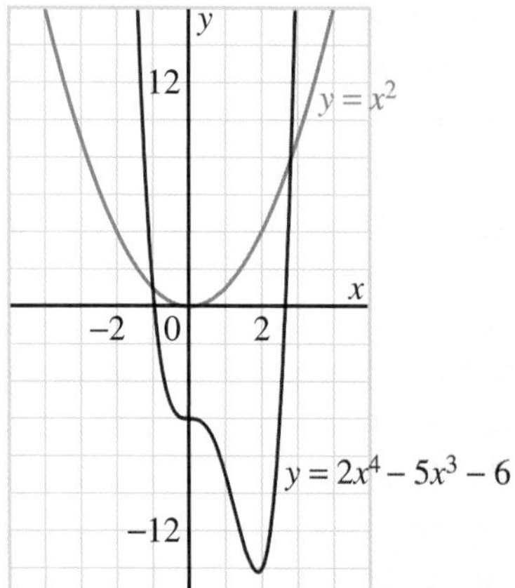

When the leading coefficient is positive, the graph extends from the 2nd quadrant to the 1st quadrant, as the graph of $y = x^2$ does.

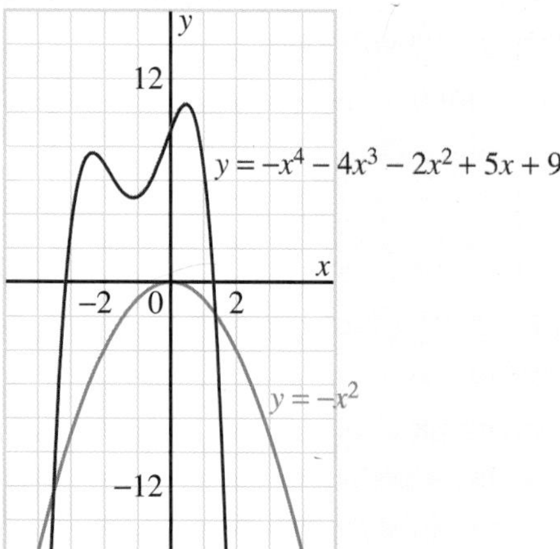

When the leading coefficient is negative, the graph extends from the 3rd quadrant to the 4th quadrant, as the graph of $y = -x^2$ does.

Symmetry

Another feature of many graphs of polynomial functions is *symmetry*. Recall that the graph of any quadratic function has *line symmetry* about the vertical line through the vertex.

Here is the graph of $f(x) = (x - 3)^2 - 4$. The parabola has line symmetry. $x = 3$ is the axis of symmetry for the parabola.

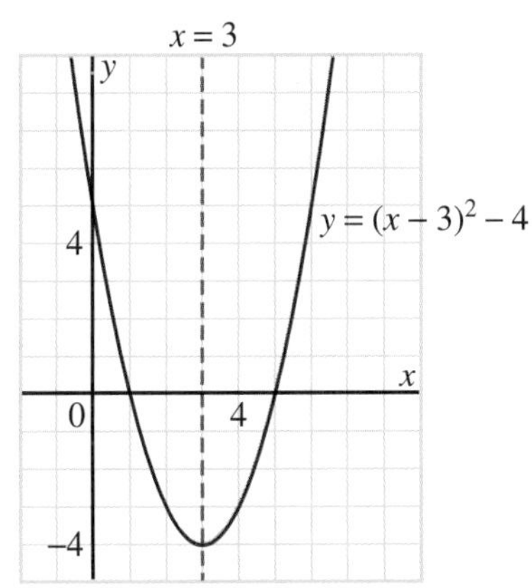

The graph of $f(x) = x^3 - 7x$ is shown at the right. Suppose we make a tracing of the graph and rotate the tracing 180° about the origin. The tracing will coincide with the original graph. Any line that passes through the origin and another point on the graph intersects the graph again at a third point. The origin is the midpoint of the line segment joining the two other points. We say this graph has *point symmetry* about the origin. The graphs of all cubic functions have point symmetry. However, this symmetry may not be about the origin.

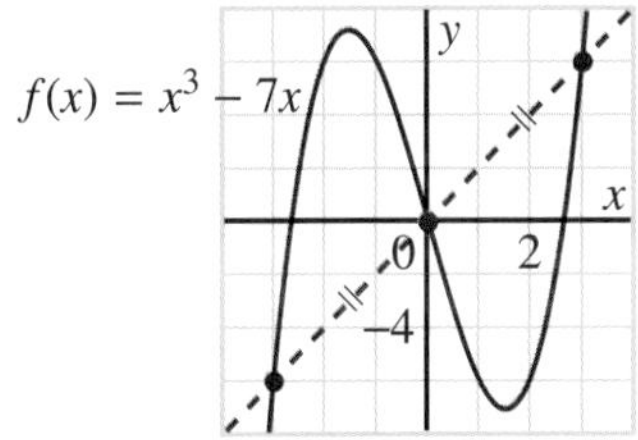

Finite Differences

We shall investigate how the differences in the table of values for a function illustrate how the graph of the function changes.

In a graphing calculator, we input $y = -x^3$.
Press [2nd] [GRAPH] for TABLE, then press [▲] or [▼] to display the table below left.
The graph of the function is shown below right.

X	Y1
-3	27
-2	8
-1	1
0	0
1	-1
2	-8
3	-27

X=-3

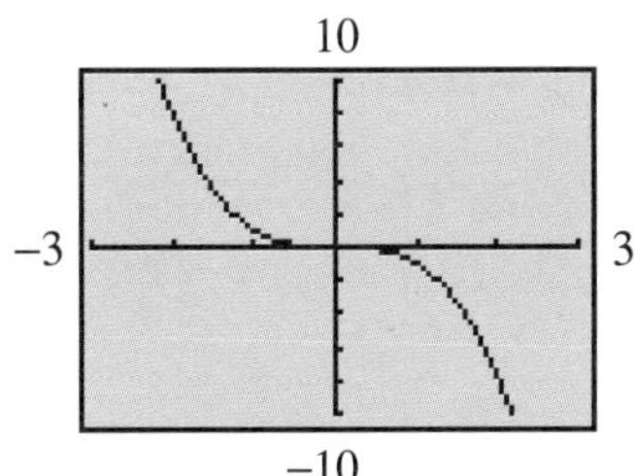

We use the table of values on the screen to construct a table of finite differences:

x	y	First differences	Second differences	Third differences
−3	27			
		−19		
−2	8		12	
		−7		−6
−1	1		6	
		−1		−6
0	0		0	
		−1		−6
1	−1		−6	
		−7		−6
2	−8		−12	
		−19		−6
3	−27		−18	
		−37		
4	−64			

The values of x are consecutive integers.
The values of y are decreasing.
When the values of x are consecutive integers, the first differences are the slopes of the secants that join the corresponding points.

The first differences are negative for all values of x. This corresponds to the fact that the y-values of the function are decreasing as x increases. Recall, from Chapter 1, that the slope of a secant represents the average rate of change from one point to the next. Look at the graph of $y = -x^3$. Each secant on this graph is drawn through the adjacent points whose coordinates are given in the table above.

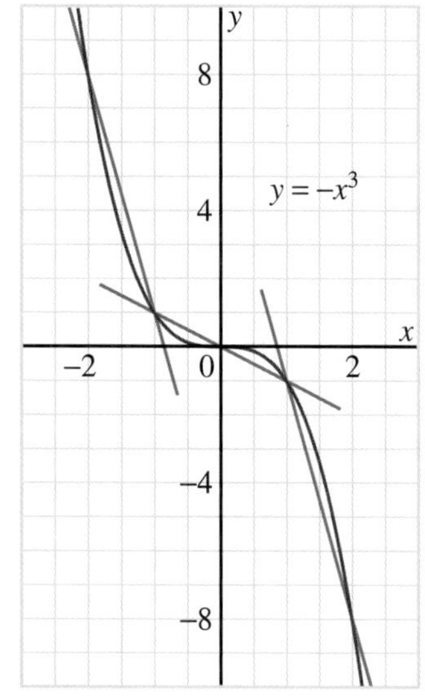

The second differences are positive, then negative. This means that the first differences are increasing, then decreasing. Thus, the graph is concave up where the second differences are positive and concave down where the second differences are negative. The point where the curve changes concavity is the *point of inflection*.

The third differences are constant and the degree of the function is 3. This is true in general. The nth differences of a function of degree n are constant. You will verify this for different functions in exercises 15 to 17.

Example

Identify the function that corresponds to each graph. Justify your choices.

$g(x) = x^3 - x^2$ $\quad$ $h(x) = x^4 - 3x^3 + x - 1$

$f(x) = -3x^3 + 8x^2 + 7$ $\quad$ $k(x) = -x^4 - x^3 + 11x^2 + 9x - 3$

a)

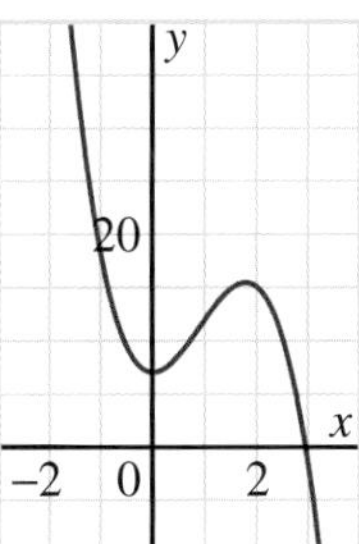

b)

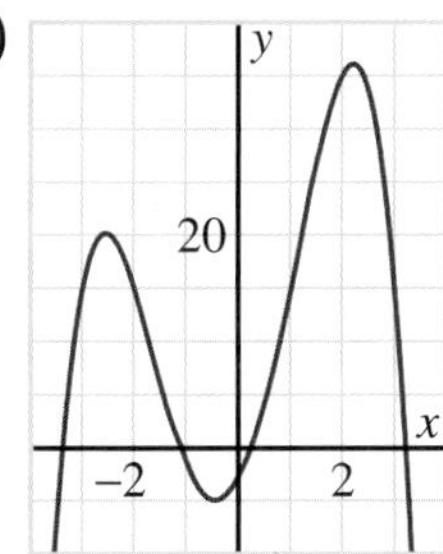

c)

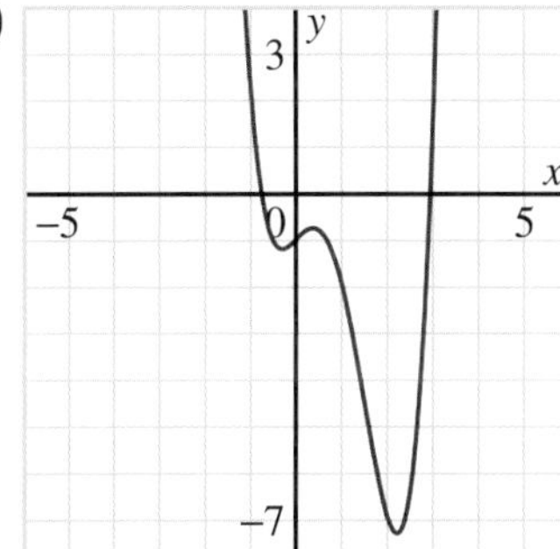

d)

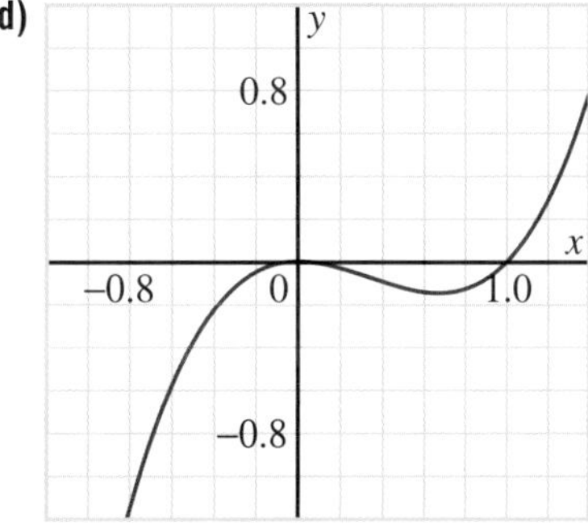

Solution

a) As x increases, this graph extends from the 2nd quadrant to the 4th quadrant, like the line $y = -x$. Hence, its equation is an odd-degree polynomial with a negative leading coefficient. The only possibility among those listed is $f(x) = -3x^3 + 8x^2 + 7$.

b) As x increases, this graph extends from the 3rd quadrant to the 4th quadrant, like the parabola $y = -x^2$. Its equation is an even-degree polynomial with a negative leading coefficient, $k(x) = -x^4 - x^3 + 11x^2 + 9x - 3$.

c) As x increases, this graph extends from the 2nd quadrant to the 1st quadrant, like the parabola $y = x^2$. Its equation is an even-degree polynomial with a positive leading coefficient, $h(x) = x^4 - 3x^3 + x - 1$.

d) As x increases, this graph extends from the 3rd quadrant to the 1st quadrant, like the line $y = x$. Its equation is an odd-degree polynomial with a positive leading coefficient, $g(x) = x^3 - x^2$.

Discuss

Identify the points about which the graphs in parts a and d have point symmetry. Estimate the coordinates of these points.

From the examples in this section, we see that the graphs of polynomial functions change in different ways, often increasing to form a hill, or decreasing to form a valley. Beginning in Section 3.3, we will use the derivative to determine: the values of x for which the function is increasing or decreasing; the values of x where there are maximum or minimum points; and the values of x where the graph changes concavity. In Section 3.2, we will examine the nature of the relationship between the graph of a polynomial function and its x-intercepts.

3.1 Exercises

A

1. Determine whether each function is a polynomial function or some other type of function. Justify your decision.

   a) $f(x) = 2x^3 + x^2 - 5$
   b) $f(x) = x^2 + 3x - 2$
   c) $y = 2x + 7$
   d) $y = \sqrt{x + 1}$
   e) $y = \dfrac{x^2 + x - 4}{x + 2}$
   f) $f(x) = x(x - 1)^2$
   g) $g(x) = 1.2x^2 + \dfrac{1}{2}x - \pi$
   h) $y = \sqrt{2x^3} + 2x - 0.5$

2. Which graph(s) could represent cubic functions? Explain.

a)
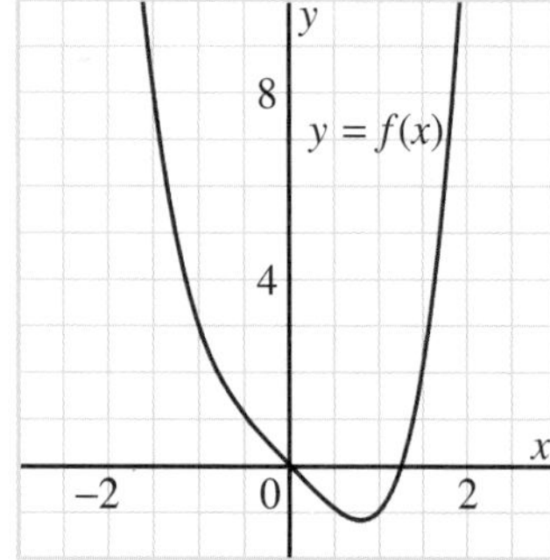

b)
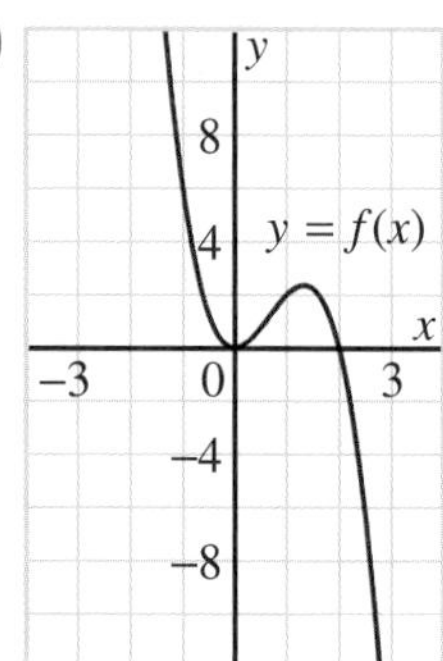

c)
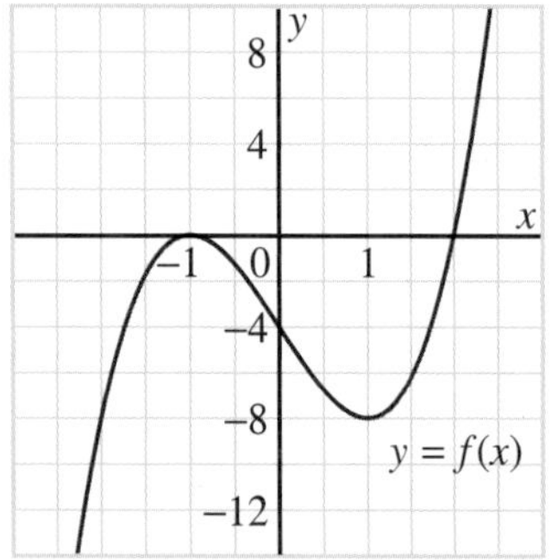

d)
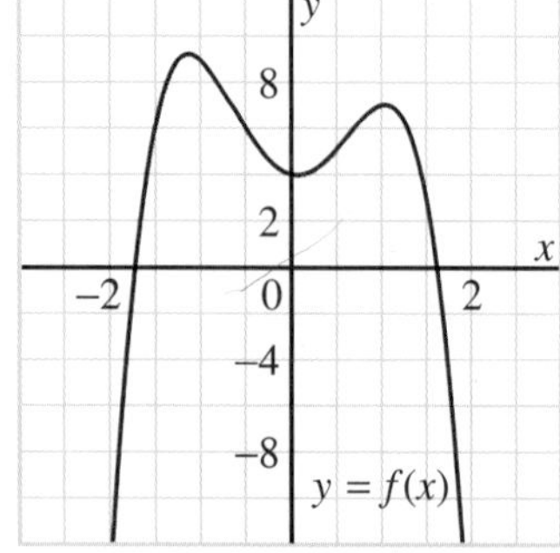

3. Which graph(s) in exercise 2 could represent quartic functions? Explain.

4. Which graphs represent functions with negative leading coefficients? Explain.

a)
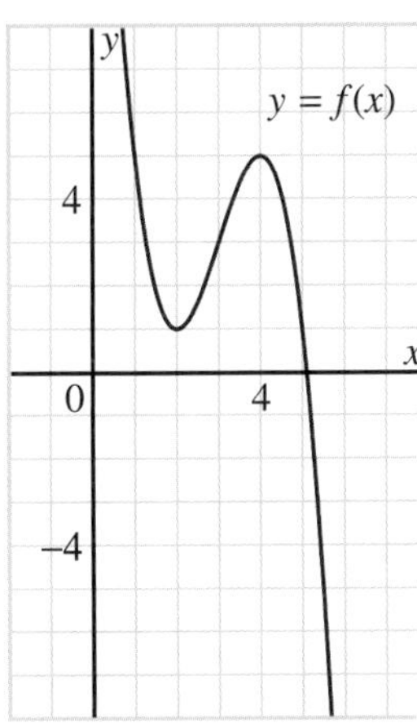

b)
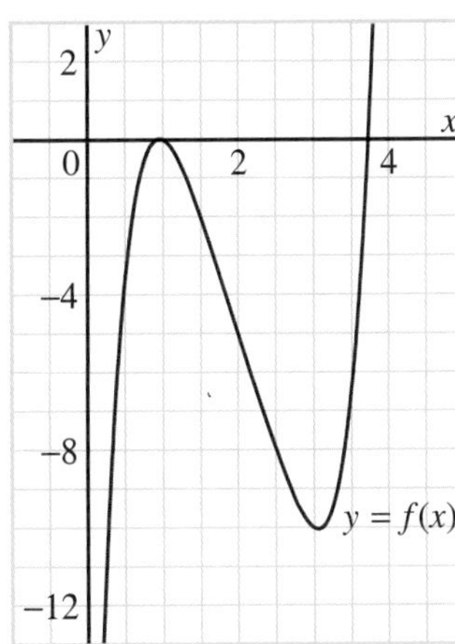

c)
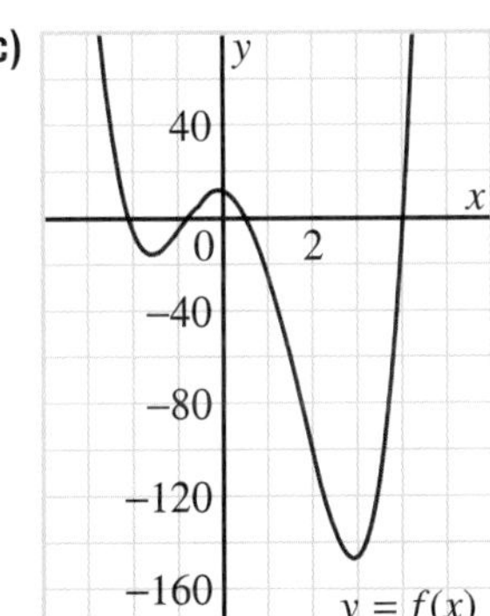

d)
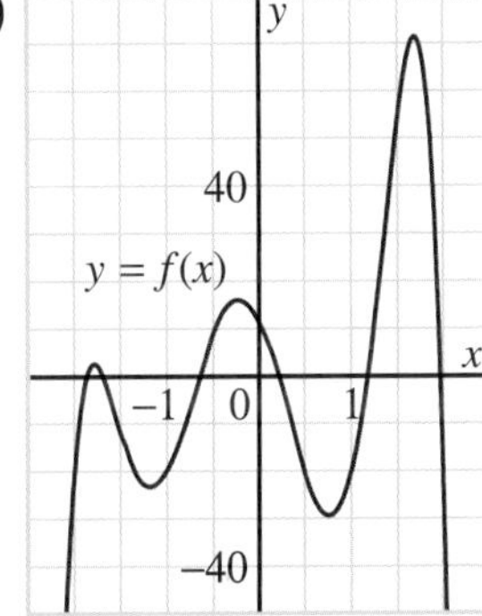

5. State the number of x-intercepts for each graph in exercise 4.

6. **Knowledge/Understanding** Identify the function that corresponds to each graph. Justify each choice.

$f(x) = x^3 - 3x^2 - 5x + 16$; $g(x) = x^4 - 10x^2 - 5x + 5$; $k(x) = 5x^4 - 14x^3$

a)
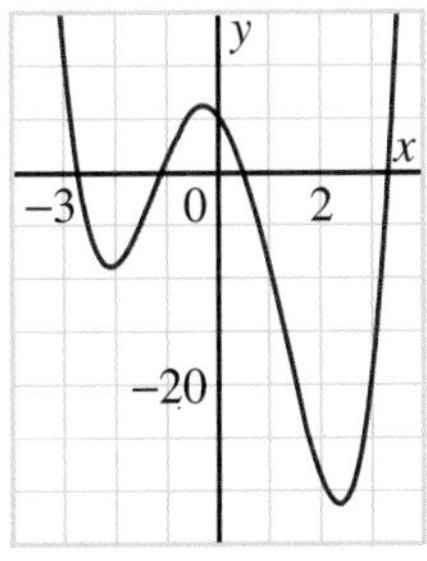

b)
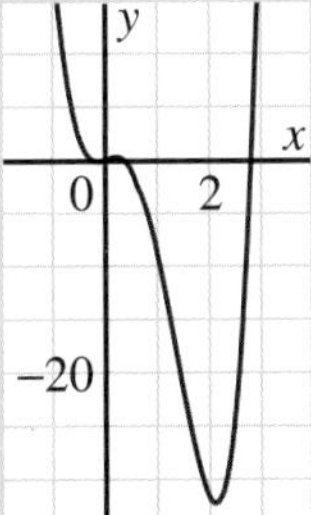

c)
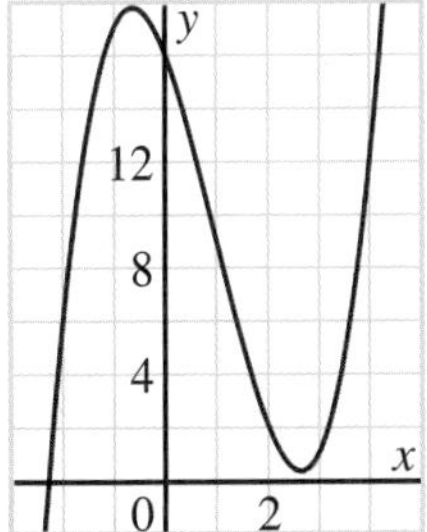

7. Sketch an example of a cubic function whose graph intersects the x-axis at each number of points.

a) only 1 point
b) 2 different points
c) 3 different points

8. Sketch an example of a quartic function whose graph intersects the x-axis at each number of points.

a) no points
b) only 1 point
c) 2 different points
d) 3 different points
e) 4 different points

9. Sketch an example of a quintic function for each possible number of points where its graph can intersect the x-axis.

10. **Communication** Suppose you are given the graph of a polynomial function.

a) How can you tell whether the degree of the function is odd or even?

b) How can you tell whether the leading coefficient is positive or negative?

c) What else can you tell about the function from its graph?

Exercise 10 is significant because we can obtain important information about a polynomial function from its graph. In particular, by examining what happens to the graph when $|x|$ is large, we can tell whether the degree of the function is odd or even, and whether the leading coefficient in its equation is positive or negative.

11. **Application** Examine these graphs. Which could be the graphs of polynomial functions? Explain.

a)
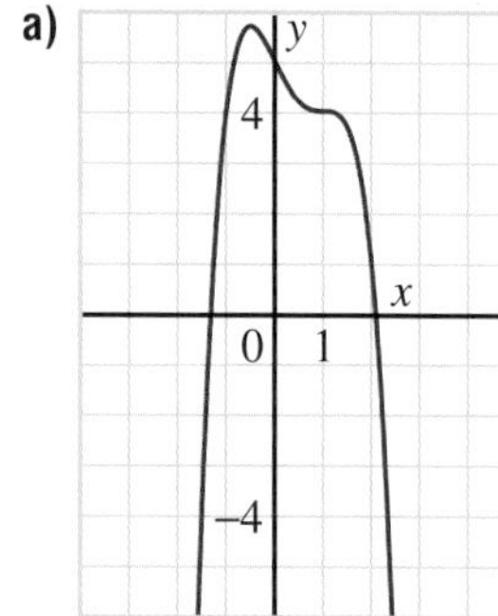

b)
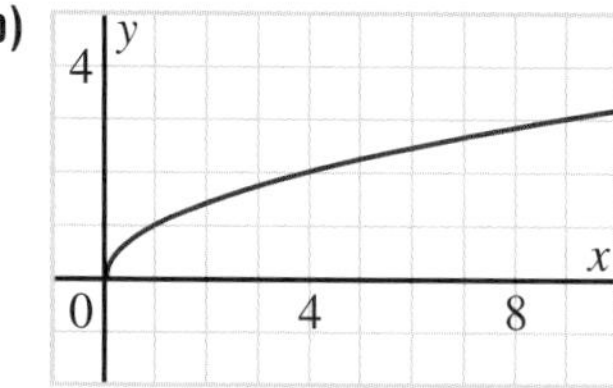

c)
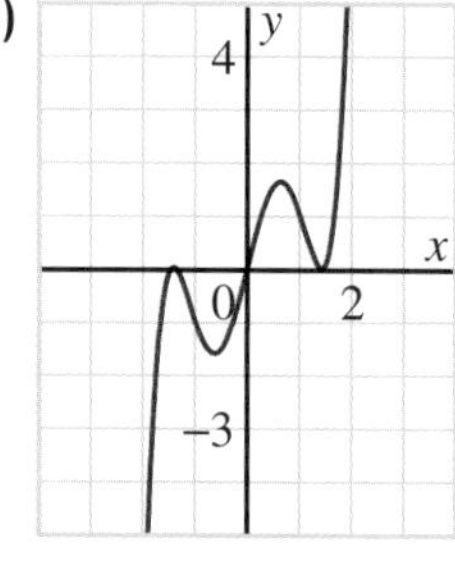

d)
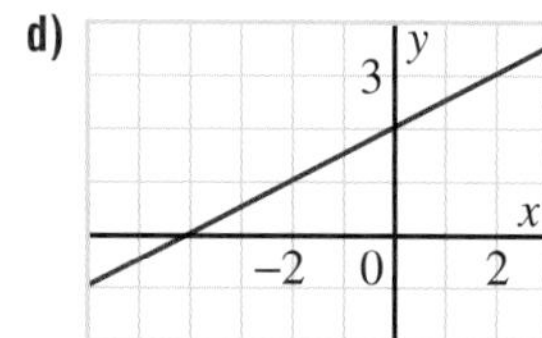

e)
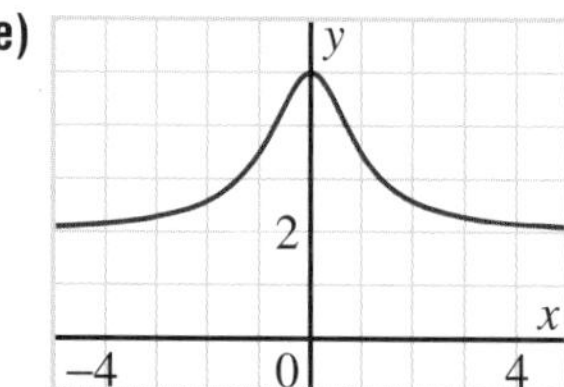

f)
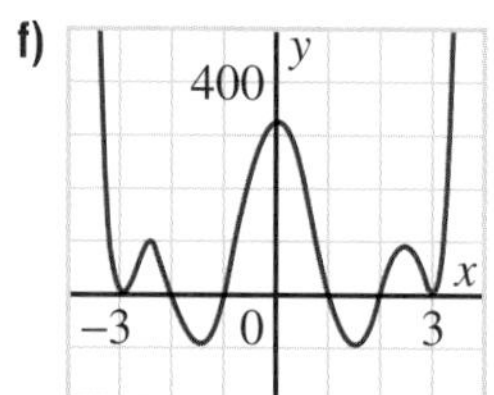

12. State whether each graph in exercise 11 has line symmetry, point symmetry, or neither. Explain.

13. Suppose you are given the graph of a polynomial function. Just by looking at the graph, can you always tell the degree of the function? Explain.

14. Determine whether the graph of each polynomial function has point or line symmetry.

a) $y = -2x^3 + 8x$
b) $y = 3x^5 - 20x^3 + x^2$
c) $y = 6x^4 - 5x^2$

15. a) Graph the linear function $y = 2x + 3$.

b) Copy this table. Complete the first two columns for integer values of x from –4 to 4.

x	y	First differences	Second differences

c) Complete the 3rd and 4th columns of the table.

d) **i)** What do you notice about the first differences?
ii) Is this true for all linear functions? Explain.

e) **i)** What do you notice about the second differences?
ii) Is this true for all linear functions? Explain.

f) What do the first differences indicate about the nature of change of this linear function?

g) How do the first differences relate to the degree of the function?

16. a) Graph the quadratic function $y = -x^2 + 4x - 3$.

b) Copy and complete a table similar to the one in exercise 15.

c) **i)** What do you notice about the first differences?
ii) Is this true for all quadratic functions? Explain.

d) **i)** What do you notice about the second differences?
ii) Is this true for all quadratic functions? Explain.

e) What do the first differences indicate about the nature of change of this quadratic function?

f) What do the second differences indicate about the nature of change of this quadratic function?

g) How do the second differences relate to the degree of the function?

17. a) Graph the quartic function $y = x^4$.

b) Copy and complete a table similar to the one in exercise 15.

c) What does each set of differences indicate about the nature of change of this quartic function?
i) first differences **ii)** second differences

d) Extend the table of differences to include third differences and fourth differences. What do you notice? Explain.

e) How do the fourth differences relate to the degree of the function?

18. a) Explain how first differences can be used to describe how a polynomial function, $y = f(x)$, changes.

b) Explain how the second differences can be used to describe how the function changes.

19. Look at the tables of values for different functions in this section. Can you identify the symmetry of a polynomial function from its table of values? Explain.

20. a) Describe how the graphs of these polynomial functions are related.

i) $y = x^3$ ii) $y = x^3 + 2$ iii) $y = (x - 2)^3$

iv) $y = 2x^3$ v) $y = -2(x + 1)^3 - 3$

b) Predict a relationship between $y = x^3$ and $y = a(x - p)^3 + q$.

c) Check your prediction by graphing three more functions of the form $y = a(x - p)^3 + q$.

21. a) Without graphing them, describe how the functions $y = x^4$ and $y = -2(x - 5)^4 + 7$ are related.

b) Graph the functions to check your description in part a.

c) Predict a relationship between $y = x^n$ and $y = a(x - p)^n + q$.

22. **Thinking/Inquiry/Problem Solving** These screens show the graphs of $y = x^2$, $y = x^4$, and $y = x^6$, respectively. The window settings are $-2 \le x \le 2$ and $-1.5 \le y \le 1.5$.

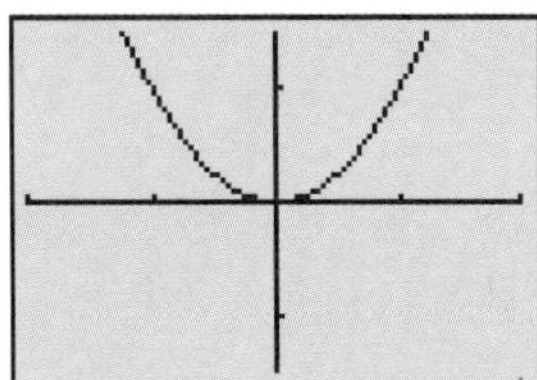
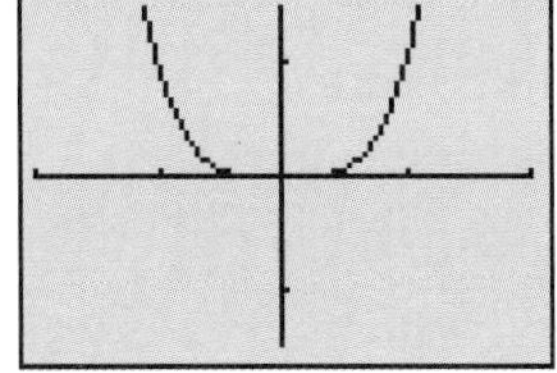
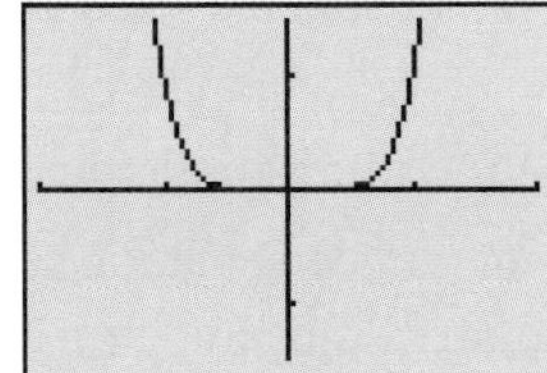

a) In what ways are the graphs similar? In what ways are they different? Explain the similarities and differences.

b) Predict what the graph of $y = x^n$ would look like if n were a much greater even number than those in part a. Use a graphing calculator to check your prediction.

c) Visualize the graphs changing from values of n that are relatively small to values of n that are relatively large. Describe what happens to the graph of $y = x^n$ as n increases through even values of n.

23. a) Use the same window settings as in exercise 22. Graph the functions $y = x^3$, $y = x^5$, and $y = x^7$.

b) Predict what the graph of $y = x^n$ would look like if n were a much greater odd number than those in part a. Use a graphing calculator to check your prediction.

c) Visualize the graphs changing from values of n that are relatively small to values of n that are relatively large. Describe what happens to the graph of $y = x^n$ as n increases through odd values of n.

3.2 Relating Polynomial Functions and Equations

Recall that when the graph of a function intersects the x-axis, the x-intercepts are the zeros of the function and the roots of the corresponding equation. For example, all the quadratic functions below have an equation of the form $f(x) = x^2 - 6x + c$. The graphs and the equations illustrate what happens as c increases from 8 to 10.

$$f(x) = x^2 - 6x + 8 = (x - 2)(x - 4)$$

zeros 2 and 4

$$x^2 - 6x + 8 = 0 \qquad (x - 2)(x - 4) = 0$$

roots 2 and 4

$$f(x) = x^2 - 6x + 9 = (x - 3)(x - 3)$$

zeros 3 and 3

$$x^2 - 6x + 9 = 0 \qquad (x - 3)(x - 3) = 0$$

roots 3 and 3

$$f(x) = x^2 - 6x + 10$$

There are no real zeros.

$$x^2 - 6x + 10 = 0$$

There are no real roots.

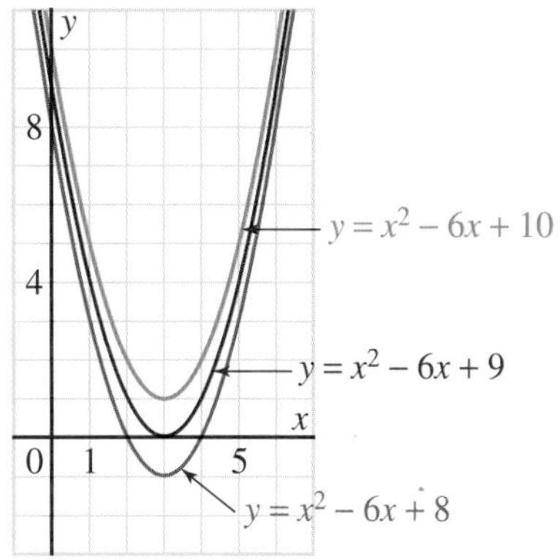

When $c < 9$, the graph intersects the x-axis at two different points. The corresponding quadratic function has two different zeros, and the corresponding quadratic equation has two different roots.

As c approaches 9, the two points of intersection with the x-axis come closer together. When $c = 9$, these points coincide at $x = 3$.

When $c > 9$, the graph does not intersect the x-axis.

We know the function $f(x) = x^2 - 6x + 8$ has zeros at $x = 2$ and $x = 4$ because the equation of the function can be written as $f(x) = (x - 2)(x - 4)$. We can tell this from the graph because it intersects the x-axis at (2, 0) and (4, 0). We know the quadratic equation $x^2 - 6x + 8 = 0$ has roots $x = 2$ and $x = 4$ because the equation can be written as $(x - 2)(x - 4) = 0$.

We say the function $f(x) = x^2 - 6x + 9$ has a *zero of order 2* at $x = 3$. We know this from the equation of the function, since it can be written as $y = (x - 3)^2$. We can tell this from the graph because it intersects the x-axis at (3, 0) but does not cross it. We know the quadratic equation $x^2 - 6x + 9 = 0$ has a *double root*, $x = 3$, because the equation can be written as $(x - 3)^2 = 0$.

The function $f(x) = x^2 - 6x + 10$ has no real zeros because its graph does not intersect the x-axis. The quadratic equation $x^2 - 6x + 10 = 0$ has no real roots. Similar results occur with all polynomial functions.

The function $y = x^2 - 6x + 8$ has zeros at $x = 2$ and $x = 4$, but this is not the only quadratic function with these zeros. Any function of the form $y = a(x^2 - 6x + 8)$ has these zeros. Some examples are shown at the right.

The graphs are vertical expansions or compressions of the graph of $y = x^2 - 6x + 8$.

The graphs of the functions with $a < 0$ are reflections of the graphs with $a > 0$. The reflection line is the x-axis.

Each graph intersects the x-axis at $x = 2$ and $x = 4$.

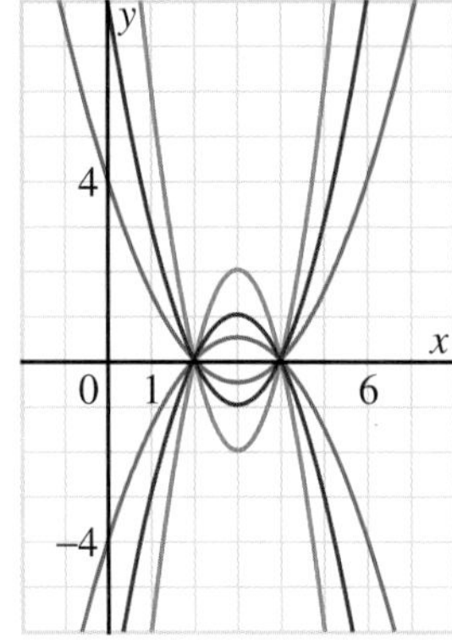

■ $y = (x - 2)(x - 4)$
■ $y = 2(x - 2)(x - 4)$
■ $y = \frac{1}{2}(x - 2)(x - 4)$

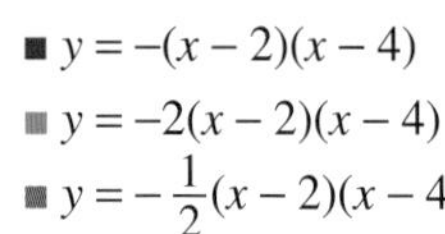

All six graphs above belong to the family of functions defined by $y = a(x^2 - 6x + 8)$.

Similarly, $y = x^2 - 6x + 9$ is not the only quadratic function with a zero of order 2 at $x = 3$. Any function of the form $y = a(x^2 - 6x + 9)$ has this zero of order 2. The graphs at the right are vertical expansions or compressions of the graph of $y = x^2 - 6x + 9$, combined with a reflection in the x-axis when $a < 0$.

Each graph intersects the x-axis at (3, 0) but does not cross it.

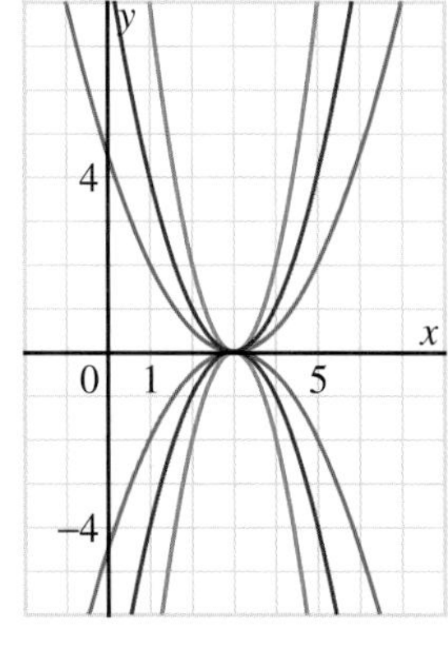

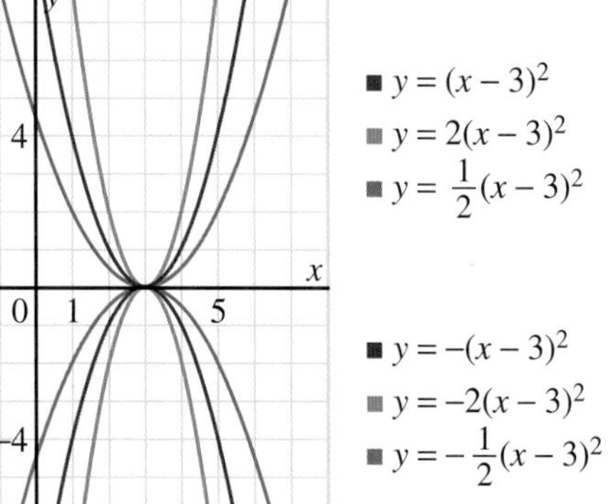

All six graphs above belong to the family of functions defined by $y = a(x^2 - 6x + 9)$.

Zeros and Roots of Polynomial Functions

The x-intercepts of the graph of a polynomial function $y = f(x)$ are the *zeros* of the function and the *roots* of the corresponding equation $f(x) = 0$.

Polynomial functions that have the same zeros and are vertical *expansions* or *compressions* of each other belong to the same *family* and have equations of the form $y = af(x)$.

Each diagram shows two polynomial functions whose graphs intersect the x-axis. Factors of the corresponding equations are indicated on the graphs.

Quadratic

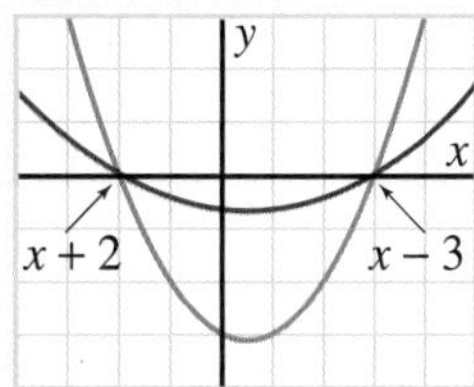

$f(x) = a(x + 2)(x - 3)$
Zeros of f: $-2, 3$
Roots of $f(x) = 0$: $-2, 3$

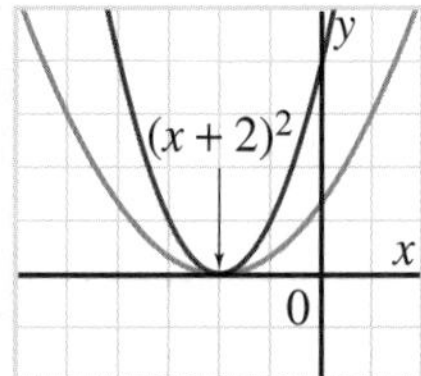

$f(x) = a(x + 2)^2$
Zero of f: -2 (of order 2)
Roots of $f(x) = 0$: -2 (double root)

Cubic

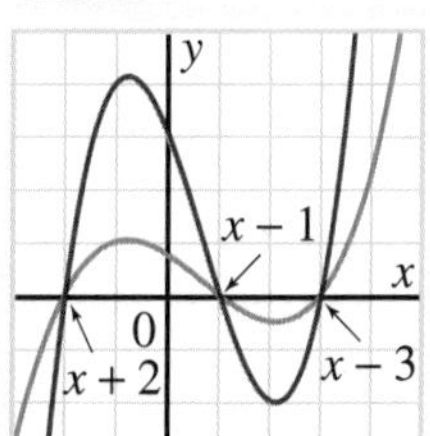

$f(x) = a(x + 2)(x - 1)(x - 3)$
Zeros of f: $-2, 1, 3$
Roots of $f(x) = 0$: $-2, 1, 3$

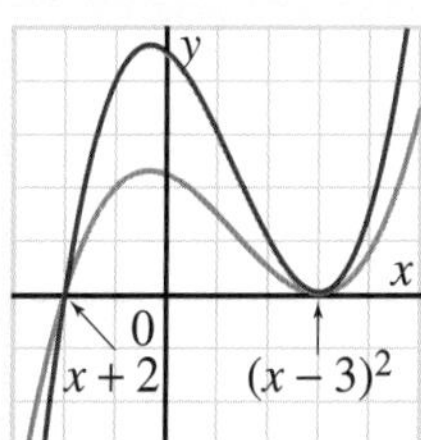

$f(x) = a(x + 2)(x - 3)^2$
Zeros of f: $-2, 3$ (of order 2)
Roots of $f(x) = 0$: $-2, 3$ (double root)

Quartic

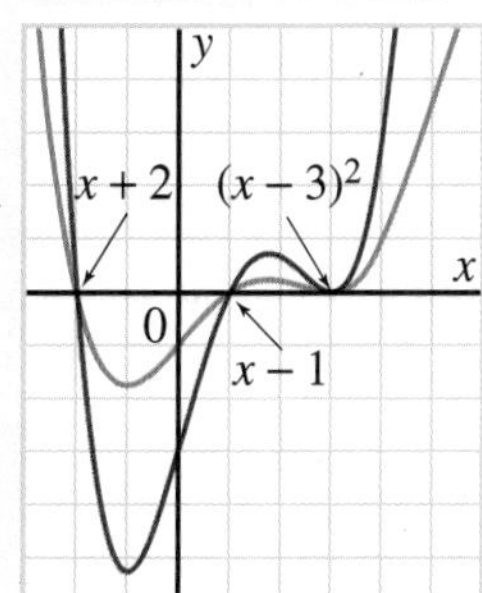

$f(x) = a(x + 2)(x - 1)(x - 3)^2$
Zeros of f: $-2, 1, 3$ (of order 2)
Roots of $f(x) = 0$: $-2, 1, 3$ (double root)

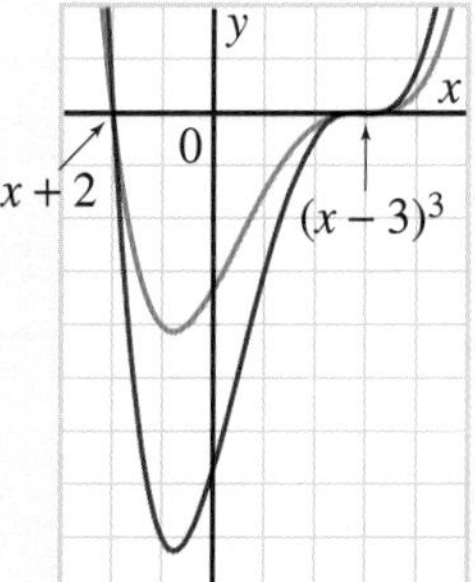

$f(x) = a(x + 2)(x - 3)^3$
Zeros of f: $-2, 3$ (of order 3)
Roots of $f(x) = 0$: $-2, 3$ (triple root)

Example 1

Consider the family of quadratic functions, where each function has zeros -2 and 3.

a) Write the equation of the family, then state two functions that belong to this family.

b) Determine the equation of the member of the family that passes through (1, 4).

c) Graph the function in part b.

Solution

a) Since the zeros are -2 and 3, the equation for this family of functions is $y = a(x+2)(x-3)$. Two other functions are found by substituting any real numbers for a.
For $a = -4$, $y = -4(x+2)(x-3)$
For $a = 5$, $y = 5(x+2)(x-3)$

b) Since the graph passes through (1, 4), the coordinates of this point satisfy the equation.
Substitute $x = 1$ and $y = 4$ in $y = a(x+2)(x-3)$.
$4 = a(1+2)(1-3)$
$4 = a(3)(-2)$
$a = -\frac{2}{3}$
The equation of the function is $y = -\frac{2}{3}(x+2)(x-3)$.

Discuss

In the equation of the function, why did we not expand the product $(x + 2)(x - 3)$?

c) The graph is concave down. Its axis of symmetry is halfway between $x = -2$ and $x = 3$; that is, $x = 0.5$. By symmetry, the point (0, 4) also lies on the graph. Draw a smooth curve through all known points.

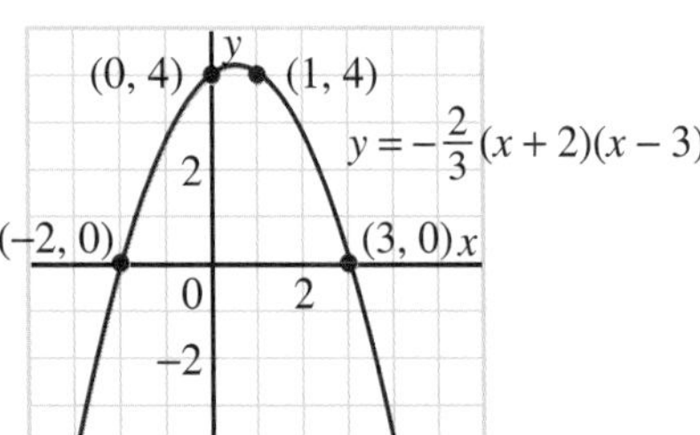

When the equation of a polynomial function is expressed in factored form, we can use the factors to graph the function.

Example 2

Sketch a graph of each function.

a) $y = (x + 2)(x - 5)^2$ **b)** $y = (x + 2)^2(x - 5)^2$

Solution

a) $y = (x + 2)(x - 5)^2$

This cubic function has a zero at $x = -2$ and a zero of order 2 at $x = 5$. Mark these points on the x-axis. Since the coefficient of x^3 is positive, the graph extends from the 3rd quadrant to the 1st quadrant. Begin in the 3rd quadrant. The curve passes up through $(-2, 0)$, back down to touch the x-axis at $(5, 0)$, then up into the 1st quadrant.

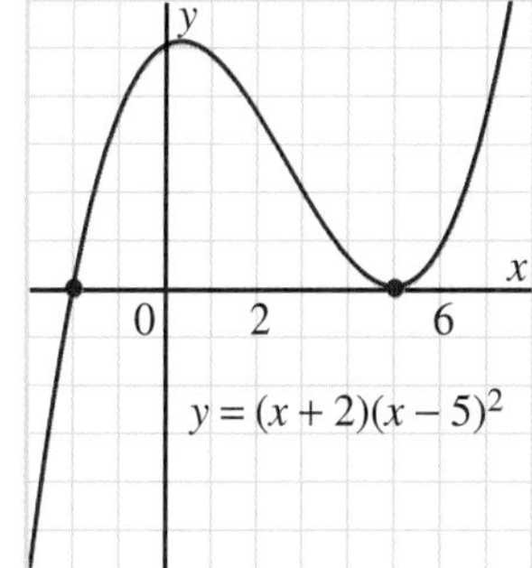

b) $y = (x + 2)^2(x - 5)^2$

This quartic function has zeros of order 2 at $x = -2$ and $x = 5$. Since the coefficient of x^4 is positive, the graph extends from the 2nd quadrant to the 1st quadrant. Begin in the 2nd quadrant. The curve passes down to touch the x-axis at $(-2, 0)$, up into the 2nd quadrant and 1st quadrant, back down to touch the x-axis at $(5, 0)$, then up into the 1st quadrant.

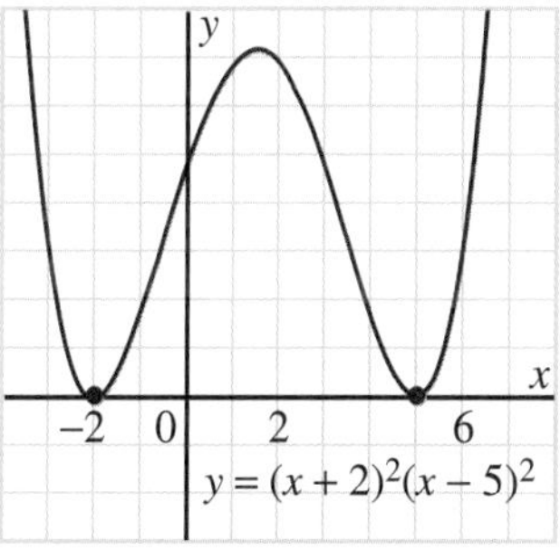

In *Example 2a*, the function $y = (x + 2)(x - 5)^2$ changes sign at the zero $x = -2$. This function does not change sign at $x = 5$, which is a zero of order 2.

Similarly, in *Example 2b*, the function $y = (x + 2)^2(x - 5)^2$ does not change sign at either zero because each zero has order 2.

In *Example 2*, the polynomial functions were given in factored form. You already know how to factor polynomials of degree 2. In Section 3.5, you will learn how to factor polynomials of degree greater than 2.

Many polynomial functions do not factor. We can use a graphing calculator to graph such functions, then use Trace or Calculate to approximate each zero.

We may use the idea of a family of functions to determine the equation of a graph of a polynomial function.

Example 3

Determine the equation of this polynomial function.

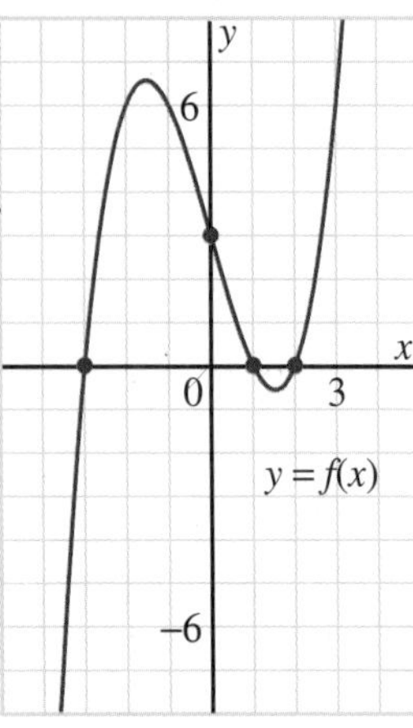

Solution

From the shape of the graph, the function is cubic.
Its equation has the form $y = a(x - b)(x - c)(x - d)$.
From the graph, the zeros of the function are –3, 1, and 2.
So, the equation is $y = a(x + 3)(x - 1)(x - 2)$.
From the graph, the y-intercept is 3.
Substitute $x = 0$, $y = 3$ in $y = a(x + 3)(x - 1)(x - 2)$.

$$3 = a(3)(-1)(-2)$$
$$3 = 6a$$
$$a = \frac{3}{6}$$
$$= \frac{1}{2}$$

The equation is $y = \frac{1}{2}(x + 3)(x - 1)(x - 2)$.

3.2 Exercises

A

1. What is the greatest number of zeros each function could have? Explain.

a) $f(x) = -2x^3 + 6x^2 - 4x + 1$

b) $g(x) = 7x^2 - 3x + 8$

c) $h(x) = -4x^2 + 2x - x^3 + 2$

d) $p(x) = -5x^4 + x^2 - 7x$

e) $f(x) = x^5 + 3x^4 - x^2 - 6$

f) $h(x) = -3x^2 + 11x^4 - 9$

2. Explain why the function $y = a(x^2 - 7x + 12)$ has the same zeros as the function $y = x^2 - 7x + 12$.

3. Which polynomial function has a graph for which the x-axis is a tangent at $x = -2$? Explain.

a) $f(x) = (x + 2)(x - 2)(x + 3)$ **b)** $f(x) = (x + 2)(x - 2)^2$

c) $f(x) = (x + 2)^2(x - 2)$

B

4. Each member of a family of quadratic functions has zeros -1 and 4.

a) Write the equation of the family, then state two functions that belong to this family.

b) Determine the equation of the member of the family that passes through (5, 9).

c) Graph the function.

5. **Knowledge/Understanding** Sketch a graph of each function.

a) $y = x(x - 3)(x - 2)$ **b)** $y = -(x - 1)(x - 2)(x - 3)$

c) $y = x(x - 1)^2$ **d)** $y = (x + 2)^3$

6. Sketch an example of a cubic function with the given zeros. Then write an equation of the function. Is the equation unique? Explain.

a) -2, 1, 4 **b)** 3 (of order 3) **c)** -2, 2 (of order 2)

7. Sketch a graph of each function.

a) $y = x(x + 1)(x - 2)(x - 4)$ **b)** $y = x^2(x + 2)^2$

c) $y = x(x + 3)^3$ **d)** $y = (x - 1)^4$

8. Sketch an example of a quartic function with the given zeros. Then write an equation of the function. Write the equations of two other functions that belong to the same family.

a) -3, -2, 1, 4 **b)** -1 (of order 2), 2 (of order 2)

c) -3, 0, 3 (of order 2) **d)** 3 (of order 4)

9. **Communication** Which statements are true and which are false? Explain.

a) Not all quadratic functions have zeros.

b) Not all cubic functions have zeros.

c) Every cubic function has at least one zero.

d) Every quartic function has at least one zero.

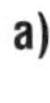
10. Determine the equation of each quadratic function.

a) 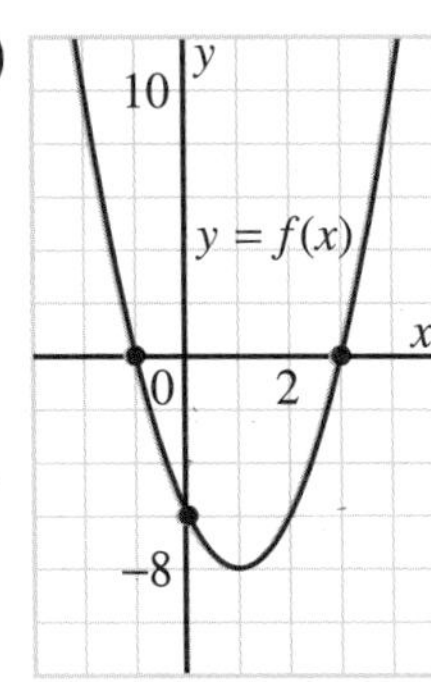

b) 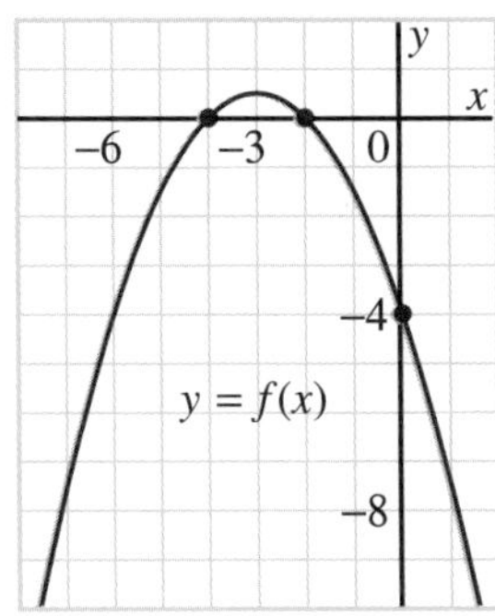

11. Sketch a graph of each function.

a) $y = (x - 2)^2(x + 1)$
b) $y = -x(x + 3)(x - 1)(x + 2)$
c) $y = -3(x + 5)(x + 4)$
d) $y = (x + 1)^2(x - 3)^2$

12. Refer to the functions in exercise 11. Which have each number of zeros given below? Explain how you know.

a) no zero of order 2
b) one zero of order 2
c) two zeros of order 2

✓ 13. A cubic function has zeros –3, –1, and 2. The y-intercept of its graph is 12.

a) Determine the equation of the function.
b) Sketch the graph of the function.

✓ 14. Determine the equation of each function, then sketch its graph.

a) quadratic function with zero 2 (of order 2); graph has y-intercept 12
b) cubic function with zeros –2, 1, 4; graph has y-intercept 24
c) cubic function with zeros –2 and 2 (of order 2); graph has y-intercept –16
d) cubic function with zeros 0, 2, 4; graph passes through (3, 9)

Exercises 13 and 14 are significant because many polynomial functions with the same degree can have the same zeros. This is illustrated in *Take Note*, page 145. To determine the equation of one particular function, additional information about it is required.

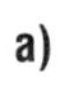
15. **Application** Determine the zeros of each function.

a) $f(x) = x^2 - 10x + 16$
b) $f(x) = x^3 - 7x^2 + 12x$
c) $f(x) = x^3 - 5x^2 - 14x$

✓ 16. Determine the equation of each cubic function.

a) 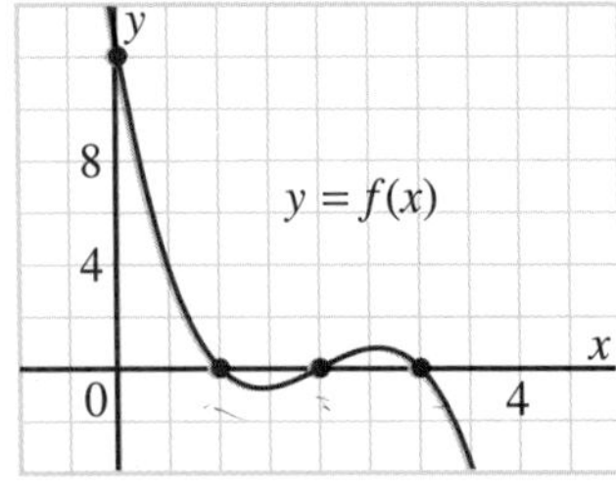

b) 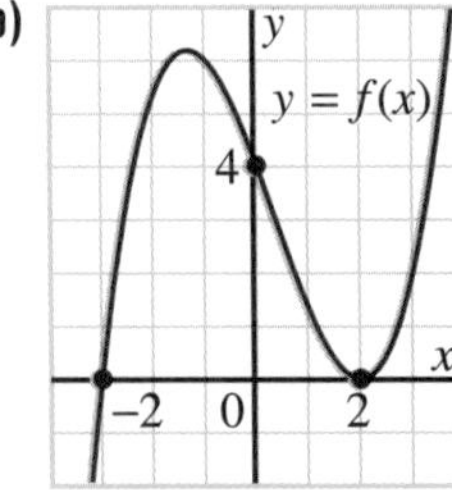

17. Determine an equation to represent the graph of each polynomial function.

a) 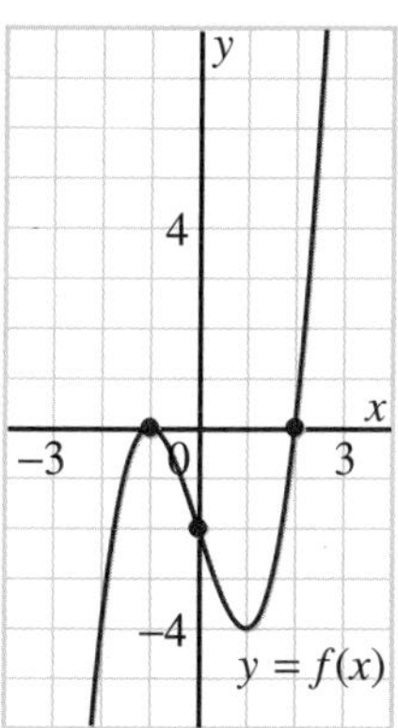

b) 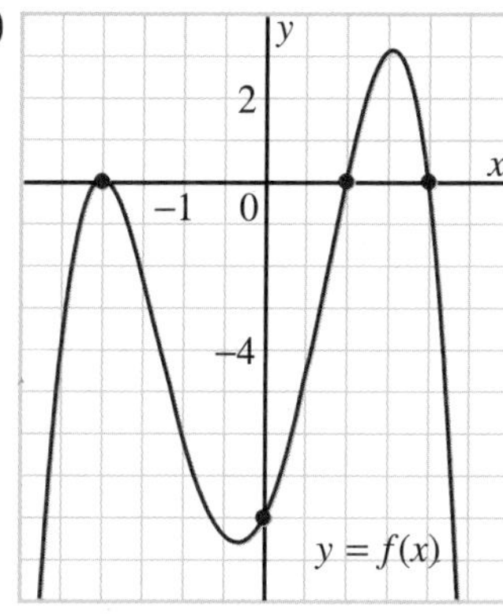

c) 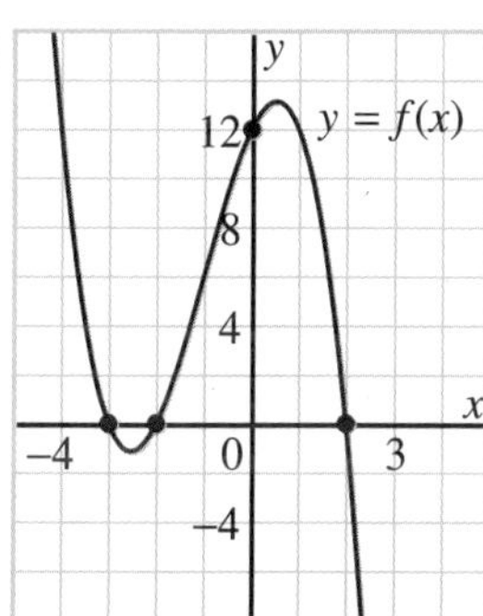

d) 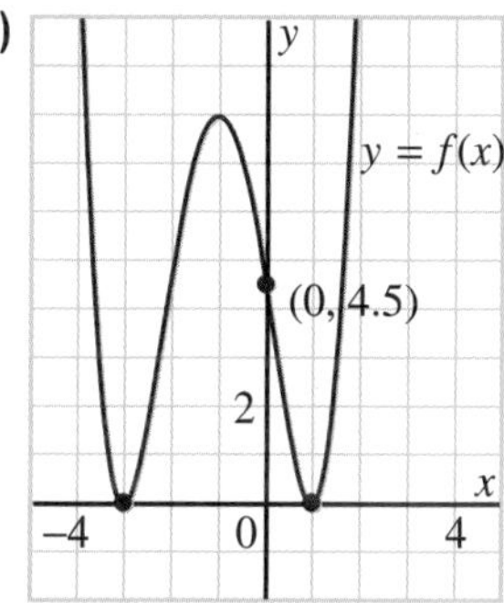

18. **Thinking/Inquiry/Problem Solving** This screen was obtained by graphing the function $y = x^3 - 250x^2 + 4000$. The function appears to have two zeros.

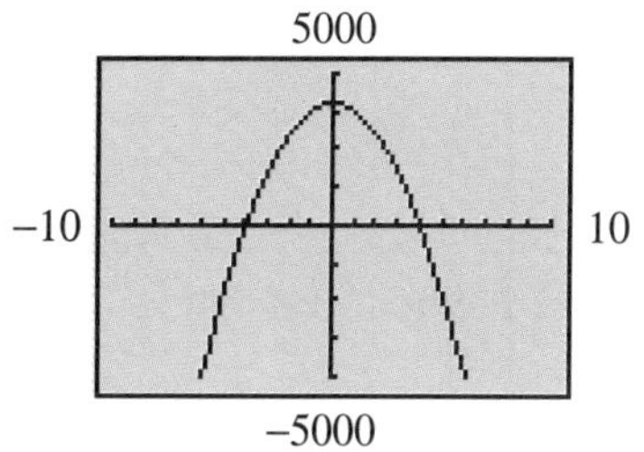

a) Explain why a third zero must exist for this function.

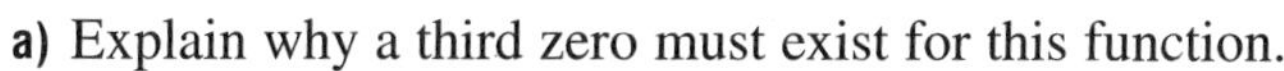

b) Is the third zero positive or negative? Explain.

c) Is the y-axis an axis of symmetry? Explain.

19. Recall that the order of a zero is the number of times its factor appears in the equation of the function. For the function $y = (x - 4)(x - 1)^2(x + 2)^3$, the zero $x = 4$ has order 1, the zero $x = 1$ has order 2, and the zero $x = -2$ has order 3. Describe a relationship between the order of the zeros and the graph of the function.

C

20. a) Consider the function $f(x) = x^3 - 2x^2 + k$, where k is a constant. Explain why $k = 0$ ensures that $f(x) = 0$ has a double root.

b) Graph $f(x) = x^3 - 2x^2 + k$ for different values of k.

c) Determine another value of k that ensures $f(x) = 0$ has a double root.

Self-Check NS, 3.1, 3.2

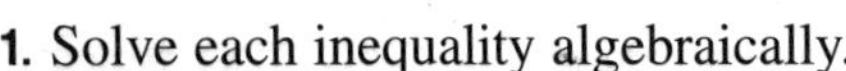

1. Solve each inequality algebraically.

   a) $(2x - 1)(x + 3) \geq 0$ b) $(5x + 2)(x - 3) < 0$

2. Solve each inequality graphically.

   a) $-3x^2 + 7x + 3 > 0$ b) $2x^2 - 3x + 4 \leq 0$

3. Determine each quotient and remainder. State the restrictions on the divisor.

   a) $(3x^2 - 5 + 2x) \div (x - 2)$ b) $(x^3 - 6x^2 + 4) \div (x + 2)$

   c) $(4x^3 - 10x^2 + 6x - 15) \div (2x - 5)$ d) $(x^5 - 1) \div (x - 1)$

4. Identify the function that corresponds to each graph. Justify each choice.

   a) $f(x) = -2x^3 + 5x$ b) $g(x) = x^4 - 3x^2 + x$ c) $p(x) = -x^6 + 3x^4 + 2x - 5$

i)

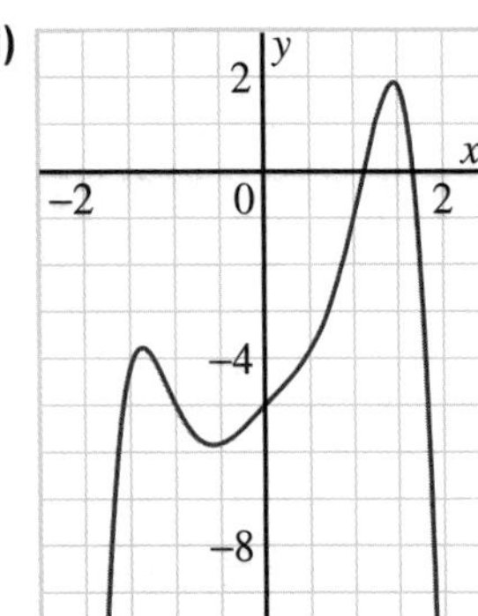

ii)

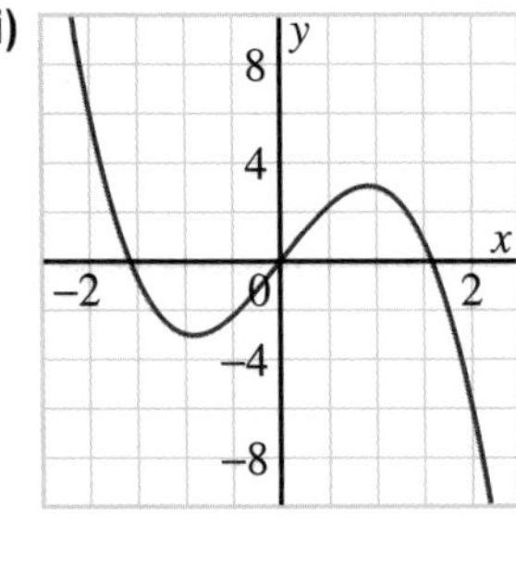

iii)

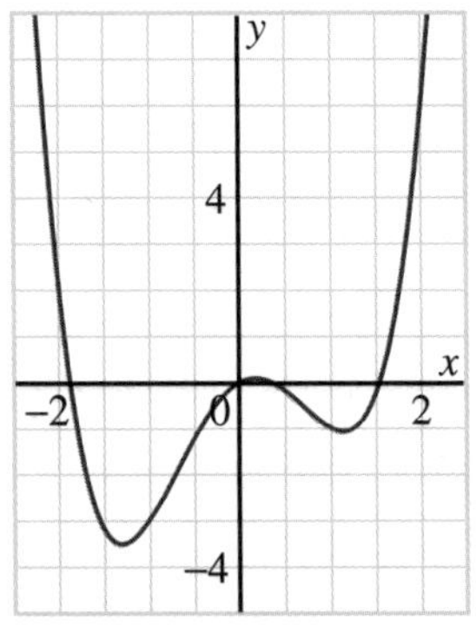

5. State whether each graph in exercise 4 has point symmetry, line symmetry, or neither. Explain.

6. Sketch a graph of each function.

   a) $y = -(x + 2)(x + 1)(x - 3)$ b) $y = (x - 1)^2(x + 4)^2$

7. Determine the zeros of each function.

   a) $y = x^4 - 16x^2$ b) $y = (x^2 - 4x - 12)(x^2 - 5x + 6)$

PERFORMANCE ASSESSMENT

8. a) Write an equation of a cubic function that has zeros –1, 2, 4 and y-intercept –3.

   b) Write an equation of a quartic function that has zeros –3, –2, 2 (of order 2).

   c) Is each equation in parts a and b unique? Explain.

3.3 Using First Derivatives to Graph Polynomial Functions

Investigation

A Cubic Function and Related Functions

The graph of the cubic function $y = x^3 - 2x^2 - 5x + 3$ is shown.

1. a) Suppose we delete the cubic term. Graph the parabola $y = -2x^2 - 5x + 3$.
   b) How is the graph of the parabola related to the graph of the cubic function?
   c) Use calculus to show that both curves have the same tangent at (0, 3).
   d) What is the equation of the common tangent?

2. a) Suppose we delete the quadratic term in the cubic function. Graph the cubic function $y = x^3 - 5x + 3$.
   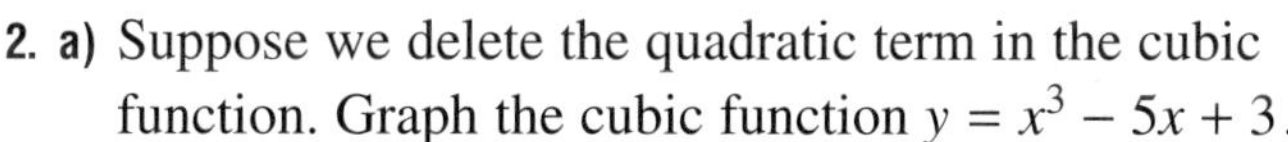
   b) How are the graphs of $y = x^3 - 2x^2 - 5x + 3$ and $y = x^3 - 5x + 3$ related?

3. a) Suppose we delete the linear term in the cubic function. Graph the cubic function $y = x^3 - 2x^2 + 3$.
   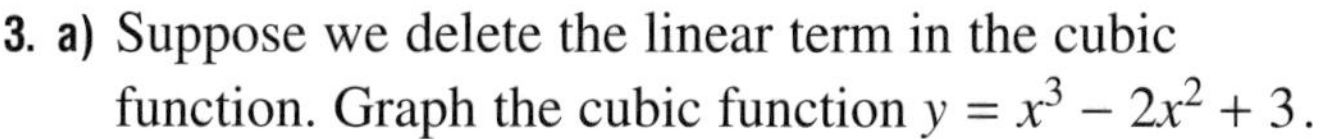
   b) How are the graphs of $y = x^3 - 2x^2 - 5x + 3$ and $y = x^3 - 2x^2 + 3$ related?

4. a) Suppose we delete the cubic and quadratic terms in the cubic function. Graph the linear function $y = -5x + 3$.
   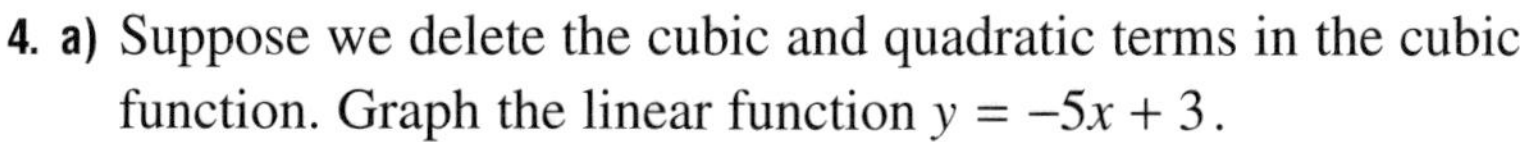
   b) How are the graphs of the linear function and the cubic function related?

5. Write a cubic function similar to that preceding exercise 1. Repeat exercises 1 to 4 for your function. Explain what you discover.

In the *Investigation*, you should have found that any cubic function $y = ax^3 + bx^2 + cx + d$ and a corresponding quadratic function $y = bx^2 + cx + d$ intersect on the y-axis at $(0, d)$. The slope of their common tangent is c, and the equation of the common tangent is $y = cx + d$.

In Section 3.1, you investigated properties of the graphs of polynomial functions. These graphs are smooth curves that usually have hills with maximum points and valleys with minimum points, as illustrated below. In Chapters 1 and 2, you used the derivative to determine the slope of a tangent. Since the tangents at maximum and minimum points are horizontal, with slope 0, the derivative is 0 at these points.

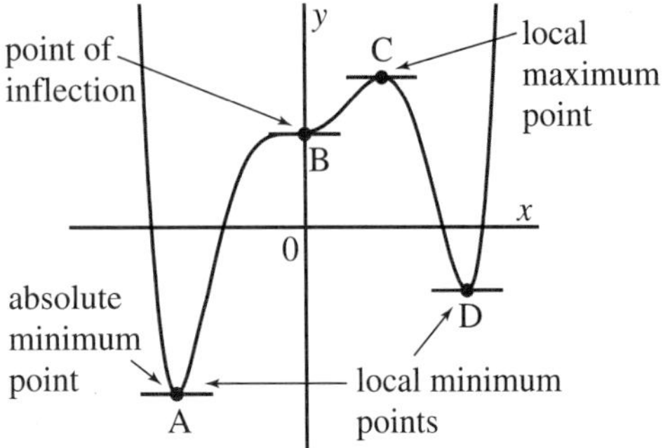

This graph shows the situations that arise when the derivative is 0. This occurs at points A, B, C, and D.

- Points A and D are called *local minimum points* because the value of y at each point is less than the values of y in a small interval containing that point. In addition, A is called an *absolute minimum point* because the value of y at that point is less than all other possible values of y.
- Point C is a *local maximum point* because the value of y at that point is greater than the values of y in a small interval containing that point. However, it is not an absolute maximum point because there are other points on the graph with greater y-values. This graph has no absolute maximum point.
- Point B is neither a local maximum point nor a local minimum point. It is a *point of inflection.*

We define a *critical point* to be a point where the derivative is 0. All four points A, B, C, and D on the graph above are critical points. The value of x at a critical point is called a *critical value*. The maximum and minimum values divide the graph of the function into intervals where the function is either increasing or decreasing.

Note that the derivative is not necessarily 0 at a point of inflection. That is, a point of inflection is a critical point only when the derivative is 0 at that point.

Consider an example, such as the quadratic function $y = x^2 - 8x + 12$, which can be written in the form $y = (x - 2)(x - 6)$. This function has the graph shown on page 155, with zeros at $x = 2$ and $x = 6$, and a minimum point at $x = 4$ with coordinates $(4, -4)$. The derivative is $y' = 2x - 8$. Its graph is a straight line with slope 2 and y-intercept -8. We compare the graph of the function and the graph of its derivative as follows:

When $x < 4$:

- The function f is decreasing.

- The derivative f' is negative.

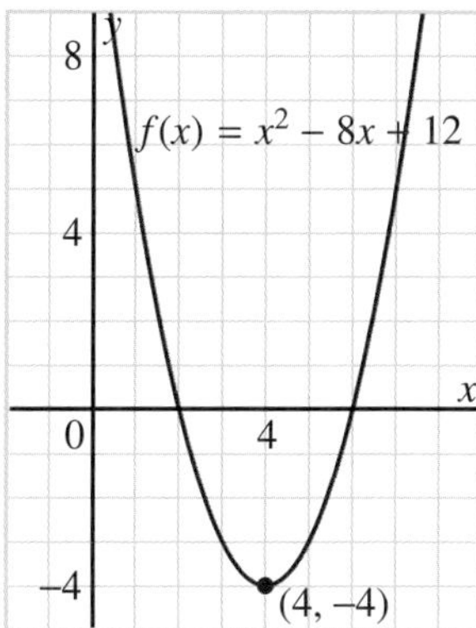

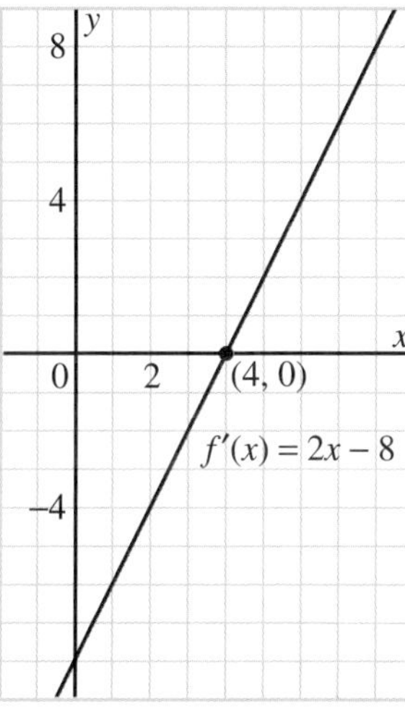

When $x > 4$:

- The function f is increasing.

- The derivative f' is positive.

We can tell from the information beside each graph that there is a local minimum point when $x = 4$. The information about the derivative is easier to use. When f' changes sign from negative to positive, there is a local minimum point on the graph of f.

The above analysis applies to any polynomial function $y = f(x)$. At a local minimum point:

- $f'(x) = 0$
- As x increases, the slope of the tangent increases as it moves around the point.
- As x increases, the derivative f' changes sign from negative to positive around the point.

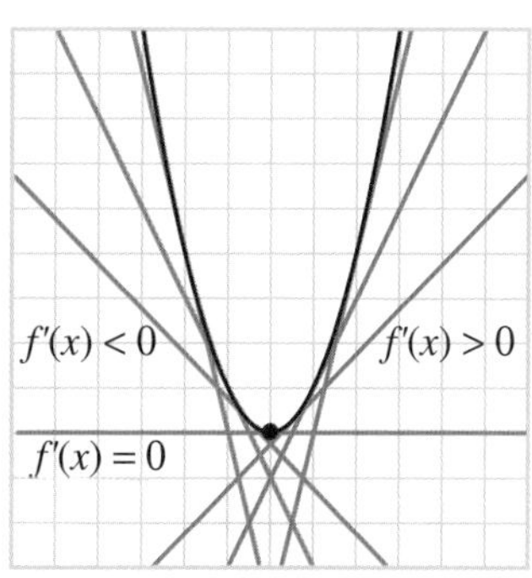

The situation for a maximum point is similar. Consider the function $y = -x^2 + 8x - 12$ (the reflection of the function $y = x^2 - 8x + 12$ in the x-axis) and its derivative $y' = -2x + 8$.

When $x < 4$:

- The function f is increasing.

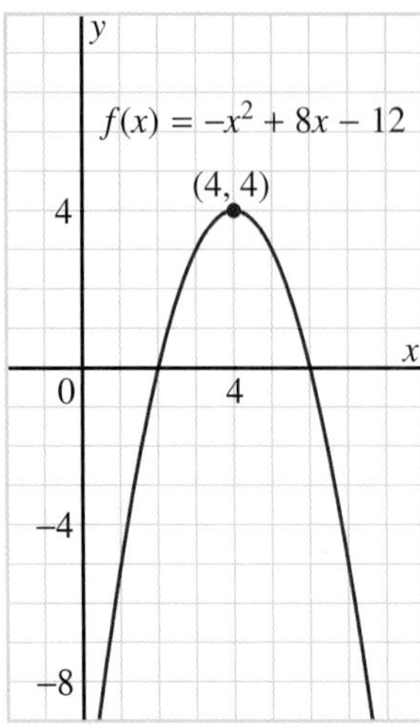

When $x > 4$:

- The function f is decreasing.

- The derivative f' is positive.

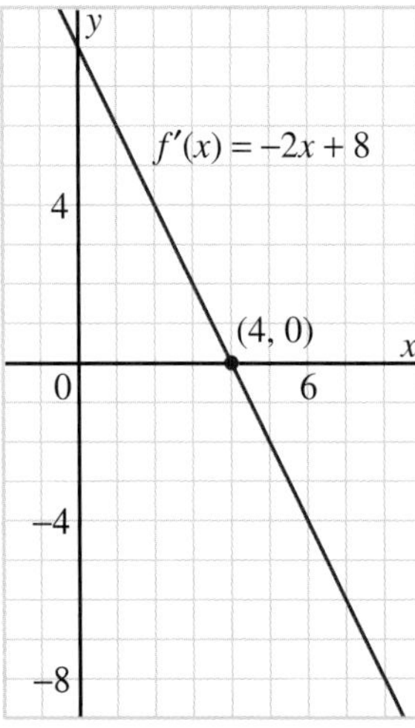

- The derivative f' is negative.

We can tell from the information beside each graph that there is a local maximum point when $x = 4$. The information about the derivative is easier to use. When f' changes sign from positive to negative, there is a local maximum point on the graph of f.

Again, the above analysis applies to any polynomial function $y = f(x)$. At a local maximum point:

- $f'(x) = 0$
- As x increases, the slope of the tangent decreases as it moves around the point.
- As x increases, the derivative f' changes sign from positive to negative around the point.

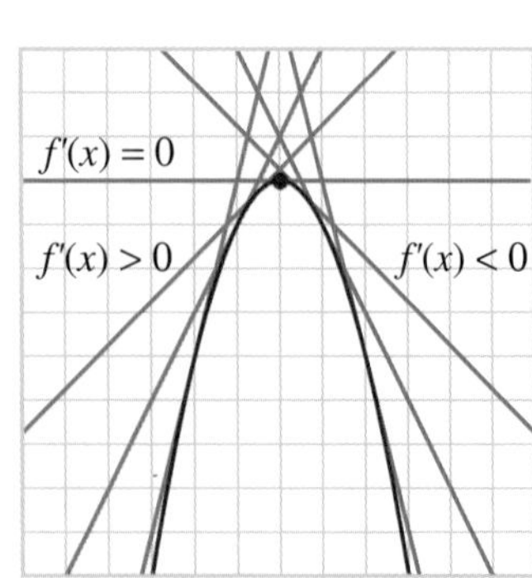

We summarize the above results on page 157.

Take Note

First Derivative Analysis of a Polynomial Function

Let f be a polynomial function.

Intervals of Increase and Decrease

- If $f'(x) > 0$ for all values of x in an interval, then f is increasing in that interval.
- If $f'(x) < 0$ for all values of x in an interval, then f is decreasing in that interval.

First Derivative Test for Local Maximum or Minimum Points

Suppose $f'(a) = 0$. Then $x = a$ is a critical value. As x increases:

- If $f'(x)$ changes sign from negative to positive at $x = a$, then f has a local minimum at $x = a$.
- If $f'(x)$ changes sign from positive to negative at $x = a$, then f has a local maximum at $x = a$.
- If $f'(x)$ does not change sign at $x = a$, then f has a point of inflection at $x = a$.

Example 1

Consider the polynomial function $f(x) = x^3 - 12x$.

a) Determine the x- and y-intercepts.

b) Determine the coordinates of the critical points.

c) Identify the nature of the critical points.

d) Graph the function.

e) State the intervals of increase and decrease.

Solution

a) $f(x) = x^3 - 12x$

The y-intercept is $f(0) = 0$.

To determine the x-intercepts, solve the equation $f(x) = 0$.

$$x^3 - 12x = 0$$
$$x(x^2 - 12) = 0$$

Either $x = 0$ or $x^2 - 12 = 0$

$$x^2 = 12$$
$$x = \pm\sqrt{12}$$

The x-intercepts are 0, $\sqrt{12}$, and $-\sqrt{12}$. In decimal form, the latter two are approximately 3.46 and –3.46.

On a sketch, plot the points that correspond to the intercepts. Since the leading coefficient of this cubic function is positive, the graph extends from the 3rd quadrant to the 1st quadrant. Sketch the curve. There appears to be a maximum point and a minimum point. We use calculus to verify this.

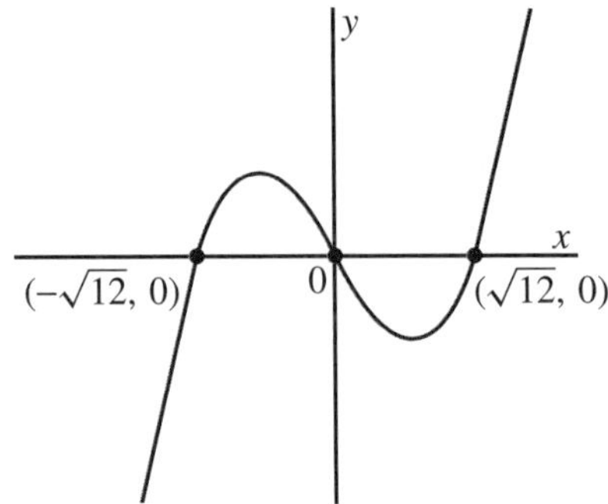

b) $f(x) = x^3 - 12x$

Differentiate.

$f'(x) = 3x^2 - 12$

Substitute $f'(x) = 0$ to determine the critical values.

$$3x^2 - 12 = 0$$
$$x^2 = 4$$
$$x = \pm 2$$

These are the critical values.

Determine the corresponding values of y.

Substitute $x = 2$ in $y = x^3 - 12x$.

$$y = 8 - 24$$
$$= -16$$

Substitute $x = -2$ in $y = x^3 - 12x$.

$$y = -8 + 24$$
$$= 16$$

The critical points are $(2, -16)$ and $(-2, 16)$. These may or may not be local maximum or local minimum points.

c) Consider each critical point and determine the values of $f'(x) = 3x^2 - 12$ at nearby points on each side of the critical point.

For the point $(-2, 16)$:

To the left, consider $x = -2.1$.

$$f'(-2.1) = 3(-2.1)^2 - 12$$
$$= 1.23$$

To the right, consider $x = -1.9$.

$$f'(-1.9) = 3(-1.9)^2 - 12$$
$$= -1.17$$

As x increases from -2.1 to -1.9, f' changes sign from positive to negative. Therefore, $(-2, 16)$ is a local maximum point.

For the point $(2, -16)$:

To the left, consider $x = 1.9$.

$$f'(1.9) = 3(1.9)^2 - 12$$
$$= -1.17$$

To the right, consider $x = 2.1$.

$$f'(2.1) = 3(2.1)^2 - 12$$
$$= 1.23$$

As x increases from 1.9 to 2.1, f' changes sign from negative to positive. Therefore, $(2, -16)$ is a local minimum point.

d) On a grid, plot the local maximum and minimum points from part c.
Plot the points corresponding to the intercepts.
Draw a smooth curve through these points, extending from the 3rd quadrant to the 1st quadrant.

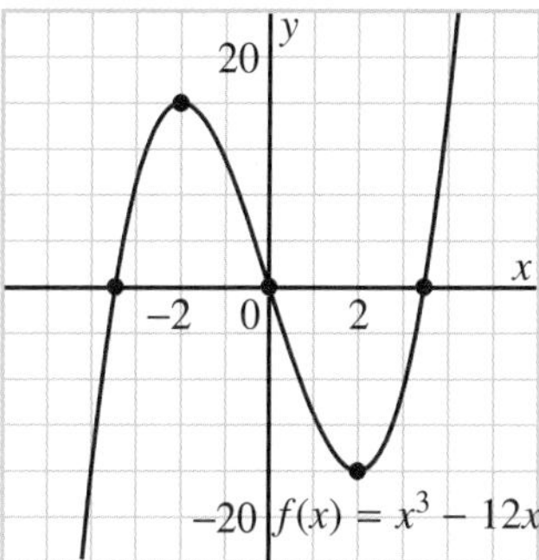

e) We can tell the intervals of increase and decrease from the graph.

The intervals of increase are $x < -2$ and $x > 2$, or $|x| > 2$.

The interval of decrease is $-2 < x < 2$, or $|x| < 2$.

In *Example 1c*, we could have illustrated the critical points on a number line. We show whether f' is negative, 0, or positive; then we use an arrow to indicate the corresponding direction of the graph. This enables us to visualize how the direction of the graph changes.

Direction of f	↗	→	↘	→	↗
x		-2		2	
Sign of f'	+	0	−	0	+

Compare the direction of the graph indicated by the arrows with the graph of $f(x) = x^3 - 12x$ above.

For the function in *Example 1*, we consider the *end behaviour* of the function to verify the shape of the graph. From the graph, when x is large, as x increases, y increases. Substitute $x = 10$ in the equation of the function to verify this.

$$\begin{aligned} f(10) &= (10)^3 - 12(10) \\ &= 880 \end{aligned}$$

Similarly, from the graph, when x is small, as x decreases, y decreases. Substitute $x = -10$ in the equation of the function to verify this.

$$\begin{aligned} f(-10) &= (-10)^3 - 12(-10) \\ &= -880 \end{aligned}$$

Example 2

Consider the polynomial function $f(x) = x^4 - 3x^3$.

a) Determine the x- and y-intercepts.

b) Determine the coordinates of the critical points.

c) Identify the nature of the critical points.

d) Graph the function.

e) State the intervals of increase and decrease.

Solution

a) $f(x) = x^4 - 3x^3$

The y-intercept is $f(0) = 0$.

To determine the x-intercepts, solve the equation $f(x) = 0$.

$x^4 - 3x^3 = 0$

$x^3(x - 3) = 0$

Either $x^3 = 0$ or $x = 3$

$x = 0$

The x-intercepts are 0 and 3.

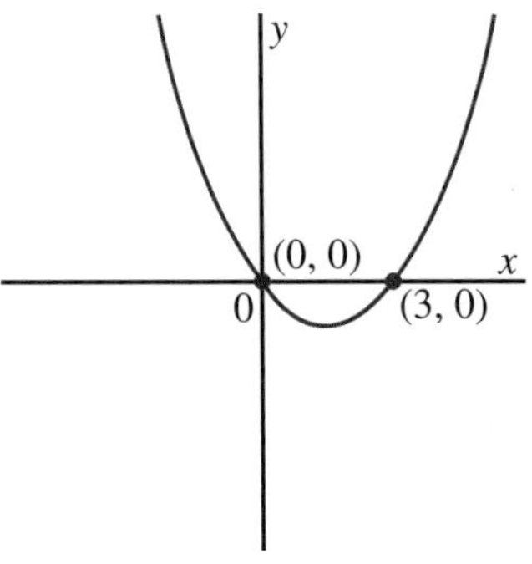

On a sketch, plot the points that correspond to the intercepts. Since the leading coefficient of this quartic function is positive, the graph extends from the 2nd quadrant to the 1st quadrant. Sketch the curve. There appears to be a minimum point. We use calculus to verify this.

b) $f(x) = x^4 - 3x^3$

Differentiate.

$f'(x) = 4x^3 - 9x^2$

Substitute $f'(x) = 0$ to determine the critical values.

$4x^3 - 9x^2 = 0$

$x^2(4x - 9) = 0$

Either $x^2 = 0$ or $4x - 9 = 0$

$x = 0$ $\qquad x = \frac{9}{4}$, or 2.25

The critical values are 0 and 2.25.

Determine the corresponding values of y.

Substitute $x = 0$ in $y = x^4 - 3x^3$.

$y = 0^4 - 3(0)^3$

$= 0$

Substitute $x = 2.25$ in $y = x^4 - 3x^3$.

$y = (2.25)^4 - 3(2.25)^3$

$\doteq -8.543$

The critical points are (0, 0) and approximately (2.25, –8.543).

c) Consider each critical point and determine the values of $f'(x) = 4x^3 - 9x^2$ on each side of the critical point.

For the point (0, 0):

To the left, consider $x = -0.1$.	To the right, consider $x = 0.1$.
$f'(-0.1) = 4(-0.1)^3 - 9(-0.1)^2$	$f'(0.1) = 4(0.1)^3 - 9(0.1)^2$
$= -0.094$	$= -0.086$

As x increases from -0.1 to 0.1, f' does not change sign. The point (0, 0) is neither a local maximum nor a local minimum. It is a point of inflection.

For the point (2.25, −8.543):

To the left, consider $x = 2.2$.	To the right, consider $x = 2.3$.
$f'(2.2) = 4(2.2)^3 - 9(2.2)^2$	$f'(2.3) = 4(2.3)^3 - 9(2.3)^2$
$= -0.968$	$= 1.058$

As x increases from 2.2 to 2.3, f' changes sign from negative to positive. Hence, (2.25, −8.543) is a local minimum point.
Illustrate the critical points on a number line.

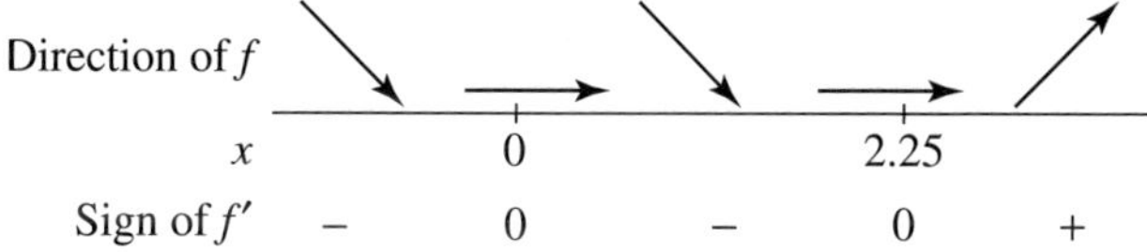

d) On a grid, plot the local minimum point and the point of inflection from part c.
Plot the points corresponding to the intercepts.
Draw a smooth curve through these points, extending from the 2nd quadrant to the 1st quadrant. Draw a point of inflection at (0, 0) and a local minimum point at approximately (2.25, −8.543).

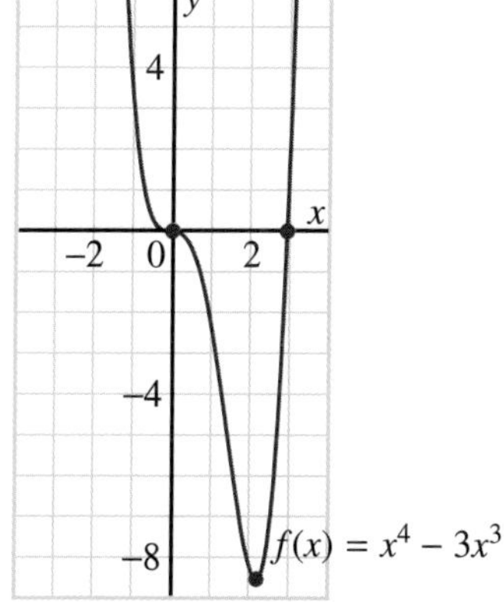

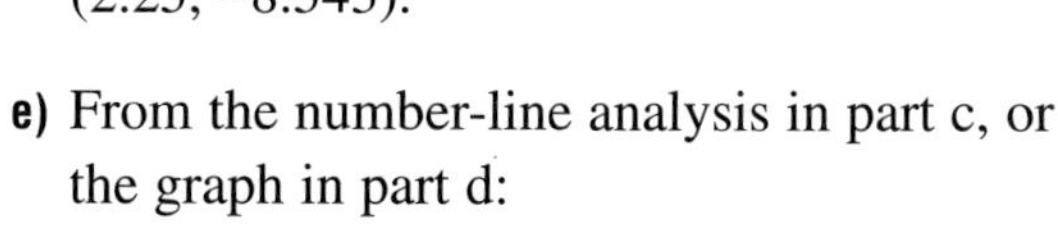

e) From the number-line analysis in part c, or the graph in part d:

The interval of increase is $x > 2.25$.

The interval of decrease is $x < 2.25$.

3.3 Exercises

1. For each graph of $y = f(x)$, state the coordinates of the local minimum and maximum points.

a)

b)

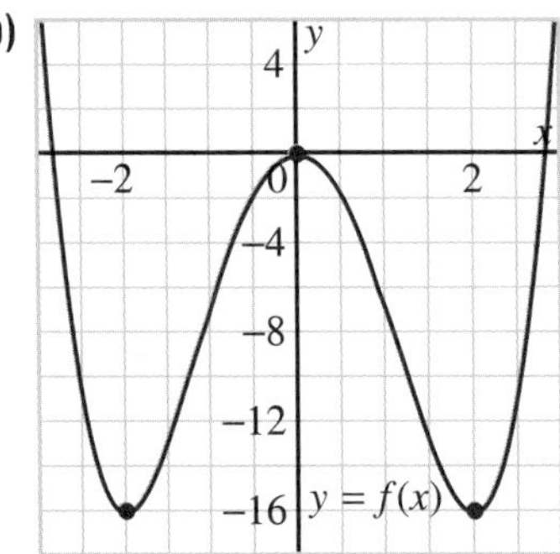

c)

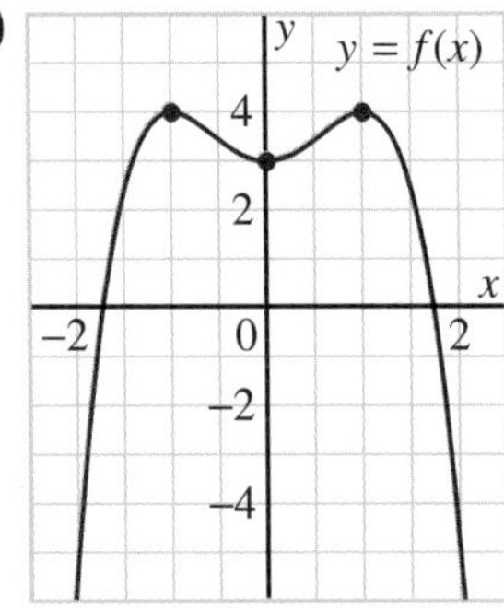

2. For each graph of $y = f(x)$ in exercise 1:
   i) State the critical values.
   ii) State the values of x for which the derivative is positive.
   iii) State the values of x for which the derivative is negative.

3. For each graph of $y = f'(x)$, state the critical values of f.

a)

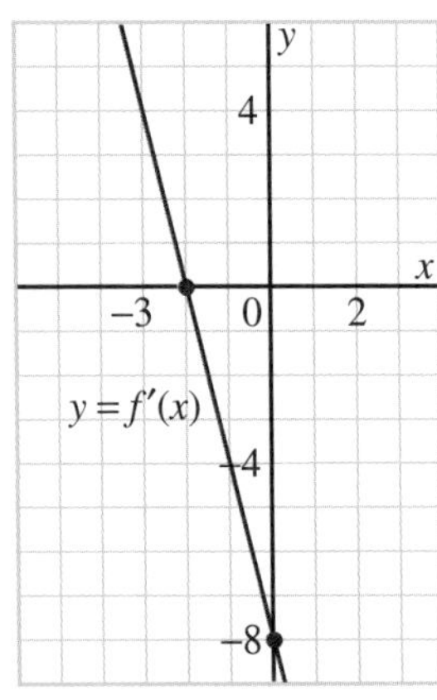

b)

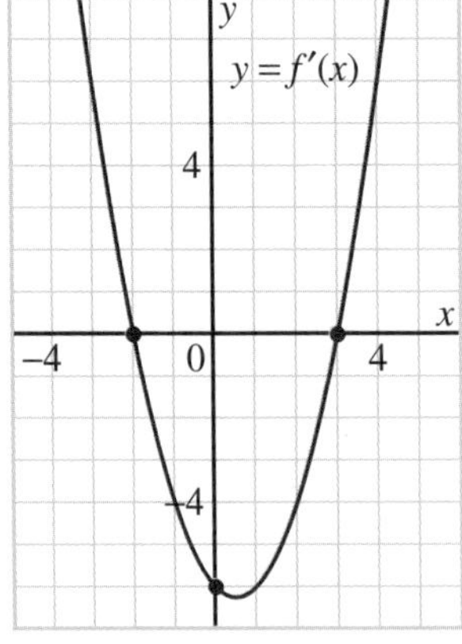

c)

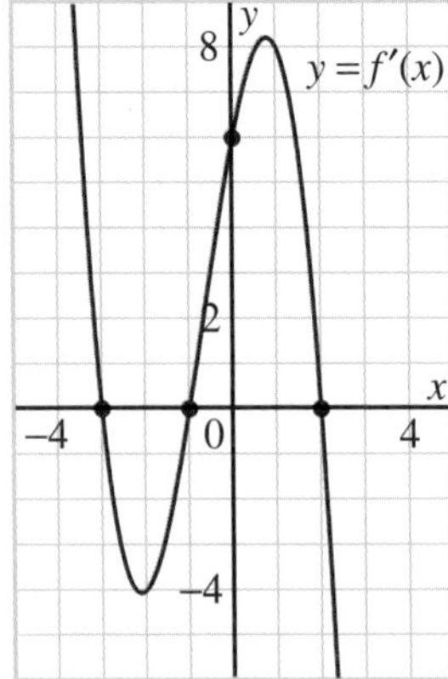

4. For each graph of $y = f'(x)$ in exercise 3:
   i) State whether f has a local minimum or a local maximum point at each critical value.
   ii) State the values of x for which f is increasing.
   iii) State the values of x for which f is decreasing.
   iv) Sketch a possible graph of $y = f(x)$.

5. For each graph of $y = f(x)$, state the intervals of increase and decrease for f.

a) 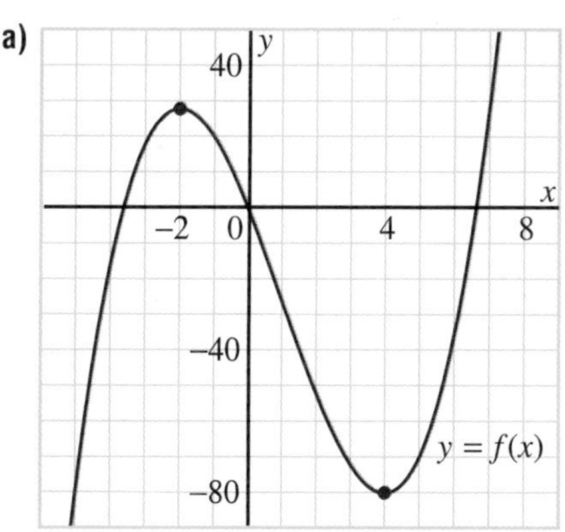

b) 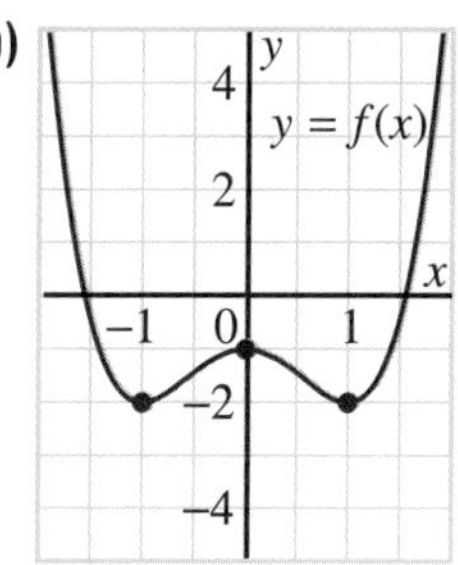

6. For each graph of $y = f(x)$ in exercise 5:

i) State the zeros of the corresponding derivative function.

ii) Does f have local or absolute maximum or minimum points? Justify your answer.

B

7. For each derivative function f', state the values of x where the graph of $y = f(x)$ has horizontal tangents.

a) $f'(x) = -5x(2x + 3)$

b) $f'(x) = 4(x + 1)(5x + 3)$

8. For each derivative function f', state the values of x where the function f may have local minimum or maximum points.

a) $f'(x) = -2(3x + 4)(x + 1)(x + 2)$

b) $f'(x) = 9x(6x + 1)(x + 7)(2x - 1)$

9. Differentiate each polynomial function.

a) $y = 3x^3 + 4x$

b) $y = x^4 - 2x^3 + x$

c) $y = -x^2 + 3x + 4$

d) $y = -2x^4 + 5x^2 - 3$

e) $y = -4x^3 - x^2 + x + 1$

f) $y = 2x^5 + 3x^3 + x + 1$

Exercises 9 and 10 are significant because they illustrate two important properties of the derivative of a polynomial function:

- The derivative of a polynomial function is another polynomial function.
- The degree of the derivative is one less than the degree of the function.

10. State as much as you can about how the derivatives in exercise 9 compare with the original functions.

11. **Communication** What properties of the graphs of polynomial functions can be determined from the derivative? Explain how the derivative is used to get this information.

12. **Application** For each derivative function f', determine the values of x for which the function f is increasing.

a) $f'(x) = -3x + 2$

b) $f'(x) = 4x(x + 3)(x - 1)$

13. Determine the coordinates of the critical points for each function.

a) $f(x) = x^3 - 6x$ b) $f(x) = x^3 - 9x$ c) $f(x) = x^4 - 2x^3$

d) $f(x) = x^4 + 2x^3$ e) $f(x) = x^4 + 3x^3$ f) $f(x) = x^4 - 4x^3$

✓ 14. Consider the polynomial function $f(x) = -3x^3 + 9x$.

a) Determine the x- and y-intercepts.

b) Determine the coordinates of the critical points.

c) Identify the nature of the critical points.

d) Graph the function.

e) State the intervals of increase and decrease.

✓ 15. **Knowledge/Understanding** Consider the polynomial function $f(x) = -4x^3 + 6x^2$.

a) Determine the x- and y-intercepts.

b) Determine the coordinates of the critical points.

c) Identify the nature of the critical points.

d) Graph the function.

e) State the intervals of increase and decrease.

16. Consider the polynomial function $f(x) = x^4 - 8x^2$.

a) Determine the x- and y-intercepts.

b) Determine the coordinates of the critical points.

c) Identify the nature of the critical points.

d) Graph the function.

e) State the intervals of increase and decrease.

✓ 17. For each graph of $y = f(x)$, sketch the graph of the derivative function $y = f'(x)$.

a)

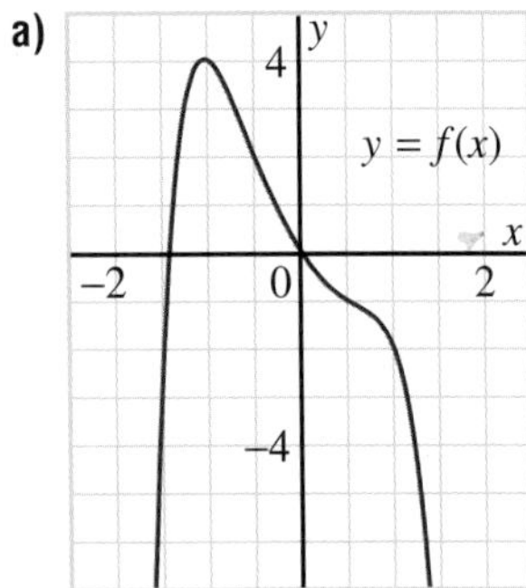

b)

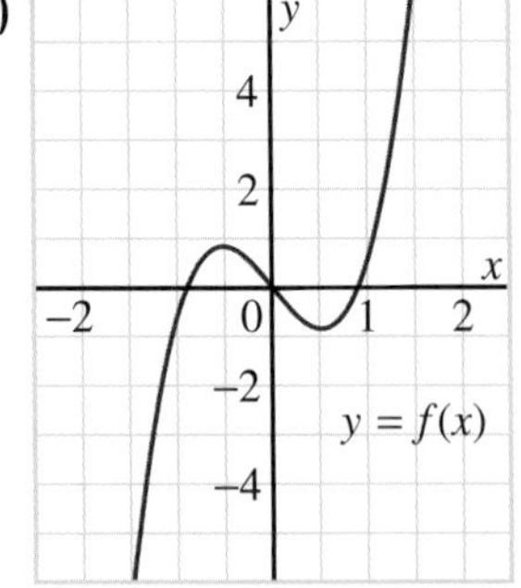

c)

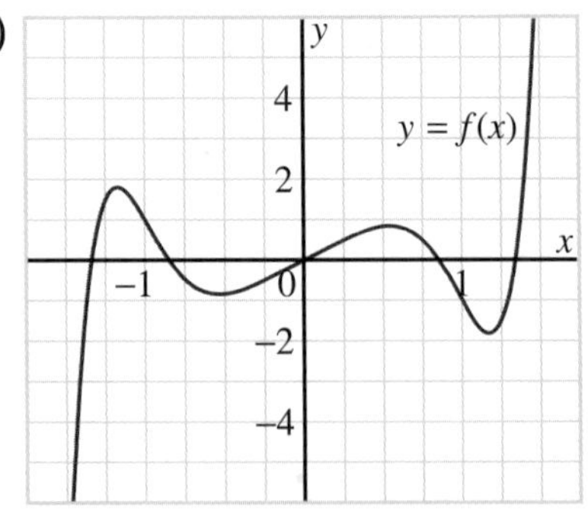

18. a) Determine the value of m so the function $y = x^2 + mx - 3$ has a local minimum point when $x = 6$.

b) Determine the local minimum value of the function.

19. **Thinking/Inquiry/Problem Solving** Determine the values of m and n so the polynomial function $y = 2x^3 + mx^2 + nx - 10$ has a local maximum when $x = -2$ and a local minimum when $x = 3$. Justify the solution.

20. The graph of $f(x) = x^3(4 - x)$ is shown, below left. There appears to be a maximum point at R(3, 27). Verify algebraically that R(3, 27) is a maximum point.

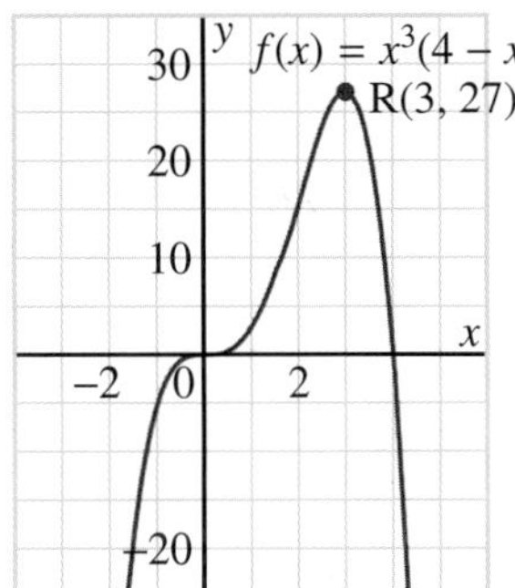

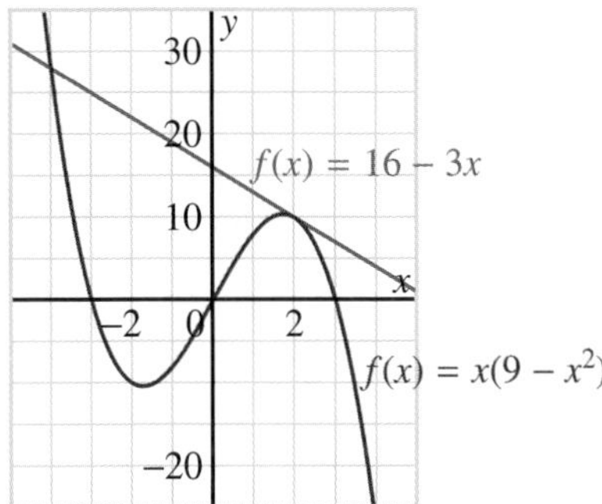

21. The graphs of $f(x) = x(9 - x^2)$ and $f(x) = 16 - 3x$ are shown, above right. Show algebraically that the line is a tangent to the curve at the point where $x = 2$.

22. The graph of the function $f(x) = 2x^3 - 25x$ is shown. The local maximum and local minimum points are marked. These appear to occur at $x = \pm 2$. Determine whether these are the correct x-coordinates for these points.

y
40
x
−2 0 2
−40 $f(x) = 2x^3 - 25x$

23. Show that the polynomial function $y = x^3 + 3x$ does not have any maximum or minimum points.

24. Consider the polynomial function $f(x) = -x^5 + 15x^3$.

a) Determine the x- and y-intercepts.

b) Determine the coordinates of the critical points.

c) Identify the nature of the critical points.

d) Graph the function.

e) State the intervals of increase and decrease.

25. In *Example 1*, we determined the coordinates of the critical points for $f(x) = x^3 - 12x$. In exercise 13, you determined the coordinates of these points for $f(x) = x^3 - 6x$ and $f(x) = x^3 - 9x$. These functions are graphed, below left.

   a) Verify that all these critical points lie on the graph of $y = -2x^3$.

   b) Determine whether this pattern continues. Can you find other cubic functions whose critical points lie on the graph? Explain.

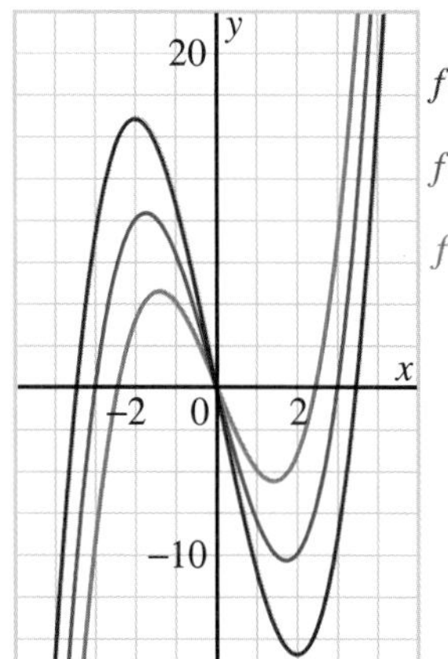

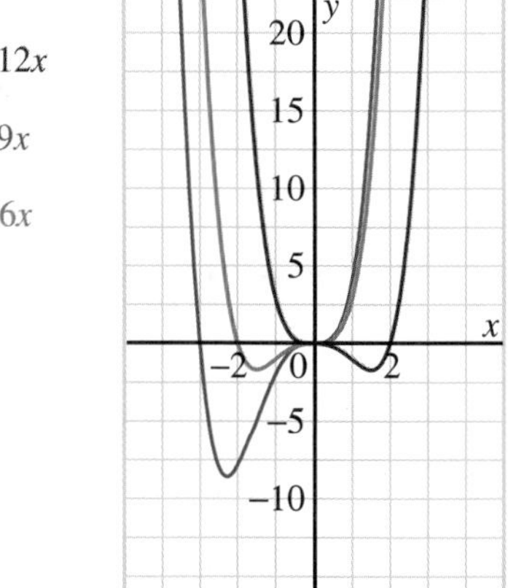

26. In *Example 2*, we determined the coordinates of the critical points for $f(x) = x^4 - 3x^3$. In exercise 13, you determined the coordinates of the critical points for 4 other quartic functions. Three of these functions are graphed, above right.

   a) Verify that all these critical points lie on the graph of $y = -\frac{x^4}{3}$.

   b) Identify the pattern in the equations of the 5 quartic functions whose critical points you determined. Can you find other quartic functions whose critical points lie on the graph of $y = -\frac{x^4}{3}$? Explain.

27. Sketch the graph of a function f that satisfies these conditions. Explain your solution.

   $f(-1) = -5, f(0) = 0$, and $f(5) = -12$

   $f'(-1) = f'(5) = 0$

   $f'(x) < 0$ when $x < -1$ and $0 < x < 5$

   $f'(x) > 0$ when $-1 < x < 0$ and $x > 5$

28. In exercise 27, is the function f unique? Explain.

29. Determine the coordinates of the points on the graph of $y = x^4 - 3x^2 + 5x - 1$ where the tangent is horizontal.

3.4 Using Second Derivatives to Graph Polynomial Functions

In Section 3.3, we used first derivatives to locate the critical points on the graphs of polynomial functions. To classify each point as a local maximum or minimum point, we determined whether the sign of the derivative changed at each point. This involved two steps because we had to calculate the value of the derivative on either side of the critical point. A more efficient method is to use the second derivative, which is the derivative of the derivative. The second derivative also provides information about a point of inflection.

Take Note

Second Derivative

The *second derivative* of a function f is the derivative of f', and is written as f''. The derivative f' is called the *first derivative.*

For example, for the function $y = x^3 - 12x$, the first derivative is $y' = 3x^2 - 12$ and the second derivative is $y'' = 6x$. The second derivative is the instantaneous rate of change of the first derivative. It tells us how the first derivative is changing.

We graphed the function $y = x^3 - 12x$ in Section 3.3, *Example 1*. The graph is repeated below.

This function has:

a local maximum point at $x = -2$
a local minimum point at $x = 2$
a point of inflection at $x = 0$

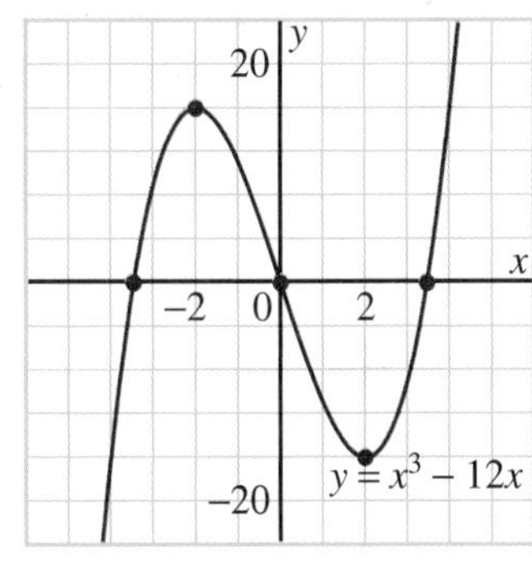

The explanations on pages 168 and 169 show how the first and second derivatives of this function are related to these points. The illustrated relationships apply to all polynomial functions. Make sure that you understand these relationships completely.

Obtaining information about local maxima and minima from f' and f''

When $x = -2$:

- The function f has a local maximum point.

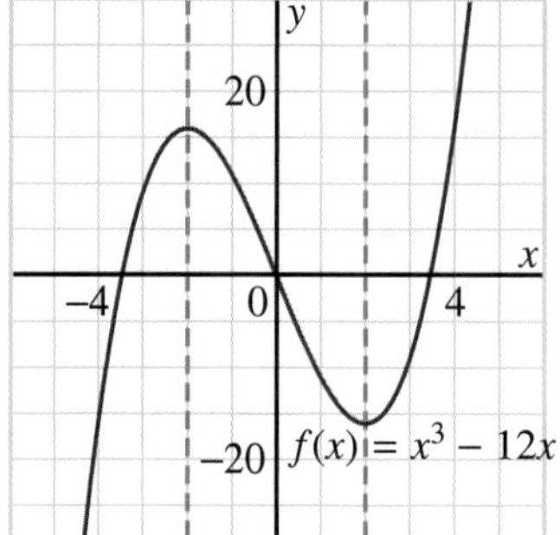

When $x = 2$:

- The function f has a local minimum point.

- The first derivative f' is equal to 0.

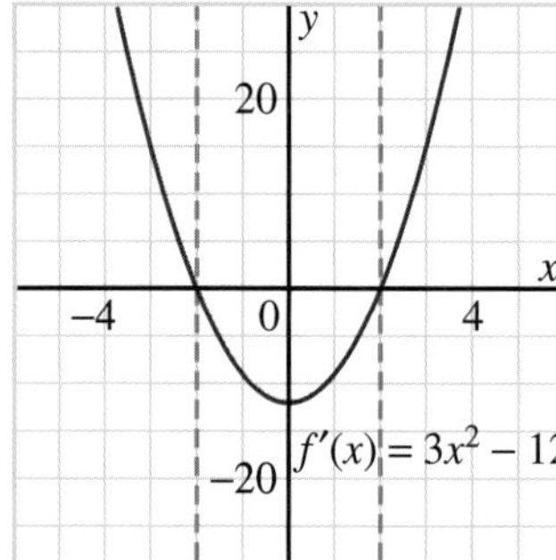

- The first derivative f' is equal to 0.

- The second derivative f'' is negative.

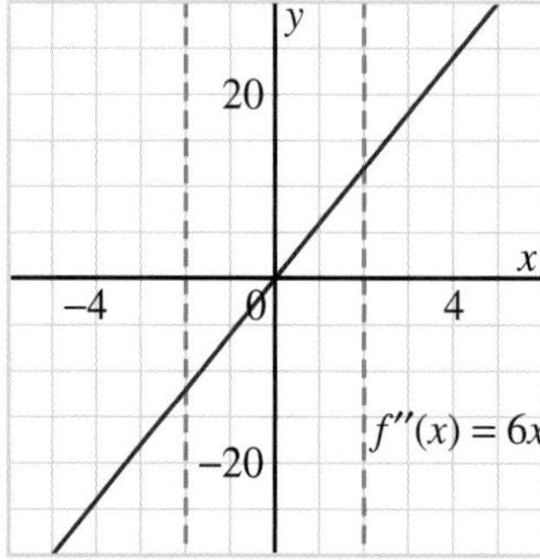

- The second derivative f'' is positive.

Although the explanation above is only one example, the information beside the graphs also applies to other polynomial functions. We use this information to modify the first derivative test in Section 3.3, as follows:

Take Note

First and Second Derivative Test for Local Maximum and Minimum Points

Let $y = f(x)$ be a polynomial function.

- At a local maximum point, $f'(x) = 0$ and $f''(x) < 0$
- At a local minimum point, $f'(x) = 0$ and $f''(x) > 0$

Obtaining information about a point of inflection from f' and f''

When $x < 0$:

- The slopes of the tangents to the graph of f are decreasing. The graph is concave down.

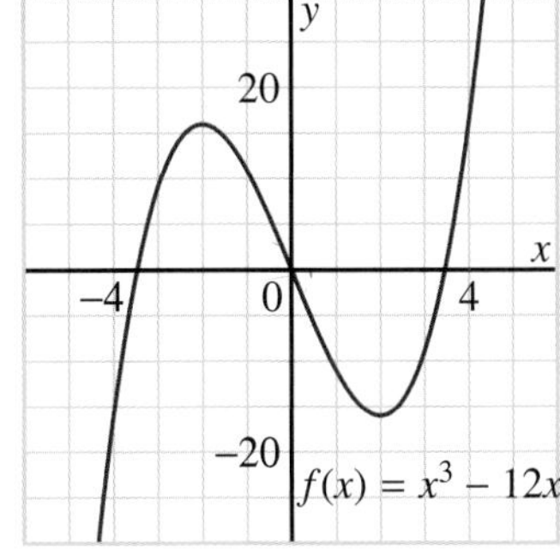

When $x > 0$:

- The slopes of the tangents to the graph of f are increasing. The graph is concave up.

When $x < 0$:

- The first derivative f' is decreasing.

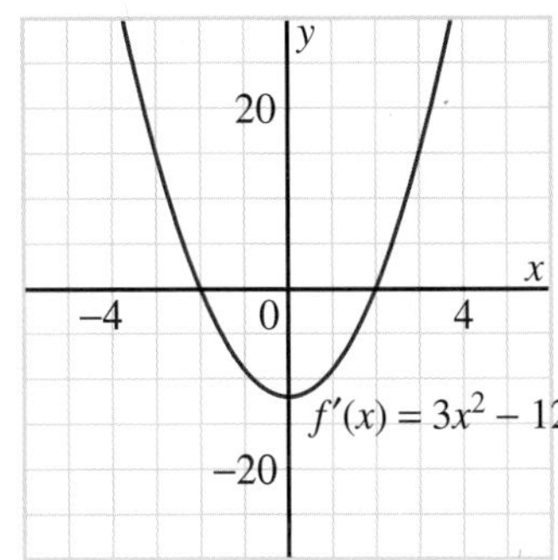

When $x > 0$:

- The first derivative f' is increasing.

When $x < 0$:

- The second derivative f'' is negative.

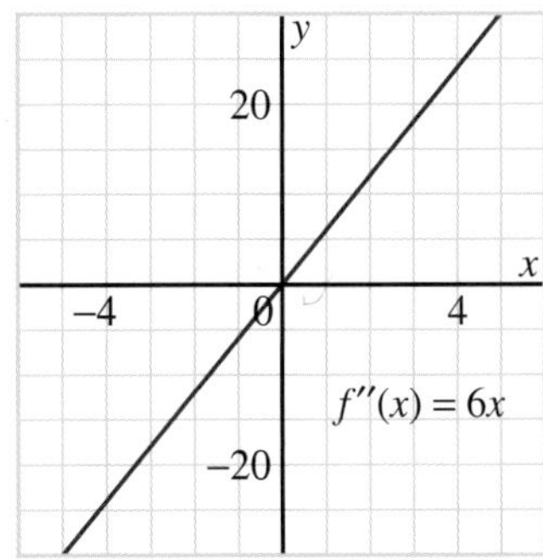

When $x > 0$:

- The second derivative f'' is positive.

We can tell from the information beside any of these graphs that there is a point of inflection on the graph of f at $x = 0$. The information about the second derivative is easiest to use. When f'' changes sign, there is a point of inflection; the graph of f changes from concave down to concave up, or vice versa.

The analysis on page 169 applies to any polynomial function $y = f(x)$. At a point of inflection:

- $f''(x) = 0$
- As x increases, the slopes of the tangents change from decreasing to increasing, or vice versa.
- As x increases, the second derivative f'' changes sign from negative to positive, or vice versa.

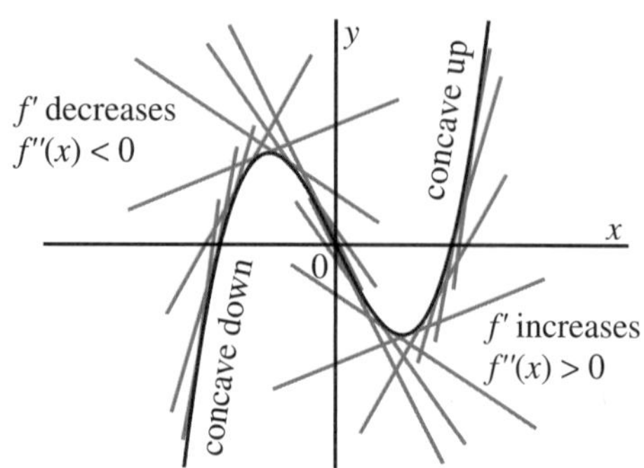

Take Note

Second Derivative Tests for Concavity and Points of Inflection

Let $y = f(x)$ be a polynomial function.

Intervals of Concavity

- If $f''(x) > 0$ for all values of x in an interval, then the graph of f is concave up in that interval.
- If $f''(x) < 0$ for all values of x in an interval, then the graph of f is concave down in that interval.

Points of Inflection

- At a point of inflection, $f''(x) = 0$ and, as x increases, f'' changes sign from positive to negative, or vice versa.
- At a point of inflection, the graph of f changes from concave up to concave down, or vice versa.

Example

Consider the polynomial function $f(x) = -x^4 + 9x^2$.

a) Determine the x- and y-intercepts.

b) Determine the coordinates of the critical points.

c) Identify the nature of the critical points.

d) Determine the coordinates of the points of inflection.

e) Graph the function.

f) State the intervals of increase and decrease.

g) State the intervals of concavity.

Solution

a) $f(x) = -x^4 + 9x^2$

The y-intercept is $f(0) = 0$.

To determine the x-intercepts, solve the equation $f(x) = 0$.

$$-x^4 + 9x^2 = 0$$
$$x^2(-x^2 + 9) = 0$$

Either $x^2 = 0$ or $-x^2 + 9 = 0$

$x = 0$ $\qquad x^2 = 9$

$x = \pm 3$

The x-intercepts are 0 and ± 3.

On a sketch, plot the points that correspond to the intercepts. Since $x = 0$ has order 2, the function does not change sign at that point. Since the leading coefficient of this quartic function is negative, the graph extends from the 3rd quadrant to the 4th quadrant. Sketch the curve. There appears to be two maximum points and one minimum point. We use calculus to verify this.

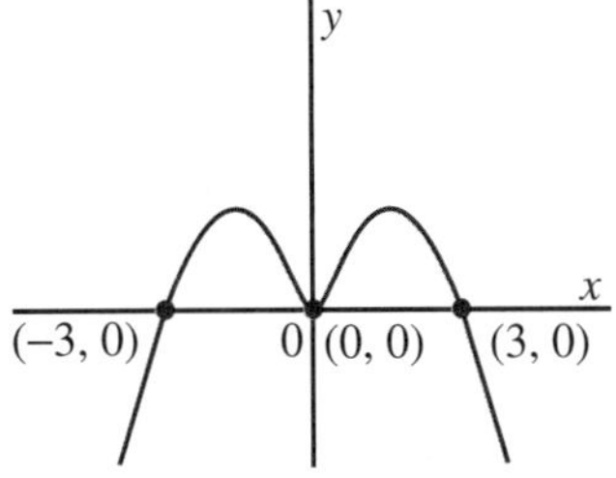

b) $f(x) = -x^4 + 9x^2$

Differentiate.

$f'(x) = -4x^3 + 18x$

Substitute $f'(x) = 0$ to determine the critical values.

$$-4x^3 + 18x = 0$$
$$2x(-2x^2 + 9) = 0$$

Either $2x = 0$ or $-2x^2 + 9 = 0$

$x = 0$ $\qquad 2x^2 = 9$

$x^2 = 4.5$

$x = \pm\sqrt{4.5}$

$\doteq \pm 2.1$

The critical values are 0 and $\pm\sqrt{4.5}$.

Determine the corresponding values of y.

Substitute $x = 0$ in $y = -x^4 + 9x^2$.

$y = 0$

Substitute $x = \sqrt{4.5}$ in $y = -x^4 + 9x^2$.

$$y = -\left(\sqrt{4.5}\right)^4 + 9\left(\sqrt{4.5}\right)^2$$
$$= 20.25$$

Substitute $x = -\sqrt{4.5}$ in $y = -x^4 + 9x^2$.

$$y = -\left(-\sqrt{4.5}\right)^4 + 9\left(-\sqrt{4.5}\right)^2$$
$$= 20.25$$

The critical points are $(0, 0)$, $\left(\sqrt{4.5}, 20.25\right)$, and $\left(-\sqrt{4.5}, 20.25\right)$.

c) To determine whether each critical point is a local maximum, a local minimum, or a point of inflection, use the second derivative.

$f'(x) = -4x^3 + 18x$

Differentiate.

$f''(x) = -12x^2 + 18$

Substitute each critical value in f'' to determine the sign of f''.
Substitute $x = -\sqrt{4.5}$ in $f''(x) = -12x^2 + 18$.

$$\begin{aligned} f''\left(-\sqrt{4.5}\right) &= -12\left(-\sqrt{4.5}\right)^2 + 18 \\ &= -36 \end{aligned}$$

Since $f'\left(-\sqrt{4.5}\right) = 0$ and $f''\left(-\sqrt{4.5}\right) < 0$, then $\left(-\sqrt{4.5}, 20.25\right)$ is a local maximum point.

Substitute $x = 0$ in $f''(x) = -12x^2 + 18$.

$$\begin{aligned} f''(0) &= -12(0)^2 + 18 \\ &= 18 \end{aligned}$$

Since $f'(0) = 0$ and $f''(0) > 0$, then $(0, 0)$ is a local minimum point.

Substitute $x = \sqrt{4.5}$ in $f''(x) = -12x^2 + 18$.

$$\begin{aligned} f''\left(\sqrt{4.5}\right) &= -12\left(\sqrt{4.5}\right)^2 + 18 \\ &= -36 \end{aligned}$$

Since $f'\left(\sqrt{4.5}\right) = 0$ and $f''\left(\sqrt{4.5}\right) < 0$, then $\left(\sqrt{4.5}, 20.25\right)$ is a local maximum point.

d) $f''(x) = -12x^2 + 18$

Substitute $f''(x) = 0$ to determine the points of inflection.

$$\begin{aligned} -12x^2 + 18 &= 0 \\ 12x^2 &= 18 \\ x^2 &= 1.5 \\ x &= \pm\sqrt{1.5} \\ &\doteq \pm 1.2 \end{aligned}$$

Consider each value of x and determine the values of $f''(x) = -12x^2 + 18$ on each side of this point.

For $x = -\sqrt{1.5}$:

To the left, consider $x = -1.3$.

$$\begin{aligned} f''(-1.3) &= -12(-1.3)^2 + 18 \\ &= -2.28 \end{aligned}$$

To the right, consider $x = -1.1$.

$$\begin{aligned} f''(-1.1) &= -12(-1.1)^2 + 18 \\ &= 3.48 \end{aligned}$$

As x increases from -1.3 to -1.1, f'' changes sign from negative to positive. The graph changes from concave down to concave up. Hence, there is a point of inflection at $x = -\sqrt{1.5}$.

For $x = \sqrt{1.5}$:

To the left, consider $x = 1.1$.

$f''(1.1) = -12(1.1)^2 + 18$
$= 3.48$

To the right, consider $x = 1.3$.

$f''(1.3) = -12(1.3)^2 + 18$
$= -2.28$

As x increases from 1.1 to 1.3, f'' changes sign from positive to negative. The graph changes from concave up to concave down. Hence, there is a point of inflection at $x = \sqrt{1.5}$.

Determine the corresponding values of y.

Substitute $x = -\sqrt{1.5}$ in $y = -x^4 + 9x^2$.

$y = -(-\sqrt{1.5})^4 + 9(-\sqrt{1.5})^2$
$= 11.25$

Substitute $x = \sqrt{1.5}$ in $y = -x^4 + 9x^2$.

$y = -(\sqrt{1.5})^4 + 9(\sqrt{1.5})^2$
$= 11.25$

The points of inflection are $(-\sqrt{1.5}, 11.25)$ and $(\sqrt{1.5}, 11.25)$.

We use the results of the second derivative test to illustrate the critical points, points of inflection, and concavity on a number line.

Shape of f	∩	Point of inflection	∪	Point of inflection	∩
x	$-\sqrt{4.5}$	$-\sqrt{1.5}$	0	$\sqrt{1.5}$	$\sqrt{4.5}$
Sign of f''	$-$	0	+	0	$-$

e) On a grid, plot the local maximum and the local minimum points, and the points of inflection.

Plot the points corresponding to the intercepts.

Draw a smooth curve through these points, extending from the 3rd quadrant to the 4th quadrant.

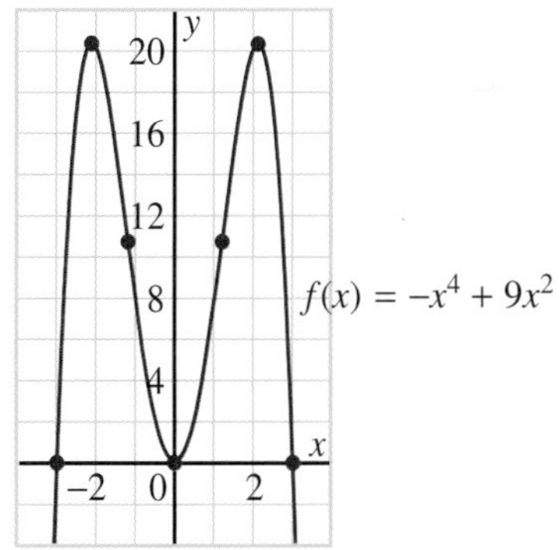

f) From the number-line analysis in part d or the graph in part e:

The intervals of increase are $x < -\sqrt{4.5}$ and $0 < x < \sqrt{4.5}$.

The intervals of decrease are $-\sqrt{4.5} < x < 0$ and $x > \sqrt{4.5}$.

g) The graph changes concavity at the points of inflection.
From the number-line analysis or the graph:

The graph of f is concave down for $x < -\sqrt{1.5}$ and $x > \sqrt{1.5}$; that is, $|x| > \sqrt{1.5}$.

The graph of f is concave up for $-\sqrt{1.5} < x < \sqrt{1.5}$; that is, $|x| < \sqrt{1.5}$.

3.4 Exercises

1. For each function f, state the value of x where a point of inflection may occur.
   a) $f'(2) = 0, f''(-3) = 0$
   b) $f'(-1) = 0, f''(7) = 0$
   c) $f(-4) = 6, f''(-4) = 0$
   d) $f(5) = 8, f'(1) = 0, f''(3) = 0$

2. Determine the second derivative of each polynomial function.
   a) $y = -x^3$
   b) $y = 2x^2 + 3x - 1$
   c) $y = 2x^3 + 3x^2 + x$
   d) $y = 3x^4 + 5x^2$
   e) $y = -x^4 + x^3 - 3x^2$
   f) $y = -2x^5 + 4x^4 - 3$

3. The second derivative of a polynomial function f is given. Determine the values of x where points of inflection may occur on the graph of f.
   a) $f''(x) = x(2x + 3)$
   b) $f''(x) = -3(4x - 5)(x + 2)(3x - 1)$
   c) $f''(x) = x^2 - 5x + 6$
   d) $f''(x) = 2x^4 - 32$

4. Determine the second derivative of each polynomial function.
   a) $f(x) = 3x^3 - 4x + 7$
   b) $g(x) = 5x^4 - 2x^3 + 4x - 1$
   c) $y = -2(x - 4)^2 + 11$
   d) $h(x) = 2(x^2 - 1)(x + 2)$
   e) $y = -4(x + 2)(2x - 1)^2$
   f) $p(x) = (2x + 1)^3$

5. A polynomial function f and its second derivative f'' are graphed on the same grid. Which graph is f and which graph is f''? Explain.

a)

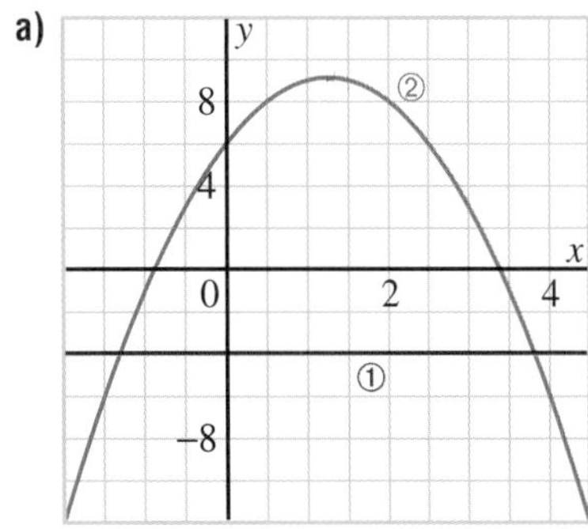

b)

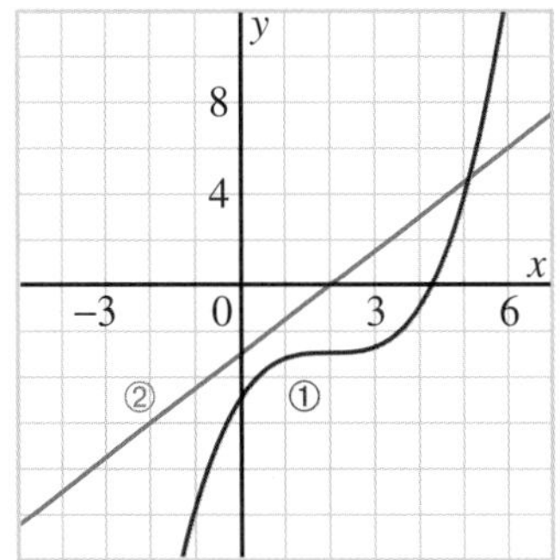

c)

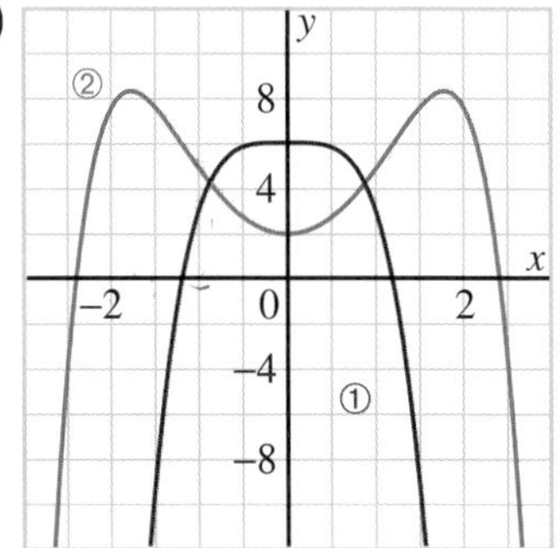

6. For each function f in exercise 5, explain how you would use the graph of f'' in each case.
   i) to determine the points of inflection on the graph of f
   ii) to determine the intervals of concavity on the graph of f

7. For each function f in exercise 5:

   i) Estimate the coordinates of the local minimum and local maximum points.

   ii) Estimate the intervals of increase and decrease.

   iii) Estimate the coordinates of the points of inflection.

   iv) Estimate the intervals of concavity.

8. For each function f in exercise 5:

   i) Determine the degree of f and the degree of f''.

   ii) Determine the sign of the leading coefficient of f.

9. A polynomial function, f, and its derivatives, f' and f'', are graphed on the same grid. Which curve is f, f', and f''? Give reasons for your answer.

a)

b) 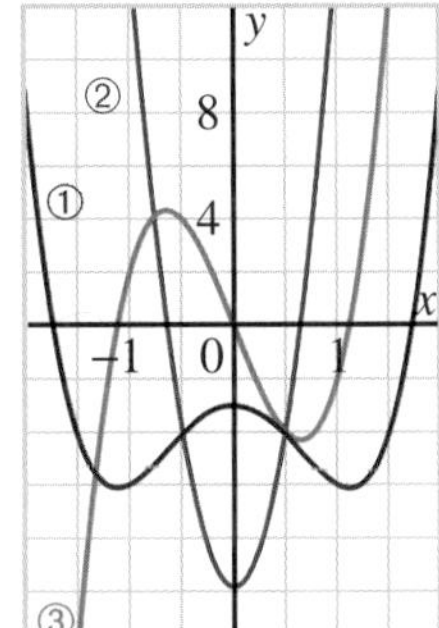

c) 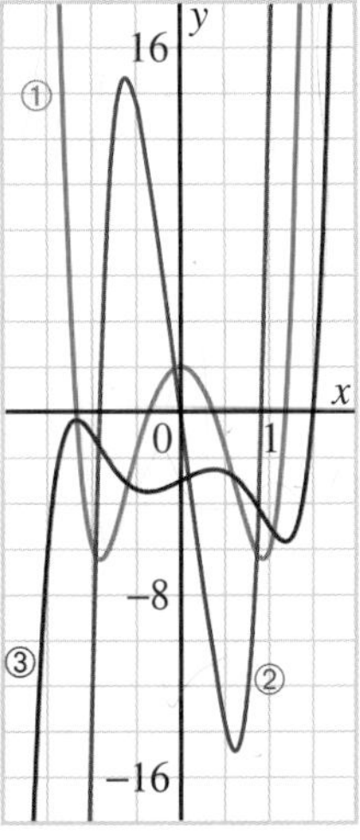

10. For each derivative function f', determine the x-coordinate(s) of the point(s) of inflection of the function f.

   a) $f'(x) = -3x^4 + 2x^3$ b) $f'(x) = x^5 + 20x^2$ c) $f'(x) = x^4 + 6x$

11. **Knowledge/Understanding** Consider the polynomial function $f(x) = -x^3 + 4x$.

   a) Determine the x- and y-intercepts.

   b) Determine the coordinates of the critical points.

   c) Identify the nature of the critical points.

   d) Determine the coordinates of the points of inflection.

   e) Graph the function.

   f) State the intervals of increase and decrease.

   g) State the intervals of concavity.

12. Consider the polynomial function $f(x) = -2x^4 + x^2$.

a) Determine the x- and y-intercepts.

b) Determine the coordinates of the critical points.

c) Identify the nature of the critical points.

d) Determine the coordinates of the points of inflection.

e) Graph the function.

f) State the intervals of increase and decrease.

g) State the intervals of concavity.

13. Consider the function $f(x) = (x^2 - 9)(5x + 2)$.

a) Determine the x- and y-intercepts.

b) Determine the coordinates of the critical points.

c) Identify the nature of the critical points.

d) Determine the coordinates of the points of inflection.

e) Graph the function.

f) State the intervals of increase and decrease.

g) State the intervals of concavity.

14. Consider the function $y = -2x^4 + 16x^2$.

a) Determine the x- and y-intercepts.

b) Determine the coordinates of the critical points.

c) Identify the nature of the critical points.

d) Determine the coordinates of the points of inflection.

e) Graph the function.

f) State the intervals of increase and decrease.

g) State the intervals of concavity.

15. As x increases, the graph of a polynomial function may change direction from going up to going down, or vice versa.

a) How many times is it possible for the graph of each polynomial function to change direction? Sketch examples to illustrate each answer.

i) linear function

ii) quadratic function

iii) cubic function

iv) quartic function

b) How many times is it possible for the graph of a polynomial function with degree n to change direction? Explain.

Exercise 15 is significant because counting the number of times a graph changes direction is a useful way to think about the graph of a polynomial function. For any given degree, different numbers of direction changes are possible, up to a certain maximum number. The maximum is always 1 less than the degree of the polynomial.

✓ **16. Communication** Determine whether each statement is true or false. Explain.

a) The second derivative of a cubic function is a linear function.

b) The first derivative is used to determine a point of inflection.

c) The second derivative is used to determine intervals of concavity.

d) The second derivative is positive at a maximum point.

e) The second derivative is positive at a minimum point.

✓ **17. Application** The following information is taken from the graph of a polynomial function f.

$f(3) = 5, f(7) = -2, f(5) = 2$
$f'(3) = f'(7) = 0$
$f'(x) > 0$ for $x < 3$ and $x > 7$
$f'(x) < 0$ for $3 < x < 7$
$f''(5) = 0$
$f''(x) < 0$ for $x < 5$ and $f''(x) > 0$ for $x > 5$

For the graph of f:

a) Determine the coordinates of its local minimum and local maximum point(s).

b) Determine the intervals of increase and decrease.

c) Determine the coordinates of its point of inflection.

d) Graph the function.

e) Determine the intervals of concavity.

✓ **18.** For each function f, sketch the graphs of the first and second derivative functions.

a)
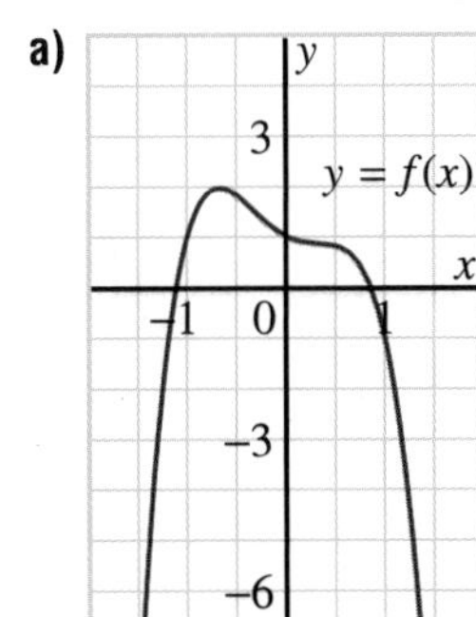

b)
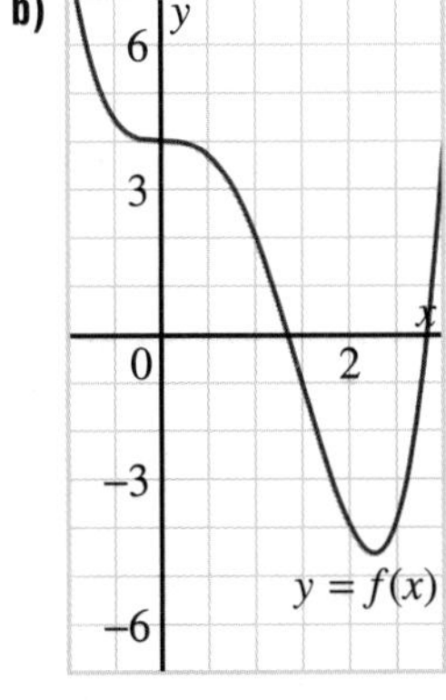

c)
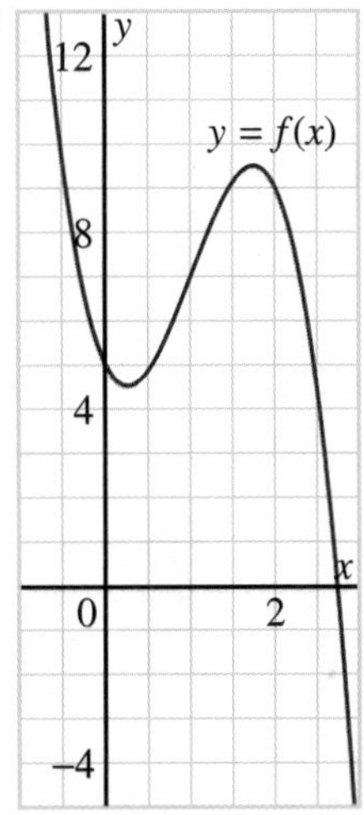

✓ **19. Thinking/Inquiry/Problem Solving** The polynomial function $f(x) = x^3 + ax^2 + x + b$ has a point of inflection at $(-2, 9)$.

a) Determine the values of a and b.

b) Determine the equation of the tangent at the point of inflection.

C

20. a) Suppose a polynomial function $y = f(x)$ has a local minimum at $x = c$. What is the sign of $f''(c)$? Explain.

b) Suppose a polynomial function $y = f(x)$ has a local maximum at $x = c$. What is the sign of $f''(c)$? Explain.

c) State a relationship that describes the results of parts a and b.

21. Use the relationship in exercise 20 to determine whether the polynomial function $y = f(x)$ has a local minimum point, local maximum point, or neither, at each x-value.

a) $x = 3$; $f'(3) = 0$, $f''(3) > 0$

b) $x = -4$; $f'(-4) = 0$, $f''(-4) = 0$

c) $x = -1$; $f'(-1) = 2$, $f''(-1) < 0$

d) $x = -6$; $f'(-6) = 0$, $f''(-6) < 0$

e) $x = 0.5$; $f'(0.5) = -3$, $f''(0.5) > 0$

f) $x = 0$; $f'(0) = 0$, $f''(0) > 0$

22. Consider the function $f(x) = -3x^5 + 5x^3$.

a) Determine the intervals of increase and decrease.

b) Determine the coordinates and nature of the critical points.

c) Determine the coordinates of the points of inflection.

d) Determine the intervals of concavity.

e) Graph the function.

23. In exercise 18, page 151, a screen showed a small portion of the graph of the cubic function $y = x^3 - 250x^2 + 4000$. A different screen for this function is shown at the right.

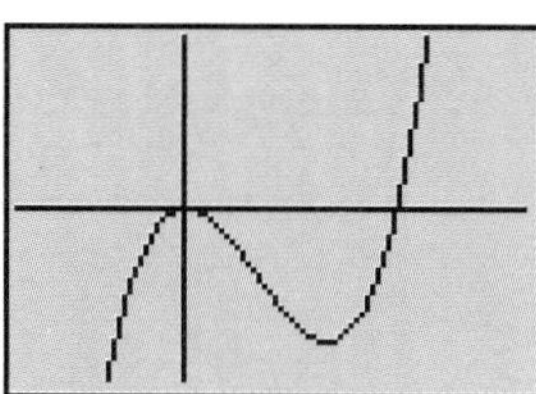

a) Determine the coordinates of the local minimum point.

b) Determine the coordinates of the point of inflection.

c) The window settings for this screen are not shown. Use the results of parts a and b to estimate the window settings that were used to make this screen.

24. Determine the point of intersection of the tangents at the points of inflection of the curve $f(x) = x^4 - 24x^2 - 1$.

3.5 The Factor Theorem

The factored form of a polynomial function is used to determine information about the graph of the function. In Section 3.2, the zeros of a polynomial function were determined by solving the corresponding polynomial equation.

Conversely, when we know the zeros of a polynomial function f, we may write an equation to represent the family of polynomial functions with those zeros. For example, a 4th degree polynomial function with zeros 0, –2, 1, 3, has the equation $f(x) = ax(x + 2)(x - 1)(x - 3)$. We need more information about the function to determine the value of a.

In Sections 3.3 and 3.4, we factored polynomial functions using common factors, difference of squares, and factoring trinomials. In this section, we will factor a polynomial of degree greater than 2. We shall use long division.

Recall from *Necessary Skills*, page 125, the division statement:

dividend = (divisor)(quotient) + remainder

In some problems involving division, only the remainder is needed. Consider the polynomial $f(x) = x^3 + 5x^2 + 7x - 1$. We shall determine the remainder when $f(x)$ is divided by $x + 2$.

$$\begin{array}{r} x^2 + 3x + 1 \\ x + 2 \overline{) x^3 + 5x^2 + 7x - 1} \\ \underline{x^3 + 2x^2} \\ 3x^2 + 7x \\ \underline{3x^2 + 6x} \\ x - 1 \\ \underline{x + 2} \\ -3 \end{array}$$

The remainder is –3.

The division statement is $x^3 + 5x^2 + 7x - 1 = (x + 2)(x^2 + 3x + 1) - 3$.

Notice what happens when we substitute $x = -2$ in each side of the division statement.

$$\begin{aligned} \text{L.S.} &= x^3 + 5x^2 + 7x - 1 \\ &= (-2)^3 + 5(-2)^2 + 7(-2) - 1 \\ &= -8 + 20 - 14 - 1 \\ &= -3 \end{aligned}$$

$$\begin{aligned} \text{R.S.} &= (x + 2)(x^2 + 3x + 1) - 3 \\ &= (-2 + 2)((-2)^2 + 3(-2) + 1) - 3 \\ &= 0(-1) - 3 \\ &= -3 \end{aligned}$$

Each side of the division statement is an expression for $f(x)$.
So, when we substitute $x = -2$, we get the value of $f(-2)$.
Hence, $f(-2) = -3$

This illustrates that when the polynomial $x^3 + 5x^2 + 7x - 1$ is divided by $x + 2$, the remainder is $f(-2)$.

This is an example of a general result called the *Remainder Theorem.* It is true for any polynomial.

Take Note

Remainder Theorem

When a polynomial, $f(x)$, is divided by $x - k$, the remainder is equal to $f(k)$, the number obtained by substituting k for x in the polynomial.

Suppose a polynomial is divided by a binomial. We can use the Remainder Theorem to determine the remainder without doing the division.

Example 1

Determine the remainder when $x^3 + 2x^2 + 3x - 1$ is divided by each binomial.

a) $x - 1$

b) $x + 3$

Solution

a) Let $f(x) = x^3 + 2x^2 + 3x - 1$.

According to the Remainder Theorem, the remainder when $f(x)$ is divided by $x - 1$ is $f(1)$.

$$\begin{aligned} f(1) &= 1^3 + 2(1)^2 + 3(1) - 1 \\ &= 5 \end{aligned}$$

The remainder is 5.

b) The remainder when $f(x)$ is divided by $x + 3$ is $f(-3)$.

$$\begin{aligned} f(-3) &= (-3)^3 + 2(-3)^2 + 3(-3) - 1 \\ &= -27 + 18 - 9 - 1 \\ &= -19 \end{aligned}$$

The remainder is -19.

We can use the Remainder Theorem to solve a problem involving a polynomial and its remainder after division by a binomial.

Example 2

When $2x^3 - 3x^2 + cx + 6$ is divided by $x - 2$, the remainder is 4. Determine the value of c.

Solution

Let $f(x) = 2x^3 - 3x^2 + cx + 6$.
According to the Remainder Theorem, the remainder when $f(x)$ is divided by $x - 2$ is $f(2)$.

$f(2) = 2(2)^3 - 3(2)^2 + 2c + 6$
$f(2) = 16 - 12 + 2c + 6$
$f(2) = 10 + 2c$

Since the remainder is 4, $f(2) = 4$. Substitute $f(2) = 4$.

$4 = 10 + 2c$
$2c = -6$
$c = -3$

Discuss

How could you check the result?

According to the Remainder Theorem, if a number, k, is substituted for x in a polynomial in x, the result is the remainder when the polynomial is divided by $x - k$. If this remainder is 0, then $x - k$ is a factor of the polynomial.

This special case of the Remainder Theorem is called the *Factor Theorem.*

Take Note

Factor Theorem

If k is substituted for x in $f(x)$, and $f(k)$ is 0, then $x - k$ is a factor of $f(x)$.

We use the Factor Theorem to determine whether a binomial of the form $x - k$ is a factor of a given polynomial.

Example 3

Determine which binomials are factors of $x^3 - x^2 - 10x - 8$.

a) $x + 2$ **b)** $x - 1$ **c)** $x + 1$ **d)** $x - 4$

Solution

Let $f(x) = x^3 - x^2 - 10x - 8$.
If the binomial is a factor, then there is no remainder when the polynomial is divided by the binomial.

a) For the binomial $x + 2$, substitute $x = -2$.

$$\begin{aligned} f(-2) &= (-2)^3 - (-2)^2 - 10(-2) - 8 \\ &= -8 - 4 + 20 - 8 \\ &= 0 \end{aligned}$$

Since $f(-2) = 0$, $x + 2$ is a factor of $x^3 - x^2 - 10x - 8$.

b) For the binomial $x - 1$, substitute $x = 1$.

$$\begin{aligned} f(1) &= 1^3 - 1^2 - 10 - 8 \\ &= 1 - 1 - 10 - 8 \\ &= -18 \end{aligned}$$

Since $f(1) \neq 0$, $x - 1$ is not a factor of $x^3 - x^2 - 10x - 8$.

c) For the binomial $x + 1$, substitute $x = -1$.

$$\begin{aligned} f(-1) &= (-1)^3 - (-1)^2 - 10(-1) - 8 \\ &= -1 - 1 + 10 - 8 \\ &= 0 \end{aligned}$$

Since $f(-1) = 0$, $x + 1$ is a factor of $x^3 - x^2 - 10x - 8$.

d) For the binomial $x - 4$, substitute $x = 4$.

$$\begin{aligned} f(4) &= 4^3 - 4^2 - 10(4) - 8 \\ &= 64 - 16 - 40 - 8 \\ &= 0 \end{aligned}$$

Since $f(4) = 0$, $x - 4$ is a factor of $x^3 - x^2 - 10x - 8$.

In *Example 3*, we found three factors of $x^3 - x^2 - 10x - 8$.
The product of these three factors is $x^3 - x^2 - 10x - 8$.
Hence, we can write this polynomial in factored form:
$x^3 - x^2 - 10x - 8 = (x + 2)(x + 1)(x - 4)$
The product of the constant terms in the factors is $(+2)(+1)(-4) = -8$. This is also the constant term in the polynomial. This suggests the following property of the factors of a polynomial.

Take Note

Factor Property

A polynomial $f(x)$ has integer coefficients with leading coefficient 1.
If $f(x)$ has any factor of the form $x - k$, then k is a factor of the constant term of $f(x)$.

We can use the Factor Theorem and the Factor Property to factor a polynomial. We must find a value of x that results in 0 when it is substituted in the polynomial. The Factor Property helps us decide which values to test.

Example 4

Factor $x^3 + 3x^2 - 6x - 8$.

Solution

Let $f(x) = x^3 + 3x^2 - 6x - 8$.
Find a value of x so that $f(x) = 0$.
According to the Factor Property, the numbers to test are the factors of –8; that is, 1, 2, 4, 8, –1, –2, –4, and –8.

Try $x = 1$.

$$\begin{aligned} f(1) &= 1^3 + 3(1)^2 - 6(1) - 8 \\ &= 1 + 3 - 6 - 8 \\ &\neq 0 \end{aligned}$$

So, $x - 1$ is not a factor of $x^3 + 3x^2 - 6x - 8$.
Try $x = -1$.

$$\begin{aligned} f(-1) &= (-1)^3 + 3(-1)^2 - 6(-1) - 8 \\ &= -1 + 3 + 6 - 8 \\ &= 0 \end{aligned}$$

So, $x + 1$ is one factor of $x^3 + 3x^2 - 6x - 8$.

To determine the other factors, use long division.

$$\begin{array}{r} x^2 + 2x - 8 \\ x + 1 \overline{\big) x^3 + 3x^2 - 6x - 8} \\ \underline{x^3 + x^2 } \\ 2x^2 - 6x \\ \underline{2x^2 + 2x } \\ -8x - 8 \\ \underline{-8x - 8} \\ 0 \end{array}$$

The other factor is $x^2 + 2x - 8$. This can be factored as $(x + 4)(x - 2)$.
Therefore, $x^3 + 3x^2 - 6x - 8 = (x + 1)(x + 4)(x - 2)$

Discuss

We could have determined the other factors by repeatedly using the Factor Theorem to check ±2, ±4, ±8. Which method is easier? Explain.

Here is an alternative way to determine the other factor, when one factor is known.

From *Example 4*, one factor of $x^3 + 3x^2 - 6x - 8$ is $x + 1$.

Let the other factor be $x^2 + bx + c$.
Then, $(x + 1)(x^2 + bx + c) = x^3 + 3x^2 - 6x - 8$ ... ①

This equation is an identity. It is true for all values of x. It must be true when any number is substituted for x. Substitute $x = 0$ in ①.

$$\begin{aligned} (0 + 1)(0^2 + b(0) + c) &= 0^3 + 3(0)^2 - 6(0) - 8 \\ c &= -8 \end{aligned}$$

Hence, equation ① becomes $(x + 1)(x^2 + bx - 8) = x^3 + 3x^2 - 6x - 8$ ... ②

This equation is also an identity. It is true for all values of x. It must be true when any number is substituted for x. Substitute $x = 1$ in ②.

$$\begin{aligned}(1 + 1)(1^2 + 1b - 8) &= 1^3 + 3(1)^2 - 6(1) - 8\\ 2(b - 7) &= -10\\ 2b - 14 &= -10\\ 2b &= 4\\ b &= 2\end{aligned}$$

Discuss
Why did we substitute $x = 1$ in equation ②? Could we have substituted $x = 0$? $x = -1$? $x = 2$? Explain.

The other factor of $x^3 + 3x^2 - 6x - 8$ is $x^2 + 2x - 8$.

Therefore, $x^3 + 3x^2 - 6x - 8 = (x + 1)(x^2 + 2x - 8)$
$= (x + 1)(x + 4)(x - 2)$

3.5 Exercises

1. Determine the remainder when $x^3 + 3x^2 - 5x - 10$ is divided by each binomial.
   a) $x - 1$ **b)** $x - 2$ **c)** $x - 5$
   d) $x + 1$ **e)** $x + 2$ **f)** $x + 5$

2. Which binomial in exercise 1 is a factor of $x^3 + 3x^2 - 5x - 10$?

3. Determine the remainder when each polynomial is divided by $x + 3$.
   a) $7x^2 - 11x + 1$ **b)** $x^3 + 4x^2 + 2$ **c)** $x^4 + x^3 + x - 9$

4. Suppose $x - 5$ is a factor of a polynomial $f(x)$. What is the value of $f(5)$?

5. For each table of values for the function $y = f(x)$, identify as many binomial factors as possible.

a)

x	f(x)
–4	28
–3	22
–2	0
–1	–20
0	–20
1	28
2	112

b)

x	f(x)
–3	–28
–2	0
–1	6
0	2
1	0
2	12
3	50

c)

x	f(x)
–2	–12
–1	0
0	2
1	0
2	0
3	8
4	30

6. Which polynomials have $x - 2$ as a factor?

a) $2x^3 - 5x - 6$
b) $3x^3 + x^2 - 1$
c) $x^4 - 7x^2 + 12$
d) $2x^4 - x^3 + 4$

7. **Communication** For the polynomial $x^3 - 5x^2 - 8x + 12$, which values of x should be chosen to test for factors of this polynomial? Explain.

B

8. Determine each value of k.

a) When $x^3 + kx^2 - 5x + 7$ is divided by $x + 1$, the remainder is 8.

b) When $x^4 + kx^3 - 3x^2 + 2x - 6$ is divided by $x - 2$, the remainder is -14.

c) When $-2x^3 - 5x^2 + kx - 9$ is divided by $x + 3$, the remainder is -15.

9. a) Show that both $x + 3$ and $x - 3$ are factors of $2x^3 - x^2 - 18x + 9$.

b) Determine another factor of $2x^3 - x^2 - 18x + 9$.

10. a) Show that $x + 2$ is a factor of $x^3 - 3x^2 - 6x + 8$.

b) Determine the remaining factors without using long division.

11. Factor each polynomial.

a) $x^3 - 2x^2 - x + 2$
b) $x^3 + 2x^2 - 5x - 6$
c) $x^3 + 4x^2 + x - 6$

12. Factor each polynomial.

a) $x^3 + 5x^2 + 2x - 8$
b) $x^3 + 9x^2 + 23x + 15$
c) $x^3 + 3x^2 - 4x - 12$

13. **Knowledge/Understanding** Factor each polynomial.

a) $x^3 + x^2 - 16x + 20$
b) $x^3 - 5x^2 - 4x + 20$

14. Factor each polynomial. Use the Factor Theorem.

a) $x^3 + 1$
b) $x^3 - 1$

15. Factor each polynomial.

a) $2x^3 - x^2 - 5x - 2$
b) $2x^3 + 5x^2 - x - 6$
c) $3x^3 + 4x^2 - 35x - 12$

16. Factor each polynomial.

a) $3x^3 + 2x^2 - 7x + 2$
b) $x^3 - 5x^2 - 2x + 24$
c) $5x^3 + 11x^2 - 13x - 3$

17. Factor the polynomial $x^4 - 2x^3 - 13x^2 + 14x + 24$.

18. When the polynomial $2x^2 + bx - 5$ is divided by $x - 3$, the remainder is 7.

a) Determine the value of b.

b) Determine the remainder when the polynomial is divided by $x - 2$.

✓ **19.** When the polynomial $ax^3 + 2x^2 - x + 3$ is divided by $x + 1$, the remainder is 4.

a) Determine the value of a.

b) What is the remainder when the polynomial is divided by $x - 1$?

20. When the polynomial $x^3 + bx^2 + 2x + 9$ is divided by $x - 2$, the remainder is 1.

a) Determine the value of b.

b) What is the remainder when the polynomial is divided by $x + 2$?

21. Determine the value of k in each case.

a) $2x + 3$ is a factor of $2x^3 + kx^2 - 4x + 3$.

b) $3x - 1$ is a factor of $kx^3 - 2x^2 + x - 1$.

c) $4x + 5$ is a factor of $8x^3 + 2x^2 + kx + 10$.

✓ **22. Application** For what values of a does the polynomial $-x^3 + 3x^2 + ax + 1$ have the same remainder when divided by $x + 2$ or by $x - 2$?

✓ **23.** For what values of b does the polynomial $x^4 - 4x^3 + bx^2 - x + 4$ have the same remainder when divided by $x + 1$ or by $x - 2$?

24. Consider the function $f(x) = 2x^3 - 7x^2 - 5x + 4$.

a) Use division to show that $2x - 1$ is a factor of $f(x)$.

b) What value of x should be substituted to give a value of 0 for this polynomial? Show that your answer is correct.

25. Without dividing, determine each remainder.

a) $(6x^2 - 10x + 7) \div (3x + 1)$ **b)** $(-8a^2 - 2a - 3) \div (4a - 1)$

✓ **26. Thinking/Inquiry/Problem Solving**
Here is the graph of the polynomial function $y = x^4 - 11x^3 + 40x^2 - 55x + 25$. It intersects the x-axis at $x = 1$, $x = 5$, and at two other points. Determine these other two x-intercepts.

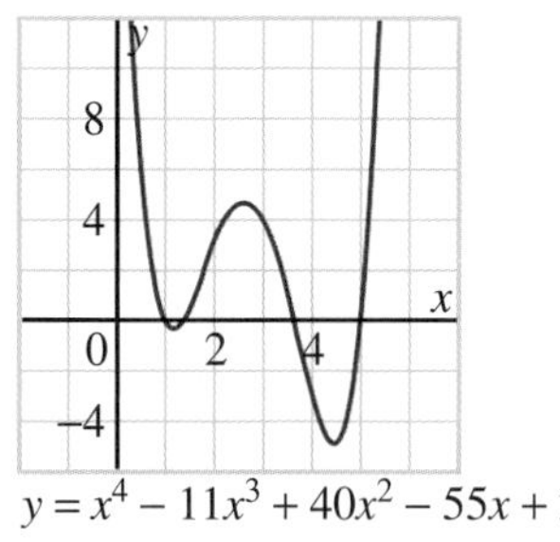

$y = x^4 - 11x^3 + 40x^2 - 55x + 25$

C

27. a) Is $3x + 2$ a factor of $6x^3 + 7x^2 - 16x - 12$? Explain.

b) Determine the factors of the polynomial in part a.

28. When $kx^3 + px^2 + 5x - 4$ is divided by $x + 2$, the remainder is -4. When this polynomial is divided by $x - 2$, the remainder is 4. Determine the values of k and p.

Self-Check 3.3 – 3.5

1. Determine the coordinates of all local maximum points, local minimum points, and points of inflection for each polynomial function.

   a) $f(x) = -2x^4 + 6x^2$ b) $g(x) = x^6 - 3.6x^5 + 3x^4 - 7$

   c) $h(x) = 6x^4 + 11x^3 - 15x^2 + 7$

2. For each polynomial function:

   i) Determine the x- and y-intercepts.

   ii) Determine the coordinates of the critical points.

   iii) Identify the nature of the critical points.

   iv) Determine the coordinates of the points of inflection.

   v) Graph the function.

   vi) State the intervals of increase and decrease.

   vii) State the intervals of concavity.

   a) $f(x) = -2x^3 + 3x^2$ b) $h(x) = x^4 - 6x^2$

3. Determine the remainder when $x^3 + 3x^2 - 5x + 4$ is divided by each binomial.

   a) $x - 1$ b) $x - 2$ c) $x - 3$

   d) $x + 1$ e) $x + 2$ f) $x + 3$

4. Factor.

   a) $x^3 - 3x^2 - 6x + 8$ b) $x^3 + 4x^2 - 7x - 10$ c) $x^3 - 8x^2 + 20x - 16$

   d) $x^3 + x^2 - 8x - 12$ e) $x^3 - 2x^2 - 11x + 12$ f) $x^3 + 3x^2 - 6x - 8$

5. When the polynomial $2x^2 + bx - 5$ is divided by $x - 3$, the remainder is 7.

   a) Determine the value of b.

   b) What is the remainder when the polynomial is divided by $x - 2$?

PERFORMANCE ASSESSMENT

6. Here is the graph of the derivative f' of a function f. The maximum point, minimum point, and intercepts are labelled.

   a) Determine the intervals of increase and decrease of f.

   b) Determine the critical values of f.

   c) Determine the intervals of concavity of f.

   d) Sketch a possible graph of $y = f(x)$.

   Explain how you completed each part.

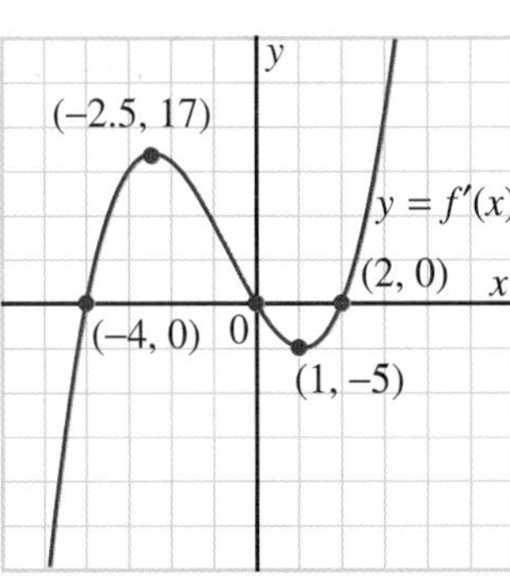

3.6 Solving Polynomial Equations and Inequalities

Recall, from Section 3.2, that when a polynomial function is expressed in factored form, the zeros of the function can be determined by inspection.
In this section, we shall apply the Factor Theorem to solve polynomial equations and inequalities of degree greater than two.
We shall then solve polynomial equations and inequalities that do not factor.

Example 1

Solve for x.

a) $x^3 - 4x^2 - 12x = 0$ **b)** $x^3 + 1 = 0$

Solution

a) $x^3 - 4x^2 - 12x = 0$

Factor. $x(x^2 - 4x - 12) = 0$

$x(x + 2)(x - 6) = 0$

Then $x = 0$ or $x + 2 = 0$ or $x - 6 = 0$

$x = -2$ $x = 6$

The roots are -2, 0, 6.

b) $x^3 + 1 = 0$

Use the Factor Theorem.

Let $f(x) = x^3 + 1$.

Find a value of x so that $f(x) = 0$.

According to the factor property, the numbers to test are the factors of 1; that is, ±1.

Try $x = 1$.

$f(1) = (1)^3 + 1$

$= 2$

Thus, $x - 1$ is not a factor of $f(x)$.

Try $x = -1$.

$f(-1) = (-1)^3 + 1$

$= 0$

Therefore, $x + 1$ is a factor of $f(x)$.

Use long division to determine the other factors.

$$\begin{array}{r} x^2 - x + 1 \\ x+1 \overline{\smash{)}\, x^3 + 0x^2 + 0x + 1} \\ \underline{x^3 + x^2} \\ -x^2 + 0x \\ \underline{-x^2 - x} \\ x + 1 \\ \underline{x + 1} \\ 0 \end{array}$$

Therefore, the given equation can be written in the form
$(x + 1)(x^2 - x + 1) = 0$.
Either $x + 1 = 0$ or $x^2 - x + 1 = 0$
Thus $x = -1$ is one root of the given equation.
The quadratic equation $x^2 - x + 1 = 0$ cannot be factored.

Use the quadratic formula to determine the roots.

$$x = \frac{-b \pm \sqrt{b^2 - 4ac}}{2a}$$

$$\begin{aligned} x &= \frac{-(-1) \pm \sqrt{(-1)^2 - 4(1)(1)}}{2(1)} \\ &= \frac{1 \pm \sqrt{1 - 4}}{2} \\ &= \frac{1 \pm \sqrt{-3}}{2} \\ &= \frac{1}{2} \pm \frac{i\sqrt{3}}{2} \end{aligned}$$

Recall that $\sqrt{-1} = i$.

There is only one root, $x = -1$, in the set of real numbers.

In the set of complex numbers, there are two additional roots,
$x = \frac{1}{2} \pm \frac{i\sqrt{3}}{2}$.

In *Example 1b*, the equation $x^3 + 1 = 0$ has only one real root. Therefore, the function $f(x) = x^3 + 1$ has only one real zero. The complex numbers $\frac{1}{2} \pm \frac{i\sqrt{3}}{2}$ are the complex zeros of $f(x) = x^3 + 1$.

Example 2

Solve the inequality $x^3 - 2x^2 - 9x + 18 < 0$.

Solution

$x^3 - 2x^2 - 9x + 18 < 0$

Factor by grouping to determine the roots of the equation
$x^3 - 2x^2 - 9x + 18 = 0$.

$$x^2(x-2) - 9(x-2) = 0$$
$$(x-2)(x^2-9) = 0$$
$$(x-2)(x+3)(x-3) = 0$$

The equation $(x-2)(x+3)(x-3) = 0$ has roots 2, −3, and 3.
Use these roots to solve the inequality $(x-2)(x+3)(x-3) < 0$.

The roots divide a number line into 4 intervals. Mark a point at each root.

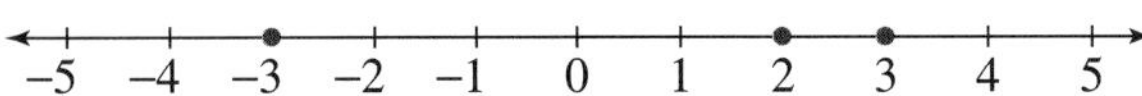

We use the roots on the number line to determine the sign of the polynomial in each interval.

Choose a value of x in each interval, and substitute it into the polynomial $(x-2)(x+3)(x-3)$. Consider the sign of each linear factor.
For $x < -3$, choose $x = -4$: $(-4-2)(-4+3)(-4-3)$
Each factor is negative, so the polynomial is negative.

For $-3 < x < 2$, choose $x = 0$: $(0-2)(0+3)(0-3)$
Two factors are negative, one is positive, so the polynomial is positive.

For $2 < x < 3$, choose $x = 2.5$: $(2.5-2)(2.5+3)(2.5-3)$
Two factors are positive, one is negative, so the polynomial is negative.

For $x > 3$, choose $x = 4$: $(4-2)(4+3)(4-3)$
Each factor is positive, so the polynomial is positive.

Record these signs on the number line.

x −5 −4 −3 −2 −1 0 1 2 3 4 5

Sign of polynomial − + − +

The solution of the inequality $x^3 - 2x^2 - 9x + 18 < 0$ is the intervals where the polynomial is negative. The solution is $x < -3$ and $2 < x < 3$.

In *Example 2*, each linear factor $(x-a)$ changes sign at the zero from negative for $x < a$ to positive for $x > a$.

$(x-a)$ is a factor.

a

When $x < a$, $(x-a)$ is negative.

When $x > a$, $(x-a)$ is positive.

The sign of the polynomial in each interval can be determined directly from the number line by determining the sign of the polynomial to the left of the first zero, then alternating signs after each zero.

An exception to this rule is when a polynomial has 2 equal factors. That is, if two factors of a polynomial are $(x - a)^2$, then the signs of the polynomial are the same for $x < a$ and $x > a$.

$(x - a)^2$ is a factor.

When $x < a$, $(x - a)^2$ is positive.	a	When $x > a$, $(x - a)^2$ is positive.

In *Example 2*, the solution of the inequality $(x - 2)(x + 3)(x - 3) < 0$ corresponds to the intervals where the graph of the function $f(x) = (x - 2)(x + 3)(x - 3)$ is negative; that is, below the x-axis. From the graph of $f(x) = (x - 2)(x + 3)(x - 3)$, the solutions are $x < 3$ and $2 < x < 3$.

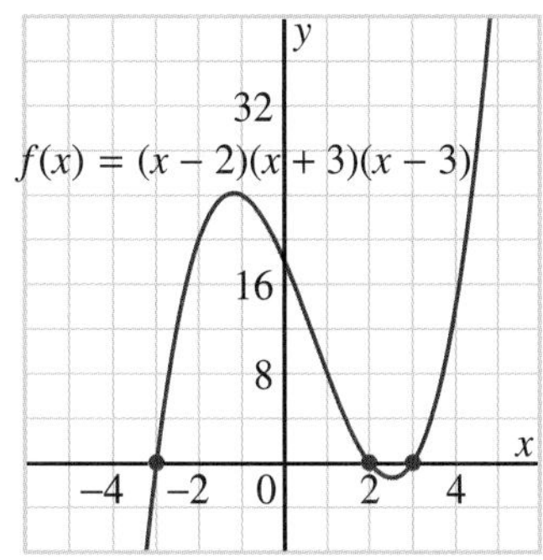

Similarly, the solution of the inequality $(x - 2)(x + 3)(x - 3) \geq 0$ corresponds to the intervals where the graph of $f(x) = (x - 2)(x + 3)(x - 3)$ is 0 or positive; that is, on or above the x-axis. The solutions are $-3 \leq x \leq 2$ and $x \geq 3$.

Example 3 illustrates this graphical method to solve an inequality once the roots of the corresponding equation have been determined.

Example 3

Solve the inequality $x^4 + 2x^3 - 4x^2 - 8x > 0$.

Solution

$x^4 + 2x^3 - 4x^2 - 8x > 0$

Factor by grouping.

$$x^3(x + 2) - 4x(x + 2) > 0$$
$$(x + 2)(x^3 - 4x) > 0$$
$$x(x + 2)(x^2 - 4) > 0$$
$$x(x + 2)(x + 2)(x - 2) > 0$$
$$x(x - 2)(x + 2)^2 > 0$$

The solution of the inequality $x(x - 2)(x + 2)^2 > 0$ is the set of values of x where the polynomial is positive.

Consider the function $f(x) = x(x - 2)(x + 2)^2$ and the set of values of x where $f(x)$ is positive.

Discuss

How do we know when we can factor by grouping, and when we must use the Factor Theorem?

The zeros of the function $f(x) = x(x-2)(x+2)^2$ are −2, 0, and 2. There is a zero of order 2 at −2, so the graph touches the x-axis at −2.

This quartic function has a positive leading coefficient, so the graph extends from the 2nd quadrant to the 1st quadrant.

Mark points at the zeros on the x-axis. Sketch a curve through the points, extending from the 2nd quadrant to the 1st quadrant.

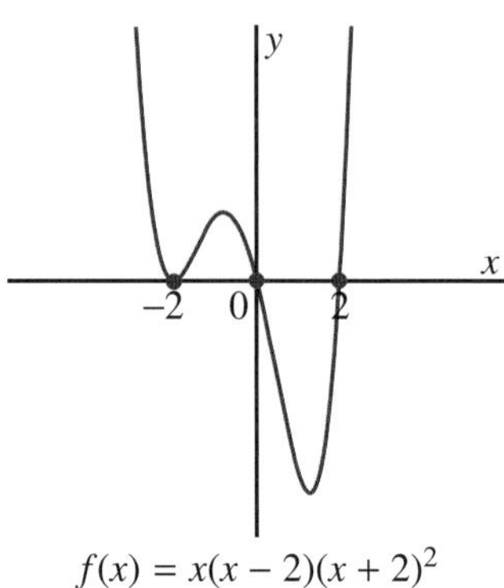

$f(x) = x(x-2)(x+2)^2$

From the graph, $f(x)$ is positive when $x < -2$, $-2 < x < 0$, and $x > 2$.

Thus, the solution of the inequality $x^4 + 2x^3 - 4x^2 - 8x > 0$ is $x < -2$, $-2 < x < 0$, and $x > 2$.

In *Example 3*, we could have used a number-line analysis to solve the inequality $x(x-2)(x+2)^2 > 0$. The roots are −2, 0, and 2. The root −2 is a double root, so the polynomial does not change sign at this root.

For $x < -2$, substitute $x = -3$ in $x(x-2)(x+2)^2$: $-3(-3-2)(-3+2)^2$
The polynomial is positive.
So, for $-2 < x < 0$, the polynomial is still positive since the polynomial does not change sign at the double root.
At $x = 0$, the polynomial becomes negative and at $x = 2$, the polynomial becomes positive.

x		−2		0		2	
Sign of polynomial	+		+		−		+

From the number line, the solution is $x < -2$, $-2 < x < 0$, and $x > 2$.

Not all equations and inequalities can be solved by factoring. We can use a graphing calculator to estimate the real roots of the equation and the zeros of the corresponding function.

Example 4

a) Solve the equation $x^3 + x = 20$. **b)** Solve the inequality $x^3 + x < 20$.

Solution

a) $x^3 + x = 20$

Write the equation as $x^3 + x - 20 = 0$.
The expression $x^3 + x - 20$ cannot be factored.
The roots of the equation $x^3 + x - 20 = 0$ may be determined from the graph of the corresponding function $f(x) = x^3 + x - 20$.

Input $y = x^3 + x - 20$.
Adjust the window so that all points where the graph intersects the x-axis are visible.
Press [2nd] [TRACE] for CALC.
Then press **2** for zero.
There is one real zero on the screen.
Use the arrow keys to move the cursor to the left of the zero.
Press [ENTER].
Move the cursor to the right of the zero.
Press [ENTER] [ENTER].
$x \doteq 2.59$ is displayed.

The real root of the equation $x^3 + x = 20$ is $x \doteq 2.59$.

b) $x^3 + x < 20$

Write the inequality as $x^3 + x - 20 < 0$.
From the screen in part a, the graph is below the x-axis when $x < 2.59$.
The solution of the inequality $x^3 + x < 20$ is approximately $x < 2.59$.

Example 5

Consider the equation $2x^2 + mx + 9 = 0$. For what value(s) of m is one root double the other root?

Solution

$2x^2 + mx + 9 = 0$ ... ①

Let the roots be r and $2r$. Then, the equation can be written in the form:

$2(x - r)(x - 2r) = 0$

Expand the left side.

$2(x^2 - 3rx + 2r^2) = 0$

$2x^2 - 6rx + 4r^2 = 0$ ... ②

The constant terms are equal in equations ① and ②.

Hence, $4r^2 = 9$

$r^2 = \frac{9}{4}$

$r = \pm\frac{3}{2}$

The coefficients of x are equal.

Hence, $m = -6r$

Substitute $r = \frac{3}{2}$.

$m = -6\left(\frac{3}{2}\right)$
$= -9$

Substitute $r = -\frac{3}{2}$.

$m = -6\left(-\frac{3}{2}\right)$
$= 9$

Therefore, when $m = \pm 9$, one root of the equation is double the other root.

Discuss
How could we check the result?

A polynomial equation or inequality may be solved algebraically if it is in factored form. However, the Factor Theorem can only be used if the equation has one or more integral (or possibly rational) roots. Many polynomial equations do not have integral or rational roots.

A quadratic equation that cannot be factored may be solved using the quadratic formula. There are complicated formulas for solving cubic and quartic equations, but there are no formulas for solving higher-degree equations.

There are limitations to the algebraic methods for solving a polynomial equation or inequality. However, all polynomial equations and inequalities can be solved using technology. A graphing calculator may be used to estimate the roots of the equation.

If a polynomial equation has real roots, the corresponding polynomial function has real zeros. If a polynomial equation has complex roots, the corresponding polynomial function has complex zeros. A polynomial function $y = f(x)$ can change sign only at its real zeros.

3.6 Exercises

Use a graphing calculator, where necessary, to complete the exercises.

A

1. Determine the roots of each equation.

   a) $(x - 4)(x + 3) = 0$
   b) $(x - 5)(2x + 1)(3x - 2) = 0$
   c) $(x - 3)(x + 1)(x - 4)(x + 5) = 0$
   d) $x(x + 2)(4x - 5) = 0$

2. Use the graph to estimate the root(s) of each equation.

   a) $x^3 + 2x^2 - 10 = 0$
   b) $-x^3 - 3x^2 + 5x + 16 = 0$

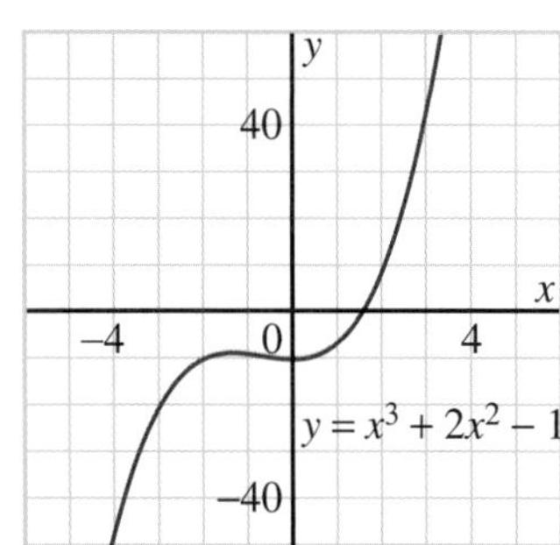

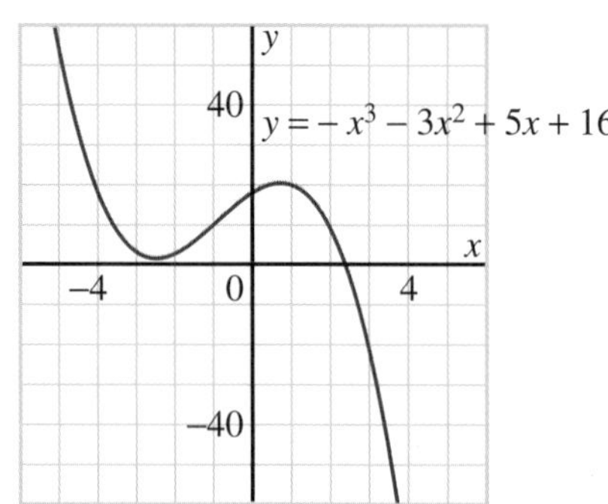

c) $x^4 - 10x^2 - 5x + 5 = 0$

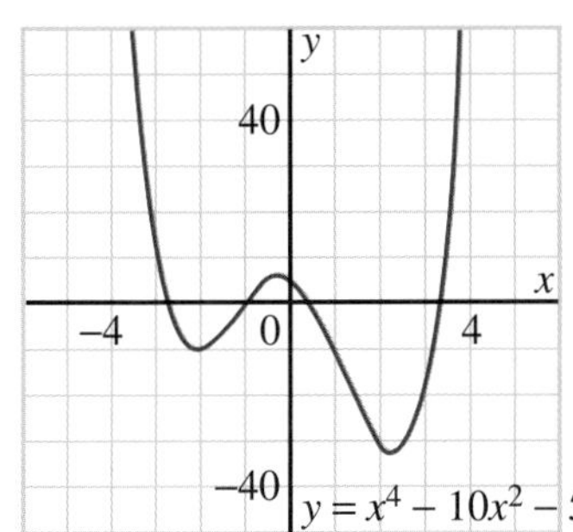

d) $x^5 - 10x^3 + 15x = 0$

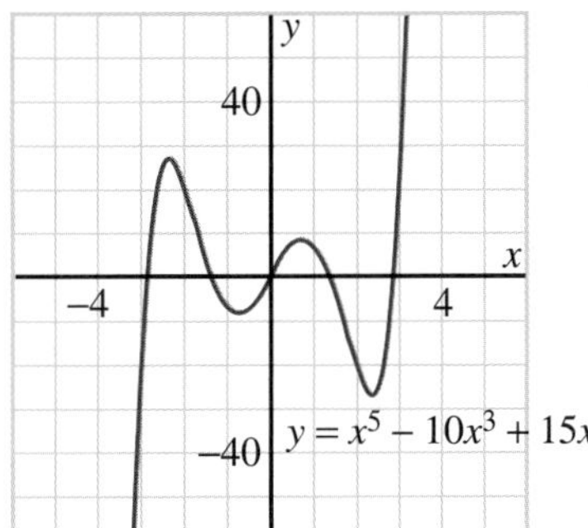

✓ **3.** Refer to the graphs in exercise 2.

a) Which equation has a double root? Explain.

b) Which equation has a complex root? Explain.

✓ **4.** Use the graphs and your estimates from exercise 2 to approximate the solution of each inequality.

a) $x^3 + 2x^2 - 10 < 0$

b) $-x^3 - 3x^2 + 5x + 16 > 0$

c) $x^4 - 10x^2 - 5x + 5 \geq 0$

d) $x^5 - 10x^3 + 15x \leq 0$

B

5. Write a polynomial equation that has each set of roots.

a) 3, –2

b) –1, 0, 3

c) 4, –5, 6

✓ **6. Knowledge/Understanding** Solve for x.

a) $x^3 - x^2 - 20x = 0$

b) $x^4 - x^3 - 13x^2 + x + 12 = 0$

c) $x^3 - x^2 - 20x > 0$

d) $x^4 - x^3 - 13x^2 + x + 12 \leq 0$

✓ **7.** Solve for x.

a) $x^3 + 10x^2 + 21x = 0$

b) $x^3 - 3x^2 - 4x + 12 = 0$

c) $2x^3 - x^2 - 15x + 18 = 0$

d) $x^4 - x^3 - 11x^2 + 9x + 18 = 0$

✓ **8.** Use your results from exercise 7 to solve each inequality.

a) $x^3 + 10x^2 + 21x < 0$

b) $x^3 - 3x^2 - 4x + 12 > 0$

c) $2x^3 - x^2 - 15x + 18 \geq 0$

d) $x^4 - x^3 - 11x^2 + 9x + 18 \leq 0$

✓ **9.** Solve graphically.

a) $x^3 + 2x = 10$

b) $x^3 - 1 = 3x$

c) $x^3 + 2x < 10$

d) $x^3 - 1 \geq 3x$

10. Determine the real and/or complex roots of each equation.

a) $x^2 + 2x - 5 = 0$

b) $x^2 + x + 6 = 0$

c) $x^2 + 2x + 2 = 0$

d) $x^2 + x - 6 = 0$

11. Determine the real and/or complex roots of each equation.

a) $x^3 - 2x^2 + x - 2 = 0$ b) $x^3 + 64 = 0$

c) $x^3 - 1 = 0$ d) $x^3 - 3x^2 + 4 = 0$

12. Determine the real and/or complex roots of each equation.

a) $x^4 + x^2 = 0$ b) $x^4 - 4 = 0$

c) $x^4 + 5x^2 + 4 = 0$ d) $x^4 + 2x^3 - x - 2 = 0$

13. Look at the results of exercises 10 to 12. How many possible complex roots may each type of equation have? Explain.

a) a quadratic equation b) a cubic equation c) a quartic equation

14. Solve each equation.

a) $x^3 - 3x^2 - 6x + 8 = 0$ b) $x^3 + 4x^2 - 7x - 10 = 0$

c) $x^3 - 8x^2 + 20x - 16 = 0$ d) $x^3 + x^2 - 8x - 12 = 0$

e) $x^3 - x^2 - 3x + 2 = 0$ f) $x^3 - 4x^2 + 8x - 8 = 0$

15. Solve each equation. Give the results to 1 decimal place.

a) $x^3 - 15x - 10 = 0$ b) $x^3 + x - 15 = 0$

c) $x^4 - 15x^2 + 20 = 0$ d) $x^4 - 5x^2 - 10x - 25 = 0$

16. Solve each inequality. Give the results to 1 decimal place where necessary.

a) $x^3 + 10x - 20 > 0$ b) $x^3 - 3x^2 + x - 10 \leq 0$

c) $x^4 - 10x^2 + 5x + 7 < 0$ d) $x^4 - 4x^3 + 16x - 25 \geq 0$

e) $2x^3 - 7x^2 + 2x + 3 > 0$ f) $3x^3 - 4x^2 - 5x + 2 < 0$

17. Solve. Give the results to 1 decimal place where necessary.

a) $x^3 + 2x \geq 5x - 10$ b) $6x^3 - 5x^2 - 12x - 4 \leq 0$

c) $2x^3 + 7x^2 + x - 10 \geq 0$ d) $x^3 + 2x < 10x - 5$

e) $x^3 - 5x^2 + 3x + 4 > 0$ f) $2x^3 + 5x^2 - 4 \leq 0$

18. **Communication** Explain how you would show, without solving, that $\frac{1}{2}$, $\frac{1}{3}$, and -2 are the roots of the polynomial equation $6x^3 + 7x^2 - 9x + 2 = 0$.

19. Determine the real and/or complex roots of each equation.

a) $3(x^2 - 4) = -x(x^2 - 4)$ b) $8x + 12 = x(x^2 + x)$

c) $3x^2 + 4 = x(3 - x^2)$ d) $12 = 3x^2(5 - x^2)$

20. Determine the value(s) of k in each equation so that one root is double the other root.

a) $x^2 - kx + 50 = 0$ b) $2x^2 - 3x + k = 0$

21. **Application** One root of the polynomial equation $3x^3 - 15x^2 + mx - 4 = 0$ is 2. Determine the value of m, then find the other roots. Justify your reasoning.

22. Determine the value(s) of m in each equation so that the two roots are equal.

a) $x^2 - 6x + m = 0$ b) $4x^2 + mx + 25 = 0$

23. **Thinking/Inquiry/Problem Solving** Both equations $x^3 - 12x + 16 = 0$ and $x^3 - 12x - 16 = 0$ have a double real root, and one other real root that is different from the double root.

a) Use this information. Determine which of the equations at the right have:
i) three different real roots ii) only one real root

$x^3 - 12x + 20 = 0$
$x^3 - 12x + 10 = 0$
$x^3 - 12x - 20 = 0$

b) Determine the values of k for which the equation $x^3 - 12x + k = 0$ has:
i) three different real roots
ii) two different real roots
iii) only one real root

24. a) Write a polynomial equation that has each set of roots.
i) $\sqrt{3}, -\sqrt{3}$ ii) $2 + \sqrt{3}, 2 - \sqrt{3}$ iii) $4 - \sqrt{3}, 4 + \sqrt{3}$

b) Write the equation of a family of polynomial functions that has each set of zeros in part a.

25. Determine the value(s) of n in each equation so that one root is triple the other root.

a) $3x^2 - 4x + n = 0$ b) $4x^2 + nx + 27 = 0$

26. Find the points on the graph of $y = x^3 - 9x^2$ where the tangent is horizontal.

27. Show that the graph of $y = 2x^3 + 5x - 7$ has no tangents with slope 4.

28. At what point on the graph of $y = x^4 - 20x + 3$ is the tangent to the graph parallel to the line $12x - y + 7 = 0$? Explain.

Exercises 29 and 30 are significant because they involve the products of complex conjugates. These exercises illustrate that the results on page 145, concerning the number of roots of a polynomial equation and the number of zeros of the corresponding function, also apply when some or all of the roots (or zeros) are complex.

29. a) Write a polynomial equation that has each set of roots.
i) $i\sqrt{3}, -i\sqrt{3}$ ii) $i + \sqrt{3}, -i + \sqrt{3}$ iii) $1 + i\sqrt{3}, 1 - i\sqrt{3}$

b) Write the equation of a family of polynomial functions that has each set of zeros in part a.

c) Describe the graph of any member from each family in part b. What do the zeros of the function indicate about its graph?

30. a) Write a polynomial equation that has each set of roots.
i) $i, -i, 1$ ii) $3, -5, 2i, -2i$ iii) $2, 3 + i\sqrt{5}, 3 - i\sqrt{5}$

b) Write the equation of a family of polynomial functions that has each set of zeros in part a.

c) Describe the graph of any member from each family in part b. What do the zeros of the function indicate about its graph?

31. Suppose the roots of the quadratic equation $2x^2 - 7x - 15 = 0$ are represented by a and b. Determine the equation whose roots are $a + b$ and ab.

3.7 Graphing Polynomial Functions

In Sections 3.3 and 3.4, we used derivatives to determine the coordinates of the local minimum and local maximum points, and the points of inflection of the graph of a polynomial function. We used the graph to determine the intervals of increase and decrease, and the intervals of concavity. The polynomial functions in Sections 3.3 and 3.4 were factored without the Factor Theorem.

In Section 3.5, we learned how to use the Factor Theorem. In Section 3.6, we applied this theorem to solve polynomial equations and inequalities.

In this section, we combine the methods of the previous sections to determine properties of the graphs of polynomial functions.

Example 1

Consider the polynomial function $f(x) = x^3 + 6x^2 + 9x + 4$.

a) Determine the x- and y-intercepts.

b) Determine the coordinates and nature of the critical points.

c) Determine the coordinates of the points of inflection.

d) Graph the function.

e) State the intervals of increase and decrease.

f) State the intervals of concavity.

Solution

a) $f(x) = x^3 + 6x^2 + 9x + 4$

The y-intercept is $f(0) = 4$.

To determine the x-intercepts, solve the equation $f(x) = 0$.

$x^3 + 6x^2 + 9x + 4 = 0$

Use the Factor Theorem.

Test the factors of 4; that is, $\pm 1, \pm 2, \pm 4$.

Since every coefficient in the equation is positive, none of $x = 1$, 2, or 4 will make $f(x)$ zero.

Try $x = -1$.

$$\begin{aligned} f(-1) &= (-1)^3 + 6(-1)^2 + 9(-1) + 4 \\ &= -1 + 6 - 9 + 4 \\ &= 0 \end{aligned}$$

Therefore, one factor of $x^3 + 6x^2 + 9x + 4$ is $x + 1$.

To determine the other factors, use long division.

$$\begin{array}{r} x^2 + 5x + 4 \\ x+1 \overline{\smash{)}\, x^3 + 6x^2 + 9x + 4} \\ \underline{x^3 + x^2} \qquad\qquad \\ 5x^2 + 9x \quad \\ \underline{5x^2 + 5x} \quad \\ 4x + 4 \\ \underline{4x + 4} \\ 0 \end{array}$$

The other factor is $x^2 + 5x + 4$. This is factored as $(x + 4)(x + 1)$.
Therefore, $x^3 + 6x^2 + 9x + 4 = (x + 1)^2(x + 4)$
The x-intercepts are -1 and -4.

On a sketch, plot the points that correspond to the intercepts. Since $x = -1$ is a double root, the function does not change sign at that point. Since the leading coefficient of the cubic function is positive, the graph extends from the 3rd quadrant to the 1st quadrant. Sketch the curve. There appears to be one maximum point and one minimum point. We use calculus to verify this.

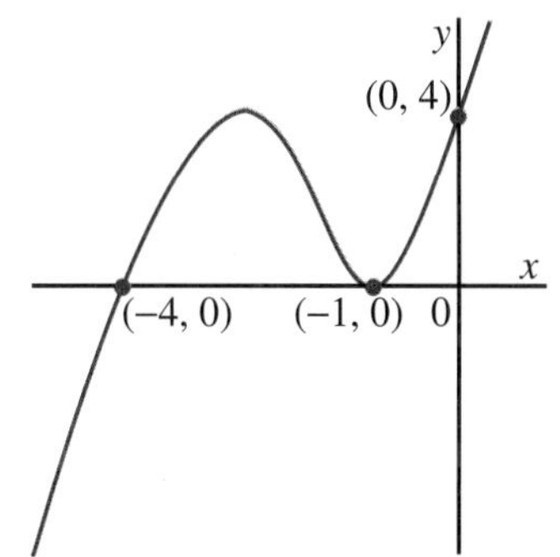

b) $f(x) = x^3 + 6x^2 + 9x + 4$

Differentiate.

$f'(x) = 3x^2 + 12x + 9$

Substitute $f'(x) = 0$ to determine the critical values.

$3x^2 + 12x + 9 = 0$

Factor.

$3(x^2 + 4x + 3) = 0$
$3(x + 3)(x + 1) = 0$

Either $x + 3 = 0$ or $x + 1 = 0$
$x = -3$ $\qquad$ $x = -1$

The critical values are -3 and -1.

Determine the corresponding values of y.

Substitute $x = -3$.

$$\begin{aligned} y &= x^3 + 6x^2 + 9x + 4 \\ &= (-3)^3 + 6(-3)^2 + 9(-3) + 4 \\ &= 4 \end{aligned}$$

Substitute $x = -1$.

$$\begin{aligned} y &= x^3 + 6x^2 + 9x + 4 \\ &= (-1)^3 + 6(-1)^2 + 9(-1) + 4 \\ &= 0 \end{aligned}$$

The critical points are $(-3, 4)$ and $(-1, 0)$.

To determine whether each critical point is a local maximum, a local minimum, or a point of inflection, use the second derivative.

$f'(x) = 3x^2 + 12x + 9$

Differentiate.

$f''(x) = 6x + 12$

Substitute $x = -3$.

$$\begin{aligned} f''(x) &= 6x + 12 \\ f''(-3) &= 6(-3) + 12 \\ &= -6 \end{aligned}$$

Since $f'(-3) = 0$ and $f''(-3) < 0$, then $(-3, 4)$ is a local maximum point.

Substitute $x = -1$.

$$\begin{aligned} f''(x) &= 6x + 12 \\ f''(-1) &= 6(-1) + 12 \\ &= 6 \end{aligned}$$

Since $f'(-1) = 0$ and $f''(-1) > 0$, then $(-1, 0)$ is a local minimum point.

c) To determine whether there are points of inflection, solve $f''(x) = 0$.

$$\begin{aligned} f''(x) &= 6x + 12 \\ 6x + 12 &= 0 \\ x &= -2 \end{aligned}$$

Consider the values of $f''(x)$ on each side of $x = -2$.

To the left, consider $x = -2.1$.

$$\begin{aligned} f''(-2.1) &= 6(-2.1) + 12 \\ &= -0.6 \end{aligned}$$

To the right, consider $x = -1.9$.

$$\begin{aligned} f''(-1.9) &= 6(-1.9) + 12 \\ &= 0.6 \end{aligned}$$

As x increases from -2.1 to -1.9, f'' changes sign from negative to positive. Hence, there is a point of inflection at $x = -2$. Determine the corresponding value of y.

Substitute $x = -2$ in $y = x^3 + 6x^2 + 9x + 4$.

$$\begin{aligned} y &= (-2)^3 + 6(-2)^2 + 9(-2) + 4 \\ &= 2 \end{aligned}$$

The point of inflection is $(-2, 2)$.

Illustrate the critical points, point of inflection, and concavity on a number line.

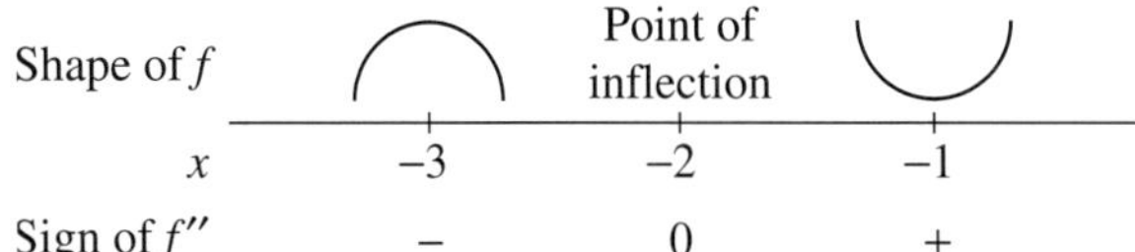

d) On a grid, plot the local maximum and local minimum points, and the point of inflection. Plot the points corresponding to the intercepts. Draw a smooth curve through the points, extending from the 3rd quadrant to the 1st quadrant.

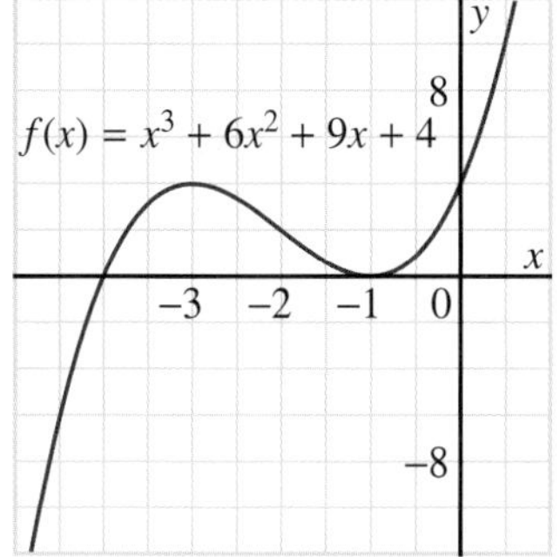

e) From the graph in part d:
The intervals of increase are $x < -3$ and $x > -1$.
The interval of decrease is $-3 < x < -1$.

f) The graph of f is concave down for $x < -2$.
The graph of f is concave up for $x > -2$.

Discuss

In part a, we used long division to determine two factors. Could we have used the Factor Theorem instead? Explain.

Example 2

Consider the polynomial function $f(x) = -3x^4 - 2x^3 + 15x^2 - 12x + 2$.

a) Determine the x- and y-intercepts.

b) Determine the coordinates and nature of the critical points.

c) Determine the coordinates of the points of inflection.

d) Graph the function.

e) State the intervals of increase and decrease.

f) State the intervals of concavity.

Solution

a) $f(x) = -3x^4 - 2x^3 + 15x^2 - 12x + 2$

The y-intercept is $f(0) = 2$.

To determine the x-intercepts, solve the equation $f(x) = 0$. Use the Factor Theorem. Test the factors of 2; that is, ± 1, ± 2.

Try $x = 1$.

$$\begin{aligned} f(1) &= -3(1)^4 - 2(1)^3 + 15(1)^2 - 12(1) + 2 \\ &= 0 \end{aligned}$$

Therefore, one factor of $-3x^4 - 2x^3 + 15x^2 - 12x + 2$ is $x - 1$.

To determine the other factors, use long division.

$$\begin{array}{r}
-3x^3 - 5x^2 + 10x - 2 \\
x - 1 \enclose{longdiv}{-3x^4 - 2x^3 + 15x^2 - 12x + 2} \\
\underline{-3x^4 + 3x^3} \\
-5x^3 + 15x^2 \\
\underline{-5x^3 + 5x^2} \\
10x^2 - 12x \\
\underline{10x^2 - 10x} \\
-2x + 2 \\
\underline{-2x + 2} \\
0
\end{array}$$

The other factor is $-3x^3 - 5x^2 + 10x - 2$. Use the Factor Theorem again.

Let $g(x) = -3x^3 - 5x^2 + 10x - 2$. Test ±1, ±2.

Try $x = 1$.

$g(1) = -3 - 5 + 10 - 2$

$\quad = 0$

So, $x - 1$ is a factor of $g(x) = -3x^3 - 5x^2 + 10x - 2$.

Use long division.

$$\begin{array}{r}
-3x^2 - 8x + 2 \\
x - 1 \enclose{longdiv}{-3x^3 - 5x^2 + 10x - 2} \\
\underline{-3x^3 + 3x^2} \\
-8x^2 + 10x \\
\underline{-8x^2 + 8x} \\
2x - 2 \\
\underline{2x - 2} \\
0
\end{array}$$

The other factor is $-3x^2 - 8x + 2$, which cannot be factored further.

Therefore, $f(x) = -3x^4 - 2x^3 + 15x^2 - 12x + 2$

$\qquad\qquad = (x - 1)^2(-3x^2 - 8x + 2)$

Use the quadratic formula to solve $-3x^2 - 8x + 2 = 0$ to determine the other two zeros.

$$x = \frac{-b \pm \sqrt{b^2 - 4ac}}{2a}$$

$$x = \frac{-(-8) \pm \sqrt{(-8)^2 - 4(-3)(2)}}{2(-3)}$$

$$= \frac{8 \pm \sqrt{88}}{-6}$$

$$x = \frac{8 + \sqrt{88}}{-6} \quad \text{or} \quad x = \frac{8 - \sqrt{88}}{-6}$$

$$\doteq -2.90 \qquad\qquad \doteq 0.23$$

The x-intercepts are 1 and approximately -2.9 and 0.2.

On a sketch, plot the points that correspond to the intercepts. Since $x = 1$ is a double root, the function does not change sign at that point. Since the leading coefficient of the quartic function is negative, the graph extends from the 3rd quadrant to the 4th quadrant. Sketch the curve. There appear to be 2 maximum points and 1 minimum point. We use calculus to verify this.

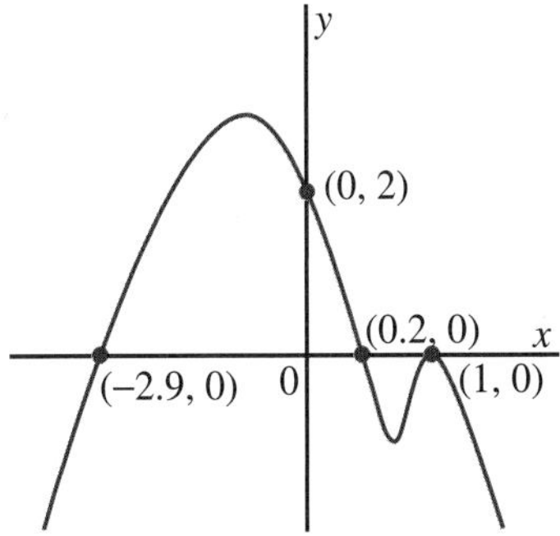

b) The critical values occur when $f'(x) = 0$.

$f(x) = -3x^4 - 2x^3 + 15x^2 - 12x + 2$

Differentiate.

$f'(x) = -12x^3 - 6x^2 + 30x - 12$

Substitute $f'(x) = 0$ to determine the critical values.

$-12x^3 - 6x^2 + 30x - 12 = 0$

Factor.

$-6(2x^3 + x^2 - 5x + 2) = 0$

$2x^3 + x^2 - 5x + 2 = 0$

Use the Factor Theorem. Let $g(x) = 2x^3 + x^2 - 5x + 2$.

Test the factors of 2; that is, ±1, ±2.

Try $x = 1$.

$$\begin{aligned} g(1) &= 2(1)^3 + (1)^2 - 5(1) + 2 \\ &= 0 \end{aligned}$$

Therefore, one factor of $2x^3 + x^2 - 5x + 2$ is $x - 1$.

To determine the other factors, use long division.

$$\begin{array}{r} 2x^2 + 3x - 2 \\ x - 1 \overline{)\, 2x^3 + x^2 - 5x + 2} \\ \underline{2x^3 - 2x^2} \\ 3x^2 - 5x \\ \underline{3x^2 - 3x} \\ -2x + 2 \\ \underline{-2x + 2} \\ 0 \end{array}$$

The other factor is $2x^2 + 3x - 2$. This is factored as $(2x - 1)(x + 2)$.

Therefore, $2x^3 + x^2 - 5x + 2 = (x - 1)(2x - 1)(x + 2)$

$$\begin{aligned} f'(x) &= -12x^3 - 6x^2 + 30x - 12 \\ &= -6(2x^3 + x^2 - 5x + 2) \\ &= -6(x - 1)(2x - 1)(x + 2) \end{aligned}$$

The critical values are 1, 0.5, and −2.

Determine the corresponding values of y. Substitute each value of x into $y = -3x^4 - 2x^3 + 15x^2 - 12x + 2$.

When $x = -2$, $y = -3(-2)^4 - 2(-2)^3 + 15(-2)^2 - 12(-2) + 2$
$= 54$

When $x = 0.5$, $y = -3(0.5)^4 - 2(0.5)^3 + 15(0.5)^2 - 12(0.5) + 2$
$= -0.6875$

When $x = 1$, $y = -3 - 2 + 15 - 12 + 2$
$= 0$

The critical points are $(-2, 54)$, $(0.5, -0.6875)$, and $(1, 0)$.

To determine whether each critical point is a local maximum, a local minimum, or a point of inflection, use the second derivative.

$f'(x) = -12x^3 - 6x^2 + 30x - 12$

Differentiate.

$f''(x) = -36x^2 - 12x + 30$

Substitute each value of x in $f''(x) = -36x^2 - 12x + 30$.

When $x = -2$, $f''(-2) = -36(-2)^2 - 12(-2) + 30$
$= -90$

Since $f'(-2) = 0$ and $f''(-2) < 0$, then $(-2, 54)$ is a local maximum point.

When $x = 0.5$, $f''(0.5) = -36(0.5)^2 - 12(0.5) + 30$
$= 15$

Since $f'(0.5) = 0$ and $f''(0.5) > 0$, then $(0.5, -0.6875)$ is a local minimum point.

When $x = 1$, $f''(1) = -36(1)^2 - 12(1) + 30$
$= -18$

Since $f'(1) = 0$ and $f''(1) < 0$, then $(1, 0)$ is a local maximum point.

c) To determine whether there are points of inflection, solve $f''(x) = 0$.

$f''(x) = -36x^2 - 12x + 30$

$-36x^2 - 12x + 30 = 0$

$-6(6x^2 + 2x - 5) = 0$

$6x^2 + 2x - 5 = 0$

This equation does not factor.

Use the quadratic formula.

$$x = \frac{-b \pm \sqrt{b^2 - 4ac}}{2a}$$

$$x = \frac{-2 \pm \sqrt{2^2 - 4(6)(-5)}}{2(6)}$$

$$= \frac{-2 \pm \sqrt{124}}{12}$$

$$x = \frac{-2 + \sqrt{124}}{12} \quad \text{or} \quad x = \frac{-2 - \sqrt{124}}{12}$$

$$\doteq 0.7613 \qquad\qquad \doteq -1.0946$$

The roots are approximately 0.8 and –1.1.

Check that $f''(x)$ changes sign at each x-value.
Consider each point of inflection, and determine the sign of $f''(x) = -36x^2 - 12x + 30$ on each side of the point.

When $x \doteq -1.1$

To the left, consider $x = -1.2$.

$$f''(-1.2) = -36(-1.2)^2 - 12(-1.2) + 30$$
$$= -7.44$$

To the right, consider $x = -1$.

$$f''(-1) = -36(-1)^2 - 12(-1) + 30$$
$$= 6$$

As x increases from –1.2 to –1, f'' changes sign. So, $x \doteq -1.1$ is a point of inflection.

When $x \doteq 0.8$

To the left, consider $x = 0.7$.

$$f''(0.7) = -36(0.7)^2 - 12(0.7) + 30$$
$$= 3.96$$

To the right, consider $x = 0.9$.

$$f''(0.9) = -36(0.9)^2 - 12(0.9) + 30$$
$$= -9.96$$

As x increases from 0.7 to 0.9, f'' changes sign. So, $x \doteq 0.8$ is a point of inflection.

Determine the corresponding values of y. Substitute each value of x into $y = -3x^4 - 2x^3 + 15x^2 - 12x + 2$.

When $x \doteq -1.09$

$$y \doteq -3(-1.09)^4 - 2(-1.09)^3 + 15(-1.09)^2 - 12(-1.09) + 2$$
$$\doteq 31.26$$

When $x \doteq 0.76$

$$y \doteq -3(0.76)^4 - 2(0.76)^3 + 15(0.76)^2 - 12(0.76) + 2$$
$$\doteq -0.33$$

The points of inflection are approximately (–1.1, 31.3) and (0.8, –0.3).

Illustrate the critical points, points of inflection, and concavity on a number line.

Shape of f	$\frown$	Point of inflection	$\smile$	Point of inflection	$\frown$
x	-2	-1.1	0.5	0.8	1
Sign of f''	$-$	0	$+$	0	$-$

d) On a grid, plot the local maximum and local minimum points, and the points of inflection. Plot the points corresponding to the intercepts. Draw a smooth curve through the points, extending from the 3rd quadrant to the 4th quadrant.

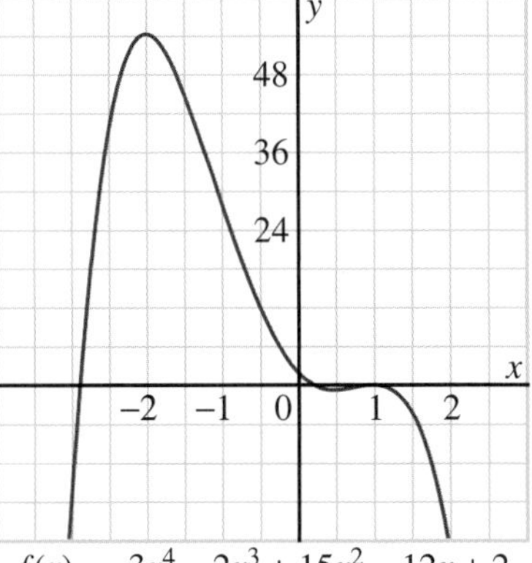

$f(x) = -3x^4 - 2x^3 + 15x^2 - 12x + 2$

e) From the graph in part d:
The intervals of increase are $x < -2$ and $0.5 < x < 1$.
The intervals of decrease are $-2 < x < 0.5$ and $x > 1$.

f) The graph of f is concave down for $x < -1.1$ and $x > 0.8$. The graph of f is concave up for $-1.1 < x < 0.8$.

THE BIG PICTURE

Using Polynomial Functions to Approximate Sinusoidal Functions

The graphs of polynomial functions have hills and valleys, as do the graphs of sinusoidal functions. These hills and valleys can be made to coincide. For example, the graphs of the 4th-degree function $y = 2\left(\left(\frac{x}{\pi}\right)^2 - 1\right)^2 - 1$ and the function $y = \cos x$ are shown. For values of x between $-\pi$ and π, the approximation is quite close. If more and more terms are used, and if the coefficients of the polynomial function are chosen carefully, the graph of a polynomial function can come closer and closer to the graph of $y = \cos x$.

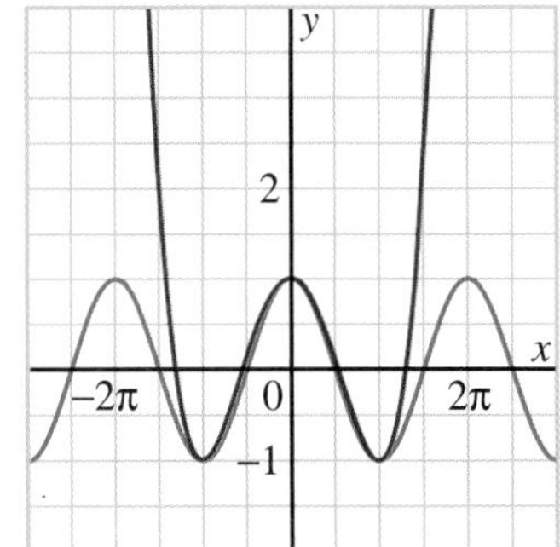

Polynomial functions are used to approximate other functions in applications involving periodic phenomena such as electronics, mechanical vibrations, complex musical tones, and economic cycles.

3.7 Exercises

1. State the x-values where the function f may have a local minimum or local maximum value.

   a) $f'(x) = (2x + 3)(x - 5)(x + 1)$ b) $f'(x) = -x(x + 2)(3x - 1)(x + 1)$

2. State the x-values where the graph of f may have a point of inflection.

   a) $f''(x) = (2x - 1)(x + 2)(x + 6)$ b) $f''(x) = (4x - 3)(x + 1)(x - 1)(3x + 5)$

3. State the coordinates of the local minimum or local maximum point on the graph of f.

   a) $f''(2) = -3,\ f'(2) = 0,\ f(2) = -1$

   b) $f''(-5) = 1,\ f'(-5) = 0,\ f(-5) = 3$

   c) $f''(-1) = 4,\ f'(-1) = 0,\ f(-1) = 8$

   d) $f''(2.5) = -6,\ f'(2.5) = 0,\ f(2.5) = 10$

4. Which graphs of f may have a point of inflection? State the coordinates of each point.

   a) $f''(-4) = 0,\ f'(-4) = -1,\ f(-4) = 3$

   b) $f''(2) = 3,\ f'(2) = 0,\ f(2) = -6$

   c) $f''(-1.5) = 0,\ f'(-1.5) = 0,\ f(-1.5) = 2$

   d) $f''(3) = 0,\ f'(3) = -7,\ f(3) = 12$

B

5. For the graph of each function f, determine the x-coordinates of the local minimum and local maximum points.

   a) $f'(x) = 6x^2 - 7x - 3$ b) $f'(x) = 2x^3 - 3x^2 - 2x + 3$

6. Determine the local minimum and local maximum values of the polynomial function $f(x) = x^3 - 3x^2 - 5x + 2$.

7. For the graph of each function f, determine the coordinates of the local minimum and local maximum points.

   a) $f(x) = 2x^3 - 3x^2 - 12x + 4$ b) $f(x) = x^5 - 3x^3 + 4$

8. For the graph of each function f, determine the x-coordinates of any point(s) of inflection.

   a) $f''(x) = 3x^3 + 2x^2 - 12x - 8$ b) $f''(x) = -2x^4 + 2x^3 + 20x^2 + 16x$

9. Consider the polynomial function $f(x) = x^3 - 4x^2 - x + 4$.

a) Determine the x- and y-intercepts.

b) Determine the coordinates and nature of the critical points.

c) Determine the coordinates of the points of inflection.

d) Graph the function.

e) State the intervals of increase and decrease.

f) State the intervals of concavity.

✓ 10. Consider the polynomial function $f(x) = -2x^3 + 3x^2 + 3x - 2$. Repeat exercise 9a to f for this function.

✓ 11. **Knowledge/Understanding** Consider the polynomial function $f(x) = 4x^3 + 12x^2 - 16$. Repeat exercise 9a to f for this function.

12. Consider the polynomial function $f(x) = x^4 - 5x^3 + x^2 + 21x - 18$. Repeat exercise 9a to f for this function.

✓ 13. **Application** Consider the function $f(x) = x^4 + 5x^2 + 3$.

a) State the maximum number of minimum or maximum values this function may have.

b) State the maximum number of points of inflection this function may have.

c) Determine the coordinates of the minimum and maximum points, and the points of inflection.

d) Compare your findings in part c with your statements in parts a and b. Give reasons for these results.

Exercise 13 is significant because it illustrates the relationship between the number of real and complex zeros of f' and f'', and the number of possible minimum, maximum, and points of inflection that the graph of f may have.

14. **Thinking/Inquiry/Problem Solving**

a) Determine the coordinates of the absolute and local minimum and maximum points of the graph of each f on the given interval.

i) $f(x) = 2x^3 - 3x^2 - 72x + 7, \quad -2 \leq x \leq 5$

ii) $f(x) = x^4 - 8x^2 + 7, \quad -3 \leq x \leq 3$

b) Suppose the domain of f was not restricted. What are the coordinates of the absolute and local minimum and maximum points on the graph of f?

c) Compare the answers to parts a and b. Explain any similarities and differences.

 **15. Communication**

a) Explain how the degree of a polynomial function is related to the maximum number of minimum and maximum points the graph of the function may have.

b) What may cause the polynomial function to have fewer minimum and maximum points? Give examples to support your answers.

C

16. For each graph of a polynomial function f, state a possible expression for f'.

a)

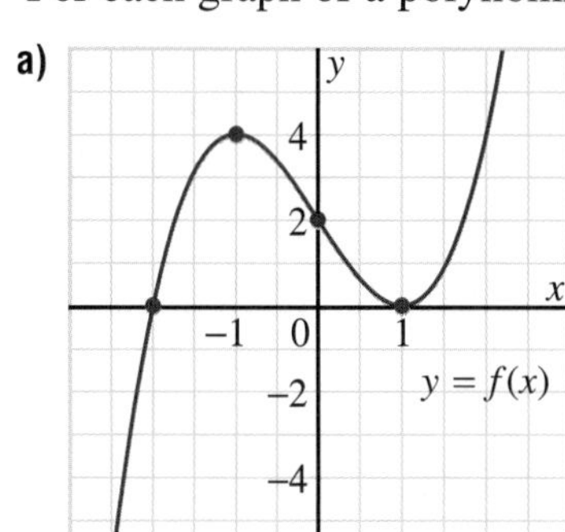

b)

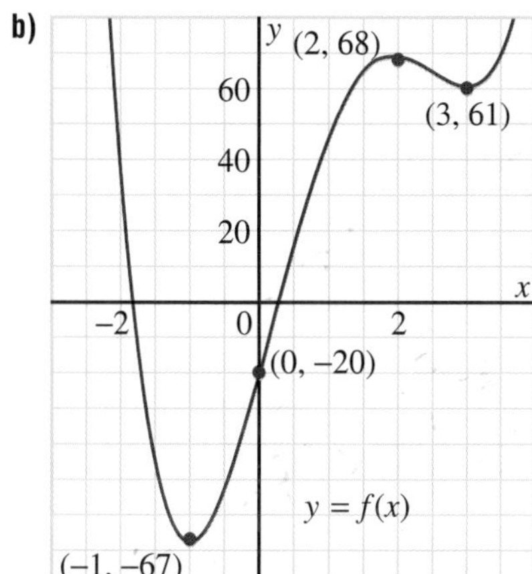

17. Sketch a graph of a function f that satisfies these conditions. Explain how you drew the graph.

$f(-1) = -5$, $f(0) = 0$, and $f(5) = -12$

$f'(-1) = f'(0) = f'(5) = 0$

$f'(x) < 0$ when $x < -1$ or $0 < x < 5$

$f'(x) > 0$ when $-1 < x < 0$ or $x > 5$

18. The point $(-1, 5)$ is a point of inflection on the graph of $f(x) = 2x^3 + mx^2 - 3x + n$.

a) Determine the values of m and n.

b) Determine the intervals of concavity and the coordinates of the minimum and maximum points on the graph of f.

19. a) Determine the value of m so the function $f(x) = x^2 + mx - 3$ has a local minimum value when $x = 6$.

b) What are the coordinates of the minimum point?

20. Determine the values of m and n so the function $f(x) = 2x^3 + mx^2 + nx - 10$ has a local maximum value at $x = -2$ and a local minimum value at $x = 3$.

3.8 Applications of Polynomial Functions

Velocity and Acceleration

The motion of an object that moves in a straight line can be represented by a polynomial function. Some examples are: a car travelling along a straight road; a train moving on a straight track; an arrow that is shot upward; and the launching of a rocket or missile. The position function, $y = s(t)$, is used to represent the displacement, or position, of the object on a line from the origin at time, t.

For an object that moves horizontally, the positive direction is conventionally used when the object moves to the right and the negative direction when the object moves to the left. For an object that moves vertically, the positive direction is conventionally up and the negative direction is down.

The velocity of an object that moves in a straight line is the rate of change of displacement with respect to time. It is the derivative of the displacement:

$v(t) = s'(t)$

The acceleration of an object that moves in a straight line is the rate of change of velocity with respect to time. It is the derivative of the velocity:

$a(t) = v'(t)$

Since velocity is the rate of change of displacement, and acceleration is the rate of change of velocity, then acceleration is the rate of change of the rate of change of displacement. The acceleration is the second derivative of the displacement:

$a(t) = s''(t)$

Take Note

Velocity and Acceleration

The *position function* of an object moving along a straight line is represented by $y = s(t)$, where s is the distance of the object from the origin at time t.
Velocity is the rate of change of displacement with respect to time: $v(t) = s'(t)$

An object is moving in a positive direction (to the right) when $v(t) > 0$ and in a negative direction (to the left) when $v(t) < 0$.
Acceleration is the rate of change of velocity with respect to time: $a(t) = s''(t)$ or $a(t) = v'(t)$

When the acceleration is negative; that is, $a(t) < 0$, the velocity is decreasing.
When the acceleration is positive; that is, $a(t) > 0$, the velocity is increasing.

In grade 10, you used the graph of a quadratic function to solve a problem similar to *Example 1*. You now have the tools to solve this type of problem algebraically.

Example 1

A ball is thrown upward with an initial velocity of 29.4 m/s from a balcony 10 m above the ground. The height, h metres, of the ball at time, t seconds, after being thrown is represented by the quadratic function $h(t) = -4.9t^2 + 29.4t + 10,\ t \geq 0$.

a) Determine the velocity of the ball after 1 s.

b) When does the ball reach its maximum height?

c) What is the maximum height of the ball?

d) How long before the ball hits the ground?

e) What is the velocity of the ball as it hits the ground?

f) What is the acceleration of the ball at time t?

Solution

a) Differentiate the height function to determine the velocity function.

$h(t) = -4.9t^2 + 29.4t + 10$

$v(t) = h'(t)$

$\quad\ = -9.8t + 29.4$

Substitute $t = 1$.

$v(1) = -9.8(1) + 29.4$

$\quad\ = 19.6$

After 1 s, the ball is travelling at 19.6 m/s upward.

b) The ball reaches its maximum height at the time when the first derivative is 0. To determine this time, solve $v(t) = 0$.

$v(t) = -9.8t + 29.4$

Substitute $v(t) = 0$.

$0 = -9.8t + 29.4$

$9.8t = 29.4$

$t = 3$

The ball reaches its maximum height after 3 s.

c) Since the maximum height occurs after 3 s, substitute $t = 3$ into the height equation to determine the maximum height.

$$\begin{aligned} h(t) &= -4.9t^2 + 29.4t + 10 \\ h(3) &= -4.9(3)^2 + 29.4(3) + 10 \\ &= 54.1 \end{aligned}$$

The maximum height of the ball is approximately 54 m.

d) When the ball hits the ground, its height is 0.

To determine the time when this occurs, solve $h(t) = 0$.

$h(t) = -4.9t^2 + 29.4t + 10$

Substitute $h(t) = 0$.

$0 = -4.9t^2 + 29.4t + 10$

Use the quadratic formula.

$$\begin{aligned} t &= \frac{-29.4 \pm \sqrt{29.4^2 - 4(-4.9)(10)}}{2(-4.9)} \\ &= \frac{-29.4 \pm \sqrt{1060.36}}{-9.8} \\ &= \frac{29.4 \pm \sqrt{1060.36}}{9.8} \end{aligned}$$

Since $t \geq 0$, consider only the positive value of t.

$$\begin{aligned} t &\doteq \frac{29.4 + \sqrt{1060.36}}{9.8} \\ &\doteq 6.323 \end{aligned}$$

It takes the ball approximately 6.3 s to hit the ground.

e) To determine the velocity of the ball as it hits the ground, substitute $t = 6.323$ in $v(t) = -9.8t + 29.4$.

$$\begin{aligned} v(6.323) &= -9.8(6.323) + 29.4 \\ &= -32.57 \end{aligned}$$

The ball hits the ground with a velocity of approximately 32.6 m/s downward.

f) To determine the acceleration, differentiate $v(t) = -9.8t + 29.4$.

$$\begin{aligned} a(t) &= v'(t) \\ &= -9.8 \end{aligned}$$

The acceleration is -9.8 m/s^2.

Discuss

In part a, how do we know the ball is moving up?

In *Example 1e,* the velocity is -32.57 m/s. We say that the speed is 32.57 m/s. That is, velocity indicates speed and direction.

To explain the results of *Example 1*, we look at the graph of the height function, below left. The function, $h(t) = -4.9t^2 + 29.4t + 10$, represents a parabola that is concave down. The maximum occurs at the vertex (3, 54.1). This point represents the maximum height of the ball: when $t = 3$ s, $h = 54.1$ m.

The t-intercept represents the time when the ball hits the ground: when $t \doteq 6.3$ s, $h = 0$. The h-intercept is the initial height of the ball (10 m). We only consider values of the function such that $t \geq 0$ and $h \geq 0$.

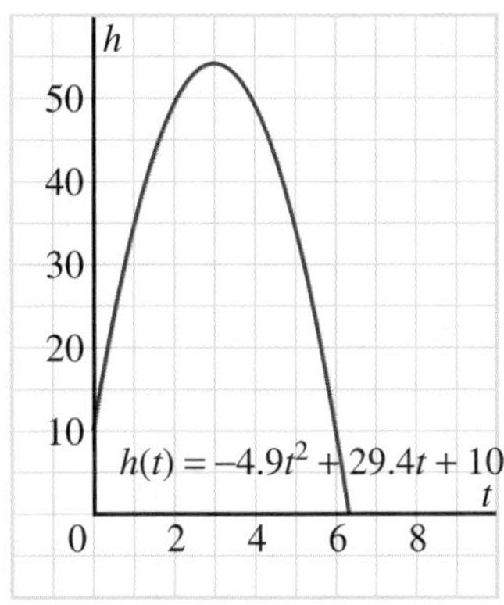

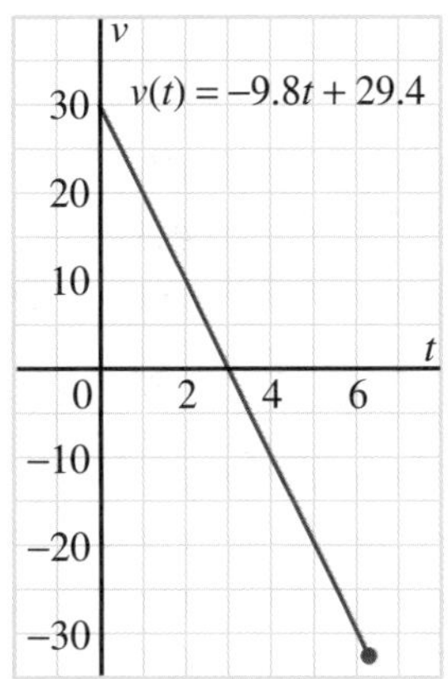

The velocity function, $v(t) = -9.8t + 29.4$, has a linear graph with constant slope, above right. The initial velocity (29.4 m/s) is the vertical intercept of the velocity graph. The acceleration (-9.8 m/s^2) is the slope of the velocity graph. The ball hits the ground when $t \doteq 6.3$ and $v \doteq -32.6$, so the graph does not extend beyond this point.

An object that moves vertically experiences an acceleration of -9.8 m/s^2 due to the force of gravity. The acceleration is constant; that is, it does not depend on t.

In $h(t)$ and $v(t)$, the acceleration is negative because it acts downward against the initial motion of the object. The effect of a negative acceleration is to slow down the object as it rises (the velocity and speed decrease) and to speed up the object as it falls (the velocity continues to decrease while the speed increases).

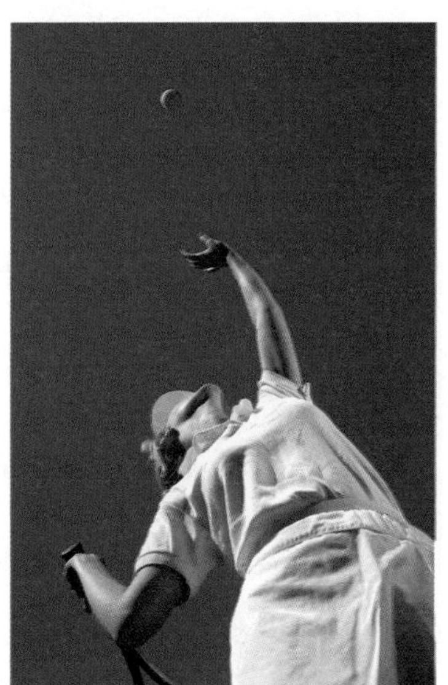

In general, the equation $h(t) = 0.5at^2 + v_0t + c_0$ represents the vertical motion of an object with an initial height of c_0, an initial velocity of v_0, and constant acceleration a.

In *Example 1*, $a = -9.8$ m/s^2, $v_0 = 29.4$ m/s, and $c_0 = 10$ m

Example 2 illustrates the motion of an object along a straight line.

Example 2

The motion of an object along a straight path is given by the polynomial function $s(t) = t^3 - 12t^2 + 36t + 5$, where s is the displacement in metres and t is the time in seconds, $t \geq 0$.

a) Determine the velocity and acceleration of the object after 1 s.

b) When is the object at rest? What is its displacement at these times?

c) When is the object moving in a positive direction? Draw a diagram to illustrate the motion of the object.

d) When is the acceleration 0? Determine the velocity and displacement of the object at this time.

e) Calculate the total distance travelled by the object during the first 7 s.

Solution

a) Use the first derivative to determine velocity. Use the second derivative to determine acceleration.

$s(t) = t^3 - 12t^2 + 36t + 5$

$$\begin{aligned} v(t) &= s'(t) \\ &= 3t^2 - 24t + 36 \end{aligned}$$

Substitute $t = 1$.

$$\begin{aligned} v(1) &= 3 - 24 + 36 \\ &= 15 \end{aligned}$$

$$\begin{aligned} a(t) &= s''(t) \\ &= 6t - 24 \end{aligned}$$

Substitute $t = 1$.

$$\begin{aligned} a(1) &= 6 - 24 \\ &= -18 \end{aligned}$$

After 1 s, the object is travelling at a velocity of 15 m/s to the right. The acceleration is negative, so the object is decelerating (or slowing down) at 18 m/s^2.

b) The object is at rest when the velocity is zero.

$v(t) = 3t^2 - 24t + 36$

Substitute $v(t) = 0$.

$$\begin{aligned} 3t^2 - 24t + 36 &= 0 \\ 3(t^2 - 8t + 12) &= 0 \\ 3(t - 6)(t - 2) &= 0 \\ t = 6 \text{ or } t &= 2 \end{aligned}$$

The object is at rest at 2 s and 6 s.

To determine the displacement at these times, substitute $t = 2$ and $t = 6$ in $s(t) = t^3 - 12t^2 + 36t + 5$.

$$\begin{aligned} s(2) &= 2^3 - 12(2)^2 + 36(2) + 5 \\ &= 8 - 48 + 72 + 5 \\ &= 37 \end{aligned}$$

$$\begin{aligned} s(6) &= 6^3 - 12(6)^2 + 36(6) + 5 \\ &= 216 - 432 + 216 + 5 \\ &= 5 \end{aligned}$$

After 2 s, the displacement is 37 m.
After 6 s, the displacement is 5 m.

c) The object is moving in a positive direction when the velocity is positive; that is, when $v(t) > 0$.
Solve $3t^2 - 24t + 36 > 0$.
Consider $3t^2 - 24t + 36 = 0$.
From part b, the equation factors as $3(t - 6)(t - 2) = 0$, and the solution is $t = 2$ and $t = 6$.

To solve the inequality, use the number line below. The roots $t = 2$ and $t = 6$ divide the number line into 3 intervals.
For $0 < t < 2$, substitute $t = 1$ in $3(t - 6)(t - 2)$: $3(1 - 6)(1 - 2)$
The polynomial is positive.
The polynomial changes sign at $t = 2$.
So, for $2 < t < 6$, the polynomial is negative.
The polynomial changes sign at $t = 6$.
So, for $t > 6$, the polynomial is positive.

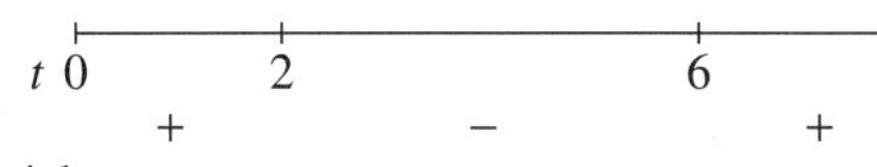

From the number line, the object moves in the positive direction when $0 < t < 2$ s and when $t > 6$ s.
When $t = 0$, the displacement is $s(0) = 5$.
The displacement is 5 m when $t = 0$.
On a displacement line, mark the origin, $s = 0$ and the initial displacement $s(0) = 5$. Mark the displacements when $v = 0$; that is, $s(2) = 37$ and $s(6) = 5$. At these displacements, the object changes direction.
Connect the points using a smooth curve. Use arrows to indicate the direction of the motion of the object.

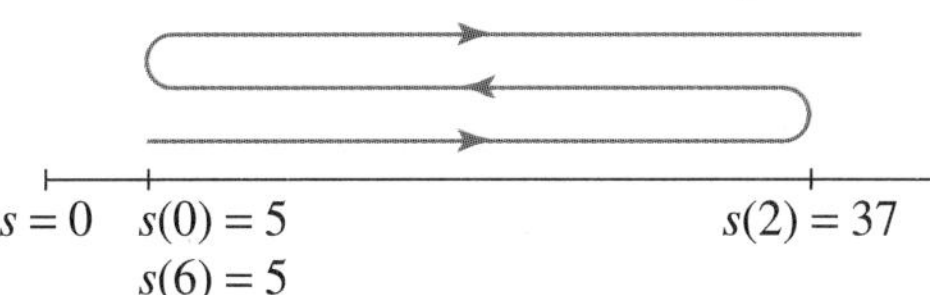

d) The acceleration is zero when $a(t) = 0$. Substitute $a(t) = 0$ in $a(t) = 6t - 24$.

$$6t - 24 = 0$$
$$6t = 24$$
$$t = 4$$

The acceleration is 0 after 4 s.
To determine the velocity after 4 s, substitute $t = 4$ in $v(t) = 3t^2 - 24t + 36$.

$$\begin{aligned} v(4) &= 3(4)^2 - 24(4) + 36 \\ &= 48 - 96 + 36 \\ &= -12 \end{aligned}$$

To determine the displacement of the object after 4 s, substitute $t = 4$ in $s(t) = t^3 - 12t^2 + 36t + 5$.

$s(4) = 4^3 - 12(4)^2 + 36(4) + 5$

$= 21$

When the acceleration is 0, the displacement is 21 m and the velocity is 12 m/s to the left.

e) From the diagram in part c, the total distance travelled is the sum of the distances travelled for $0 \le t < 2, 2 < t < 6$, and $6 < t < 7$.

From part c, $s(0) = 5, s(2) = 37$, and $s(6) = 5$

Calculate $s(7)$.

$s(7) = 7^3 - 12(7)^2 + 36(7) + 5$

$= 12$

From $t = 0$ s to $t = 2$ s, the distance travelled is

$|s(2) - s(0)| = |37 - 5| = 32$.

From $t = 2$ s to $t = 6$ s, the distance travelled is

$|s(6) - s(2)| = |5 - 37| = 32$.

From $t = 6$ s to $t = 7$ s, the distance travelled is

$|s(7) - s(6)| = |12 - 5| = 7$.

The total distance is $32 + 32 + 7 = 71$.

The total distance travelled in the first 7 s is 71 m.

Other applications of polynomial functions involve rates of change of quantities such as population and prices. You will learn about these applications in the exercises that follow.

3.8 Exercises

1. The graph of a position function $y = s(t)$ is given. Sketch the graphs of $y = v(t)$ and $y = a(t)$.

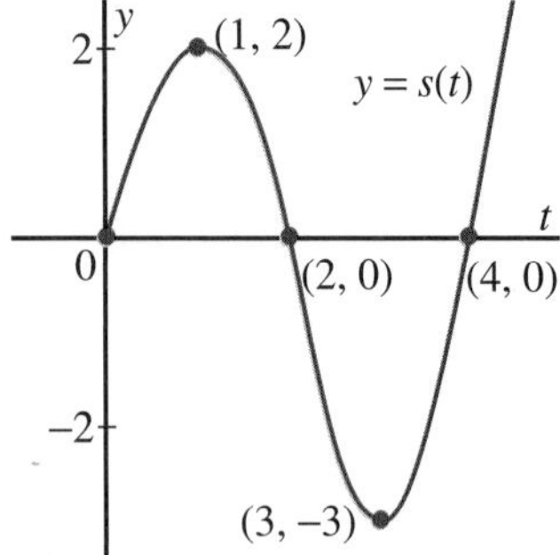

2. Each graph represents a position function $y = s(t)$. Describe $y = v(t)$ as increasing or decreasing, and $y = a(t)$ as positive or negative.

a)

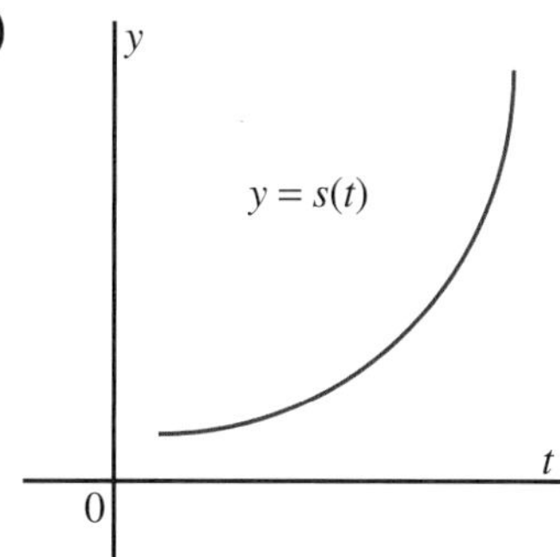

b)

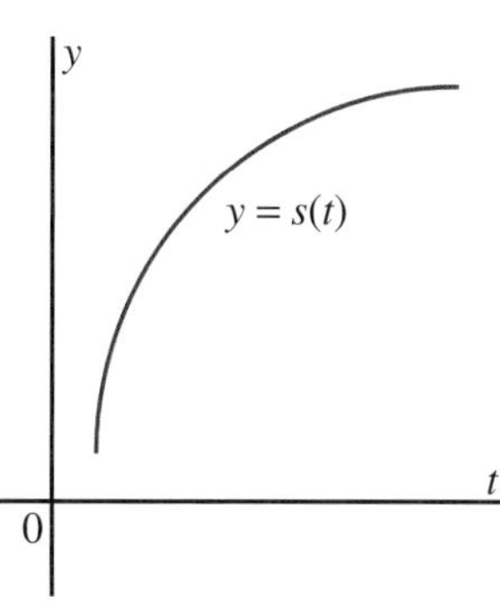

c)

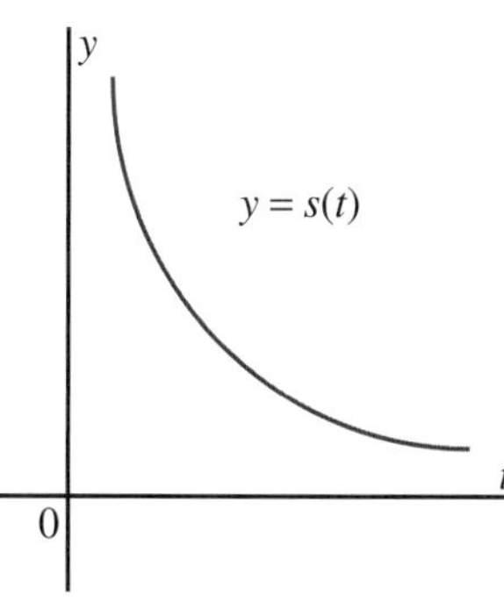

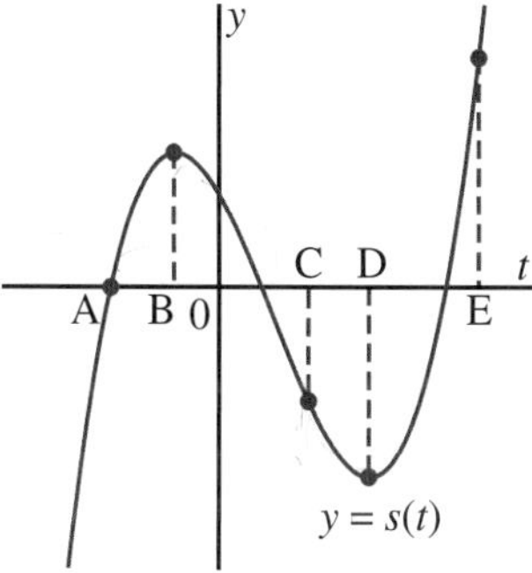

3. For the graph of the position function $y = s(t)$, determine the interval(s) in each case.

a) Both the displacement and velocity are increasing.

b) The displacement is increasing while the velocity is decreasing.

c) The displacement is decreasing while the velocity is increasing.

d) Both the displacement and velocity are decreasing.

B

4. Communication

a) Consider the motion of a car that is braking while moving forward.

i) Is the velocity positive or negative? Explain.

ii) Is the acceleration positive or negative? Explain.

iii) Is the velocity increasing or decreasing? Explain.

b) Suppose the car is accelerating while moving forward. Answer each question in part a for this situation.

5. For each position function s, determine s', and the velocity after 3 s. The displacement is measured in metres.

a) $s(t) = -5t^3 + 8t^2 - 1$ **b)** $s(t) = 3t^4 - 2t^3 + 5t - 1$

c) $s(t) = 2t^2 - 7t + 5$ **d)** $s(t) = 8 - 11t$

6. For each position function s, determine s'', and the acceleration after 2 s. The displacement is measured in metres.

a) $s(t) = t^2(t - 1)$ **b)** $s(t) = -5t^3 + 4t^2 - t + 3$

7. Each function describes the displacement, in metres, of an object that moves along a straight line. Determine whether the object is moving toward or away from the origin at time $t = 1$ s and at time $t = 5$ s.

a) $s(t) = t(t + 2)(t - 1)$ **b)** $s(t) = -t^2 + 5t + 4$ **c)** $s(t) = 2t^3 - 6t^2 + 3$

8. Refer to exercise 7c. For the position function $s(t) = 2t^3 - 6t^2 + 3$:

a) Determine when the object is at rest.

b) Determine when the object is moving to the right, and to the left.

c) Draw a diagram to illustrate the motion of the object.

d) Determine the total distance travelled in the first 8 s.

9. A bicycle rider travelling along a straight path applies the brakes. The rider's position is then modelled by the quadratic function $s(t) = -0.8t^2 + 6.4t$, where s is measured in metres and t is measured in seconds.

a) What is the rider's initial velocity?

b) How long does it take the rider to stop?

10. Knowledge/Understanding The position function of an object travelling in a straight line is given by $s(t) = t^3 - 6t^2 + 9t + 2$, where s is measured in metres and t is measured in seconds, $t \geq 0$.

a) Calculate the velocity after 3 s.

b) Calculate the acceleration after 4 s.

c) When is the object travelling to the right? To the left?

d) Draw a diagram to illustrate the motion of the object.

e) Calculate the total distance travelled in the first 5 s.

11. A stone is thrown upward with an initial velocity of 12 m/s from a height of 45 m. Its height, h metres, after t seconds is represented by $h(t) = -4.9t^2 + 12t + 45$.

a) Determine the velocity of the stone after 1 s and after 3 s.

b) When will the stone hit the ground?

c) When does the stone reach its maximum height?

d) What is the maximum height?

12. Application The graph represents the position function of a motorcycle travelling along a highway.

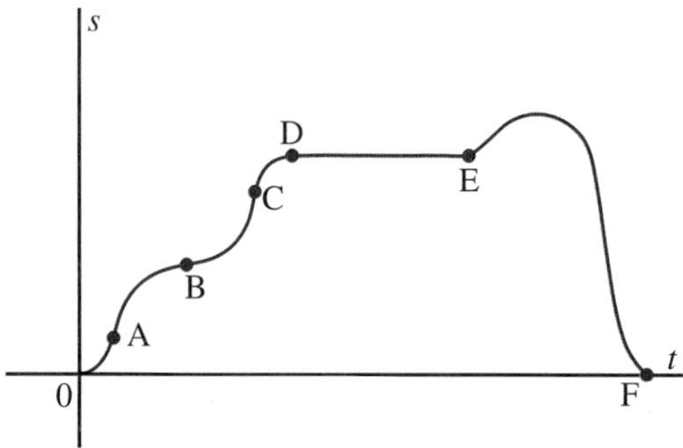

a) Was the motorcycle travelling faster at B or at C? Explain.

b) Was the motorcycle accelerating or decelerating at A, B, and C? Explain.

c) What happened from D to E? Explain.

d) What happened at F? Explain.

13. A demographer developed a function to model the population, P, of a town x years from today: $P(x) = 6500 - 0.25x^2 + 235x$.

a) Determine the present population of the town.

b) Predict the population in 5 years and in 10 years.

c) What is the rate of growth of the population of the town?

d) When will the population reach 10 000?

e) When will the rate of growth of the population be 225 people per year?

14. A clinical test was conducted of a new drug designed to eliminate microbes. The results of a patient's blood samples showed that the number of microbes per millilitre, M, at t hours after taking the drug is $M(t) = 40\,000(36 - t)^2, t \geq 10$.

a) How long will it take for all the microbes to be eliminated?

b) Determine the rate of decrease in the number of microbes per millilitre 15 h after the drug is taken.

✓ 15. **Thinking/Inquiry/Problem Solving** The revenue of a company that produces CDs is modelled by the function $R(x) = x(100 - 2x)$, where R is in thousands of dollars and x dollars is the price charged per box of CDs, $0 \leq x \leq 30$.

a) Determine the rate of change of the revenue with respect to the price when the price per box is \$10, and when the price per box is \$15.

b) Explain what the values of the rates of change in part a mean to the company.

c) For what prices do revenues increase?

d) Determine the price that produces a \$0 rate of change of revenue.

e) Explain the significance of the price in part d.

16. After the driver of a sports utility vehicle applies the brakes, the position of the vehicle is modelled by $s(t) = 2t(14 - 0.1t^2)$, where t is measured in seconds and s in metres.

a) Determine the acceleration of the vehicle after 2 s.

b) How far does the vehicle travel before it comes to a complete stop?

✓ 17. An object in free fall travels a distance of $s(t) = 0.5gt^2$, where s is measured in metres, t is measured in seconds, and $g = 9.8$ m/s^2.

a) How far does the object fall in the first 5 s?

b) Determine the velocity of the object at time t seconds.

c) What is the velocity after 6 s?

d) Suppose the object falls from a cliff 400 m high. What is the velocity of the object as it hits the ground?

Self-Check 3.6 – 3.8

1. Solve for x.

   a) $x^3 - 5x^2 + 2x + 8 = 0$

   b) $x^3 - 2x^2 - 5x + 6 = 0$

2. Solve each inequality.

   a) $4x^3 - 8x^2 + x + 3 < 0$

   b) $x^4 + 3x^3 - x^2 - 6x > 0$

3. Determine the value(s) of m such that $x = 1$ is a root of $x^4 + 2mx^3 = -2 + m^2x$.

4. Suppose a, b, and c are the roots of the equation $x^3 - 3x^2 - 4x + 12 = 0$. Determine an equation that has roots $a + b + c$ and $\frac{bc}{a}$.

5. Write a polynomial equation with each set of roots.

   a) $-1, \pm i$

   b) $2 + \sqrt{3}, 2 - \sqrt{3}, 1$

6. Consider the polynomial function $f(x) = 3x^3 + 7x^2 + 3x - 1$.

   a) Determine the x- and y-intercepts.

   b) Determine the coordinates and nature of the critical points.

   c) Determine the coordinates of the points of inflection.

   d) Graph the function.

   e) State the intervals of increase and decrease.

   f) State the intervals of concavity.

7. For each position function $y = s(t)$, calculate the velocity and acceleration after 2 s.

   a) $s(t) = -4t^3 + 8t^2 + 5t - 7$

   b) $s(t) = (2t - 1)^2(3t + 5)$

8. The driver of a car is braking while reversing.

   a) Is the velocity positive or negative? Explain.

   b) Is the acceleration positive or negative? Explain.

   c) Is the velocity increasing or decreasing? Explain.

PERFORMANCE ASSESSMENT

9. The position function of an object that moves along a straight path is given by $s(t) = t^3 - 9t^2 + 15t + 4$, where s is measured in metres and t in seconds, $t \geq 0$.

   a) When is the object at rest? When is it moving in the positive direction?

   b) Calculate the velocity and displacement when the acceleration is zero.

   c) Draw a diagram to illustrate the motion of the object.

   d) How far does the object travel in the first 7 s?

Review Exercises

Mathematics Toolkit

Polynomial Tools

Polynomial Functions and Equations

The x-intercepts of the graph of a polynomial function $y = f(x)$ are the real zeros of the function and the real roots of the corresponding polynomial equation $f(x) = 0$.

When the graph of $y = f(x)$ intersects the x-axis at $(a, 0)$, then $(x - a)$ occurs as a factor in the equation $y = f(x)$.

When the x-axis is a tangent to the graph of $y = f(x)$ at $(a, 0)$, then $(x - a)$ occurs more than once as a factor in the equation $y = f(x)$.

Remainder Theorem
When a polynomial, $f(x)$, is divided by $x - k$, the remainder is $f(k)$, the number obtained by substituting k for x in the polynomial.

Factor Theorem
If k is substituted for x in $f(x)$, and $f(k) = 0$, then $x - k$ is a factor of $f(x)$.

Factor Property
A polynomial $f(x)$ has integer coefficients with leading coefficient 1. If $f(x)$ has any factor of the form $x - k$, then k is a factor of the constant term of $f(x)$.

Derivatives and Graphs of Polynomial Functions

Let $y = f(x)$ be a polynomial function.

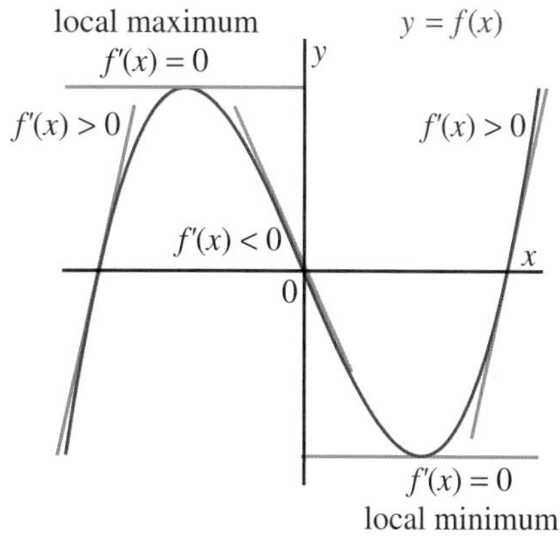

Intervals of Increase and Decrease

- If $f'(x) > 0$ for all values of x in an interval, then f is increasing in that interval.
- If $f'(x) < 0$ for all values of x in an interval, then f is decreasing in that interval.

First Derivative Test for Local Maximum or Minimum Points
Suppose $f'(a) = 0$. Then $x = a$ is a critical value.
As x increases:

- If $f'(x)$ changes sign from negative to positive at $x = a$, then f has a local minimum at $x = a$.
- If $f'(x)$ changes sign from positive to negative at $x = a$, then f has a local maximum at $x = a$.

- If $f'(x)$ does not change sign at $x = a$, then f has a point of inflection at $x = a$.

The second derivative of a function f is the derivative of f', and it is written f''. The second derivative f'' is the rate of change of the first derivative f'.

First and Second Derivative Test for Local Maximum and Minimum Points

- At a local maximum point, $f'(x) = 0$ and $f''(x) < 0$
- At a local minimum point, $f'(x) = 0$ and $f''(x) > 0$

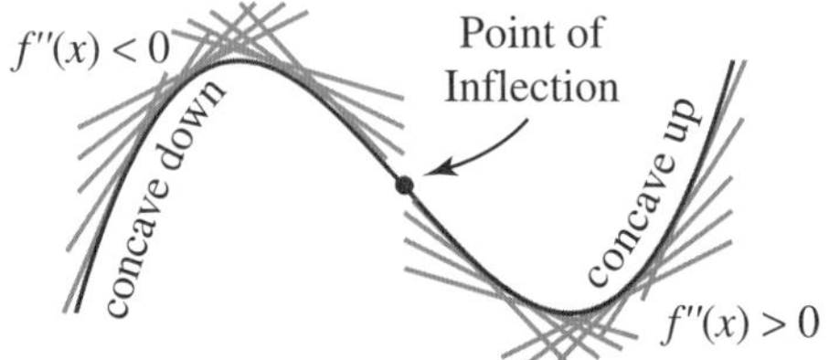

Intervals of Concavity

- If $f''(x) > 0$ for all values of x in an interval, then the graph of f is concave up in that interval.
- If $f''(x) < 0$ for all values of x in an interval, then the graph of f is concave down in that interval.

Points of Inflection

- At a point of inflection, $f''(x) = 0$ and, as x increases, f'' changes sign from positive to negative, or vice versa.
- At a point of inflection, the graph of f changes from concave up to concave down, or vice versa.

Velocity and Acceleration

The position function of an object moving along a straight line is represented by a polynomial function $y = s(t)$, where s is the displacement of the object at time t.

Velocity is the rate of change of displacement with respect to time: $v(t) = s'(t)$
Velocity indicates speed and direction. An object is moving in a positive direction when $v(t) > 0$ and is moving in a negative direction when $v(t) < 0$.

Acceleration is the rate of change of velocity with respect to time: $a(t) = s''(t)$ or $a(t) = v'(t)$
When acceleration is negative ($a(t) < 0$), the velocity is decreasing.
When acceleration is positive ($a(t) > 0$), the velocity is increasing.

NS **1.** Solve each inequality. Show each solution on a number line.

a) $x^2 + 7x + 12 > 0$ **b)** $x^2 + 6x - 27 < 0$

c) $2x^2 + 9x + 10 \leq 0$ **d)** $10x^2 + x - 2 \geq 0$

2. Determine each quotient and remainder. State the restrictions on the divisor.

a) $(2x^2 + 5x - 4) \div (2x - 1)$ **b)** $(3x^3 - x^2 - x - 2) \div (x + 1)$

c) $(3x^3 - x - 3) \div (x - 1)$ **d)** $(6x^3 + x^2 - 4x - 3) \div (2x + 3)$

3. When a certain polynomial is divided by $3x + 4$, the quotient is $x^3 + 2x - 5$ and the remainder is -2. What is the polynomial?

3.1 **4.** Identify the function that corresponds to each graph below. Justify your choices.

$h(x) = -x^4 - 4x^3 + 10$ $f(x) = x^3 - 3x^2 + 1$

$k(x) = 2x^4 - 3x^2 - 21$ $g(x) = -2x^3 + 4x^2 + 7x - 3$

a)
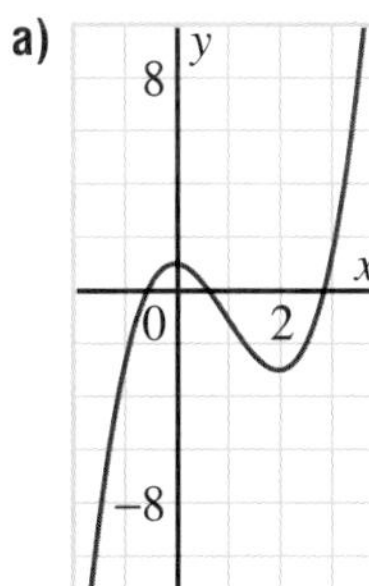

b)
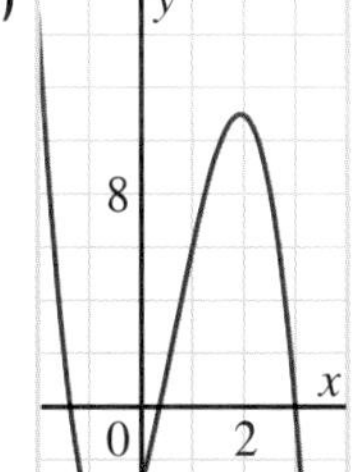

c)
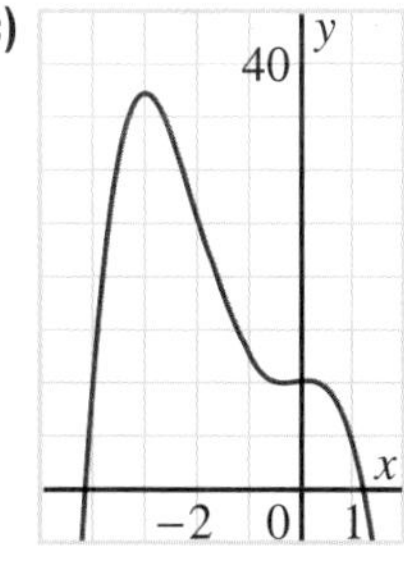

d)
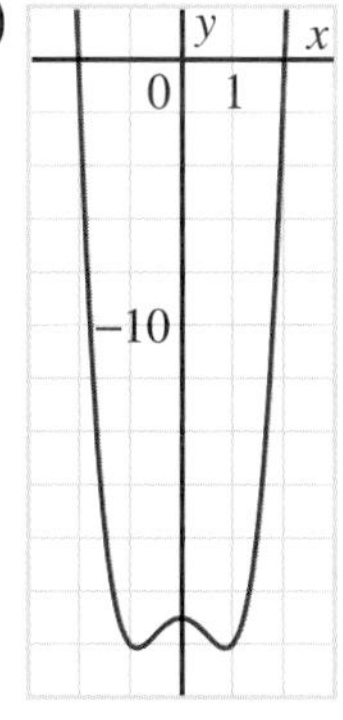

5. For each polynomial function below:

i) Describe the nature of change of the function using first and second differences.

ii) Graph the function.

iii) Compare your results in parts i and ii.

a) $f(x) = -2x^3 - x$ **b)** $y = 2x^4 - 5x$ **c)** $g(x) = -2x^4 + 4x^3 + 47x^2 - 13x - 12$

6. Refer to the graphs on page 224.

i) Which functions have an odd degree? Explain.

ii) Which functions have an even degree? Explain.

iii) Which functions have a negative leading coefficient? Explain.

iv) Which functions have a positive leading coefficient? Explain.

v) Which graphs have symmetry? Describe the symmetry.

a) 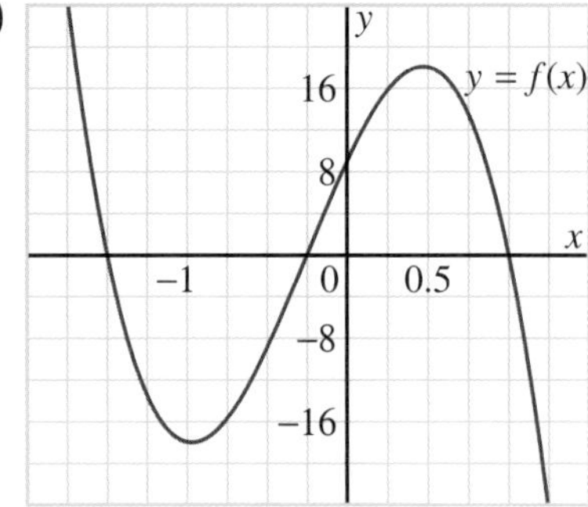

b) 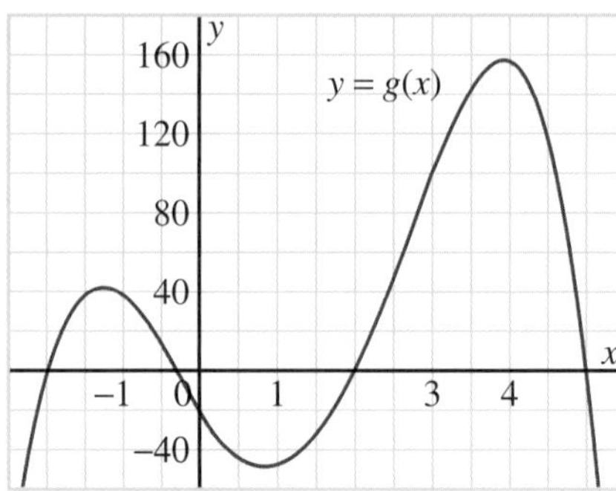

c) 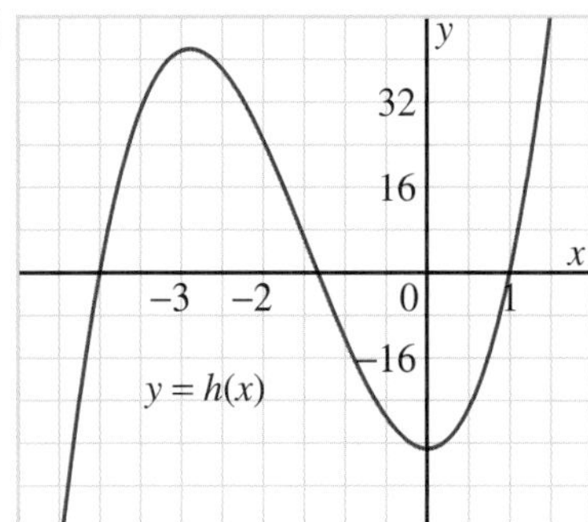

d) 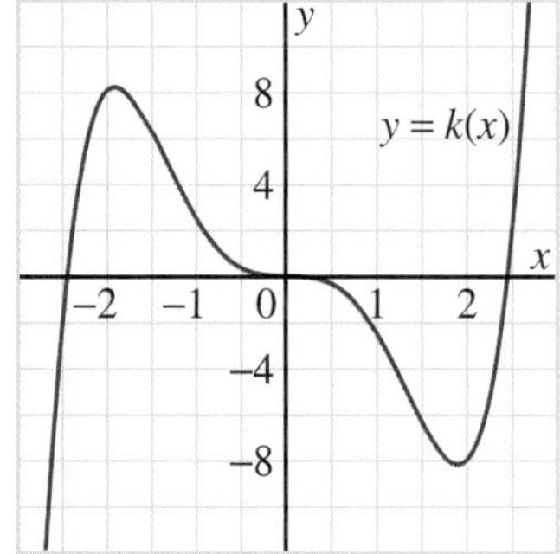

3.2 **7.** Determine the zeros of each polynomial function.

a) $g(x) = (x - 3)(2x + 5)(x + 2)^2$ **b)** $f(x) = 2x^2(3x + 4)(x + 5)(x - 1)$

8. The zeros of four functions are given below.

i) Which function has a zero of order 2?

ii) Which function has a zero of order 3?

iii) Write a polynomial function that has each set of zeros.

a) 3, –2, 1 **b)** 0, 0, 1 **c)** –4, 5, –1 **d)** –3, –3, –3

9. State the zeros of each function, then sketch its graph.

a) $f(x) = (x + 3)(x - 4)$

b) $f(x) = -(x + 3)^2(x - 4)$

c) $f(x) = (x - 1)(x + 2)(x - 3)$

10. Determine the equation of each function, then sketch its graph.

a) a cubic function with zeros –3, 4, 4; graph passes through (5, 12)

b) a quartic function with zeros –2, 0, 0, 1; graph passes through (–3, –12)

11. Explain the relationship between a polynomial function and its corresponding polynomial equation. Give examples.

12. i) Sketch an example of each cubic function with the given zeros.
 ii) State the defining equation for the family of cubic functions represented by each set of zeros.

 a) $-2, 1, 3$ b) $-1, 0, 1$ c) $-4, -4, -4$

3.3 13. The graphs of functions f are given. Sketch the graph of each derivative f'. Explain your sketches.

 a)

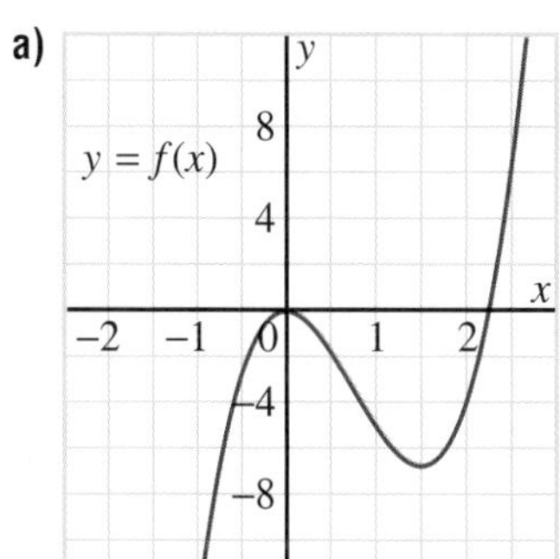

 b)

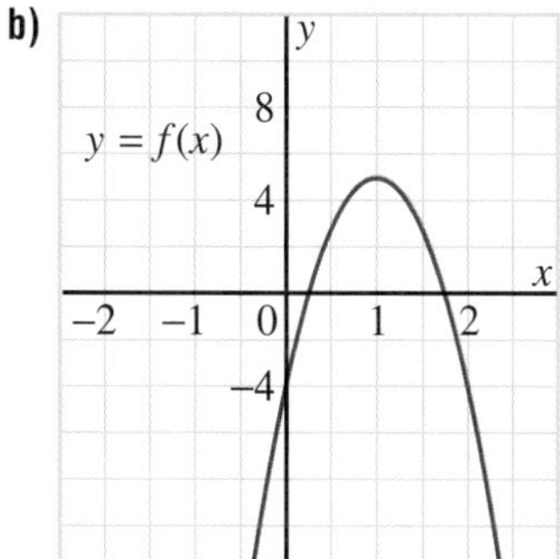

14. For each graph in exercise 13:
 i) State the approximate coordinates of the local maximum and local minimum points.
 ii) State the intervals of increase and decrease for f.

15. Determine the critical values of each polynomial function.

 a) $y = x^4 - 3x^3 + 5$ b) $y = -3x^3 + 9x^2 - 11$
 c) $y = -2x^4 - 6x^2 - 9$ d) $y = x(x - 2)^2$

16. For each function in exercise 15, determine the coordinates of the local minimum and local maximum points, then use these points to sketch a graph.

17. For each function below:
 i) Determine the x- and y-intercepts.
 ii) Determine the coordinates of the critical points.
 iii) Identify the nature of the critical points.
 iv) Sketch the graph.
 v) Identify the intervals of increase and decrease.

 a) $y = -4x^3 - 3x^2$ b) $y = x^4 - 2x^2$

Review Exercises

3.4 **18.** For each function $y = f(x)$, sketch the graphs of the first and second derivative functions. Explain each sketch.

a)

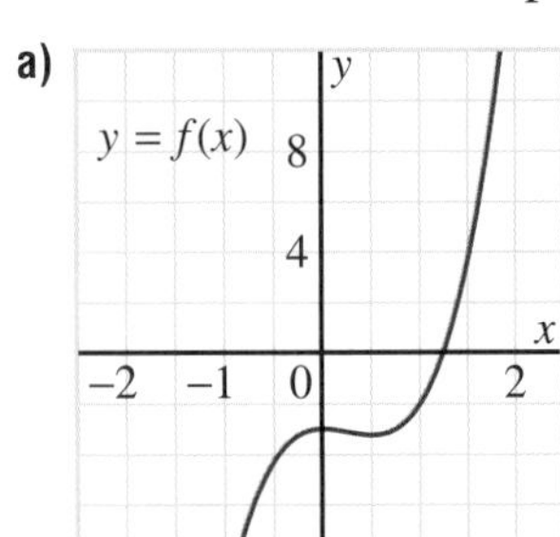

b)

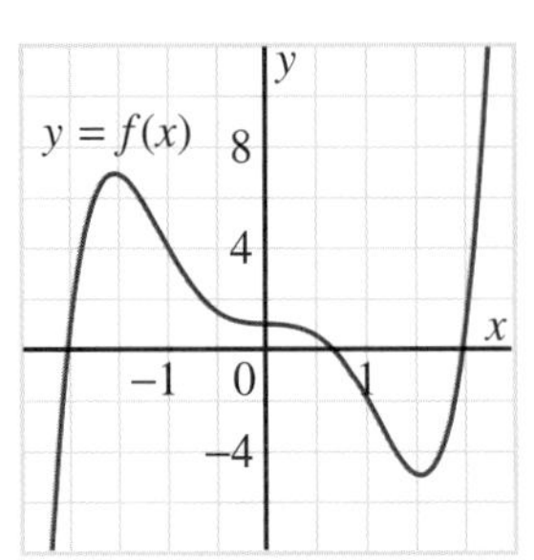

c)

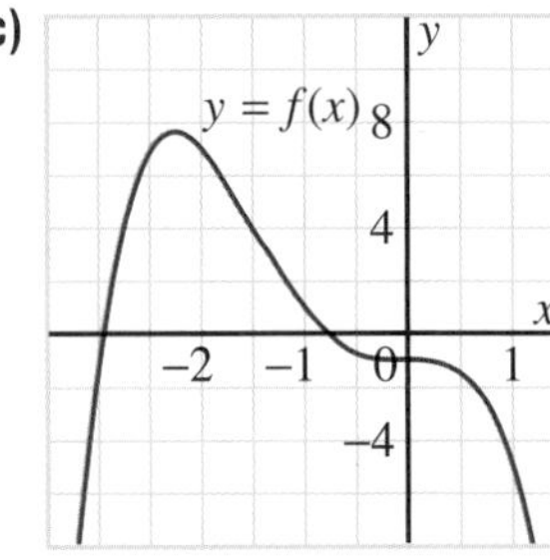

19. Determine the second derivative of each polynomial function.

a) $f(x) = x^4 - 2x^3 + 16x^2 - 7$ **b)** $f(x) = 3x(x - 2)(3x + 1)$

c) $f(x) = (3x + 2)^2(x - 1)(x + 1)$ **d)** $f(x) = -2x(x + 5) - 3x(4x^2 + 1)$

20. For each function in exercise 19, determine the coordinates of the points of inflection.

21. Consider the polynomial function $f(x) = -5x^4 + 4x^3$.

a) Determine the x- and y-intercepts.

b) Determine the coordinates of the critical points.

c) Identify the nature of the critical points.

d) Determine the coordinates of the points of inflection.

e) Graph the function.

f) State the intervals of increase and decrease.

g) State the intervals of concavity.

22. Consider the polynomial function $f(x) = (3x - 3)(x + 2)(x - 1)$. Repeat exercise 21a to g for this function.

23. Sketch a graph of the polynomial function that satisfies this set of conditions:

$f(-3) = -8,\ f(0) = 6,\ f(4) = -10,\ f(-1) = -1,\ f(2) = -4$

$f'(-3) = f'(0) = f'(4) = 0$

$f'(x) < 0$ for $x < -3$ and $0 < x < 4$; $f'(x) > 0$ for $-3 < x < 0$ and $x > 4$

$f''(-1) = f''(2) = 0$

$f''(x) < 0$ for $-1 < x < 2$; $f''(x) > 0$ for $x < -1$ and $x > 2$

3.5 **24. a)** Determine the remainder when $x^3 - 4x^2 + x + 6$ is divided by each binomial.

i) $x - 1$ **ii)** $x - 2$ **iii)** $x - 3$ **iv)** $x - 4$

b) What does a zero remainder tell about the polynomial $x^3 - 4x^2 + x + 6$?

25. Which polynomial has $x - 1$ as a factor? Explain.

a) $x^3 + x^2 - x - 1$ **b)** $2x^3 - 4x^2 - 7x + 8$ **c)** $4x^3 - 2x^2 + 7$

d) $x^4 - 2x^3 + 3x^2 + 4x + 5$ **e)** $3x^3 - 2x^2 + 7x - 6$ **f)** $x^3 - 5x^2 + 4x - 8$

26. Factor each polynomial.

a) $x^3 - 4x^2 + x + 6$ **b)** $x^3 + 2x^2 - 19x - 20$ **c)** $x^3 + 4x^2 - 7x - 10$

27. Factor each polynomial.

a) $6x^3 - 17x^2 + 15x - 4$ **b)** $2x^4 + 5x^3 - 14x^2 - 5x + 12$

c) $-2x^4 - 3x^3 + 12x^2 + 7x - 6$

28. Determine the value of m so that each polynomial is divisible by the given binomial.

a) $x^3 - 2x^2 + 5x + m;\ x + 1$ **b)** $x^3 - 3x^2 + mx + 6;\ x - 2$

3.6 **29.** Solve for x.

a) $2x^3 - 5x^2 - 4x + 3 = 0$ **b)** $x^4 - 3x^3 - 7x^2 + 27x - 18 = 0$

30. Use the results of exercise 29 to solve each inequality.

a) $2x^3 - 5x^2 - 4x + 3 < 0$ **b)** $x^4 - 3x^3 - 7x^2 + 27x - 18 > 0$

31. Determine the real and/or complex roots of each polynomial equation.

a) $x^3 - 27 = 0$ **b)** $x^2 + 3x - 8 = 0$

c) $x^4 - x^3 - x + 1 = 0$ **d)** $x^3 + 8 = 0$

e) $2x^5 + x^3 = 0$ **f)** $x^4 - 16 = 0$

32. Determine the value(s) of k in each equation so that one root is double the other root.

a) $x^2 - kx + 7 = 0$ **b)** $4x^2 + kx + 4 = 0$

33. Determine the value(s) of k in each equation for which one root is triple the other root.

a) $x^2 - kx + 3 = 0$ **b)** $x^2 + 8x + k = 0$

c) $kx^2 - 8x + 3 = 0$ **d)** $3x^2 + kx + 4 = 0$

e) $x^2 - 16x + k = 0$ **f)** $16x^2 + kx + 27 = 0$

34. Determine the value(s) of m in each equation so that the two roots are equal.

a) $9x^2 - mx + 1 = 0$ **b)** $9x^2 - 42x + m = 0$

35. Solve each inequality.

a) $x^3 + x^2 + x - 3 < 0$ **b)** $3x^4 + x^3 \geq 13x^2 - 4x + 4$ **c)** $3x^5 \leq 27x$

Review Exercises

36. Write a polynomial equation that has each set of roots. Is the equation unique? Explain.

    a) $-2, 4, \frac{2}{5}, -1$ b) $\pm 1, 2 - \sqrt{3}, 2 + \sqrt{3}$ c) $-2, -3 + 4i, -3 - 4i$

3.7 37. Determine the coordinates of the local minimum and local maximum points of the graph of each polynomial function.

    a) $f(x) = 3x^4 - 8x^3 - 6x^2 + 24x - 9$ b) $f(x) = -3x^3 + 9x^2 - 9x + 11$

38. Determine the coordinates of the points of inflection of the graph of each function.

    a) $h(x) = 2x^3 - 12x^2 + 18x - 1$ b) $g(x) = x^4 - 6x^3 + 9x^2 + 3x$

39. Consider the polynomial function $f(x) = 2x^4 - 26x^2 + 72$.

    a) Determine the x- and y-intercepts.

    b) Determine the coordinates and the nature of the critical points.

    c) Determine the coordinates of the points of inflection.

    d) Graph the function.

    e) Determine the intervals of increase and decrease.

    f) Determine the intervals of concavity.

3.8 40. A ball is thrown vertically with a velocity of 18 m/s. The ball is h metres above the ground after t seconds, where $h(t) = -5t^2 + 18t + 12$.

    a) What is the height of the ball after 2 s?

    b) What is the velocity of the ball after 2 s?

    c) What is the maximum height of the ball?

    d) What is the velocity of the ball as it hits the ground?

41. The position function of an object moving along a straight path is given by $s(t) = 3t^4 - 16t^3 + 24t^2$, where s is measured in metres and t in seconds, $t \geq 0$.

    a) Calculate the velocity after 3 s.

    b) Calculate the acceleration after 3 s.

    c) When is the object at rest?

    d) When is the object travelling to the right?

    e) Draw a diagram to illustrate the motion of the object.

    f) Calculate the total distance travelled in the first 5 s.

Self-Test

1. **Knowledge/Understanding** For each graph of a polynomial function:
   - **i)** Determine the sign of the leading coefficient.
   - **ii)** Determine the degree of the function.
   - **iii)** Determine the real zeros of the function.
   - **iv)** Determine the y-intercept of the function.
   - **v)** Determine the equation of the function.

a)

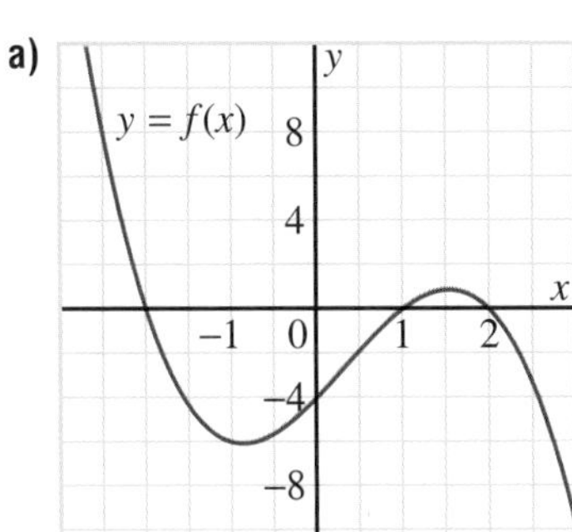

b)

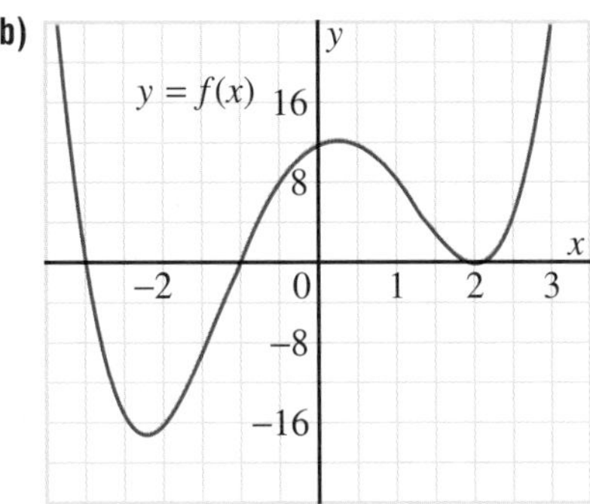

2. Sketch the graphs of the first and second derivative functions for each function in exercise 1.

3. Consider the polynomial function $y = 2x^3 + 3x^2$.
   - **a)** Determine the x- and y-intercepts.
   - **b)** Determine the coordinates and nature of the critical points.
   - **c)** Determine the coordinates of the points of inflection.
   - **d)** Graph the function.
   - **e)** Determine the intervals of increase and decrease.
   - **f)** Determine the intervals of concavity.

4. When $x^4 - 5x^3 + mx^2 + 3x - 11$ is divided by $x + 2$, the remainder is -1. Determine the value of m.

5. **Communication** Explain how the first and second derivatives can be used to identify properties of the graph of a polynomial function. Use the polynomial function $y = 3x^4 - 8x^3 - 30x^2 + 72x + 50$ in your discussion.

6. **a)** Determine all the real and complex zeros of the polynomial function $f(x) = 3x^4 + 4x^3 - 2x^2 - 31x - 10$.

   b) Use the results of part a to solve the inequality $3x^4 + 4x^3 - 2x^2 - 31x - 10 > 0$.

7. **Application** The general position function of an object moving along a straight line is modelled by $s(t) = s_0 + v_0t + 0.5gt^2$, where s is measured in metres, t in seconds, and s_0, v_0, and g are constants.

   a) What is the position function for a rocket fired vertically from the deck of a ship, 30 m high, with an initial velocity of 55 m/s?

   b) How long does it take for the rocket to hit the sea?

   c) What is the velocity of the rocket as it hits the sea?

   d) What is the maximum height of the rocket? How long does it take to reach this height?

8. Determine the equation of a quartic function that passes through $(1, -8)$ and has roots $-1, \frac{2}{3}, 2 + i, 2 - i$.

9. **Thinking/Inquiry/Problem Solving** A tangent is drawn at each point of inflection on the graph of $f(x) = x^4 - 6x^2 + 2$. Determine the coordinates of the point of intersection of these tangents.

PERFORMANCE ASSESSMENT

10. When $f(x)$ is a polynomial with $f(2) = f(-1) = 0$, then $x - 2$ and $x + 1$ are factors of $f(x)$. Suppose $g(x)$ is a polynomial with $g(2) = g(-1) = 4$. What conclusions can you draw? Explain.

11. When we differentiate the function $y = x^3 + 3x^2 + 6x + 6$, we get $y' = 3x^2 + 6x + 6$. The derivative is the part of the function without the cubic term.

    a) How are the graphs of the function and its derivative related?

    b) Make up a different cubic function so that its derivative is the part of the function without the cubic term. Explain how you did this. Repeat part a for your function. Describe what you notice.

    c) Make up a quartic function so that its derivative is the part of the function without the quartic term. Repeat part a for your quartic function. Describe what you notice.

    d) For a function of any degree, is it always possible to construct a function so that its derivative is the part of the function without the highest degree term? Explain.

Cumulative Review (Chapters 1–3)

1. Draw the graph illustrated by the data below.

a) Determine the average rate of change between $x = 2$ and $x = 4$.

b) Estimate the instantaneous rate of change of y with respect to x when $x = 4$.

Illustrate these rates of change on your sketch.

x	0	2	4	6	8
y	0	2	10	14	16

2. Consider the function $y = \sqrt{8 - x}$.

a) Graph the function for $-1 \leq x \leq 8$.

b) Calculate the average rate of change of the function between $x = 1$ and $x = 4$.

c) Estimate the instantaneous rate of change of y with respect to x when $x = 4$.

d) Explain geometrically the meaning of your answers in parts b and c.

3. For each function $y = f(x)$, sketch the derivative of y as a function of x.

a)

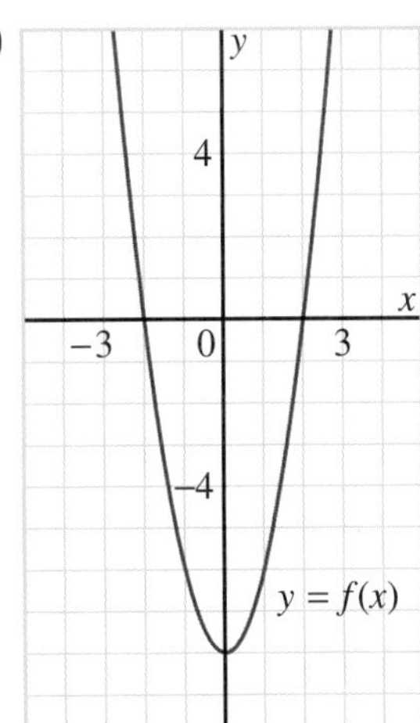

b)

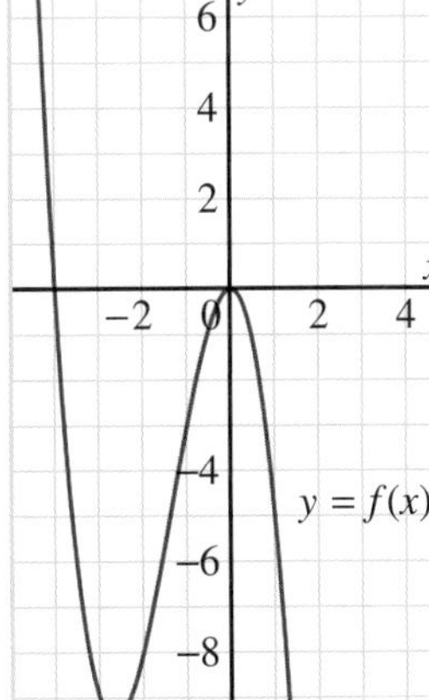

c)

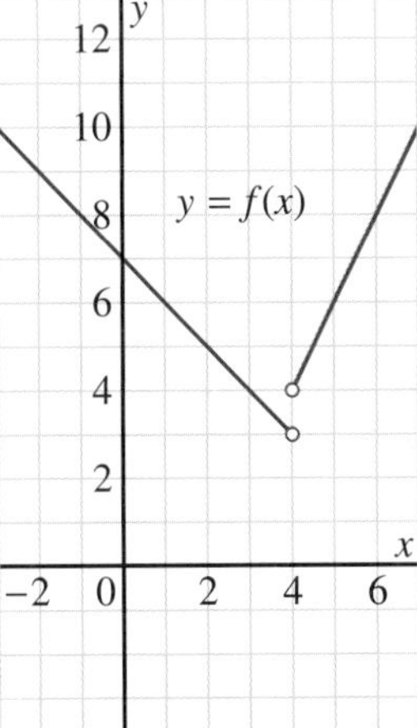

4. Determine the equation of the tangent to the graph of $y = \frac{3}{x}$ at the point where $x = -1$.

5. Use the first-principles definition to determine the derivative of each function.

a) $f(x) = 4x^2$ **b)** $f(x) = \frac{x}{3x - 1}$ **c)** $f(x) = \sqrt{1 - 4x}$

6. At what point(s) on the graph of $y = 4x^3 - x^2$ does the tangent have a slope of 4?

7. Use the graphical meaning of the derivative to explain how to determine the derivative of $y = 4x^3 - 8$ from the derivative of $y = x^3$.

8. For each derivative function $y = f'(x)$, sketch a possible graph of the function $y = f(x)$.

a) 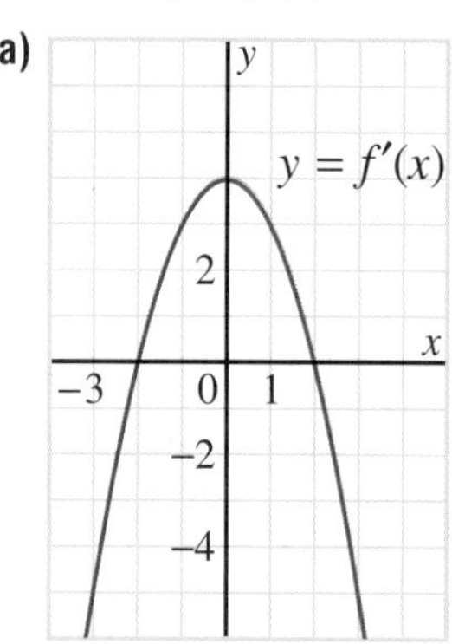

b) 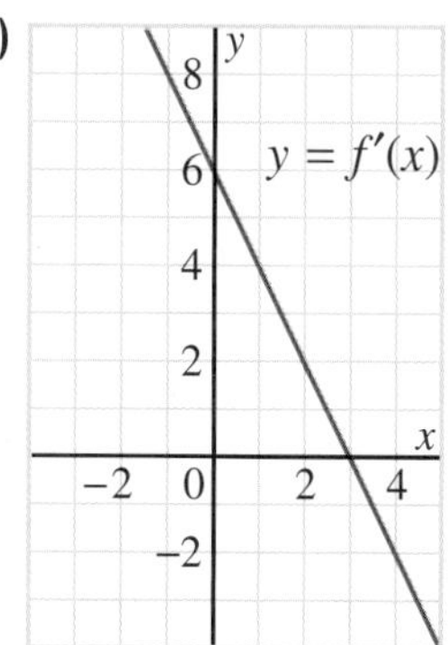

c) 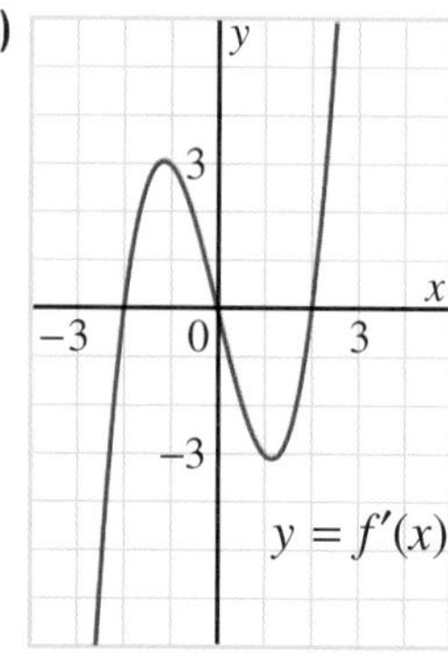

9. A toy rocket is fired vertically into the air from a launching pad on level ground. Its height, h metres, above the ground at any time, t seconds, is given by $h(t) = -4.9t^2 + 49t + 1$.

a) Determine $\frac{dh}{dt}$ and explain what it means.

b) Determine the maximum height reached by the rocket.

c) Determine the average speed of the rocket during the first 3 s of motion.

10. Solve the inequality $x^4 - 9x^2 - 4x + 12 < 0$.

11. Determine the equation of a polynomial function that has roots $3, 2 - \sqrt{3}i, 2 + \sqrt{3}i$.

12. When $2x^6 - 4kx^3 - 3$ is divided by $x + 1$, the remainder is 1. Determine the value of k.

13. For the graph of each polynomial function, determine the intercepts, critical points, points of inflection, and intervals of increase, decrease, and concavity; then sketch the graph.

a) $f(x) = x^3 + 3x^2 - 9x - 27$

b) $f(x) = -x^4 + 18x^2 - 32$

14. A particle moves along the x-axis so that its position is given by $s(t) = t^3 - 5t^2 + 3t - 2$, where s is measured in metres and t in seconds.

a) Determine the velocity and acceleration at any time t.

b) When is the particle at rest?

c) When is the particle accelerating? Give reasons for your answer.

d) Determine the total distance travelled during the first 5 s of motion.

e) What is the average velocity of the particle during the first 4 s of motion?

Calculating Derivatives Using Differentiation Rules

4

Curriculum Expectations

By the end of this chapter, you will:

- Identify composition as an operation in which two functions are applied in succession.
- Demonstrate an understanding that the composition of two functions exists only when the range of the first function overlaps the domain of the second.
- Determine the composition of two functions expressed in function notation.
- Decompose a given composite function into its constituent parts.
- Describe the effect of the composition of inverse functions.
- Justify the constant, power, sum and difference, product, quotient, and chain rules for determining derivatives.
- Determine the derivatives of polynomial and rational functions, using the constant, power, sum and difference, product, quotient, and chain rules for determining derivatives.
- Determine derivatives, using implicit differentiation in simple cases.
- Determine the equation of the tangent to the graph of a function, or of a conic.
- Solve problems of rates of change drawn from a variety of applications involving polynomial and rational functions.

Necessary Skills

1. Review: Algebraic Operations

In this chapter, you will simplify algebraic expressions. The following examples and exercises illustrate the types of simplification required.

Example 1

Simplify.

a) $2x(5x - 4) + (x^2 + 1)(5)$

b) $(6x)(x^2 + 1) + (3x^2 + 2)(2x)$

c) $(3x^2)(x + 1)^2 + 2x^3(x + 1)$

Solution

a) $2x(5x - 4) + (x^2 + 1)(5)$

Expand.

$$\begin{aligned} 2x(5x - 4) + (x^2 + 1)(5) &= 10x^2 - 8x + 5x^2 + 5 \\ &= 15x^2 - 8x + 5 \end{aligned}$$

b) $(6x)(x^2 + 1) + (3x^2 + 2)(2x)$

Remove $2x$ as a common factor from each term.

$$\begin{aligned} (6x)(x^2 + 1) + (3x^2 + 2)(2x) &= 2x[3(x^2 + 1) + (3x^2 + 2)] \\ &= 2x[3x^2 + 3 + 3x^2 + 2] \\ &= 2x(6x^2 + 5) \end{aligned}$$

c) $(3x^2)(x + 1)^2 + 2x^3(x + 1)$

Remove $x^2(x + 1)$ as a common factor.

$$\begin{aligned} (3x^2)(x + 1)^2 + 2x^3(x + 1) &= x^2(x + 1)[3(x + 1) + 2x] \\ &= x^2(x + 1)(3x + 3 + 2x) \\ &= x^2(x + 1)(5x + 3) \end{aligned}$$

In some exercises, you will simplify rational expressions.

Example 2

Simplify $\frac{(x^3)(-6x) - (4 - 3x^2)(2x^3)}{x^6}$.

Solution

$$\frac{(x^3)(-6x) - (4 - 3x^2)(2x^3)}{x^6}$$

Remove $2x^3$ as a common factor in the numerator.

$$\begin{aligned}\frac{(x^3)(-6x) - (4 - 3x^2)(2x^3)}{x^6} &= \frac{2x^3[(-3x) - (4 - 3x^2)]}{x^6} \\ &= \frac{2x^3(-3x - 4 + 3x^2)}{x^6} \\ &= \frac{2x^3(3x^2 - 3x - 4)}{x^6}\end{aligned}$$

Divide the numerator and denominator by the common factor x^3, where $x \neq 0$.

$$\frac{(x^3)(-6x) - (4 - 3x^2)(2x^3)}{x^6} = \frac{2(3x^2 - 3x - 4)}{x^3}$$

Exercises

1. Factor, then simplify.

a) $(x^2 + 2x)(2) + (2x + 1)(2x + 3)$

b) $(1 + 4x)(3 - 2x) + (3x - x^2)(4)$

c) $(2x^2 - 1)(4x) + (3 + 2x^2)(4x)$

d) $(x^3 - 3)(6x^2) + (1 + 2x^3)(3x^2)$

e) $(4x^3)(x - 1) + (x - 1)^2(6x^2)$

f) $(2 - x)^2(2x) - (2x^2)(2 - x)$

2. Factor, then simplify.

a) $(2x - x^2)(x + 1) + (x + 1)^2(1 - x)$

b) $(3 - 2x)^2(2x) - (4x^2)(3 - 2x)$

c) $(2x - x^2)(2x - 3) + (x^2 - 3x)(2 - 2x)$

d) $(3x^2 + 2x + 1)(-2x - 3) + (-x^2 - 3x + 5)(6x + 2)$

e) $(2x^2 - x + 2)(1 - 6x) + (x - 3x^2)(4x - 1)$

f) $(2x^2 - 3x)(4x - 3) + (2x^2 - 3x)(4x - 3)$

3. Simplify.

a) $\frac{(2x^2)(3x^2) - (3 + x^3)(4x)}{4x^4}$

b) $\frac{(2 + x^2)(2x + 1) - (x^2 + x)(2x)}{(2 + x^2)^2}$

c) $\frac{(x - 1)(2x) - (3 + x^2)}{(x - 1)^2}$

d) $\frac{(x^2 - 2x)(2x) - (x^2 + 3)(2x - 2)}{(x^2 - 2x)^2}$

2. Review: Inverse of a Function

Remember how to determine the inverse of a function $y = f(x)$.

- Solve the equation of the function for x in terms of y.
- Interchange x and y.

On a graph, the inverse of the function $y = f(x)$ is obtained by reflecting its graph in the line $y = x$. The result is not necessarily a function, because it might not satisfy the vertical line test. When the result is a function, it is represented by $y = f^{-1}(x)$.

Example 1

Determine the equation of the inverse of each function. Then graph each function and its inverse. Use both the equation and the graph to explain whether the inverse is a function.

a) $y = 4x - 2$ **b)** $y = x^2$

Solution

a) $y = 4x - 2$

Solve the equation for x.

$x = \dfrac{y+2}{4}$

Interchange x and y.

$y = \dfrac{x+2}{4}$

The inverse is a function because there is only one value of y for each value of x. The graph of $y = \dfrac{x+2}{4}$ satisfies the vertical line test.

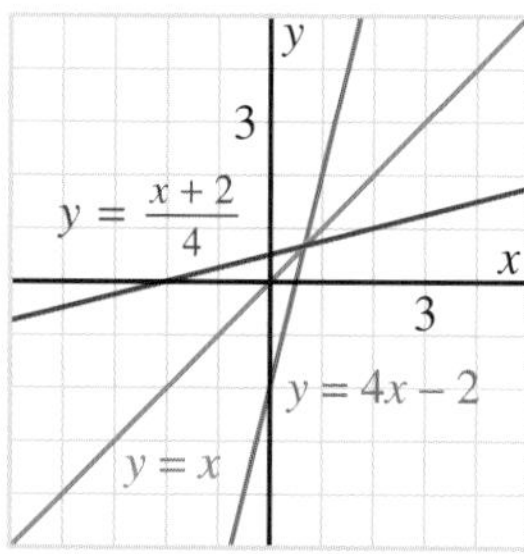

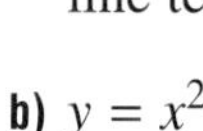

b) $y = x^2$

Solve the equation for x.

$x = \pm\sqrt{y}$

Interchange x and y.

$y = \pm\sqrt{x}, x \geq 0$

The inverse is not a function because there is more than one value of y for each value of $x > 0$. The graph of $y = \pm\sqrt{x}$ does not satisfy the vertical line test.

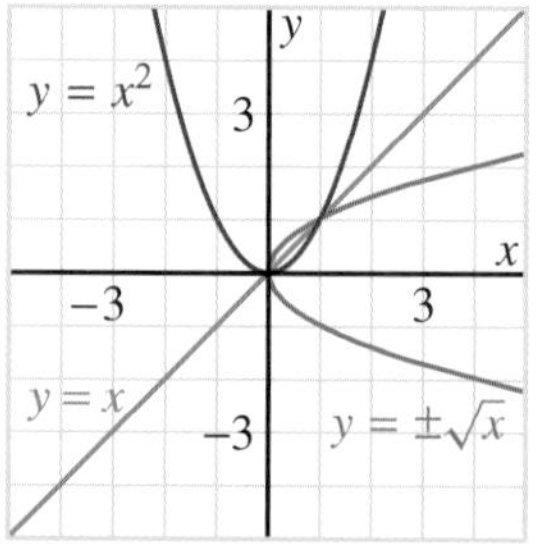

Example 2

Consider $f(x) = \frac{2x+1}{x-1}$. Determine $f^{-1}(x)$.

Solution

Let $y = \frac{2x+1}{x-1}$, where $x \neq 1$.

Solve the equation for x.

$$xy - y = 2x + 1$$
$$xy - 2x = y + 1$$
$$x(y - 2) = y + 1$$
$$x = \frac{y+1}{y-2}$$

Interchange x and y.

$y = \frac{x+1}{x-2}, x \neq 2$

The inverse is a function because there is only one value of y for each value of x for which the expression is defined.

Therefore, $f^{-1}(x) = \frac{x+1}{x-2}$

Exercises

1. Determine the inverse of each function. Is the inverse a function? Explain.

a) $y = 3 - x$ **b)** $y = \frac{1}{2}x + 5$ **c)** $y = x^2 - 4$

d) $y = \frac{1}{x+1}$ **e)** $y = \frac{x+1}{x-1}$ **f)** $y = \frac{3}{x^2+1}$

2. For each function in exercise 1a, b, and c:

i) Graph the function and its inverse on the same grid.

ii) Use the graph to explain whether the inverse is a function.

3. For each function, determine $f^{-1}(x)$. Explain why f^{-1} is a function.

a) $f(x) = 2 - 3x$ **b)** $f(x) = \frac{4}{x} + 2$ **c)** $f(x) = \frac{x}{2x-1}$

4.1 The Product Rule

Recall that the derivative of a function f is a function, f', that represents the instantaneous rate of change of f at x, when f' exists. Since the derivative also represents the slope of the tangent to the graph of a function, it is a powerful tool for analysing functions. We developed some differentiation rules in Chapter 2, which we used in Chapter 3 to analyse polynomial functions. To analyse other kinds of functions, we shall develop more rules for differentiating functions in this chapter.

From Chapter 2, we know that the derivative of the sum or the difference of two functions is the sum or the difference of their derivatives. That is,

$$\frac{d}{dx}[f(x) \pm g(x)] = f'(x) \pm g'(x)$$

We shall now consider whether there is a similar rule for the product of two functions. That is, we shall investigate whether the derivative of the product of two functions is equal to the product of their derivatives.

Consider the functions $f(x) = x^2 + 1$ and $g(x) = 2x - 1$. Their product is:

$$f(x)g(x) = (x^2 + 1)(2x - 1)$$

We can expand the product on the right side, then differentiate the resulting polynomial.

$$f(x)g(x) = 2x^3 - x^2 + 2x - 1$$

$$\frac{d}{dx}[f(x)g(x)] = 6x^2 - 2x + 2$$

The derivative of the product of the two functions is $6x^2 - 2x + 2$.

The derivatives of the two functions are:

$$f'(x) = 2x \quad \text{and} \quad g'(x) = 2$$

The product of the derivatives is $(2x)(2) = 4x$.

The derivative of the product is *not* the product of the derivatives.

If we can determine a rule for differentiating the product of two functions, we could use it to differentiate $(x^2 + 1)(2x - 1)$ without expanding the product.

We can use the first-principles definition to determine a rule for differentiating the product of two functions.

Let the functions be $f(x)$ and $g(x)$.
According to the first-principles definition,

$$\frac{d}{dx}[f(x)g(x)] = \lim_{h \to 0} \frac{f(x + h)g(x + h) - f(x)g(x)}{h}$$

We add and subtract $f(x)g(x + h)$ in the numerator.

$$\frac{d}{dx}[f(x)g(x)] = \lim_{h \to 0} \frac{f(x+h)g(x+h) + f(x)g(x+h) - f(x)g(x+h) - f(x)g(x)}{h}$$

Rearrange the numerator.

$$\frac{d}{dx}[f(x)g(x)] = \lim_{h \to 0} \frac{f(x+h)g(x+h) - f(x)g(x+h) + f(x)g(x+h) - f(x)g(x)}{h}$$

We remove a common factor $g(x + h)$ from the first two terms and a common factor $f(x)$ from the last two terms.

$$\frac{d}{dx}[f(x)g(x)] = \lim_{h \to 0}\left[\frac{f(x+h) - f(x)}{h}g(x+h) + f(x)\frac{g(x+h) - g(x)}{h}\right]$$

As h gets closer and closer to 0, the expression in the square brackets gets closer and closer to $f'(x)g(x) + f(x)g'(x)$.

Therefore, $\frac{d}{dx}[f(x)g(x)] = f'(x)g(x) + f(x)g'(x)$

This is the *product rule*.

Take Note

Product Rule

$$\frac{d}{dx}[f(x)g(x)] = f'(x)g(x) + f(x)g'(x)$$

We can use the product rule to differentiate the product of the functions $f(x)$ and $g(x)$ on page 238.

Example 1

Determine $\frac{dy}{dx}$ for $y = (x^2 + 1)(2x - 1)$.

Solution

$y = (x^2 + 1)(2x - 1)$

This is a product of functions $f(x) = x^2 + 1$ and $g(x) = 2x - 1$.

Use the product rule: $\frac{d}{dx}[f(x)g(x)] = f'(x)g(x) + f(x)g'(x)$

$$\begin{aligned}\frac{dy}{dx} &= \left[\frac{d}{dx}(x^2 + 1)\right](2x - 1) + (x^2 + 1)\left[\frac{d}{dx}(2x - 1)\right]\\ &= (2x)(2x - 1) + (x^2 + 1)(2)\\ &= 2[x(2x - 1) + x^2 + 1]\\ &= 2(2x^2 - x + x^2 + 1)\\ &= 2(3x^2 - x + 1)\end{aligned}$$

The derivative in *Example 1* is the same as the derivative we obtained on page 238 by expanding the product, then differentiating the polynomial.

Example 2

Differentiate $y = (3x + 4)(x^2 + 5x + 2)$.

Solution

$y = (3x + 4)(x^2 + 5x + 2)$

This is a product of functions $f(x) = 3x + 4$ and $g(x) = x^2 + 5x + 2$.

Use the product rule.

$$\begin{aligned}\frac{dy}{dx} &= \left[\frac{d}{dx}(3x + 4)\right](x^2 + 5x + 2) + (3x + 4)\left[\frac{d}{dx}(x^2 + 5x + 2)\right] \\ &= 3(x^2 + 5x + 2) + (3x + 4)(2x + 5) \\ &= 3x^2 + 15x + 6 + 6x^2 + 23x + 20 \\ &= 9x^2 + 38x + 26\end{aligned}$$

Example 3

Determine the derivative of $y = (x^3 + x^2)^2$.

Solution

$y = (x^3 + x^2)^2$

Rewrite as a product.

$y = (x^3 + x^2)(x^3 + x^2)$

This is a product of functions $f(x) = x^3 + x^2$ and $g(x) = x^3 + x^2$.

Use the product rule.

$$\begin{aligned}\frac{dy}{dx} &= \left[\frac{d}{dx}(x^3 + x^2)\right](x^3 + x^2) + (x^3 + x^2)\left[\frac{d}{dx}(x^3 + x^2)\right] \\ &= (3x^2 + 2x)(x^3 + x^2) + (x^3 + x^2)(3x^2 + 2x) \\ &= 2(3x^2 + 2x)(x^3 + x^2)\end{aligned}$$

In *Example 3*, we could have expanded the product $2(3x^2 + 2x)(x^3 + x^2)$ and obtained $\frac{dy}{dx} = 6x^5 + 10x^4 + 4x^3$. Or, we could have factored the product and obtained $\frac{dy}{dx} = 2x^3(3x + 2)(x + 1)$. However, we left the derivative as $\frac{dy}{dx} = 2(3x^2 + 2x)(x^3 + x^2)$ because the given function $y = (x^3 + x^2)^2$ is not in expanded form or in fully factored form.

Example 4

Differentiate $y = x^3 \sin x$.

Solution

$y = x^3 \sin x$

This is a product of functions $f(x) = x^3$ and $g(x) = \sin x$.

Use the product rule.

$$\frac{dy}{dx} = \left[\frac{d}{dx}(x^3)\right](\sin x) + (x^3)\left[\frac{d}{dx}(\sin x)\right]$$
$$= 3x^2 \sin x + x^3 \cos x$$
$$= x^2(3 \sin x + x \cos x)$$

Discuss

Describe another way to differentiate the functions in *Examples 2* and *3*. Can we differentiate the function in *Example 4* another way? Explain.

Here is an example of the product rule applied to a problem from the field of economics.

Example 5

An orange grove contains 150 trees producing an average yield of 300 oranges per tree. The owner is expanding the grove at the rate of 13 trees per year. Due to improved farming methods, the average annual yield is increasing at a rate of 10 oranges per tree.

a) Write an equation to represent the annual production, P, as a function of n, the number of years from now.

b) Determine the current rate of increase in the annual production of oranges.

Solution

a) This year, year 0, the annual production is (150)(300).
Next year, year 1, the annual production will be (150 + 13)(300 + 10).
Next year, year 2, the annual production will be (150 + 13(2))(300 + 10(2)).
⋮
In year n, the annual production will be $(150 + 13n)(300 + 10n)$.
The function is $P = (150 + 13n)(300 + 10n)$.

b) The rate of increase in the annual production is $\frac{dP}{dn}$.

Since $(150 + 13n)(300 + 10n)$ is a product, use the product rule to differentiate.

$$\begin{aligned}\frac{dP}{dn} &= 13(300 + 10n) + (150 + 13n)(10)\\ &= 3900 + 130n + 1500 + 130n\\ &= 5400 + 260n\end{aligned}$$

The current rate of increase is the value of $\frac{dP}{dn}$ when $n = 0$.

So, the current rate of increase is 5400 oranges per year.

In *Example 5*, n can only take whole number values, but we can replace $P(n)$ by a smooth approximating function so that the derivative, $\frac{dP}{dn}$, exists.

4.1 Exercises

1. Determine the derivative of each product.

   a) $y = 3x(2x + 1)$ b) $y = x^2(3x + 2)$ c) $y = x(3x^2 + 1)$

   d) $y = 5x(1 - 2x)$ e) $y = x^2(3 - x)$ f) $y = 2x(3x^2 + 6x + 5)$

2. Differentiate each function.

   a) $y = (x + 2)(x + 3)$ b) $y = (2x - 1)(3x + 2)$

   c) $y = (1 - 3x)(2x + 6)$ d) $y = (3x^2 + 2)(2x + 1)$

   e) $y = (6x - 5)(5x - 6)$ f) $y = (x^2 - 2x + 1)(x^2 - 1)$

3. a) Determine the derivative of each product.

   i) $y = x(x + 1)$ ii) $y = (x + 1)(x + 2)$

   iii) $y = (x + 2)(x + 3)$ iv) $y = (x + 3)(x + 4)$

   b) Use the pattern in part a to predict the derivative of the product $y = (x + k)(x + k + 1)$, where k is a positive integer.

4. **Communication** Make up an example, different from that on page 238, to show that the derivative of the product of two functions is not the product of their derivatives. Justify your choice of example.

5. **Knowledge/Understanding** Differentiate each function.

a) $y = 5x^2(2x^2 + 3)$
b) $y = x^2(1 - x^2)$
c) $y = (x^2 + 1)(2x^2 - x)$
d) $y = (3x^2 + 5)(2x^3 + 3x^2 + x - 5)$
e) $y = (5 - 3x - 4x^2)(3x^2 + 6x + 5)$
f) $y = x(x + 1)(x - 1)$

6. Determine the derivative of each product.

a) $y = (x^2 + 3x)^2$
b) $y = (2x - x^2)^2$
c) $y = (5x^3 - 2x^2)^2$
d) $y = (3x^2 + 2x - 1)^2$

7. To differentiate $y = x^5$, you can write x^5 as a product of two powers of x, then use the product rule. Show that you obtain $y' = 5x^4$ regardless of which two powers of x are used.

8. Consider $P(x) = (ax + b)(cx + d)$, where a, b, c, and d are constants. Show that $P'(x) = 2acx + ad + bc$ in two ways:

a) by using the product rule
b) by expanding the product first

9. Consider $P(x) = l(x)q(x)$, where $l(x) = mx + n$ and $q(x) = ax^2 + bx + c$. Show that $P'(x) = 3amx^2 + 2(bm + an)x + (bn + cm)$ in two ways:

a) by using the product rule
b) by expanding the product first

10. Determine the equation of the tangent to the graph of $y = x^2(1 - 2x^2)$ at the point C(1, –1).

11. Determine the coordinates of the points on the graph of $y = (2x^2 - 1)^2$ where the tangent is horizontal.

12. The reaction, r, of the body to a dose of medicine may be represented by an equation such as $r = m^2\left(\frac{k}{2} - \frac{m}{3}\right)$, where k is a positive constant, and m is the volume of medicine in millilitres absorbed into the blood. Determine $\frac{dr}{dm}$, which is the sensitivity of the body to the medicine.

13. a) Use the product rule to differentiate each function.

i) $y = 5x$
ii) $y = 5x^2$
iii) $y = 5\left(\frac{1}{x}\right)$

b) Use the product rule to show that $\frac{d}{dx}[cf(x)] = cf'(x)$.

14. Use the product rule to differentiate each function. Explain the results.

a) $y = (x - 1)(x + 1)$
b) $y = (x - 1)(x^2 + x + 1)$
c) $y = (x - 1)(x^3 + x^2 + x + 1)$
d) $y = (x - 1)(x^4 + x^3 + x^2 + x + 1)$

15. Use the product rule to differentiate each function. Explain the results.

a) $y = (1 + x)(1 - x)$
b) $y = (1 + x)(1 - x + x^2)$
c) $y = (1 + x)(1 - x + x^2 - x^3)$
d) $y = (1 + x)(1 - x + x^2 - x^3 + x^4)$

16. **a)** Use the product rule to differentiate $y = (x + 1)(x^2 + 2)$.

b) Use the result of part a and the product rule to differentiate $y = (x + 1)(x^2 + 2)(x^3 + 3)$.

c) Use the result of part b to differentiate $y = (x + 1)(x^2 + 2)(x^3 + 3)(x^4 + 4)$.

17. **Thinking/Inquiry/Problem Solving** The product rule can be expressed as $(fg)' = f'g + fg'$.

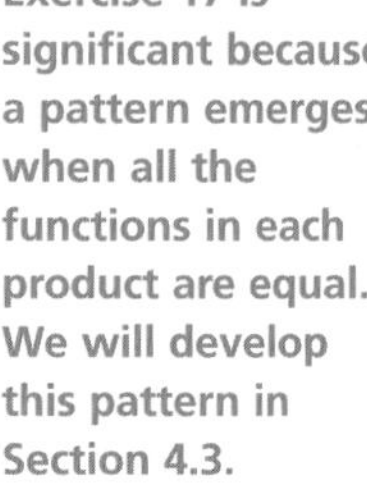

Exercise 17 is significant because a pattern emerges when all the functions in each product are equal. We will develop this pattern in Section 4.3.

a) By writing $fgh = (fg)h$, and using the product rule twice, determine a similar expression for the derivative of the product of 3 functions.

b) Describe any patterns in the result.

c) Write a similar expression for the derivative of the product of 4 functions, $fghk$, then show that your expression is correct.

d) Describe how you could determine the derivative of the product of any number of differentiable functions.

18. **Application** The volume of a crop, V bushels, is changing according to the function $V(t) = 65t(20 - t)$, where t is the time in weeks, and $0 \le t \le 20$. The price of the crop, P dollars per bushel, changes according to the function $P(t) = 8 - 0.5t$.

a) Write an expression for the value of the crop in dollars, $C(t)$, after t weeks.

b) During which time interval is the value of the crop increasing? Explain.

19. Consider $f(x) = x^2 - 3$ and $g(x) = x + 1$. Determine all values of x for which $f'(x) \times g'(x) = (f(x) \times g(x))'$.

20. Determine the derivative of each product.

a) $y = \sin x \cos x$ **b)** $y = x^2 \sin x$ **c)** $y = -x^3 \cos x$

d) $y = \sin^2 x$ **e)** $y = 4x \sin x$ **f)** $y = 2 \sin x \cos x$

21. Determine the derivative of each product.

a) $y = x\sqrt{x}$ **b)** $y = (x + 1)\sqrt{x}$ **c)** $y = -x^2\sqrt{x}$

22. On page 239, we developed the product rule by adding and subtracting the term $f(x)g(x + h)$. Use the development on page 239 as a guide. Develop the product rule by adding and subtracting the term $f(x + h)g(x)$.

23. A *normal* is a line that is perpendicular to a tangent. Determine the coordinates of the point(s) on the graph of $y = \left(\frac{1}{8}x - 4\right)^3$ where the normal has slope $-\frac{1}{6}$.

4.2 The Quotient Rule

The rule for the derivative of a quotient of functions follows from the product rule.

Consider the function $Q(x) = \frac{f(x)}{g(x)}$, where $g(x) \neq 0$.

Thus, $Q(x)g(x) = f(x)$

To find a formula that relates $Q'(x)$ to $f'(x)$ and $g'(x)$, differentiate each side. Use the product rule on the left side.
$Q'(x)g(x) + Q(x)g'(x) = f'(x)$

Solve for $Q'(x)$.

$$Q'(x)g(x) = f'(x) - Q(x)g'(x)$$
$$Q'(x) = \frac{f'(x) - Q(x)g'(x)}{g(x)}$$

Rewrite $Q(x)$ as $\frac{f(x)}{g(x)}$.

$$Q'(x) = \frac{f'(x) - \frac{f(x)}{g(x)}g'(x)}{g(x)}$$

Simplify by multiplying the numerator and denominator by $g(x)$.

$$Q'(x) = \frac{f'(x) - \frac{f(x)}{g(x)}g'(x)}{g(x)} \times \frac{g(x)}{g(x)}$$
$$= \frac{f'(x)g(x) - f(x)g'(x)}{[g(x)]^2}$$

We have shown that if $Q(x) = \frac{f(x)}{g(x)}$, then $Q'(x) = \frac{f'(x)g(x) - f(x)g'(x)}{[g(x)]^2}$.

This is the *quotient rule*. As with a product, the derivative of a quotient of functions is *not* the quotient of the derivatives of the functions.

Take Note

Quotient Rule

$$\frac{d}{dx}\left[\frac{f(x)}{g(x)}\right] = \frac{f'(x)g(x) - f(x)g'(x)}{[g(x)]^2}$$

Discuss

How is the quotient rule similar to the product rule? How is it different from the product rule? Explain.

We could have proved the quotient rule using the first-principles definition. The proof is similar to that for the product rule and has been left to the exercises. (See exercise 20, page 249).

Example 1

Determine the derivative of the function $y = \frac{2x^2 - 3}{x^2}$.

Solution

$y = \frac{2x^2 - 3}{x^2}$

This is a quotient of functions $f(x) = 2x^2 - 3$ and $g(x) = x^2$.

Use the quotient rule: $\frac{d}{dx}\left[\frac{f(x)}{g(x)}\right] = \frac{f'(x)g(x) - f(x)g'(x)}{[g(x)]^2}$

$$\frac{dy}{dx} = \frac{\left[\frac{d}{dx}(2x^2 - 3)\right](x^2) - (2x^2 - 3)\left[\frac{d}{dx}(x^2)\right]}{(x^2)^2}$$

$$= \frac{(4x)(x^2) - (2x^2 - 3)(2x)}{x^4}$$

$$= \frac{4x^3 - 4x^3 + 6x}{x^4}$$

$$= \frac{6x}{x^4} \qquad \text{Divide by } x\text{, where } x \neq 0.$$

$$= \frac{6}{x^3}$$

Discuss

We used the quotient rule to determine the derivative of $y = \frac{2x^2 - 3}{x^2}$.

Describe two other ways to differentiate $y = \frac{2x^2 - 3}{x^2}$.

Which of the three methods is easiest? Explain.

Example 2

Differentiate $y = \frac{x^2 - 1}{x^2 + 1}$.

Solution

$y = \frac{x^2 - 1}{x^2 + 1}$

This is a quotient of functions $f(x) = x^2 - 1$ and $g(x) = x^2 + 1$.
Use the quotient rule.

$$\frac{dy}{dx} = \frac{\left[\frac{d}{dx}(x^2 - 1)\right](x^2 + 1) - (x^2 - 1)\left[\frac{d}{dx}(x^2 + 1)\right]}{(x^2 + 1)^2}$$

$$= \frac{(2x)(x^2 + 1) - (x^2 - 1)(2x)}{(x^2 + 1)^2}$$

$$= \frac{2x(x^2 + 1 - x^2 + 1)}{(x^2 + 1)^2}$$

$$= \frac{4x}{(x^2 + 1)^2}$$

Discuss

Consider this line from the second step in the solution:

$\frac{(2x)(x^2 + 1) - (x^2 - 1)(2x)}{(x^2 + 1)^2}$

Why can we not divide the first term in the numerator and the denominator by $x^2 + 1$?

A function that has an equation of the form $y = \frac{1}{f(x)}$ is a *reciprocal function.*

We can use the quotient rule to differentiate a reciprocal function.

Example 3

Determine the derivative of the function $y = \dfrac{1}{3x^2 + 4}$.

Solution

$y = \dfrac{1}{3x^2 + 4}$

This is a quotient of functions $f(x) = 1$ and $g(x) = 3x^2 + 4$.
Use the quotient rule.

$$\frac{dy}{dx} = \frac{\left[\frac{d}{dx}(1)\right](3x^2 + 4) - (1)\left[\frac{d}{dx}(3x^2 + 4)\right]}{(3x^2 + 4)^2}$$

$$= \frac{(0)(3x^2 + 4) - (1)(6x)}{(3x^2 + 4)^2}$$

$$= \frac{-6x}{(3x^2 + 4)^2}$$

4.2 Exercises

A

1. Determine the derivative of each quotient.

a) $y = \dfrac{x}{x + 1}$ **b)** $y = \dfrac{x^2}{x + 1}$ **c)** $y = \dfrac{x^3}{x + 1}$

d) $y = \dfrac{-x}{x + 1}$ **e)** $y = \dfrac{-x^2}{x + 1}$ **f)** $y = \dfrac{-x^3}{x + 1}$

2. Differentiate each function.

a) $y = \dfrac{x - 1}{x}$ **b)** $y = \dfrac{x - 1}{x^2}$ **c)** $y = \dfrac{x - 1}{x^3}$

d) $y = \dfrac{x - 1}{-x}$ **e)** $y = \dfrac{x - 1}{-x^2}$ **f)** $y = \dfrac{x - 1}{-x^3}$

3. Determine the derivative of each function.

a) $y = \dfrac{x}{x + 1}$ **b)** $y = \dfrac{x}{x + 3}$ **c)** $y = \dfrac{x}{3x + 1}$

d) $y = \dfrac{x + 2}{x + 3}$ **e)** $y = \dfrac{x - 1}{x - 2}$ **f)** $y = \dfrac{x}{5x + 2}$

✓ 4. Differentiate each function.

a) $y = \dfrac{3x}{2x + 1}$ **b)** $y = \dfrac{x}{3x + 2}$ **c)** $y = \dfrac{-x}{3x + 1}$

d) $y = \dfrac{x + 2}{x - 3}$ **e)** $y = \dfrac{2x - 1}{3x + 2}$ **f)** $y = \dfrac{6x - 5}{5x - 6}$

5. **a)** Determine the derivative of each quotient.

   i) $y = \frac{x}{x+1}$ **ii)** $y = \frac{x+1}{x+2}$ **iii)** $y = \frac{x+2}{x+3}$ **iv)** $y = \frac{x+3}{x+4}$

   b) Use the pattern in part a to predict the derivative of the quotient $y = \frac{x+k}{x+k+1}$, where k is a positive integer.

6. **Communication** Look at the pattern in the derivatives in exercise 5. Is there a similar pattern for the derivative of $y = \frac{x+k}{x+k+1}$, where k is a negative integer? Investigate and explain what you discovered.

7. Determine the derivative of each reciprocal function.

   a) $y = \frac{1}{x^2+2}$ **b)** $y = \frac{1}{1-x^2}$ **c)** $y = -\frac{1}{x-1}$

   d) $y = \frac{1}{x^2+x}$ **e)** $y = \frac{1}{x-x^2}$ **f)** $y = \frac{1}{x^2+2x+3}$

8. Differentiate each rational function.

   a) $y = \frac{x}{x^2+2}$ **b)** $y = \frac{2}{1-x^2}$ **c)** $y = \frac{-x^2}{x-1}$

   d) $y = \frac{-3x}{x^2+x}$ **e)** $y = \frac{x^3}{x-x^2}$ **f)** $y = \frac{2x^2}{x^2+2x+3}$

9. **Knowledge/Understanding** Differentiate each quotient.

   a) $y = \frac{1}{x^2+x^3}$ **b)** $y = \frac{-3x}{5-x}$ **c)** $y = \frac{x^2}{1-2x^2}$

10. Make up an example to show that the derivative of the quotient of two functions is not equal to the quotient of their derivatives.

11. **a)** Write the power x^3 as a quotient of two powers in three different ways.

    b) Differentiate each quotient in part a. Describe any patterns in the solutions.

12. **Application**

    a) Differentiate each reciprocal function.

    i) $y = \frac{1}{x}$ **ii)** $y = \frac{1}{x^2}$ **iii)** $y = \frac{1}{x^3}$ **iv)** $y = \frac{1}{x^4}$ **v)** $y = \frac{1}{x^5}$

    b) Explain how you could use the results of part a to differentiate each function.

    i) $y = x^{-1}$ **ii)** $y = x^{-2}$ **iii)** $y = x^{-3}$ **iv)** $y = x^{-4}$ **v)** $y = x^{-5}$

    c) Use the method of part b to differentiate each function.

    i) $y = 2x^{-2}$ **ii)** $y = \frac{1}{2}x^{-3}$ **iii)** $y = -5x^{-4}$ **iv)** $y = -\frac{3}{2}x^{-6}$

13. Consider the function $y = \frac{1}{f(x)}$, where $f(x) \neq 0$. Determine an expression for $\frac{dy}{dx}$ in terms of $f(x)$ and $f'(x)$.

Exercise 13 is significant because it provides a rule for the derivative of the reciprocal of a function. The rule is a special case of a more general rule, called the chain rule. **See *The Big Picture*, page 271.**

✓ **14.** Determine the equation of the tangent to the graph of $y = \dfrac{x}{x+1}$ (below left) at the point $\left(1, \frac{1}{2}\right)$.

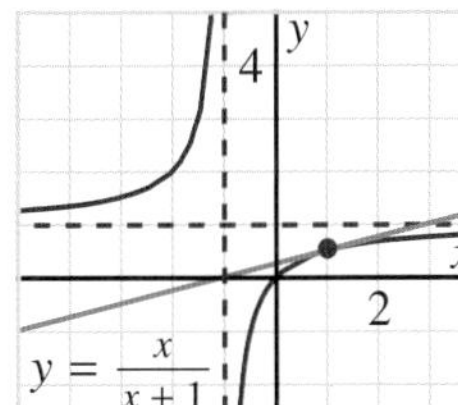

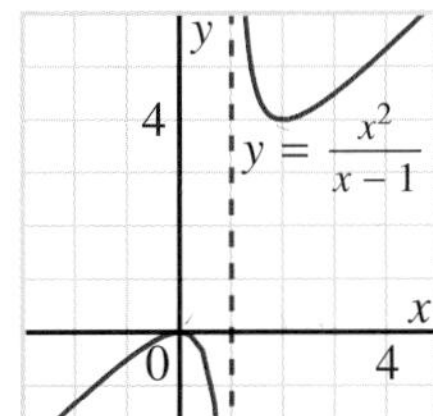

✓ **15.** Determine the coordinates of the point(s) on the graph of $y = \dfrac{x^2}{x-1}$ (above right) where the tangent(s) have slope 0.

✓ **16. Thinking/Inquiry/Problem Solving**

a) Determine the equation of the tangent to the graph of $y = \dfrac{x}{x-1}$ at the point $\left(a, \dfrac{a}{a-1}\right)$.

b) Where does this tangent intersect the x-axis? Explain.

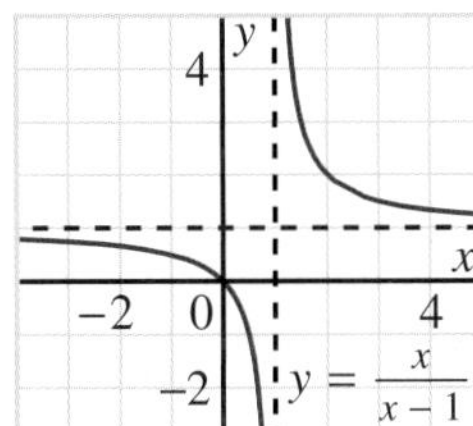

17. Consider $P = fg$ and $Q = \dfrac{f}{g}$, where f and g are functions of x. According to the product and quotient rules, $P' = f'g + fg'$ and $Q' = \dfrac{f'g - fg'}{g^2}$.

The expressions for P' and Q' are similar. The numerator in Q' contains the same terms as P' but the signs connecting the terms are opposite. Q' contains a denominator that is not present in P'. This suggests that there may be a way to write these rules symmetrically.

a) Show that $\dfrac{P'}{P} = \dfrac{f'}{f} + \dfrac{g'}{g}$.

b) Show that $\dfrac{Q'}{Q} = \dfrac{f'}{f} - \dfrac{g'}{g}$.

18. Determine the derivative of each function.

a) $y = \dfrac{3x}{\sin x}$ **b)** $y = \dfrac{\cos x}{3x}$ **c)** $y = \dfrac{\sin x}{x}$

d) $y = \dfrac{\sin x}{\cos x}$ **e)** $y = \dfrac{\cos x}{\sin x}$ **f)** $y = \dfrac{1}{\cos x}$

19. Differentiate each function.

a) $y = \dfrac{\sqrt{x}+1}{2\sqrt{x}-1}$ **b)** $y = \dfrac{\sqrt{x}}{x-2}$ **c)** $y = \dfrac{\sin x}{3-\sqrt{x}}$

20. Use the first-principles definition of a derivative to show that the derivative of $\dfrac{f(x)}{g(x)}$ is $\dfrac{f'(x)g(x) - f(x)g'(x)}{[g(x)]^2}$.

4.3 The Power of a Function Rule

The differentiation rules we have developed so far allow us to differentiate many polynomial and rational functions. These rules cannot be used to differentiate powers of these functions, such as $y = (3x + 1)^7$ or $y = \frac{1}{(x^3 + 1)^5}$.

To differentiate $y = (3x + 1)^7$, we would either have to expand $(3x + 1)^7$ or repeatedly apply the product rule. Both methods are tedious. We need a more efficient method to differentiate a power of a function.

In Section 4.1, we developed the product rule for a product of two factors. In 4.1 Exercises, exercises 16 and 17, we extended the product rule for more than two factors. The results are summarized below:

$(fg)' = f'g + fg'$

$(fgh)' = f'gh + fg'h + fgh'$

$(fghk)' = f'ghk + fg'hk + fgh'k + fghk'$

This pattern continues.

When the factors in a product are equal, the function is a power. Thus,

$$\begin{aligned}(f^2)' &= f'f + ff' \\ &= 2ff' \\ (f^3)' &= f'ff + ff'f + fff' \\ &= 3f^2f' \\ (f^4)' &= f'fff + ff'ff + fff'f + ffff' \\ &= 4f^3f'\end{aligned}$$

These results form a pattern that suggests a general result.
For any positive integer n,

$$\frac{d}{dx}[f(x)]^n = n[f(x)]^{n-1} \times f'(x)$$

We now have an efficient way to determine the derivative of a power of a function.

Take Note

Power of a Function Rule

For any positive integer n, $\frac{d}{dx}[f(x)]^n = n[f(x)]^{n-1} \times f'(x)$

Example 1

Differentiate each function.

a) $y = (3x + 1)^7$ **b)** $y = -(1 - 3x^2)^3$

Solution

a) $y = (3x + 1)^7$

This is a power of the form $[f(x)]^n$, where $f(x) = 3x + 1$ and $n = 7$.

Use the power of a function rule: $\frac{d}{dx}[f(x)]^n = n[f(x)]^{n-1} \times f'(x)$

$$\begin{aligned}\frac{dy}{dx} &= 7(3x + 1)^6 \frac{d}{dx}(3x + 1)\\ &= 7(3x + 1)^6(3)\\ &= 21(3x + 1)^6\end{aligned}$$

b) $y = -(1 - 3x^2)^3$

This is a power of the form $-[f(x)]^n$, where $f(x) = 1 - 3x^2$ and $n = 3$.

Use the power of a function rule.

$$\begin{aligned}\frac{dy}{dx} &= -3(1 - 3x^2)^2 \frac{d}{dx}(1 - 3x^2)\\ &= -3(1 - 3x^2)^2(-6x)\\ &= 18x(1 - 3x^2)^2\end{aligned}$$

To determine some derivatives, we use both the power of a function rule and the product or quotient rule.

Example 2

Determine the derivative of each function.

a) $y = \left(\frac{x}{x+2}\right)^3$

b) $y = (4x - 1)^2(3x + 1)^3$

Solution

a) $y = \left(\frac{x}{x+2}\right)^3$

This is a power of the form $[f(x)]^n$, where $f(x) = \frac{x}{x+2}$ and $n = 3$.

Use the power of a function rule.

$$\frac{dy}{dx} = 3\left(\frac{x}{x+2}\right)^2 \frac{d}{dx}\left(\frac{x}{x+2}\right)$$

Use the quotient rule to determine $\frac{d}{dx}\left(\frac{x}{x+2}\right)$.

Discuss

Describe another method to determine the derivative in part a.

$$\frac{d}{dx}\left(\frac{x}{x+2}\right) = \frac{\left[\frac{d}{dx}(x)\right](x+2) - (x)\left[\frac{d}{dx}(x+2)\right]}{(x+2)^2}$$
$$= \frac{(1)(x+2) - (x)(1)}{(x+2)^2}$$
$$= \frac{2}{(x+2)^2}$$

Thus, $\frac{dy}{dx} = 3\left(\frac{x}{x+2}\right)^2\left(\frac{2}{(x+2)^2}\right)$

$$= \frac{6x^2}{(x+2)^4}$$

b) $y = (4x-1)^2(3x+1)^3$

This is a product of functions $f(x) = (4x-1)^2$ and $g(x) = (3x+1)^3$.

Use the product rule.

$$\frac{dy}{dx} = \left[\frac{d}{dx}(4x-1)^2\right](3x+1)^3 + (4x-1)^2\left[\frac{d}{dx}(3x+1)^3\right]$$

Use the power of a function rule to determine $\frac{d}{dx}(4x-1)^2$ and $\frac{d}{dx}(3x+1)^3$.

$$\frac{dy}{dx} = [2(4x-1)(4)](3x+1)^3 + (4x-1)^2[3(3x+1)^2(3)]$$
$$= 8(3x+1)^3(4x-1) + 9(3x+1)^2(4x-1)^2$$

To simplify, remove the greatest common factor: $(3x+1)^2(4x-1)$

$$\frac{dy}{dx} = (3x+1)^2(4x-1)[8(3x+1) + 9(4x-1)]$$
$$= (3x+1)^2(4x-1)[24x + 8 + 36x - 9]$$
$$= (3x+1)^2(4x-1)(60x-1)$$

Example 3

Determine the derivative of each function.

a) $y = \sin^2 x$

b) $y = \sin^3 x \cos x$

Solution

a) $y = \sin^2 x$

Write $\sin^2 x$ as $(\sin x)^2$.

Use the power of a function rule.

$$\frac{dy}{dx} = \frac{d}{dx}(\sin x)^2$$
$$= (2\sin x)\left[\frac{d}{dx}(\sin x)\right]$$
$$= 2\sin x \cos x$$

b) $y = \sin^3 x \cos x$

Use the product rule, then the power of a function rule.

$$\frac{dy}{dx} = \left[\frac{d}{dx}(\sin^3 x)\right](\cos x) + (\sin^3 x)\left[\frac{d}{dx}(\cos x)\right]$$
$$= (3\sin^2 x)\left(\frac{d}{dx}(\sin x)\right)(\cos x) + (\sin^3 x)\left[\frac{d}{dx}(\cos x)\right]$$
$$= (3\sin^2 x)(\cos x)(\cos x) + (\sin^3 x)(-\sin x)$$
$$= 3\sin^2 x\cos^2 x - \sin^4 x$$
$$= \sin^2 x(3\cos^2 x - \sin^2 x)$$

We can use the quotient rule to determine the derivative of the reciprocal of a power function, such as $y = \frac{1}{(x^3+1)^5}$.

Example 4

Differentiate $y = \frac{1}{(x^3+1)^5}$.

Solution

$$y = \frac{1}{(x^3+1)^5}$$

This is a quotient of the form $\frac{f(x)}{g(x)}$, where $f(x) = 1$ and $g(x) = (x^3+1)^5$.
Use the quotient rule.

$$\frac{dy}{dx} = \frac{(0)(x^3+1)^5 - (1)\left[5(x^3+1)^4(3x^2)\right]}{\left[(x^3+1)^5\right]^2}$$
$$= \frac{-15x^2(x^3+1)^4}{(x^3+1)^{10}}$$
$$= \frac{-15x^2}{(x^3+1)^6}$$

In *Example 4*, if we write $y = \frac{1}{(x^3+1)^5}$ as $y = (x^3+1)^{-5}$, then use the power of a function rule to differentiate, we get $\frac{dy}{dx} = -5(x^3+1)^{-6}(3x^2) = -15x^2(x^3+1)^{-6}$. This agrees with the result of *Example 4*.

This illustrates that the power of a function rule $\frac{d}{dx}[f(x)]^n = n[f(x)]^{n-1} \times f'(x)$ applies to negative exponents n.

4.3 Exercises

1. Determine the derivative of each power.

   a) $y = (2x + 2)^6$ b) $y = (2x - 5)^{10}$

   c) $y = (9 - 3x)^5$ d) $y = (5 + 3x)^2$

2. Differentiate each power.

   a) $y = (2x + 1)^3$ b) $y = (1 - 5x)^7$

   c) $y = (5 - 3x)^4$ d) $y = (7x - 5)^5$

3. Determine the derivative of each power.

   a) $y = (4x + 1)^{-2}$ b) $y = (3 - 2x)^{-6}$

   c) $y = (3x + 5)^{-4}$ d) $y = (2 + 5x)^{-3}$

4. Differentiate each power.

   a) $y = (4x + 2)^{-6}$ b) $y = (1 - 2x)^{-4}$

   c) $y = (-2 - 3x)^{-5}$ d) $y = (6x - 5)^{-3}$

5. Determine the derivative of each function.

   a) $y = (x^2 + 1)^3$ b) $y = -(2x^2 + 3)^5$

   c) $y = (3x^2 + 5x - 1)^2$ d) $y = -(x^2 - 2x + 5)^4$

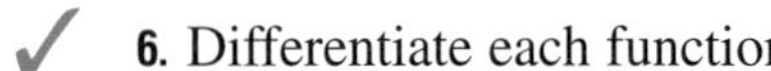

6. Differentiate each function.

   a) $y = -(1 - x^2)^{-2}$ b) $y = (3 - x - 3x^2)^{-5}$

   c) $y = -(2 - 3x^2)^{-7}$ d) $y = (2x^2 + 3x + 1)^{-4}$

7. Determine the derivative of each reciprocal function.

   a) $y = \dfrac{1}{(1 + x^2)^4}$ b) $y = \dfrac{-1}{(2 - 3x^3)^2}$

   c) $y = \dfrac{1}{(x^2 + x + 1)^3}$ d) $y = \dfrac{1}{(2x^2 + x^4)^2}$

8. Differentiate each function.

   a) $y = \left(\dfrac{x}{x - 1}\right)^3$ b) $y = \left(\dfrac{x - 2}{2x}\right)^2$

   c) $y = \left(\dfrac{-4 + x^2}{x}\right)^5$ d) $y = \left(\dfrac{x^2}{x^3 + 9}\right)^4$

9. Differentiate each function.

   a) $y = (x - 1)^2(x + 3)^3$ b) $y = -(2x^2 + 3)^3(1 - x)^4$

   c) $y = (2x - x^2)^4(x + 5)^2$ d) $y = -(x^2 - 2)^3(x^2 + 3)^3$

10. Differentiate each function.

a) $y = \frac{(x+1)^2}{-2x}$ b) $y = \frac{(x^2+1)^3}{x^2}$ c) $y = \frac{-(x+2)^3}{(3x+2)^2}$

d) $y = \frac{(1-x)^2}{-x}$ e) $y = \frac{(1-x^2)^3}{x^3}$ f) $y = \frac{(1-x-x^2)^3}{(x+2)^2}$

11. **Communication** Describe how the power of a function rule is similar to, and different from, the power rule on page 83.

12. **Knowledge/Understanding** Differentiate each function.

a) $y = (2x^2 - 3x + 6)^5$ b) $y = \frac{-1}{(2-3x^2)^3}$

c) $y = (x^4 - x^2)^2(3x^2 - 1)^3$ d) $y = \frac{(x^2-x)^5}{(3x+4)^2}$

13. Differentiate each function.

a) $y = \sin^2 x \cos x$ b) $y = \cos^2 x \sin x$

c) $y = \cos^2 x \sin^2 x$ d) $y = \cos^3 x \sin x$

14. **Application** A cylindrical tube is 100 cm tall, and has a cross-sectional area of 1 cm^2. The tube is full of water. At time $t = 0$ seconds, water begins to flow out the bottom of the tube. The volume, V cubic centimetres, of water in the tube at any time, t seconds, is given by $V = \frac{4}{25}(t-25)^2$.

a) What is the volume of water in the tube after 5 s?

b) How long does it take to empty the tube?

c) What is the rate of water flow when $t = 10$ s?

d) i) When is the volume of water in the tube 64 cm^3?

ii) What is the rate of water flow at this time?

e) At what rate would water have to be poured into the tube to maintain the volume at 36 cm^3? Explain.

15. Consider $y = [f(x)]^{-3}$.

a) Write $y = \frac{1}{[f(x)]^3}$, and use the quotient rule together with the power of a function rule to determine $\frac{dy}{dx}$.

b) Show that $\frac{dy}{dx} = -3[f(x)]^{-4} \times f'(x)$.

16. Let $y = [f(x)]^{-n}$, where n is a positive integer.

a) Write $y = \frac{1}{[f(x)]^n}$, and determine $\frac{dy}{dx}$.

b) Show that $\frac{dy}{dx} = -n[f(x)]^{-n-1} \times f'(x)$.

Exercise 16 is significant because it shows that the power of a function rule, $\frac{d}{dx}[f(x)]^n = n[f(x)]^{n-1} \times f'(x)$, applies to all integral exponents n.

17. Determine the equation of the tangent to the graph of $y = (x-1)^2(3x-2)^2$ at the point (2, 16).

18. Determine the equation of the tangent to the graph of $y = \frac{-1}{(x^2 + 1)^3}$ (below left) at the point $(1, -\frac{1}{8})$.

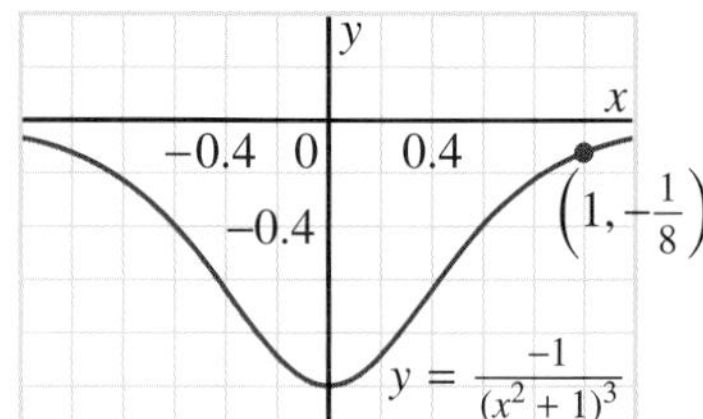

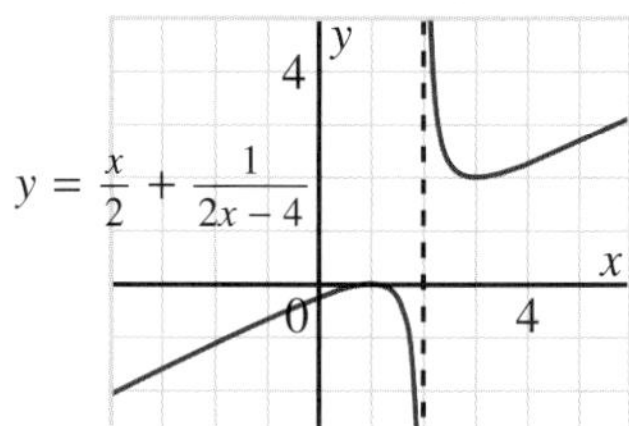

19. Determine whether there are any points on the graph of $y = \frac{x}{2} + \frac{1}{2x - 4}$ (above right) where the slope of the tangent is $-\frac{3}{2}$.

20. An object starts at $t = 0$ seconds and moves in a straight line. Its distance, d metres, from its starting point at time t seconds is given by $d = \frac{4t}{1 + t^2}$.

 a) How far is the object from the starting point when its velocity is 0?

 b) Determine the acceleration of the object when its velocity is 0. Explain what this represents.

21. **Thinking/Inquiry/Problem Solving** Consider the function $y = \frac{(x + 1)(x^2 - 3)}{x^4}$.

 a) Describe 3 different ways to differentiate this function.

 b) Choose one way from part a. Differentiate the function this way.

22. The graph of $y = \frac{-1}{(2x - 1)^2}$ is shown below left. Determine the coordinates of the point on the graph where the slope of the tangent is $\frac{1}{2}$.

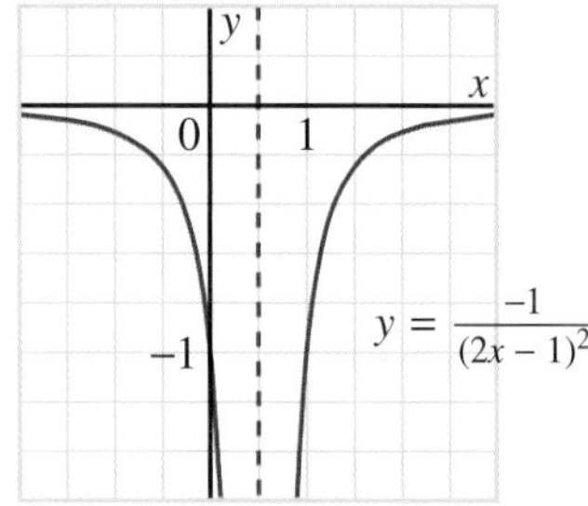

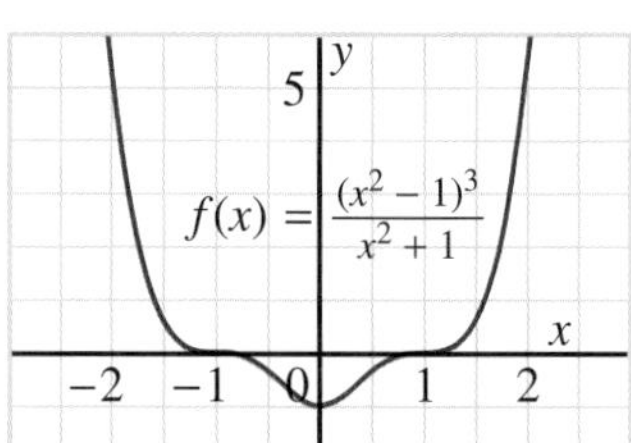

23. The graph of $f(x) = \frac{(x^2 - 1)^3}{x^2 + 1}$ is shown above right.

 a) Determine the derivative of $y = f(x)$.

 b) Sketch the graph of the derivative of $y = f(x)$.

Self-Check 4.1 – 4.3

1. Determine the derivative of each product.

   a) $y = x(x^2 + 3)$
   b) $y = 2x^3(x - 1)$
   c) $y = 5x^3(2x^2 + 3x - 3)$
   d) $y = (1 - x^2)(2 + x)$

2. Determine the derivative of each quotient.

   a) $y = \dfrac{x^2}{x + 3}$
   b) $y = \dfrac{2 - x}{2 + x}$
   c) $y = \dfrac{x + 2}{x - 1}$
   d) $y = \dfrac{2x}{x - 2}$

3. Differentiate each function.

   a) $y = \dfrac{1}{3x^2 + x}$
   b) $y = \dfrac{4x^3}{x^2 + 1}$
   c) $y = (2x^2 + 3)(1 - x)$
   d) $y = (x + 1)(-3x^2 - 2x + 5)$

4. Determine the derivative of each function.

   a) $y = \sin x(1 + \cos x)$
   b) $y = x^3(2 + \sin x)$
   c) $y = \dfrac{x}{1 + \sin x}$
   d) $y = \dfrac{\cos x}{2 \sin x}$

5. Differentiate each power function.

   a) $y = (4 + 3x^2)^3$
   b) $y = -(x - 2x^2)^4$
   c) $y = (7x^3 + x)^{-5}$
   d) $y = -(1 + x - 6x^2)^{-2}$

6. The graph of $y = \dfrac{x + 1}{x - 1}$ is shown below left. Determine the equation of the tangent to the graph at the point on the graph where $x = 3$.

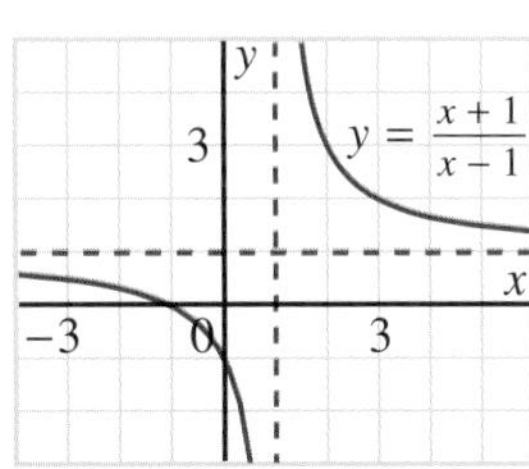

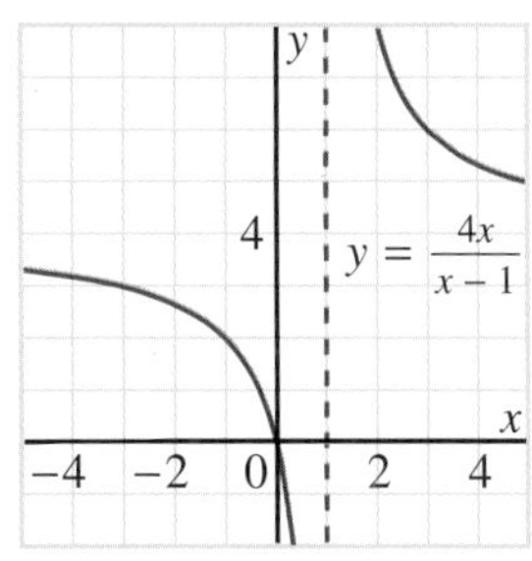

PERFORMANCE ASSESSMENT

7. The graph of $y = \dfrac{4x}{x - 1}$ is shown above right.
   Determine the coordinates of the point(s) on the graph where the slope of the tangent is $-\dfrac{1}{4}$.

4.4 The Composition of Functions

In Section 4.3, we differentiated powers of functions such as $h(x) = (2x + 1)^3$. Inside the brackets, there is a first-degree polynomial that represents a linear function. The function h is a combination of a linear function and a cubic function; h is a function of a function. A function, such as h, is called a *composite function*. In Section 4.5, we will present a general rule to differentiate a composite function.

A composite function arises in a situation when a quantity is given as a function of one variable that, in turn, can be written as a function of another variable. For example, consider how to express the cost of fuel, when taking a trip by car, as a function of the distance driven. The cost of fuel is a function of the amount of fuel consumed. In turn, the amount of fuel consumed is a function of the distance driven.

When fuel costs 70 ¢/L, the cost, C cents, for n litres of fuel is given by the function:

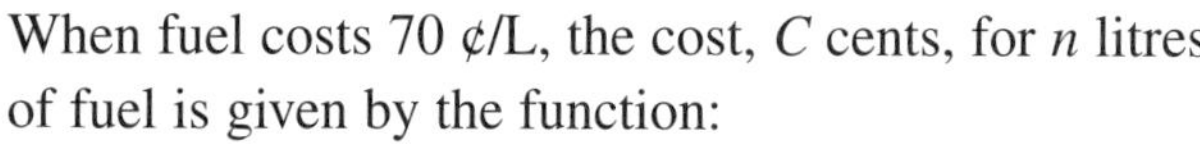

$$C(n) = 70n$$

When the car consumes fuel at the rate of 8.0 L/100 km, the amount of fuel consumed to travel x kilometres is given by the function:

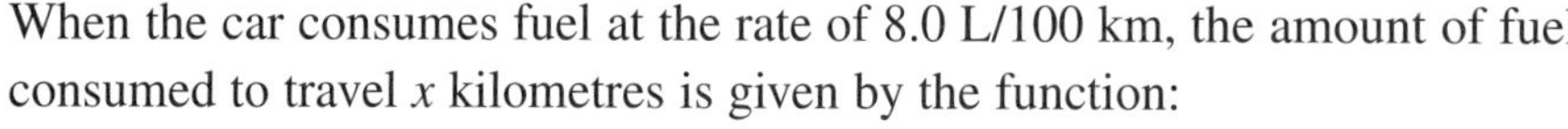

$$n(x) = 0.080x$$

Thus, C is a function of n, and n is a function of x. The cost of fuel as a function of the distance driven is given by the function $C(n(x))$. Substitute $n(x) = 0.080x$ for n in $C(n)$.

$$\begin{aligned} C(n(x)) &= 70(0.080x) \\ &= 5.6x \end{aligned}$$

The cost of fuel is $5.6x$ cents; that is, the product of 5.6 cents and the distance driven.

When two functions are applied in succession, the resulting function is called the *composite* of the two given functions. In the above example, the function $y = C(n(x))$ is the composite of the functions C and n. It is the function that results when the function C is applied on the function n.

$$C \circ n$$

$$x \xrightarrow{n} 0.080x \xrightarrow{C} 70(0.080x) = 5.6x$$

The composition of $C(n)$ and $n(x)$ is written $C \circ n(x)$ or $C(n(x))$. We say "C of n at x," or "C at n at x," or "C following n."

Take Note

Function Composition

Given two functions f and g, the composition of f and g is defined by $y = f(g(x))$.

$f(g(x))$ may be written as $f \circ g(x)$.

In the composition $y = f(g(x))$, the function f is applied on the function g. We can also calculate the composition $y = g(f(x))$, where the function g is applied on the function f.

Example 1

Consider $f(x) = x^2 - x$ and $g(x) = 4x + 1$.

a) Determine $f(g(x))$.

b) Determine $g(f(x))$.

Solution

$f(x) = x^2 - x$ and $g(x) = 4x + 1$

a) For $f(g(x))$, function f is applied on function g.
To determine $f(g(x))$, substitute $g(x)$ for x in $f(x)$.

$$\begin{aligned} f(x) &= x^2 - x \\ f(g(x)) &= (g(x))^2 - g(x) \\ &= (4x + 1)^2 - (4x + 1) \\ &= 16x^2 + 8x + 1 - 4x - 1 \\ &= 16x^2 + 4x \end{aligned}$$

b) For $g(f(x))$, function g is applied on function f.
To determine $g(f(x))$, substitute $f(x)$ for x in $g(x)$.

$$\begin{aligned} g(x) &= 4x + 1 \\ g(f(x)) &= 4f(x) + 1 \\ &= 4(x^2 - x) + 1 \\ &= 4x^2 - 4x + 1 \end{aligned}$$

Example 1 illustrates that, in general, $f(g(x))$ does not equal $g(f(x))$.

In *Example 1*, we can substitute any real number x into the equations for $f(g(x))$ and $g(f(x))$. That is, $f(g(x))$ and $g(f(x))$ are defined for all real numbers x. However, this may not always be the case, as illustrated in *Example 2*.

Example 2

Consider $f(x) = \sqrt{x}$ and $g(x) = x - 5$.

a) Evaluate $f(g(9))$.

b) Evaluate $f(g(1))$.

Solution

a) To evaluate $f(g(9))$, evaluate $g(9)$ then substitute this result in the function f.

$$\begin{aligned} g(x) &= x - 5 \\ g(9) &= 9 - 5 \\ &= 4 \end{aligned}$$

$$\begin{aligned} f(x) &= \sqrt{x} \\ f(g(9)) &= \sqrt{g(9)} \\ &= \sqrt{4} \\ &= 2 \end{aligned}$$

Thus, $f(g(9)) = 2$

b) To evaluate $f(g(1))$, evaluate $g(1)$ then substitute this result in the function f.

$$\begin{aligned} g(x) &= x - 5 \\ g(1) &= 1 - 5 \\ &= -4 \end{aligned}$$

$$\begin{aligned} f(x) &= \sqrt{x} \\ f(g(1)) &= \sqrt{g(1)} \\ &= \sqrt{-4} \end{aligned}$$

There is no real number whose square is –4.
Thus, $f(g(1))$ is undefined.

Discuss

Describe another way to evaluate these composite functions.

In *Example 2b*, the composite function $y = f(g(x))$ was not defined when $x = 1$. This suggests that we should consider function composition more closely.

In the function $y = f(g(x))$, the functions g and f are applied in succession. First, g acts upon x, then f acts upon $g(x)$.

$$x \xrightarrow{g} g(x) \xrightarrow{f} f(g(x)) \quad (f \circ g : x \mapsto f(g(x)))$$

If x is not a valid input for the function g, then $g(x)$ does not exist and the composite function $y = f(g(x))$ cannot be formed. Thus, for $f(g(x))$ to be defined, x must be in the domain of g.

If x is in the domain of g, we can evaluate $g(x)$. Then $g(x)$ becomes the input for the function f. For $f(g(x))$ to be defined, $g(x)$ must be a suitable input for f. Therefore, for $f(g(x))$ to be defined, $g(x)$ must be in the domain of f.

That is, $f(g(x))$ is defined only when these two conditions are met:

1. x is in the domain of g.
2. $g(x)$ is in the domain of f.

Consider these conditions applied to the functions $f(x) = \sqrt{x}$ and $g(x) = x - 5$ from *Example 2*.

The domain of g is all real numbers. This means there is no restriction on the domain of $f(g(x))$ from condition 1.

Next, we look at condition 2. The domain of f is $x \geq 0$. This means that only non-negative numbers are valid inputs for f. For $g(x)$ to be in the domain of f, $g(x) \geq 0$. That is,

$$x - 5 \geq 0$$
$$x \geq 5$$

Conditions 1 and 2 imply that $y = f(g(x))$ is only defined when $x \geq 5$. This explains why $y = f(g(x))$ was undefined at $x = 1$ but defined at $x = 9$ in *Example 2b*.

Take Note

Domain of the Composite Function $y = f(g(x))$

The domain of the composite function $y = f(g(x))$ is the set of all x such that:

- x is in the domain of g, and
- $g(x)$ is in the domain of f.

Example 3

Consider $f(x) = x^2 - 9$, $g(x) = \sqrt{x}$, and $h(x) = \frac{1}{x}$. Determine the domain of each composite function.

a) $f \circ g$ **b)** $h \circ f$ **c)** $h \circ g$

Solution

a) $f \circ g$

$f(x) = x^2 - 9$ and $g(x) = \sqrt{x}$

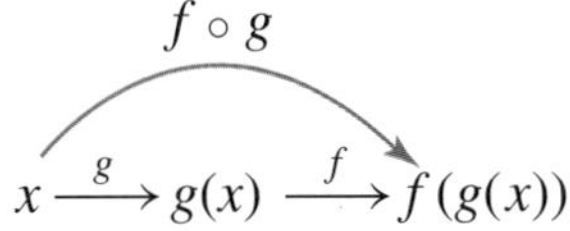

- The domain of g is $x \geq 0$. This means that negative numbers are excluded from the domain of $y = f(g(x))$.
- The domain of f is all real numbers. This means there are no further restrictions on the domain of $y = f(g(x))$.

  Therefore, the domain of $y = f(g(x))$ is $x \geq 0$; in set notation, $D = \{x \mid x \in \Re, x \geq 0\}$.

b) $h \circ f$

$h(x) = \frac{1}{x}$ and $f(x) = x^2 - 9$

$$x \xrightarrow{f} f(x) \xrightarrow{h} h(f(x)) \quad (h \circ f)$$

- The domain of f is all real numbers. This means that $f(x)$ is defined for all values of x.
- The domain of h is all real numbers except $x = 0$. This means any number except 0 is a valid input for h. For $f(x)$ to be in the domain of h, $f(x) \neq 0$; that is,

  $$x^2 - 9 \neq 0$$
  $$x^2 \neq 9$$
  $$x \neq \pm 3$$

  Therefore, the domain of $y = h(f(x))$ is all real numbers except ± 3; in set notation, $D = \{x \mid x \in \Re, x \neq \pm 3\}$.

c) $h \circ g$

$h(x) = \frac{1}{x}$ and $g(x) = \sqrt{x}$

$$x \xrightarrow{g} g(x) \xrightarrow{h} h(g(x)) \quad (h \circ g)$$

- The domain of g is $x \geq 0$. This means that negative numbers are excluded from the domain of $y = h(g(x))$.
- The domain of h is all real numbers except $x = 0$. For $g(x)$ to be in the domain of h, $g(x) \neq 0$; that is,

  $$\sqrt{x} \neq 0$$
  $$x \neq 0$$

  Therefore, all negative numbers and 0 are excluded from the domain of $h(g(x))$. Thus, the domain of $y = h(g(x))$ is $x > 0$; in set notation, $D = \{x \mid x \in \Re, x > 0\}$.

In Section 4.5, we shall develop a rule to calculate the derivative of the composite function $y = f(g(x))$ in terms of the derivatives of f and g, the functions that make up the composition. To use this rule, we must recognize a composite function and determine the functions used to construct it. For example, suppose we are given the function $y = \sqrt{x+1}$. We must be able to *decompose* $y = \sqrt{x+1}$ into functions f and g so that $y = f(g(x))$.

In the function $y = f(g(x))$, g first acts upon x, then f acts upon $g(x)$.

$$f \circ g$$

$$x \xrightarrow{g} g\,(x) \xrightarrow{f} f(g(x))$$

To evaluate $\sqrt{x+1}$ at a particular value of x, we add 1 to that value then take the square root. Therefore, in the composition $y = f(g(x))$, the first function adds 1 and the second function takes the square root.

$$f \circ g$$

$$x \xrightarrow{g\text{: add 1}} x + 1 \xrightarrow{f\text{: square root}} \sqrt{x+1}$$

The function g adds 1 to its input.
Thus, the defining equation of g is $g(x) = x + 1$.
The function f takes the square root of its input.
Thus, the defining equation of f is $f(x) = \sqrt{x}$.

We can verify our choices for f and g by finding the composite $y = f(g(x))$.
$g(x) = x + 1$ and $f(x) = \sqrt{x}$

$$\begin{aligned} f(x) &= \sqrt{x} \\ f(g(x)) &= \sqrt{g(x)} \\ &= \sqrt{x+1} \end{aligned}$$

Thus, $y = \sqrt{x+1}$ is the composite $y = f(g(x))$ of the functions $f(x) = \sqrt{x}$ and $g(x) = x + 1$.

Example 4

For each function, determine functions f and g so that $y = f(g(x))$.

a) $y = \frac{1}{x^2}$ **b)** $y = (2x^3 + 1)^5$

Solution

a) To evaluate $y = \frac{1}{x^2}$, we square x then take the reciprocal.

$$f \circ g$$

$$x \xrightarrow{g:\ square} x^2 \xrightarrow{f:\ reciprocal} \frac{1}{x^2}$$

Let $f(x) = \frac{1}{x}$ and $g(x) = x^2$.

Then, $f(g(x)) = \frac{1}{g(x)}$

$= \frac{1}{x^2}$

Thus, $y = \frac{1}{x^2}$ is the composite $y = f(g(x))$ of the functions $f(x) = \frac{1}{x}$ and $g(x) = x^2$.

b) $(2x^3 + 1)^5$ is the fifth power of $2x^3 + 1$. To evaluate this function at a particular value of x, calculate $2x^3 + 1$ first, then take the fifth power.

$$f \circ g$$

$$x \xrightarrow{g:\ cube,\ multiply\ by\ 2,\ add\ 1} 2x^3 + 1 \xrightarrow{f:\ 5th\ power} (2x^3 + 1)^5$$

Let $f(x) = x^5$ and $g(x) = 2x^3 + 1$.

Then, $f(g(x)) = [g(x)]^5$

$= (2x^3 + 1)^5$

Thus, $y = (2x^3 + 1)^5$ is the composite $y = f(g(x))$ of the functions $f(x) = x^5$ and $g(x) = 2x^3 + 1$.

The solution of *Example 4b* shows one way to decompose the function $y = (2x^3 + 1)^5$ into two functions f and g such that $y = f(g(x))$. This decomposition can be done in other ways (see exercise 23).

4.4 Exercises

A

1. Consider $f(x) = x - 1$ and $g(x) = 4x + 3$. Determine each value.

   a) $f(2)$ **b)** $g(f(2))$ **c)** $g(2)$ **d)** $f(g(2))$

2. For each function in exercise 1, determine $g(f(x))$ and $f(g(x))$.

3. Consider $f(x) = 2 - x^2$ and $g(x) = -3x$. Determine each value.

   a) $f(-1)$ **b)** $g \circ f(-1)$ **c)** $g(-1)$ **d)** $f \circ g(-1)$

4. For each function in exercise 3, determine $f \circ g(x)$ and $g \circ f(x)$.

5. Determine $y = f(g(x))$ and $y = g(f(x))$ for each pair of functions.

a) $f(x) = -x^2 + 1;\ g(x) = 3 - x$

b) $f(x) = \frac{1}{2}x - x^3;\ g(x) = -4x$

c) $f(x) = 2 - x + 2x^2;\ g(x) = -2x^2 + x - 2$

B

6. The circumference, C, of a circle is a function of its diameter, d, where $C = \pi d$. Express the circumference as a function of the radius, r.

7. The volume, V, of a cylinder that has a height equal to its radius, r, is given by $V = \pi r^3$. Express the volume as a function of the diameter, d.

8. The perimeter, P, and area, A, of a square are functions of its side length, s.

a) Express the perimeter as a function of the area.

b) Express the area as a function of the perimeter.

9. The volume, V, and surface area, A, of a cube are functions of its edge length, s.

a) Express the volume as a function of the surface area.

b) Express the surface area as a function of the volume.

10. **Knowledge/Understanding** Consider $f(x) = 3x + 2$, $g(x) = x - 4$, and $h(x) = \frac{1}{x+1}$. Determine each composite function.

a) $f(g(x))$ b) $g(f(x))$ c) $f(h(x))$

d) $h(f(x))$ e) $g(h(x))$ f) $h(g(x))$

11. Determine the domain of each composite function in exercise 10.

12. Consider $f(x) = 2 - x$, $g(x) = \frac{2}{5-x}$, and $h(x) = \frac{1}{x-1}$. Determine each composite function.

a) $f \circ g$ b) $g \circ f$ c) $f \circ h$

d) $h \circ f$ e) $g \circ h$ f) $h \circ g$

13. Determine the domain of each composite function in exercise 12.

14. **Application** When determining the composition of two functions, the two functions need not be different. That is, we can determine the composition of a function with itself. Determine $f(f(x))$ for each function.

a) $f(x) = x + 5$ b) $f(x) = 2x + 1$ c) $f(x) = x + x^2$ d) $f(x) = \sqrt{x+1}$

15. Consider $f(x) = 2x + 3$.

a) Determine $f^{-1}(x)$.

b) i) Show that $f(f^{-1}(x)) = x$. ii) Show that $f^{-1}(f(x)) = x$.

✓ **16.** For each function, determine $f^{-1}(x)$. Then verify that $f(f^{-1}(x)) = x$ and $f^{-1}(f(x)) = x$.

a) $f(x) = 1 - 2x$ **b)** $f(x) = \sqrt{x^2 + 1}$ **c)** $f(x) = \frac{1}{x + 5}$

✓ **17.** Compare your results in exercises 15 and 16.

a) Describe the effect of the composition of two inverse functions.

b) Given that $f(x)$ and $f^{-1}(x)$ are inverse functions, explain why $f(f^{-1}(x)) = x$ and $f^{-1}(f(x)) = x$.

✓ **18.** For each pair of functions, determine:

i) the domain of f and of g **ii)** the range of f and of g

iii) $f \circ g$ **iv)** the domain of $f \circ g$ **v)** the range of $f \circ g$

a) $f(x) = 5x + 2$; $g(x) = 2x + 1$ **b)** $f(x) = x^2 + 2$; $g(x) = 3x + 5$

c) $f(x) = 4 - x^2$; $g(x) = \sqrt{x}$ **d)** $f(x) = \frac{1}{x^2}$; $g(x) = 5 - x^2$

✓ **19. Communication** Use the results of exercise 18. For the composition of two functions to exist, describe how the domains and ranges of the two functions are related.

20. For each function, determine functions f and g so that $y = f(g(x))$.

a) $y = (x + 1)^2$ **b)** $y = -(x + 1)^2$ **c)** $y = \frac{1}{x + 1}$

d) $y = \frac{-1}{x + 1}$ **e)** $y = \sqrt{x + 1}$ **f)** $y = -2\sqrt{x + 1}$

✓ **21.** For each function, determine functions f and g so that $y = f(g(x))$.

a) $y = (x^2 + 3)^4$ **b)** $y = \sqrt{2x - 1}$

c) $y = \frac{1}{3x + 2}$ **d)** $y = -(4x - 5)^3$

e) $y = (2x + 3)^2 + 3(2x + 3) - 4$ **f)** $y = -2\sqrt{x - 1}$

22. For each function, determine functions f and g so that $y = f(g(x))$.

a) $y = (x + 1)^2 - 2(x + 1) + 3$ **b)** $y = \sqrt{4x - 5}$

c) $y = \frac{1}{2x + 3}$ **d)** $y = x^4 - 4x^2 + 4$

23. Determine two different ways in which each function is the composition of two functions f and g.

a) $y = \sqrt{1 - x^2}$ **b)** $y = (3x^2 + 4)^2$ **c)** $y = \frac{1}{2x - 4}$

24. The velocity, v metres per second, of a ball thrown vertically upward is given by $v(t) = 49 - 9.8t$. The kinetic energy of the ball, K joules, is a function of its velocity, where $K(v) = 0.4v^2$. Express the kinetic energy as a function of time.

25. Thinking/Inquiry/Problem Solving Find two non-inverse functions f and g such that $f(g(x)) = g(f(x))$.

4.5 Differentiating Composite Functions

In this section, we will determine a formula for the derivative of a function, $f(g(x))$; that is, the composite of two functions, $f(x)$ and $g(x)$. To do this, we will examine the derivatives of some functions we differentiated previously, and compare the results.

Power of a Function

In Section 4.3, *Example 1a*, we differentiated the function $y = (3x + 1)^7$ and obtained:

$$\frac{dy}{dx} = 7(3x + 1)^6 \times 3 \qquad \ldots ①$$

The power function $y = (3x + 1)^7$ is a composite function $y = f(g(x))$, where $g(x) = 3x + 1$ and $f(x) = x^7$.
Observe that $f'(x) = 7x^6$ and $g'(x) = 3$.
Also, $f'(g(x)) = 7(3x + 1)^6$
Therefore, expression ① is the product of $f'(g(x))$ and $g'(x)$.
Hence, in this example, we may write:

$$\frac{d}{dx}[f(g(x))] = f'(g(x)) \times g'(x)$$

We would have obtained the same result using any other power of a function.

Reciprocal of a Function

In Section 4.2, *Example 3*, we differentiated the function $y = \frac{1}{3x^2 + 4}$ and obtained:

$$\frac{dy}{dx} = \frac{-6x}{(3x^2 + 4)^2} \qquad \ldots ②$$

The reciprocal function, $y = \frac{1}{3x^2 + 4}$, is a composite function $y = f(g(x))$, where $g(x) = 3x^2 + 4$ and $f(x) = \frac{1}{x}$.

Observe that $f'(x) = -\frac{1}{x^2}$ and $g'(x) = 6x$.

Also, $f'(g(x)) = \frac{-1}{(3x^2 + 4)^2}$

Therefore, expression ② is the product of $f'(g(x))$ and $g'(x)$.
Hence, in this example, we may write:

$$\frac{d}{dx}[f(g(x))] = f'(g(x)) \times g'(x)$$

We would have obtained the same result using any other reciprocal of a function.

The Chain Rule

In both situations above, we obtained the same result for $\frac{d}{dx}[f(g(x))]$. It can be proved that this result is true for any differentiable functions $f(x)$ and $g(x)$, but the proof is not required in this course.

Take Note

The Chain Rule

If $f(x)$ and $g(x)$ are differentiable functions, then
$\frac{d}{dx}[f(g(x))] = f'(g(x)) \times g'(x)$.

To determine the derivative of $f(g(x))$, determine $f'(x)$, replace x with $g(x)$, then multiply by $g'(x)$.

Example 1

Differentiate each function.

a) $y = (x^2 + 1)^4$ **b)** $y = \frac{1}{x^2 + 1}$

Solution

a) $y = (x^2 + 1)^4$

This is the composition $y = f(g(x))$ of the functions $f(x) = x^4$ and $g(x) = x^2 + 1$; that is, $f(g(x)) = (x^2 + 1)^4$.

Use the chain rule: $\frac{d}{dx}[f(g(x))] = f'(g(x)) \times g'(x)$

$f'(x) = 4x^3$

$f'(g(x)) = 4(x^2 + 1)^3$

$g'(x) = 2x$

So, $\frac{d}{dx}[f(g(x))] = 4(x^2 + 1)^3 \times 2x$

Therefore, $\frac{dy}{dx} = 8x(x^2 + 1)^3$

b) $y = \frac{1}{x^2 + 1}$

This is the composition $y = f(g(x))$ of the functions $f(x) = \frac{1}{x}$ and $g(x) = x^2 + 1$; that is, $f(g(x)) = \frac{1}{x^2 + 1}$.

Use the chain rule.

$f'(x) = -\frac{1}{x^2}$

$f'(g(x)) = -\frac{1}{(x^2 + 1)^2}$

$g'(x) = 2x$

So, $\frac{d}{dx}[f(g(x))] = -\frac{1}{(x^2 + 1)^2} \times 2x$

Therefore, $\frac{dy}{dx} = -\frac{2x}{(x^2 + 1)^2}$

Example 2

Determine the derivative of $y = \sqrt{x^2 + 1}$.

Solution

$y = \sqrt{x^2 + 1}$

This is the composition $y = f(g(x))$ of the functions $f(x) = \sqrt{x}$ and $g(x) = x^2 + 1$; that is, $f(g(x)) = \sqrt{x^2 + 1}$.

Use the chain rule.

$f'(x) = \dfrac{1}{2\sqrt{x}}$

$f'(g(x)) = \dfrac{1}{2\sqrt{x^2 + 1}}$

$g'(x) = 2x$

So, $\dfrac{d}{dx}[f(g(x))] = \dfrac{1}{2\sqrt{x^2 + 1}} \times 2x$

Therefore, $\dfrac{dy}{dx} = \dfrac{x}{\sqrt{x^2 + 1}}$

We can use the chain rule to differentiate a trigonometric function.

Example 3

Determine the derivative of each function.

a) $y = \sin x^3$

b) $y = \cos \dfrac{1}{2x}$

Solution

a) $y = \sin x^3$

This is the composition $y = f(g(x))$ of the functions $f(x) = \sin x$ and $g(x) = x^3$; that is, $f(g(x)) = \sin x^3$.

Use the chain rule.

$f'(x) = \cos x$

$f'(g(x)) = \cos x^3$

$g'(x) = 3x^2$

So, $\dfrac{d}{dx}[f(g(x))] = \cos x^3 \times 3x^2$

Therefore, $\dfrac{dy}{dx} = 3x^2 \cos x^3$

b) $y = \cos \frac{1}{2x}$

This is the composition $y = f(g(x))$ of the functions $f(x) = \cos x$ and $g(x) = \frac{1}{2x}$; that is, $f(g(x)) = \cos \frac{1}{2x}$.

Use the chain rule.

$f'(x) = -\sin x$

$f'(g(x)) = -\sin \frac{1}{2x}$

$g'(x) = -\frac{1}{2x^2}$

So, $\frac{d}{dx}[f(g(x))] = -\sin \frac{1}{2x} \times \left(-\frac{1}{2x^2}\right)$

Therefore, $\frac{dy}{dx} = \frac{1}{2x^2} \sin \frac{1}{2x}$

We can use the chain rule to differentiate the power of a quotient.

Example 4

Determine the derivative of $y = \left(\frac{x+2}{x-2}\right)^4$.

Solution

$y = \left(\frac{x+2}{x-2}\right)^4$

This is the composition $y = f(g(x))$ of the functions $f(x) = x^4$ and $g(x) = \frac{x+2}{x-2}$; that is, $f(g(x)) = \left(\frac{x+2}{x-2}\right)^4$.

Use the chain rule.

$f'(x) = 4x^3$

$f'(g(x)) = 4\left(\frac{x+2}{x-2}\right)^3$

$$g'(x) = \frac{(x-2)(1) - (x+2)(1)}{(x-2)^2} = \frac{-4}{(x-2)^2}$$

$$\text{So, } \frac{d}{dx}[f(g(x))] = 4\left(\frac{x+2}{x-2}\right)^3 \times \left(\frac{-4}{(x-2)^2}\right) = \frac{-16(x+2)^3}{(x-2)^5}$$

Discuss

Use the quotient rule to differentiate the function $y = \left(\frac{x+2}{x-2}\right)^4$, then use the product rule to differentiate it. Which of the three methods is most efficient? Explain.

THE BIG PICTURE

Justifying the Chain Rule

Although a proof of the chain rule is beyond the scope of this course, we can justify the chain rule in different ways.

Differentiating the power of a function

On page 267, an example that involves the power of a function rule was used to justify the chain rule. In general, the power of a function rule states that the derivative of the function $y = [f(x)]^n$ is $\frac{dy}{dx} = n[f(x)]^{n-1} \times f'(x)$.

Differentiating the reciprocal of a function

Consider the function $y = \frac{1}{f(x)}$.

In Section 4.2, exercise 13, page 248, you used the quotient rule to show that $\frac{dy}{dx} = -\frac{1}{[f(x)]} \times f'(x)$.

Differentiating the square root of a function

Consider the function $y = \sqrt{f(x)}$.

In Section 4.6, exercise 19, page 281, you will show that $\frac{dy}{dx} = \frac{1}{2\sqrt{f(x)}} \times f'(x)$. You will do this without using the chain rule.

All three of the above results are special cases of the chain rule.

4.5 Exercises

1. Use the chain rule to differentiate each function.
   a) $y = (4x + 3)^2$　b) $y = (x^2 - 3)^5$　c) $y = (x^3 + 1)^2$
   d) $y = (3x - 2)^4$　e) $y = (x^2 - 2)^5$　f) $y = (x^2 - 2x - 3)^5$
2. Use the chain rule to differentiate each function.
   a) $y = (2 - 3x)^3$　b) $y = (1 - 2x^2)^2$　c) $y = (3 - 4x^3)^4$
   d) $y = (-5 - x^2)^2$　e) $y = (-3x^2 - x)^3$　f) $y = (-5 - 10x^4)^2$
3. Use the chain rule to differentiate each function.
   a) $y = (1 + 2x)^{-2}$　b) $y = (-x^2 + 3x)^{-1}$　c) $y = -(x - 2)^{-3}$
   d) $y = (-x^2 + x^3)^{-3}$　e) $y = (2 + x + 3x^2)^{-4}$　f) $y = -(3x^2 + 2x^3)^{-1}$

B

4. Use the chain rule to differentiate each function.

a) $y = \dfrac{1}{2x - 1}$ b) $y = \dfrac{1}{(2 - 4x)^3}$ c) $y = \dfrac{1}{x^2 - 2x}$

d) $y = \dfrac{1}{(x^3 + 3)^2}$ e) $y = \dfrac{-1}{2x + 6x^2}$ f) $y = \dfrac{-1}{(x^2 - x - 1)^4}$

5. Use the chain rule to differentiate each function.

a) $y = \sqrt{1 + x}$ b) $y = \sqrt{x^2 - x}$ c) $y = -\sqrt{x^3 - x}$

d) $y = \sqrt{2 - x + x^2}$ e) $y = -\sqrt{3 - 2x^2}$ f) $y = \sqrt{2x + x^4}$

6. Differentiate each function.

a) $y = \left(\dfrac{x + 1}{x - 1}\right)^3$ b) $y = \left(\dfrac{x^2 + 1}{x + 1}\right)^2$ c) $y = \left(\dfrac{1 - 2x}{1 + 2x^2}\right)^3$

d) $y = \left(\dfrac{x^2 + x}{x + x^3}\right)^2$ e) $y = \left(\dfrac{1 + x + x^2}{1 - x - x^2}\right)^4$ f) $y = \left(\dfrac{2 - x^2}{2 + x^3}\right)^3$

7. **Knowledge/Understanding** Differentiate each function.

a) $y = (2x^2 - 3x)^5$ b) $y = (x - x^2 + x^3)^{-1}$

c) $y = \left(\dfrac{4x - 3}{4 - 3x}\right)^3$ d) $y = \dfrac{1}{(x^2 - 3)^4}$

8. Use the chain rule to differentiate each function.

a) $y = \sin 2x$ b) $y = \cos 3x$ c) $y = \sin x^2$

d) $y = \cos x^3$ e) $y = \sin \dfrac{1}{x}$ f) $y = \cos \dfrac{1}{x^2}$

9. **Communication** Explain how the power of a function rule is a special case of the chain rule. Give two examples of functions that can be differentiated using the chain rule but not by using the power of a function rule.

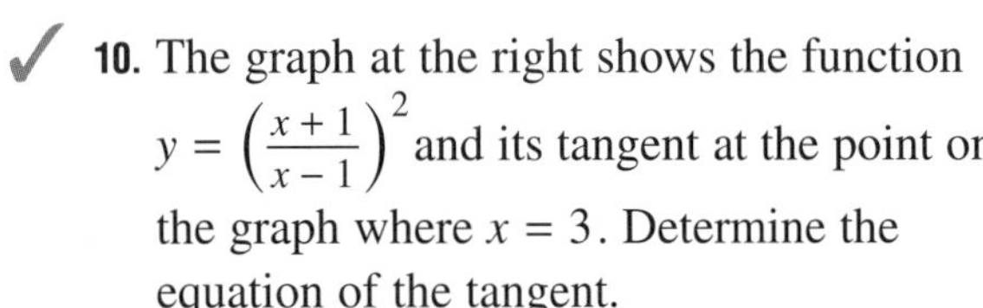

10. The graph at the right shows the function $y = \left(\dfrac{x + 1}{x - 1}\right)^2$ and its tangent at the point on the graph where $x = 3$. Determine the equation of the tangent.

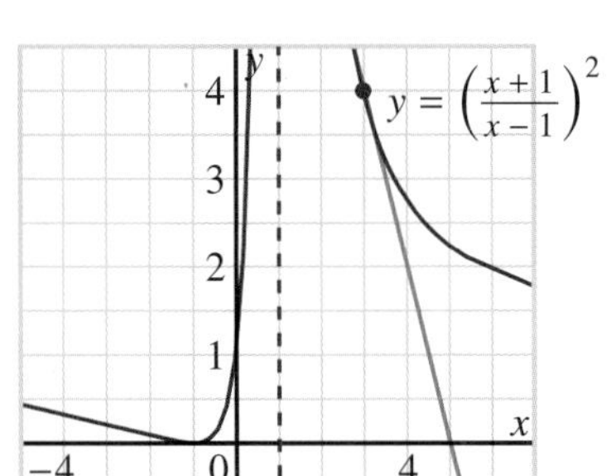

11. Two tangents to the graph of $y = (4x^2 - 1)^2$ are perpendicular to the line $x + 5 = 0$. Determine the equations of the tangents.

12. The temperature of Earth's crust increases by about 1°C for every 100 m below the surface. The temperature at the top of a mine shaft is 20°C. An elevator moves down the shaft at 5 m/s. The temperature, T degrees Celsius, at a depth of d metres, is $T = 20 + 0.01d$. The depth of the elevator, d metres, after t seconds, is $d = 5t$. For people in the elevator, determine the rate of change of temperature with time, $\dfrac{dT}{dt}$.

✓ 13. **Application** An iron bar has a temperature of 20°C. The bar is heated and its temperature increases at a rate of 3°C per minute. The temperature, C degrees Celsius, at any time, t minutes, after heat is first applied, is given by $C = 20 + 3t$. To convert from C degrees Celsius to F degrees Fahrenheit, the equation is $F = 1.8C + 32$. Calculate the rate of change of the temperature of the bar in degrees Fahrenheit per minute.

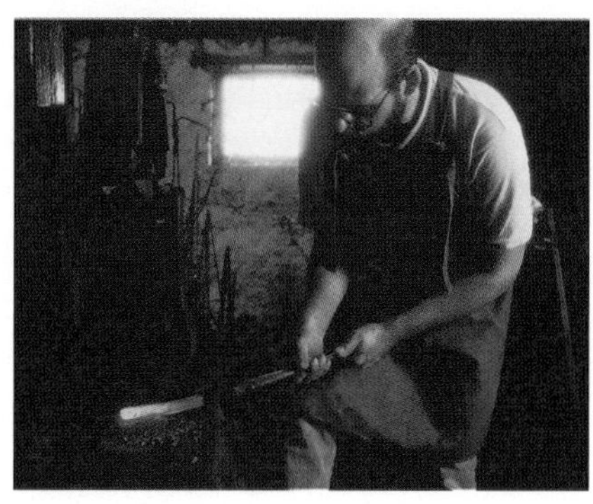

✓ 14. The square-root function $y = \sqrt{5x + 1}$ is a composite function $y = f(g(x))$, where $g(x) = 5x + 1$ and $f(x) = \sqrt{x}$. In this exercise, you will determine $\frac{dy}{dx}$ without using the chain rule. Consider the product $5x + 1 = (\sqrt{5x + 1})(\sqrt{5x + 1})$.

Exercise 14 is significant because it provides further justification that the chain rule is correct.

a) Determine the derivative of the expression on the left side.

b) Use the product rule to differentiate the right side. Use y' for the derivative of $\sqrt{5x + 1}$, where required.

c) Use the results of parts a and b to determine y'.

d) Show that $y' = f'(g(x)) \times g'(x)$.

✓ 15. **Thinking/Inquiry/Problem Solving** Consider the function $y = \frac{1}{x - 1}$.

a) Determine y' and y''.

b) The third and fourth derivatives are represented by y''' and y'''', respectively. Determine y''' and y''''.

c) The nth derivative is represented by $y^{(n)}$. Conjecture a formula for $y^{(n)}$.

16. **a)** Determine the equation of the tangent to the graph of $y = x\sqrt{1 + 3x^2}$ at the point on the graph where $x = 1$.

b) The tangent in part a intersects the graph of $y = x\sqrt{1 + 3x^2}$ at a second point. Determine the coordinates of this point.

17. Two tangents are drawn from the point (2, 9) to the graph of $y = -x^2 + 4$. Determine the coordinates of the points where the tangents touch the graph.

4.6 Implicit Differentiation

On pages 3–5, you reviewed several ways to write the equation of a line.

Slope y-intercept form:	$y = mx + b$	... ①
Point-slope form:	$y = m(x - p) + q$	... ②
Standard form:	$Ax + By + C = 0$	... ③

In equations ① and ②, the dependent variable y is isolated and expressed in terms of the independent variable x. An equation written in this form is called an *explicit* equation because y is expressed explicitly in terms of x. An equation, such as ③, in which the dependent variable is not isolated, is called an *implicit* equation.

To determine the slope of the tangent to the graph of a relation defined by an implicit equation, we can sometimes solve for y to obtain an explicit equation. We can then determine the derivative $\frac{dy}{dx}$.

For example, consider the equation $x^2 + y^2 = 9$, which represents a circle with centre (0, 0) and radius 3. We can solve this equation explicitly for y:

$$x^2 + y^2 = 9$$
$$y^2 = 9 - x^2$$
$$y = \pm\sqrt{9 - x^2}$$

The implicit equation $x^2 + y^2 = 9$ defines *two* explicit equations: $y = \sqrt{9 - x^2}$ and $y = -\sqrt{9 - x^2}$.

The graph of $y = \sqrt{9 - x^2}$ is the upper half of the circle (since $y \geq 0$), while the graph of $y = -\sqrt{9 - x^2}$ is the lower half of the circle (since $y \leq 0$).

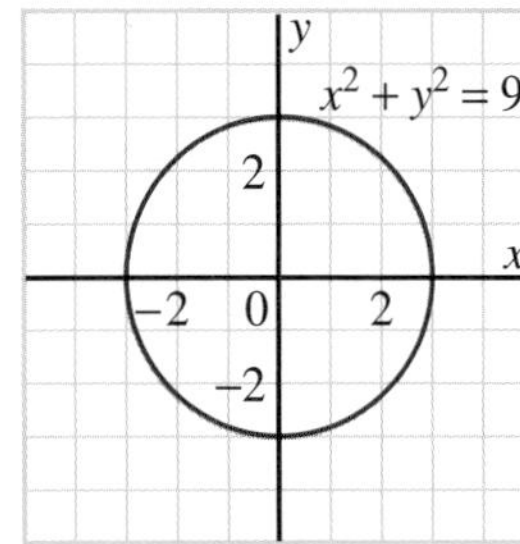

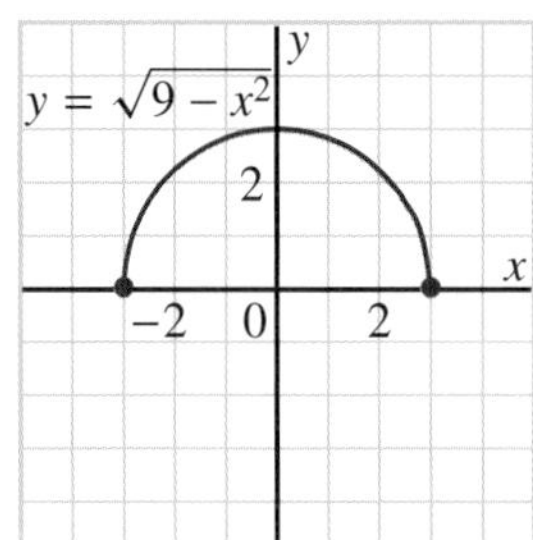

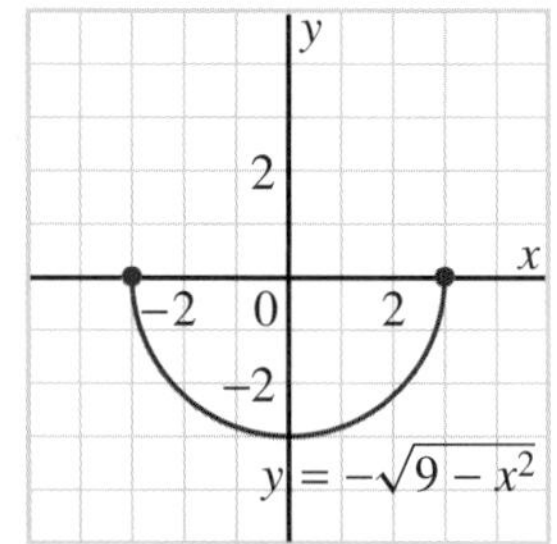

The equation $x^2 + y^2 = 9$ does not represent a function since a vertical line intersects its graph at more than one point. However, both $y = \sqrt{9 - x^2}$ and $y = -\sqrt{9 - x^2}$ satisfy the vertical line test and, hence, represent functions. We can therefore determine their derivatives $\frac{dy}{dx}$. We use the chain rule.

$$y = \sqrt{9 - x^2}$$
$$\frac{dy}{dx} = \frac{1}{2\sqrt{9 - x^2}} \times \frac{d}{dx}(9 - x^2)$$
$$= \frac{1}{2\sqrt{9 - x^2}} \times (-2x)$$
$$= -\frac{x}{\sqrt{9 - x^2}}, x \neq \pm 3$$

$$y = -\sqrt{9 - x^2}$$
$$\frac{dy}{dx} = \frac{-1}{2\sqrt{9 - x^2}} \times \frac{d}{dx}(9 - x^2)$$
$$= \frac{-1}{2\sqrt{9 - x^2}} \times (-2x)$$
$$= \frac{x}{\sqrt{9 - x^2}}, x \neq \pm 3$$

Each $\frac{dy}{dx}$ is not defined when $x = \pm 3$. At these points, the tangent is vertical and the slope of the graph is undefined.

We can determine $\frac{dy}{dx}$ more easily by differentiating each side of the implicit equation $x^2 + y^2 = 9$ with respect to x. We treat y as a differentiable function of x.

$$x^2 + y^2 = 9$$
$$\frac{d}{dx}(x^2 + y^2) = \frac{d}{dx}(9)$$
$$\frac{d}{dx}(x^2) + \frac{d}{dx}(y^2) = \frac{d}{dx}(9)$$
$$2x + \frac{d}{dx}(y^2) = 0$$

Consider the expression $\frac{d}{dx}(y^2)$. Since y is implicitly defined as a function of x, we differentiate y^2 using the power of a function rule. We know that $\frac{d}{dx}[f(x)]^2 = 2f(x)\,f'(x)$, so $\frac{d}{dx}(y^2) = 2y\frac{dy}{dx}$.

Therefore,

$$2x + 2y\frac{dy}{dx} = 0$$
$$2y\frac{dy}{dx} = -2x$$
$$\frac{dy}{dx} = -\frac{2x}{2y}$$
$$= -\frac{x}{y}, y \neq 0$$

The derivative is expressed in terms of both x and y. This is because $x^2 + y^2 = 9$ defines two explicit functions of x, and the derivative applies to each function.

When $y = \sqrt{9 - x^2}$,

$$\frac{dy}{dx} = -\frac{x}{y}$$
$$= -\frac{x}{\sqrt{9 - x^2}}, \ x \neq \pm 3$$

This agrees with the result above.

Similarly, when $y = -\sqrt{9 - x^2}$,

$$\frac{dy}{dx} = -\frac{x}{y}$$
$$= -\frac{x}{-\sqrt{9 - x^2}}$$
$$= \frac{x}{\sqrt{9 - x^2}}, \ x \neq \pm 3$$

This agrees with the result above.

$\frac{dy}{dx} = -\frac{x}{y}$ is not defined when $y = 0$. For the equation $x^2 + y^2 = 9$, $y = 0$ when $x = \pm 3$. This agrees with the results above.

The process of differentiating an equation that implicitly defines one or more functions is called *implicit differentiation*. To determine $\frac{dy}{dx}$, we treat y as a differentiable function of x. We differentiate each term of the equation with respect to x, then solve for $\frac{dy}{dx}$.

Example 1

Use implicit differentiation to determine $\frac{dy}{dx}$ for the parabola $3x^2 + 2y = 6$.

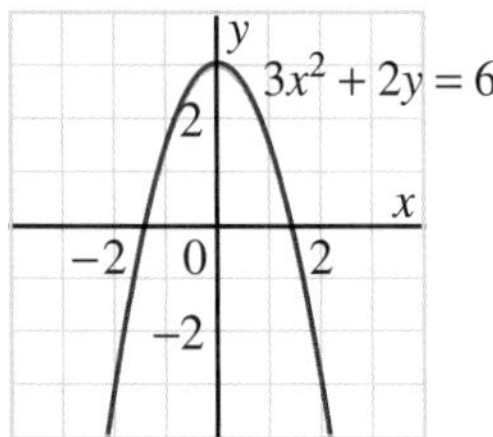

Solution

$3x^2 + 2y = 6$

Take the derivative of each term. Since y is implicitly defined as a function of x, use the chain rule to differentiate the term $2y$.

$$3x^2 + 2y = 6$$

$$\frac{d}{dx}(3x^2) + \frac{d}{dx}(2y) = \frac{d}{dx}(6)$$

$$6x + 2\frac{dy}{dx} = 0$$

$$2\frac{dy}{dx} = -6x$$

$$\frac{dy}{dx} = \frac{-6x}{2}$$

$$= -3x$$

Discuss

Describe another way to determine the derivative of the function.

You may have studied conics in grade 11. Most conics have equations that implicitly define one or more functions of x. We can use implicit differentiation to determine the slopes of tangents to the graphs of these conics.

Example 2

The graph of the ellipse $x^2 + 5y^2 = 36$ is shown.

a) Determine $\frac{dy}{dx}$.

b) Determine the slope of the tangent to the graph at the point R(4, 2).

c) Determine the equation of the tangent at R.

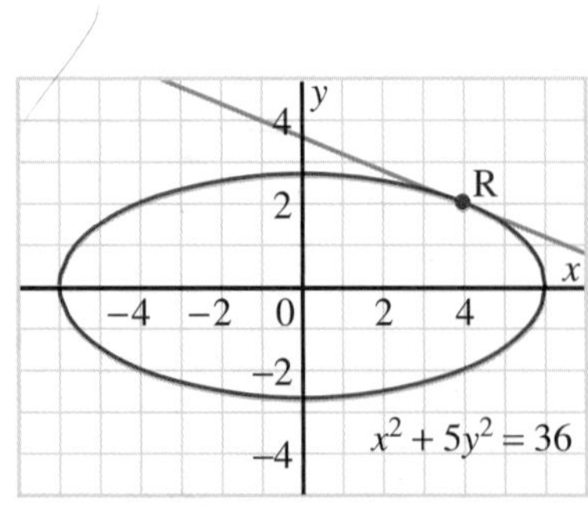

Solution

a) Use implicit differentiation.

$$x^2 + 5y^2 = 36$$

$$\frac{d}{dx}(x^2) + \frac{d}{dx}(5y^2) = \frac{d}{dx}(36)$$

$$2x + 10y\frac{dy}{dx} = 0$$

$$\frac{dy}{dx} = -\frac{2x}{10y}$$

$$= -\frac{x}{5y}$$

b) The slope of the tangent at any point (x, y) is $\frac{dy}{dx}$.

To determine the slope of the tangent at the point R(4, 2), substitute these coordinates into the expression for $\frac{dy}{dx}$.

$$\frac{dy}{dx} = -\frac{x}{5y}$$

$$= -\frac{4}{5(2)}$$

$$= -\frac{2}{5}$$

The slope of the tangent at R is $-\frac{2}{5}$.

c) To determine the equation of the line with slope $-\frac{2}{5}$ that passes through R(4, 2), use the point-slope form $y = m(x - p) + q$.

So, $y = -\frac{2}{5}(x - 4) + 2$

$$y = -\frac{2}{5}x + \frac{8}{5} + 2$$

$$y = -\frac{2}{5}x + \frac{18}{5}$$

Some equations in x and y are difficult or impossible to write in explicit form. We use implicit differentiation to determine $\frac{dy}{dx}$.

Example 3

Determine $\frac{dy}{dx}$ for the equation $x^3 + y^3 = 3xy$.

Solution

$x^3 + y^3 = 3xy$

Differentiate each term with respect to x.

$\frac{d}{dx}(x^3) + \frac{d}{dx}(y^3) = \frac{d}{dx}(3xy)$

On the left side, use the power of a function rule to differentiate y^3. On the right side, use the product rule.

$$3x^2 + 3y^2\frac{dy}{dx} = 3x\frac{d}{dx}(y) + y\frac{d}{dx}(3x)$$
$$3x^2 + 3y^2\frac{dy}{dx} = 3x\frac{dy}{dx} + 3y$$

Solve for $\frac{dy}{dx}$.

$$3y^2\frac{dy}{dx} - 3x\frac{dy}{dx} = 3y - 3x^2$$
$$(3y^2 - 3x)\frac{dy}{dx} = 3y - 3x^2$$
$$\frac{dy}{dx} = \frac{3y - 3x^2}{3y^2 - 3x}$$
$$= \frac{3(y - x^2)}{3(y^2 - x)}$$
$$= \frac{y - x^2}{y^2 - x}$$

We can use implicit differentiation to determine the derivative of a power of x, when the exponent is a rational number.

Example 4

a) Determine the derivative of $y = x^{\frac{2}{3}}$.

b) Express the derivative in terms of x.

Solution

a) $y = x^{\frac{2}{3}}$

Raise each side to the power 3.

$$y^3 = \left(x^{\frac{2}{3}}\right)^3$$
$$y^3 = x^2$$

Take the derivative of each side.

$$3y^2\frac{dy}{dx} = 2x$$
$$\frac{dy}{dx} = \frac{2x}{3y^2}$$

b) To express $\frac{dy}{dx} = \frac{2x}{3y^2}$ in terms of x, substitute $y = x^{\frac{2}{3}}$, or $y^2 = x^{\frac{4}{3}}$.

$$\frac{dy}{dx} = \frac{2x}{3x^{\frac{4}{3}}}$$
$$= \frac{2}{3}x^{1-\frac{4}{3}}$$
$$= \frac{2}{3}x^{-\frac{1}{3}}$$

Look at the derivative in *Example 4*. Suppose we assume the power of a function rule is true for rational exponents and differentiate $y = x^{\frac{2}{3}}$ using that rule.

$$y = x^{\frac{2}{3}}$$

Then $\frac{dy}{dx} = \frac{2}{3}x^{\frac{2}{3}-1}$

$$= \frac{2}{3}x^{-\frac{1}{3}}$$

This is the result we obtained in *Example 4*.
Example 4 illustrates that the power of a function rule is true for rational exponents.

Take Note

Power of a Function Rule for Rational Exponents

For any rational number n,

$$\frac{d}{dx}[f(x)]^n = n[f(x)]^{n-1} \times f'(x)$$

This means that we can differentiate functions such as $y = (3 - x)^{\frac{1}{2}}$ and $y = (x^2 + x)^{-\frac{1}{2}}$. You will do this in the exercises that follow.

4.6 Exercises

A

1. Classify each equation as implicit or explicit.

   a) $y = 3x + 5$ b) $x = 3y + 5$ c) $y = \frac{1}{x}$

   d) $xy = 1$ e) $4x^2 + 9y^2 = 36$ f) $\frac{x^2}{9} + \frac{y^2}{4} = 1$

2. Use implicit differentiation to determine $\frac{dy}{dx}$.

   a) $3x + 5y = 4$ b) $2x - 7y = 14$ c) $2x - 3y + 4 = 0$

3. Write each equation in exercise 2 as an explicit equation. Determine $\frac{dy}{dx}$.

B

4. a) Differentiate $2x + 3y + 6 = 0$ implicitly.

   b) Differentiate $2x + 3y + 6 = 0$ explicitly.

   c) Compare the answers to parts a and b. Explain what you notice.

✓ 5. Determine $\frac{dy}{dx}$ for each relation.

   a) $xy = 1$ b) $x^2y = 1$ c) $xy^2 = 1$

   d) $xy = x$ e) $x^2y^2 = 1$ f) $x^2y^2 = x$

✓ 6. **Knowledge/Understanding** Determine $\frac{dy}{dx}$ for each relation.

   a) $y^2 = 5x^2$ b) $x^2 + y = 1$ c) $x + y^2 = 1$

   d) $x^3 + y = 1$ e) $4y^2 - 2x^2 = 8$ f) $x^3 + y^3 = 1$

✓ 7. Differentiate.

   a) $y = x^{\frac{1}{3}}$ b) $y = x^{\frac{3}{4}}$ c) $y = x^{-\frac{2}{3}}$

8. Consider the function $y = x^{\frac{m}{n}}$, where m and n are positive integers. Raise each side to the nth power to obtain $y^n = x^m$. Then use implicit differentiation to show that $\frac{dy}{dx} = \frac{m}{n}x^{\frac{m}{n}-1}$.

Exercise 8 is significant because it proves that the power rule $\frac{d}{dx}(x^n) = nx^{n-1}$ applies to rational exponents.

9. Determine $\frac{dy}{dx}$ for each function.

a) $y = (3x-2)^{\frac{1}{2}}$ b) $y = (x^2+x)^{-\frac{1}{2}}$ c) $y = (4-x^2)^{\frac{3}{2}}$

d) $y = (x^2+x-2)^{\frac{1}{3}}$ e) $y = (-5-x^3)^{-\frac{1}{3}}$ f) $y = (6x-2x^4)^{-\frac{3}{2}}$

10. Determine $\frac{dy}{dx}$ for each ellipse.

a) $4x^2 + 8y^2 = 32$ b) $10x^2 + 2y^2 = 20$ c) $x^2 + 16y^2 = 16$

11. Each equation that follows represents a hyperbola. Determine $\frac{dy}{dx}$ for each equation.

a) $x^2 - y^2 = 5$ b) $2x^2 - 3y^2 = 6$

c) $4x^2 - 3y^2 = -12$ d) $9x^2 - 4y^2 = -36$

12. a) Determine $\frac{dy}{dx}$ for each parabola. Include any values of y for which $\frac{dy}{dx}$ is undefined.

i) $y^2 = x$ ii) $y^2 = x + 1$ iii) $(y-2)^2 = x + 1$

b) Sketch each parabola in part a. Describe how $\frac{dy}{dx}$ relates to each graph.

13. a) Determine $\frac{dy}{dx}$ for each circle.

i) $x^2 + y^2 = 25$ ii) $2x^2 + 2y^2 = 5$ iii) $5x^2 + 5y^2 = 12$

b) What do you notice about the values of $\frac{dy}{dx}$ in part a? Explain the results in terms of the graphs of the circles.

14. The ellipse $3x^2 + 4y^2 = 12$ is shown below left.

a) Determine $\frac{dy}{dx}$ for the ellipse.

b) Determine the equation of the tangent to the ellipse at the point (1, 1.5).

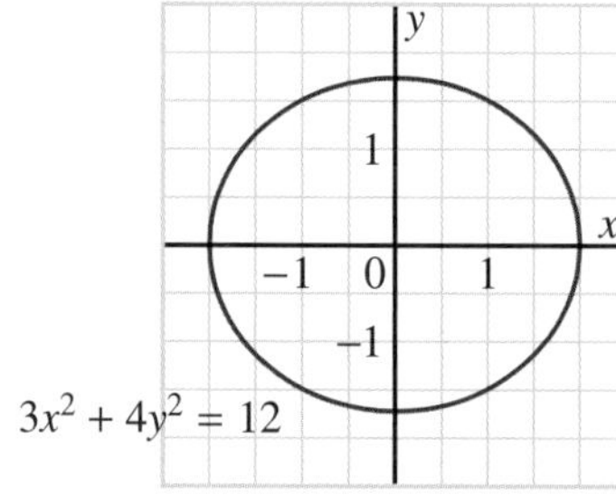

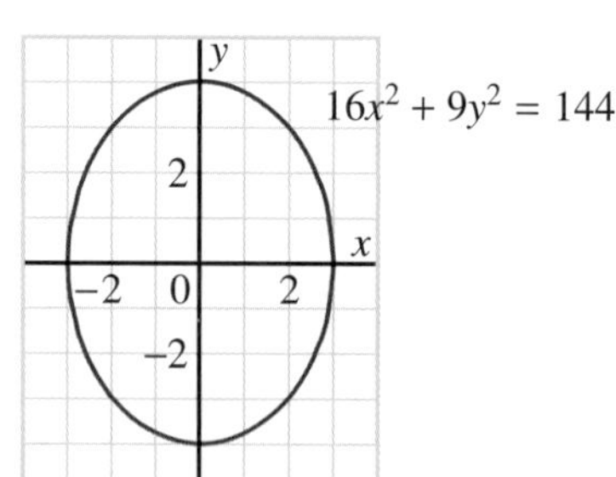

15. The ellipse $16x^2 + 9y^2 = 144$ is shown above right.

a) Determine $\frac{dy}{dx}$ for the ellipse.

b) Determine the equation of the tangent to the ellipse at the point $(-1.5, 2\sqrt{3})$.

✓ **16.** The hyperbola $x^2 - 4y^2 = -16$ is shown below left.

a) Determine $\frac{dy}{dx}$ for the hyperbola.

b) Determine the equation of the tangent to the hyperbola at the point $(2, -\sqrt{5})$.

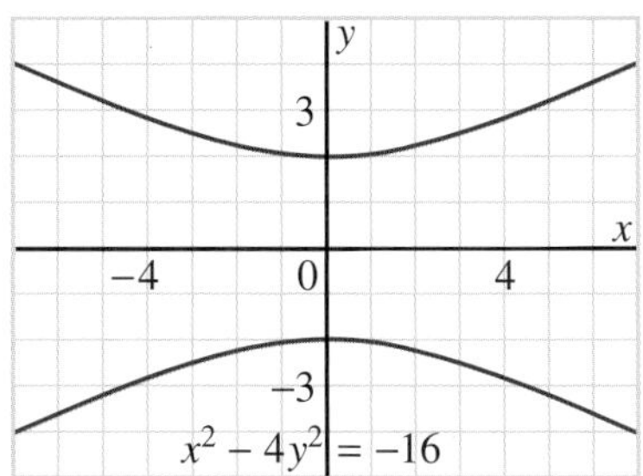

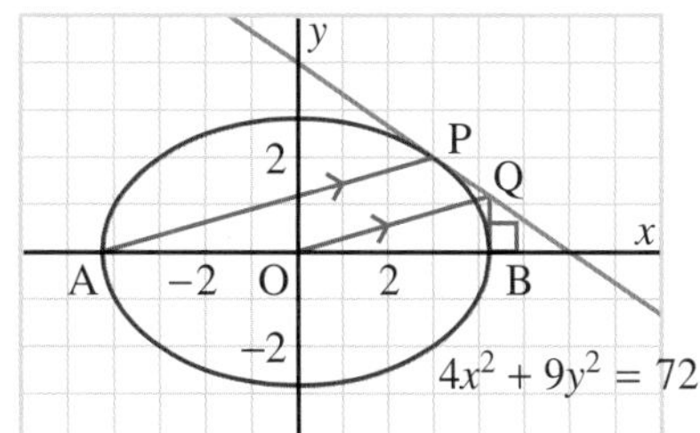

✓ **17. Thinking/Inquiry/Problem Solving** Point P(3, 2) lies on the ellipse $4x^2 + 9y^2 = 72$, above right. Points A and B are at the x-intercepts. Segment OQ is parallel to AP. Segment QB is perpendicular to the x-axis. Show that PQ is a tangent to the ellipse at P.

✓ **18. Communication** Consider y implicitly defined as a function of x. When a term that contains y is differentiated, the result includes $\frac{dy}{dx}$. Any term that contains only a constant or a power of x does not have $\frac{dy}{dx}$ in its derivative. Explain.

✓ **19.** Consider the function $y = \sqrt{f(x)}$, where $f(x) \geq 0$. Square each side, then use implicit differentiation to determine an expression for $\frac{dy}{dx}$ in terms of $f(x)$ and $f'(x)$.

Exercise 19 is significant because it provides a rule for the derivative of the square root of a function. This rule is a special case of the chain rule. See *The Big Picture*, page 271.

20. a) Show that the point (2, 3) lies on the graph of $x^2 + xy - y^2 = 1$.

b) Determine the equation of the tangent to the graph of $x^2 + xy - y^2 = 1$ at the point (2, 3).

✓ **21. Application** Show that the two circles $(x - 6)^2 + (y - 3)^2 = 20$ and $x^2 + y^2 + 2x + y = 10$ have a common tangent at the point (2, 1). Illustrate your solution geometrically.

✓ **22.** Determine $\frac{dy}{dx}$ for each relation.

a) $y = x^{\frac{7}{8}}$ **b)** $y = (4x + 3)^{\frac{3}{4}}$ **c)** $x^{\frac{1}{3}} + y^{\frac{1}{3}} = 1$

23. Determine $\frac{dy}{dx}$ for the relation $(x + y)^2 - (x - y)^2 = x^4 + y^4$.

24. Determine $\frac{dy}{dx}$ for each relation.

a) $x^2y^2 = x^2 + y^2$ **b)** $x^2y - xy^2 + x^2 + y^2 = 0$ **c)** $x^2 - xy + y^2 = 0$

25. Determine $\frac{d^2y}{dx^2}$ for each relation.

a) $xy + y^2 = 1$ **b)** $x^2 - xy + y^2 = 1$

Self-Check 4.4 – 4.6

1. Determine $y = f(g(x))$ and $y = g(f(x))$ for each pair of functions.

   a) $f(x) = 5x - 1;\ g(x) = 3x + 2$ b) $f(x) = \frac{1}{3 - x};\ g(x) = \frac{3x - 1}{x}$

2. Determine the domain of each composite function in exercise 1.

3. For the pair of functions, $f(x) = \frac{15}{x^2 + 1}$ and $g(x) = x^2 - 1$, determine:

   a) the domain and range of $y = f(x)$ and $y = g(x)$

   b) $y = f(g(x))$

   c) the domain of $y = f(g(x))$

   d) the range of $y = f(g(x))$

4. For each function, determine functions f and g so that $y = f(g(x))$.

   a) $y = \sqrt{\frac{3x + 1}{2}}$ b) $y = (3x + 2)^{-2}$ c) $y = (x^2 + 1)^2 + 1$

5. For each function f, determine $f^{-1}(x)$. Then show that $f(f^{-1}(x)) = f^{-1}(f(x)) = x$.

   a) $f(x) = \frac{3}{2x + 1}$ b) $f(x) = \frac{x + 1}{x - 2}$

6. Differentiate each function.

   a) $y = (x^2 + 2)^6$ b) $y = (x + 1)^3(2x - 1)^2$ c) $y = (3x + 5)^{-4}$

7. Differentiate each function.

   a) $y = \frac{1}{(x^2 + 3x)^4}$ b) $y = \left(\frac{2x + 1}{3x}\right)^3$ c) $y = \left(\frac{1 - 3x^2}{3x + 2}\right)^4$

8. Determine the equations of the tangents to the hyperbola $8x^2 - y^2 = 7$ at the points on the graph where $x = -2$.

9. The equation $x^2 + 2xy + y^2 = 1$ can be written $(x + y)^2 = 1$.

   a) Determine $\frac{dy}{dx}$ for each form of the equation.

   b) Interpret the results of part a in terms of the graph of the relation.

10. a) Show that the point $(-1, 3)$ lies on the graph of $x^2y^2 = 9$.

    b) Determine the equation of the tangent to the graph of $x^2y^2 = 9$ at the point $(-1, 3)$.

PERFORMANCE ASSESSMENT

11. Determine $\frac{dy}{dx}$ for each relation.

    a) $x^2 + 3y^2 = 3$ b) $4x^2 - 5y^2 = 20$ c) $x^2y = y^2x$

    d) $y = x^{\frac{3}{4}}$ e) $y = (7x - 2x^2)^{\frac{1}{2}}$ f) $y^2 + xy = y^3$

Review Exercises

Mathematics Toolkit

Derivative and Function Tools

- Product rule: $\frac{d}{dx}[f(x)(g(x))] = f'(x)g(x) + f(x)g'(x)$
- Quotient rule: $\frac{d}{dx}\left[\frac{f(x)}{g(x)}\right] = \frac{f'(x)g(x) - f(x)g'(x)}{[g(x)]^2}$
- Power of a function rule: for any rational number n, $\frac{d}{dx}[f(x)]^n = n[f(x)]^{n-1} \times f'(x)$
- Chain rule: $\frac{d}{dx}[f(g(x))] = f'(g(x)) \times g'(x)$
- Implicit differentiation: to differentiate an implicit equation, differentiate each term with respect to x, then solve for $\frac{dy}{dx}$.
- In the composition $y = f(g(x))$ or $f \circ g$, the function f is applied on the function g.
- The domain of $y = f(g(x))$ is the set of all x such that x is in the domain of g, and $g(x)$ is in the domain of f.
- For inverse functions $f(x)$ and $f^{-1}(x)$, $f(f^{-1}(x)) = f^{-1}(f(x)) = x$

4.1

1. Determine the derivative of each product.

   a) $y = 2x(1 - 3x)$ b) $y = -2x^2(3 + x)$ c) $y = (1 + 3x^2)(2 - 5x)$

   d) $y = x^3(x^2 - 5x + 7)$ e) $y = (x^3 + 1)(-4x + 2)$ f) $y = (2x^2 + x)(6 - 2x^3)$

2. Determine $\frac{d^2y}{dx^2}$ for each function in exercise 1.

3. Determine the equation of the tangent to the graph of $y = x^3(3 - x)$ at the point $(-1, -4)$.

4. An apple orchard contains 250 trees. Each tree produces an average annual yield of 200 apples. The farmer is expanding her orchard at the rate of 10 trees per year. The average annual yield is increasing at 7 apples per tree.

   a) Write an equation to represent the annual production, P, as a function of t, the number of years from now.

   b) Determine the current rate of increase in the annual production of apples.

Review Exercises

5. Determine $\frac{dy}{dx}$ for each function.

a) $y = -\sin x \cos x$

b) $y = x^3 \cos x$

c) $y = x^2\sqrt{x}$

d) $y = x\sqrt{x+1}$

4.2 6. Differentiate each quotient.

a) $y = \frac{x^2}{x-3}$

b) $y = \frac{-x^3}{x^2+2}$

c) $y = \frac{5-x^2}{x+1}$

d) $y = \frac{4x-x^2}{x^3-1}$

e) $y = \frac{x^2+x+3}{x^2-x-3}$

f) $y = \frac{x^3-3x}{2x^2+x+2}$

7. Determine the derivative of each reciprocal function.

a) $y = \frac{1}{x}$

b) $y = \frac{1}{x^2+2}$

c) $y = \frac{1}{5-x^2}$

d) $y = \frac{1}{x^3+2x}$

e) $y = \frac{-1}{x-2x^3}$

f) $y = \frac{-1}{3x^3+2x}$

8. Differentiate each function.

a) $y = \frac{1}{\sqrt{x}}$

b) $y = \frac{1}{\sin x}$

c) $y = \frac{-\cos x}{\sin x}$

d) $y = \frac{x}{\sqrt{x}}$

e) $y = \frac{-\sqrt{x}}{x+1}$

f) $y = \frac{5x}{\cos x}$

9. The graph of $y = \frac{x^2}{x-2}$ is shown below left. Determine the equation of the tangent to the graph at the point on the graph where $x = -2$.

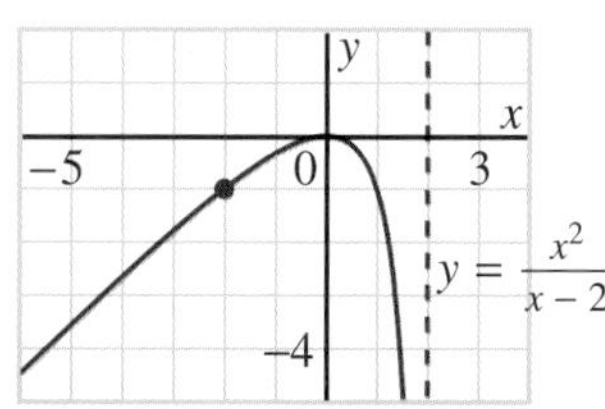

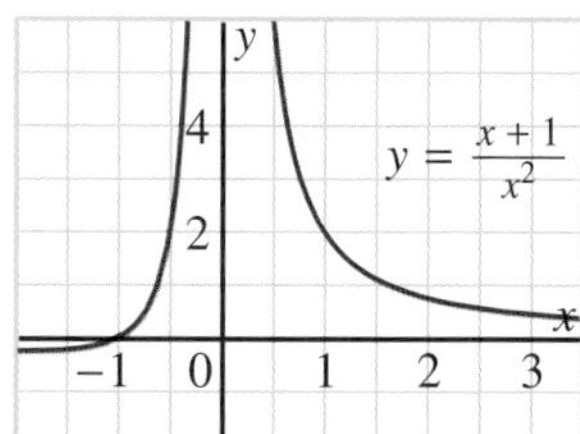

10. The graph of $y = \frac{x+1}{x^2}$ is shown above right. A tangent to this graph has slope –3. Determine the equation of the tangent.

4.3 11. Determine the derivative of each power.

a) $y = (3x-5)^4$

b) $y = (2-7x)^3$

c) $y = -(x^2+8x)^5$

d) $y = (3+4x)^{-2}$

e) $y = (x^2+x)^{-3}$

f) $y = (x^3-5x)^{-1}$

12. Determine the derivative of each reciprocal function.

a) $y = \frac{1}{(1+3x)^5}$

b) $y = \frac{-1}{(x^2-x^3)^3}$

c) $y = \frac{1}{(5-2x+3x^2)^2}$

13. Differentiate each function.

a) $y = \left(\frac{x}{x+1}\right)^6$

b) $y = \left(\frac{x-3}{x+5}\right)^4$

c) $y = \left(\frac{2x}{1-3x}\right)^2$

d) $y = \left(\frac{x - 2x^2}{x + 1}\right)^3$ e) $y = \left(\frac{5x^2 - x^3}{x + 1}\right)^2$ f) $y = \left(\frac{x^3 + 1}{x^3 - x}\right)^3$

14. Differentiate each function.

a) $y = \frac{-x}{(x + 1)^2}$ b) $y = \frac{(2 + 5x)^2}{x}$ c) $y = \frac{-(x + 1)^2}{(3 + x^2)^3}$

15. Differentiate each function.

a) $y = -\sin^2 x \cos^2 x$ b) $y = \frac{-\sin^2 x}{\cos^2 x}$ c) $y = \frac{\sin^3 x}{\cos x}$

4.4 16. Determine $y = f(g(x))$ and $y = g(f(x))$ for each pair of functions.

a) $f(x) = x + 3$; $g(x) = x^2 - 4$ b) $f(x) = \frac{-1}{x + 1}$; $g(x) = 1 - x^3$

c) $f(x) = \sqrt{x - 1}$; $g(x) = 2x + 3$ d) $f(x) = \frac{1}{1 - x}$; $g(x) = \sqrt{1 + x}$

17. Consider $f(x) = 4 - 3x$, $g(x) = \frac{1}{\sqrt{x + 1}}$, and $h(x) = \sqrt{x + 2}$. Determine the domain of each composite function.

a) $f \circ g$ b) $g \circ f$ c) $f \circ h$ d) $h \circ f$ e) $g \circ h$ f) $h \circ g$

18. The volume, V, and surface area, A, of a sphere are functions of its radius, r: $V = \frac{4}{3}\pi r^3$ and $A = 4\pi r^2$

a) Express the volume as a function of the diameter, d.

b) Express the surface area as a function of the diameter, d.

c) Express the volume as a function of the surface area.

d) Express the surface area as a function of the volume.

19. For each pair of functions, determine:

i) the domain of $y = f(x)$ and of $y = g(x)$

ii) the range of $y = f(x)$ and of $y = g(x)$

iii) $y = f(g(x))$ iv) the domain of $y = f(g(x))$ v) the range of $y = f(g(x))$

a) $f(x) = 1 + x^2$; $g(x) = \sqrt{x + 1}$ b) $f(x) = \frac{-1}{x^2}$; $g(x) = 1 - x^2$

20. For each function, determine $y = f(f(x))$.

a) $f(x) = 2x - x^2$ b) $f(x) = \sqrt{2 + x}$ c) $f(x) = \frac{1}{x^2 - 3}$

21. For each function, determine functions f and g so that $y = f(g(x))$.

a) $y = \frac{1}{(x + 1)^2}$ b) $y = \frac{-1}{x - 2}$ c) $y = (x + 3)^2 + 2(x + 3)$

d) $y = \sqrt{2x + 5}$ e) $y = (4 - 3x)^2$ f) $y = x^4 + 2x^2 + 1$

22. For each function f, determine a function g so that $f(g(x)) = g(f(x))$.

a) $f(x) = 5 - 2x^2$ b) $f(x) = \frac{2}{x + 1}$ c) $f(x) = \sqrt{x^2 - 1}$

Review Exercises

23. The graphs of $y = f(g(x))$ and $y = g(f(x))$ are shown below left, where $f(x) = \frac{1}{x+1}$ and $g(x) = x^2$. Identify each graph. Explain your answer.

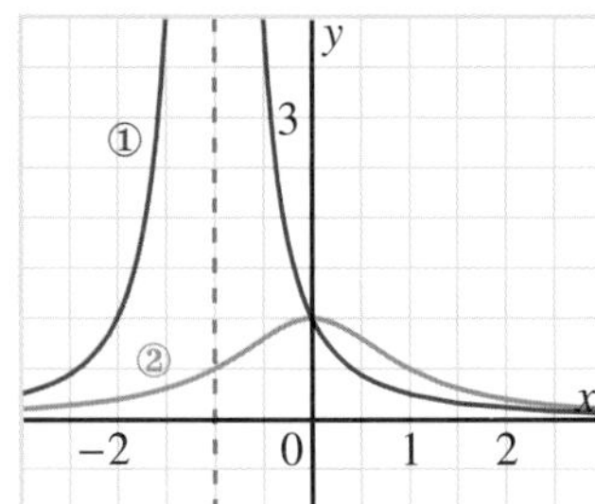

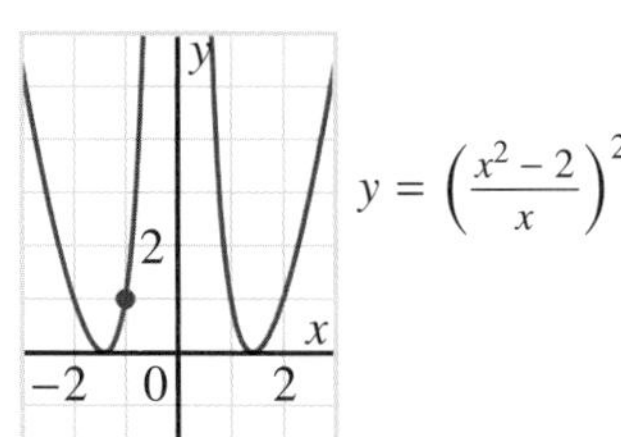

4.5 24. The graph of $y = \left(\frac{x^2-2}{x}\right)^2$ is shown above right. Determine the equation of the tangent to the graph at the point on the graph where $x = -1$.

25. Differentiate each function.

a) $y = (2x - 6)^3$ **b)** $y = -(x^2 - 5)^4$ **c)** $y = (-2x + 3x^2)^5$

d) $y = (-x^2 + 2)^{-2}$ **e)** $y = -(x^2 - 2)^{-3}$ **f)** $y = -(x - 2x^3)^{-4}$

26. Differentiate each function.

a) $y = \frac{1}{(x^2 + 3x)^3}$ **b)** $y = \frac{-1}{(5 - x^3)^2}$ **c)** $y = \frac{1}{(-2x + x^4)^4}$

27. Differentiate each function.

a) $y = \left(\frac{3-x}{2+x}\right)^5$ **b)** $y = \left(\frac{3-x^2}{2-x}\right)^3$ **c)** $y = \left(\frac{3x-x^2}{2+x^3}\right)^4$

4.6 28. Determine $\frac{dy}{dx}$ for each relation.

a) $y^2 = 3x$ **b)** $y^2 + x = 2y$ **c)** $x^2 + y = 2x$

d) $y^3 + x = 1$ **e)** $2y^2 - 2x^2 = x$ **f)** $3y^2 - x^2 = y$

29. **a)** Determine $\frac{dy}{dx}$ for the hyperbola $x^2 - y^2 = 9$.

b) Determine the equation of the tangent to the hyperbola at the point $(-5, 4)$.

30. **a)** Determine $\frac{dy}{dx}$ for the ellipse $2x^2 + y^2 = 6$.

b) Determine the equations of the tangents to the ellipse at the two points on the ellipse where $x = 1$.

31. Determine $\frac{dy}{dx}$ for each relation.

a) $y = (2x^2 + 1)^{\frac{1}{3}}$ **b)** $y = (x - x^3)^{-\frac{1}{2}}$ **c)** $x^{\frac{2}{3}} + y^{\frac{2}{3}} = 1$

32. Determine $\frac{dy}{dx}$ for each relation.

a) $x^3y^3 = x^3 + y^3$ **b)** $x^2 + 2xy + y^2 = 1$

Self-Test

1. **Knowledge/Understanding** Determine $\frac{dy}{dx}$ for each relation.

   a) $y = 2x^2(3x - 5)$
   b) $y = \frac{x^2 + 2x - 1}{x^2 - 5x + 7}$
   c) $y = (5x^2 - 1)^4$
   d) $y = \sqrt{\frac{x}{x - 5}}$
   e) $y^2 - 9x^2 = 36$
   f) $xy^2 + 2y^3 = x$

2. Determine $\frac{d^2y}{dx^2}$ for the functions in exercise 1a, b, and c.

3. Determine the equations of the tangents to the ellipse $4x^2 + y^2 = 100$ at the two points where $y = 8$.

4. Determine the coordinates of the point(s) on the graph of $y = (1 - 2x^2)^2$ where the slope of the tangent is –8.

5. **Thinking/Inquiry/Problem Solving** Consider the functions $f(x) = x^2 + 2$ and $g(x) = \sqrt{k - x^2}$, where k is a constant. For which values of k will the composition $y = f(g(x))$ exist? Explain.

6. **Communication** "The composition of two functions exists only when the range of the first function overlaps the domain of the second function." Explain the meaning of this statement. Include two functions and their composite in your explanation.

7. Determine two functions f and g for which $f(g(x)) = g(f(x))$. Are these functions unique? Explain.

8. **Application** The volume of water, V litres, in a tank t minutes after the tank has started to drain is given by $V(t) = 1000(25 - t)^2,\ 0 \le t \le 25$. How fast is the water draining after 8 min? Explain.

9. The Body Mass Index (BMI) is a measure of a person's physical fitness. An equation for BMI is $B = \frac{m}{h^2}$, where m is a person's mass in kilograms, and h is that person's height in metres. The desirable range for BMI is between 20 and 25. Suppose a person has a mass of 55 kg.

   a) What is the rate of change of BMI with respect to height? Explain.

   b) Calculate $\frac{dB}{dh}$ for a height of 1.75 m. Explain what it means.

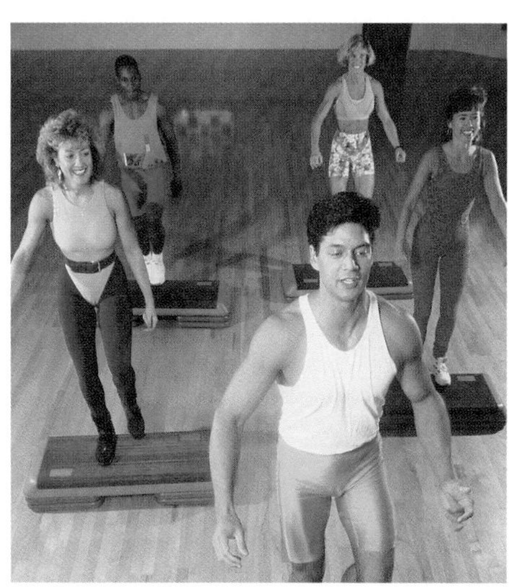

Self-Test

10. a) Differentiate each function.

i) $y = x(x + 1)(x + 2)$

ii) $y = (x + 1)(x + 2)(x + 3)$

iii) $y = (x + 2)(x + 3)(x + 4)$

iv) $y = (x + 3)(x + 4)(x + 5)$

b) Describe any patterns in part a. Predict $\frac{dy}{dx}$ for $y = (x + k)(x + k + 1)(x + k + 2)$, where k is a positive integer. Check your prediction by differentiating.

11. Consider the function $f(x) = \frac{x+1}{x+2}$.

a) Determine each composition.

i) $y = f(f(x))$ **ii)** $y = f(f(f(x)))$ **iii)** $y = f(f(f(f(x))))$

b) Describe any patterns in part a. Predict the results of further compositions of $f(x)$ with itself.

PERFORMANCE ASSESSMENT

12. Consider the linear function $f(x) = 2x + 1$.

a) Show that $f(f(x)) = 4x + 3$.

b) Find another linear function g so that $g(g(x)) = 4x + 3$.

13. The function $y = \sqrt{\frac{1}{2x+3}}$ can be considered as the composition of three functions.

a) Determine functions f, g, and h so that $y = f(g(h(x)))$.

b) Describe how the chain rule can be used to determine $\frac{dy}{dx}$.

14. At what points on the ellipse $16x^2 + 9y^2 = 144$ is the tangent parallel to the line $y = x$? Explain.

15. Points A(4, 4) and B(1, –2) lie on the parabola $y^2 = 4x$. The midpoint of AB is C. A line is drawn through C parallel to the x-axis to meet the parabola at D. Show that the tangent at D is parallel to AB.

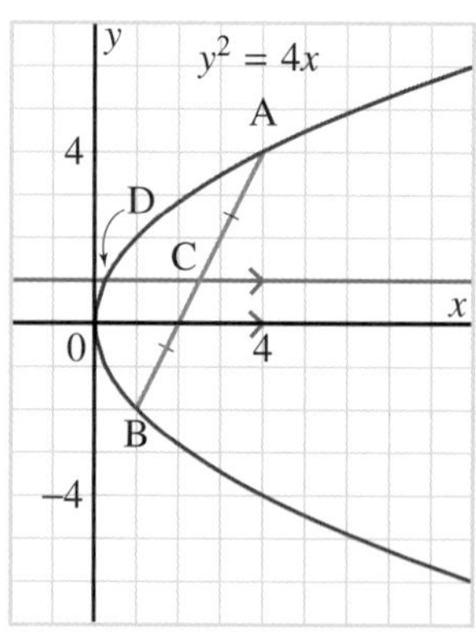

Mathematical Modelling

Optimal Profits

There is an old saying that you have to spend money to make money. In most businesses, we have to make an initial investment to get started, then pay expenses such as salaries and supplies. A typical problem is to find the right level of investment. If this is either too small or too large, your profit will be less than it could be.

Problems in economics and management are often of this type. To make a profit we have to make an investment, and the greater the investment the greater the profit. But, on the other hand, a greater investment means a greater cost. Somewhere there should be an investment level that maximizes the profit.

Here is an example. Suppose you have converted an old warehouse into a widget-making factory. You need to decide how many workers to hire. You need at least 20 workers to operate the machinery, but the space and equipment you have suggests that it should be profitable to have many more than that. However, too many workers will crowd the facilities and profitability will decrease. You need to know how the number of workers is related to the daily revenue. There is no simple formula for this, but there is an empirical graph for operations of this type and size, and it is shown at the right.

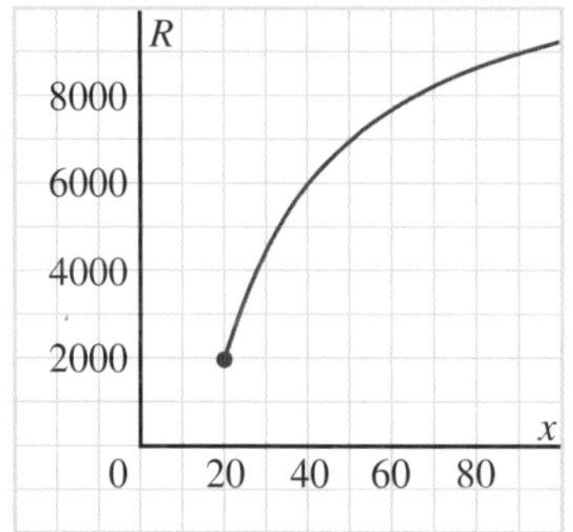

x workers produce a daily revenue, R dollars.

Formulating a Hypothesis

From the graph, when $x = 20$ workers (the minimum viable number), the revenue is \$2000 per day. The revenue quickly increases with extra workers. For example, 40 workers provide a revenue of \$6000 per day. But, as x increases, the rate of return (the slope of the graph) decreases, and 100 workers provide a revenue of less than \$10 000 per day. We want to maximize the profit. However, more workers will incur higher wage costs and that will reduce the profit. We make the following hypothesis:

Somewhere between 40 and 100 workers, there should be an optimum number that maximizes the daily profit.

Developing a Mathematical Model

Profit is revenue minus cost.

$P = R - C$

In this model, R is the *net* revenue, which takes account of the cost of materials. So the cost, C, that remains is the fixed costs and workers' wages. These are as follows:

Fixed costs: $1000/day
Wage per worker: $80/day

To understand net revenue, consider the graph on page 289. According to the graph, 20 workers produce a net revenue of $2000/day. Twenty workers can make 100 widgets per day. Each widget sells for $30, but consumes $10 in raw materials. The net revenue is:
100(30 – 10) = 2000

For x workers, the cost is given by:

$$C = 1000 + 80x$$

The daily profit is given by:

$$P = R - C$$
$$P = R - (1000 + 80x) \qquad \ldots ①$$

This is what we need to maximize. We cannot do that yet because there are two variables, R and x. But R is a function of x, as shown on the graph on page 289. If we had an equation for this function, we could use it to simplify equation ① and express P as a function of x.

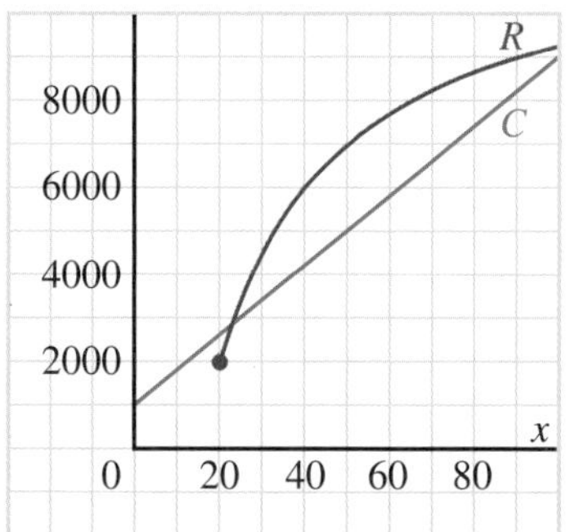

Since we have no equation for $R(x)$, we use the graph of $R(x)$ and equation ①. At the right, we plot revenue, R, and cost, C, on the same grid. Since C is a linear function of x, it has a straight-line graph.

Since P is the difference between R and C, it is represented as the vertical distance between the two graphs.

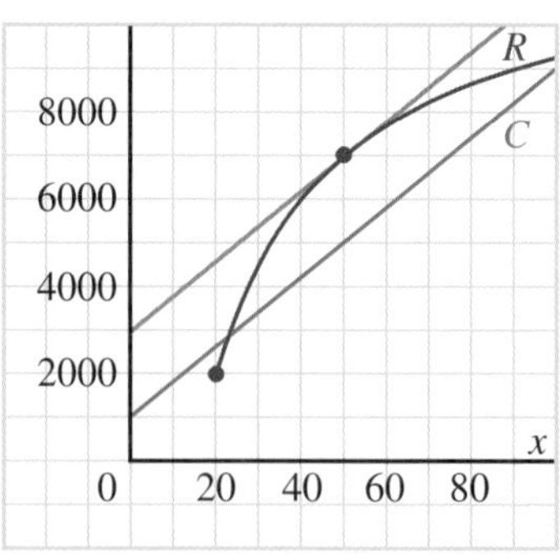

Hence, we choose x to make the vertical difference between the graphs as large as possible. This happens at the value of x where the R-graph is as far as possible above the C-line. Visualize translating the C-line vertically until it just loses contact with the R-graph. That last point of contact is the highest point and will give us the optimum value of x. This is the point where the tangent to the graph of R has the same slope as the line C. On the graph at the right, it appears to occur near $x = 50$.

To maximize your profit, you should employ 50 workers. With this number, the graph shows that $P = R - C = 2000$, which represents a maximum profit of $2000 per day.

Key Features of the Model

Our model for solving the optimal profits problem consists of the graph of $R(x)$ and equation ①. This model is significant because the profit is not represented on the axes of the graph. It is represented by the vertical distance between the curve and the line. We can see how the profit changes as x changes. For example, with $x = 20$ workers, the graph of R is below the graph of C and the profit is negative (we lose money). We need at least 23 workers to make any profit. Then, as the number of workers increases, the profit increases. The profit is a maximum near $x = 50$. For more than 50 workers, the profit decreases.

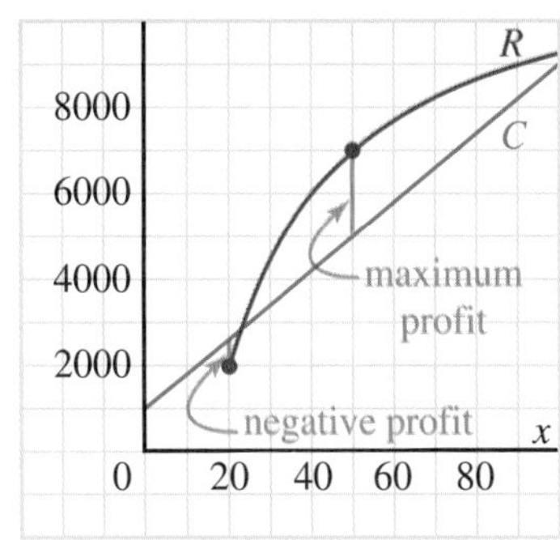

The equation of the revenue graph is:

$$R = \frac{12\,000(x - 15)}{x + 10} \text{ for } x \geq 20$$

Since the equation contains the same information as the graph, we can use it to solve the optimal profits problem.

In applications of mathematics to the social sciences, the functions we work with are not usually given by equations. All we know is something about the general shape and position of the graphs. This is why the graphical model for solving the optimal profits problem is so important. We need to be able to use the graph to solve the problem.

1. a) Substitute the expression for R in equation ① to obtain an equation that expresses the profit, P, as a function of the number of workers, x.

b) Determine the derivative, $\frac{dP}{dx}$.

c) Determine the value of x that produces the maximum profit.

d) How can you be certain that the point you found is a maximum, and not a minimum or a point of inflection?

Making Inferences from the Models

We now have two models for solving the optimal profits problem:

Graphical model: the graph and equation ①

Algebraic model: the equation in exercise 1a

Use both models in exercises 2, 3, and 4. To use the graphical model, you will need a copy of the graph on page 289.

2. Suppose a worker's wage was \$100/day instead of \$80/day.

a) Express the profit, P, as a function of x.

b) Determine the number of workers who should be employed to produce the maximum profit.

c) How high could a worker's wages be before the business becomes unprofitable?

3. Suppose the fixed costs were \$2000/day instead of \$1000/day.

   a) Determine the number of workers who should be employed to produce the maximum profit.

   b) How high could a worker's wages be before the business becomes unprofitable?

4. Suppose you wanted to maximize the net revenue, R, produced per worker. How many workers would you hire? Explain.

Related Problems

5. Use the equation $P = R - (1000 + 80x)$, where R is a function of x.

   a) Determine an expression for $\frac{dP}{dx}$. Do not substitute for R.

   b) Use the result of part a to show that the maximum profit occurs when the graph of R has slope 80.

   c) Explain why the result in part b is the same as the result on page 290.

For exercise 6, you need a copy of the graph at the right.

6. You want to know how much to spend to advertise the school concert on the local radio station. The more you advertise, the more tickets you will sell (up to a point), but the higher your advertising costs will be. A graph that shows how the expected revenue, R dollars, from ticket sales depends on the number of advertising minutes, x, is shown at the right. A ticket costs \$10, so $R = 10T$, where T is the number of tickets sold.

   a) Do you think the shape of the graph is reasonable? Explain.

   b) Suppose advertising costs \$30/min. Determine the value of x that produces a maximum profit, P (ticket revenue minus advertising costs).

   c) Let c denote the advertising rate in dollars per minute. How high can c be before you should decide not to advertise?

   d) The graph of R against x is a parabola with vertex at $x = 50$. Determine the equation of the parabola, then use it for an algebraic verification of your answers to parts b and c.

Revenue from Ticket Sales

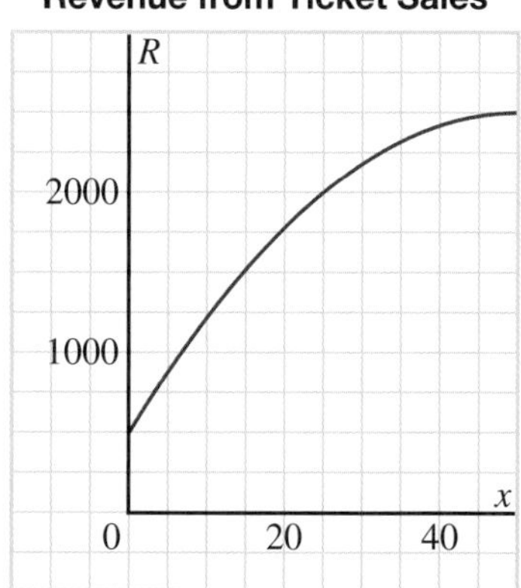

With no advertising ($x = 0$), you will sell 50 tickets at \$10/ticket, for a revenue of \$500.

7. Write a report to describe what you have learned about optimal profits. In your report, include an answer to this question: How much should be invested to produce a maximum profit? Include some equations and graphs in your report.

Rational Functions 5

Curriculum Expectations

By the end of this chapter, you will:

- Describe the key features of a given graph of a function, including intervals of increase and decrease, critical points, points of inflection, and intervals of concavity.
- Identify the nature of the rate of change of a given function, and the rate of change of the rate of change, as they relate to the key features of the graph of that function.
- Sketch, by hand, the graph of the derivative of a given graph.
- Determine the limit of a rational function.
- Demonstrate an understanding that limits can give information about some behaviours of graphs of functions.
- Identify examples of discontinuous functions and the types of discontinuities they illustrate.
- Determine the derivatives of rational functions, using the constant, power, sum and difference, product, quotient, and chain rules for determining derivatives.
- Determine second derivatives.
- Determine the equation of the tangent to the graph of a rational function.
- Determine, from the equation of a rational function, the intercepts and the positions of the vertical and horizontal or oblique asymptotes to the graph of the function.
- Determine, from the equation of a rational function, the key features of the graph of the function, using the techniques of differential calculus, and sketch the graph by hand.
- Sketch the graph of the first and second derivative functions, given the graph of the original function.
- Sketch the graph of a function, given the graph of its derivative function.

1. Review: Rational Expressions

Recall that a rational expression is an expression that can be written in the form $\frac{p(x)}{q(x)}$, where $p(x)$ and $q(x)$ are polynomials, and $q(x) \neq 0$. A rational expression is not defined when its denominator is 0. A value of x that makes the denominator 0 is a *non-permissible value*.

Example

State the non-permissible value(s) for each rational expression.

a) $\frac{x}{x-4}$ **b)** $\frac{x^3+1}{x^2+5x}$ **c)** $\frac{x-3}{x^2-9}$ **d)** $\frac{x}{x^2+2}$

Solution

A non-permissible value is the value of the variable that makes the denominator 0. Therefore, equate each denominator to 0, then solve the resulting equation.

a) $\frac{x}{x-4}$

Let $x - 4 = 0$.

$x = 4$

Therefore, $x = 4$ is a non-permissible value.

b) $\frac{x^3+1}{x^2+5x}$

Let $x^2 + 5x = 0$.

$x(x + 5) = 0$

Either $x = 0$ or $x + 5 = 0$

$x = -5$

Therefore, $x = 0$ and $x = -5$ are non-permissible values.

c) $\frac{x-3}{x^2-9}$

Let $x^2 - 9 = 0$.

$(x - 3)(x + 3) = 0$

$x = \pm 3$

Therefore, $x = 3$ and $x = -3$ are non-permissible values.

d) $\frac{x}{x^2+2}$

Let $x^2 + 2 = 0$.

$x^2 = -2$

Since the square of any real number is never negative, the expression $x^2 + 2$ is never equal to 0.

Therefore, all real values of x are permissible.

Exercises

1. Evaluate each rational expression for $x = 2$.

a) $\frac{x + 3}{x^2 + 3x}$ **b)** $\frac{x - 2}{x + 1}$ **c)** $\frac{x}{x - 2}$ **d)** $\frac{x - 2}{x^2 - 4}$ **e)** $\frac{1}{x^2 + 4}$

2. Suppose you evaluate each rational expression in exercise 1 for $x = 0$. Which expression(s) are not defined? Explain.

3. Determine the non-permissible values of x for each rational expression.

a) $\frac{x + 1}{2x}$ **b)** $\frac{-3}{x + 5}$ **c)** $\frac{x^3}{x^2 - 1}$

d) $\frac{x^2 - 4}{x^2 + 4}$ **e)** $\frac{x - 6}{x^2 - 2x - 24}$ **f)** $\frac{x^2 - 81}{x - 9}$

4. For which value(s) of x is each rational expression not defined?

a) $\frac{x^2 + 2x}{3x^3}$ **b)** $\frac{6 - x^3}{3 - x}$ **c)** $\frac{5x}{4x^2 - 9}$

d) $\frac{x^3}{2x^2 + 1}$ **e)** $\frac{x - 1}{x^2 - 1}$ **f)** $\frac{x^4}{x^3 - 25x}$

2. New: Equivalent Forms of Rational Expressions

To change the form of a rational expression, we divide.

Dividing a Polynomial by a Monomial

Dividing a polynomial by a monomial is similar to dividing in arithmetic. Compare the steps in these two examples.

$$\frac{17}{6} = \frac{12 + 5}{6} = \frac{12}{6} + \frac{5}{6} = 2 + \frac{5}{6}$$

The rational number $\frac{17}{6}$ is equivalent to the expression $2 + \frac{5}{6}$.

$$\frac{x^2 + 2x + 3}{x} = \frac{x^2}{x} + \frac{2x}{x} + \frac{3}{x} = x + 2 + \frac{3}{x}, \text{ where } x \neq 0$$

The rational expression $\frac{x^2 + 2x + 3}{x}$ is equivalent to the expression $x + 2 + \frac{3}{x}$, where $x \neq 0$. This is the sum of a polynomial and a rational expression.

Necessary Skills

Example 1

Use division to write the rational expression $\frac{4x^2 - 3x + 6}{2x}$ as the sum of a polynomial and another rational expression.

Solution

$$\frac{4x^2 - 3x + 6}{2x} = \frac{4x^2}{2x} - \frac{3x}{2x} + \frac{6}{2x}$$
$$= 2x - \frac{3}{2} + \frac{3}{x}$$

Therefore, $\frac{4x^2 - 3x + 6}{2x} = 2x - \frac{3}{2} + \frac{3}{x}, x \neq 0$

Dividing a Polynomial by a Binomial

Recall, from Chapter 3, the steps in the division $(6x^2 + 7x + 9) \div (2x + 1)$.

$$\begin{array}{r} 3x + 2 \\ 2x + 1 \overline{)\, 6x^2 + 7x + 9} \\ \underline{6x^2 + 3x} \\ 4x + 9 \\ \underline{4x + 2} \\ 7 \end{array}$$

The rational expression $\frac{6x^2 + 7x + 9}{2x + 1}$ is equivalent to $3x + 2 + \frac{7}{2x + 1}$.

Example 2

Use division to write the rational expression $\frac{4x + 6x^3 - 33}{2x - 4}$ as the sum of a polynomial and another rational expression.

Solution

Recall that the divisor and dividend are written in descending powers of x, with any missing powers inserted with zero coefficients. Rearrange, then rewrite $4x + 6x^3 - 33$ as $6x^3 + 0x^2 + 4x - 33$.

$$\begin{array}{r} 3x^2 + 6x + 14 \\ 2x - 4 \overline{)\, 6x^3 + 0x^2 + 4x - 33} \\ \underline{6x^3 - 12x^2} \qquad\qquad \\ 12x^2 + 4x \qquad \\ \underline{12x^2 - 24x} \qquad \\ 28x - 33 \\ \underline{28x - 56} \\ 23 \end{array}$$

Therefore, $\frac{6x^3 + 4x - 33}{2x - 4} = 3x^2 + 6x + 14 + \frac{23}{2x - 4}, x \neq 2$

Exercises

1. Use division to write each rational expression as the sum of a polynomial and another rational expression.

   a) $\frac{x+3}{x}$ b) $\frac{x^2-5x+4}{x}$ c) $\frac{4x^3+10x^2-5}{2x}$

2. Use division to write each rational expression as the sum of a polynomial and another rational expression.

   a) $\frac{2x^2+5x-1}{x+1}$ b) $\frac{3x^2+2x-5}{x-2}$ c) $\frac{x^3-5x^2+10x-15}{x-3}$

   d) $\frac{x^2-2}{x+4}$ e) $\frac{6x^3}{2x+1}$ f) $\frac{6x^2-3}{2x+4}$

3. Use division to write each rational expression as the sum of a polynomial and another rational expression.

   a) $\frac{x^2-9}{x-3}$ b) $\frac{x^2-9}{x+3}$ c) $\frac{x^2+9}{x-3}$ d) $\frac{x^2+9}{x+3}$

4. In exercise 3, the remainder was 0 for two rational expressions. How are the dividend and the divisor related in each case?

5. a) Divide $4x^3+x^2-2x+1$ by x^2-3.

   b) Use the result from part a to write the rational expression $\frac{4x^3+x^2-2x+1}{x^2-3}$ in the form $Q+\frac{R}{D}$, where Q is the quotient, R is the remainder, and D is the divisor.

   c) For a polynomial, P, when the rational expression $\frac{P}{D}$ is written in the form $Q+\frac{R}{D}$, must R always be a constant? If not, how are the remainder, R, and the divisor, D, related?

3. Review: Graphs of Polynomial Functions

Recall the key features of the graph of a polynomial function $y=f(x)$. The methods that follow were developed in Chapter 3 using polynomial functions, but these methods apply to all functions. In this chapter, you will use these methods with rational functions.

Necessary Skills

Intercepts

To determine the y-intercept, calculate $f(0)$.
To determine the x-intercepts, solve $f(x) = 0$.

Critical points

At a critical point, the graph of a polynomial function has a horizontal tangent. To find the critical values, determine the derivative f'. The critical values are the solutions of the equation $f'(x) = 0$. A critical point can be a local maximum, a local minimum, or a point of inflection.

Intervals of increase and decrease

When $f'(x) > 0$, the function f is increasing as x increases.
When $f'(x) < 0$, the function f is decreasing as x increases.

Points of inflection

At a point of inflection, $(a, f(a))$, the graph of the function changes concavity. To find the points of inflection, determine the second derivative f''. The x-coordinates of possible points of inflection are solutions of the equation $f''(x) = 0$. The sign of f'' on each side of a possible point of inflection must be examined to determine the concavity. If the sign changes, then the concavity of the graph of the function $y = f(x)$ changes, and there is a point of inflection.

Intervals of concavity

When $f''(x) > 0$, the graph of the function $y = f(x)$ is concave up.
When $f''(x) < 0$, the graph of the function $y = f(x)$ is concave down.

Example

Analyse the key features of the graph of $f(x) = x^3 - 27x$, then sketch the graph.

Solution

Intercepts

$f(x) = x^3 - 27x$

The y-intercept is $f(0) = 0$.

Solve $f(x) = 0$ to determine the x-intercepts.

$$x^3 - 27x = 0$$
$$x(x^2 - 27) = 0$$
$$x = 0 \text{ or } x = \pm\sqrt{27}$$

Therefore, the x-intercepts are 0 and $\pm\sqrt{27}$, or 0 and approximately ± 5.2.

Critical points

Consider the first derivative of $f(x)$.

$f(x) = x^3 - 27x$

Differentiate.

$f'(x) = 3x^2 - 27$

Solve $f'(x) = 0$ to determine the critical values.

$$3x^2 - 27 = 0$$
$$3x^2 = 27$$
$$x^2 = 9$$
$$x = \pm 3$$

Therefore, there are critical points at $x = \pm 3$.

Intervals of increase and decrease

Use a number-line analysis to determine the sign of f' in each interval.

Consider $f'(x) = 3x^2 - 27$ and the critical values.

When $x < -3$, say $x = -3.1$, $f'(x) = 3(-3.1)^2 - 27$
$= 1.83$, which is positive

When $-3 < x < 3$, say $x = 0$, $f'(x) = 3(0)^2 - 27$
$= -27$, which is negative

When $x > 3$, say $x = 3.1$, $f'(x) = 3(3.1)^2 - 27$
$= 1.83$, which is positive

Direction of f	↗	→	↘	→	↗
x		-3		3	
Sign of f'	+	0	−	0	+

Therefore, the function f is increasing when $x < -3$ or $x > 3$; that is, when $|x| > 3$.

The function is decreasing when $-3 < x < 3$; that is, when $|x| < 3$.

Points of inflection

Consider the second derivative of $f(x)$.

$f'(x) = 3x^2 - 27$

Differentiate.

$f''(x) = 6x$

Solve $f''(x) = 0$ to determine the possible points of inflection.

$6x = 0$

$x = 0$

There is a possible point of inflection when $x = 0$.

Intervals of concavity

Use a number-line analysis to determine the sign of f'' in each interval.

Use $f''(x) = 6x$.

When $x < 0$, say $x = -1$, f'' is negative.

When $x > 0$, say $x = 1$, f'' is positive.

The graph of the function $y = f(x)$ is concave down when $x < 0$ and concave up when $x > 0$.

Shape of f	⌒	Point of inflection	⌣
x		0	
Sign of f''	−	0	+

Since f'' changes sign at $x = 0$, the graph of the function $y = f(x)$ has a point of inflection at $x = 0$. When $x = 0$, $f(0) = 0$; so, the coordinates of the point of inflection are (0, 0).

Use the information to sketch the graph.

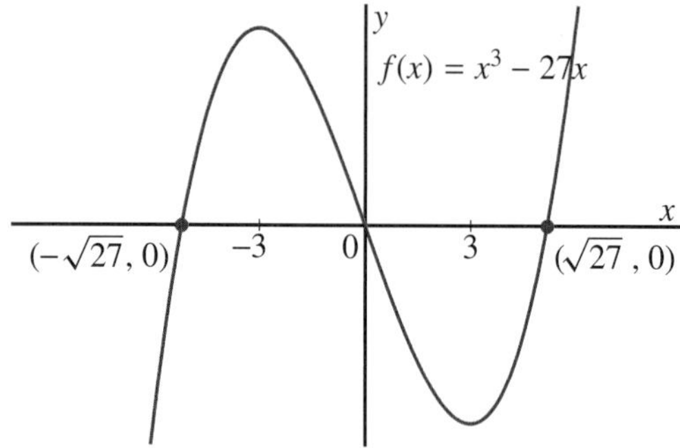

To draw a more accurate graph, we would determine the y-coordinates of the critical points.

Exercises

1. Use the graph of the polynomial function $y = f(x)$, below left, to estimate the intervals of increase, decrease, and concavity.

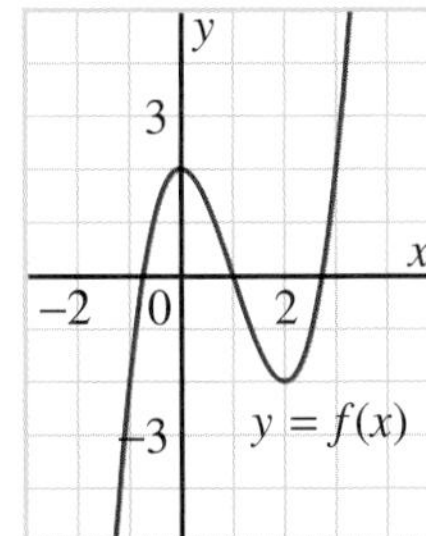

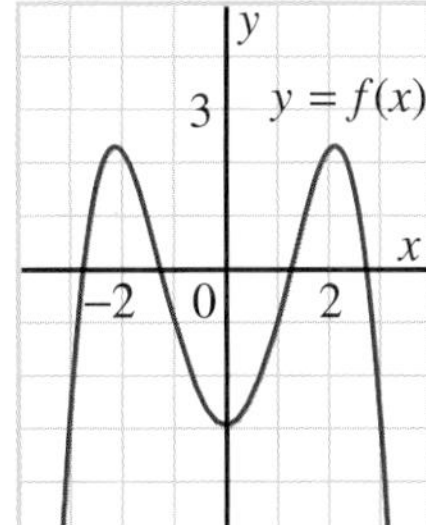

2. Use the graph of the polynomial function $y = f(x)$, above right, to estimate where f' and f'' are 0, positive, and negative. Explain.

3. Sketch a graph of a polynomial function $y = f(x)$ that satisfies these conditions:

$f(0) = 0, f(3) = -2, f(4) = 0$
$f'(0) = 0, f'(3) = 0$
$f'(x) < 0$ when $x < 3$; $f'(x) > 0$ when $x > 3$
$f''(x) < 0$ when $0 < x < 2$; $f''(x) > 0$ when $x < 0$ or $x > 2$

4. Analyse the key features of the graph of each polynomial function, then sketch the graph.

a) $y = x^2 + 2x - 3$ **b)** $y = x^3 - 3x^2 + 20$

c) $y = x^4 - 6x^2 + 9$ **d)** $y = 3x^4 + 16x^3 + 24x^2$

4. Review: Differentiation Tools

Recall the differentiation tools you have used throughout this text. Two of these rules are important when determining the derivative of a rational function.

The Power of a Function Rule

$$\frac{d}{dx}[f(x)]^n = n[f(x)]^{n-1} \times f'(x)$$

Quotient Rule

$$\frac{d}{dx}\left[\frac{f(x)}{g(x)}\right] = \frac{f'(x)g(x) - f(x)g'(x)}{[g(x)]^2}$$

Example 1

Determine the first and second derivative of the rational function $y = \frac{x+1}{x-1}$.

Solution

$y = \frac{x+1}{x-1}$

Use the quotient rule.

$$\begin{aligned} y' &= \frac{1(x-1)-(x+1)(1)}{(x-1)^2} \\ &= \frac{x-1-x-1}{(x-1)^2} \\ &= \frac{-2}{(x-1)^2} \end{aligned}$$

Therefore, $y' = \frac{-2}{(x-1)^2}$

To determine y'', write y' using negative exponents.

$y' = -2(x-1)^{-2}$

Use the power of a function rule.

$$\begin{aligned} y'' &= -2(-2(x-1)^{-3}) \\ &= 4(x-1)^{-3} \\ &= \frac{4}{(x-1)^3} \end{aligned}$$

Therefore, $y'' = \frac{4}{(x-1)^3}$

Example 2

Determine the first derivative of each rational function.

a) $y = \frac{x^3}{x^2-x-6}$

b) $y = \frac{x+5}{x^2-25}$

Solution

a) $y = \frac{x^3}{x^2-x-6}$

Use the quotient rule.

$$\begin{aligned} y' &= \frac{3x^2(x^2-x-6)-(x^3)(2x-1)}{(x^2-x-6)^2} \\ &= \frac{x^2(3x^2-3x-18-2x^2+x)}{(x^2-x-6)^2} \\ &= \frac{x^2(x^2-2x-18)}{(x^2-x-6)^2} \end{aligned}$$

Therefore, $y' = \frac{x^2(x^2 - 2x - 18)}{(x^2 - x - 6)^2}$

b) $y = \frac{x + 5}{x^2 - 25}$

Factor, then simplify before differentiating.

$$\begin{aligned} y &= \frac{x + 5}{x^2 - 25} \\ &= \frac{x + 5}{(x + 5)(x - 5)} \\ &= \frac{1}{x - 5}, \ x \neq -5 \end{aligned}$$

Use the power of a function rule.

$$\begin{aligned} y &= \frac{1}{x - 5} \\ &= (x - 5)^{-1} \end{aligned}$$

$$y' = -(x - 5)^{-2}$$

Therefore, $y' = \frac{-1}{(x - 5)^2}, \ x \neq -5$

Exercises

1. Determine the first and second derivative of each rational function.

a) $y = \frac{1}{x}$

b) $y = \frac{1}{x + 2}$

c) $y = \frac{1}{x^2 - 4}$

d) $y = \frac{1}{(x - 3)^2}$

e) $y = \frac{x}{x + 6}$

f) $y = \frac{x + 2}{x - 5}$

2. Determine the first derivative of each rational function.

a) $y = \frac{x^2 - 2}{x^2 + 3}$

b) $y = \frac{x^3}{x^2 - 25}$

c) $y = \frac{x^3 - 1}{x^2}$

d) $y = \frac{x}{x^2 - x - 12}$

e) $y = \frac{x^2}{(x^2 + 1)^3}$

f) $y = \frac{x}{(x^3 - 5)^2}$

3. Determine the first derivative of each rational function.

a) $y = \frac{x^2 - 4}{x - 2}$

b) $y = \frac{x - 2}{x^2 - 4}$

c) $y = \frac{x^3 + 3x}{x}$

d) $y = \frac{x}{x^2 + 3x}$

5.1 Rational Functions and Their Derivatives

We introduced a new tool, the derivative, in Chapters 1 and 2 then used it in Chapter 3 to study polynomial functions. In Chapter 4, we developed some rules for differentiating more complicated functions. In particular, the quotient rule was used to differentiate rational functions. A function whose defining equation contains a rational expression is called a *rational function*. Thus, a rational function has an equation of the form $f(x) = \frac{p(x)}{q(x)}$, where $p(x)$ and $q(x)$ are polynomial functions, and $q(x) \neq 0$. We now study these rational functions in more detail using the differentiation tools from Chapter 4.

Investigation

Rational Functions

1. The graphs of six rational functions are shown. Their equations are listed on the next page. Match each graph with its equation. Consider intercepts, domain restrictions, and what happens when x is large.

a)

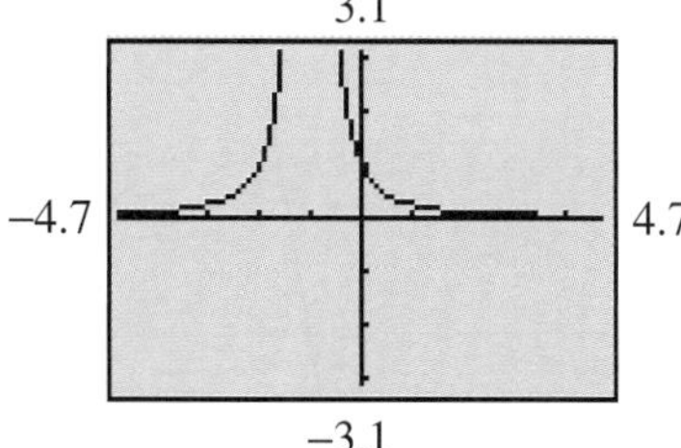

b)

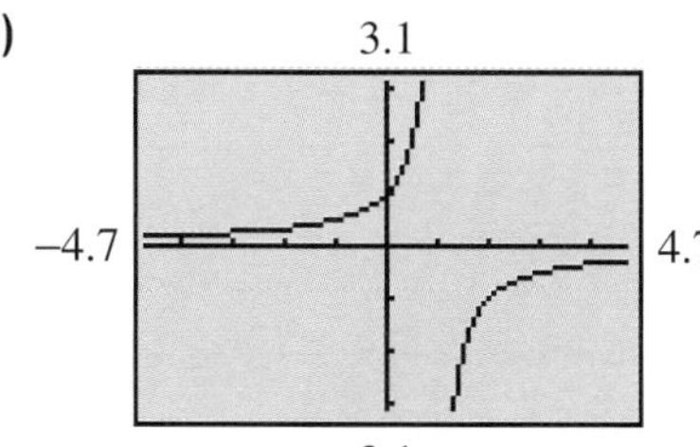

c)

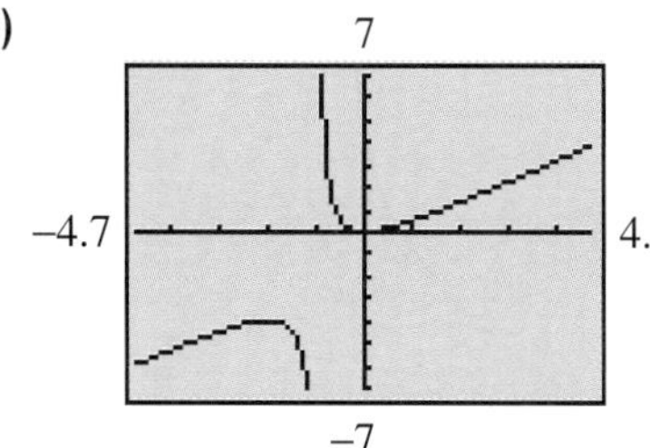

d)

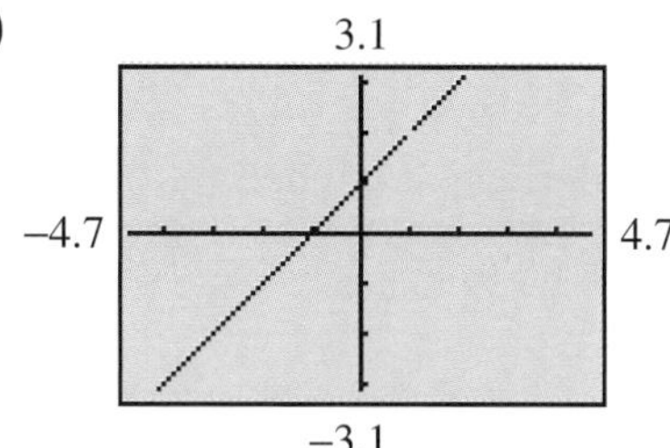

e)

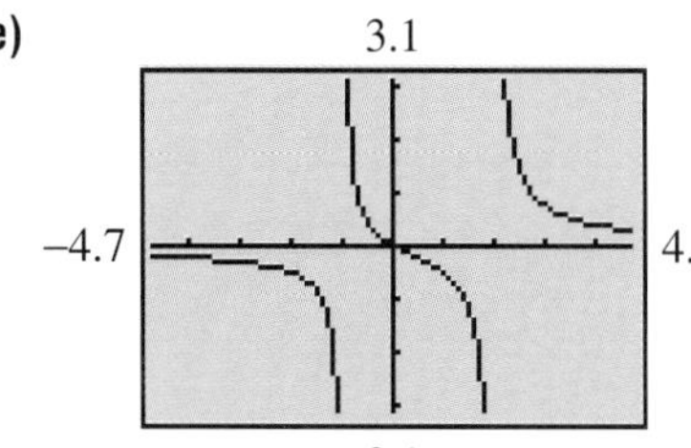

f)

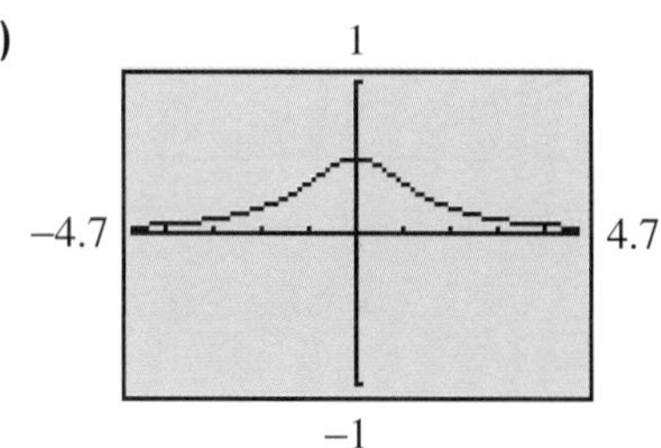

i) $y = \dfrac{x^2 - 1}{x - 1}$ **ii)** $y = \dfrac{1}{(x + 1)^2}$ **iii)** $y = \dfrac{1}{x^2 + 2}$

iv) $y = \dfrac{-1}{x - 1}$ **v)** $y = \dfrac{x}{(x + 1)(x - 2)}$ **vi)** $y = \dfrac{x^2}{x + 1}$

2. Consider the rational function $f(x) = \dfrac{p(x)}{q(x)}$. Describe what happens to its graph for any value of $x = a$ when the denominator $q(a) = 0$.

3. When the graph of a function can be drawn without lifting the pencil, we say that the function is *continuous* for all values of x. When the graph of a function cannot be drawn without lifting the pencil, we say that the function is *discontinuous*. The functions in exercise 1b and d are discontinuous when $x = 1$.

a) Describe the two different types of discontinuities that occur at $x = 1$.

b) Examine the corresponding equations, then make a conjecture about how you could predict each type of discontinuity.

4. The graphs in exercise 1a and c have a vertical asymptote at $x = -1$, yet each graph behaves differently as x gets close to -1. For each graph:

a) Describe what happens to y when x gets close to -1.

b) What happens to y when x is very large?

5. a) Do all rational functions have discontinuous graphs? Explain.

b) If your answer to part a is no, give examples of rational functions that have continuous graphs. Verify your assertions using a graphing calculator.

6. The graphs of some rational functions have horizontal asymptotes.

a) Which graphs in exercise 1 have a horizontal asymptote?

b) What are the equations of the asymptotes?

7. In what ways can the graph of a rational function differ from the graph of a polynomial function?

From the *Investigation*, you learned that the graphs of rational functions may not be continuous. The graph of a rational function $y = f(x)$ is discontinuous at any x-value where the denominator is equal to 0. There are two types of discontinuities for rational functions. We use the graphs in the *Investigation* to illustrate these discontinuities.

Infinite Discontinuity

When the graph of a rational function has a vertical asymptote, we say that the function has an *infinite discontinuity* at that point. For example, the graph in exercise 1e has infinite discontinuities at $x = -1$ and $x = 2$. That is, the graph gets closer and closer to the lines $x = -1$ and $x = 2$, but never intersects these lines. Therefore, the corresponding function $y = \frac{x}{(x+1)(x-2)}$ has infinite discontinuities at $x = -1$ and $x = 2$.

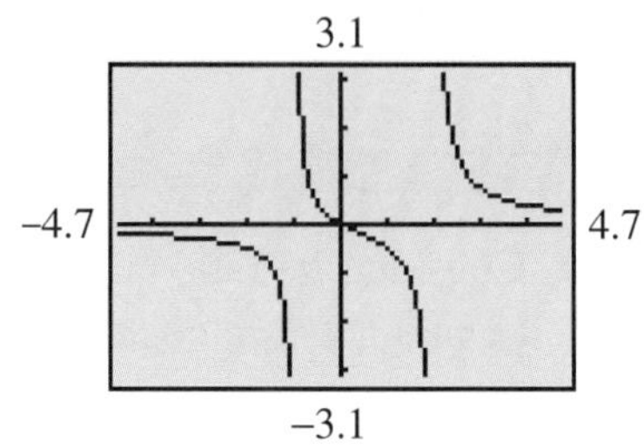

Point Discontinuity

When the graph of a rational function has a "hole", we say that the function has a *point discontinuity*. For example, the graph in exercise 1d has a point discontinuity at $x = 1$. Therefore, the corresponding function $y = \frac{x^2 - 1}{x - 1}$ has a point discontinuity at $x = 1$.

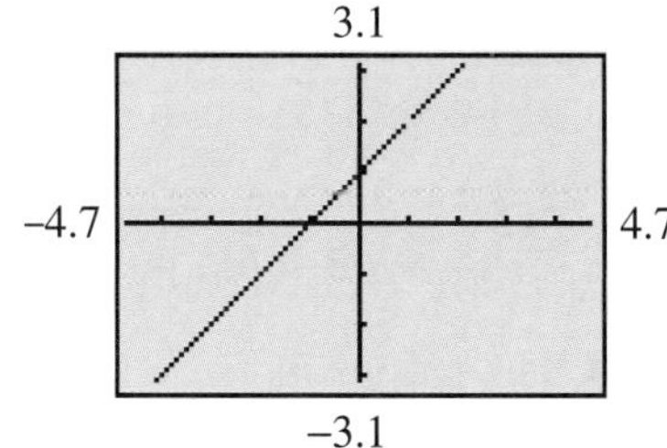

In the *Investigation*, you examined what happened to the y-value when x became very large. A rational function may have an oblique asymptote or a horizontal asymptote. For example, the graph in exercise 1c, below left, has an oblique asymptote. That is, the graph appears to get closer and closer to an oblique line but never intersects the line. The graph in exercise 1f, below right, has a horizontal asymptote. That is, the graph gets closer and closer to the x-axis, but never intersects the x-axis.

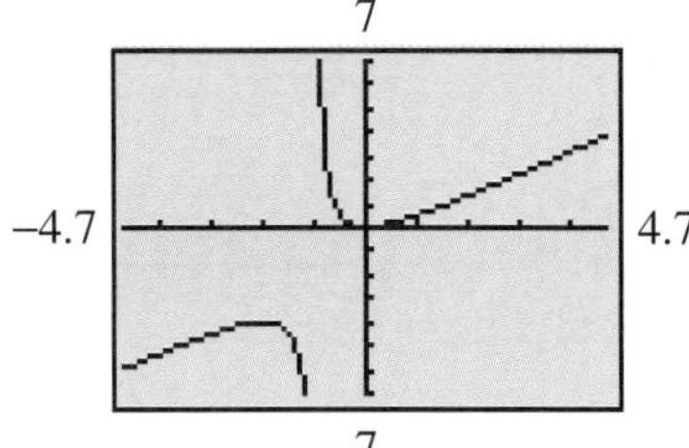

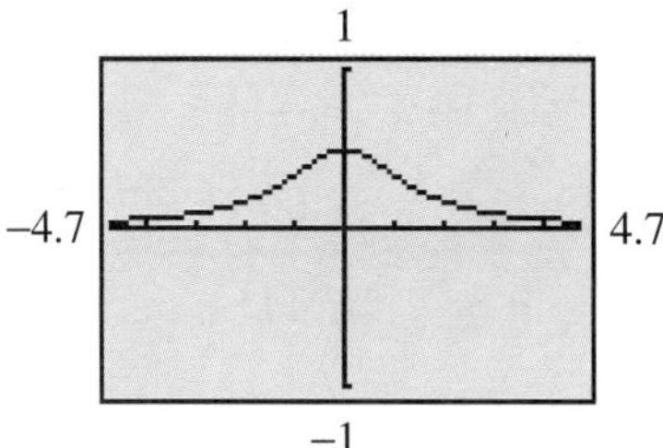

Rational functions may have critical points, intervals of increase or decrease, points of inflection, and so on. Therefore, we can use the derivative to learn more about rational functions. We can differentiate a rational function if we know its equation. If we have the graph of a rational function, we can sketch the graph of the derivative function by analysing the slope of the tangent at different points on the graph.

Example 1

For the rational function $y = f(x)$:

a) Estimate the slope of the tangent at different points on its graph, then sketch the graph of the derivative function.

b) Use the graph of the first derivative function to sketch the graph of the second derivative function.

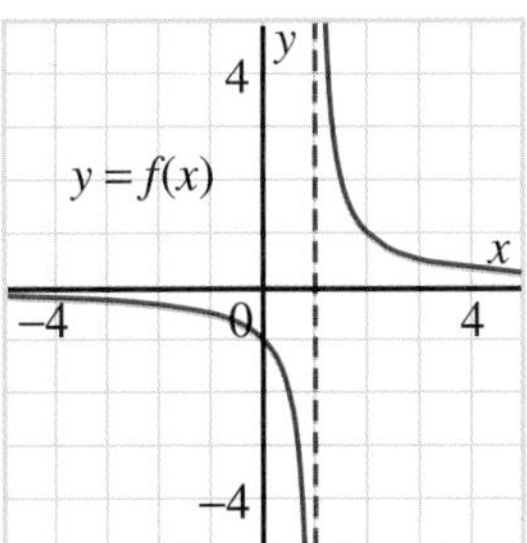

Discuss

What happens to the graph of the derivative function when the original function has a discontinuity?

Solution

a) Move from left to right on the graph of the function. Start at the left. The slope of the tangent is negative and close to zero. As x increases, the slopes of the tangents decrease until $x = 1$, where the slope is undefined. When $x > 1$, the slopes of the tangents increase from a negative number with a large absolute value to a negative number close to zero.

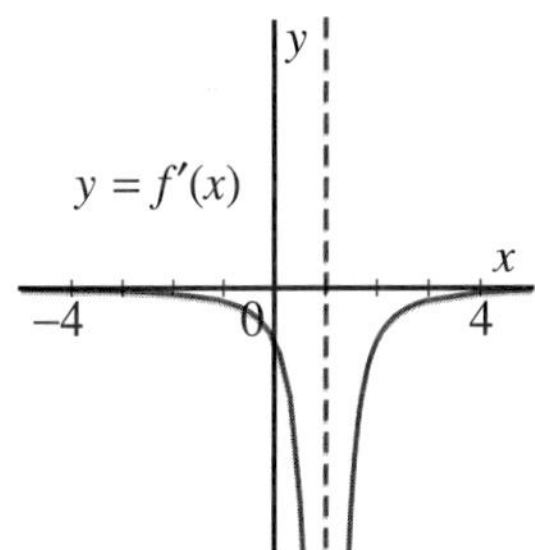

b) Move from left to right on the graph of the derivative function. Start at the left. The slope of the tangent is negative and close to zero. As x increases, the slopes of the tangents decrease until $x = 1$, where the slope is undefined. When $x > 1$, the slopes of the tangents decrease from a large positive number to a positive number close to zero.

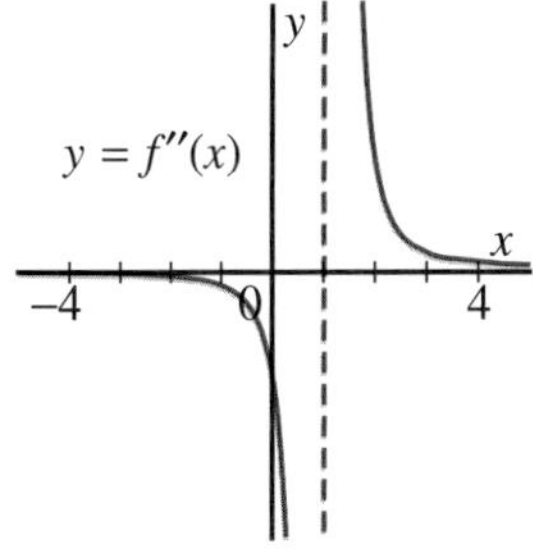

Example 2

The graphs of a rational function and its derivative are given. Which graph is the function and which graph is the derivative? Explain.

Graph 1

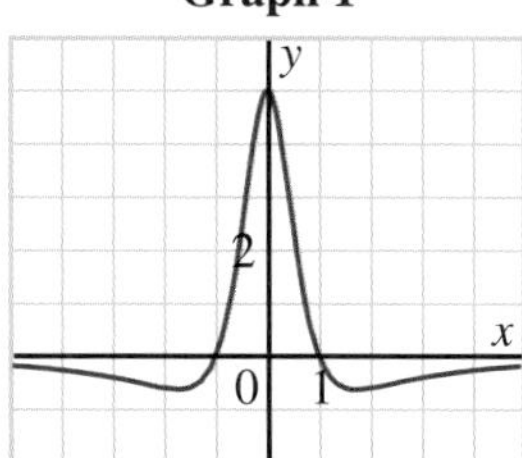

Graph 2

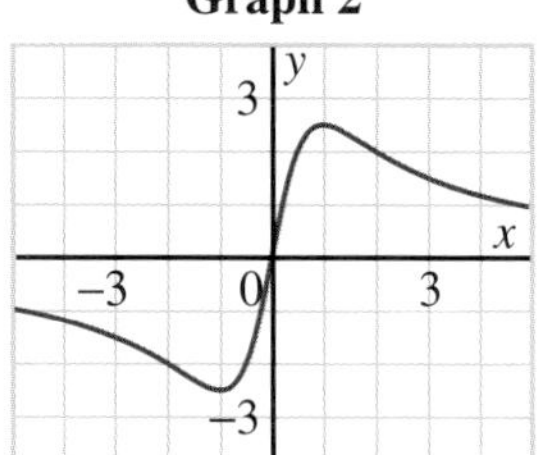

Solution

Look at the points where each graph intersects the x-axis.
Graph 1 has x-intercepts at $x = -1$ and $x = 1$.
Graph 2 has horizontal tangents at these x-values.
This suggests that graph 2 is the function and graph 1 is the derivative.
Graph 2 intersects the x-axis at $x = 0$. If this graph were the derivative, then graph 1 would have a horizontal tangent only at this point. Graph 1 has horizontal tangents at three points: $x = 0$, $x \doteq 2$, and $x \doteq -2$.
Therefore, graph 2 cannot be the derivative.
Graph 2 is the function and graph 1 is the derivative.

5.1 Exercises

A

1. Which equations represent rational functions? Explain.

   a) $y = x^2 - x - 12$ **b)** $y = \frac{x}{x^2 - x - 12}$ **c)** $y = \frac{1}{x^2 + x}$

   d) $y = (x - 1)^{-1}$ **e)** $y = \frac{x^3 - 10x + 6}{2}$ **f)** $y = 3^{-x} + 1$

2. Examine each screen. Which could be a graph of a rational function, and which could be a graph of a polynomial function? Explain.

a)

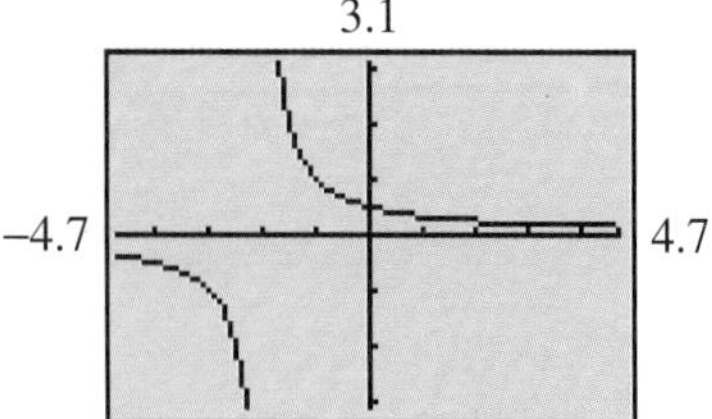

b)

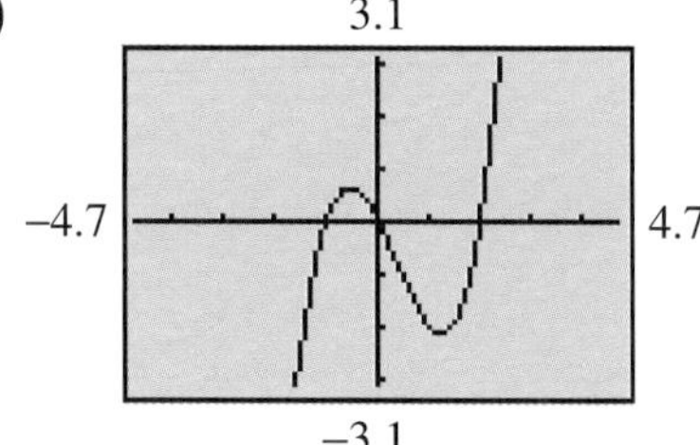

c)

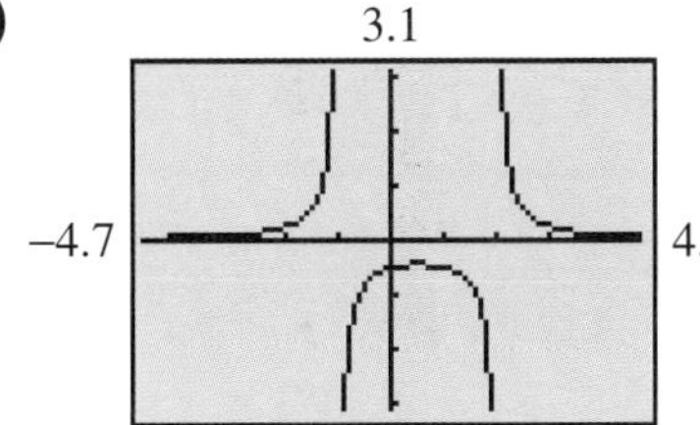

d)

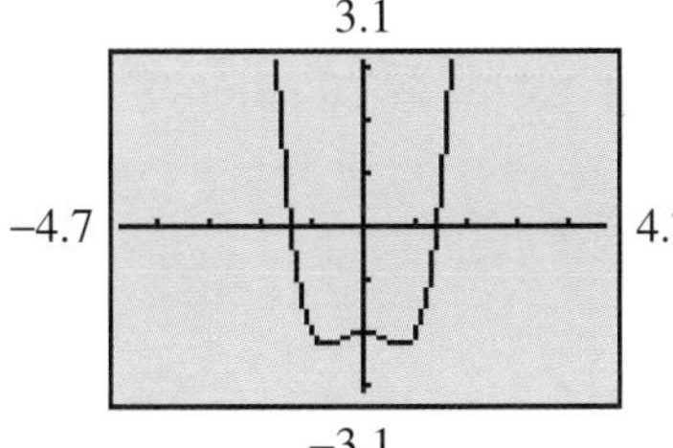

e)

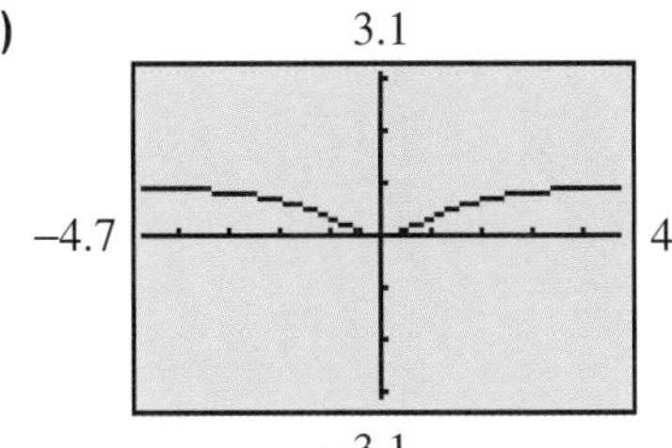

f)

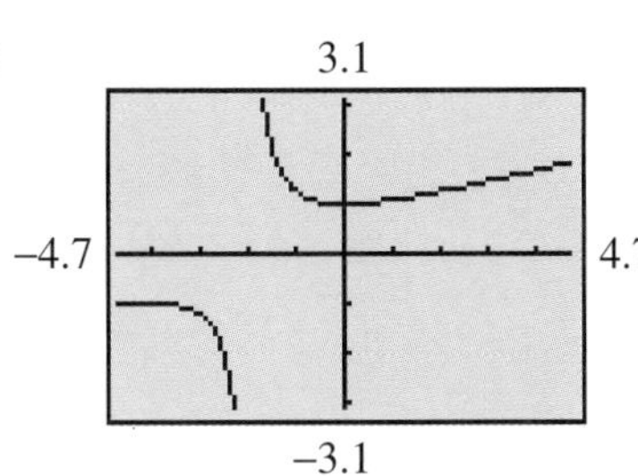

B

3. The graphs of four rational functions are shown. Their equations are listed below. Match each graph with its equation.

a)

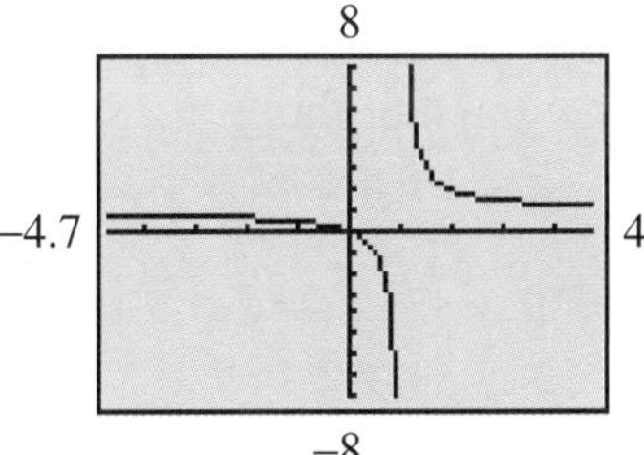

b)

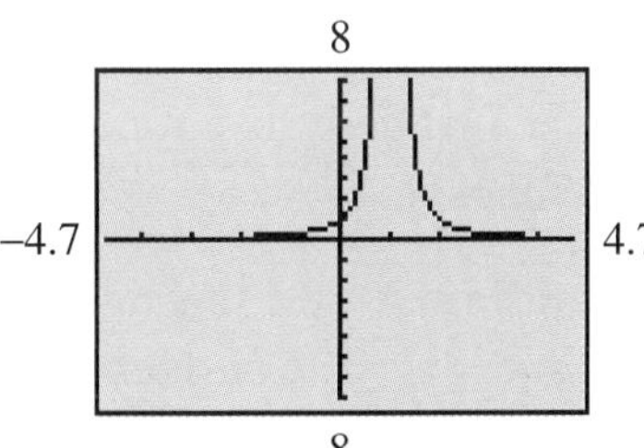

c)

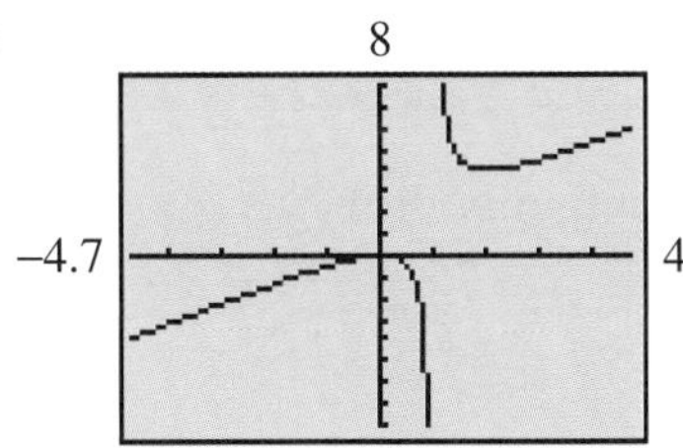

d)

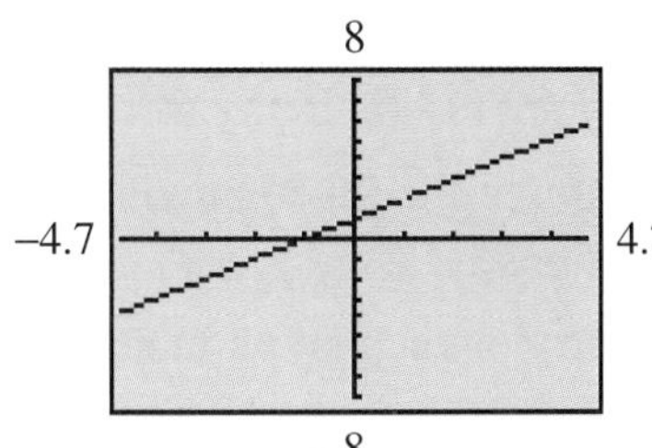

i) $y = \frac{x^2}{x - 1}$ **ii)** $y = \frac{x^2 - 1}{x - 1}$ **iii)** $y = \frac{x}{x - 1}$ **iv)** $y = \frac{1}{(x - 1)^2}$

4. All the functions in exercise 3 are discontinuous at $x = 1$. Consider these functions.

a) What is similar about all the equations that produce this discontinuity?

b) Which graphs have a vertical asymptote at $x = 1$?

c) Which graph has a point discontinuity at $x = 1$?

d) How does the equation of the function with the point discontinuity differ from the other equations?

5. The graphs of four rational functions are shown. Their equations are listed below. Match each graph with its equation.

a) 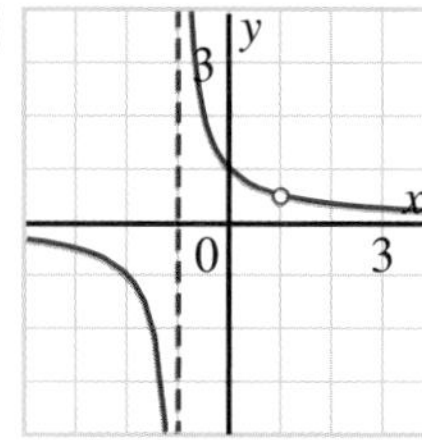

b) 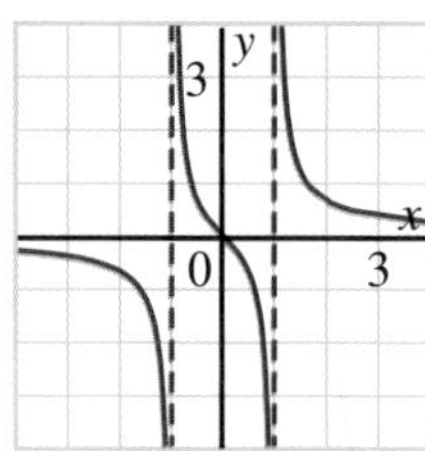

c) 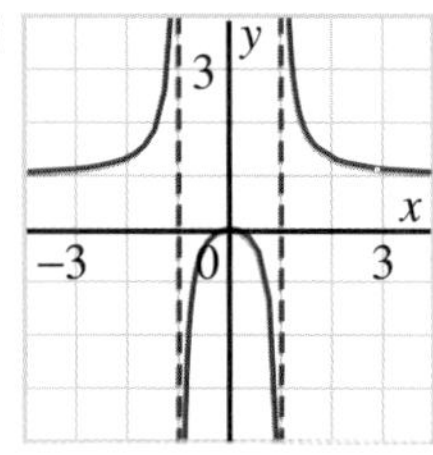

d) 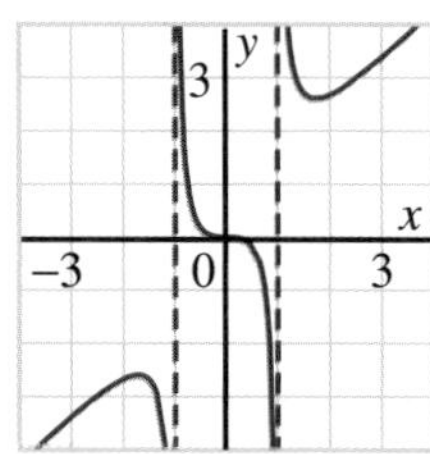

i) $y = \frac{x^3}{x^2 - 1}$ **ii)** $y = \frac{x - 1}{x^2 - 1}$ **iii)** $y = \frac{x}{x^2 - 1}$ **iv)** $y = \frac{x^2}{x^2 - 1}$

6. For each graph in exercise 5:
   i) Estimate the slope of the tangent at different points, then sketch the graph of the derivative function.
   ii) Use the graph of the first derivative function to sketch the graph of the second derivative function.

7. **Knowledge/Understanding** All the functions in exercise 5 are discontinuous at $x = 1$ and $x = -1$. Consider these functions.
   a) What is similar about all the equations that produce this discontinuity?
   b) Which graphs have vertical asymptotes at $x = 1$?
   c) Which graph has a point discontinuity at $x = 1$?
   d) How does the equation of the function with the point discontinuity differ from the other equations?
   e) Describe the symmetry of each graph in exercise 5.

8. **a)** For each rational function, estimate the slope of the tangent at different points on its graph, then sketch the graph of the derivative function.

i) 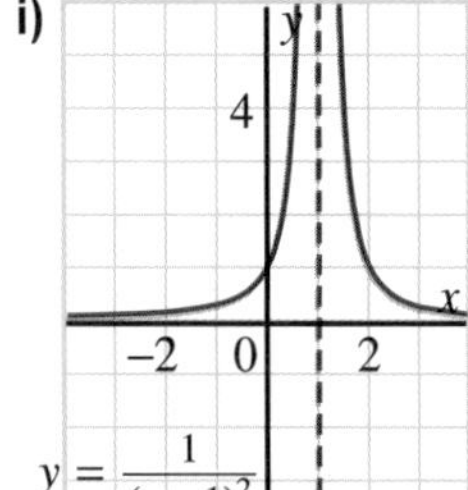

ii) 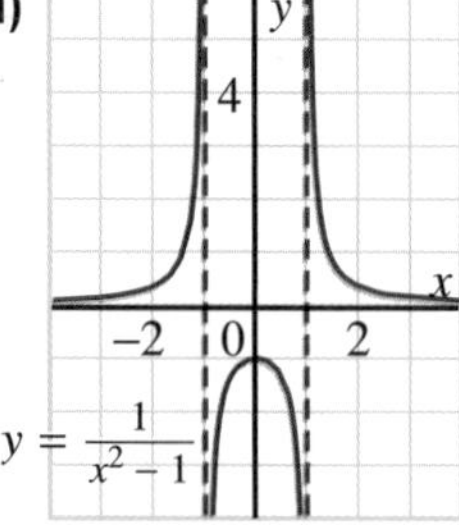

iii) 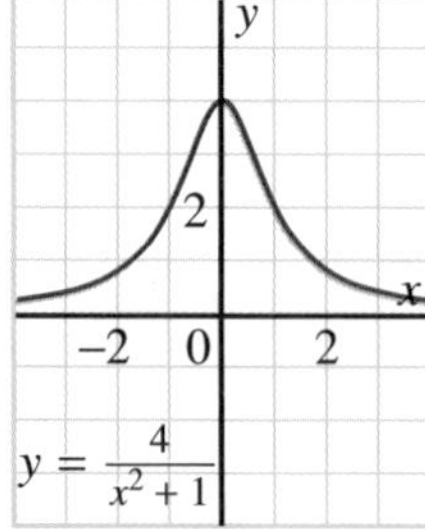

b) Verify the graph of each derivative function in part a by determining the derivative of each function, then use a graphing calculator to graph it.

c) In two graphs in part a, there is a critical point at $x = 0$. Use the equations in part a to determine the value of each function at $x = 0$.

d) In one graph in part a, there appear to be two points of inflection where the concavity of the graph changes. Use the second derivative to determine the exact coordinates of these points of inflection.

✓ **9.** A function of the form $y = \frac{1}{f(x)}$ is the reciprocal of the function $y = f(x)$. For each function, write the equation of the corresponding reciprocal function.

a) $y = x - 2$ b) $y = x^2 - 1$ c) $y = \sqrt{x + 1}$

d) $y = x^2 - x - 6$ e) $y = 2^x$ f) $y = x^3 + x$

✓ **10.** Which reciprocal functions in exercise 9 are rational functions? Explain.

Exercise 11 is significant because it considers how a rational function is related to a polynomial function and how the reciprocal transformation, $\frac{1}{f(x)}$, can produce a rational function if $y = f(x)$ is a non-constant polynomial function.

✓ **11. Communication** Use the results of exercises 9 and 10.

a) Is the reciprocal of every polynomial function a rational function? Explain.

b) Is the reciprocal of every rational function a polynomial function? Explain.

✓ **12.** Graph each function on grid paper. Graph its reciprocal function on the same grid, then identify the vertical asymptotes (if they exist).

a) $y = x + 1$ b) $y = x^2 - 4$ c) $y = x^2 + 1$

✓ **13.** Determine the equation of the tangent to the graph of each reciprocal function in exercise 12 at the point on the graph where $x = 1$.

14. Thinking/Inquiry/Problem Solving Suppose $y = f(x)$ is a non-constant polynomial function. Explain your answer to each question.

a) Is the graph of $y = \frac{1}{f(x)}$ always non-linear?

b) Does the graph of $y = \frac{1}{f(x)}$ always have a vertical asymptote?

c) Does $y = \frac{1}{f(x)}$ always have value(s) of x for which the function is not defined?

d) When $y = \frac{1}{f(x)}$ is undefined at a value $x = a$, what is true about this value $x = a$ for the function $y = f(x)$?

15. The degree of the derivative of a polynomial function is 1 less than the degree of the function. So, the graph of the derivative of a polynomial function is "simpler" than the graph of the function. Is this generally true for all functions? Explain.

16. A rational function f and its derivative f' are graphed below. Which graph is the function and which graph is the derivative? Explain.

Graph 1

Graph 2

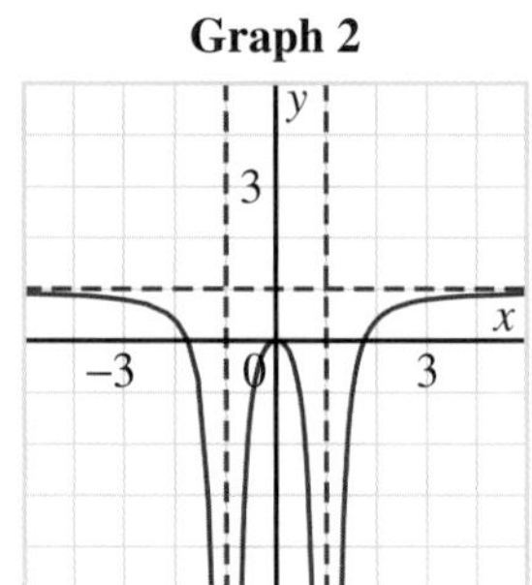

17. Describe these features of graph 1 in exercise 16: intervals of increase and decrease; critical points; points of inflection; and intervals of concavity.

18. Each graph below is the graph of the derivative of a function. Sketch a possible graph of the function.

a)

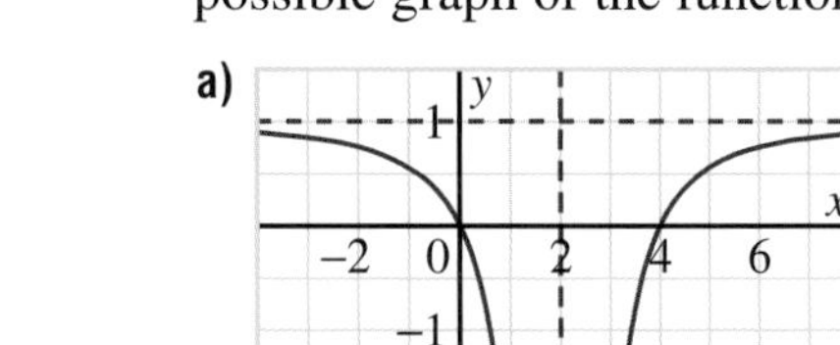

b)

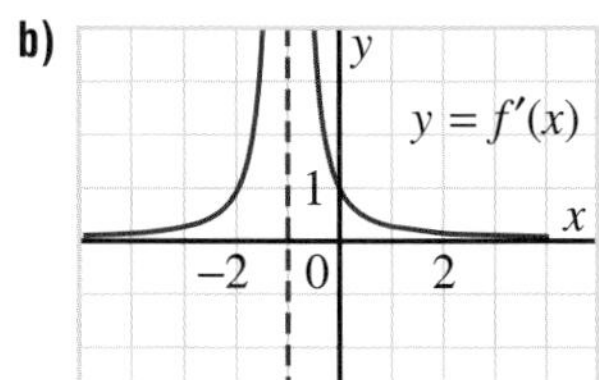

19. **Application** Graph each function, then complete parts i to iii.

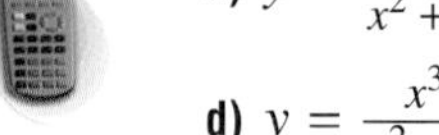

a) $y = \frac{1}{x^2 + 1}$ b) $y = \frac{x}{x^2 + 1}$ c) $y = \frac{x^2}{x^2 + 1}$

d) $y = \frac{x^3}{x^2 + 1}$ e) $y = \frac{x^4}{x^2 + 1}$ f) $y = \frac{x^5}{x^2 + 1}$

i) All the functions are continuous. Explain why.

ii) Identify which graphs have either line or point symmetry.

iii) Zoom in closer to $x = 0$ on each graph. What does each graph look like for very small values of x? Explain.

20. For very small values of x (that is, as x approaches 0), what new function does the graph of each rational function approximate? Explain.

a) $y = \frac{1}{x^2 - 1}$ b) $y = \frac{x}{x^2 - 1}$ c) $y = \frac{x^2}{x^2 - 1}$

d) $y = \frac{x^3}{x^2 - 1}$ e) $y = \frac{x^4}{x^2 - 1}$ f) $y = \frac{x^5}{x^2 - 1}$

21. Consider a non-polynomial function, such as $f(x) = \sin x$. Does the reciprocal function $y = \frac{1}{f(x)}$ have vertical asymptotes? Explain.

5.2 The Limit of a Rational Function

We can learn more about the graph of a rational function $y = f(x)$ by observing what happens to the value of y as x approaches a fixed value. We use the word *limit* when the value of y also approaches a fixed value. Recall that in Chapter 2 the derivative was defined as a limit. Limits are particularly useful for studying rational functions.

Investigation 1

Limits as $x \to a$

1. The graph of the function $y = \dfrac{x^2 - 1}{x - 1}$ and the corresponding table of values are shown. The graph is a straight line with a point discontinuity when $x = 1$. The reason is that the function is not defined when $x = 1$.

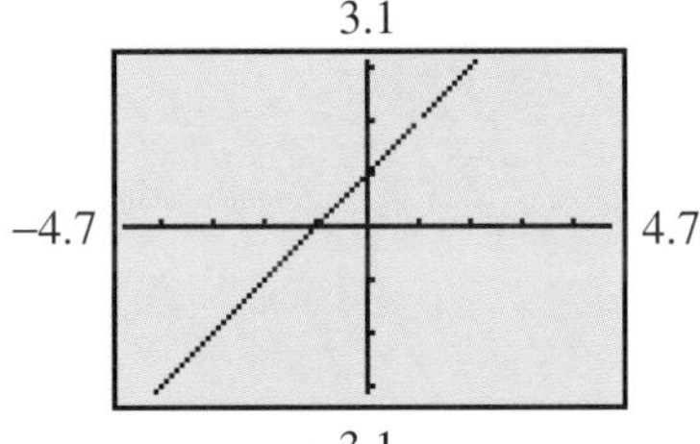

X	Y1
.7	1.7
.8	1.8
.9	1.9
1	ERROR
1.1	2.1
1.2	2.2
1.3	2.3

Y1=ERROR

   a) Use the screens to determine what happens to the values of y as x gets close to 1.

   b) Determine $\lim\limits_{x \to 1} \dfrac{x^2 - 1}{x - 1}$.

2. a) Graph the function $y = \dfrac{x^2 - 4}{x + 2}$.

   b) Trace close to $x = -2$ on either side of $x = -2$ on the graph. What happens to the values of y as x gets close to -2?

   c) Determine $\lim\limits_{x \to -2} \dfrac{x^2 - 4}{x + 2}$.

3. The graph of the function $y = \dfrac{1}{x + 1}$ and the corresponding table of values are shown. The function is not defined when $x = -1$.

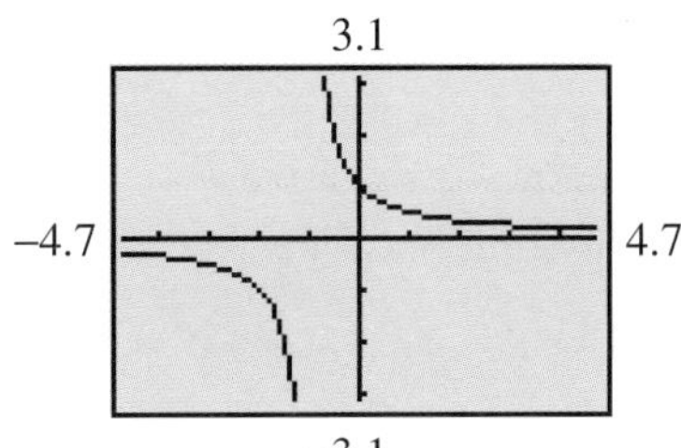

X	Y1
-1.3	-3.333
-1.2	-5
-1.1	-10
-1	ERROR
-.9	10
-.8	5
-.7	3.3333

Y1=ERROR

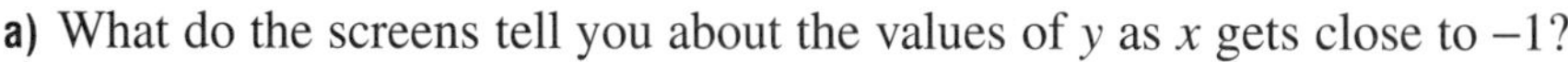

a) What do the screens tell you about the values of y as x gets close to -1?

b) What does the result of part a tell you about $\lim_{x\to -1} \frac{1}{x+1}$?

4. a) Graph the function $y = \frac{5}{x-2}$.

b) Trace close to $x = 2$ on either side of $x = 2$ on the graph. What happens to the values of y as x gets close to 2?

c) What does the result of part b tell you about $\lim_{x\to 2} \frac{5}{x-2}$?

Limits that Exist as $x \to a$

Look at the graph in *Investigation 1*, exercise 1. If you trace closer and closer to $x = 1$ on the graph of $y = \frac{x^2-1}{x-1}$, the value of y becomes closer and closer to 2. We say that the limit of $\frac{x^2-1}{x-1}$ as x approaches 1 is 2. We write: $\lim_{x\to 1} \frac{x^2-1}{x-1} = 2$

The statement, $\lim_{x\to 1} \frac{x^2-1}{x-1} = 2$, means that we can make $\frac{x^2-1}{x-1}$ as close to 2 as we like by substituting values of x that are close enough to 1. For example, we get 1.99 by substituting $x = 0.99$. We get 2.0001 by substituting $x = 1.0001$, and so on.

We can show algebraically that $\lim_{x\to 1} \frac{x^2-1}{x-1} = 2$, as follows. We simplify the expression to get a meaningful result when x approaches 1.

$$\begin{aligned}\lim_{x\to 1} \frac{x^2-1}{x-1} &= \lim_{x\to 1} \frac{(x+1)(x-1)}{x-1} \\ &= \lim_{x\to 1}(x+1) \\ &= 2\end{aligned}$$

Recall from Chapter 2 that similar steps occurred when we determined the derivatives of functions from first principles.

Limits that Do Not Exist as $x \to a$

Look at the graph in *Investigation 1*, exercise 3. If you trace closer and closer to $x = -1$ from the right side on the graph of $y = \frac{1}{x+1}$, the value of y becomes larger and larger, and does not approach any number. We say that the limit of $\frac{1}{x+1}$ as x approaches -1 from the right does not exist.

We write: $\lim_{x\to -1^+} \frac{1}{x+1}$ does not exist.

If you trace closer and closer to $x = -1$ on the left side, the value of y is negative with a larger and larger absolute value, and does not approach any number. The limit of $\frac{1}{x+1}$ as x approaches -1 from the left does not exist either.

We write: $\lim_{x\to -1^-} \frac{1}{x+1}$ does not exist.

To show that $\lim_{x \to -1^+} \frac{1}{x+1}$ and $\lim_{x \to -1^-} \frac{1}{x+1}$ do not exist, without using the graph, consider the expression $\frac{1}{x+1}$. As x gets closer and closer to -1 (whether greater or less than -1), the denominator gets closer and closer to 0 and the value of this expression does not approach any number. Therefore, $\lim_{x \to -1^+} \frac{1}{x+1}$ and $\lim_{x \to -1^-} \frac{1}{x+1}$ do not exist.

Take Note

Definition of $\lim_{x \to a} f(x)$

$\lim_{x \to a} f(x) = L$ means that we can make $f(x)$ as close to L as we like by substituting values of x close enough to a but not equal to a.

If there is no such number L, this limit does not exist.

If the limit expression contains $x \to a^+$ or $x \to a^-$, the values of x that are substituted must be greater than a or less than a, respectively.

Example 1

Determine each limit.

a) $\lim_{x \to 5} \frac{x^2 - 25}{x - 5}$ **b)** $\lim_{x \to 0} \frac{x^2 - 3x}{x^2 + 2x}$ **c)** $\lim_{x \to 2} \frac{1}{x+1}$ **d)** $\lim_{x \to 1} \frac{1}{(x-1)^2}$

Solution

a)
$$\begin{aligned}\lim_{x \to 5} \frac{x^2 - 25}{x - 5} &= \lim_{x \to 5} \frac{(x-5)(x+5)}{x-5} \\ &= \lim_{x \to 5}(x+5) \\ &= 10\end{aligned}$$

b)
$$\begin{aligned}\lim_{x \to 0} \frac{x^2 - 3x}{x^2 + 2x} &= \lim_{x \to 0} \frac{x(x-3)}{x(x+2)} \\ &= \lim_{x \to 0} \frac{x-3}{x+2} \\ &= -\frac{3}{2}\end{aligned}$$

c)
$$\begin{aligned}\lim_{x \to 2} \frac{1}{x+1} &= \frac{1}{2+1} \\ &= \frac{1}{3}\end{aligned}$$

d) $\lim_{x \to 1} \frac{1}{(x-1)^2}$

When $x \to 1, (x-1)^2 \to 0$, so

$\lim_{x \to 1} \frac{1}{(x-1)^2}$ does not exist.

Example 1c illustrates that it is not always necessary to change the form of the function to determine its limit. Since the denominator is not equal to 0 at $x = 2$, this value can be substituted directly into the expression.

In *Example 1d*, $\lim_{x \to 1} \frac{1}{(x-1)^2}$ does not exist. A shorthand way to write this is $\lim_{x \to 1} \frac{1}{(x-1)^2} = \infty$. This is not an equation in the usual sense. An equation is a statement that two numbers or algebraic expressions are equal. But ∞ is neither of these. In particular, ∞ is not a number.

$\lim_{x \to 1} \frac{1}{(x-1)^2} = \infty$ means that $\lim_{x \to 1} \frac{1}{(x-1)^2}$ does not exist.

Investigation 2

Limits as $x \to \infty$

1. The graph of the function $y = \frac{5}{x^2 + 1}$ and the corresponding table of values are shown.

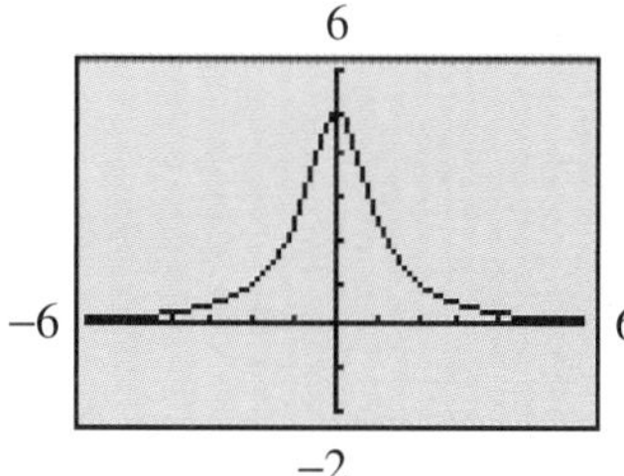

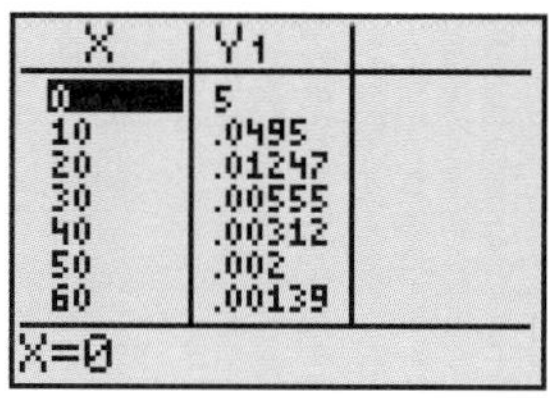

   a) Use the screens to determine what happens to the values of y as x gets larger and larger.

   b) Suppose the values of x get larger and larger. If the values of $f(x)$ get closer and closer to a particular number, we use the expression $\lim_{x \to \infty} f(x)$ to represent this number. Determine $\lim_{x \to \infty} \frac{5}{x^2 + 1}$.

   c) Explain what the expression $\lim_{x \to -\infty} f(x)$ represents. Determine $\lim_{x \to -\infty} \frac{5}{x^2 + 1}$, and explain the result.

2. a) Graph the function $y = \frac{-2x^2}{x^2 + x - 6}$ for $-9.4 \le x \le 9.4$ and $-10 \le y \le 10$.

   b) Use the trace or table feature to determine each limit, if it exists. If the limit does not exist, explain.

   i) $\lim_{x \to \infty} \frac{-2x^2}{x^2 + x - 6}$ ii) $\lim_{x \to -\infty} \frac{-2x^2}{x^2 + x - 6}$

3. a) Graph the function $y = \frac{x^2}{x - 2}$ for $-9.4 \le x \le 9.4$ and $-15 \le y \le 15$.

   b) Use the trace or table feature to determine each limit, if it exists. If the limit does not exist, explain.

   i) $\lim_{x \to \infty} \frac{x^2}{x - 2}$ ii) $\lim_{x \to -\infty} \frac{x^2}{x - 2}$

Limits that Exist as $x \to \infty$

Look at the graph in *Investigation 2*, exercise 1. You should have found that as x gets larger and larger, the values of $\frac{5}{x^2 + 1}$ become closer and closer to 0. We say that the limit of $\frac{5}{x^2 + 1}$ as x approaches infinity is 0.

We write: $\lim\limits_{x \to \infty} \frac{5}{x^2 + 1} = 0$

The statement $\lim\limits_{x \to \infty} \frac{5}{x^2 + 1} = 0$ means that we can make $\frac{5}{x^2 + 1}$ as close to 0 as we like by substituting values of x that are large enough. For example, we get approximately 0.0005 by substituting $x = 100$. We get approximately 0.5×10^{-6} by substituting $x = 1000$, and so on.

Similarly, if x is negative and has a larger and larger absolute value, the values of $\frac{5}{x^2 + 1}$ become closer and closer to 0. We say that the limit of $\frac{5}{x^2 + 1}$ as x approaches negative infinity is 0.

We write: $\lim\limits_{x \to -\infty} \frac{5}{x^2 + 1} = 0$

This means that we can make $\frac{5}{x^2 + 1}$ as close to 0 as we like by substituting values of x that are negative and have a large enough absolute value.

Limits that Do Not Exist as $x \to \infty$

In *Investigation 2*, exercise 3, you should have found that as x becomes larger and larger, the values of $\frac{x^2}{x - 2}$ become larger and larger, and do not approach any number. Therefore, $\lim\limits_{x \to \infty} \frac{x^2}{x - 2}$ does not exist. Similarly, $\lim\limits_{x \to -\infty} \frac{x^2}{x - 2}$ does not exist.

Take Note

Definition of $\lim\limits_{x \to \infty} f(x)$ ***and*** $\lim\limits_{x \to -\infty} f(x)$

$\lim\limits_{x \to \infty} f(x) = L$ means that we can make $f(x)$ as close to L as we like by substituting positive values of x that are large enough.

$\lim\limits_{x \to -\infty} f(x) = L$ means that we can make $f(x)$ as close to L as we like by substituting negative values of x with large enough absolute values.

If there is no such number L, these limits do not exist.

There is an algebraic method to determine limits as $x \to \infty$ or $x \to -\infty$. It involves an important principle from *Necessary Skills*, page 295; that is, sometimes it is helpful to change the form of a rational expression. As well as dividing, we can also change the form by multiplying or dividing the numerator and denominator by the same expression.

For example, consider *Investigation 2*, exercise 2. You should have found that $\lim_{x \to \infty} \frac{-2x^2}{x^2 + x - 6} = -2$. To determine this limit algebraically, divide the numerator and denominator of the rational expression by the highest power of x in the denominator.

$$\begin{aligned}\lim_{x \to \infty} \frac{-2x^2}{x^2 + x - 6} &= \lim_{x \to \infty} \frac{\frac{-2x^2}{x^2}}{\frac{x^2}{x^2} + \frac{x}{x^2} - \frac{6}{x^2}} \\ &= \lim_{x \to \infty} \frac{-2}{1 + \frac{1}{x} - \frac{6}{x^2}}\end{aligned}$$

Look at the numerator and denominator separately and take the limit as $x \to \infty$. The numerator is constant. In the denominator, as x gets larger, $\frac{1}{x}$ and $\frac{6}{x^2}$ get smaller and approach 0.

$$\begin{aligned}\text{So, } \lim_{x \to \infty} \frac{-2x^2}{x^2 + x - 6} &= \frac{-2}{1} \\ &= -2\end{aligned}$$

Look at the same analysis for *Investigation 2*, exercise 3: $\lim_{x \to -\infty} \frac{x^2}{x - 2}$. The highest power of x in the denominator is x. Therefore, divide each term in the numerator and denominator by x.

$$\begin{aligned}\lim_{x \to -\infty} \frac{x^2}{x - 2} &= \lim_{x \to -\infty} \frac{\frac{x^2}{x}}{\frac{x}{x} - \frac{2}{x}} \\ &= \lim_{x \to -\infty} \frac{x}{1 - \frac{2}{x}}\end{aligned}$$

In the numerator, $\lim_{x \to -\infty} x = -\infty$

In the denominator, $\lim_{x \to -\infty} \left(1 - \frac{2}{x}\right) = 1$

Therefore, $\lim_{x \to -\infty} \frac{x^2}{x - 2} = -\infty$; that is, the limit does not exist.

The above analysis illustrates that when we have changed the form of the rational function, we determine the limit of the quotient by considering the limits of the numerator and denominator.

Example 2

Determine each limit, if it exists.

a) $\lim_{x \to \infty} \frac{3x^2 - 2x + 1}{5x^2 + 4}$ **b)** $\lim_{x \to \infty} \frac{4x^2}{x^3 - 2x + 3}$ **c)** $\lim_{x \to \infty} \frac{3 - x^4}{x^3 - x^2 + 1}$

Solution

a)
$$\begin{aligned}\lim_{x \to \infty} \frac{3x^2 - 2x + 1}{5x^2 + 4} &= \lim_{x \to \infty} \frac{\frac{3x^2}{x^2} - \frac{2x}{x^2} + \frac{1}{x^2}}{\frac{5x^2}{x^2} + \frac{4}{x^2}} \\ &= \lim_{x \to \infty} \frac{3 - \frac{2}{x} + \frac{1}{x^2}}{5 + \frac{4}{x^2}}\end{aligned}$$

In the numerator, $\lim_{x \to \infty} \frac{2}{x} = 0$ and $\lim_{x \to \infty} \frac{1}{x^2} = 0$

In the denominator, $\lim_{x \to \infty} \frac{4}{x^2} = 0$

So, $\lim_{x \to \infty} \frac{3x^2 - 2x + 1}{5x^2 + 4} = \frac{3}{5}$

b) $$\lim_{x \to \infty} \frac{4x^2}{x^3 - 2x + 3} = \lim_{x \to \infty} \frac{\frac{4x^2}{x^3}}{\frac{x^3}{x^3} - \frac{2x}{x^3} + \frac{3}{x^3}}$$

$$= \lim_{x \to \infty} \frac{\frac{4}{x}}{1 - \frac{2}{x^2} + \frac{3}{x^3}}$$

In the numerator, $\lim_{x \to \infty} \frac{4}{x} = 0$

In the denominator, $\lim_{x \to \infty} \frac{2}{x^2} = 0$ and $\lim_{x \to \infty} \frac{3}{x^3} = 0$

So, $\lim_{x \to \infty} \frac{4x^2}{x^3 - 2x + 3} = \frac{0}{1}$, or 0

c) $$\lim_{x \to \infty} \frac{3 - x^4}{x^3 - x^2 + 1} = \lim_{x \to \infty} \frac{\frac{3}{x^3} - \frac{x^4}{x^3}}{\frac{x^3}{x^3} - \frac{x^2}{x^3} + \frac{1}{x^3}}$$

$$= \lim_{x \to \infty} \frac{\frac{3}{x^3} - x}{1 - \frac{1}{x} + \frac{1}{x^3}}$$

In the numerator, $\lim_{x \to \infty} \frac{3}{x^3} = 0$ and $\lim_{x \to \infty} x = \infty$

In the denominator, $\lim_{x \to \infty} \frac{1}{x} = 0$ and $\lim_{x \to \infty} \frac{1}{x^3} = 0$

So, $\lim_{x \to \infty} \frac{3 - x^4}{x^3 - x^2 + 1} = -\infty$

That is, the limit does not exist.

5.2 Exercises

1. Each screen shows a table of values close to $x = 3$ for a rational function f. Determine $\lim_{x \to 3} f(x)$, if it exists.

a)

X	Y1
2.97	-33.33
2.98	-50
2.99	-100
3	ERROR
3.01	100
3.02	50
3.03	33.333

X=2.97

b)

X	Y1
2.97	1111.1
2.98	2500
2.99	10000
3	ERROR
3.01	10000
3.02	2500
3.03	1111.1

X=2.97

c)

X	Y1
2.97	5.97
2.98	5.98
2.99	5.99
3	ERROR
3.01	6.01
3.02	6.02
3.03	6.03

X=2.97

2. Consider the graph of the rational function $y = f(x)$, below left. Determine each value, if possible.

a) $\lim_{x \to -1} f(x)$ b) $f(-1)$ c) $\lim_{x \to \infty} f(x)$ d) $\lim_{x \to -\infty} f(x)$

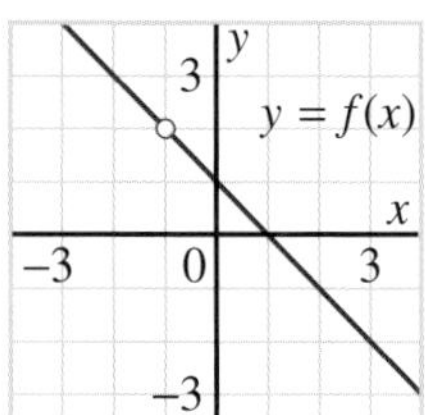

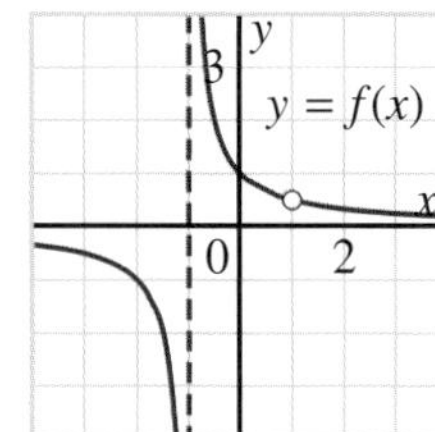

3. Consider the graph of the rational function $y = f(x)$, above right. Determine each value, if it exists.

a) $\lim_{x \to 1} f(x)$ b) $f(1)$ c) $\lim_{x \to -1^+} f(x)$ d) $\lim_{x \to -1^-} f(x)$

e) $\lim_{x \to -1} f(x)$ f) $f(-1)$ g) $\lim_{x \to \infty} f(x)$ h) $\lim_{x \to -\infty} f(x)$

4. For the graph of each rational function $y = f(x)$ below, determine each limit.

i) $\lim_{x \to 1^+} f(x)$ ii) $\lim_{x \to 1^-} f(x)$ iii) $\lim_{x \to 1} f(x)$ iv) $\lim_{x \to -1^+} f(x)$

v) $\lim_{x \to -1^-} f(x)$ vi) $\lim_{x \to -1} f(x)$ vii) $\lim_{x \to \infty} f(x)$ viii) $\lim_{x \to -\infty} f(x)$

a)

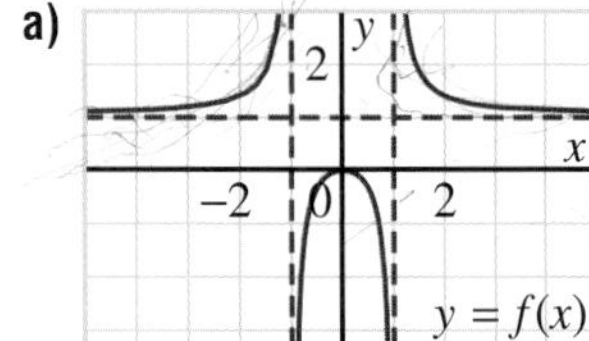

b)

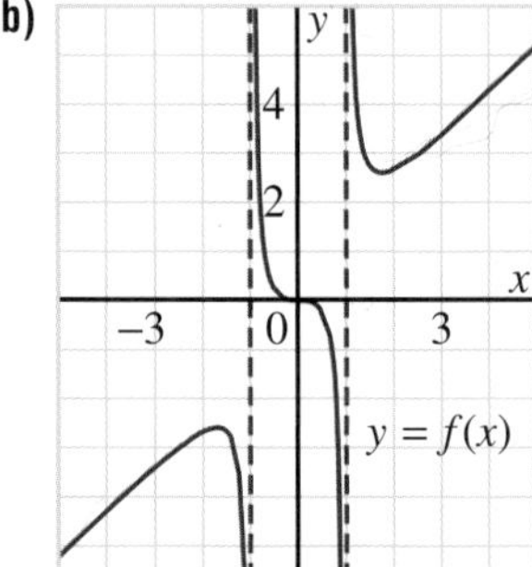

5. Evaluate each function at each value of x. Use the result to determine each limit.

a) $f(x) = \dfrac{9 - x^2}{3 - x}$; $x = 2.9, 2.99, 2.999, 3.1, 3.01, 3.001$; $\lim_{x \to 3} \dfrac{9 - x^2}{3 - x}$

b) $f(x) = \dfrac{1}{3 - x}$; $x = 2.9, 2.99, 2.999, 3.1, 3.01, 3.001$; $\lim_{x \to 3} \dfrac{1}{3 - x}$

c) $f(x) = \dfrac{1}{(3 - x)^2}$; $x = 2.9, 2.99, 2.999, 3.1, 3.01, 3.001$; $\lim_{x \to 3} \dfrac{1}{(3 - x)^2}$

6. **Knowledge/Understanding** Evaluate each limit.

a) $\lim_{x \to 3} \dfrac{x^2 - 9}{x - 3}$ b) $\lim_{x \to 3} \dfrac{x - 3}{x^2 - 9}$

c) $\lim_{x \to 2} \frac{x^2 - 4}{x - 2}$

d) $\lim_{x \to 2} \frac{x - 2}{x^2 - 4}$

e) $\lim_{x \to -1} \frac{x^2 - 2x - 3}{x + 1}$

f) $\lim_{x \to -3} \frac{x^2 + 7x + 12}{x^2 + 2x - 3}$

g) $\lim_{x \to 0} \frac{x}{x^2 + 2x}$

h) $\lim_{x \to 0} \frac{x^3 - 5x}{x^2 + 2x}$

7. Evaluate each limit.

a) $\lim_{x \to 3} \frac{4}{x + 1}$

b) $\lim_{x \to -1} \frac{4}{x + 1}$

c) $\lim_{x \to 2} \frac{x - 2}{x + 2}$

d) $\lim_{x \to -2} \frac{x - 2}{x + 2}$

e) $\lim_{x \to -1} \frac{x}{x + 1}$

f) $\lim_{x \to -3} \frac{x - 1}{x^2 + 2x - 3}$

g) $\lim_{x \to 3} \frac{1}{(x - 3)^2}$

h) $\lim_{x \to -2} \frac{5}{(x + 2)^2}$

8. Evaluate each limit.

a) $\lim_{x \to \infty} \frac{1}{x + 1}$

b) $\lim_{x \to \infty} \frac{x}{x + 1}$

c) $\lim_{x \to \infty} \frac{3x^2}{x^2 + 2x}$

d) $\lim_{x \to \infty} \frac{2x^3 - 5}{7x^3 + x - 1}$

e) $\lim_{x \to \infty} \frac{x^3}{x^2 - 4x}$

f) $\lim_{x \to \infty} \frac{3x^2 - 4x + 1}{x - 2}$

g) $\lim_{x \to \infty} \frac{5x^2 + x}{x^3 + 2x + 3}$

h) $\lim_{x \to \infty} \frac{3 - x}{x^2 + 2x}$

9. Communication In each part of *Example 2*, pages 318, 319, compare the degree of the numerator to the degree of the denominator. Compare the limits in *Example 2* with the limits in exercise 8. Describe the limits to infinity in terms of these three cases:

- The degrees of the numerator and the denominator are equal.
- The degree of the numerator is less than the degree of the denominator.
- The degree of the numerator is greater than the degree of the denominator.

10. a) Explain what this limit means: $\lim_{x \to \infty} \frac{3x}{x + 2} = 3$.

b) What values of x will produce a value for $\frac{3x}{x + 2}$ within 0.001 of 3?

11. a) Explain what this limit means: $\lim_{x \to -2} \frac{1}{(x + 2)^2} = \infty$.

b) What values of x will produce a value for $\frac{1}{(x + 2)^2}$ equal to 10 000?

12. a) Explain what this limit means: $\lim_{x \to 2} \frac{x^2 - 4}{x - 2} = 4$.

b) What values of x will produce a value for $\frac{x^2 - 4}{x - 2}$ within 0.001 of 4?

13. **a)** Explain what this limit means: $\lim_{x \to 0^+} \frac{1}{x} = \infty$.

b) What value of x will produce a value for $\frac{1}{x}$ equal to 10^{10}?

14. Evaluate each limit.

a) $\lim_{x \to 1^+} \frac{1}{x - 1}$

b) $\lim_{x \to -1^+} \frac{x}{x + 1}$

c) $\lim_{x \to -2^-} \frac{1}{x + 2}$

d) $\lim_{x \to -2^+} \frac{1}{x + 2}$

e) $\lim_{x \to -\infty} \frac{1}{x - 4}$

f) $\lim_{x \to -\infty} \frac{x^2}{x - 4}$

g) $\lim_{x \to \infty} \frac{5x^2}{x - 3x^2}$

h) $\lim_{x \to \infty} \frac{2x^3}{x - 7x^2}$

15. Consider the function $f(x) = \frac{x^2 - 9}{x - 3}$.

a) Determine each value.

i) $f(3)$

ii) $\lim_{x \to 3} f(x)$

b) Use the results of part a to describe the graph of $y = f(x)$ near $x = 3$.

Exercise 15 is significant because it challenges you to consider what the limit of a rational function indicates about the graph of the function.

16. **Application** Until now, a function has been defined with a single formula. However, sometimes we define a function by applying different formulas to different parts (or pieces) of its domain. For example, we could define the function $y = f(x)$ to be equal to $y = \frac{x^2 - 9}{x - 3}$ for all values of x other than $x = 3$ and to be equal to $y = 6$ when $x = 3$. When a function is defined *in pieces*, it is called a *piecewise function* and a brace is used as shown:

$$f(x) = \begin{cases} \frac{x^2 - 9}{x - 3} & x \neq 3 \\ 6 & x = 3 \end{cases}$$

a) Determine each value.

i) $f(3)$

ii) $\lim_{x \to 3} f(x)$

b) Use the results of part a to describe the graph of $y = f(x)$.

17. Evaluate each limit.

a) $\lim_{x \to -2} \frac{x^2 + 2x}{x^2 + x - 2}$

b) $\lim_{x \to -2} \frac{x^2 + x - 2}{x^2 + 2x}$

18. Determine a possible equation for a rational function $y = f(x)$ that has a point discontinuity at $x = -4$, but the limit exists at $x = -4$.

19. **Thinking/Inquiry/Problem Solving** Consider the rational function $y = f(x)$ and $\lim_{x \to 2} f(x) = \frac{5}{3}$. Would it always be true that $\lim_{x \to 2} \frac{1}{f(x)} = \frac{3}{5}$? Explain.

Self-Check 5.1, 5.2

1. For the rational function at the right:

   a) Estimate the slope of the tangent at different points on its graph, then sketch the graph of the derivative function.

   b) Use the graph of the first derivative function to sketch the graph of the second derivative function.

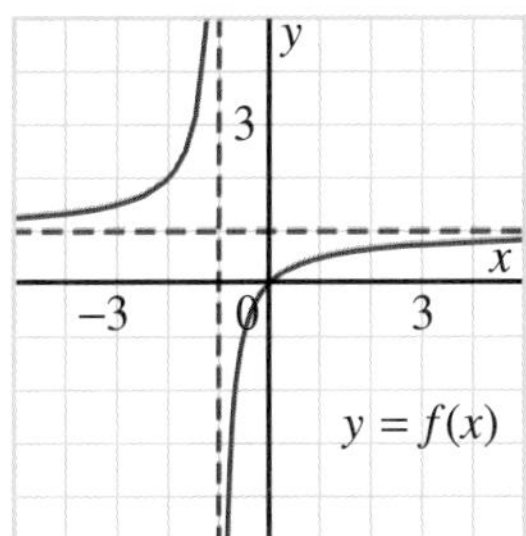

2. The equation of the function in exercise 1 is $y = \frac{x}{x+1}$. Verify the graph of the derivative function in exercise 1: differentiate the function, then use a graphing calculator to graph the derivative function.

3. Describe these features of the graph in exercise 1: intervals of increase and decrease; critical points; points of inflection; and intervals of concavity.

4. Determine the equation of the tangent to the graph of the function in exercise 1 at the point (–2, 2).

5. Evaluate each limit.

   a) $\lim_{x \to -2} \frac{x^2 - 4}{x + 2}$ b) $\lim_{x \to 5} \frac{x - 5}{x^2 - 25}$ c) $\lim_{x \to 0} \frac{x}{x^2 + 7x}$

   d) $\lim_{x \to 4} \frac{x^2 - x - 12}{x - 4}$ e) $\lim_{x \to 1} \frac{x}{x - 1}$ f) $\lim_{x \to -3} \frac{1}{x + 3}$

   g) $\lim_{x \to 5} \frac{1}{(x - 5)^2}$ h) $\lim_{x \to -1} \frac{x}{(x + 1)^2}$ i) $\lim_{x \to 3} \frac{1}{x + 3}$

6. Evaluate each limit.

   a) $\lim_{x \to \infty} \frac{1}{x + 6}$ b) $\lim_{x \to \infty} \frac{x^3}{x^2 - 4x}$ c) $\lim_{x \to \infty} \frac{5x^2 + 4x - 12}{3 - x - 2x^2}$

7. Determine a possible equation for a rational function $y = f(x)$ in each case.

   a) There is a point discontinuity at $x = -2$ and the limit exists at $x = -2$.

   b) There is an infinite discontinuity at $x = 1$ and the limit does not exist at $x = 1$.

PERFORMANCE ASSESSMENT

8. a) Evaluate each limit, if it exists.

   i) $\lim_{x \to 2^+} \frac{1}{x - 2}$ ii) $\lim_{x \to 2^-} \frac{1}{x - 2}$ iii) $\lim_{x \to 2} \frac{1}{x - 2}$

   b) Which limits in part a exist? Explain.

   c) Use the limits in part a to describe the graph of the rational function $y = \frac{1}{x - 2}$ near $x = 2$.

5.3 Features of the Graph of a Rational Function

We can analyse the equation of a rational function to determine features of its graph. Recall that a rational function may have a domain restriction since the denominator in a rational expression can never equal 0. In the following *Investigation*, you will determine what effect domain restrictions have on the graph of a rational function.

Investigation

Point Discontinuity and Vertical Asymptote

The equation and the corresponding graph of each rational function are shown.

a) $f(x) = \frac{1}{x-1}$

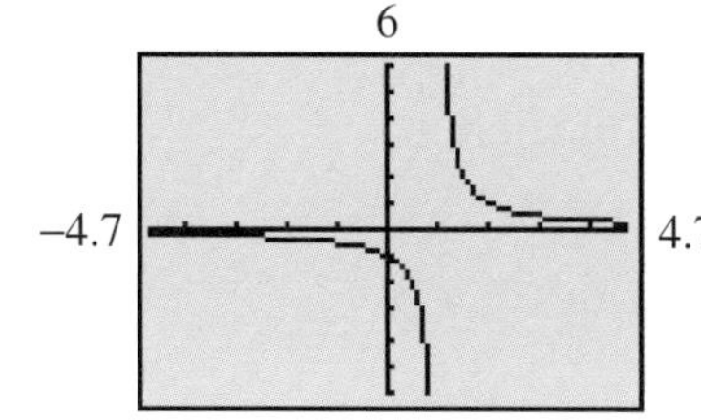

b) $f(x) = \frac{-1}{(x-1)^2}$

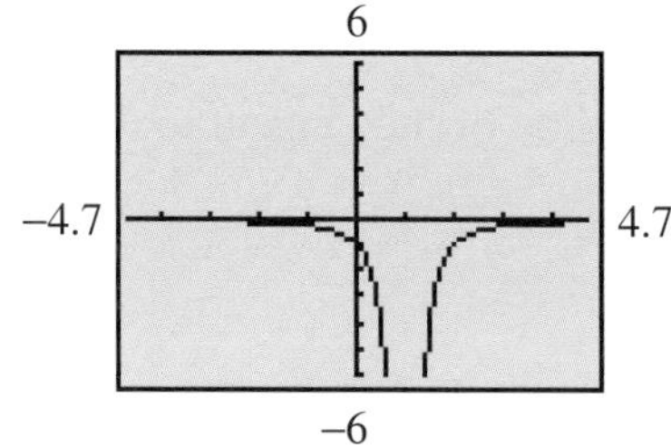

c) $f(x) = \frac{x^2-1}{x-1}$

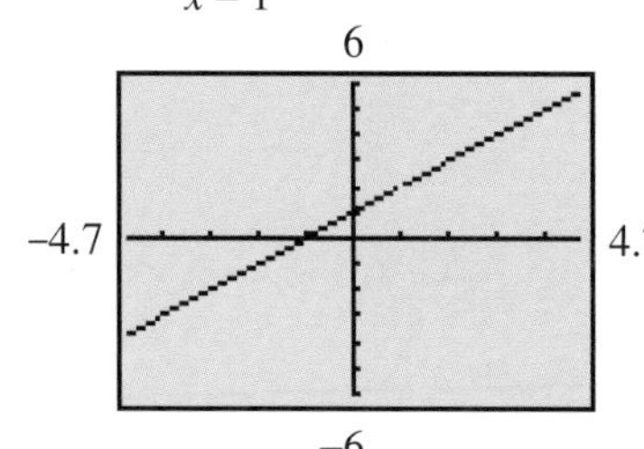

d) $f(x) = \frac{x^2}{x-1}$

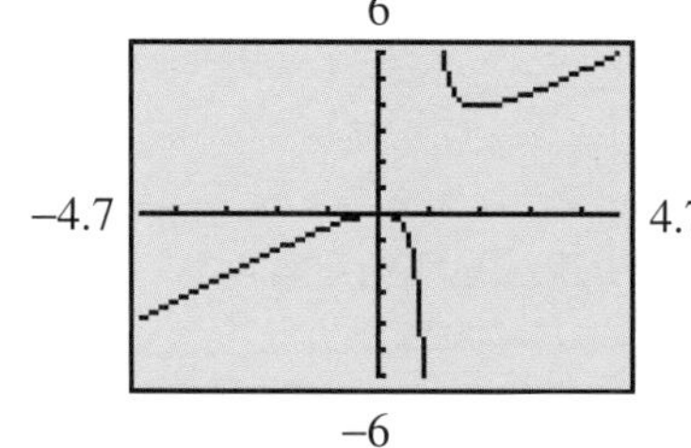

1. All the functions have the same domain restriction. Identify the domain restriction. Explain how this restriction is indicated in each graph.

2. For each function above, evaluate each limit. What do you notice?

i) $\lim_{x \to 1^+} f(x)$ **ii)** $\lim_{x \to 1^-} f(x)$ **iii)** $\lim_{x \to 1} f(x)$

3. The graph of $f(x) = \frac{x^2-1}{x-1}$ has a point discontinuity at $x = 1$. From the results of exercise 2, make a conjecture about how you can tell whether the graph of a rational function $y = f(x)$ has a point discontinuity at $x = a$.

4. Each of the graphs $f(x) = \frac{1}{x-1}$, $f(x) = \frac{-1}{(x-1)^2}$, and $f(x) = \frac{x^2}{x-1}$ has a vertical asymptote at $x = 1$. From the results of exercise 2, make a conjecture about how you can tell whether the graph of a rational function $y = f(x)$ has a vertical asymptote at $x = a$.

5. When $\lim_{x \to a^+} f(x) = \lim_{x \to a^-} f(x)$, does the rational function $y = f(x)$ have a point discontinuity at $x = a$? Explain.

From the *Investigation*, you should have determined that the domain restriction at $x = 1$ could result in either a vertical asymptote or a point discontinuity at that value. Three graphs in exercise 1 have a vertical asymptote at $x = 1$ and one graph has a point discontinuity at $x = 1$. When you examined the limit as x approached 1, you should have discovered that the function with the point discontinuity was the only one that had a finite limit. That is, for $f(x) = \frac{x^2 - 1}{x - 1}$, $\lim_{x \to 1} f(x) = 2$. The discontinuity of the function could be "removed" by defining $f(1) = 2$.

For a function $y = f(x)$, if the denominator equals 0 for $x = a$, there is usually a vertical asymptote at $x = a$. However, when the numerator also equals 0 for $x = a$, there may be a point discontinuity at $x = a$.

Take Note

Point Discontinuity and Vertical Asymptote

For a rational function $f(x) = \frac{p(x)}{q(x)}$:

- If $q(a) = p(a) = 0$, then the graph of $y = f(x)$ may have a point discontinuity at $x = a$. If $\lim_{x \to a} f(x) = L$, where L is a real number, a new continuous function $y = F(x)$ can be created by defining $F(x) = f(x)$ for $x \neq a$ and $F(a) = L$.
- If $q(a) = 0$ and $p(a) \neq 0$, then the graph of $y = f(x)$ has a vertical asymptote at $x = a$.

Example 1

Determine the equation(s) of any vertical asymptotes, or the x-value of any point discontinuity for the graph of each function. Use a graphing calculator to illustrate the results.

a) $y = \frac{x}{x-3}$ **b)** $y = \frac{1}{x^2-4}$ **c)** $y = \frac{x^2-25}{x-5}$ **d)** $y = \frac{x}{x^2-2x}$

Solution

a) $y = \dfrac{x}{x-3}$

Determine where the denominator equals 0.

Let $x - 3 = 0$.

$x = 3$

Therefore, the graph has either a vertical asymptote or a point discontinuity at $x = 3$.

When $x = 3$, the numerator is 3.

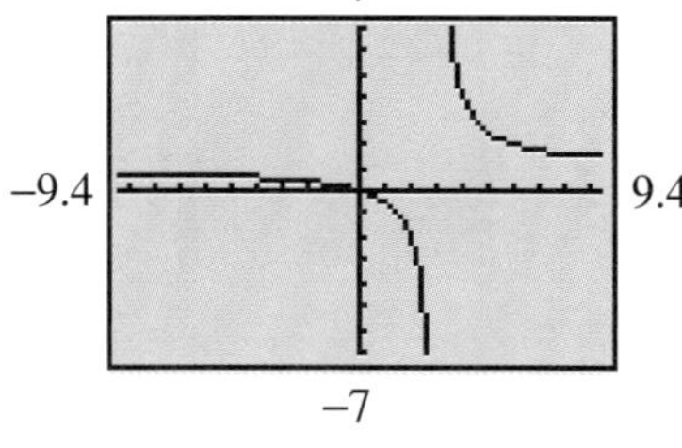

Since the numerator $\neq 0$, there is a vertical asymptote at $x = 3$. The graph of $y = \dfrac{x}{x-3}$ illustrates the vertical asymptote.

b) $y = \dfrac{1}{x^2 - 4}$

Determine where the denominator equals 0.

Let $x^2 - 4 = 0$.

$x = \pm 2$

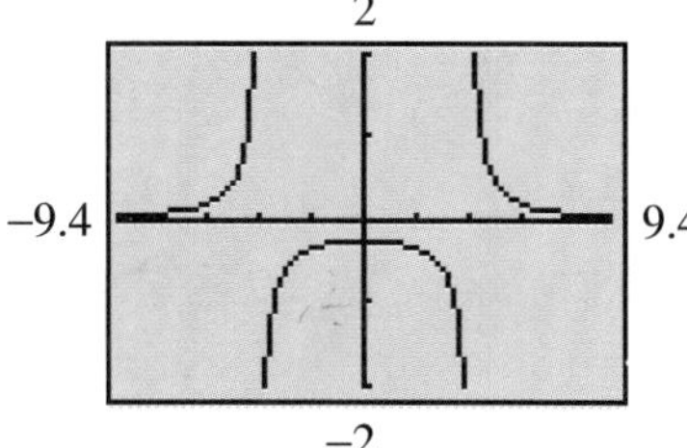

Therefore, the graph has either a vertical asymptote or a point discontinuity at $x = 2$ and $x = -2$. Since the numerator $\neq 0$, there are vertical asymptotes at $x = 2$ and $x = -2$. The graph of $y = \dfrac{1}{x^2 - 4}$ illustrates the vertical asymptotes.

c) $y = \dfrac{x^2 - 25}{x - 5}$

Determine where the denominator equals 0.

Let $x - 5 = 0$.

$x = 5$

Therefore, the graph has either a vertical asymptote or a point discontinuity at $x = 5$.

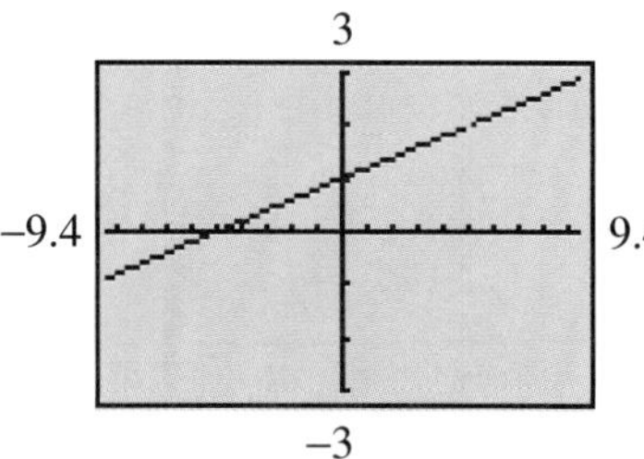

When $x = 5$, the numerator $= 0$; so, there is a point discontinuity at $x = 5$. Look carefully at the screen. The graph of $y = \dfrac{x^2 - 25}{x - 5}$ illustrates the point discontinuity.

d) $y = \dfrac{x}{x^2 - 2x}$

Determine where the denominator equals 0.

Let $x^2 - 2x = 0$.

$x(x - 2) = 0$

$x = 0$ or $x = 2$

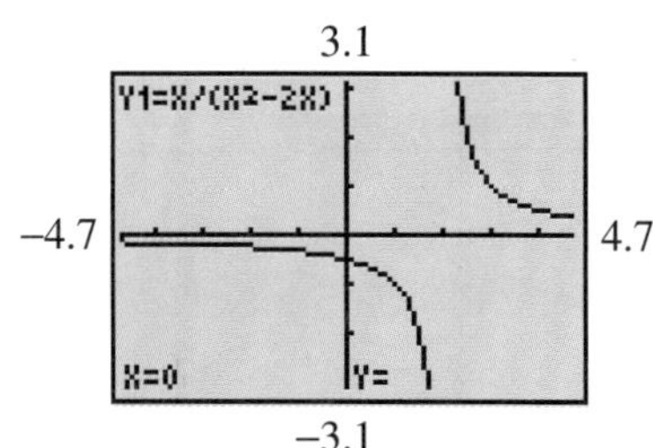

Therefore, the graph has either a vertical asymptote or a point discontinuity at $x = 0$ and $x = 2$.

When $x = 0$, the numerator $= 0$; so, there is a point discontinuity at $x = 0$.

When $x = 2$, the numerator $\neq 0$; so, there is a vertical asymptote at $x = 2$.

The graph of $y = \frac{x}{x^2 - 2x}$ illustrates the vertical asymptote and point discontinuity.

The point discontinuity is not seen on the screen because it coincides with the y-axis. So, we trace to $x = 0$ to check that there is no corresponding y-value.

Horizontal Asymptote

Here is the graph of the function $f(x) = \frac{1}{x}$.

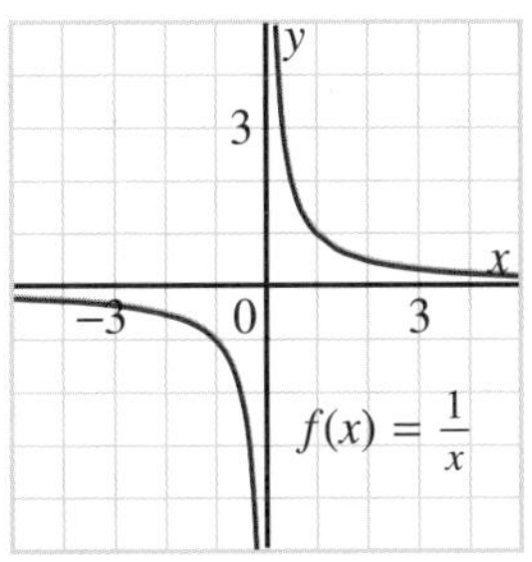

From the graph, the x-axis appears to be a horizontal asymptote.

That is, as $x \to \infty$ or as $x \to -\infty$, $f(x)$ gets closer and closer to 0.

We say that the line $y = 0$ is the horizontal asymptote to the graph of $f(x) = \frac{1}{x}$ because $\lim\limits_{x \to \infty} \frac{1}{x} = 0$. This result is true in general.

We define a horizontal asymptote as follows:

Take Note

Horizontal Asymptote

The line $y = k$ is a *horizontal asymptote* to the graph of the function $y = f(x)$ if either $\lim\limits_{x \to \infty} f(x) = k$ or $\lim\limits_{x \to -\infty} f(x) = k$.

For a rational function, it is not possible for $\lim\limits_{x \to \infty} f(x)$ and $\lim\limits_{x \to -\infty} f(x)$ to be different. That is, if $\lim\limits_{x \to \infty} f(x) = k$, then $\lim\limits_{x \to -\infty} f(x)$ is also equal to k (see exercise 24). Therefore, to determine the equation of any horizontal asymptote to the graph of a rational function, it is only necessary to determine $\lim\limits_{x \to \infty} f(x)$. The graph of a rational function can only have one horizontal asymptote.

Example 2

Determine the equation of any horizontal asymptote to the graph of each function. Use a graphing calculator to illustrate the results.

a) $y = \frac{3x}{x^2 - 9}$ **b)** $y = \frac{6x^2 - x + 2}{3x^2 - 5}$ **c)** $y = \frac{x^2}{x - 1}$

Solution

For each function, take the limit as $x \to \infty$ to determine the horizontal asymptote. Use the technique of dividing the numerator and denominator by the highest power of x in the denominator.

a) 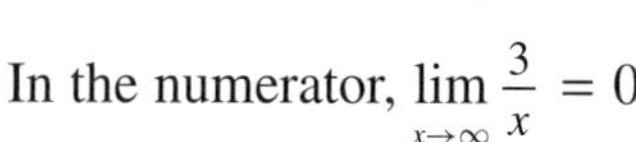

$$\lim_{x\to\infty}\frac{3x}{x^2-9} = \lim_{x\to\infty}\frac{\frac{3x}{x^2}}{\frac{x^2}{x^2}-\frac{9}{x^2}}$$

$$= \lim_{x\to\infty}\frac{\frac{3}{x}}{1-\frac{9}{x^2}}$$

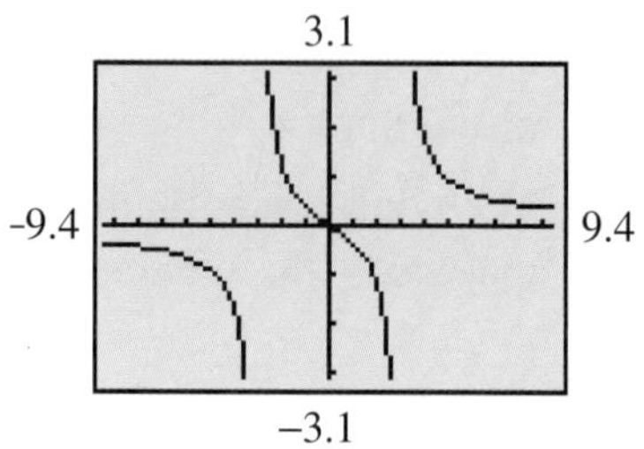

In the numerator, $\lim_{x\to\infty}\frac{3}{x} = 0$

In the denominator, $\lim_{x\to\infty}\frac{9}{x^2} = 0$

So, $\lim_{x\to\infty}\frac{3x}{x^2-9} = \frac{0}{1}$, or 0

Therefore, the horizontal asymptote is $y = 0$.

The graph of $y = \frac{3x}{x^2-9}$ illustrates this horizontal asymptote.

b) 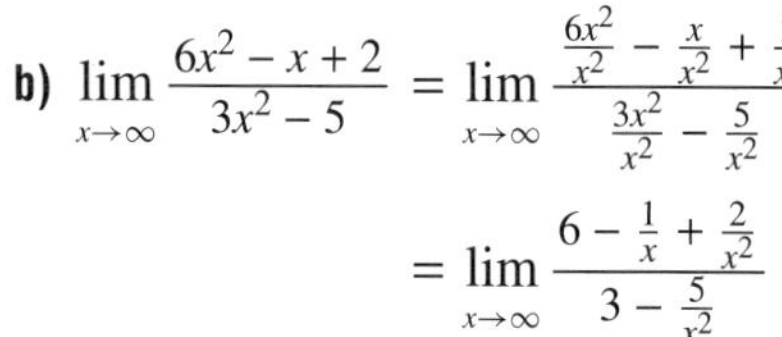

$$\lim_{x\to\infty}\frac{6x^2-x+2}{3x^2-5} = \lim_{x\to\infty}\frac{\frac{6x^2}{x^2}-\frac{x}{x^2}+\frac{2}{x^2}}{\frac{3x^2}{x^2}-\frac{5}{x^2}}$$

$$= \lim_{x\to\infty}\frac{6-\frac{1}{x}+\frac{2}{x^2}}{3-\frac{5}{x^2}}$$

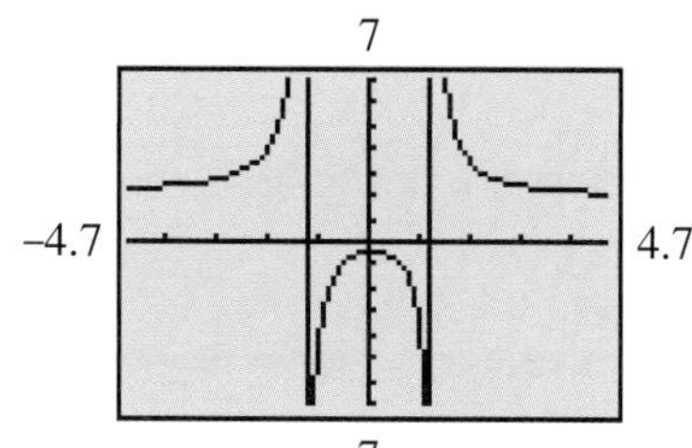

In the numerator, $\lim_{x\to\infty}\frac{1}{x} = 0$ and $\lim_{x\to\infty}\frac{2}{x^2} = 0$

In the denominator, $\lim_{x\to\infty}\frac{5}{x^2} = 0$

So, $\lim_{x\to\infty}\frac{6x^2-x+2}{3x^2-5} = \frac{6}{3}$

$= 2$

Therefore, the horizontal asymptote is $y = 2$.

The graph of $y = \frac{6x^2-x+2}{3x^2-5}$ illustrates this horizontal asymptote.

c) 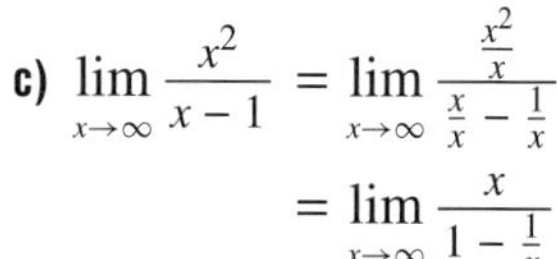

$$\lim_{x\to\infty}\frac{x^2}{x-1} = \lim_{x\to\infty}\frac{\frac{x^2}{x}}{\frac{x}{x}-\frac{1}{x}}$$

$$= \lim_{x\to\infty}\frac{x}{1-\frac{1}{x}}$$

7

−4.7 4.7

−7

In the numerator, $\lim_{x\to\infty} x = \infty$

In the denominator, $\lim_{x\to\infty}\frac{1}{x} = 0$

So, $\lim_{x\to\infty}\frac{x^2}{x-1} = \infty$

The limit does not exist.

Therefore, there is no horizontal asymptote.

The graph of $y = \frac{x^2}{x-1}$ illustrates that there is no horizontal asymptote.

Intercepts

Another feature of the graph of a function is its intercepts. Recall that the intercepts are the values of x and y where the graph intersects the x- and y-axes, respectively. To determine the x-intercept, substitute $y = 0$. To determine the y-intercept, substitute $x = 0$.

Example 3

Determine the x- and y-intercepts of the graph of each function.

a) $y = \frac{1}{x + 2}$ **b)** $y = \frac{x}{x - 4}$ **c)** $y = \frac{x^2 - 2x - 3}{x - 3}$

Solution

a) $y = \frac{1}{x + 2}$

Substitute $y = 0$, then solve for x.

$0 = \frac{1}{x + 2}$

This equation has no solution since there is no value of x that will make the expression $\frac{1}{x + 2}$ equal to 0. Therefore, there is no x-intercept.

Substitute $x = 0$.

$$y = \frac{1}{0 + 2}$$
$$= \frac{1}{2}$$

The y-intercept is $\frac{1}{2}$.

b) $y = \frac{x}{x - 4}$

Substitute $y = 0$, then solve for x.

$0 = \frac{x}{x - 4}$

For the rational expression $\frac{x}{x - 4}$ to equal 0, the numerator must be 0; that is, $x = 0$.

The x-intercept is 0.

Substitute $x = 0$.

$$y = \frac{0}{0 - 4}$$
$$= 0$$

The y-intercept is 0.

c) $y = \frac{x^2 - 2x - 3}{x - 3}$

Factor.

$$y = \frac{(x - 3)(x + 1)}{x - 3}$$

If $x \neq 3$, divide numerator and denominator by $x - 3$.

$y = x + 1$

Substitute $y = 0$, then solve for x.

$0 = x + 1$

$x = -1$

The x-intercept is -1.

Substitute $x = 0$.

$y = 0 + 1$

$= 1$

The y-intercept is 1.

5.3 Exercises

1. Determine the x- and y-intercepts of the graph of each function.

   a) $y = \frac{1}{x - 3}$ b) $y = \frac{x}{x - 3}$ c) $y = \frac{1}{x}$ d) $y = \frac{x^2 - 4x + 4}{x - 2}$

   e) $y = \frac{x^2 - x}{x - 1}$ f) $y = \frac{x - 1}{x^2 - x}$ g) $y = \frac{x^2 - x - 6}{x - 3}$ h) $y = \frac{x^2 + 6x + 9}{x + 3}$

2. Match each equation with the graph of the corresponding function below.

   i) $y = \frac{x}{x - 1}$ ii) $y = \frac{2x^2}{x^2 - 4}$ iii) $y = \frac{1}{x - 1}$ iv) $y = \frac{x^2 - 4}{x - 2}$

a)

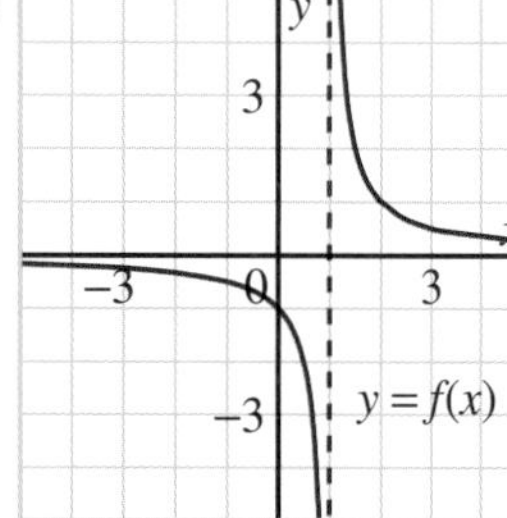

b)

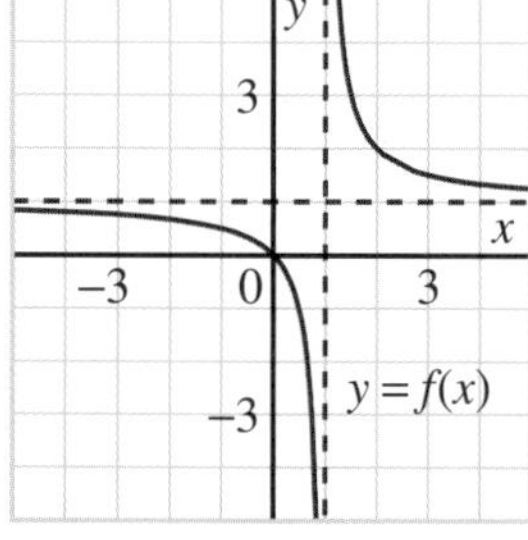

c)

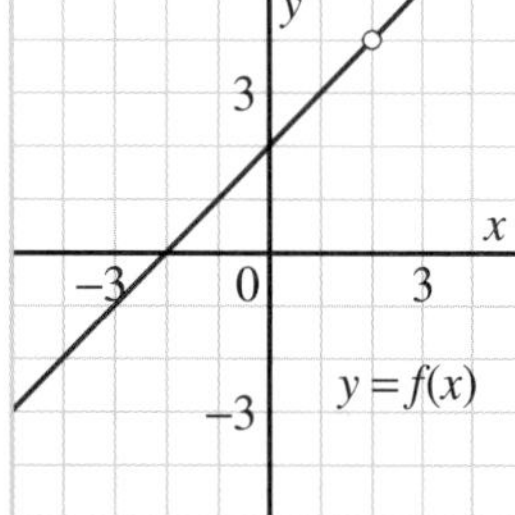

d)

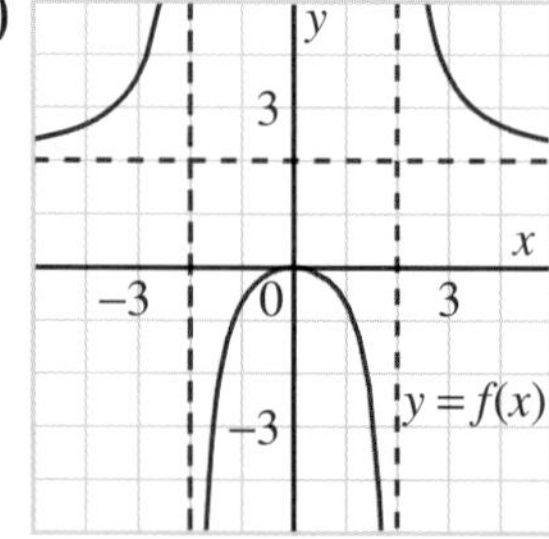

3. For each graph in exercise 2, determine, if possible, the x- and y-intercepts, the equations of any vertical or horizontal asymptotes, and the x-values of any point discontinuities.

4. **Communication** Only one graph in exercise 2 has a point discontinuity. What is different about its equation compared with the other equations? Explain.

5. Determine the equation(s) of any vertical asymptotes or the x-value of any point discontinuity for the graph of each function. Use a graphing calculator to illustrate.

   a) $y = \frac{1}{x+4}$ b) $y = \frac{x}{x-2}$ c) $y = \frac{x^2-2x}{x}$

   d) $y = \frac{x^2-9}{x-3}$ e) $y = \frac{x-1}{x^2-1}$ f) $y = \frac{x+2}{x^2-x-6}$

6. Determine the equation of any horizontal asymptote to the graph of each function. Use a graphing calculator to illustrate.

   a) $y = \frac{1}{x-2}$ b) $y = \frac{5}{x}$ c) $y = \frac{x^3-1}{x^3}$

   d) $y = \frac{3x}{2x-1}$ e) $y = \frac{x}{x^2+3}$ f) $y = \frac{x^2+3}{x}$

7. For each polynomial function $y = f(x)$, determine the equation(s) of any vertical asymptote(s) to the graph of the reciprocal function $y = \frac{1}{f(x)}$.

   a) $f(x) = x + 5$ b) $f(x) = 3x - 1$ c) $f(x) = x^2 - 25$

   d) $f(x) = x^2 + 4$ e) $f(x) = x^2 - x - 12$ f) $f(x) = x^3 + 8$

8. A polynomial function $y = f(x)$ has zeros at $x = -1$ and $x = 4$. Determine the equations of the vertical asymptotes to the graph of the reciprocal function $y = \frac{1}{f(x)}$.

9. A polynomial function $y = f(x)$ has no zeros. Determine the equations of the vertical asymptotes to the graph of the reciprocal function $y = \frac{1}{f(x)}$.

10. **Application** A large tank contains 1000 L of pure water. Salt water that contains 20 g of salt per litre is pumped into the tank at a rate of 10 L/min. The concentration of salt, $C(t)$ grams per litre, after t minutes, is given by the equation $C(t) = \frac{20t}{100+t}$.

    a) Determine the concentration of salt in the tank after each time.

       i) 1 h ii) 5 h iii) 15 h

    b) Determine $\lim_{t \to \infty} \frac{20t}{100+t}$.

    c) Interpret the limit in part b in terms of the concentration of salt in the tank.

 **11.** For the graph of each function, determine the equations of any vertical or horizontal asymptotes and the x-value of any point discontinuity.

a) $y = \frac{1}{x^2 - 4}$ **b)** $y = \frac{x}{x^2 - 4}$ **c)** $y = \frac{x - 2}{x^2 - 4}$

d) $y = \frac{x^2}{x^2 - 4}$ **e)** $y = \frac{x^3}{x^2 - 4}$ **f)** $y = \frac{x^2 - 16}{x - 4}$

 **12. Knowledge/Understanding** Consider $f(x) = \frac{x^2 - 1}{x^2 + 1}$.

a) Determine the x- and y-intercepts of the graph of $y = f(x)$.

b) Determine the equations of any vertical or horizontal asymptotes to the graph of $y = f(x)$.

c) Graph $y = f(x)$ to verify parts a and b.

13. Consider $f(x) = \frac{x^2 - 25}{x^2 + 25}$.

a) Determine the x- and y-intercepts of the graph of $y = f(x)$.

b) Determine the equations of any vertical or horizontal asymptotes to the graph of $y = f(x)$.

c) Graph $y = f(x)$ to verify parts a and b.

14. Consider any function of the form $f(x) = \frac{x^2 - a^2}{x^2 + a^2}$.

a) Determine the x- and y-intercepts of the graph of $y = f(x)$.

b) Determine the equations of any vertical or horizontal asymptotes to the graph of $y = f(x)$.

15. For each function in exercises 12 to 14, consider the reciprocal function $y = \frac{1}{f(x)}$. For the graph of each reciprocal function:

a) Determine the x- and y-intercepts.

b) Determine the equations of any vertical or horizontal asymptotes.

 **16.** Draw a possible graph of each rational function $y = f(x)$ in the neighbourhood of $x = 2$ for each set of conditions.

a) $f(2)$ does not exist.
$\lim_{x \to 2^+} f(x) = 3$
$\lim_{x \to 2^-} f(x) = 3$

b) $f(2)$ does not exist.
$\lim_{x \to 2^+} f(x) = \infty$
$\lim_{x \to 2^-} f(x) = -\infty$

c) $f(2)$ does not exist.
$\lim_{x \to 2^+} f(x) = -\infty$
$\lim_{x \to 2^-} f(x) = -\infty$

Exercise 16 is significant because it considers how the graph of a rational function is related to the right side and left side limits at a particular value, $x = 2$.

✓ **17.** The graph of a rational function $y = f(x)$ has a vertical asymptote at $x = a$. Which limit(s) could be true (where c is a constant real number)? Explain.

a) $\lim_{x \to a^+} f(x) = c$ **b)** $\lim_{x \to a^-} f(x) = c$ **c)** $\lim_{x \to a^+} f(x) = \infty$ **d)** $\lim_{x \to a^-} f(x) = \infty$

18. A rational function $y = f(x)$ has a point discontinuity at $x = a$. Which limit(s) must be true (where c is a constant real number)? Explain.

a) $\lim_{x \to a^+} f(x) = c$ **b)** $\lim_{x \to a^-} f(x) = c$ **c)** $\lim_{x \to a^+} f(x) = \infty$ **d)** $\lim_{x \to a^-} f(x) = \infty$

✓ 19. The graph of a rational function $y = f(x)$ has a horizontal asymptote at $y = a$. Which limit(s) must be true (where c is a constant real number)? Explain.

a) $\lim_{x \to \infty} f(x) = c$ **b)** $\lim_{x \to \infty} f(x) = 0$ **c)** $\lim_{x \to a} f(x) = \infty$ **d)** $\lim_{x \to \infty} f(x) = a$

20. For each pair of polynomial functions $y = p(x)$ and $y = q(x)$:

i) Determine the equation of the rational function $f(x) = \frac{p(x)}{q(x)}$.

ii) Determine the equations of any vertical or horizontal asymptotes.

iii) Determine the x-value of any point discontinuity for the graph of $y = f(x)$.

a) $p(x) = x + 5$, $q(x) = x^2 - 25$ **b)** $p(x) = x$, $q(x) = x^2 - 3x$

c) $p(x) = x^2$, $q(x) = x^2 + 3$ **d)** $p(x) = x^3$, $q(x) = x^3 - 27$

e) $p(x) = x^2$, $q(x) = x + 7$ **f)** $p(x) = x^4$, $q(x) = x^3 - 4x$

21. Use the polynomial functions $y = p(x)$ and $y = q(x)$ from each part of exercise 20 to form a new rational function $g(x) = \frac{q(x)}{p(x)}$. Determine the equations of any vertical or horizontal asymptotes and the x-value of any point discontinuity for the graph of each function $y = g(x)$.

✓ 22. **Thinking/Inquiry/Problem Solving** Consider the function $f(x) = \frac{x}{x + 1}$.

a) Determine the equations of the vertical and horizontal asymptotes to the graph of $y = f(x)$.

b) Determine the derived function $y = f'(x)$, then determine the equations of the vertical and horizontal asymptotes to its graph.

c) Determine the second derivative function $y = f''(x)$, then determine the equations of the vertical and horizontal asymptotes to its graph.

d) Which asymptote remains the same for f, f', and f''? Will this always be true for any function of the form $f(x) = \frac{x}{x + k}$, where k is a constant? Explain.

C

23. Suppose $f(x)$ satisfies these conditions: $f(a)$ does not exist, $\lim_{x \to a^+} f(x) = \infty$, and $\lim_{x \to a^-} f(x) = \infty$. Determine $\lim_{x \to a^+} f'(x)$ and $\lim_{x \to a^-} f'(x)$.

24. Let $f(x)$ be a rational function such that $\lim_{x \to \infty} f(x) = k$. Explain why $\lim_{x \to -\infty} f(x) = k$.

5.4 Oblique Asymptotes

The graph of the rational function $y = \frac{x^2}{x+1}$ appears to have an oblique asymptote. To find the equation of the oblique asymptote, we consider the limit as $x \to \infty$, just as we did with a horizontal asymptote. However, we change the form of the rational expression first before we take the limit.

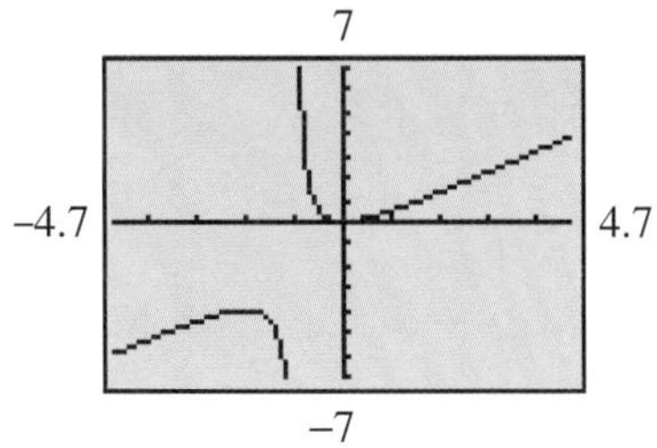

Recall, from *Necessary Skills*, how we changed a rational expression to the sum of a polynomial and another rational expression. For the function $y = \frac{x^2}{x+1}$, we use long division to change the expression $\frac{x^2}{x+1}$ to an equivalent form:

$$
\begin{array}{r}
x - 1 \\
x + 1 \overline{) \; x^2 + 0x + 0} \\
\underline{x^2 + \;\; x} \quad\;\; \\
-x + 0 \\
\underline{-x - 1} \\
1
\end{array}
$$

Therefore, $\frac{x^2}{x+1} = x - 1 + \frac{1}{x+1}, x \neq -1$

As $x \to \infty$, the term $\frac{1}{x+1} \to 0$; that is, the term $\frac{1}{x+1}$ becomes insignificant when x is large. Therefore, as x increases, the graph gets closer and closer to the line $y = x - 1$. So, $y = x - 1$ is the oblique asymptote.

There is a dynamic way to see what happens to the graph as x gets larger and larger. Use a graphing calculator. Input $y = \frac{x^2}{x+1}$, then press ZOOM **3** ENTER. Continue to press ENTER several times. The screens below show how the graph behaves more and more like $y = x - 1$, as x gets very large.

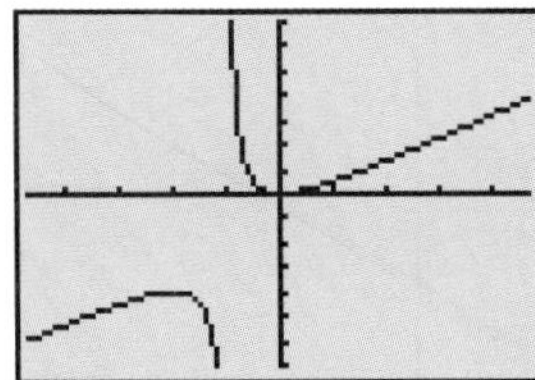

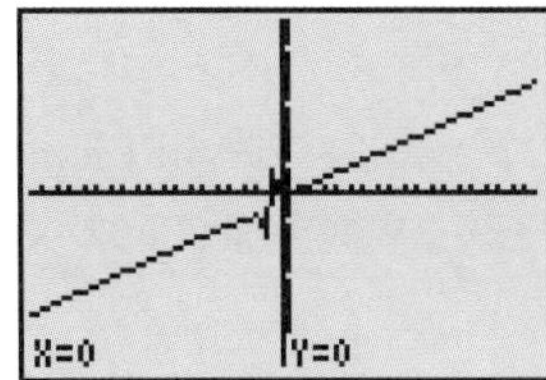

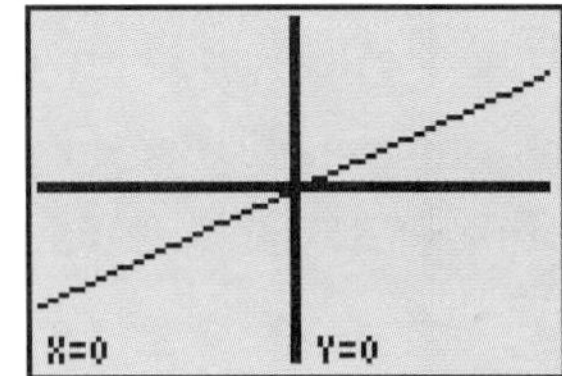

Example 1

Determine the equation of the oblique asymptote to the graph of each function. Use a graphing calculator to illustrate the results.

a) $y = \dfrac{x^2 - 5}{x + 4}$ **b)** $y = \dfrac{x^3}{x^2 - 4}$

Solution

Use long division to change each expression to an equivalent form.

a) $y = \dfrac{x^2 - 5}{x + 4}$

$$\begin{array}{r} x - 4 \\ x + 4 \overline{)\, x^2 + 0x - 5} \\ \underline{x^2 + 4x} \\ -4x - 5 \\ \underline{-4x - 16} \\ 11 \end{array}$$

Therefore,

$\dfrac{x^2 - 5}{x + 4} = x - 4 + \dfrac{11}{x + 4}, x \neq -4$

As $x \to \infty$, the term $\dfrac{11}{x + 4} \to 0$

Therefore, the function $y = \dfrac{x^2 - 5}{x + 4}$ has the oblique asymptote $y = x - 4$.

The graphs of $y = \dfrac{x^2 - 5}{x + 4}$ and $y = x - 4$ are shown.

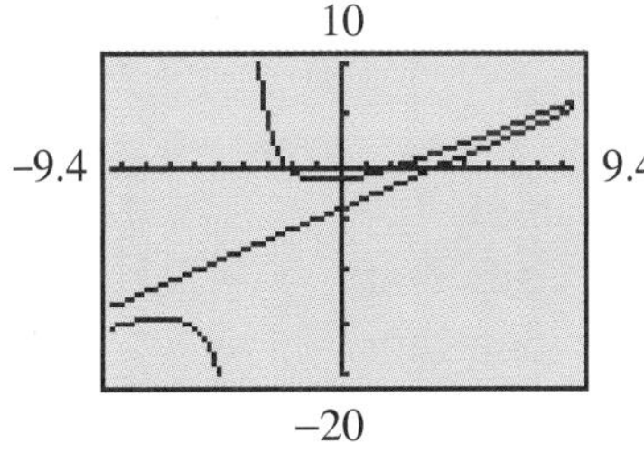

b) $y = \dfrac{x^3}{x^2 - 4}$

$$\begin{array}{r} x \\ x^2 - 4 \overline{)\, x^3 + 0x^2 + 0x} \\ \underline{x^3 \qquad - 4x} \\ 4x \end{array}$$

Therefore,

$\dfrac{x^3}{x^2 - 4} = x + \dfrac{4x}{x^2 - 4}, x \neq \pm 2$

As $x \to \infty$, the term $\dfrac{4x}{x^2 - 4} \to 0$

Therefore, the function $y = \dfrac{x^3}{x^2 - 4}$ has the oblique asymptote $y = x$.

The graphs of $y = \dfrac{x^3}{x^2 - 4}$ and $y = x$ are shown.

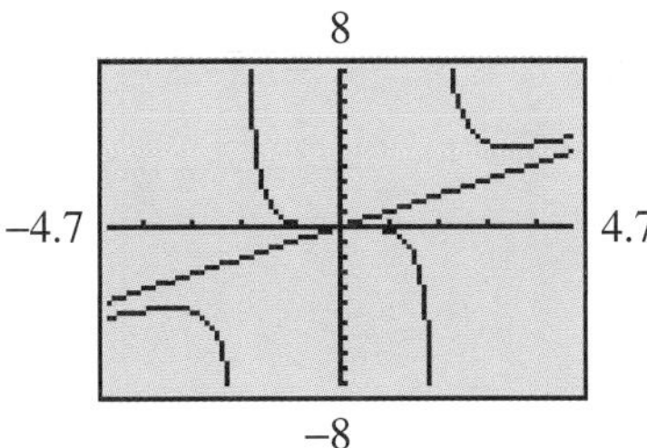

In each equation in *Example 1*, the degree of the numerator is 1 more than the degree of the denominator. This is why we get the sum of a linear expression and a rational expression when we divide.

We can use the table feature of a graphing calculator to justify that a line is an oblique asymptote. Consider the function in *Example 1a*. The line $y = x - 4$ is an oblique asymptote to the graph of $y = \frac{x^2 - 5}{x + 4}$, or $y = x - 4 + \frac{11}{x + 4}$.

Input the rational function and its asymptote. Set the table to start at 0 with increments of 10.

```
Plot1 Plot2 Plot3
\Y1=(X²-5)/(X+4)
\Y2=X-4
\Y3=
\Y4=
\Y5=
\Y6=
```

X	Y1	Y2
0	-1.25	-4
10	6.7857	6
20	16.458	16
30	26.324	26
40	36.25	36
50	46.204	46
60	56.172	56

X=0

From the table, we can see that as x becomes larger and larger, the y-coordinates of the points on the graph of $y = \frac{x^2 - 5}{x + 4}$ and the y-coordinates of the points on the line $y = x + 4$ get closer and closer. This illustrates that the line $y = x - 4$ is an oblique asymptote to the graph of $y = \frac{x^2 - 5}{x + 4}$.

In each part of *Example 1*, we noted that the degree of the numerator is 1 greater than the degree of the denominator. We shall now consider different cases concerning the relative degrees of the numerator and the denominator. Consider the four cases below. All the denominators are $x + 2$. The degree of the numerator changes in each equation.

Case I. The degree of the numerator is less than the degree of the denominator.

$$y = \frac{1}{x + 2}$$

The x-axis is a horizontal asymptote. This makes sense if we consider what is happening to the values of y as $x \to \infty$.

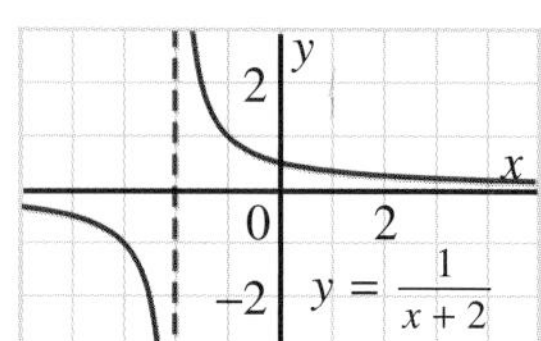

Case II. The degree of the numerator equals the degree of the denominator.

$$y = \frac{2x}{x + 2}$$

There is a horizontal asymptote, but it is not the x-axis. It is $y = 2$, since $\lim_{x \to \infty} \frac{2x}{x + 2} = 2$.

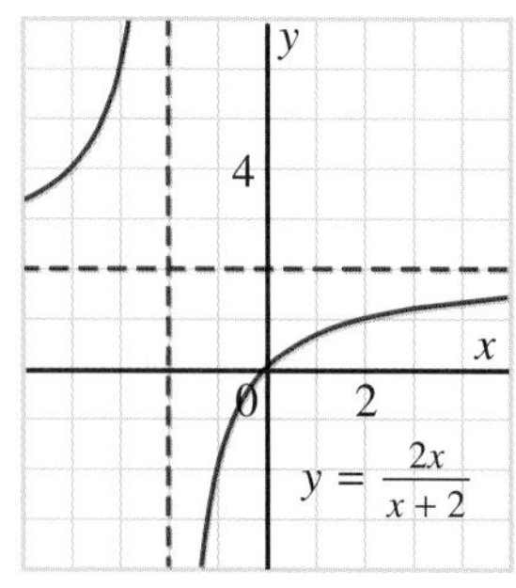

Case III. The degree of the numerator is 1 greater than the degree of the denominator.

$y = \dfrac{x^2}{x+2}$

There is an oblique asymptote.

$y = \dfrac{x^2}{x+2}$ can be written as $y = x - 2 + \dfrac{4}{x+2}$.

As $x \to \infty$, $y = x - 2 + \dfrac{4}{x+2} \to y = x - 2$

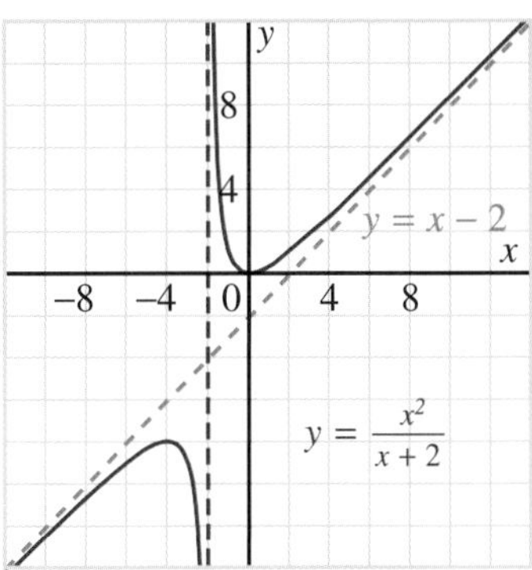

Case IV. The degree of the numerator is more than 1 greater than the degree of the denominator.

$y = \dfrac{x^3}{x+2}$

There are no horizontal or oblique asymptotes.

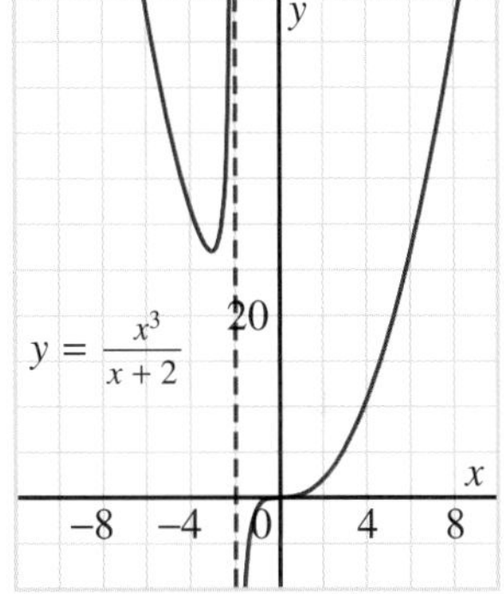

There is a horizontal asymptote when the degree of the numerator equals the degree of the denominator, or is less than the degree of the denominator. There is an oblique asymptote when the degree of the numerator is 1 greater than the degree of the denominator.

Take Note

Asymptotes

For a rational function $f(x) = \dfrac{p(x)}{q(x)}$:

- When $q(a) = 0$ and $p(a) \neq 0$, there is a vertical asymptote $x = a$.
- When $\lim_{x \to \infty} f(x) = k$, there is a horizontal asymptote $y = k$.
  When the degree of $p(x) <$ the degree of $q(x)$, the horizontal asymptote is $y = 0$.
  When the degree of $p(x) =$ the degree of $q(x)$, the horizontal asymptote is $y = k$, $k \neq 0$.
- When the degree of $p(x)$ is 1 greater than the degree of $q(x)$, there is an oblique asymptote.
- When the degree of $p(x)$ is more than 1 greater than the degree of $q(x)$, there is no horizontal or oblique asymptote.

Example 2

For each rational function, determine whether its graph has a horizontal asymptote or an oblique asymptote. Explain. If the graph of the function has one of these asymptotes, write its equation.

a) $y = \frac{5x}{3x - 1}$ **b)** $y = \frac{2x^2}{x - 3}$ **c)** $y = \frac{1}{x^2 - 9}$ **d)** $y = \frac{x^4}{x^2 + 3}$

Solution

a) Since the degree of the numerator is the same as the degree of the denominator, the graph of $y = \frac{5x}{3x - 1}$ has a horizontal asymptote.

As $x \to \infty$, $\frac{5x}{3x - 1} \to \frac{5x}{3x} = \frac{5}{3}$

The equation of the horizontal asymptote is $y = \frac{5}{3}$.

b) Since the degree of the numerator is 1 greater than the degree of the denominator, the graph of $y = \frac{2x^2}{x - 3}$ has an oblique asymptote. Divide to determine its equation.

$$
\begin{array}{r}
2x + \ \ 6 \\
x - 3 \enclose{longdiv}{2x^2 + 0x + \ \ 0} \\
\underline{2x^2 - 6x} \\
6x + \ \ 0 \\
\underline{6x - 18} \\
18
\end{array}
$$

$\frac{2x^2}{x - 3} = 2x + 6 + \frac{18}{x - 3}$

As $x \to \infty$, $2x + 6 + \frac{18}{x - 3} \to 2x + 6$

The equation of the oblique asymptote is $y = 2x + 6$.

c) Since the degree of the numerator is less than the degree of the denominator, the graph of $y = \frac{1}{x^2 - 9}$ has a horizontal asymptote.
The equation of the horizontal asymptote is $y = 0$.

d) Since the degree of the numerator is more than 1 greater than the degree of the denominator, the graph of $y = \frac{x^4}{x^2 + 3}$ has no horizontal or oblique asymptote.

THE BIG PICTURE

Changing the Scale of the Graphs of Rational Functions

The graphs of rational functions have many different shapes. For example, the graph of $y = \frac{x^5 - 10x^3 + 20}{x^2 - 9}$ is shown below. There are vertical asymptotes at $x = \pm 3$. If we change the scale to show the graph for much larger values of x, it becomes virtually indistinguishable from the graph of $y = x^3$. The reason is that as x becomes larger and larger, the lower-degree terms have less effect on the values of y, and the function becomes closer and closer to $y = \frac{x^5}{x^2}$, or $y = x^3$. There are still vertical asymptotes at $x = \pm 3$, but they cannot be seen on this scale.

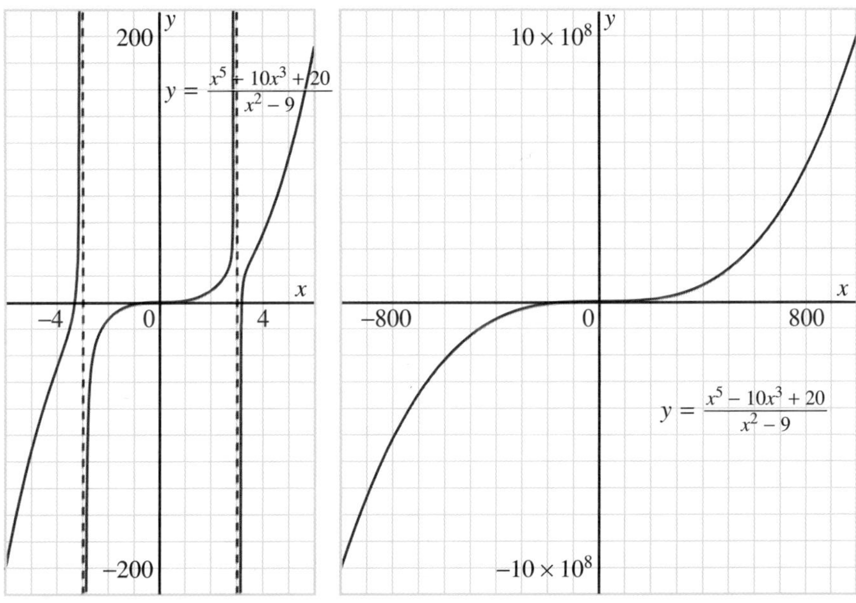

When we graph a rational function, we usually choose a scale to show the fine detail of the graph. This shows how the function behaves for relatively small values of x. However, we can also choose a scale to show how the function behaves for larger values of x. If the degree of the numerator is greater than that of the denominator, a rational function approximates a power function for very large values of x.

5.4 Exercises

A

1. For each function, the asymptotes are not drawn. Estimate the equations of any vertical, horizontal, or oblique asymptotes. Use a ruler to determine the oblique asymptote.

a) 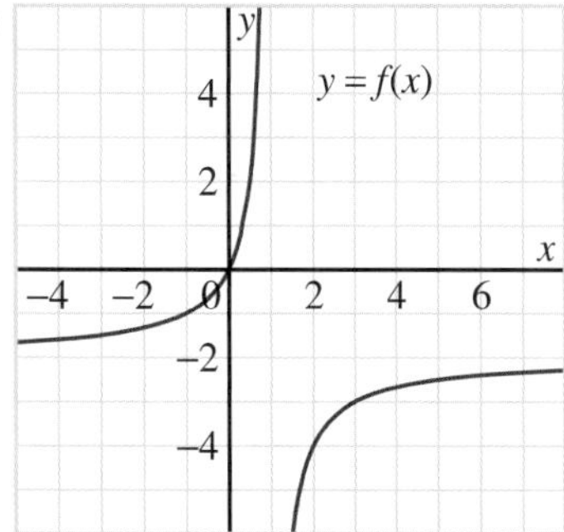

b) 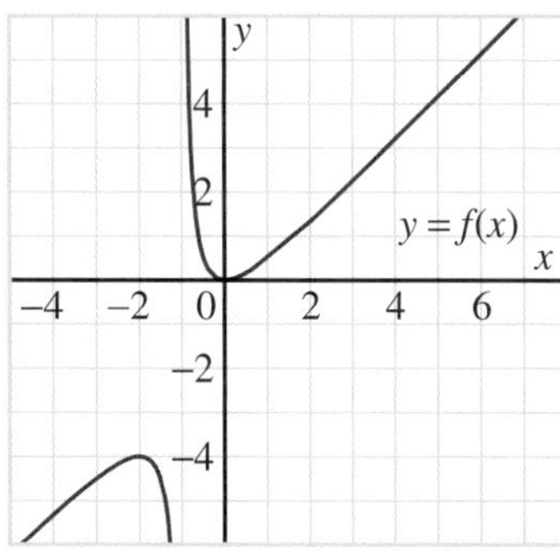

c) 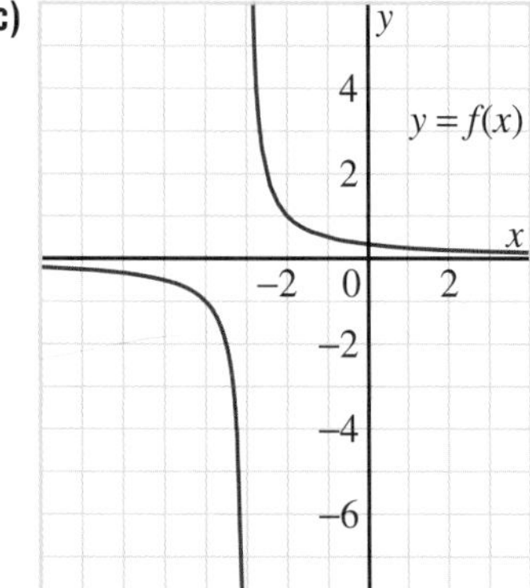

d) 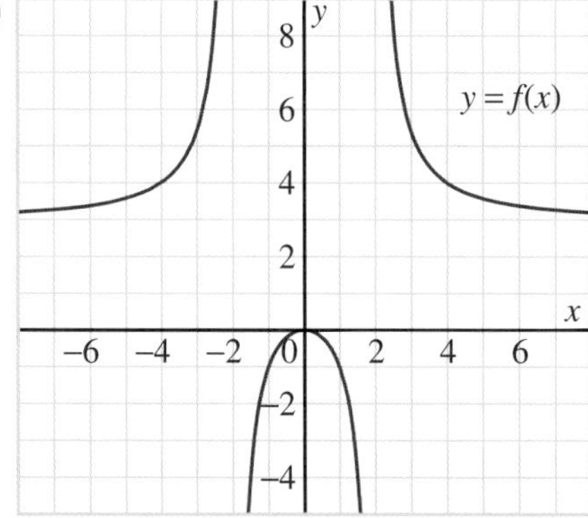

2. For each rational function, determine whether its graph has a horizontal asymptote, an oblique asymptote, or neither. Explain.

a) $y = \frac{1}{x+5}$ b) $y = \frac{x}{x+5}$ c) $y = \frac{x^2}{x+5}$ d) $y = \frac{x^3}{x+5}$

B

3. For the graph of each function in exercise 2, determine the equation of the horizontal or oblique asymptote.

4. **Knowledge/Understanding** Determine the equation of the oblique asymptote to the graph of each function. Use a graphing calculator to illustrate the results.

a) $y = \frac{x^2}{x-4}$ b) $y = \frac{x^2-16}{x-3}$ c) $y = \frac{4x^2}{x-1}$

d) $y = \frac{8x^2}{2x-1}$ e) $y = \frac{x^3}{x^2+2}$ f) $y = \frac{x^3-2}{x^2+1}$

5. a) Determine the equation of the oblique asymptote to the graph of each function.

i) $y = \dfrac{x^2}{x+6}$ ii) $y = \dfrac{2x^2}{x+6}$ iii) $y = \dfrac{3x^2}{x+6}$

b) Use the pattern in part a. Without using long division, predict the equation of the oblique asymptote to the graph of each function.

i) $y = \dfrac{4x^2}{x+6}$ ii) $y = \dfrac{10x^2}{x+6}$ iii) $y = \dfrac{kx^2}{x+6}$ (where k is a constant)

✓ 6. a) Determine the equation of the oblique asymptote to the graph of each function.

i) $y = \dfrac{x^2}{x-3}$ ii) $y = \dfrac{2x^2}{x-3}$ iii) $y = \dfrac{3x^2}{x-3}$

b) **Communication** Use the pattern in part a. Without using long division, predict the equation of the oblique asymptote to the graph of each function. Explain.

i) $y = \dfrac{4x^2}{x-3}$ ii) $y = \dfrac{10x^2}{x-3}$ iii) $y = \dfrac{kx^2}{x-3}$ (where k is a constant)

7. Determine the equation of the horizontal asymptote to the graph of each function.

a) $y = \dfrac{5x}{2x-1}$ b) $y = \dfrac{x}{2x-3}$ c) $y = \dfrac{12x^2}{3x^2-9}$

✓ 8. **Thinking/Inquiry/Problem Solving** Consider the rational function $f(x) = \dfrac{p(x)}{q(x)}$.

a) Suppose the degree of $p(x)$ is 1 greater than the degree of $q(x)$. What is the slope of the oblique asymptote? Explain.

b) Suppose the degree of $p(x)$ is the same as the degree of $q(x)$. What is the slope of the oblique asymptote? Explain.

✓ 9. Daniel discovered a new way to obtain the equation of the oblique asymptote to the graph of the rational function $y = \dfrac{x^2}{x-2}$. By adding and subtracting the same number in the numerator, he wrote the rational expression as a sum, without dividing. He did this:

$$\begin{aligned} y &= \frac{x^2}{x-2} \\ &= \frac{x^2-4+4}{x-2} \\ &= \frac{x^2-4}{x-2} + \frac{4}{x-2} \\ &= \frac{(x-2)(x+2)}{x-2} + \frac{4}{x-2} \\ &= x+2+\frac{4}{x-2} \end{aligned}$$

Daniel concluded that the equation of the oblique asymptote is $y = x + 2$.

a) Explain how Daniel's method works.

b) Use Daniel's method to determine the equation of the oblique asymptote to the graph of each function.

i) $y = \dfrac{x^2}{x-5}$ ii) $y = \dfrac{x^2}{x+6}$ iii) $y = \dfrac{x^2}{x-a}$ (where a is a constant)

 **10. Application** Carla says, "But your method in exercise 9 only works for certain rational functions, Daniel. What about finding the equation of the oblique asymptote to the graph of $y = \frac{x^2}{2x - 6}$ or $y = \frac{x^2 + 1}{x + 3}$, or even $y = \frac{5x^2}{6x - 12}$?" Daniel replies, "My method works with some factoring and algebraic manipulation." Use Daniel's method to determine the equation of the oblique asymptote to the graph of each function in this exercise.

11. a) Use any method to determine the equation of the oblique asymptote to the graph of each function.

i) $y = \frac{x^2}{x + 10}$ **ii)** $y = \frac{3x^2}{2x - 8}$ **iii)** $y = \frac{x^3}{x^2 - 9}$

iv) $y = \frac{x^3 + 1}{x^2 - 9}$ **v)** $y = \frac{x^3 + x}{x^2 - 9}$ **vi)** $y = \frac{x^4 - x^2}{x^3 + 8}$

b) For which functions in part a could you not use the method from exercise 9? Explain.

12. Determine the equation of the vertical asymptote to the graph of each function in exercise 11.

 **13. a)** The graph of the function $f(x) = \frac{x^2}{x - 1}$ is shown. It appears that the function f has a maximum point at (0, 0) and a minimum point at (2, 4). Determine f'. Use the derivative to verify these coordinates.

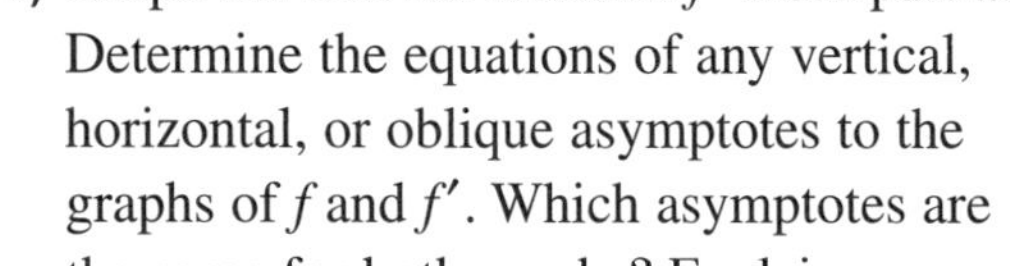

b) Graph the derived function f' from part a. Determine the equations of any vertical, horizontal, or oblique asymptotes to the graphs of f and f'. Which asymptotes are the same for both graphs? Explain.

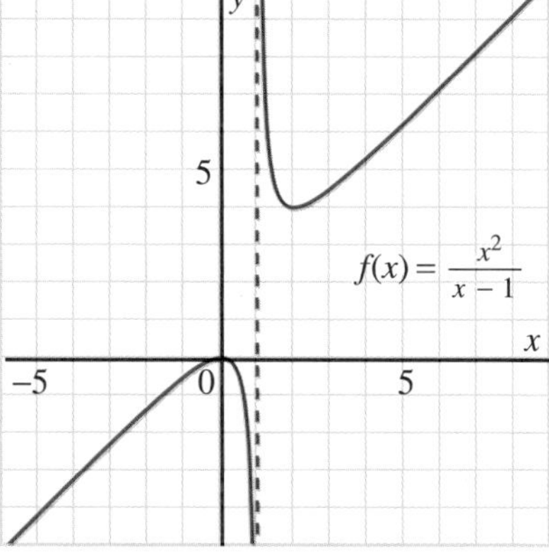

14. The sum of a positive number, n, and its reciprocal can be modelled by the rational function $S(n) = n + \frac{1}{n}$.

a) Graph the function $S(n)$. Determine the oblique asymptote to the graph.

b) Interpret the meaning of this oblique asymptote for the function as $n \to \infty$.

c) Determine $S'(n)$. Determine the value of n that produces the least sum.

5.5 Sketching the Graph of a Rational Function

Each section in this chapter has been concerned with a specific aspect of a rational function. You now know how to determine the intercepts, the equation of a vertical, horizontal, or oblique asymptote, and any point discontinuity. In this section, you will combine these skills with the techniques of calculus you learned in earlier chapters to analyse the key features of the graph of a rational function, and to sketch the graph. The key features of the graph of a function are: intercepts, domain, critical points, intervals of increase and decrease, points of inflection, intervals of concavity, asymptotes, and point discontinuities.

Example 1

Analyse the key features of the graph of $y = \frac{3x^2}{x^2 - 4}$, then sketch the graph.

Solution

$y = \frac{3x^2}{x^2 - 4}$, $x \neq \pm 2$

Intercepts

For the y-intercept, substitute $x = 0$, then solve for y.

$$y = \frac{0}{0 - 4}$$
$$= 0$$

The y-intercept is 0.

For the x-intercept, substitute $y = 0$, then solve for x.

$$\frac{3x^2}{x^2 - 4} = 0$$
$$3x^2 = 0$$
$$x = 0$$

The x-intercept is 0.

Domain

Since the denominator cannot be zero, $x^2 - 4 \neq 0$

$$x \neq \pm 2$$

The domain is all real numbers except 2 and –2.

$D = \{x | x \in \Re, x \neq \pm 2\}$

Critical points

Determine the derivative, y'.

Since $y = \frac{3x^2}{x^2 - 4}$ is a quotient of two polynomials, use the quotient rule to differentiate.

$$y' = \frac{6x(x^2 - 4) - (3x^2)(2x)}{(x^2 - 4)^2}$$
$$= \frac{6x^3 - 24x - 6x^3}{(x^2 - 4)^2}$$
$$= \frac{-24x}{(x^2 - 4)^2}, \ x \neq \pm 2$$

For a critical point, solve $y' = 0$.

$$\frac{-24x}{(x^2 - 4)^2} = 0$$
$$-24x = 0$$
$$x = 0$$

So, there is a critical point at $x = 0$.

Intervals of increase and decrease

Consider the sign of y' in each interval.

$$y' = \frac{-24x}{(x^2 - 4)^2}, \ x \neq \pm 2$$

When $x < 0$, $\frac{-24x}{(x^2 - 4)^2} > 0$, so $y' > 0$

When $x > 0$, $\frac{-24x}{(x^2 - 4)^2} < 0$, so $y' < 0$

Use a number-line analysis for the first derivative.

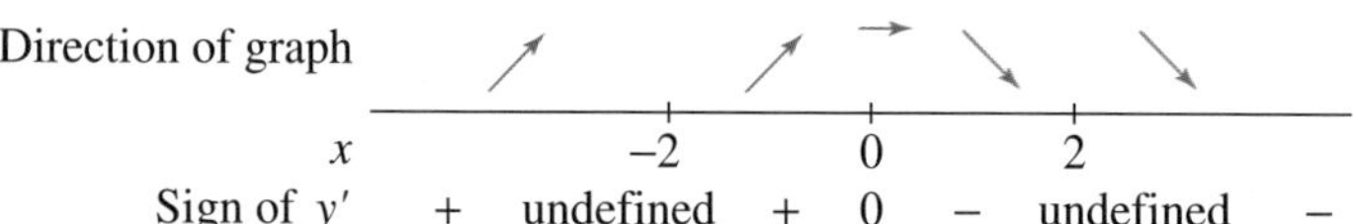

Therefore, the function is increasing when $x < 0$, $x \neq -2$; and decreasing when $x > 0$, $x \neq 2$.

Points of inflection

Determine the second derivative, y''.

$$y' = \frac{-24x}{(x^2 - 4)^2}, \ x \neq \pm 2$$

Use the quotient rule to differentiate.

$$y'' = \frac{-24(x^2 - 4)^2 - (-24x)(2)(x^2 - 4)(2x)}{(x^2 - 4)^4}$$
$$= \frac{-24(x^2 - 4)(x^2 - 4 - 4x^2)}{(x^2 - 4)^4}$$
$$= \frac{24(3x^2 + 4)}{(x^2 - 4)^3}, \ x \neq \pm 2$$

For a point of inflection, solve $y'' = 0$.

$$\frac{24(3x^2 + 4)}{(x^2 - 4)^3} = 0$$

Since the numerator can never be 0, there is no solution.

So, there are no points of inflection.

Intervals of concavity

Determine the sign of y'' in each interval.

$y'' = \frac{24(3x^2 + 4)}{(x^2 - 4)^3}$, $x \neq \pm 2$

When $x < -2$, say $x = -2.1$, y'' is positive, and the graph is concave up.
When $-2 < x < 2$, say $x = 0$, y'' is negative, and the graph is concave down.
When $x > 2$, say $x = 2.1$, y'' is positive, and the graph is concave up.

Use a number-line analysis for the second derivative.

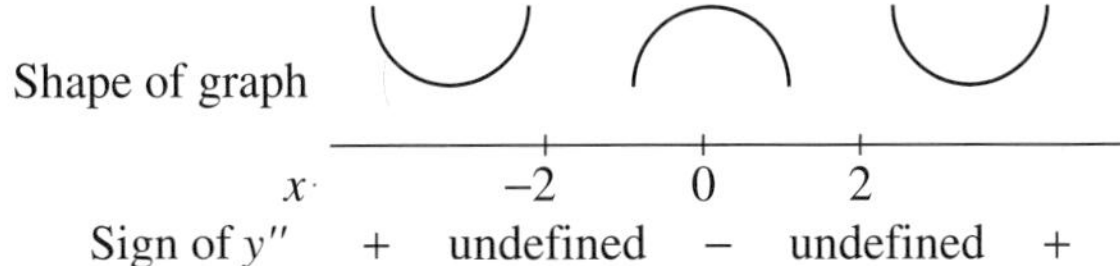

Therefore, the graph of $y = \frac{3x^2}{x^2 - 4}$ is concave up when $x < -2$ or $x > 2$ and concave down when $-2 < x < 2$.

Asymptotes and point discontinuities

There are vertical asymptotes or point discontinuities at values of x for which $y = \frac{3x^2}{x^2 - 4}$ is undefined; that is, at $x = \pm 2$.
When $x = -2$, the numerator $\neq 0$; so, there is a vertical asymptote $x = -2$.
When $x = 2$, the numerator $\neq 0$; so, there is a vertical asymptote $x = 2$.
There are no point discontinuities.

For the horizontal asymptote, determine:

$\lim_{x \to \infty} \frac{3x^2}{x^2 - 4} = 3$

Therefore, there is a horizontal asymptote $y = 3$.

Use the information to sketch the graph.

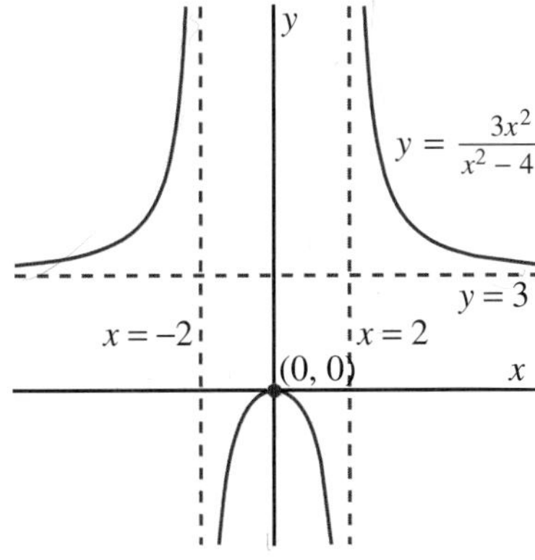

Example 2

Analyse the key features of the graph of $y = \frac{4}{x^2 + 2}$, then sketch the graph.

Solution

$$y = \frac{4}{x^2 + 2}$$

Intercepts

For the y-intercept, substitute $x = 0$, then solve for y.

$$\begin{aligned} y &= \frac{4}{0 + 2} \\ &= 2 \end{aligned}$$

So, the y-intercept is 2.

For the x-intercept, substitute $y = 0$, then solve for x.

$$\frac{4}{x^2 + 2} = 0$$

Since the numerator can never be 0, there is no solution.
So, there is no x-intercept.

Domain

The denominator, $x^2 + 2$, is never equal to zero, so the domain is all real numbers.
$D = \Re$

Critical points

Determine the derivative, y'.
Since $y = \frac{4}{x^2 + 2}$ or $y = 4(x^2 + 2)^{-1}$, use the chain rule to differentiate.

$$\begin{aligned} y' &= -4(x^2 + 2)^{-2}(2x) \\ &= \frac{-8x}{(x^2 + 2)^2} \end{aligned}$$

For a critical point, solve $y' = 0$.

$$\begin{aligned} \frac{-8x}{(x^2 + 2)^2} &= 0 \\ -8x &= 0 \\ x &= 0 \end{aligned}$$

Therefore, there is a critical point at $x = 0$.

Intervals of increase and decrease

Consider the sign of y' in each interval.

$$y' = \frac{-8x}{(x^2 + 2)^2}$$

When $x < 0$, $y' > 0$
When $x > 0$, $y' < 0$

Use a number-line analysis for the first derivative.

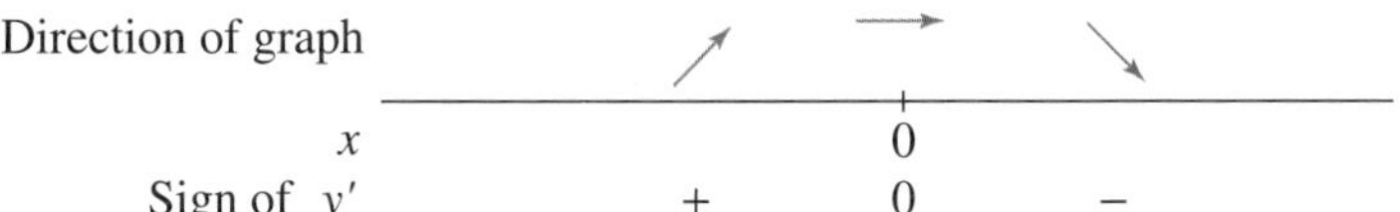

Therefore, the function is increasing when $x < 0$ and decreasing when $x > 0$.

Points of inflection

Determine the second derivative, y''.

$$y' = \frac{-8x}{(x^2+2)^2}$$

Differentiate. Use the quotient rule.

$$\begin{aligned} y'' &= \frac{-8(x^2+2)^2 - (-8x)(2)(x^2+2)(2x)}{(x^2+2)^4} \\ &= \frac{-8(x^2+2)[(x^2+2) - 2(2x)(x)]}{(x^2+2)^4} \\ &= \frac{-8[(x^2+2) - 4x^2]}{(x^2+2)^3} \\ &= \frac{-8(2-3x^2)}{(x^2+2)^3} \end{aligned}$$

For a point of inflection, solve $y'' = 0$.

$$\begin{aligned} \frac{-8(2-3x^2)}{(x^2+2)^3} &= 0 \\ 2 - 3x^2 &= 0 \\ x^2 &= \frac{2}{3} \\ x &= \pm\sqrt{\frac{2}{3}}, \text{ or approximately } \pm 0.82 \end{aligned}$$

Since $y'' = 0$ when $x = \pm\sqrt{\frac{2}{3}}$, there are 2 possible points of inflection. Check the concavity to determine whether these points of inflection exist.

Intervals of concavity

Determine the sign of $y'' = \frac{-8(2-3x^2)}{(x^2+2)^3}$ in each interval.

For $x < -\sqrt{\frac{2}{3}}$, say $x = -1$, y'' is positive, and the graph is concave up.

For $-\sqrt{\frac{2}{3}} < x < \sqrt{\frac{2}{3}}$, say $x = 0$, y'' is negative, and the graph is concave down.

For $x > \sqrt{\frac{2}{3}}$, say $x = 1$, y'' is positive, and the graph is concave up.

Use a number-line analysis for the second derivative.

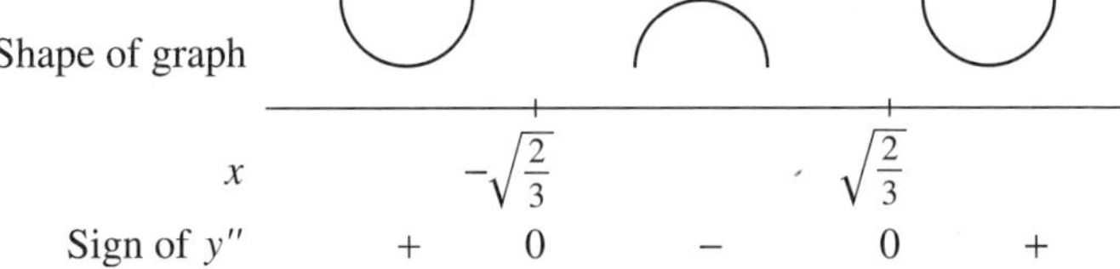

Since the graph of $y = \frac{4}{x^2 + 2}$ changes concavity at $x = \pm\sqrt{\frac{2}{3}}$, there are points of inflection where x has these values.

Asymptotes and point discontinuities

There are vertical asymptotes or point discontinuities at values of x for which $y = \frac{4}{x^2 + 2}$ is undefined.

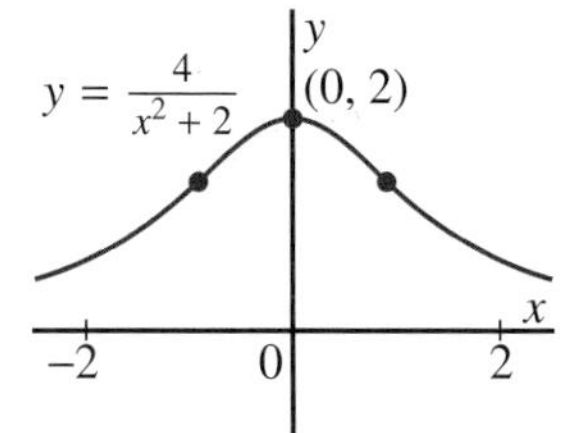

Since $x^2 + 2$ is always positive, y is always defined. So, there are no vertical asymptotes or point discontinuities.

For the horizontal asymptote, determine:

$\lim_{x \to \infty} \frac{4}{x^2 + 2} = 0$

Since the limit is 0, there is a horizontal asymptote $y = 0$.

Use the information to sketch the graph.

5.5 Exercises

A

1. Look at the conditions below for a rational function $y = f(x)$. Which conditions *must* be true in each case?

   a) The graph of $y = f(x)$ has a vertical asymptote at $x = 2$.

   b) The graph of $y = f(x)$ has a horizontal asymptote at $y = 2$.

   c) The graph of $y = f(x)$ has a point discontinuity at $x = 2$.

   i) $\lim_{x \to \infty} f(x) = 2$ ii) $\lim_{x \to 2} f(x) = \infty$

   iii) $f(2)$ is undefined. iv) $\lim_{x \to 2} f(x) = c$ (where c is a constant)

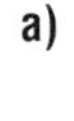

2. For each function $y = f(x)$, estimate where f' and f'' are 0, positive, and negative.

a)

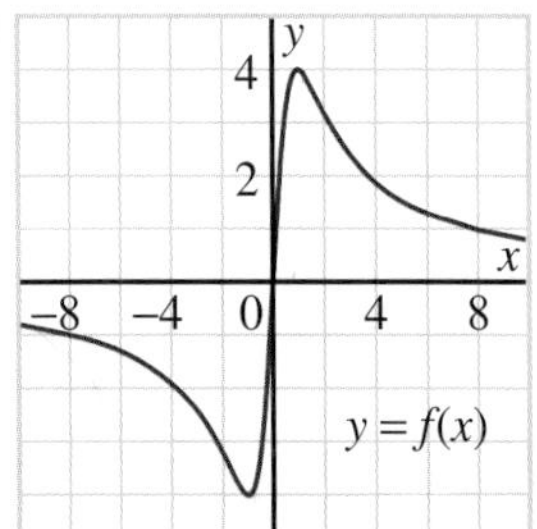

b)

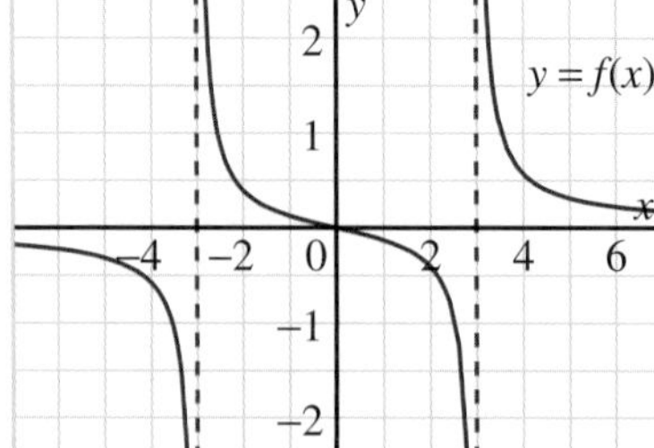

B

3. Use the graph of each function $y = f(x)$ in exercise 2 to sketch the graphs of $y = f'(x)$ and $y = f''(x)$.

4. Analyse the key features of the graph of each function, then sketch the graph of the function.

a) $y = \dfrac{x+5}{x^2-25}$ **b)** $y = \dfrac{x-2}{x^2-4}$

5. **Knowledge/Understanding** Analyse the key features of the graph of each function, then sketch the graph of the function.

a) $y = \dfrac{x^2}{x^2-4}$ **b)** $y = \dfrac{3x^2}{x^2-1}$

6. Analyse the key features of the graph of each function, then sketch the graph of the function.

a) $y = \dfrac{x^2-1}{x}$ **b)** $y = \dfrac{x^2+3}{2x}$

7. Analyse the key features of the graph of each function, then sketch the graph of the function.

a) $y = \dfrac{8}{x^2+4}$ **b)** $y = \dfrac{4x}{x^2+4}$

8. Analyse the key features of the graph of each function, then sketch the graph of the function.

a) $y = \dfrac{x^2-25}{x+5}$ **b)** $y = \dfrac{1+x^2}{1-x^2}$

c) $y = \dfrac{x^2+4}{x}$ **d)** $y = \dfrac{3}{x^2+1}$

9. Analyse the key features of the graph of each function, then sketch the graph of the function.

a) $y = \dfrac{x^2-4}{x-2}$ **b)** $y = \dfrac{x^2}{1-x^2}$

c) $y = \dfrac{x^2+1}{x}$ **d)** $y = \dfrac{2}{x^2+2}$

10. **a)** The graph of $y = f'(x)$, the derivative of the rational function $y = f(x)$, is shown.

i) Determine the intervals where f is increasing and where f is decreasing.

ii) Determine the x-values where f has a local maximum value or a local minimum value.

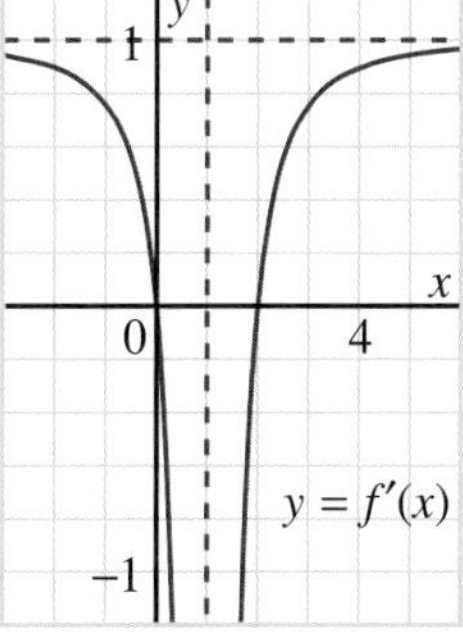

b) Use the results of part a to sketch the graph of a possible function $y = f(x)$.

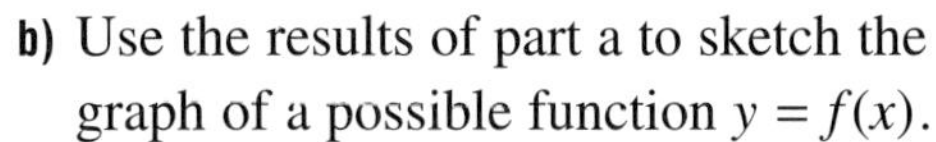

11. An equation of the function in exercise 10 is $f(x) = \dfrac{x^2}{x-1}$.

a) Determine the derived function $y = f'(x)$.

b) Graph $y = f'(x)$. Check that the graph matches that in exercise 10.

✓ **12. Application** Sketch the graph of a rational function $y = f(x)$ that satisfies these conditions:
$f(1) = 0$; $f(3) = 2$; $f(2)$ is undefined; $f'(x) < 0$ for all x except $x = 2$

Exercise 12 is significant because it demonstrates what it means for a function $y = f(x)$ to be undefined at a particular value of x. The graph of the rational function has either a point discontinuity or a vertical asymptote at this value.

13. Sketch the graph of a rational function $y = f(x)$ that satisfies these conditions:
$f(1) = 3$; $f(3) = 3$; $f(2)$ is undefined; $f'(x) < 0$ when $x > 0$ and $f'(x) > 0$ when $x < 0$

14. Sketch the graph of a rational function $y = f(x)$ that satisfies these conditions:
$f(4) = 3$; $f(0) = -1$; $f(-x) = f(x)$; $\lim_{x \to 2^+} f(x) = \infty$, $\lim_{x \to 2^-} f(x) = -\infty$;
$f'(0) = 0$; $f'(x) < 0$ when $x > 0$ and $f'(x) > 0$ when $x < 0$

15. Communication All the functions in this section could have been graphed using a graphing calculator. Explain why the analysis to determine the key features is useful. What can you learn from the analysis that you could not learn from the graph on a calculator screen?

✓ **16.** Analyse the key features of the graph of the function $f(x) = \dfrac{x^2 + 4x - 2}{x^2}$, then sketch the graph.

✓ **17. Thinking/Inquiry/Problem Solving** The three graphs below are functions of the form $f(x) = \dfrac{x^2 + 4x + k}{x^2}$, where k is a constant. Determine the possible values of k for each graph. For each screen, $-4.7 < x < 4.7$ and $-7 < y < 7$

a)
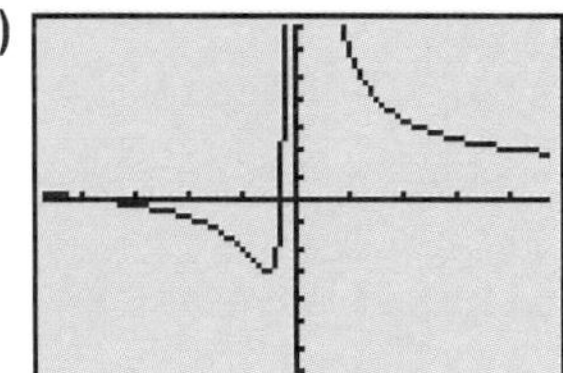

b)
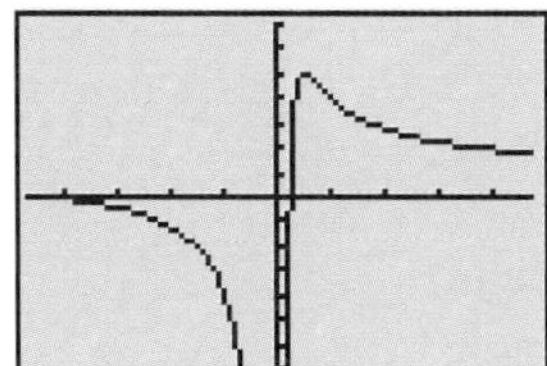

c)
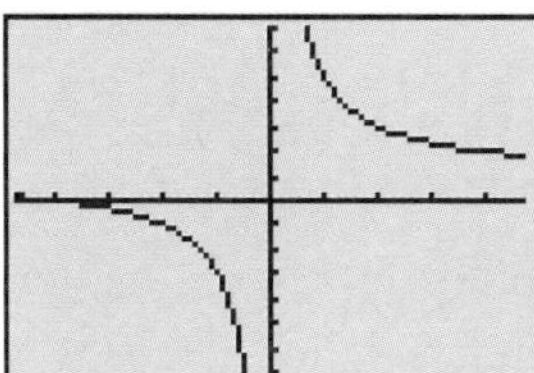

C

18. The three graphs below are functions of the form $f(x) = \dfrac{1}{x^2 + 2x + k}$, where k is a constant. As k changes, the graph of the function has two vertical asymptotes, one vertical asymptote, or no vertical asymptotes. Determine the possible values of k for each graph. For each screen, $-4.7 < x < 4.7$ and $-3.1 < y < 3.1$

a)
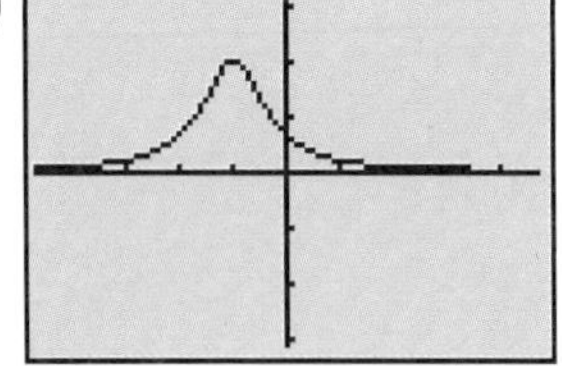

b)
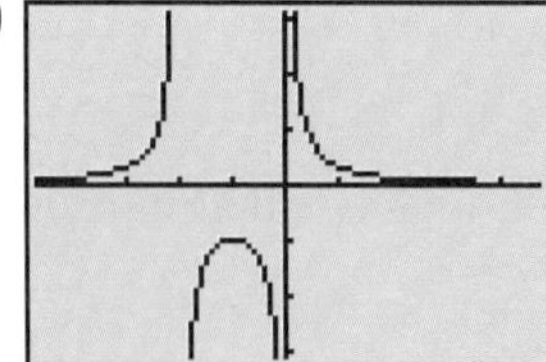

c)
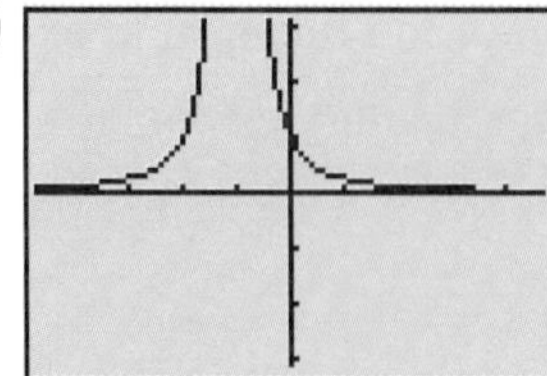

Self-Check 5.3 – 5.5

1. Determine the x- and y-intercepts of the graph of each function.

   a) $y = \frac{1}{x - 2}$ b) $y = \frac{x^2 - 1}{x - 1}$ c) $y = \frac{x + 1}{x^2 - x - 6}$

2. Determine the equation(s) of any vertical asymptotes or the x-value of any point discontinuity for the graph of each function. Use a graphing calculator to illustrate.

   a) $y = \frac{x - 1}{x - 2}$ b) $y = \frac{x^2 - 81}{x + 9}$ c) $y = \frac{x^2 - 2x}{2x^2 + 3x}$

3. Determine the equation of any horizontal asymptote to the graph of each function. Use a graphing calculator to illustrate.

   a) $y = \frac{1}{x^2 - 9}$ b) $y = \frac{x^2}{x^2 + 3}$ c) $y = \frac{5x^3 - 1}{3x^3}$

4. For each rational function:

   i) Determine whether its graph has a horizontal asymptote, an oblique asymptote, or neither. Explain.

   ii) Determine the equations of the asymptotes.

   a) $y = \frac{x}{x^2 + 5}$ b) $y = \frac{x^2}{x + 5}$ c) $y = \frac{x^2 - x + 3}{x + 5}$

5. Determine the equations of the oblique and vertical asymptotes to the graph of each function.

   a) $y = \frac{x^2}{x - 2}$ b) $y = \frac{x^2 + x}{x - 1}$ c) $y = \frac{3x^2 + 2x - 1}{x}$

6. Analyse the key features of the graph of each function, then sketch the graph. Recall that the key features are: intercepts; domain; critical points; intervals of increase, decrease, and concavity; points of inflection; asymptotes; and point discontinuities.

   a) $y = \frac{x}{x - 5}$ b) $y = \frac{x^2}{x^2 - 25}$ c) $y = \frac{x - 5}{x^2 - 25}$

   d) $y = \frac{10}{x^2 + 5}$ e) $y = \frac{2x^2 + 3}{x}$ f) $y = \frac{x^2 + 2}{x - 1}$

PERFORMANCE ASSESSMENT

7. Draw a possible graph of each rational function $y = f(x)$ in the neighbourhood of $x = 1$ for each set of conditions. Explain how you did this.

   a) $f(1)$ does not exist.
   $\lim_{x \to 1^+} f(x) = -2$
   $\lim_{x \to 1^-} f(x) = -2$

   b) $f(1)$ does not exist.
   $\lim_{x \to 1^+} f(x) = \infty$
   $\lim_{x \to 1^-} f(x) = \infty$

Review Exercises

Mathematics Toolkit

Algebra and Calculus Tools

- A rational function has an equation that can be expressed in the form $f(x) = \frac{p(x)}{q(x)}$, where $p(x)$ and $q(x)$ are polynomials, and $q(x) \neq 0$.
- A rational function $y = f(x)$ may have a limit, L, at $x = a$.
  - The limit exists if $\lim_{x \to a} f(x) = L$.
  - The limit does not exist if $\lim_{x \to a} f(x) = \infty$.
- For a rational function $f(x) = \frac{p(x)}{q(x)}$, there are 3 types of limits as $x \to \infty$:
  - When the degrees of $p(x)$ and $q(x)$ are equal, then $\lim_{x \to \infty} f(x) = \frac{a}{b}$, where a and b are the leading coefficients of $p(x)$ and $q(x)$, respectively.
  - When the degree of $p(x)$ is greater than the degree of $q(x)$, then $\lim_{x \to \infty} f(x)$ does not exist.
  - When the degree of $p(x)$ is less than the degree of $q(x)$, $\lim_{x \to \infty} f(x) = 0$.
- The graph of a rational function $y = f(x)$ may have a point discontinuity at $x = a$ when $\lim_{x \to a} f(x) = L$.
- The graph of a rational function $y = f(x)$ has a vertical asymptote $x = a$ when $\lim_{x \to a} f(x)$ does not exist.
- The graph of a rational function $y = f(x)$ has a horizontal asymptote $y = k$ when $\lim_{x \to \infty} f(x) = k$.
- The graph of a rational function $f(x) = \frac{p(x)}{q(x)}$ has an oblique asymptote when the degree of $p(x)$ is 1 greater than the degree of $q(x)$.
- These are the key features of the graph of a rational function: intercepts; domain; critical points; intervals of increase and decrease; points of inflection; intervals of concavity; asymptotes; and point discontinuities.

5.1

1. Which equations represent rational functions? Explain.

a) $y = \frac{x - 5}{x^2 - 10x + 25}$ **b)** $y = x^2 - 10x + 25$ **c)** $y = \frac{x^2 - 10x + 25}{x - 5}$

d) $y = (x + 3)^{-2}$ **e)** $y = 2^{x-1} + 3$ **f)** $y = \frac{x^3 - 15x + 4}{3}$

2. The graphs of four rational functions are shown. Their equations are listed below. Match each graph with its equation.

a)
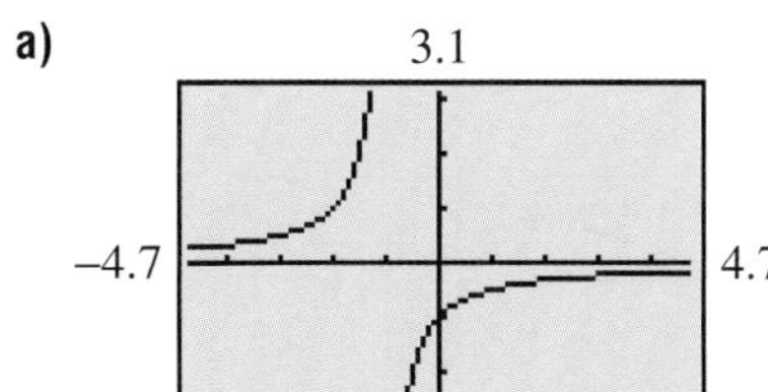

b)
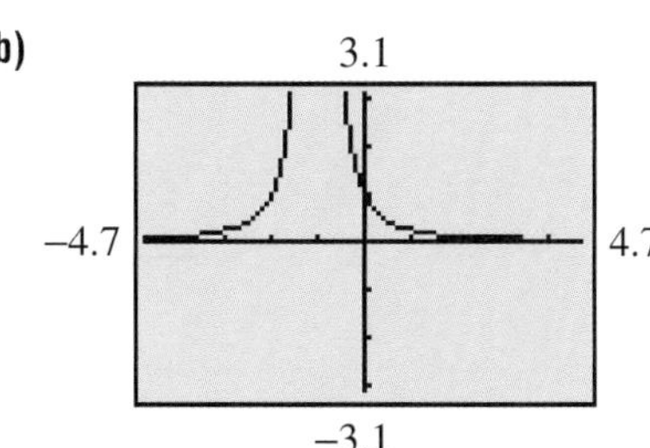

c)
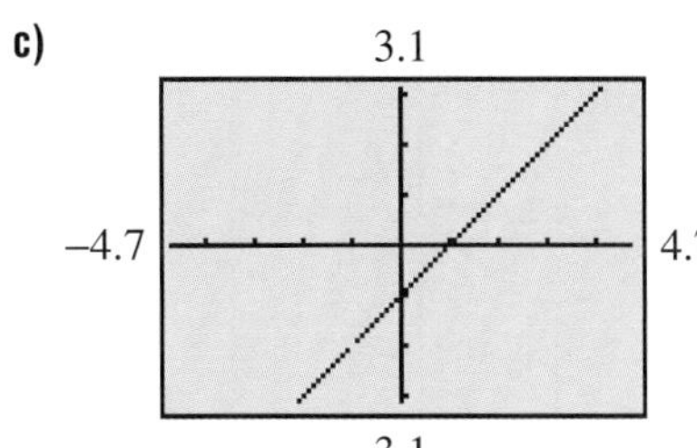

d)
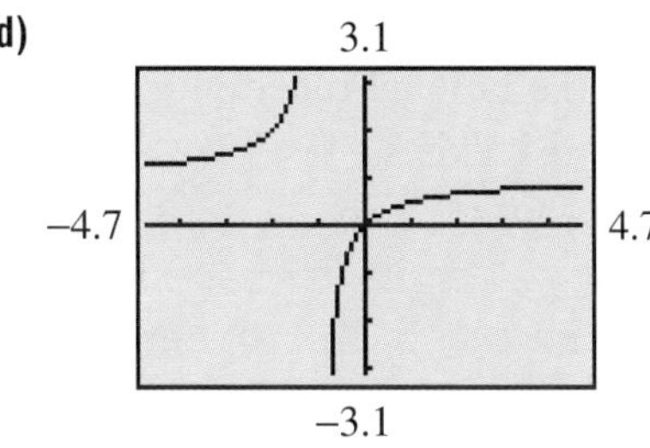

i) $y = \frac{x}{x+1}$ ii) $y = \frac{-1}{x+1}$ iii) $y = \frac{1}{(x+1)^2}$ iv) $y = \frac{x^2-1}{x+1}$

3. Determine the equation of the tangent to the graph of each function in exercise 2 at the point where $x = 2$.

4. All the functions in exercise 2 are discontinuous at $x = -1$.

a) What is similar about all the equations that produce this discontinuity?

b) Which graphs have a vertical asymptote at $x = -1$?

c) Which graph has a point discontinuity at $x = -1$?

d) How does the equation of the function with the point discontinuity differ from the other equations?

e) Determine the equation of the horizontal asymptote for any appropriate graph in exercise 2.

5. For the rational function at the right:

a) Estimate the slope of the tangent at different points on its graph, then sketch the graph of the derivative function.

b) Use the graph of the first derivative function to sketch the graph of the second derivative function.

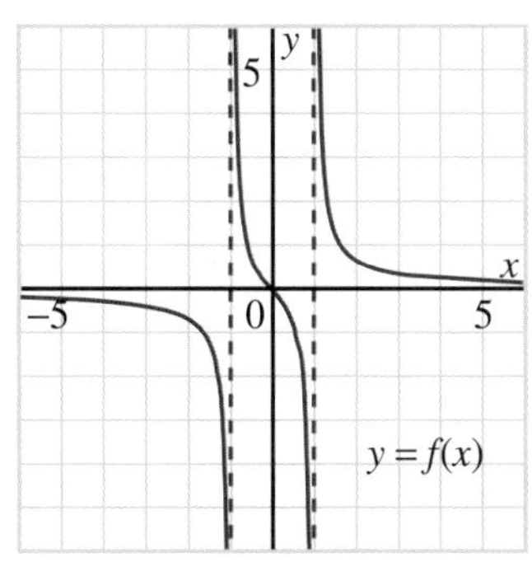

6. The graph, below left, is the graph of the derivative of a function. Sketch a possible graph of the function.

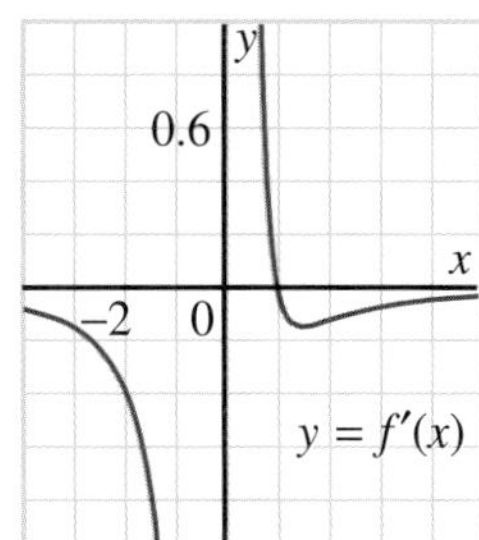

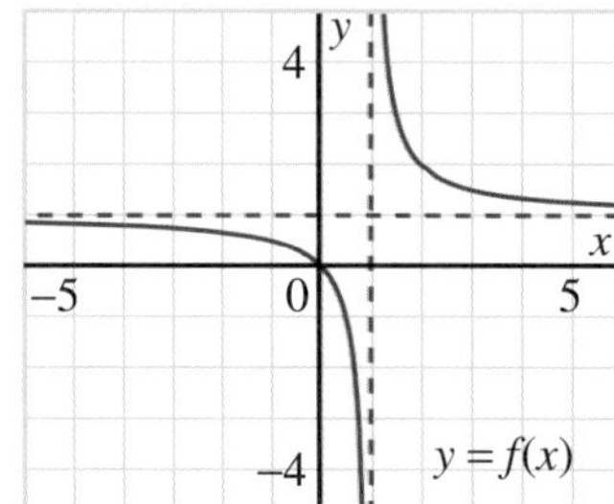

5.2 7. Consider the graph of the rational function $y = f(x)$, above right. Determine each value, if it exists.

a) $\lim_{x\to1} f(x)$ b) $f(1)$ c) $\lim_{x\to-1^+} f(x)$

d) $\lim_{x\to-1^-} f(x)$ e) $\lim_{x\to\infty} f(x)$ f) $\lim_{x\to-\infty} f(x)$

8. Evaluate each limit.

a) $\lim_{x\to-4} \frac{x^2-16}{x+4}$

b) $\lim_{x\to-4} \frac{x+4}{x^2-16}$

c) $\lim_{x\to0} \frac{x}{x^2-2x}$

d) $\lim_{x\to2} \frac{x-2}{x^2+x-2}$

e) $\lim_{x\to5} \frac{x}{x-5}$

f) $\lim_{x\to-2} \frac{1}{x+2}$

g) $\lim_{x\to3} \frac{1}{(x-3)^2}$

h) $\lim_{x\to5} \frac{x}{x^2-10x+25}$

9. Evaluate each limit.

a) $\lim_{x\to\infty} \frac{x^2-x}{x^2+2}$

b) $\lim_{x\to\infty} \frac{x^3+1}{x^3-2x^2+x-16}$

c) $\lim_{x\to\infty} \frac{x}{x^2-5x}$

d) $\lim_{x\to\infty} \frac{x^2+x+3}{x^3-2}$

e) $\lim_{x\to\infty} \frac{x^2}{x-5}$

f) $\lim_{x\to\infty} \frac{x^3+2x}{x+4}$

g) $\lim_{x\to\infty} \frac{3x^2-2x}{4x^2+5x+1}$

h) $\lim_{x\to\infty} \frac{2x^3-3x^2+5x}{x^2-4x^3}$

10. a) Evaluate each limit.

i) $\lim_{x\to6^+} \frac{x^2-36}{x-6}$ ii) $\lim_{x\to6^-} \frac{x^2-36}{x-6}$ iii) $\lim_{x\to6} \frac{x^2-36}{x-6}$

b) Use the limits in part a to describe the graph of the rational function $y = \frac{x^2-36}{x-6}$ for values of x near $x = 6$.

5.3 **11.** Determine the x- and y-intercepts of the graph of each function.

a) $y = \frac{1}{x+2}$ **b)** $y = \frac{x^2 - 9}{x - 3}$ **c)** $y = \frac{x+2}{x-1}$

12. Determine the equation(s) of any vertical asymptotes or the x-value of any point discontinuity for the graph of each function. Use a graphing calculator to illustrate.

a) $y = \frac{1}{x-1}$ **b)** $y = \frac{x}{x+3}$ **c)** $y = \frac{x^2 + 4x}{x}$

d) $y = \frac{x^2 - 4}{x+2}$ **e)** $y = \frac{x-2}{x^2 - 4}$ **f)** $y = \frac{1}{x^2 + 9}$

13. Determine the equation of any horizontal asymptote to the graph of each function. Use a graphing calculator to illustrate.

a) $y = \frac{x^2 - 1}{x^2}$ **b)** $y = \frac{6x}{x+7}$ **c)** $y = \frac{4}{x^2}$

d) $y = \frac{3x}{2x^2 + 1}$ **e)** $y = \frac{4x^2 - 1}{x+3}$ **f)** $y = \frac{x^3 + x + 1}{x^2 - 4}$

14. The graph of a rational function $y = f(x)$ has a vertical asymptote $x = b$. Which limits could be true (where k is a constant real number)?

a) $\lim_{x \to b^+} f(x) = k$ **b)** $\lim_{x \to b^-} f(x) = k$

c) $\lim_{x \to b^+} f(x) = \infty$ **d)** $\lim_{x \to b^-} f(x) = \infty$

5.4 **15.** For each rational function, determine whether its graph has a horizontal asymptote, an oblique asymptote, or neither. Explain.

a) $y = \frac{5x}{x+6}$ **b)** $y = \frac{x^2}{x+6}$ **c)** $y = \frac{x^2 + 2}{x}$ **d)** $y = \frac{x^3}{2x^4 + 5}$

16. For the graph of each function in exercise 15, determine the equation of the horizontal or oblique asymptote.

17. Determine the equation of the oblique asymptote for the graph of each function. Use a graphing calculator to illustrate the results.

a) $y = \frac{x^3}{x^2 + 2}$ **b)** $y = \frac{x^2 - 5}{x - 1}$ **c)** $y = \frac{4x^2 + 2x + 1}{x}$ **d)** $y = \frac{3x^2}{5x + 1}$

18. Predict the slope of the oblique asymptote to the graph of each function. Explain.

a) $y = \frac{5x^2}{x+4}$ **b)** $y = \frac{6x^2}{5x+1}$ **c)** $y = \frac{12x^2 + 3}{2x}$

19. The rational function $y = \frac{f(x)}{g(x)}$ can be written in the form $y = ax + b + \frac{c}{g(x)}$, where a, b, and c are constants. Determine the equation of the oblique asymptote to the graph of the rational function.

5.5 **20.** For each rational function $y = f(x)$ graphed below, estimate where f' and f'' are 0, positive, and negative.

a)

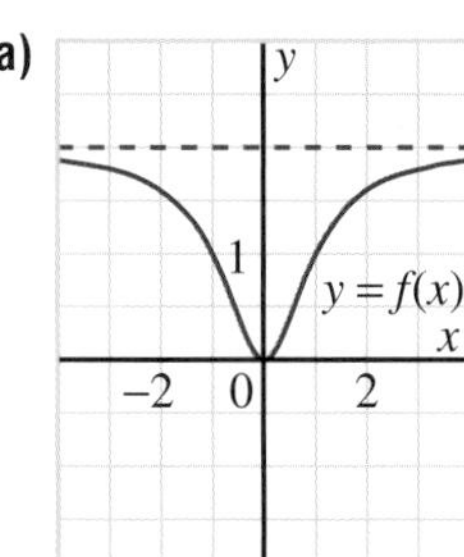

b)

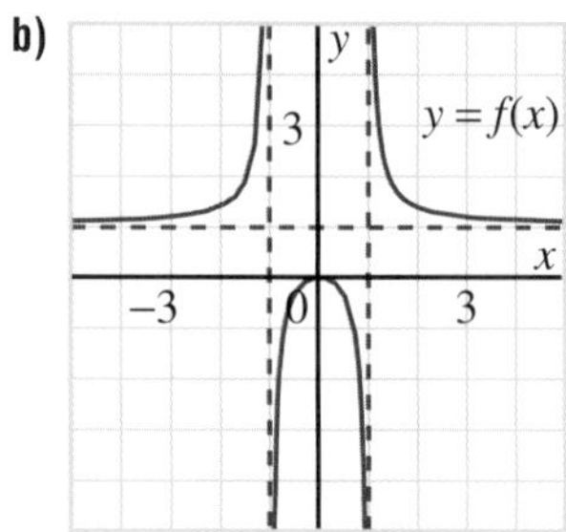

21. For each function in exercise 20:

i) Estimate the equation of any horizontal or vertical asymptote.

ii) Estimate the coordinates of any critical points.

iii) Estimate the coordinates of any points of inflection.

22. For each function in exercise 20, sketch the graphs of $y = f'(x)$ and $y = f''(x)$.

23. The graphs of $y = f'(x)$, the derivatives of the rational functions $y = f(x)$, are shown. For each graph of $y = f'(x)$:

i) Determine the intervals where f is increasing and where f is decreasing.

ii) Determine the x-values where f has a local maximum value or local minimum value.

iii) Sketch the graph of a possible function $y = f(x)$.

a)

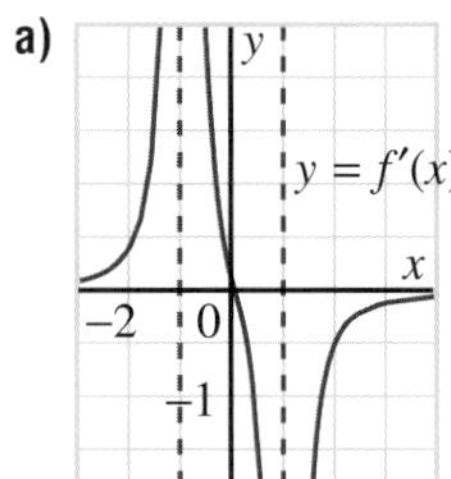

b)

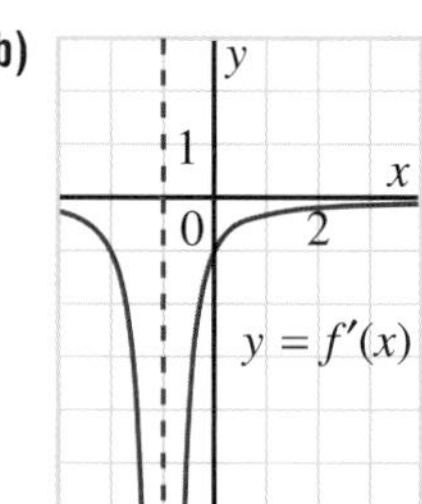

24. The equation of the function whose derivative is graphed in exercise 23b is $f(x) = \frac{1}{x+1}$. Determine the derived function $y = f'(x)$, then use a graphing calculator to graph this derived function. Check that your graph matches that in exercise 23b.

25. Analyse the key features of the graph of each function, then sketch the graph.

a) $y = \frac{1}{x+3}$ **b)** $y = \frac{x^2}{x^2+2}$ **c)** $y = \frac{x-3}{x^2-x-6}$ **d)** $y = \frac{x^2-2}{x+1}$

Self-Test

1. **Knowledge/Understanding** Evaluate each limit.

 a) $\lim_{x\to 5} \frac{x^2 - 25}{x - 5}$ b) $\lim_{x\to 0} \frac{x}{x^2 - 3x}$ c) $\lim_{x\to 4} \frac{x}{x - 4}$ d) $\lim_{x\to -2} \frac{x + 2}{(x + 2)^2}$

 e) $\lim_{x\to \infty} \frac{1}{x}$ f) $\lim_{x\to \infty} \frac{x}{x + 3}$ g) $\lim_{x\to -\infty} \frac{x^2}{x - 2}$ h) $\lim_{x\to \infty} \frac{3x^2 + x + 1}{7 - 5x^2}$

2. For the rational function $y = f(x)$, sketch the graphs of $y = f'(x)$ and $y = f''(x)$.

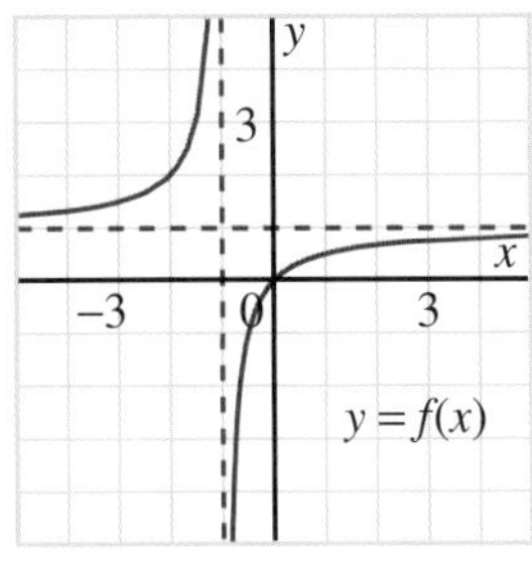

3. Analyse the key features of the graph of each function, then sketch the graph.

 a) $y = \frac{x^2 - 4}{x + 2}$ b) $y = \frac{x^2}{x^2 - 9}$

4. **Communication**

 a) The graph of one function in exercise 3 has vertical and horizontal asymptotes. Explain how you know this from an investigation of the equation of the function.

 b) The graph of one function in exercise 3 has a point discontinuity. Explain how you know this from an investigation of the equation of the function.

5. **Application** The quotient of the square of a number, n, and a number that is two less than n can be modelled by the rational function $Q(n) = \frac{n^2}{n - 2}$.

 a) Determine the equation of the oblique asymptote to the graph of $y = Q(n)$.

 b) Determine the critical points on the graph of $y = Q(n)$.

 c) For $n > 2$, determine the minimum value of the quotient.

 d) For $n < 2$, determine the maximum value of the quotient.

 e) Explain how the results of parts a to d relate to the number n.

6. Determine a possible equation for a rational function $y = f(x)$ that has a horizontal asymptote $y = 2$, vertical asymptotes $x = \pm 1$, and $f(0) = 3$.

Self-Test

7. Thinking/Inquiry/Problem Solving

a) Suppose $\lim_{x \to a} f(x) = \infty$. Which of the following statements is true about the graph of the rational function $y = f(x)$? Explain.

i) The graph has a vertical asymptote at $x = a$.
ii) The graph has a point discontinuity at $x = a$.
iii) The graph has a horizontal asymptote at $y = a$.

b) Does $\lim_{x \to a} f(x) = \infty$ mean that the limit exists? Explain.

c) Does $\lim_{x \to \infty} f(x) = a$ mean that the limit exists? Explain. How does this limit affect the graph of the rational function $y = f(x)$?

PERFORMANCE ASSESSMENT

8. a) The rational function $y = f(x)$ is such that $f(2)$ is undefined. Describe the graph of $y = f(x)$ in the neighbourhood of $x = 2$ in each case.

i) $\lim_{x \to 2} f(x) = 4$

ii) $\lim_{x \to 2} f(x) = \infty$

iii) $\lim_{x \to 2^+} f(x) = \infty$, $\lim_{x \to 2^-} f(x) = -\infty$

b) Give a possible equation for a function $y = f(x)$ that would satisfy each condition in part a.

9. Consider the rational function $f(x) = \dfrac{x^2 - 2x + 1}{x^2 - 4x + 4}$.

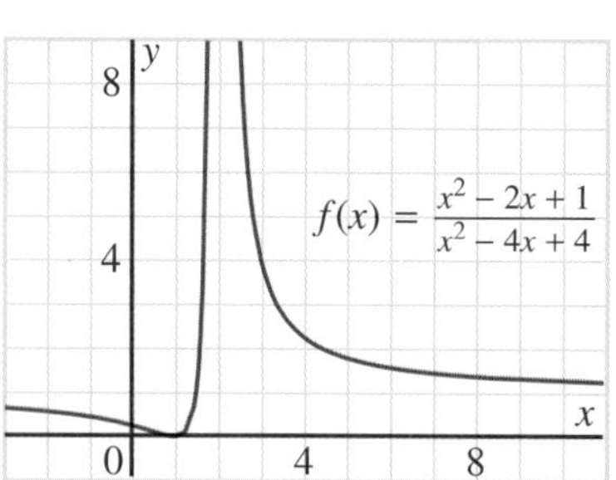

a) From the graph, estimate the equations of the vertical and horizontal asymptotes.

b) The rational function appears to have a minimum point at $x = 1$. Determine $f'(x)$, then verify that $x = 1$ is a critical value of the function $y = f(x)$.

c) Determine the equations of the vertical and horizontal asymptotes to the graph of $y = f'(x)$. Which asymptote remains the same? Is this consistent with your observation in part a? Explain.

Mathematical Modelling

Predicting Populations

We can use the world population data for the last half of the 20th century to predict the world population for the next 25 years. The population data at the right come from the U.S. Bureau of the Census.

Year	Population, P (billions)
1950	2.556
1955	2.780
1960	3.039
1965	3.345
1970	3.707
1975	4.086
1980	4.454
1985	4.851
1990	5.279
1995	5.688
2000	6.083

Developing a Mathematical Model

The table at the right shows the population data for 50 years. These data are graphed below.

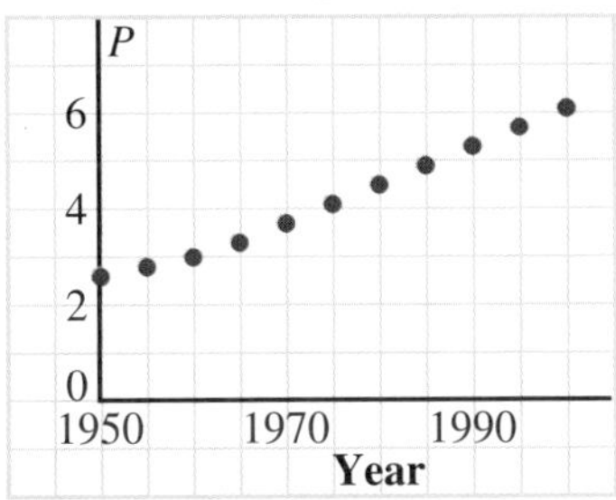

We want to understand how P changes. The data are given at 5-year intervals. We will work with the 5-year increase in population size, which we will call ΔP.

We first consider the *unit growth rate*, $\frac{\Delta P}{P}$. For example, in 1950, the ratio of population increase over the next 5 years, ΔP, to population size, P, is:

$$\frac{\Delta P}{P} = \frac{P(1955) - P(1950)}{P(1950)}$$
$$= \frac{2.780 - 2.556}{2.556}$$
$$\doteq 0.0876$$

This indicates that the population grew by approximately 0.0876, or 8.76%, over that 5-year period.

We calculate $\frac{\Delta P}{P}$ for each 5-year period. These values are shown in the table and graph that follow.

When we consider how a population grows, we expect that the growth rate will be proportional to the population. That is, twice as many parents will produce twice as many children; hence, we consider unit growth rate. However, a larger population may lead to a strain on resources, which, in turn, could lead to a decrease in the birth rate. At that time, the unit growth rate will decrease.

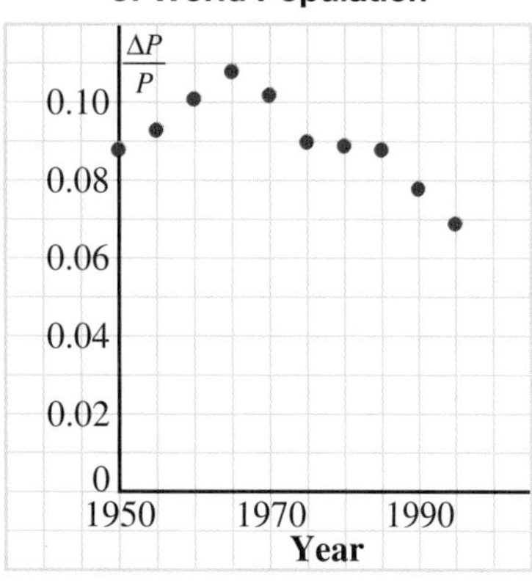

Year	Population, P (billions)	$\frac{\Delta P}{P}$
1950	2.556	
		0.0876
1955	2.780	
		0.0932
1960	3.039	
		0.1007
1965	3.345	
		0.1082
1970	3.707	
		0.1022
1975	4.086	
		0.0901
1980	4.454	
		0.0891
1985	4.851	
		0.0882
1990	5.279	
		0.0775
1995	5.688	
		0.0694
2000	6.083	

Formulating a Hypothesis

From the graph above, over the last 50 years, $\frac{\Delta P}{P}$ increased for the first 15 years, and then decreased. There is a linear trend in this 35-year decrease. We make the following hypothesis: This linear trend will continue to apply over the next 25 years and we will calculate future population sizes assuming that this is true.

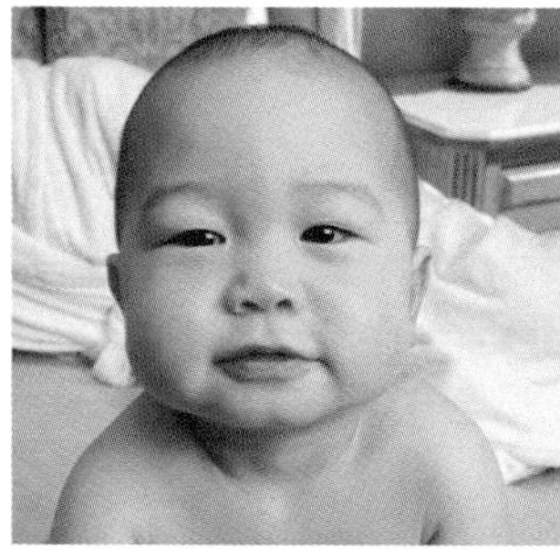

Analysing the Model

For the analysis above, we graphed $\frac{\Delta P}{P}$ against time. Now we graph $\frac{\Delta P}{P}$ against P; this is reasonable because it is the changes in population that affect $\frac{\Delta P}{P}$, rather than changes in time. We use the data in the second and third columns of the table, above right. We begin with the point (3.345, 0.1082), where the values of $\frac{\Delta P}{P}$ begin to decrease. We use a TI-83 graphing calculator to graph these data. The graph is shown top left on the next page.

Recall how to use the TI-83 graphing calculator to graph a statistical plot.

- **Press STAT 1. Clear lists L1 and L2 if necessary.**
- **Enter the population data, P, in list L1, from 3.345 to 5.688. Enter the unit growth rate data, $\frac{\Delta P}{P}$, in list L2, from 0.1082 to 0.0694.**
- **Press 2nd Y= for STATPLOT. Press 1 ENTER. Select the first plot type. Ensure that L1 is beside X list and L2 is beside Y list, respectively.**
- **Set the window for $0 < x < 7$ and $0 < y < 0.12$.**
- **Press Y=. Clear any equations on the screen. Use the arrow keys to highlight Plot 1. Press GRAPH.**

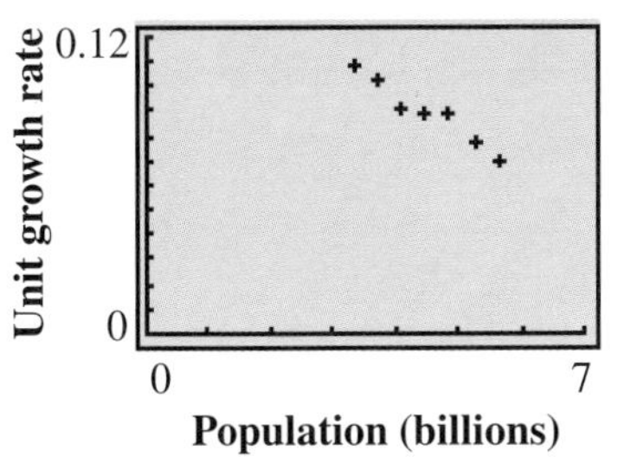

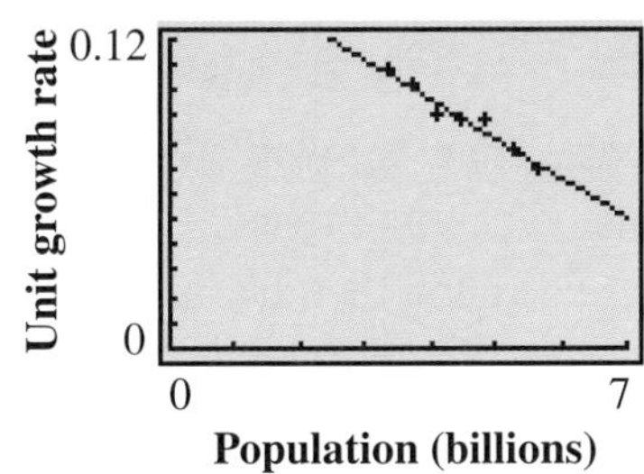

To calculate the equation of the regression line, and draw the line on the screen:

- Press STAT ▶ 4 to select LinReg(ax + b).
- Press 2nd 1 , 2nd 2 , VARS ▶ 1 1 ENTER to display the equation of the regression line.
- Press GRAPH to draw the regression line on the screen.

We then draw the regression line on the screen (above right) and display its approximate equation in the Y= list:

$y = -0.0153x + 0.1581$

So, the equation of the graph of unit growth rate against population is:

$$\frac{\Delta P}{P} = -0.0153P + 0.1581$$

Solve for ΔP.

$$\Delta P = -0.0153P^2 + 0.1581P$$

This equation is our model for population growth. It predicts the growth in population over the next 5 years, starting at any current population size, P.

Making Inferences from the Model

We use the equation above to predict the world population in 2005.

$$\Delta P = -0.0153P^2 + 0.1581P$$

Substitute the population for the year 2000; that is, $P = 6.083$.

$$\Delta P = -0.0153(6.083)^2 + 0.1581(6.083)$$
$$\doteq 0.396$$

The estimated population in 2005 will be $6.083 + 0.396 = 6.479$. In 2005, the world population will be approximately 6.5 billion.

Year	Predicted population, P (billions)	Predicted population from U.S. Bureau (billions)
2000	6.083	
2005	6.479	6.468
2010	6.861	6.849
2015	7.225	7.227
2020	7.569	7.585
2025	7.889	7.923

We repeat the calculation above to estimate the world population in each of the years 2010, 2015, 2020, 2025. These data are shown at the right, along with the estimates for the same data from the U.S. Bureau of the Census. The Bureau's data are slightly different because it uses a more complex model that takes into account different situations around the world.

Key Features of the Model

The data were given at 5-year intervals. Thus, we worked with growth increments, ΔP, rather than instantaneous growth rates, $\frac{dP}{dt}$. We plotted the unit growth rates, $\frac{\Delta P}{P}$, against P, and found that the rates were not constant. They increased, then decreased. We used the decreasing part of the graph to construct our model. We fitted a line to the data points and that gave us an equation for ΔP in terms of P. We use this equation as a predictor of future data.

Related Problems

1. In our analysis of world population growth, the trend of the last 10 years may have more effect than the trend of the last 30 years. This seems appropriate since the last 3 data points in the graph of $\frac{\Delta P}{P}$ against P seem to lie on a line.
   a) Determine the equation of the regression line for the last 3 data points.
   b) Use the equation in part a to predict the world population at 5-year intervals from 2005 to 2025.
   c) Compare the data in part b with both sets of predicted populations in the table on page 361. Describe what you notice.

2. The population of Canada, P million, at 5-year intervals from 1951 to 2001 is shown at the right.
   a) Graph the population, P, against time, t, for 1951 to 2001.
   b) Calculate $\frac{\Delta P}{P}$ for each 5-year period from 1951 to 1996.
   c) Graph $\frac{\Delta P}{P}$ against P.
   d) Determine the equation of the regression line.
   e) Use this equation of the regression line to predict the population of Canada in 2006 and 2011. Use the population in 2001 as the starting point.
   f) Write a report to describe what you have learned about predicting populations. Include a description of your predictions for the Canadian population. Describe some factors that should be taken into account when making predictions about future populations. Include equation(s) and graph(s) in your report.

Year	Population of Canada, P (millions)
1951	13.648
1956	16.081
1961	18.238
1966	20.015
1971	21.568
1976	23.518
1981	24.900
1986	26.204
1991	28.111
1996	29.959
2001	31.593

Exponential and Logarithmic Functions

6

Curriculum Expectations

By the end of this chapter, you will:

- Identify, through investigations, using graphing calculators or graphing software, the key properties of exponential functions of the form a^x $(a > 0,\ a \neq 1)$ and their graphs.
- Describe the graphical implications of changes in the parameters *a, b,* and *c* in the equation $y = ca^x + b$.
- Compare the rates of change of the graphs of exponential and non-exponential functions.
- Describe the significance of exponential growth or decay within the context of applications presented by various mathematical models.
- Pose and solve problems related to models of exponential functions drawn from a variety of applications, and communicate the solutions with clarity and justification.
- Define the logarithmic function $\log_a x$ $(a > 1)$ as the inverse of the exponential function a^x, and compare the properties of the two functions.
- Express logarithmic equations in exponential form, and vice versa.
- Simplify and evaluate expressions containing logarithms.
- Solve exponential and logarithmic equations, using the laws of logarithms.
- Solve simple problems involving logarithmic scales.

Necessary Skills

1. Review: Exponent Laws for Integer Exponents

Remember that a^n is defined as follows, where n is a positive integer:

$$a^n = a \times a \times a \times \ldots \times a \text{ (} n \text{ factors)}$$

Recall the exponent laws that follow.

Take Note

Exponent Laws for Integer Exponents

The bases a and b are real numbers.
The exponents m and n are integers.

Multiplication	$a^m \times a^n = a^{m+n}$	
Division	$\dfrac{a^m}{a^n} = a^{m-n}$	$(a \neq 0)$
Power of a power	$(a^m)^n = a^{mn}$	
Power of a product	$(ab)^n = a^n b^n$	
Power of a quotient	$\left(\dfrac{a}{b}\right)^n = \dfrac{a^n}{b^n}$	$(b \neq 0)$

Remember how a power of a is defined when the exponent is zero or a negative integer. In the following definitions, n is a positive integer.

$$a^0 = 1 \qquad (a \neq 0)$$

$$a^{-n} = \frac{1}{a^n} \qquad (a \neq 0)$$

Example 1

Evaluate each expression without using a calculator.

a) 5^{-3} **b)** 0.4^{-2}

Solution

a) $5^{-3} = \dfrac{1}{5^3}$

$= \dfrac{1}{125}$

b) $0.4^{-2} = \left(\dfrac{2}{5}\right)^{-2}$

$= \left(\dfrac{5}{2}\right)^2$

$= \dfrac{25}{4}$

Example 2

Simplify each expression.

a) $a^5 \times a^{-3}$

b) $(2a^3)^{-2}$

Solution

a) $a^5 \times a^{-3} = a^{5 + (-3)}$
$= a^{5 - 3}$
$= a^2$

b) $(2a^3)^{-2} = 2^{-2} \times (a^3)^{-2}$
$= \frac{1}{4}a^{-6}$

In *Example 2b*, we could have expressed the answer as $\frac{1}{4a^6}$, but it is not necessary, or even desirable, to do so. In this chapter, we will write similar expressions in the form ca^x, in which the exponent can be positive or negative.

Exercises

1. Evaluate each expression without using a calculator.

a) 6^0 **b)** 3^{-1} **c)** 5^{-2}

d) 0.4^2 **e)** $\left(\frac{2}{3}\right)^3$ **f)** $\left(\frac{3}{4}\right)^{-1}$

g) $\left(\frac{3}{4}\right)^{-2}$ **h)** 0.75^{-2} **i)** 0.5^{-3}

2. Simplify.

a) $m^3 \times m^{-7}$ **b)** $b^{-2} \times b^6$ **c)** $4a^{-3} \times 2a^{-1}$

d) $\frac{x^5}{x^3}$ **e)** $\frac{3x^2}{x^4}$ **f)** $\frac{a^3}{a^{-1}}$

g) $(c^3)^2$ **h)** $(3x^{-2})^2$ **i)** $(a^{-2})^{-3}$

2. Review: Exponent Laws for Rational Exponents

Remember how a power of a is defined when the exponent is a rational number. In the following definitions, m is a positive integer or 0, n is a positive integer, and a is a real number.

$a^{\frac{1}{n}} = \sqrt[n]{a}$ $a^{\frac{m}{n}} = \left(\sqrt[n]{a}\right)^m$ or $\sqrt[n]{a^m}$ (When n is even, $a \geq 0$.)

$a^{-\frac{1}{n}} = \frac{1}{\sqrt[n]{a}}$ $a^{-\frac{m}{n}} = \frac{1}{\left(\sqrt[n]{a}\right)^m}$ or $\frac{1}{\sqrt[n]{a^m}}$ (When n is even, $a > 0$.)

With these definitions, the exponent laws on page 364 still apply.

Necessary Skills

Take Note

Exponent Laws for Rational Exponents

The bases a and b are real numbers.
The exponents m and n are rational numbers.

Multiplication	$a^m \times a^n = a^{m+n}$	
Division	$\frac{a^m}{a^n} = a^{m-n}$	$(a \neq 0)$
Power of a power	$(a^m)^n = a^{mn}$	
Power of a product	$(ab)^n = a^n b^n$	
Power of a quotient	$\left(\frac{a}{b}\right)^n = \frac{a^n}{b^n}$	$(b \neq 0)$

Example 1

Evaluate each expression without using a calculator.

a) $9^{1.5}$

b) $8^{-\frac{2}{3}}$

Solution

a) $$\begin{aligned} 9^{1.5} &= 9^{\frac{3}{2}} \\ &= \left(\sqrt{9}\right)^3 \\ &= 3^3 \\ &= 27 \end{aligned}$$

b) $$\begin{aligned} 8^{-\frac{2}{3}} &= \frac{1}{8^{\frac{2}{3}}} \\ &= \frac{1}{(\sqrt[3]{8})^2} \\ &= \frac{1}{2^2} \\ &= \frac{1}{4} \end{aligned}$$

In *Example 1a,* the power has a decimal exponent. In this chapter, you will frequently write powers with decimal exponents. The definition of a power with a rational exponent is used to explain what these powers mean.

Example 2

a) Evaluate $2.15^{1.327}$ to the nearest thousandth.

b) Explain the meaning of the result.

Solution

a) Use the $\boxed{\wedge}$ key on a calculator.

$2.15^{1.327} \doteq 2.762$

b) To explain the meaning of this result, write the exponent in the form $\frac{m}{n}$.

$1.327 = \frac{1327}{1000}$

Use the definition of a power with a rational exponent.

$$2.15^{1.327} = 2.15^{\frac{1327}{1000}}$$
$$= \left(\sqrt[1000]{2.15}\right)^{1327}$$

Hence, $2.15^{1.327}$ means $\left(\sqrt[1000]{2.15}\right)^{1327}$.

Example 3

Simplify each expression.

a) $b^{1.5} \times b^{-2.2}$

b) $(2t^{1.5})^{-3}$

Solution

a) $b^{1.5} \times b^{-2.2} = b^{1.5-2.2}$
$= b^{-0.7}$

b) $(2t^{1.5})^{-3} = 2^{-3} \times t^{-4.5}$
$= \frac{1}{8}t^{-4.5}$

Exercises

1. Evaluate each expression without using a calculator.

a) $8^{\frac{1}{3}}$ **b)** $16^{\frac{1}{4}}$ **c)** $16^{\frac{3}{4}}$ **d)** $4^{1.5}$

e) $9^{-\frac{1}{2}}$ **f)** $27^{-\frac{2}{3}}$ **g)** $25^{-\frac{3}{2}}$ **h)** $4^{-2.5}$

2. Use a calculator to evaluate to the nearest thousandth.

a) $1.15^{0.328}$ **b)** $1.08^{0.633}$ **c)** $1.56^{-2.84}$ **d)** $2.45^{-0.037}$

3. Choose one part of exercise 2. Explain the meaning of the result.

4. Simplify.

a) $a^{1.8} \times a^{2.7}$ **b)** $c^{-2.3} \times c^{2.5}$ **c)** $s^{-0.25} \times s^{-0.5}$

d) $\frac{x^{3.2}}{x^{1.8}}$ **e)** $\frac{m^{1.5}}{m^{-0.5}}$ **f)** $\frac{a^{-2.6}}{a^{-1}}$

g) $(x^{1.2})^{2.5}$ **h)** $(t^{2.5})^{-3}$ **i)** $(2k^{-1.5})^{-2}$

3. Review: Transformations of Graphs

When certain changes are made to the equation of a function $y = f(x)$, corresponding changes occur to its graph.

Translating Graphs of Functions

Remember how the graph of $y = (x - 3)^2 + 1$ is related to the graph of $y = x^2$.

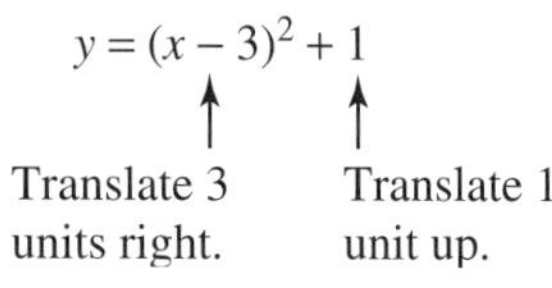

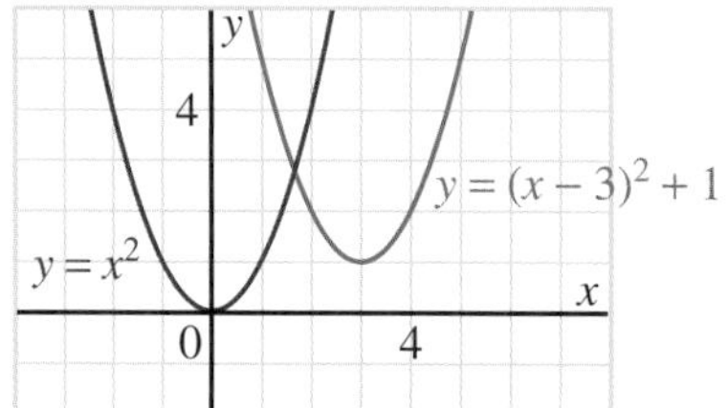

The graph is translated to the right because x must be 3 units greater than it was previously to give the same result after squaring.

Similar results occur for any function $y = f(x)$.

Expanding Graphs of Functions

Remember how the graph of $y = 3 \sin \frac{1}{2}x$ is related to the graph of $y = \sin x$.

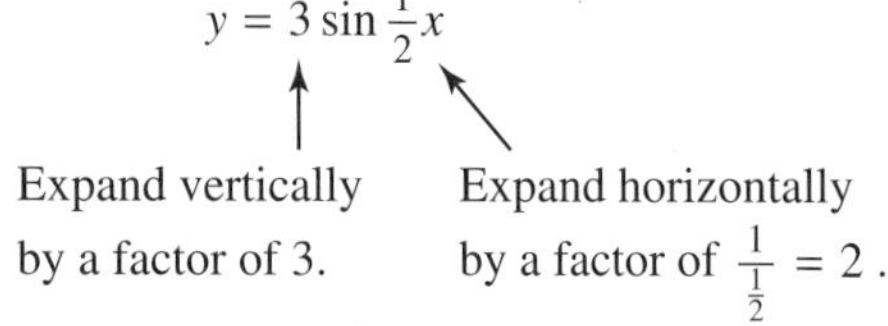

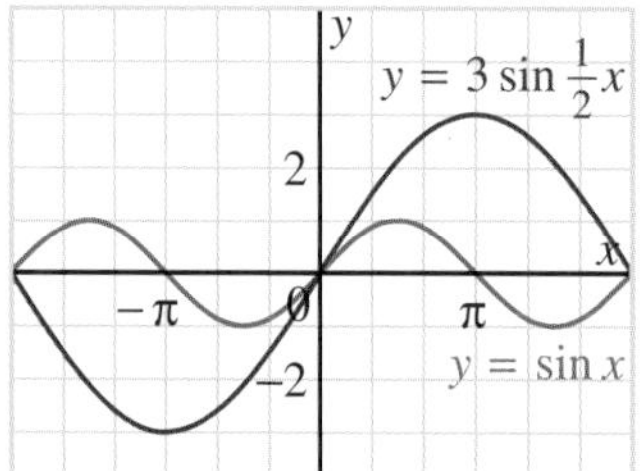

The graph is expanded horizontally because x must be 2 times its previous value to give the same result after taking the sine.

Similar results occur for any function $y = f(x)$.

Example

Use transformations to sketch a graph of the function $y = 3\sqrt{\frac{1}{5}x} + 2$.

Solution

For $y = 3\sqrt{\frac{1}{5}x} + 2$:

Start with the graph of $y = \sqrt{x}$.

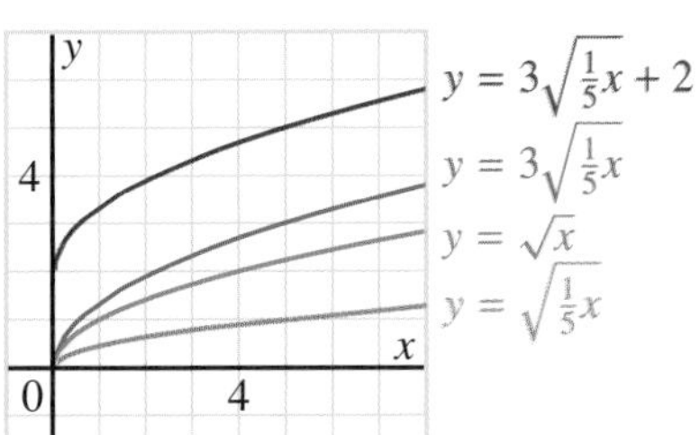

Expand this graph horizontally by a factor of 5 to obtain the graph of $y = \sqrt{\frac{1}{5}x}$.

Expand this graph vertically by a factor of 3 to obtain the graph of $y = 3\sqrt{\frac{1}{5}x}$.

Translate this graph 2 units up to obtain the graph of $y = 3\sqrt{\frac{1}{5}x} + 2$.

Exercises

1. Use the translation on page 368: translate 3 units right and 1 unit up. Write the equation of the translated graph starting with each function below instead of $y = x^2$.

a) $y = x^3$ **b)** $y = \sqrt{x}$ **c)** $y = \frac{1}{x}$

d) $y = x$ **e)** $y = 2^x$ **f)** $y = f(x)$

2. Use the second transformation on page 368: expand vertically by a factor of 3 and horizontally by a factor of 2. Write the equation of the transformed graph starting with each function below instead of $y = \sin x$.

a) $y = x^3$ **b)** $y = \sqrt{x}$ **c)** $y = \frac{1}{x}$

d) $y = x$ **e)** $y = 2^x$ **f)** $y = f(x)$

3. Use transformations to sketch the graph of each function.

a) $y = 2(x - 5)^2 + 3$ **b)** $y = 3\sin 2x$ **c)** $y = 5\sin\frac{1}{2}x$

d) $y = 3\sqrt{2x} + 1$ **e)** $y = 2\sqrt{\frac{1}{4}x} + 3$ **f)** $y = \frac{1}{2x} + 2$

6.1 The Exponential Function $y = a^x$ and Multiplicative Growth

In Chapter 2, we investigated power functions, which have equations of the form $y = x^n$, where the variable x is the base and the exponent n is constant. The basic exponential function has an equation of the form $y = a^x$, $a > 0$, where the variable x is the exponent and the base a is constant. An essential characteristic of any exponential function is that the variable is in an exponent. In this chapter, we shall examine some properties and important applications of exponential functions. We shall consider derivatives of exponential functions in Chapter 7.

Investigation

Graphing $y = a^x$, $a > 0$

Use a graphing calculator or graphing software.

1. a) Graph the function $y = 2^x$ for $-5 \le x \le 5$.

   b) Sketch the graph.

2. Use your graph in exercise 1 to explain your answer to each question.

   a) What are the x- and y-intercepts of the graph?

   b) What are the domain and range of the function?

   c) Is the function increasing or decreasing?

   d) Is the graph concave up or concave down?

   e) Does the graph have an asymptote?

3. Predict how the graph of $y = 2^x$ will change for each function below. Use a graphing calculator or graphing software to confirm your predictions. Sketch each graph.

   a) $y = 10^x$ b) $y = 3^x$ c) $y = 1.065^x$ d) $y = 1^x$

   e) $y = 0.87^x$ f) $y = 0.5^x$ g) $y = 0.1^x$

4. a) Describe how the graph of each function in exercise 3 is similar to and different from the graph of $y = 2^x$.

   b) Which, if any, of your answers in exercise 2 would change for each function in exercise 3? Explain.

The graphs of $y = a^x$ for $a = 10, 2, 1.25, 1, 0.8, 0.5$, and 0.1 are on the next page. They show how the graph of $y = a^x$ changes as a varies, and they illustrate the following properties:

Properties of the Graph of $f(x) = a^x$, $a > 0$

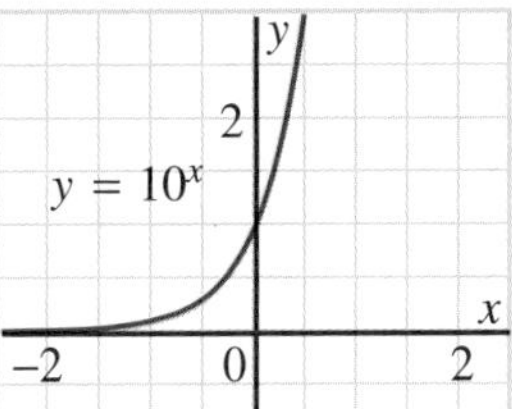

Intercepts

$f(x) = a^x$

To determine the vertical intercept, substitute $x = 0$.

Then, $f(0) = a^0$

$= 1$

The vertical intercept is 1.

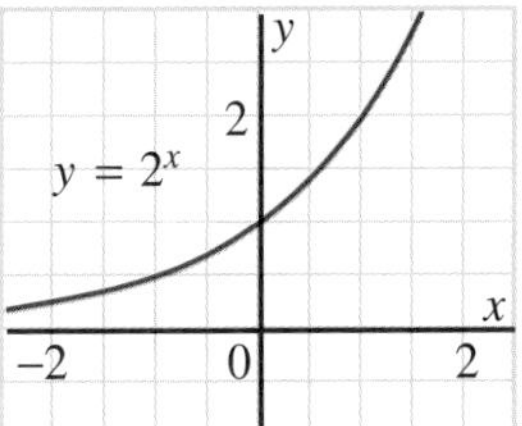

To determine horizontal intercepts, solve $f(x) = 0$.

Then, $a^x = 0$

This equation has no real solution because $a^x > 0$ for all real values of x. There are no horizontal intercepts.

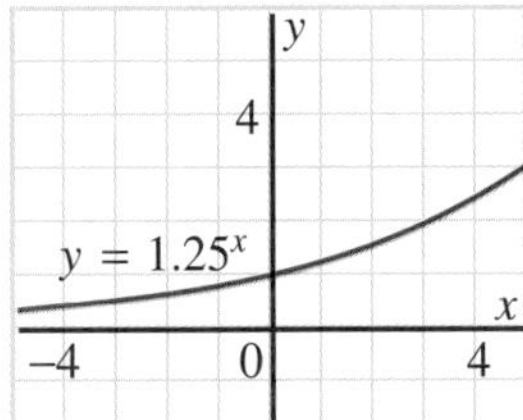

Domain and range

Since $a > 0$, we can define a^x for all real values of x, and the domain is the set of real numbers; $D = \Re$.

Since a^x is positive for all real values of x, the range is the set of positive real numbers; $R = \{y \mid y \in \Re, y > 0\}$.

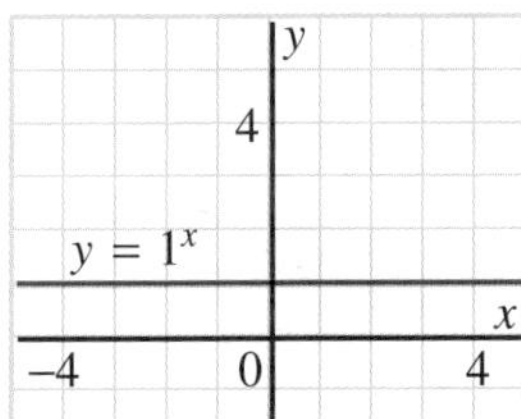

Critical points

From the graphs, there do not appear to be any critical points. We will use the derivative to prove this result in Chapter 7.

Intervals of increase and decrease

From the first 3 graphs, when $a > 1$, the function increases for all values of x.

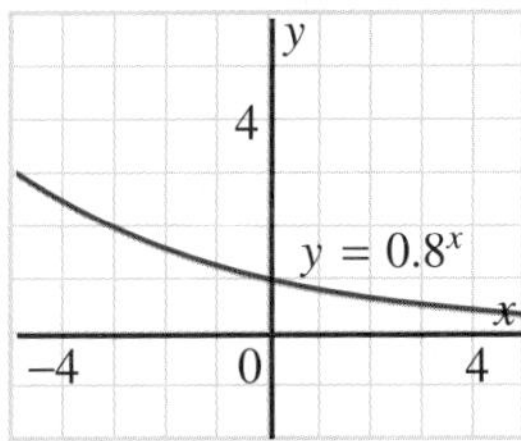

From the last 3 graphs, when $0 < a < 1$, the function decreases for all values of x.

From the 4th graph, when $a = 1$, $y = 1^x$; the function is constant for all values of x.

Points of inflection

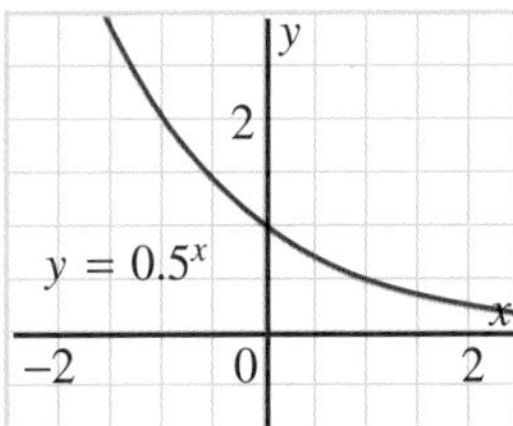

From the graphs, there do not appear to be any points of inflection. We will prove this result in Chapter 7.

Concavity

The graph is concave up for all values of a except $a = 1$. We will use the derivative to prove this result in Chapter 7.

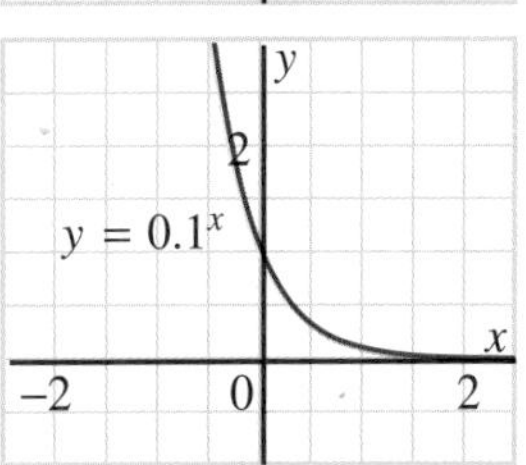

Asymptote

For $a > 1$, when x decreases through negative values, the points on the graph come closer and closer to the x-axis, but never reach it.

For $0 < a < 1$, when x increases through positive values, the points on the graph come closer and closer to the x-axis, but never reach it.

The x-axis is an asymptote.

Point discontinuities

There are no point discontinuities.

Law of exponents

The graph of $y = 2^x$ is shown.

$f(x) = 2^x$

x	y
−3	$\frac{1}{8}$
−2	$\frac{1}{4}$
−1	$\frac{1}{2}$
0	1
1	2
2	4
3	8

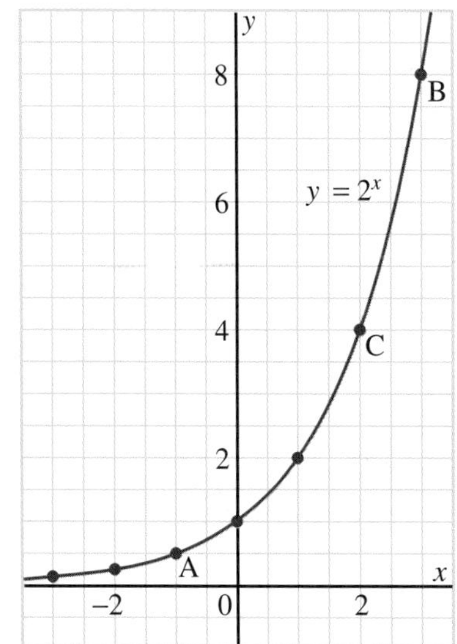

Suppose we select any two points on the graph, such as A$(-1, \frac{1}{2})$ and B(3, 8).

Add their x-coordinates.
$-1 + 3 = 2$

Multiply their y-coordinates.
$\frac{1}{2} \times 8 = 4$

The results are the coordinates of another point C(2, 4) on the graph.

Adding the x-coordinates and multiplying the y-coordinates of two points on the graph produces the coordinates of another point on the graph. This property is a consequence of the law of exponents for multiplication; that is, to multiply two powers with the same base, we add their exponents. Therefore, this law applies to the graph of $y = a^x$ for any value of a, not just $a = 2$.

Applications to Problems Involving Multiplicative Growth and Decay

Exponents were originally defined as a notation for repeated multiplication. Repeated multiplication occurs frequently in problems that involve growth and decay.

Compound Interest

Suppose you make a long-term investment of \$500 at a fixed interest rate of 6.5% compounded annually. The amount, A dollars, of your investment after n years is represented by the equation $A = 500(1.065)^n$. The amount, A, increases with time. In this equation, n is a natural number because it represents the number of years.

In the *Investigation*, you graphed the function $y = 1.065^x$. This graph is shown below, along with the graph of $A = 500(1.065)^n$. Although the scales on the two graphs are different, we can visualize how the graph of $A = 500(1.065)^n$ coincides with a vertical expansion of the graph of $y = 1.065^x$ by a factor of 500.

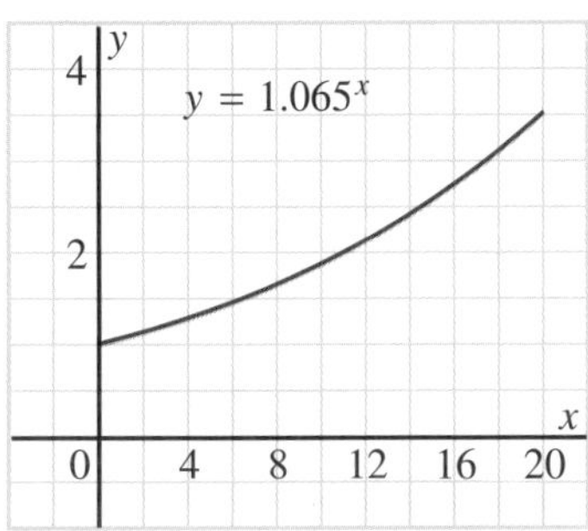

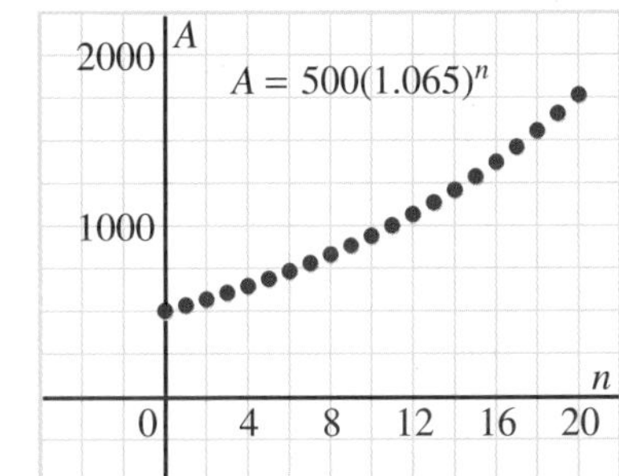

How Long Caffeine Stays in Your Bloodstream

Coffee, tea, cola, and chocolate contain caffeine. When you consume caffeine, the percent, P, left in your body can be modelled as a function of the elapsed time, n hours, by the equation $P = 100(0.87)^n$. The percent, P, decreases with time.

In the *Investigation,* you graphed the function $y = 0.87^x$. This graph is shown below, along with the graph of $P = 100(0.87)^n$. The graph of $P = 100(0.87)^n$ coincides with a vertical expansion of the graph of $y = 0.87^x$ by a factor of 100.

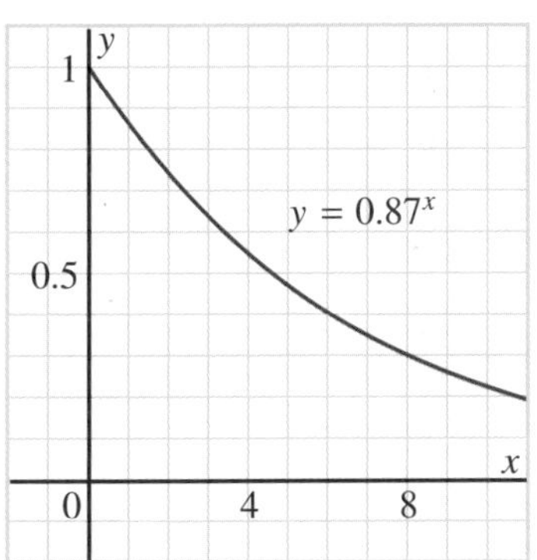

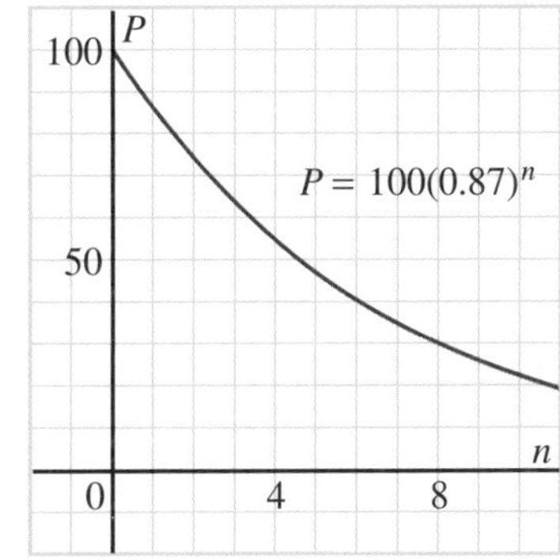

Comparing the Equations

The situations on page 373 are examples of exponential growth and exponential decay. Consider the similarities in the equations of the two functions.

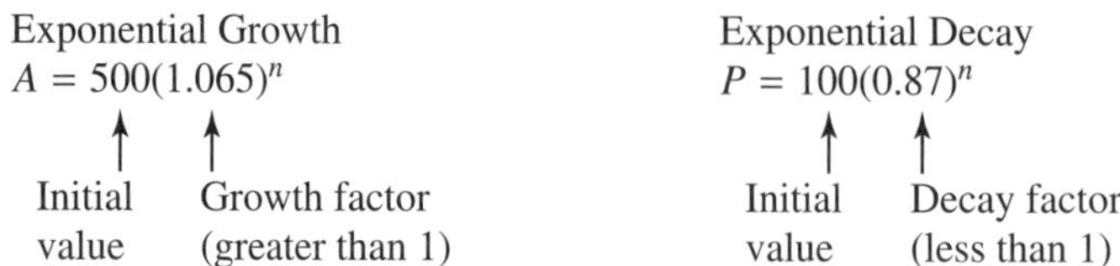

In these exponential functions, there is an underlying principle from which the equation is derived. This principle involves repeated multiplication. The first equation arises by starting with \$500 and multiplying by 1.065 for each year. The second equation arises by starting with 100% and multiplying by 0.87 for each hour. These represent multiplicative growth and decay situations, respectively.

Other examples of exponential functions arise from empirical data. The equation in the following *Example* was determined by calculating the exponential equation of best fit.

Example

The growth of the Internet can be measured by the number of computers offering information, called *hosts*. In 1995, there were about 7.4 million hosts. The number has been growing at a fairly constant rate since then, and can be modelled by the equation $h = 7.4(1.59)^n$, where h is the number of hosts in millions, and n is the number of years since 1995.

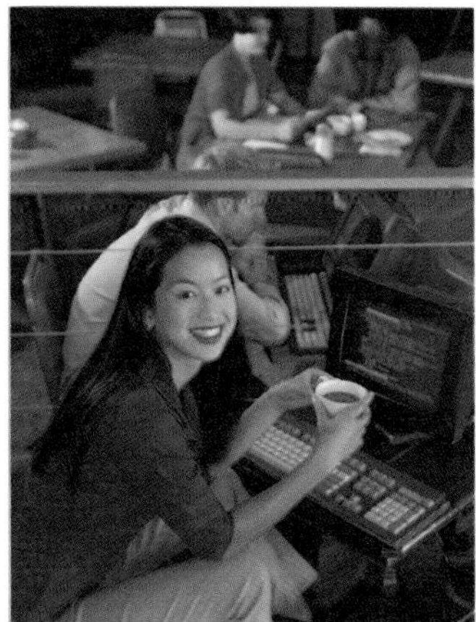

a) Graph the equation for $0 \leq n \leq 10$.

b) Estimate the number of hosts in 2001.

c) According to the model, when will the number of hosts reach 500 million?

Solution

Using paper and pencil

a) Make a table of values, then draw the graph of $h = 7.4(1.59)^n$.

n	h
0	7.4
2	18.7
4	47.3
6	119.6
8	302.3
10	764.2

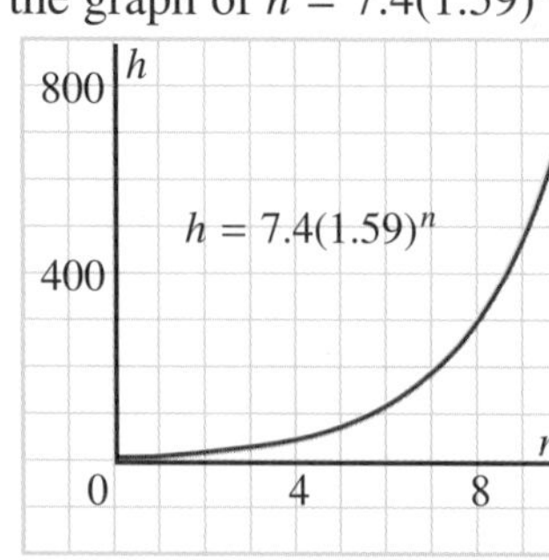

b) 2001 is 6 years from 1995. Substitute 6 for n in the equation.

$$h = 7.4(1.59)^6$$
$$\doteq 119.6$$

In 2001, there were about 120 million hosts.

c) Substitute 500 for h in the equation.

$$500 = 7.4(1.59)^n$$
$$1.59^n = \frac{500}{7.4}$$
$$1.59^n \doteq 67.6$$

Find a power of 1.59 that is close to 67.6.
Use systematic trial. Look at the table in part a.
$h = 500$ lies between $n = 8$ and $n = 10$.

Try $n = 9$: $1.59^9 \doteq 64.9$ This is close to 67.6.
The solution is $n = 9$, to the nearest whole number.
According to the model, the number of hosts will reach 500 million in $1995 + 9 = 2004$.

Using a graphing calculator

a) Input [Y=] 7.4 [(] 1.59 [)] [^] [X,T,θ,n].
Use appropriate window settings:
$0 \le x \le 10$; $-70 \le y \le 700$
Press [GRAPH].

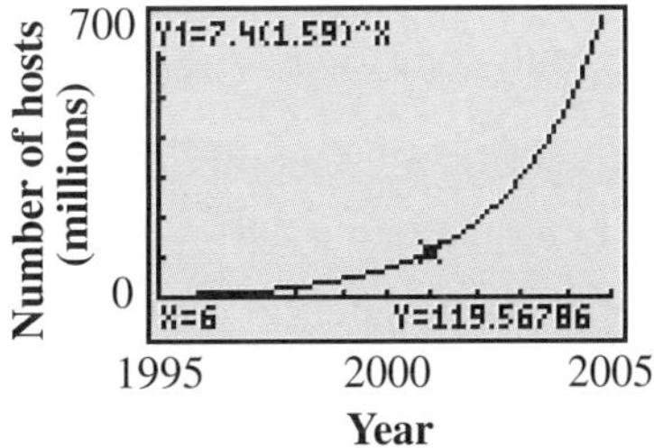

b) 2001 is 6 years from 1995. Press [TRACE] 6 [ENTER] to display X = 6, Y = 119.56786 at the bottom of the screen.
In 2001, there were about 120 million hosts.

c) Substitute 500 for h in the equation.
$500 = 7.4(1.59)^n$

Press [TRACE], then trace to the point where the y-coordinate is closest to 500. This occurs when the x-coordinate is approximately 9. According to the model, the number of hosts will reach 500 million in $1995 + 9 = 2004$.

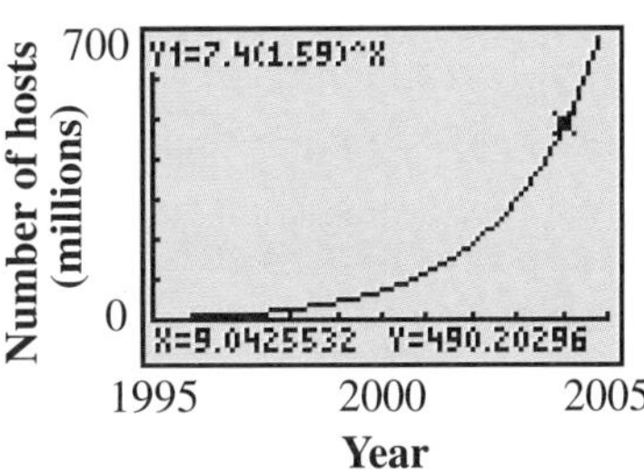

In the *Example* part c, we solved the equation $500 = 7.4(1.59)^n$ by systematic trial and by graphing, then tracing. This equation is an example of an *exponential equation.* In Sections 6.2 and 6.3, we will develop a more efficient method to solve exponential equations.

6.1 Exercises

A

1. Identify which graph best represents each function.

   a) $f(x) = 3^x$ b) $g(x) = 10^x$

   c) $h(x) = \left(\frac{3}{4}\right)^x$ d) $k(x) = \left(\frac{1}{4}\right)^x$

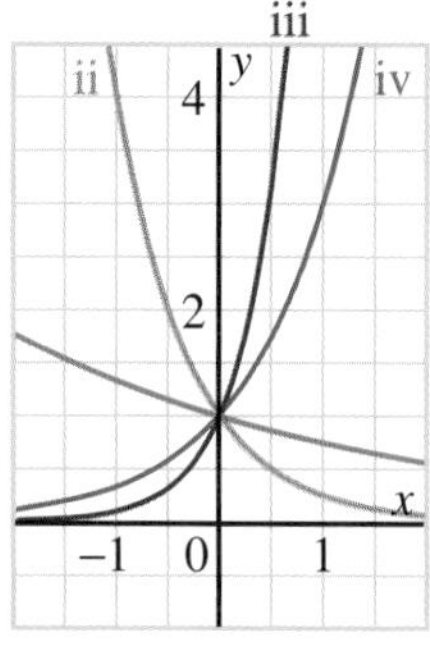

2. a) Sketch the graphs of $y = 2^x$ and $y = 6^x$ on the same grid.

   b) On the same grid as in part a, sketch the graph of $y = 3^x$.

3. Use the results of exercise 2. Sketch the graphs of $y = \left(\frac{1}{2}\right)^x$, $y = \left(\frac{1}{3}\right)^x$, and $y = \left(\frac{1}{6}\right)^x$ on the same grid.

4. Confirm your results in exercises 2 and 3.

5. The growth of $500 at 6.5% compounded annually is represented by the equation and the graph below left.

   a) Use the graph to estimate the value of the investment after 5 years.

   b) Use the graph to estimate how many years it takes for the investment to grow to $1000.

   c) Describe how both the graph and the equation change in each case.

      i) The original investment is greater than $500.

      ii) The original investment is less than $500.

      iii) The interest rate is greater than 6.5%.

      iv) The interest rate is less than 6.5%.

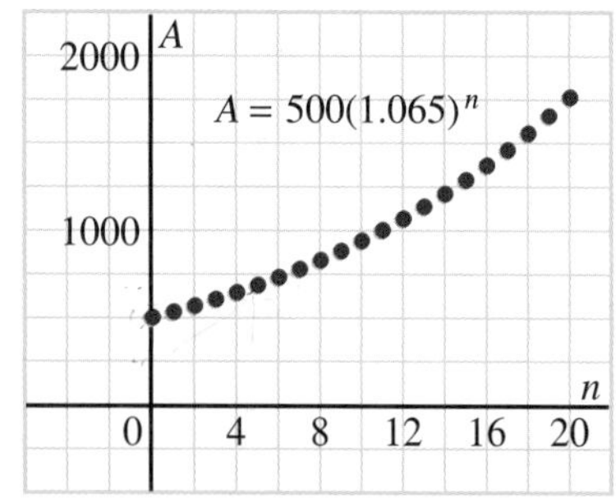

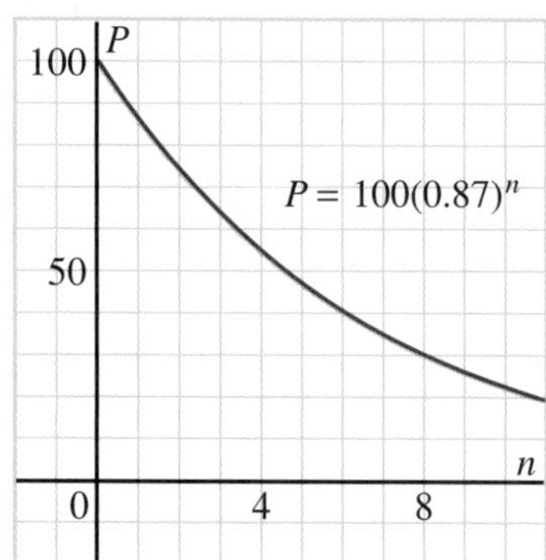

6. The time that caffeine stays in your bloodstream is represented by the equation and the graph on page 376, bottom right.

   a) Use the graph to estimate how long it takes until only 20% of the caffeine remains.

   b) Describe how both the graph and the equation would change for pregnant women, who require a much longer time to metabolize caffeine than other adults.

7. **Knowledge/Understanding** Sketch the graphs of the equations in each list.

   a) $y = 2^x$, $y = 4^x$, $y = 10^x$

   b) $y = \left(\frac{1}{2}\right)^x$, $y = \left(\frac{1}{4}\right)^x$, $y = \left(\frac{1}{10}\right)^x$

8. **Thinking/Inquiry/Problem Solving**

   a) Graph the function $y = 3^x$.

   b) Describe how the graph in part a will change for each function below. Use a graphing calculator to confirm your description.

      i) $y = 2(3)^x$ ii) $y = 0.5(3)^x$ iii) $y = -3^x$

      iv) $y = 3^x + 2$ v) $y = 3^x - 1$ vi) $y = 2(3)^x + 1$

   c) Describe the graphical implications of changes in the parameters b and c in the equation $y = c(3)^x + b$.

9. a) Graph the function $y = 2(3)^x + 1$.

   b) Describe how the graph in part a will change for each function below. Use a graphing calculator to confirm your description.

      i) $y = 2(5)^x + 1$ ii) $y = 2(0.5)^x + 1$

      iii) $y = -2(3)^x + 5$ iv) $y = 5(3)^x - 5$

   c) Describe the graphical implications of changes in the parameters a, b, and c in the equation $y = ca^x + b$.

10. The number of Canadian cell phone users, s, has grown according to the equation $s = 130\ 000(1.45)^t$, where t is the time in years since 1987.

   a) Graph the equation for $0 \le t \le 20$.

   b) Use the equation to predict the number of cell phone users in 2005. Identify any assumptions used.

   c) Estimate when the number of cell phone users reached 10 million.

   d) According to the equation, what is the average annual rate of increase in the number of cell phone users since 1987?

   e) Use parts b to d as a guide. Pose your own problem about cell phone use in Canada. Solve the problem you posed.

11. **Communication** Each graph below shows the exponential function $y = 2^x$ and a power function. On the first graph, the power function is the quadratic function $y = x^2$. On the second graph, the power function is the cubic function $y = x^3$.

a) Use the first graph. For large values of x, which function has the greater rate of increase, $y = 2^x$ or $y = x^2$? Explain.

b) Use the second graph. For large values of x, which function has the greater rate of increase, $y = 2^x$ or $y = x^3$? Explain.

c) Suppose you draw a graph to show the exponential function $y = 2^x$ and the quartic function $y = x^4$. Would you get similar results if you compare the rates of increase of these functions for large values of x? Explain.

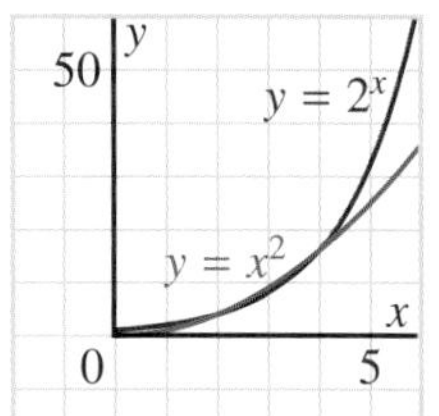

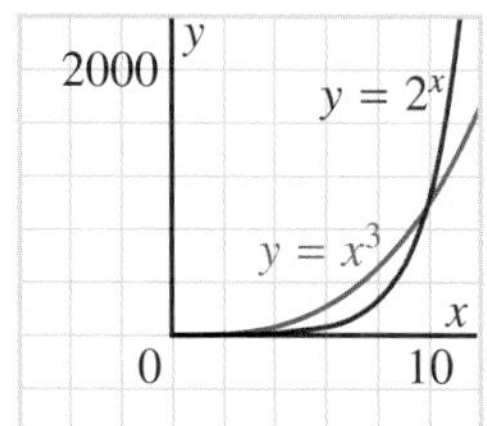

The values of polynomial functions are very large when x is large. However, if x is large enough, the values of any exponential function will eventually become greater than the values of any polynomial function. And, from that point on, the rate of change of the exponential function is greater than the rate of change of the polynomial function. Exercise 11 is significant because it illustrates this fact using particular examples.

12. **Application** The population of Canada, P million, can be modelled by the equation $P = 24.0(1.014)^t$, where t is the time in years since 1981. Use this equation.

a) Predict the population in 2011.

b) Predict when the population might reach 35 million.

c) Describe how the graph of $P = 24.0(1.014)^t$ is related to the graph of $y = 1.014^x$.

d) Use parts a and b as a guide. Pose your own problem about the population of Canada. Solve the problem you posed.

e) What assumption is made about the population growth in parts a and b?

f) Give some reasons why the population of Canada in 2011 might be different from your prediction in part a, or why the population might reach 35 million in a year different from the year you predicted in part b.

13. Suppose you make a long-term investment at a fixed interest rate of 6.5% compounded annually. The principal, P dollars, that you would invest now to have \$1000 after n years is represented by the equation $P = 1000(1.065)^{-n}$. Recall that P is the present value.

a) How much would you need to invest now to have \$1000 after 6 years?

b) Suppose you invested \$500 now at 6.5% compounded annually. How long would it take until the amount is \$1000?

✓ 14. Most chemical reactions take place more rapidly at high temperatures than at low temperatures. The time, t seconds, for a certain chemical reaction to take place can be modelled as a function of the temperature, T degrees Celsius, by the equation $t = 8.72(0.93)^T$.

a) According to this model, how long does the reaction require at 20°C?

b) At what temperature does the reaction require 30 s to take place?

c) Pose your own problem about this chemical reaction. Solve the problem you posed.

15. Compare the equations in exercises 13 and 14.

a) Both situations are exponential decay situations, yet one exponent is negative and the other is positive. Explain.

b) Rewrite each equation in the form of the other equation.

✓ 16. Several layers of glass are stacked together. Each layer reduces the light passing through it by 5%.

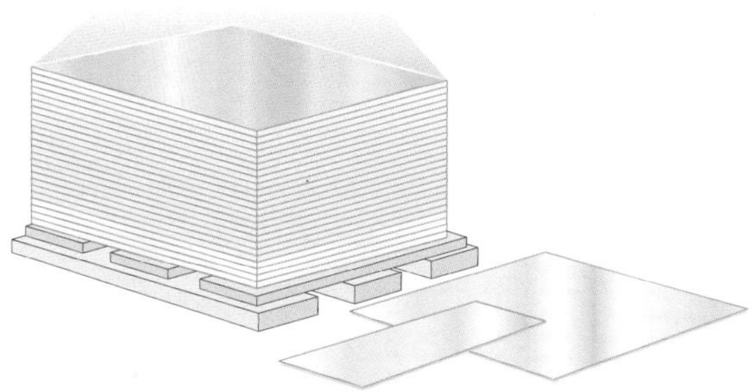

a) Write an equation to represent the percent of light, P, passing through n layers.

b) Graph the equation.

c) Estimate how many panes are needed before only 50% of the light passes through.

d) The 5% figure is for clear glass. For frosted glass, each layer reduces the light passing through it by 10%. Describe how both the graph and the equation change for frosted glass.

✓ 17. For every metre below the surface of water, the light intensity is reduced by 2.5%.

a) Write an equation to express the percent of light remaining, P, as a function of the depth, d metres, below the surface.

b) Graph P as a function of d.

c) Describe how the graph is similar to, and different from, the graph in exercise 16b.

d) Determine the light intensity at a depth of 10 m.

e) At what depth is the light intensity reduced to 50% of the intensity at the surface?

f) Pose your own problem about light intensity in water. Solve the problem you posed. Write a clear explanation of your problem and its solution.

6.2 Defining a Logarithm

Investigation

The LOG Key on a Calculator

1. Find out what the LOG key on your calculator does. Try a wide variety of numbers such as those suggested below. Record the results for later reference.

   a) Numbers selected at random; for example:
   5, 47, 329, 6388
   (To determine log 5, press LOG 5 ENTER.)

   b) Powers of 10; for example:
   10, 100, 1000, 10 000, … 1, 0.1, 0.01, 0.001, …

   c) Multiples of 10, 100, 1000; for example:
   20, 200, 2000, … 30, 300, 3000, …

   d) Numbers with the same significant digits; for example:
   5.72, 57.2, 572, 5720, … 0.572, 0.0572, 0.005 72, …

   e) Square roots of powers of 10; for example:
   $\sqrt{10}$, $\sqrt{100}$, $\sqrt{1000}$, $\sqrt{10\ 000}$, …
   $\sqrt{1}$, $\sqrt{0.1}$, $\sqrt{0.01}$, $\sqrt{0.001}$, …

   f) Zero and negative numbers; for example:
   0, –2, –5, –23.5

2. For any positive number, x, the LOG key calculates a number represented by $\log x$.

   a) Explain how $\log x$ is related to x.

   b) Which word best describes a logarithm?

In Section 6.1, you considered problems such as these:

Example, page 374

When will the number of Internet hosts reach 500 million?

In the solution, the answer was obtained by solving this equation:

$$500 = 7.4(1.59)^n \quad \ldots ①$$

Exercise 16c, page 379

How many panes of glass are needed before only 50% of the light passes through?

In your solution, you would have solved this equation:

$$50 = 100(0.95)^n \quad \ldots ②$$

Equations ① and ② can be solved using technology. However, exponential equations occur so frequently in applied problems that a way has been developed to solve them directly. This method involves logarithms. In this section, we define a logarithm and develop some of its properties. In Section 6.3, we shall solve equations ① and ②, and other similar exponential equations.

In the *Investigation,* you determined logarithms of 10, 100, and 1000, and obtained the results below, which are also shown on the calculator screen.

```
log(10)
                1
log(100)
                2
log(1000)
                3
```

$$\log 10 = 1$$
$$\log 100 = 2$$
$$\log 1000 = 3$$

These examples indicate that log x is defined as the exponent that 10 has when x is written as a power of 10. In general, log x is equal to a number b with the property that $10^b = x$. The expression log x is read "the logarithm of x."

In *Example 1,* and throughout this chapter and Chapter 7, the keystrokes and screens are for the TI-83 Plus calculator. The keystrokes work for most scientific calculators. If the keystrokes do not work for your calculator, consult its manual to determine the proper keystrokes.

Example 1

Use a calculator to determine each logarithm, if possible, to 4 decimal places. Explain the meaning of each result.

a) log 125 **b)** log 0.66 **c)** log (–4)

Solution

a) Press: [LOG] 125 [ENTER] to display 2.096910013

$\log 125 \doteq 2.0969$

This means that $10^{2.0969} \doteq 125$.

b) $\log 0.66 \doteq -0.1805$

This means that $10^{-0.1805} \doteq 0.66$.

c) Pressing [LOG] [(–)] 4 [ENTER] results in an error message.

This means that log (–4) is not defined as a real number.

There is no power of 10 that equals –4.

Discuss

How could you check the results of parts a and b?

The logarithms on page 380 and above have base 10.

The restriction to base 10 is not necessary, and logarithms can be defined with any positive base. Some examples of base-2 and base-3 logarithms are shown at the right.

$\log_2 2 = 1$	$\log_3 3 = 1$
$\log_2 4 = 2$	$\log_3 9 = 2$
$\log_2 8 = 3$	$\log_3 27 = 3$

These examples show that $\log_2 x$ is defined as the exponent that 2 has when x is written as a power of 2. And, $\log_3 x$ is defined as the exponent that 3 has when x is written as a power of 3. The expression $\log_a x$ is defined in the same way. It is read "the logarithm of x to the base a."

Take Note

Definition of $\log_a x$

The expression $\log_a x$ is an exponent. It is the exponent that a has when x is written as a power of a.

$\log_a x = y$ means $a^y = x$, where $a > 0$, $a \neq 1$, and $x > 0$.

To remember the definition of $\log_a x$, think:

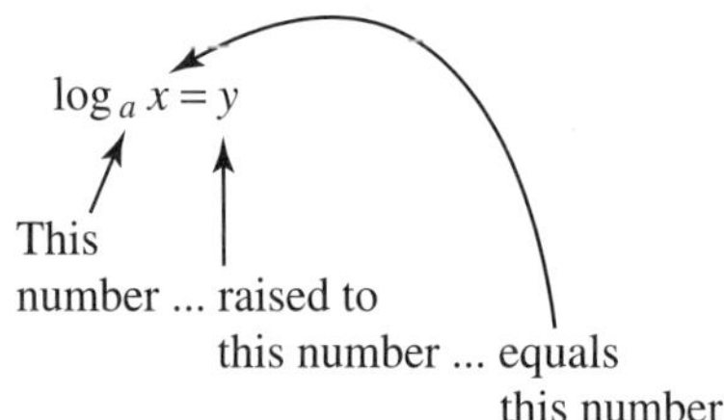

If a is omitted, the base is 10. Hence, $\log x$ means $\log_{10} x$.

Example 2

Write each expression in exponential form.

a) $\log_6 216 = 3$ **b)** $\log_5 0.2 = -1$ **c)** $\log p = q$

Solution

a) $\log_6 216 = 3$
$6^3 = 216$

b) $\log_5 0.2 = -1$
$5^{-1} = 0.2$

c) $\log p = q$
$10^q = p$

Example 3

Write each expression in logarithmic form.

a) $3^5 = 243$ **b)** $2^{-3} = \frac{1}{8}$ **c)** $a^b = c$

Solution

a) $3^5 = 243$
$\log_3 243 = 5$

b) $2^{-3} = \frac{1}{8}$
$\log_2 \left(\frac{1}{8}\right) = -3$

c) $a^b = c$
$\log_a c = b$

Example 4

Evaluate each logarithm.

a) $\log_4 16$ **b)** $\log_8 \left(\frac{1}{8}\right)$ **c)** $\log_7 1$

Solution

a) $\log_4 16$ is the power to which we raise 4 to get 16.
Since $4^2 = 16$, then $\log_4 16 = 2$

b) $\log_8 \left(\frac{1}{8}\right)$ is the power to which we raise 8 to get $\frac{1}{8}$.
Since $8^{-1} = \frac{1}{8}$, then $\log_8 \left(\frac{1}{8}\right) = -1$

c) $\log_7 1$ is the power to which we raise 7 to get 1.
Since $7^0 = 1$, then $\log_7 1 = 0$

6.2 Exercises

A

1. Knowledge/Understanding Use a calculator to determine each logarithm to 4 decimal places. Explain the meaning of each result.

a) log 8.3 **b)** log 635 **c)** log 9842

d) log 0.37 **e)** log 0.079 **f)** log 0.0045

2. Explain your answer to each question.

a) In exercise 1, why are the first three logarithms positive and the last three negative?

b) Why does a negative number not have a logarithm?

3. Evaluate the logarithms in each list.

a) log 1	**b)** $\log_5 1$	**c)** $\log_a 1$
log 10	$\log_5 5$	$\log_a a$
log 100	$\log_5 25$	$\log_a a^2$
log 1000	$\log_5 125$	$\log_a a^3$

B

4. How can you remember how to convert from exponential form to logarithmic form, and vice versa?

5. Communication Explain what is meant by a logarithm. Use some examples to illustrate your explanation.

6. Write in exponential form.

a) $\log_2 8 = 3$ b) $\log_3 81 = 4$ c) $\log 10\,000 = 4$
d) $\log 0.01 = -2$ e) $\log_5 125 = 3$ f) $\log_{16} 2 = 0.25$
g) $\log_2 \left(\frac{1}{4}\right) = -2$ h) $\log x = y$ i) $\log_a x = y$

7. Write in logarithmic form.

a) $7^2 = 49$ b) $10^3 = 1000$ c) $2^{10} = 1024$
d) $5^4 = 625$ e) $2^{-1} = 0.5$ f) $16^{\frac{1}{2}} = 4$
g) $81^{\frac{3}{4}} = 27$ h) $16^{0.25} = 2$ i) $9^{-0.5} = \frac{1}{3}$

8. Evaluate each logarithm without using a calculator.

a) $\log 1000$ b) $\log_5 25$ c) $\log_3 27$
d) $\log_2 32$ e) $\log 1$ f) $\log_6 1$
g) $\log_3 81$ h) $\log_7 49$ i) $\log 0.1$
j) $\log_2 \left(\frac{1}{2}\right)$ k) $\log_2 \left(\frac{1}{16}\right)$ l) $\log_a a$

9. Choose two logarithms in exercise 8. Explain how you evaluated each logarithm.

10. All these logarithms represent the same number:

$\log 10^2$ $\log_3 3^2$ $\log_5 5^2$ $\log_6 6^2$ $\log_a a^2$

a) What number do they represent? Explain.
b) Write a similar expression that represents this number.

11. **Application** The calculator screen shows the results of two calculations. Each calculation includes the number 1.482, and the result is also 1.482. Is this a coincidence, or is there a reason why the result is the same as the number we started with? Explain.

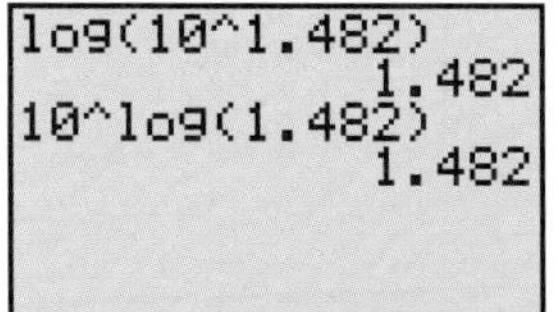

Exercises 11–13 are significant because they show that determining a logarithm and determining an exponent are inverse operations. When these operations are applied in succession, they leave a number unchanged. The following equations are true for all values of x for which the expressions are meaningful.
$\log_a a^x = x$
$a^{\log_a x} = x, x > 0$
These properties are a direct consequence of the definition of a logarithm.

12. Evaluate each logarithm without using a calculator.

a) $\log 10^3$ b) $\log_3 3^4$ c) $\log_6 6^3$
d) $\log_2 2^{-1}$ e) $\log_5 5^{-2}$ f) $\log_a a^x$

13. Simplify each expression.

a) $10^{\log 1000}$ b) $3^{\log_3 81}$ c) $6^{\log_6 216}$
d) $2^{\log_2 \left(\frac{1}{2}\right)}$ e) $5^{\log_5 \left(\frac{1}{25}\right)}$ f) $a^{\log_a x}$

14. Thinking/Inquiry/Problem Solving

a) Use a calculator to evaluate the logarithms in each list.

i) log 2	ii) log 3
log 20	log 30
log 200	log 300
log 2000	log 3000

b) Explain the patterns in your results for part a.

15. Refer to exercise 3. The numbers that follow "log" in parts a and b can be written as powers of 10 and 5, respectively, with positive exponents. Write similar lists of logarithms that involve powers of 10, 5, and *a*, with negative exponents.

16. Suppose you record the first digits of many numbers randomly selected from the pages of newspapers, stock-market prices, census data, or street addresses. Most people would think that the digits 1, 2, 3, …, 9 would occur first about the same number of times. However, this is not true. In the 1990s, it was proved that in such a set of data, the probability that the first non-zero digit of a number is *n*, is:

$$\text{P(first digit is } n) = \log\left(1 + \frac{1}{n}\right),\ n = 1,\ 2,\ 3,\ \ldots,\ 9$$

For example, the probability that the first digit is 1 is $\log\left(1 + \frac{1}{1}\right)$, or $\log 2 \doteq 0.30$.

The relationship in exercise 16 is known as Benford's Law. It has important applications in the design of computers and the detection of fraudulent accounting.

a) Calculate the probability that the first digit is each number.

i) 2 ii) 3 iii) 4 iv) 5
v) 6 vi) 7 vii) 8 viii) 9

b) Graph P(first digit is *n*) against *n*.

17. The LN key on a calculator determines logarithms of numbers to a base different from the LOG key. This base is represented by the letter *e*. Use a calculator to determine *e* to 5 decimal places.

18. Evaluate each logarithm.

a) $\log_3 9^5$ b) $\log_5 25^{-3}$ c) $\log_2 \sqrt{32}$

19. a) Evaluate each logarithm.

i) $\log_2 8$ and $\log_8 2$ ii) $\log_5 25$ and $\log_{25} 5$

b) Make a conjecture about how $\log_a b$ and $\log_b a$ are related. Prove your conjecture.

20. a) Prove that $(\log_a b)(\log_b c) = \log_a c$.

b) Explain how the equation in part a can be used to change the base of a logarithm.

6.3 Law of Logarithms for Powers

In Section 6.2, logarithms were introduced to develop a method for solving exponential equations. Since a logarithm is an exponent, we can write the laws of exponents in logarithmic form. In this section, we shall consider the law of logarithms for powers.

Suppose we use a calculator to determine the base-10 logarithms of some powers of 2. Notice the patterns in the results at the right. All the logarithms of the powers of 2 are multiples of the logarithm of 2.
For example,

$$\log 8 = 3 \times 0.30103$$

This can be written as:

$$\log 2^3 = 3 \log 2$$

$\log 2 \doteq 0.30103$
$\log 4 \doteq 0.60206$
$\log 8 \doteq 0.90309$
$\log 16 \doteq 1.20412$
$\log 32 \doteq 1.50515$
$\log 64 \doteq 1.80618$

This is an example of the law of logarithms for powers.
This law has the form:

$$\log_a x^n = n \log_a x$$

To prove this law, let $\log_a x = m$ ... ①

Step 1. Write the equation in exponential form.

$$x = a^m$$

Step 2. Raise each side to the nth power.

$$x^n = (a^m)^n$$

$$x^n = a^{mn} \qquad \text{Using the law of exponents for powers}$$

Step 3. Take the base-a logarithm of each side.

$$\log_a x^n = \log_a a^{mn}$$

$$\log_a x^n = mn$$

$$\log_a x^n = (\log_a x)n \qquad \text{Using equation ①}$$

Hence, $\log_a x^n = n \log_a x$

Take Note

Law of Logarithms for Powers

If x and n are real numbers, and $x > 0$, then

$$\log_a x^n = n \log_a x \qquad a > 0, a \neq 1$$

Two things about this law are significant. First, the way we proved the law is significant. We started with $\log_a x = m$, and expressed it in exponential form. This allowed us to use the law of exponents for powers. Then, we expressed the result in logarithmic form. Therefore, the law of logarithms for powers is simply a restatement of the law of exponents for powers in logarithmic form.

Second, the law of logarithms for powers itself is significant because n occurs as an exponent on the left side, but as a factor on the right side. This means that we can use this law to solve exponential equations.

Example 1

Solve the equation $6^x = 53$ to 4 decimal places. Check the solution.

Solution

$$6^x = 53$$

Take the base-10 logarithm of each side.

$$\log 6^x = \log 53$$

Use the law of logarithms for powers.

$$x \log 6 = \log 53$$

$$x = \frac{\log 53}{\log 6}$$

$$\doteq \frac{1.724\ 28}{0.778\ 15}$$

$$\doteq 2.2159$$

To check, determine $6^{2.2159}$. The result is approximately 53.0036.

An important example of an exponential equation occurs when we express one positive number as a power of another positive number, as *Example 2* illustrates. The significance of this example will be seen in Section 6.5 and in Chapter 7.

Example 2

Express 37 as a power of 2.

Solution

Let $2^x = 37$.

Take the base-10 logarithm of each side.

$$\log 2^x = \log 37$$

Use the law of logarithms for powers.

$$x \log 2 = \log 37$$

$$x = \frac{\log 37}{\log 2}$$

$$\doteq \frac{1.568\ 20}{0.301\ 03}$$

$$\doteq 5.2094$$

Therefore, $37 \doteq 2^{5.2094}$

At the beginning of Section 6.2, we stated that we would develop a method to solve exponential equations such as $500 = 7.4(1.59)^n$ and $50 = 100(0.95)^n$. An equation similar to these appears in *Example 3.*

Example 3

Some people invest in mutual funds. Since 1989, the number of mutual funds available in Canada has grown rapidly. Data collected from a graph in a newspaper shows that the number of mutual funds, M, can be modelled by the equation $M = 460(1.19)^n$, where n is the number of years since 1989.

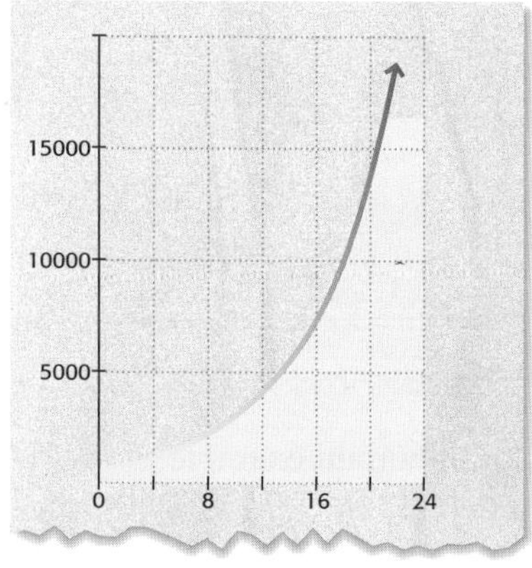

a) According to this model, when will the number of mutual funds reach 8000?

b) How many years does it take for the number of mutual funds to double?

Solution

a) $$M = 460(1.19)^n$$

Substitute 8000 for M.

$$8000 = 460(1.19)^n$$

Isolate the power.

$$1.19^n = \frac{8000}{460}$$

$$1.19^n \doteq 17.3913$$

Take the base-10 logarithm of each side.

$$\log 1.19^n \doteq \log 17.3913$$

Use the law of logarithms for powers.

$$n \log 1.19 \doteq \log 17.3913$$

$$n \doteq \frac{\log 17.3913}{\log 1.19}$$

$$\doteq 16.4180$$

There will be 8000 mutual funds approximately 16 years after 1989; that is, in 2005.

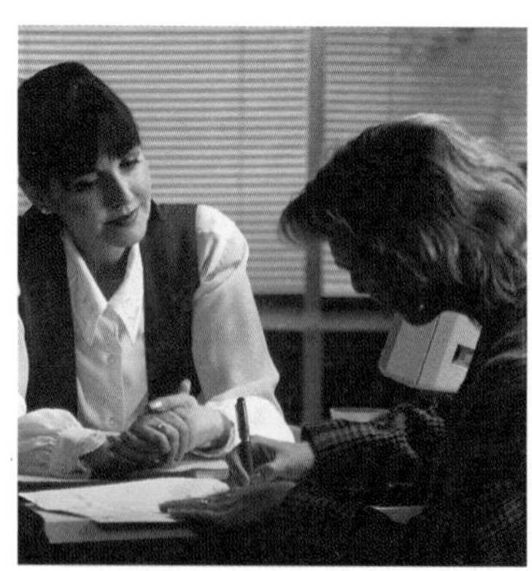

b) $M = 460(1.19)^n$

Initially, $n = 0$, so substitute this value for n in the equation.

$$\begin{aligned} M &= 460(1.19)^0 \\ &= 460 \end{aligned}$$

In 1989, there were 460 mutual funds.

Double this number is 920.

Substitute 920 for M in the equation.

$$\begin{aligned} 920 &= 460(1.19)^n \\ 2 &= 1.19^n \end{aligned}$$

Take the base-10 logarithm of each side.

$$\log 2 = \log 1.19^n$$

Use the law of logarithms for powers.

$$\begin{aligned} \log 2 &= n \log 1.19 \\ n &= \frac{\log 2}{\log 1.19} \\ &\doteq 3.9847 \end{aligned}$$

The number of mutual funds doubles in approximately 4 years.

Doubling Time for Exponential Growth

In *Example 3,* we showed that it takes approximately 4 years for the number of mutual funds to double. This doubling time applies at all times, provided the equation $M = 460(1.19)^n$ continues to be a suitable model for the growth of the mutual funds.

Suppose P and Q are points on the graph of $M = 460(1.19)^n$ so that Q is 4 units to the right of P. Then, the second coordinate of Q is double the second coordinate of P. If P and Q move along the graph, keeping 4 units apart horizontally, the second coordinate of Q is always double the second coordinate of P.

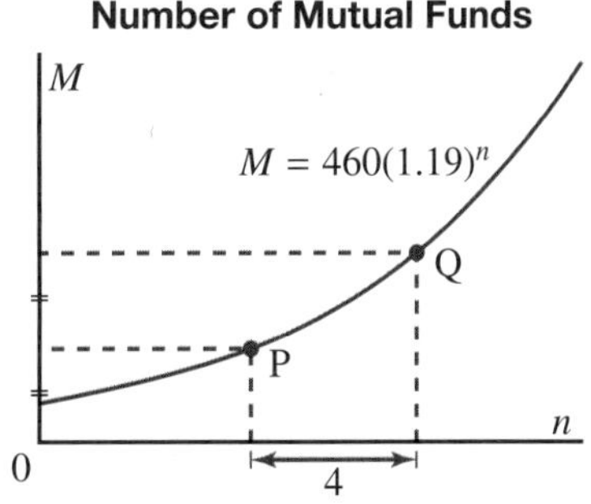

A corresponding property applies to any function $y = ca^x$ that models exponential growth ($a > 1$). When something is growing exponentially, it becomes twice as large within a fixed period of time. The doubling time depends only on the base, a, in the equation, and not on the coefficient, c. You will prove this in exercise 10.

A similar property applies to exponential decay. You will prove this property in Section 6.4.

6.3 Exercises

1. Evaluate the logarithms in each list, to 3 decimal places. Explain the pattern in the results.

   a) log 5
log 25
log 125
log 625
log 3125

   b) log 0.5
log 0.25
log 0.125
log 0.062 5
log 0.031 25

2. Evaluate the logarithms in each list and compare the results. Explain.

   a) log 2
log 4
log 8
log 16
log 32

   b) log 2
2 log 2
3 log 2
4 log 2
5 log 2

3. Create another pattern similar to each of those in exercise 2.

B

4. **Knowledge/Understanding** Solve each equation to 4 decimal places. Check the solution.

   a) $2^x = 11$ b) $3^x = 17$ c) $6^x = 5$

5. Solve the equations in each list. Give the answers to 4 decimal places.

   a) $3^x = 2$
$3^x = 4$
$3^x = 8$
$3^x = 16$

   b) $3^x = 2$
$9^x = 2$
$27^x = 2$
$81^x = 2$

   c) $2^x = 3$
$4^x = 3$
$8^x = 3$
$16^x = 3$

6. **Communication** Choose one list in exercise 5. Explain how the roots of the equations in that list are related.

7. Solve each equation to 4 decimal places.

   a) $2^x = 10$ b) $3^x = 100$ c) $10^x = 375$

   d) $2^{x-3} = 7$ e) $4^{3x} = 60$ f) $3(5)^x = 150$

8. Express:

   a) 26 as a power of 2 b) 43 as a power of 5 c) 89 as a power of 4

   d) 381 as a power of 6 e) 7 as a power of 10 f) 2 as a power of 3

✓ **9. Application** When $500 are invested at 6.5% compounded annually, the amount, A dollars, after n years is given by $A = 500(1.065)^n$.

a) Determine how many years it takes the investment of $500 to double.

b) Predict how many years it will take for the investment to double again. Check your prediction by calculation.

c) Determine the amount after 3 years. Then predict how many years it will take this amount to double. Check your prediction by calculation.

d) Your results in parts a to c illustrate a certain property of exponential growth. What is this property?

10. Thinking/Inquiry/Problem Solving Consider the exponential function $f(x) = ca^x$, where $a > 1$. Let $P(x_1, y_1)$ and $Q(x_2, y_2)$ be any two points on the graph of f.

a) Suppose $y_2 = 2y_1$. Show that $x_2 = x_1 + k$, where k depends only on a.

b) Sketch the graph of k against a.

Exercise 10 is significant because it proves that adding a certain constant to the value of x causes the value of y to double. If x represents time, then the doubling time is constant for exponential growth.

✓ **11.** Many consumers use debit cards for making purchases. According to data from the Canadian Bankers Association, the annual number of debit card transactions, D million, can be modelled by the equation $D = 328(1.384)^n$, where n is the number of years since 1994.

a) How many debit card transactions were there in 2000?

b) When will the number of debit card transactions reach 5 billion?

c) Pose your own problem about the number of debit card transactions. Solve the problem you posed.

12. In a steel mill, red-hot slabs of steel are pressed many times between heavy rollers. The drawings show two stages in rolling a slab. On all passes through the rollers, the width of the slab is constant.

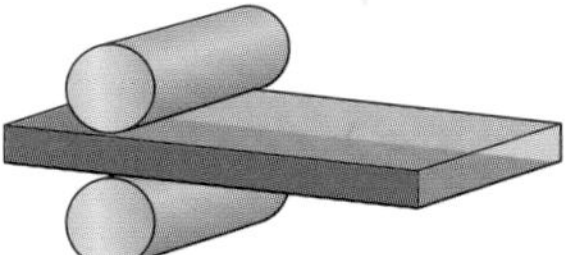

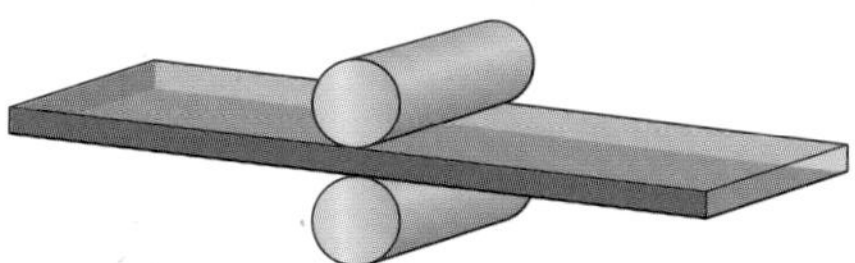

A slab is 2.00 m long and 0.50 m thick. On each pass through the rollers, its length increases by 20%, and its thickness decreases by 17%. After n passes, its length, L metres, and thickness, T metres, are modelled by the equations $L = 2.00(1.20)^n$ and $T = 0.50(0.83)^n$.

a) How long and how thick is the slab after 12 passes through the rollers?

b) How many passes are needed until the length is at least 20 m? How thick is the slab at this point?

c) How many passes are needed until the thickness is about 1 mm? How long is the slab at this point?

6.4 Laws of Logarithms for Multiplication and Division

In Section 6.3, we wrote the law of exponents for powers in logarithmic form. We can do the same with the laws of exponents for multiplication and division.

Suppose we evaluate the base-10 logarithms of 2, 20, 200, and 2000. Notice the patterns in the results at the right. Each logarithm is the sum of a natural number and the logarithm of 2.

$$\log 2 \doteq 0.30103$$
$$\log 20 \doteq 1.30103$$
$$\log 200 \doteq 2.30103$$
$$\log 2000 \doteq 3.30103$$

For example,

$$\log 2000 = 3 + 0.30103$$

This can be written as:

$$\log (1000 \times 2) = \log 1000 + \log 2$$

This is an example of the law of logarithms for multiplication. This law has the form:

$$\log_a xy = \log_a x + \log_a y$$

To prove the law, we use the same strategy as in Section 6.3 when we proved the law of logarithms for powers.

Let $\log_a x = m$ ... ①

and $\log_a y = n$ ... ②

Step 1. Write the equations above in exponential form.

$$x = a^m$$
$$y = a^n$$

Step 2. Multiply the left and right sides.

$$xy = a^m \times a^n$$
$$xy = a^{m+n}$$ Using the law of exponents for multiplication

Step 3. Take the base-a logarithm of each side.

$$\log_a xy = \log_a a^{m+n}$$
$$\log_a xy = m + n$$

Hence, $\log_a xy = \log_a x + \log_a y$ Using equations ① and ②

Take Note

Law of Logarithms for Multiplication

If x and y are positive real numbers, then

$$\log_a xy = \log_a x + \log_a y \qquad a > 0, a \neq 1$$

To prove the law, we expressed the equations $\log_a x = m$ and $\log_a y = n$ in exponential form. This allowed us to use the law of exponents for multiplication. Then, we expressed the result in logarithmic form. Therefore, the law of logarithms for multiplication is simply a restatement, in logarithmic form, of the law of exponents for multiplication.

There is a law of logarithms for division. This states that the logarithm of a quotient of two numbers is the difference of the logarithms of the two numbers. All the logarithms have the same base. This law can be proved in the same way as the law of logarithms for multiplication (see exercise 19).

Take Note

Law of Logarithms for Division

If x and y are positive real numbers, then

$$\log_a \left(\frac{x}{y}\right) = \log_a x - \log_a y \qquad a > 0, a \neq 1$$

We use the laws of exponents to simplify expressions containing exponents. Similarly, we use the laws of logarithms to simplify expressions containing logarithms.

Example 1

Simplify each expression.

a) $\log 25 + \log 4$

b) $\log 64 - \log 640$

c) $\log_3 6 + \log_3 1.5$

d) $\log_5 50 - \log_5 0.4$

Solution

a) $\log 25 + \log 4 = \log (25 \times 4)$
$= \log 100$
$= 2$

b) $\log 64 - \log 640 = \log \left(\frac{64}{640}\right)$
$= \log 0.1$
$= -1$

c) $\log_3 6 + \log_3 1.5 = \log_3 (6 \times 1.5)$
$= \log_3 9$
$= 2$

d) $\log_5 50 - \log_5 0.4 = \log_5 \left(\frac{50}{0.4}\right)$
$= \log_5 125$
$= 3$

The law of logarithms for multiplication provides an alternative way to solve an exponential equation, such as $8000 = 460(1.19)^n$, which we solved in Section 6.3, *Example 3*. Instead of isolating the power, we can take the logarithm of each side. Compare the solution of *Example 2* with the solution on page 388.

Example 2

Solve $8000 = 460(1.19)^n$ to 4 decimal places.

Solution

$8000 = 460(1.19)^n$

Take the base-10 logarithm of each side.

$\log 8000 = \log\left(460(1.19)^n\right)$

$\log 8000 = \log 460 + \log 1.19^n$ Using the law of logarithms for multiplication

$\log 8000 = \log 460 + n\log 1.19$ Using the law of logarithms for powers

Solve for n.

$n\log 1.19 = \log 8000 - \log 460$

$$n = \frac{\log 8000 - \log 460}{\log 1.19}$$

$$= \frac{\log\left(\frac{8000}{460}\right)}{\log 1.19}$$ Using the law of logarithms for division

$n \doteq 16.4180$

In *Example 2,* the step of writing $n = \frac{\log\left(\frac{8000}{460}\right)}{\log 1.19}$ is not essential, but it facilitates calculating the result.

Exponential Decay

Radioactive isotopes of certain elements decay with a characteristic half-life. This is the time required for one-half of the substance to decay to other substances. Radioactive decay can be modelled by an exponential function.

Example 3

In 1986, there was a major nuclear accident at the Chernobyl power plant in Ukraine. The atmosphere was contaminated with radioactive iodine-131, which has a half-life of 8.1 days. An equation that models the percent, P, of the original radiation that remains t days after the accident is $P = 100(0.918)^t$. How long did it take for the level of radiation to decrease to 1% of the level immediately after the accident?

Solution

Use the equation $P = 100(0.918)^t$. To determine the time when the radiation is 1% of the original radiation, substitute $P = 1$, then solve for t.

$1 = 100(0.918)^t$

Take the base-10 logarithm of each side.

$$\log 1 = \log 100 + t \log 0.918$$
$$0 = 2 + t \log 0.918$$
$$t = \frac{-2}{\log 0.918}$$
$$\doteq 53.8252$$

It took approximately 54 days for the level of radiation to decrease to 1% of the level immediately after the accident.

Half-Life for Exponential Decay

In *Example 3,* the half-life of iodine-131 is 8.1 days. This means it takes 8.1 days for the amount of radiation to decrease by 50%. This half-life applies at all times, not just for the original amount of radiation.

Suppose Q and R are points on the graph of $P = 100(0.918)^t$ so that R is 8.1 units to the right of Q. Then, the second coordinate of R is one-half the second coordinate of Q. If Q and R move along the graph, keeping 8.1 units apart horizontally, the second coordinate of R is always one-half the second coordinate of Q.

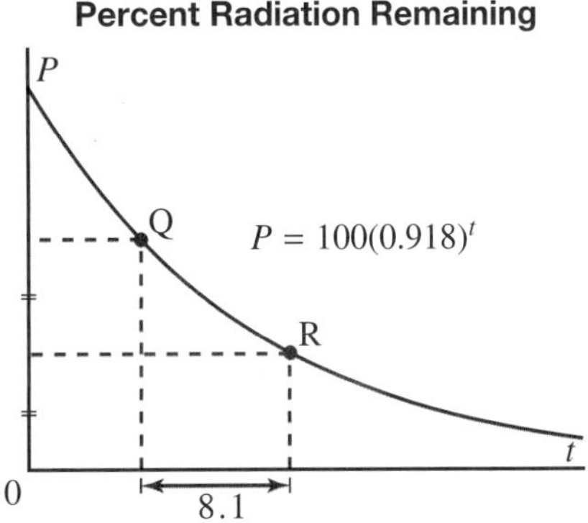

A corresponding property applies to any function $y = ca^x$ that models exponential decay $(0 < a < 1)$. When something is decaying exponentially, it becomes one-half as large within a fixed period of time. The half-life depends only on the base, *a,* in the equation, and not on the coefficient, *c*. You will prove this in exercise 13.

Just as we can use the laws of logarithms to solve exponential equations, we can also use them to solve logarithmic equations.

Example 4

Solve for *x,* then check the solution. $2 \log x = \log 3 + \log 12$

Solution

$$2 \log x = \log 3 + \log 12$$
$$\log x^2 = \log 36$$

Since the logarithms are equal,

$$x^2 = 36$$
$$x = \pm 6$$

Check.

When $x = 6$:

L.S. $= 2 \log x$
$= 2 \log 6$
$= \log 6^2$
$= \log 36$

R.S. $= \log 3 + \log 12$
$= \log 36$

Since L.S. = R.S., 6 is a root.

When $x = -6$:

L.S. $= 2 \log x$
$= 2 \log(-6)$

This is undefined.
Therefore, -6 is an extraneous root.

6.4 Exercises

1. Evaluate the logarithms in each list, to 3 decimal places. Explain the pattern in the results.

   a) log 5
   log 50
   log 500
   log 5000
   log 50 000

   b) log 0.5
   log 0.05
   log 0.005
   log 0.000 5
   log 0.000 05

2. Simplify the expressions in each list. Explain the results.

   a) log 1 + log 36
   log 2 + log 18
   log 3 + log 12
   log 4 + log 9
   log 6 + log 6

   b) log 2 − log 1
   log 4 − log 2
   log 8 − log 4
   log 16 − log 8
   log 32 − log 16

3. Create another pattern similar to each of those in exercise 2.

4. Simplify each expression.

   a) log 5 + log 2 b) log 50 + log 2 c) log 500 + log 2
   d) log 5000 + log 2 e) log 1000 + log 10 f) log 100 + log 100
   g) log 20 − log 2 h) log 200 − log 2 i) log 200 − log 20

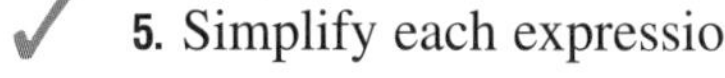

5. Simplify each expression.

   a) $\log_5 100 - \log_5 4$ b) $\log_2 96 - \log_2 6$ c) $\log_3 108 - \log_3 12$

B

6. **Knowledge/Understanding** Simplify each expression.

   a) $\log_6 9 + \log_6 4$ b) $\log_5 15 - \log_5 3$ c) $\log_4 2 + \log_4 32$
   d) $\log_2 48 - \log_2 6$ e) $\log_3 54 - \log_3 2$ f) $\log_3 9 + \log_3 9$

7. **Communication** Choose one expression in exercise 6. Write two other similar expressions that simplify to the same result. Explain why the expressions are equal.

8. Simplify.

   a) $\log_4 8 + \log_4 0.5$ b) $\log_4 8 - \log_4 0.5$

   c) $\log_6 9 + \log_6 8 - \log_6 2$ d) $\log_2 240 - \log_2 10 - \log_2 3$

9. Express as a single logarithm.

   a) $2\log 5 + \log 3$ b) $\frac{2}{5}\log 32 - \frac{1}{2}\log 4$

   c) $7\log 2 - \frac{1}{2}\log 16 + \log 3$ d) $\frac{1}{2}\log x - \frac{3}{2}\log y$

10. Solve each equation to 4 decimal places.

   a) $23(1.2)^n = 470$ b) $875(1.055)^n = 1250$

   c) $64(0.98)^n = 50$ d) $5000(0.75)^n = 2250$

11. Solve each equation, then check.

   a) $2\log x = \log 50 - \log 2$ b) $3\log x = \log 48 - \log 6$

   c) $\log x + \log 3 = \log 30$ d) $\log x - \log 5 = \log 2$

   e) $\log x + \log 8 = 2$ f) $\log x^2 - \log 2x = \log 2$

   g) $\log 4 + 9 = 2\log x + 7$ h) $\log_{12}(x-3) + \log_{12}(x+1) = 1$

12. **Application** The equation, $P = 100(0.87)^n$, models the percent of caffeine, P, left in your body as a function of time, n hours, after consumption.

   a) Determine how many hours it takes for the percent of caffeine to drop by 50%.

   b) Predict how many hours it will take for the percent of caffeine to drop by 50% again. Check your prediction by calculation.

   c) Determine the amount of caffeine present after 2 h.

   d) Predict how many hours it will take the amount you calculated in part c to drop by 50%. Check your prediction by calculation.

   e) Your results in parts a to d illustrate a certain property of exponential decay. What is this property?

13. **Thinking/Inquiry/Problem Solving** Consider the exponential function $f(x) = ca^x$, where $0 < a < 1$. Let $P(x_1, y_1)$ and $Q(x_2, y_2)$ be points on the graph of f.

   a) Suppose $y_2 = \frac{1}{2}y_1$. Show that $x_2 = x_1 + k$, where k depends only on a.

   b) Sketch the graph of k against a.

Exercise 13 is significant because it proves that adding a certain constant to the value of x causes the value of y to become one-half as large. If x represents time, then the time to become one-half as large is constant for exponential decay.

14. The table shows some substances that are present in radioactive waste from the nuclear power industry. Choose one substance.

Substance	Half-Life
iodine-131	8 days
strontium-90	28 days
tritium	12 years
nickel-59	76 000 years
iodine-129	16 000 000 years

a) Make a table of values then draw a graph to show the percent, P, that remains during the first 5 half-lives.

b) What percent of the substance remains after the first 5 half-lives?

c) The *hazardous life* of a radioactive substance is from 10 to 20 half-lives. What percent of the original radiation will be present at the end of the hazardous life?

15. Choose a different substance from the list in exercise 14.

a) Pose a problem about the decay of that substance.

b) Solve the problem you posed. Write a clear explanation of your problem and its solution.

✓ 16. Some materials used in hospitals are radioactive, generally with a half-life of 8 months or less. For example, the half-life of sodium-24 is 14.9 h. Suppose a hospital buys a 40-mg sample of sodium-24.

a) Make a table of values then draw a graph to show the mass, m milligrams, that remains during the first 5 half-lives.

b) How much of the sample will remain after 48 h?

c) How long will it be until only 1 mg remains?

✓ 17. The temperature of hot chocolate is recorded as it cools down. Analysis of the data shows that the temperature of the hot chocolate, T degrees Celsius, can be expressed as a function of time, t minutes, by the equation $T = 79(0.85)^t + 20$.

a) Graph the equation using appropriate values of t.

b) Determine the temperature of the hot chocolate after 5 min.

c) Hot chocolate is safe to drink at 40°C. How long did it take the hot chocolate to cool down to 40°C?

18. Hot objects cool down and cold objects warm up to the temperature of their surroundings. Newton's Law of Cooling states that the difference between the temperature of a hot substance and the temperature of the surroundings is an exponential function of the time the substance has been cooling.

a) What do the three numbers 20, 79, and 0.85 in the equation in exercise 17 represent?

b) Explain how the equation $T = 79(0.85)^t + 20$ illustrates Newton's Law of Cooling.

c) Use the transformation tools from *Necessary Skills*. Explain how the graph of $T = 79(0.85)^t + 20$ is related to the graph of $y = 0.85^x$.

19. Prove the law of logarithms for division:
$\log_a\left(\frac{x}{y}\right) = \log_a x - \log_a y \quad (x > 0,\ y > 0,\ a > 0,\ a \neq 1)$

20. According to data from the Canadian Bankers Association, the number of credit cards in circulation in Canada, C million, has been growing by about 7% per year since 1980. There were approximately 10.6 million credit cards in circulation that year.

a) Write an equation that expresses C as a function of t, the number of years since 1980.

b) When did the number of credit cards reach 30 million?

✓ **21.** According to data from the Canadian Bankers Association, the total credit card sales in Canada, S billion dollars, has been growing by about 13.6% per year since 1980. Credit card sales that year were approximately $10.6 billion.

a) Pose a problem about the growth of credit card sales in Canada.

b) Solve the problem you posed. Write a clear explanation of your problem and its solution.

22. Use the equations of the functions in exercises 20 and 21. Write an equation that expresses the average sales per credit card, A, as a function of the number of years, n, since 1980.

23. Recall that a prime number, such as 29, has no integral factors other than 1 and the number itself. The largest known prime number before the age of computers was calculated by hand in 1951. The number is $\frac{2^{148}+1}{17}$.

a) How many digits are there in this number?

b) What is the first digit of this number?

24. The total area of arable land in the world is about 3.2×10^9 ha. Approximately 0.4 ha of land is required to grow food for one person.

a) Assume a 2001 world population of 6.13 billion and a constant annual growth rate of 1.3%. Determine when the demand for arable land will exceed the supply.

b) Compare the effect of each scenario on the result of part a.

i) reducing the growth rate by one-half to 0.65%

ii) doubling the productivity of the land

iii) reducing the growth rate by one-half and doubling the productivity

Self-Check 6.1 – 6.4

1. Sketch graphs of each pair of functions on the same grid.

   a) $y = 2^x$ and $y = 3(2)^x + 1$ b) $y = 5^x$ and $y = -5^x - 2$

2. Write in exponential form.

   a) $\log_3 27 = 3$ b) $\log_3\left(\frac{1}{9}\right) = -2$ c) $\log_2 0.5 = -1$

3. Write in logarithmic form.

   a) $6^4 = 1296$ b) $6^{-4} = \frac{1}{1296}$ c) $10^3 = 1000$

   d) $10^{-3} = 0.001$ e) $27^{\frac{2}{3}} = 9$ f) $25^{0.5} = 5$

4. Evaluate each logarithm.

   a) $\log 100$ b) $\log\left(\frac{1}{100}\right)$

   c) $\log_3 81$ d) $\log_3\left(\frac{1}{27}\right)$

5. Solve each equation to 4 decimal places. Check the solution.

   a) $2^x = 7$ b) $3^x = 30$ c) $8^x = 6$

6. Express:

   a) 33 as a power of 2 b) 54 as a power of 7

7. Simplify without using a calculator.

   a) $\log 50 + \log 20$ b) $\log_5 1000 - \log_5 8$ c) $\frac{1}{3}\log 27 - \frac{1}{4}\log 16$

8. Solve for x. Give the answer to 4 decimal places where necessary.

   a) $350 = 200(1.05)^x$ b) $\log 3 + \log x = \log 12$ c) $\log_6 x - \log_6 (x - 1) = 1$

PERFORMANCE ASSESSMENT

9. The population of British Columbia, P million, can be modelled by the equation $P = 2.76(1.022)^t$, where t is the time in years since 1981.

   a) Predict the population in 2010.

   b) Predict when the population might reach 5 million.

   c) What assumptions are you making in parts a and b?

   d) Give some reasons why your predictions in parts a and b could be wrong.

   e) Describe how the graph of $P = 2.76(1.022)^t$ is related to the graph of $y = 1.022^x$.

   f) Pose a problem about the population in British Columbia. Solve the problem you posed.

6.5 Expressing Exponential Functions with the Same Base

In this chapter, we have encountered many examples of exponential functions that arise in real situations. The equations of these functions usually have the form $y = ca^x$, where the base a is different in each situation. In the exercises of Sections 6.3 and 6.4, you encountered problems involving the concepts of doubling time and half-life.

Stating the doubling time (or half-life) is another way to describe the rate of change of something that is growing (or decaying) exponentially. We can use the concepts of doubling time and half-life to write exponential equations with base 2.

For example, in Section 6.1, we used the equation $A = 500(1.065)^n$ to represent the amount, A, after n years when \$500 are invested at 6.5% compounded annually. The base, 1.065, shows that the amount is increasing by 6.5% per year. In an earlier exercise (Section 6.3, exercise 9), we found that the doubling time is approximately 11 years. We can use this doubling time to express the equation of the function in a different form.

The amount after:

11 years is	500×2^1	Each exponent is the number of years divided by 11.
22 years is	500×2^2	
33 years is	500×2^3	
⋮	⋮	
n years is	$500 \times 2^{\frac{n}{11}}$	

Hence, here is another equation that represents the amount after n years:

$$A = 500(2)^{\frac{n}{11}} \leftarrow \text{Doubling time}$$

↑ Initial amount

This equation is equivalent to the equation $A = 500(1.065)^n$. The only difference is that the equation is expressed with base 2 instead of base 1.065. We can explain why the equations are equivalent in two ways. One way is to show that the same numbers satisfy each equation (see exercise 1). A better way is to show that each equation can be derived from the other. Suppose we start with the equation $A = 500(1.065)^n$. To express this equation with base 2, we need to express the base, 1.065, as a power of 2.

Let $1.065 = 2^x$.

Solve for x. Take the base-10 logarithm of each side.

$$\log 1.065 = \log 2^x$$

$$\log 1.065 = x \log 2$$

$$x = \frac{\log 1.065}{\log 2}$$

$$\doteq 0.090\ 853\ 430\ 5$$

Therefore, $1.065 \doteq 2^{0.090\ 85}$

This means that the equation $A = 500(1.065)^n$ can be approximated as $A = 500(2)^{0.090\ 85n}$. Since the reciprocal of 0.090 85 (that is, $\frac{1}{0.090\ 85}$) is approximately 11, this is the same as $A = 500(2)^{\frac{n}{11}}$.

THE BIG PICTURE

Relating Exponential Functions

Consider the graph of $y = 2^x$. The graph of any other exponential function is a transformation of this graph. Some examples follow.

Suppose the graph of $y = 2^x$ is expanded horizontally by a factor of 11. Its image is the graph of $y = 2^{\frac{x}{11}}$. Since $2^{\frac{1}{11}} \doteq 1.065$, this is the graph of $y = 1.065^x$.

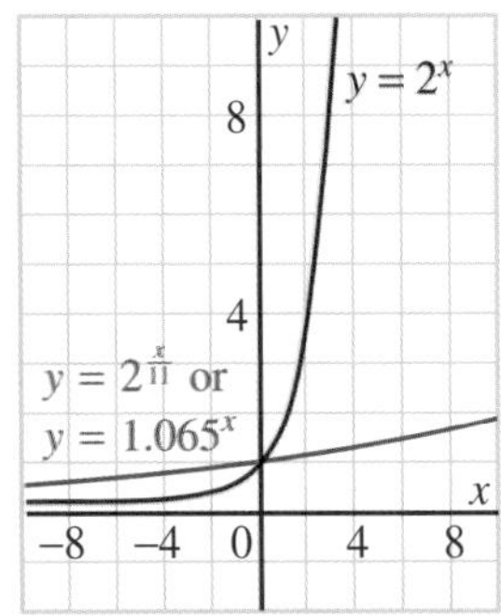

If the graph of $y = 1.065^x$ is expanded vertically by a factor of 500, the result is the graph of $y = 500(1.065)^x$. If we replace y with A and x with n, we get the graph of the amount of \$500 invested for n years at 6.5% (page 373).

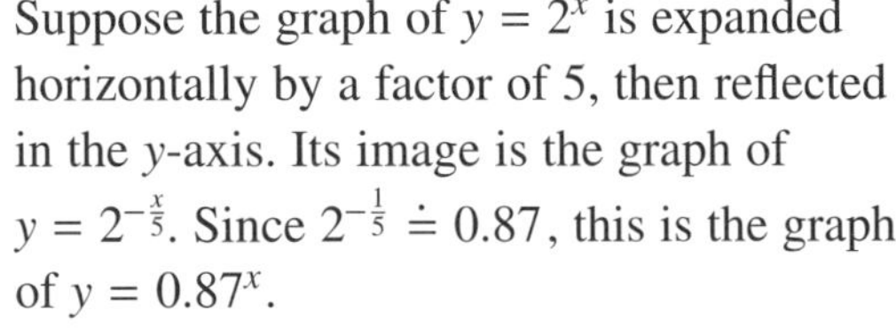

Suppose the graph of $y = 2^x$ is expanded horizontally by a factor of 5, then reflected in the y-axis. Its image is the graph of $y = 2^{-\frac{x}{5}}$. Since $2^{-\frac{1}{5}} \doteq 0.87$, this is the graph of $y = 0.87^x$.

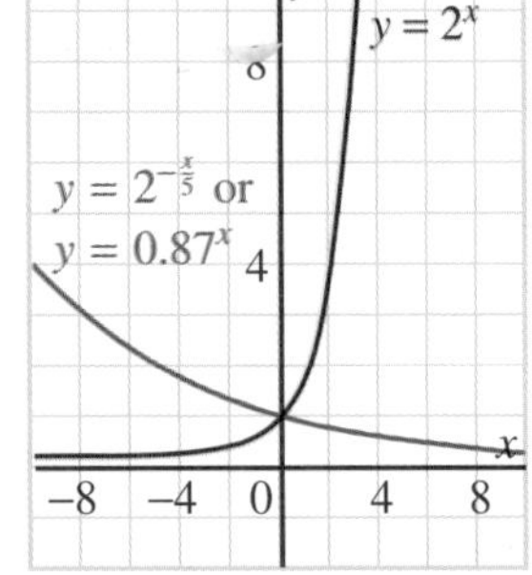

If the graph of $y = 0.87^x$ is expanded vertically by a factor of 100, the result is the graph of $y = 100(0.87)^x$. If we replace y with P and x with n, we get the graph of the percent of caffeine left in the body n hours after consumption (page 373).

We can use the method at the top of page 402 to express any exponential equation, $y = ca^x$, as an equivalent exponential equation with base 2. The new equation has the form $y = c(2)^{kx}$, where the coefficient k in the exponent compensates for the change of base.

Example 1

A cell phone battery loses 2% of its charge every day. An equation that models the percent charge, C, that remains after t days is $C = 100\,(0.98)^t$.

a) Write this equation using base 2.

b) Use the result of part a to determine the half-life of the battery.

Solution

a) Express 0.98 as a power of 2.

Let $0.98 = 2^x$.

Solve for x. Take the base-10 logarithm of each side.

$$\log 0.98 = \log 2^x$$

$$\log 0.98 = x \log 2$$

$$x = \frac{\log 0.98}{\log 2}$$

$$\doteq -0.029\ 146\ 346$$

Therefore, $0.98 \doteq 2^{-0.029\ 146\ 346}$

An equation that approximately models the percent charge remaining is $C = 100\,(2)^{-0.029t}$.

b) To determine the half-life, let $2^{-0.029\ 146\ 346t} = \frac{1}{2}$.

Then, $2^{-0.029\ 146\ 346t} = 2^{-1}$

Since the powers are equal and they have the same base, the exponents are equal.

$$-0.029\ 146\ 346t = -1$$

$$t = \frac{1}{0.029\ 146\ 346}$$

$$\doteq 34.309\ 618\ 49$$

The half-life is about 34 days. It takes approximately 34 days for the battery to lose one-half of its charge.

Discuss

How could we check that the half-life is correct?

Example 1 shows that when the equation representing an exponential decay situation is expressed using base 2, the exponent is negative. The equation can be written with a positive exponent using a base of $\frac{1}{2}$. That is, the equation $C = 100\,(2)^{-0.029t}$ is the same as $C = 100\left(\frac{1}{2}\right)^{0.029t}$.

Example 2

The astounding growth of computer technology has been described in different ways. This statement is generally regarded as being true:

"Microchips have doubled in performance power about every 18 months during the last 30 years."

In 1972, a typical computer chip could execute 60 000 instructions per second.

a) Write an equation to express the number of instructions per second, I, a chip can execute as a function of the number of years, n, since 1972.

b) Use the equation to estimate the number of instructions per second a chip could execute in 2001.

Solution

a) The doubling time is 18 months, or 1.5 years. Hence,
$I = 60\ 000(2)^{\frac{n}{1.5}}$

b) The year 2001 is 29 years after 1972.
Substitute $n = 29$ in the equation in part a.

$$I = 60\ 000(2)^{\frac{29}{1.5}}$$
$$\doteq 3.96 \times 10^{10}$$

According to the model, a chip could execute approximately 40 billion instructions per second in 2001.

We can express $y = 2^x$ as an equivalent exponential function with any base.

Example 3

Consider the function $y = 2^x$. Write an equivalent exponential function with base 3.

Solution

Express 2 as a power of 3.
Let $2 = 3^k$.
Solve for k. Take the base-10 logarithm of each side.

$$\log 2 = \log 3^k$$
$$\log 2 = k \log 3$$
$$k = \frac{\log 2}{\log 3}$$
$$\doteq 0.630\ 93$$

Therefore, $2 \doteq 3^{0.630\ 93}$
Substitute this expression for 2 in the equation $y = 2^x$.
An equivalent exponential function has the approximate equation
$y = 3^{0.630\ 93x}$.

6.5 Exercises

1. On page 401, we stated that the equations $A = 500(2)^{\frac{n}{11}}$ and $A = 500(1.065)^n$ are equivalent.

   a) Show that this is true by substituting three different values of n into each equation, then calculating A.

   b) Why is deriving one equation from the other a better way to show that the two equations are equivalent?

2. **Knowledge/Understanding** Express each function as an equivalent exponential function with base 3.

   a) $y = 5^x$ b) $y = 10^x$ c) $y = 1.5^x$

3. Express the function $y = 10^x$ as an equivalent exponential function with each base.

   a) 2   b) 5   c) 20

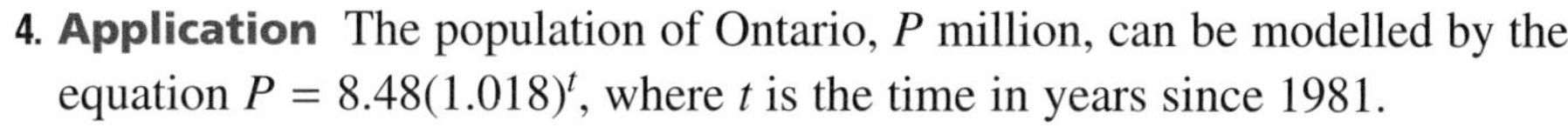

4. **Application** The population of Ontario, P million, can be modelled by the equation $P = 8.48(1.018)^t$, where t is the time in years since 1981.

   a) Write this equation using base 2.

   b) Use the result of part a to determine the number of years for the population to double.

5. The following growth situations occurred in earlier sections of this chapter. Write each equation using base 2.

   a) Hosts on the Internet, $h = 7.4(1.59)^t$

   b) Population of Canada, $P = 24.0(1.014)^t$

   c) Canadian mutual funds, $M = 460(1.19)^n$

6. The following decay situations occurred in earlier sections of this chapter. Write each equation using base 2.

a) Caffeine in the bloodstream, $P = 100(0.87)^n$

b) Chernobyl radiation, $P = 100(0.918)^t$

c) Temperature of hot chocolate, $T = 79(0.85)^t + 20$

7. **Communication** Choose one situation in exercise 5 or 6. Explain how the doubling time or the half-life can be obtained from the equation you determined. In your explanation, include an analysis of the graphs of the two equations.

8. In 1970, it was possible to put 1000 components on a computer chip. Since then, the number of components that can be put on a chip has doubled every year and a half.

a) Write an equation to express the number of components, *c*, that can be put on a chip as a function of the number of years, *n*, since 1970.

b) Write the equation in part a using base 10.

9. **Thinking/Inquiry/Problem Solving** Most laptop computers have a power meter program that can be used to monitor the battery power remaining. Data collected from one computer showed that the percent power remaining, *p*, was modelled as a function of time, *n* hours, by the equation $p = 100(0.98)^n$. When the battery power was reduced to 50%, the computer was connected to a power source and the battery power increased. Data collected using the power meter program showed that the percent battery power was modelled as a function of time by the equation $p = 100 - 50(0.98)^n$.

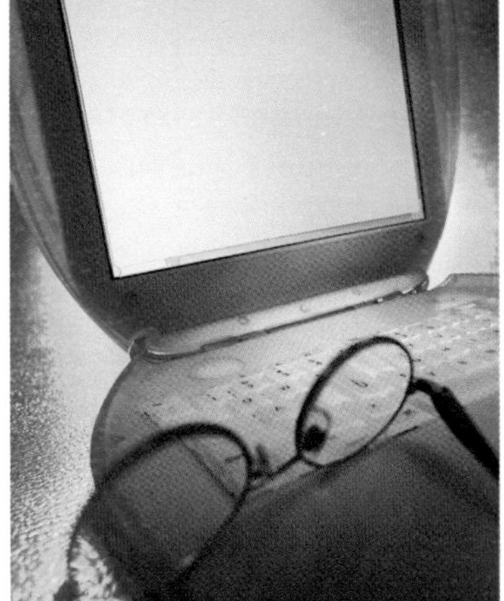

a) Sketch the graphs of these two functions on the same grid.

b) Describe how the graph of each function can be obtained by transforming the graph of $y = 0.98^x$.

c) What is a disadvantage of using $p = 100 - 50(0.98)^n$ to model the percent power remaining when the battery is being charged?

C

10. a) Show that $\log_2 9 \times \log_3 8 = \log_2 8 \times \log_3 9$.

b) The equation in part a is a special case of the following relationship. Prove the relationship.

$\log_a x \times \log_b y = \log_a y \times \log_b x$

6.6 Logarithmic Scales

In 1999, an earthquake in Turkey killed thousands of people and left hundreds of thousands homeless. This earthquake had magnitude 7.5 on the Richter scale.

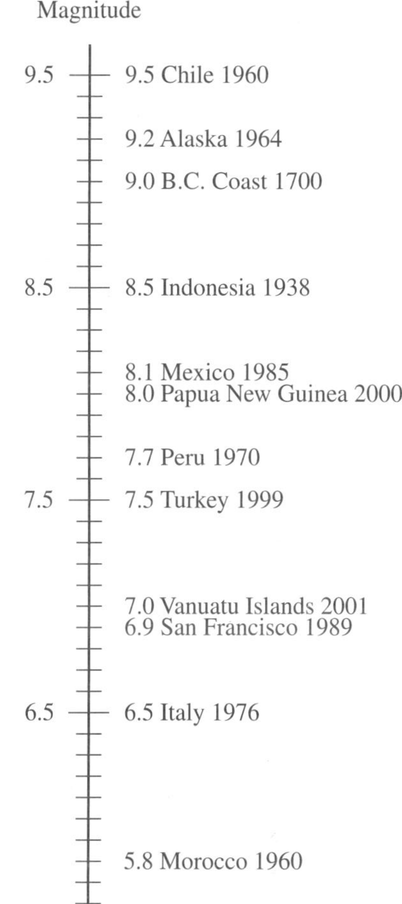

The size, or intensity, of an earthquake is determined from the amount of ground motion detected by seismographs. We use the Richter scale to compare the intensities of two earthquakes according to this principle: each increase of 1 unit in magnitude on the Richter scale represents a 10-fold increase in intensity.

On the scale at the right, there are earthquakes with magnitudes 5.5, 6.5, 7.5, 8.5, and 9.5. Since these numbers have a common difference of 1, each earthquake was 10 times as intense as the one below it. The Turkey earthquake was 10 times as intense as the 1976 Italy earthquake and 100 times as intense as the 1983 Colombia earthquake. The Turkey earthquake was $\frac{1}{10}$ as intense as the 1938 Indonesia earthquake and $\frac{1}{100}$ as intense as the 1960 Chile earthquake. The Indonesia and Chile earthquakes were, respectively, 1000 times and 10 000 times as intense as the Colombia earthquake.

Since the relationship between intensity and magnitude involves repeated multiplication, intensity is an exponential function of magnitude. But earthquake intensities are awkward to work with, and a graph of intensity against magnitude is difficult to draw. Such a graph would be similar to the graph of $y = 10^x$ shown. However, to represent the range of intensities for earthquakes with magnitudes from 5.5 to 9.5 shown above, the vertical axis would have to be 100 times as long as it is here. We avoid these difficulties by *comparing* the magnitudes and intensities of two earthquakes instead of *calculating* their intensities.

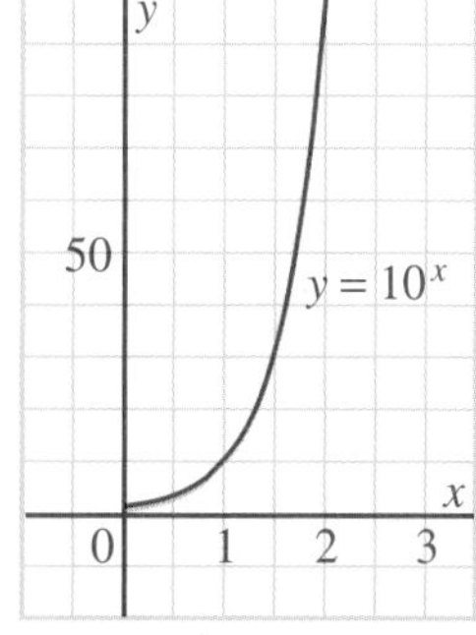

Take Note

Comparing Earthquake Intensities

Suppose I_0 represents the intensity of any earthquake with magnitude M_0. Then the intensity, I, of any other earthquake with magnitude M, is given by this formula:

$$I = I_0(10)^{M-M_0}$$

For example, suppose the 1999 Turkey earthquake had intensity I_0.
Then the intensity of the 1960 Chile earthquake was:

$$\begin{aligned} I &= I_0(10)^{9.5-7.5} \\ &= I_0(10)^2 \\ &= 100I_0 \end{aligned}$$

The 1960 Chile earthquake was 100 times as intense as the 1999 Turkey earthquake.

Example 1

How many times as intense as the 1989 San Francisco earthquake was the 1964 Alaska earthquake?

Solution

The 1989 San Francisco earthquake had magnitude 6.9.
The 1964 Alaska earthquake had magnitude 9.2.
Use $I = I_0(10)^{M-M_0}$
Let I_0 represent the intensity of the 1989 San Francisco earthquake.
Then the intensity, I, of the 1964 Alaska earthquake was:

$$\begin{aligned} I &= I_0(10)^{9.2-6.9} \\ &= I_0(10)^{2.3} \\ &\doteq 200I_0 \end{aligned}$$

The 1964 Alaska earthquake was approximately 200 times as intense as the 1989 San Francisco earthquake.

Example 2

Calculate the magnitude of an earthquake that is twice as intense as the 2000 Papua New Guinea earthquake.

Solution

The magnitude of the 2000 Papua New Guinea earthquake was 8.0.
Use the equation $I = I_0(10)^{M-8.0}$, where I_0 represents the intensity of this earthquake. For an earthquake with twice the intensity, substitute $I = 2I_0$.

$$2I_0 = I_0(10)^{M-8.0}$$
$$10^{M-8.0} = 2$$

Take the base-10 logarithm of each side.

$$\log 10^{M-8.0} = \log 2$$
$$M - 8.0 = \log 2$$
$$M = 8.0 + \log 2$$
$$\doteq 8.301$$

An earthquake that is twice as intense as the 2000 Papua New Guinea earthquake would have magnitude approximately 8.3.

The Richter scale is an example of a *logarithmic scale.* On a logarithmic scale, each increase of 1 unit on the scale corresponds to multiplication by a constant, which is usually 10. Other examples of logarithmic scales occur in exercises 10–19.

6.6 Exercises

1. How many times as intense is an earthquake with magnitude 8 than one with each magnitude?

   a) 7 b) 6 c) 5 d) 4 e) 3 f) 2

2. **Knowledge/Understanding** Use the Richter scale on page 407. Two earthquakes that hit South America were the 1960 Chile earthquake and the 1970 Peru earthquake. How many times as intense as the Peru earthquake was the Chile earthquake?

3. Explain why the term "logarithmic scale" is appropriate for a scale in which each increase of 1 unit corresponds to multiplication by a constant.

4. Most of Canada's earthquakes occur along the west coast. Three British Colombia earthquakes are listed below.

Date	Place	Magnitude
June 23, 1946	Near Campbell River	7.3
Aug 22, 1949	Queen Charlotte Islands	8.1
June 24, 1997	Southwestern B.C.	4.6

a) i) How many times as intense as the 1997 earthquake was the 1946 earthquake?

ii) How many times as intense as the 1997 earthquake was the 1949 earthquake?

iii) How many times as intense as the 1946 earthquake was the 1949 earthquake?

b) Why do earthquakes occur along Canada's west coast?

5. **Communication** Choose one part of exercise 4a. Explain how you compared the intensities of the two earthquakes.

6. From the Richter scale on page 407, find examples of two earthquakes in each case.

a) One is 2 times as intense as the other.

b) One is 4 times as intense as the other.

c) One is 8 times as intense as the other.

d) One is 16 times as intense as the other.

7. Use the results of *Examples 1* and *2*. How do the magnitudes of two earthquakes compare in each situation? Explain.

a) One earthquake is 200 times as intense as another earthquake.

b) One earthquake is 2 times as intense as another earthquake.

c) One earthquake is 20 times as intense as another earthquake.

8. For each decrease of 1 unit in magnitude, earthquakes are about 6 or 7 times as frequent. In a given year, how should the number of earthquakes with magnitudes between 4.0 and 4.9 compare with the number of earthquakes with magnitudes in each range?

a) 5.0 to 5.9 b) 6.0 to 6.9 c) 7.0 to 7.9

9. The energy released by an earthquake is a measure of its destructive power. For each increase of 1 unit in magnitude of an earthquake, there is a 31-fold increase in the amount of energy released.

a) Write an equation similar to that in *Take Note,* page 408, that you can use to compare the energy released in two earthquakes.

b) How many times as much energy was released by the 1999 Turkey earthquake as by:

i) the 1976 Italy earthquake?

ii) the 1983 Colombia earthquake?

c) How many times as much energy was released by the 1700 earthquake off the British Columbia coast as by the 1983 Colombia earthquake?

Exercises 9 and 16 are significant because they involve logarithmic scales in which the base is not 10. In exercise 9, the base is 31. In exercise 16, it is 0.5.

Loudness of Sounds

✓ **10. Application** The loudness of a sound is measured in *decibels* (dB). Each increase of 10 dB represents a 10-fold increase in loudness. For example, the increase from the sound of normal conversation to a heavy truck is 30 dB. This is 3 increases of 10 dB, so the increase in loudness is 10^3, or 1000. Hence, a heavy truck is 1000 times as loud as normal conversation. How many times as loud as:

a) normal conversation is a car horn?

b) a quiet whisper is normal conversation?

c) the rustle of a leaf is the threshold of pain?

The Decibel Scale

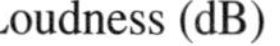

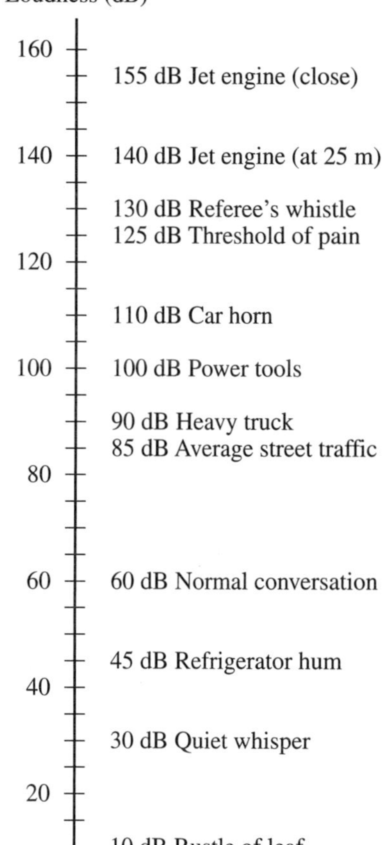

11. How many times as loud as:

a) a heavy truck is a jet engine up close?

b) a refrigerator hum is average street traffic?

12. People often wear earplugs or earmuffs when operating noisy machines or equipment. These can reduce the sound level by up to 25 dB. How many times as loud would the sound be without the earplugs or earmuffs?

13. Research was conducted to test the noise levels in school gymnasiums. Noise levels ranged from 75 dB to 115 dB. How many times as loud as a noise at 75 dB is a noise at 115 dB?

14. Let L_0 represent the loudness of a sound at 0 dB, the threshold of hearing. Write an equation to express the loudness, L, of a sound with decibel level d.

15. The loudness of a heavy snore is 69 dB.

a) How many times as loud as normal conversation is a heavy snore?

b) How many times as loud as a quiet whisper is a heavy snore?

16. **Thinking/Inquiry/Problem Solving** When a person listens to very loud sounds for prolonged periods of time, the hearing can be impaired. An 8-h exposure to a 90-dB sound is considered acceptable. For every 5-dB increase in loudness, the acceptable exposure is reduced by one-half.

   a) What is the acceptable exposure time for each noise?

      i) a rock group playing at 110 dB

      ii) a referee's whistle at 130 dB

   b) Write an equation to express the acceptable exposure time, E hours, as a function of the sound level, s decibels.

   c) Graph the equation in part b.

Acidity and Alkalinity of Solutions

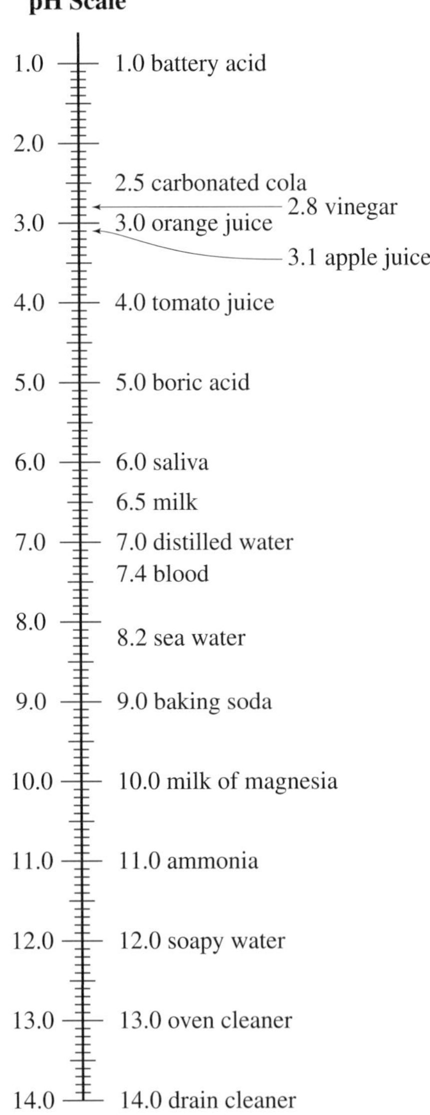

Values are approximate.

17. In chemistry, the pH scale measures the acidity or alkalinity of a solution. Most pH values range from 0 to 14, with 7 being neutral. For pH less than 7, each 1-unit decrease in pH represents a 10-fold increase in acidity. For pH greater than 7, each 1-unit increase in pH represents a 10-fold increase in alkalinity.

   a) How many times as acidic as distilled water is vinegar?

   b) How many times as acidic as milk is apple juice?

   c) How many times as alkaline as baking soda is oven cleaner?

   d) How many times as alkaline as distilled water is sea water?

18. Normal rainwater is slightly acidic (around pH 5.5) because it contains acids formed by the carbon dioxide it absorbs from the atmosphere. Determine the pH of something that is:

   a) twice as acidic as normal rainwater

   b) one-half as acidic as normal rainwater

19. According to Environment Canada's website, normal rainwater is less acidic than tomato juice, but more acidic than milk. And, in some areas of Canada, rainwater can be as acidic as vinegar.

   a) How many times as acidic as milk is normal rainwater?

   b) How many times as acidic as normal rainwater is:

      i) tomato juice? **ii)** vinegar?

6.7 The Logarithmic Function $y = \log_a x$

Many examples of exponential equations were given in earlier sections of this chapter. We can use logarithms to solve any of these equations for the variable in the exponent. Solving an equation is one of the steps to determine the equation of the inverse of a function. For example, consider again the function $f(x) = 2^x$. To determine the equation of the inverse, let $y = 2^x$ then follow these steps:

Step 1. Take the logarithm to base 2 of each side. $\quad \log_2 y = \log_2 2^x$
Step 2. Solve for x. $\quad x = \log_2 y$
Step 3. Interchange x and y. $\quad y = \log_2 x$
The equation of the inverse of $y = 2^x$ is $y = \log_2 x$.

We can graph the inverse of any function by reflecting its graph in the line $y = x$. This is equivalent to interchanging the coordinates of the points on the graph. For example, the graph below shows the function $f(x) = 2^x$. A table of values for $f(x) = 2^x$ is shown. We reflect each point on the graph of $f(x) = 2^x$ in the line $y = x$. We join these points to get the graph of the inverse function $f^{-1}(x) = \log_2 x$.

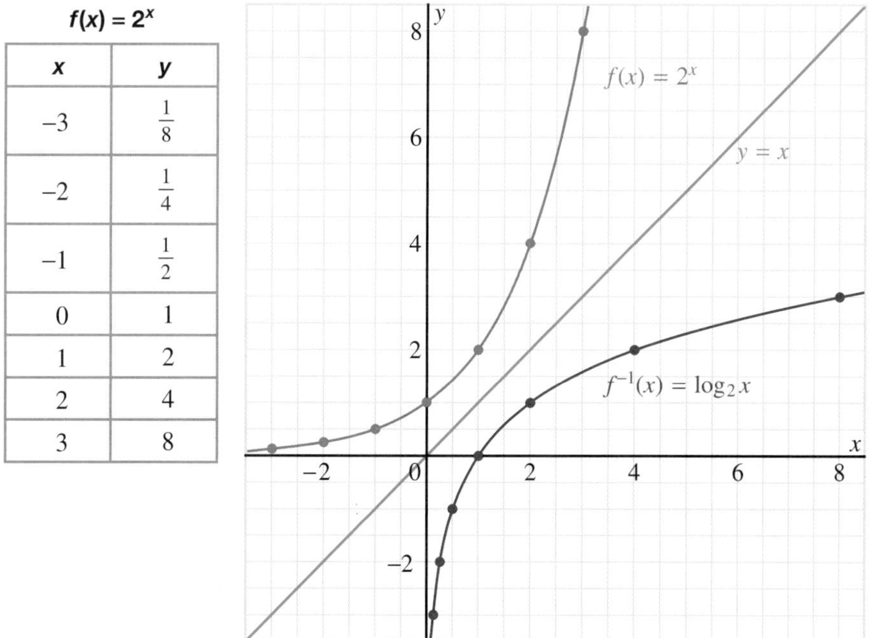

$f(x) = 2^x$

x	y
−3	$\frac{1}{8}$
−2	$\frac{1}{4}$
−1	$\frac{1}{2}$
0	1
1	2
2	4
3	8

We define the function $y = \log_a x$ to be the inverse of the function $y = a^x$, where $a > 0$, $a \neq 1$. The graphs of $y = \log_a x$ for $a = 10$, 2, 1.25, 0.8, 0.5, and 0.1 are shown on the next page. They are reflections in the line $y = x$ of the first 3 and last 3 graphs on page 371. The graphs on page 414 show how the graph of $y = \log_a x$ changes as a varies, and they illustrate the following properties:

Properties of the Graph of $f^{-1}(x) = \log_a x$, $a > 0$, $a \neq 1$

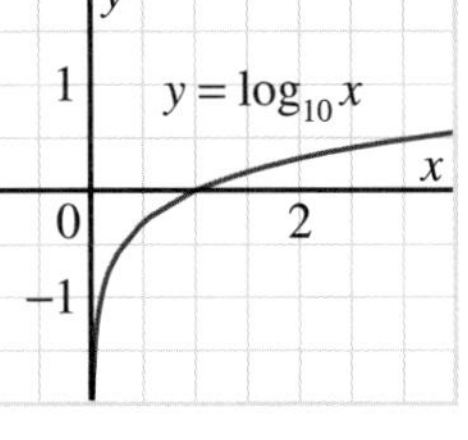

Intercepts

$f^{-1}(x) = \log_a x$
To determine the vertical intercept, substitute $x = 0$.
$f^{-1}(0) = \log_a 0$
To determine the value of $\log_a 0$, let $\log_a 0 = y$, then $a^y = 0$.
There is no value of y such that $a^y = 0$.
Hence, $\log_a 0$ is undefined.
There is no vertical intercept.

To determine horizontal intercepts, solve $f^{-1}(x) = 0$.
Then, $\log_a x = 0$, or $x = a^0$
Since $a^0 = 1$, $x = 1$
The horizontal intercept is 1.

Domain and range

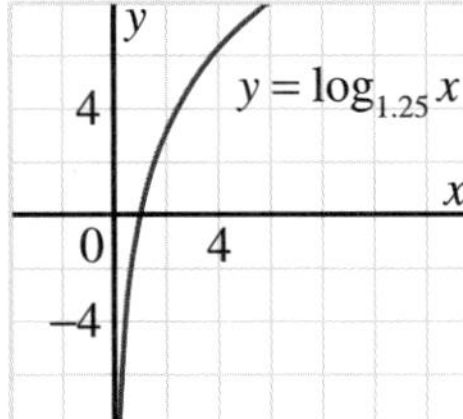

Since we can define $\log_a x$ for all positive values of x, the domain is the set of positive real numbers; $D = \{x \mid x \in \Re, x > 0\}$.
The range is the set of all real numbers; $R = \Re$.

Critical points

From the graphs, there do not appear to be any critical points. We will prove this result in Chapter 7.

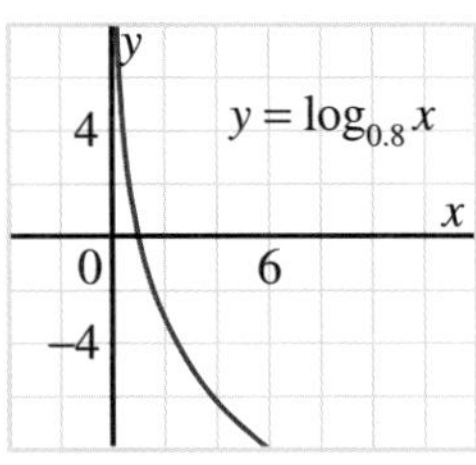

Intervals of increase and decrease

From the first 3 graphs, when $a > 1$, the function increases for $x > 0$.
From the last 3 graphs, when $0 < a < 1$, the function decreases for $x > 0$.

Points of inflection

From the graphs, there do not appear to be any points of inflection. We will prove this result in Chapter 7.

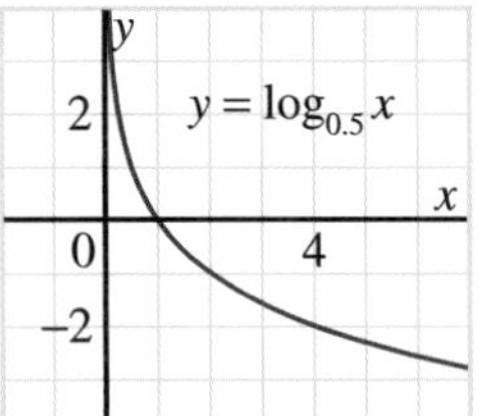

Concavity

For $a > 1$, the graph is concave down. For $0 < a < 1$, the graph is concave up. We will use the derivative to prove these results in Chapter 7.

Asymptote

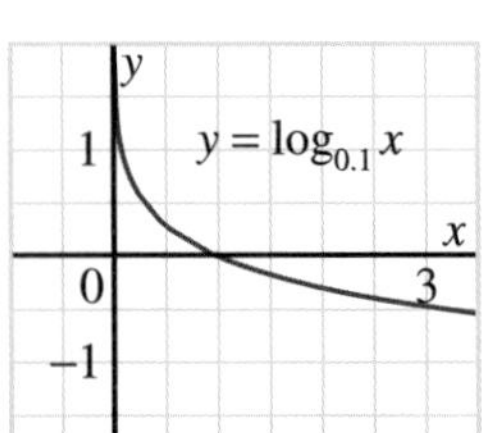

As x decreases and approaches 0, the points on the graph come closer and closer to the y-axis, but never reach it. The y-axis is an asymptote.

Point discontinuities

There are no point discontinuities.

Law of logarithms

The graph of $y = \log_2 x$ is shown.

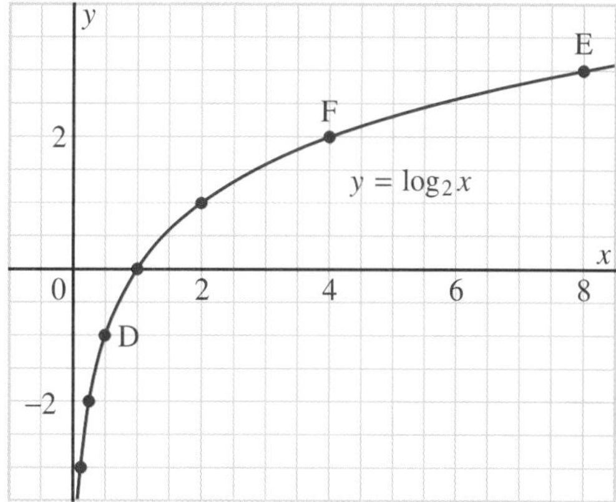

Suppose we select any two points on the graph, such as $D(\frac{1}{2}, -1)$ and E(8, 3).

Multiply their x-coordinates.

$$\frac{1}{2} \times 8 = 4$$

Add their y-coordinates.

$$-1 + 3 = 2$$

The results are the coordinates of another point F(4, 2) on the graph.

Multiplying the x-coordinates and adding the y-coordinates of two points on the graph produces the coordinates of another point on the graph. This property is a consequence of the law of logarithms for multiplication; that is, the sum of two logarithms with the same base is the logarithm of their product.

Example 1

a) Sketch a graph of the exponential function $f(x) = \left(\frac{1}{3}\right)^x$ and its inverse on the same grid.

b) Write the equation of the inverse function.

Solution

a) When x is positive, say $x = 5$,

$f(x) = \left(\frac{1}{3}\right)^5 \doteq 0.004$, which is very small and positive.

When x is negative, say $x = -5$,

$f(x) = \left(\frac{1}{3}\right)^{-5} \doteq 243$, which is very large.

Also, $f(0) = \left(\frac{1}{3}\right)^0 = 1$

Sketch the graph using this information.

Reflect the graph of $f(x) = \left(\frac{1}{3}\right)^x$ in the line $y = x$. The image is the graph of the inverse function, $f^{-1}(x)$.

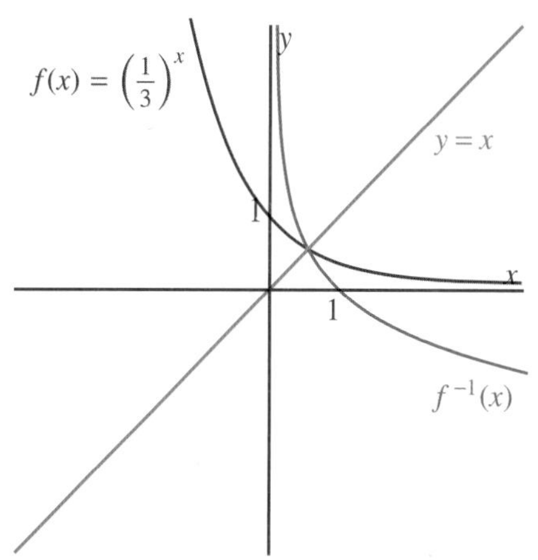

b) Let $y = \left(\frac{1}{3}\right)^x$. Take the base-$\frac{1}{3}$ logarithm of each side.

$\log_{\frac{1}{3}} y = \log_{\frac{1}{3}} \left(\frac{1}{3}\right)^x$

$\log_{\frac{1}{3}} y = x$

Interchange x and y.

$y = \log_{\frac{1}{3}} x$

The equation of the inverse function is $f^{-1}(x) = \log_{\frac{1}{3}} x$.

Logarithmic functions can arise in practical situations, but they are not as common as exponential functions.

Example 2

To determine the number of plant species in a lawn, researchers recorded the numbers of species in different areas of square patches of the lawn. The results showed that the number of plant species, s, contained in a patch of lawn with area A square metres, was modelled by the equation $s = 10.7 \log A + 8.1$, where $A > 1$.

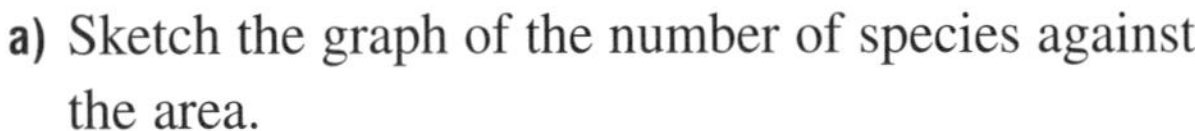

a) Sketch the graph of the number of species against the area.

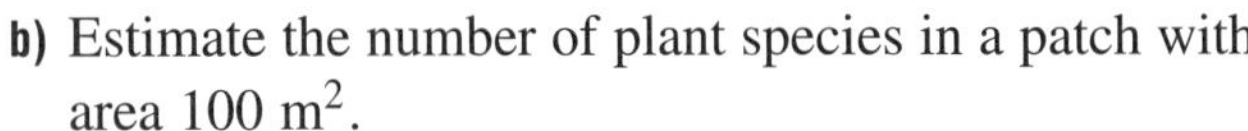

b) Estimate the number of plant species in a patch with area 100 m^2.

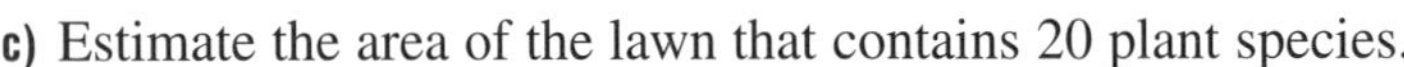

c) Estimate the area of the lawn that contains 20 plant species.

Solution

a) $s = 10.7 \log A + 8.1$

When A is small, say $A = 1.1$,

$s = 10.7 \log 1.1 + 8.1$

$\doteq 8.5$

When A is large, say $A = 200$,

$s = 10.7 \log 200 + 8.1$

$\doteq 33$

Plot these points on a sketch. Join them with a smooth curve.

b) $s = 10.7 \log A + 8.1$

Substitute $A = 100$ in the equation.

$s = 10.7 \log 100 + 8.1$

$= 10.7(2) + 8.1$

$= 29.5$

There are approximately 30 plant species in a lawn with area 100 m^2.

c) $s = 10.7\log A + 8.1$

Substitute $s = 20$ in the equation.

$$20 = 10.7\log A + 8.1$$

$$\log A = \frac{20 - 8.1}{10.7}$$

$$\log A \doteq 1.112\ 149\ 533$$

Use the definition of a logarithm.

$$A \doteq 10^{1.112\ 149\ 533}$$

$$\doteq 12.946\ 415\ 3$$

A patch that contains 20 plant species has an area approximately 13 m^2.

In *Example 2c,* we solved the logarithmic equation $20 = 10.7\log A + 8.1$. Logarithmic equations are not as common as exponential equations, but they can occur in any situation that involves exponential functions. The reason is that the equation of any exponential function can be written in logarithmic form. This form of the equation can then be used to determine values of the variables involved.

Example 3

When \$500 are invested at 6.5% compounded annually, the amount, A dollars, after n years is given by $A = 500(1.065)^n$.

a) Solve the equation for n to express n as a function of A.

b) Graph the function in part a.

c) Use the equation in part a to determine the time for the amount to double.

d) Use the equation in part a to determine the amount after 5 years.

Solution

a) $A = 500(1.065)^n$

Take the base-10 logarithm of each side.

$$\log A = \log 500 + n\log 1.065$$

Solve for n.

$$n\log 1.065 = \log A - \log 500$$

$$n\log 1.065 = \log\left(\frac{A}{500}\right)$$

$$n = \frac{1}{\log 1.065} \times \log\left(\frac{A}{500}\right)$$

$$n \doteq 36.56\log\left(\frac{A}{500}\right)$$

b) Input [Y=] 36.56 [LOG] [X,T,θ,n] [÷] 500 [)].
Set the window as shown at the right.
Press [GRAPH].

20

0 2000

c) Substitute $A = 1000$ in the equation $n \doteq 36.56 \log\left(\frac{A}{500}\right)$.

$$n \doteq 36.56 \log\left(\frac{1000}{500}\right)$$
$$\doteq 11.01$$

The amount doubles in approximately 11 years.

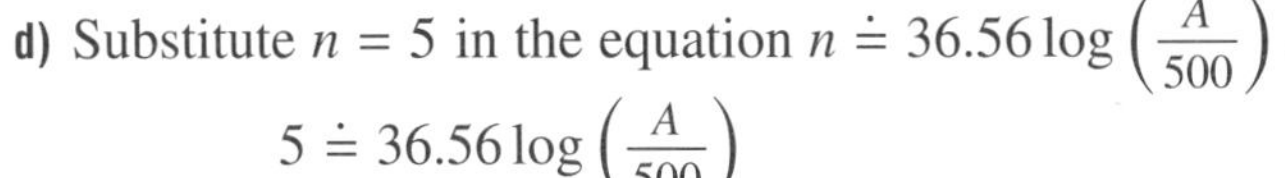

d) Substitute $n = 5$ in the equation $n \doteq 36.56 \log\left(\frac{A}{500}\right)$.

$$5 \doteq 36.56 \log\left(\frac{A}{500}\right)$$
$$\log\left(\frac{A}{500}\right) \doteq \frac{5}{36.56}$$
$$\log\left(\frac{A}{500}\right) \doteq 0.136\ 76$$
$$\frac{A}{500} \doteq 10^{0.136\ 76}$$
$$A \doteq 500(10)^{0.136\ 76}$$
$$\doteq 685.06$$

The amount after 5 years is approximately \$685.

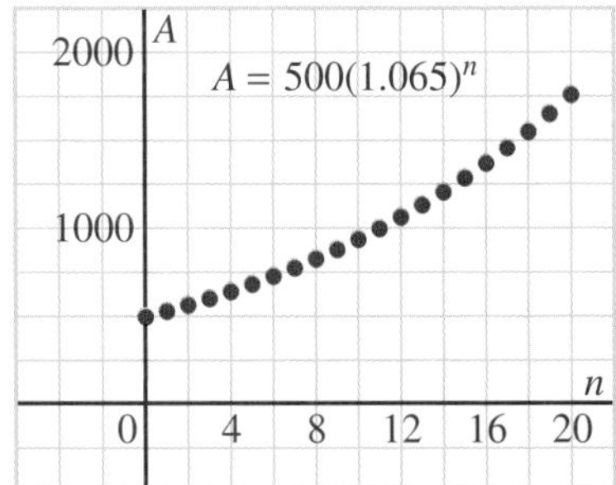

Compare the graph in the above solution with the graph of $A = 500(1.065)^n$, at the right.

The function $n = 35.56 \log\left(\frac{A}{500}\right)$ presents the same information as the function $A = 500(1.065)^n$, but in a different form. This is why we can use either equation to determine the value of one variable when the other variable is known.

6.7 Exercises

A

1. Write the equation of the inverse of each exponential function.

a) $y = 10^x$ **b)** $y = 2^x$ **c)** $f(x) = 5^x$

d) $f(x) = 0.3^x$ **e)** $g(x) = \left(\frac{3}{2}\right)^x$ **f)** $h(x) = 1.5^x$

2. Write the equation of the inverse of each logarithmic function.

a) $y = \log x$ **b)** $y = \log_3 x$ **c)** $f(x) = \log_7 x$

B

3. Compare the properties of the logarithmic function on page 414 with the properties of the exponential function on page 371. Explain why $a \neq 1$ in the properties of the logarithmic function, but $a = 1$ in the properties of the exponential function.

4. Knowledge/Understanding

a) i) Sketch the graphs of $y = 10^x$ and $y = 2^x$ on the same grid.
ii) Sketch the graph of $y = 5^x$ on the grid in part i.

b) i) Use your graph from part a as a guide. Sketch the graphs of $y = \log x$ and $y = \log_2 x$ on another grid.
ii) Sketch the graph of $y = \log_5 x$ on the grid in part b) i.

c) Explain how the graphs you drew in part b are related to those you drew in part a.

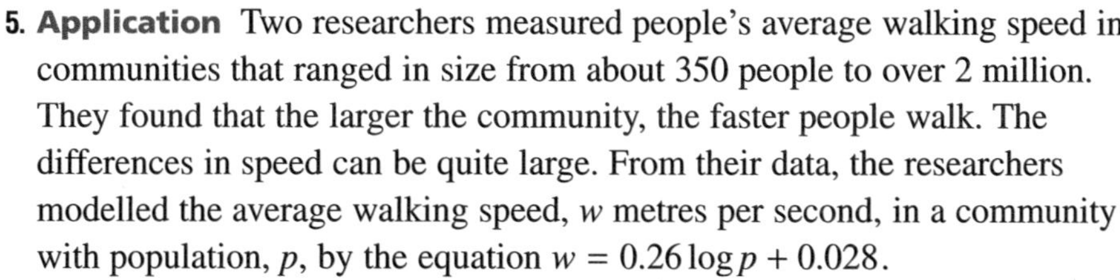

5. Application Two researchers measured people's average walking speed in communities that ranged in size from about 350 people to over 2 million. They found that the larger the community, the faster people walk. The differences in speed can be quite large. From their data, the researchers modelled the average walking speed, w metres per second, in a community with population, p, by the equation $w = 0.26 \log p + 0.028$.

a) Sketch a graph of walking speed against population.

b) What walking speed does the model predict for a community with each population?

i) 350 **ii)** 25 000 **iii)** 500 000 **iv)** 2 million

c) Predict the size of a community in which the average walking speed is 1 m/s.

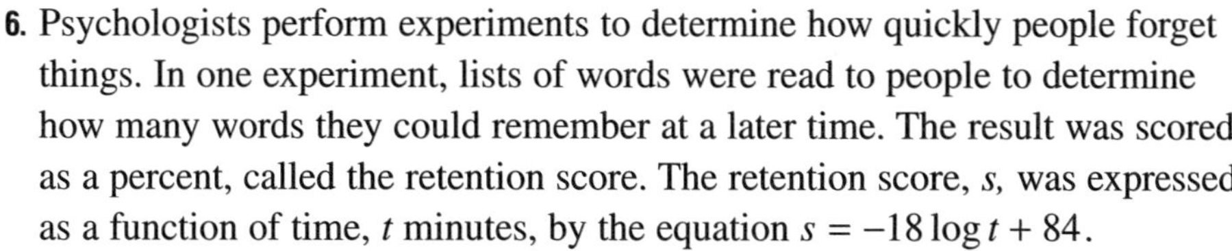

6. Psychologists perform experiments to determine how quickly people forget things. In one experiment, lists of words were read to people to determine how many words they could remember at a later time. The result was scored as a percent, called the retention score. The retention score, s, was expressed as a function of time, t minutes, by the equation $s = -18 \log t + 84$.

a) Sketch a graph of s against t.

b) What was the retention score after each time?

i) 30 min **ii)** 1 h **iii)** 24 h **iv)** 1 week

c) After how many minutes was the retention score 50%?

7. **Communication** Choose either exercise 5 or exercise 6. Explain how the graph of the equation is related to the graph of $y = \log x$.

8. The formulas $L = 2.00(1.20)^n$ and $T = 0.50(0.83)^n$ represent the length and thickness, respectively, of a slab of red-hot steel that passes n times through heavy rollers (see exercise 12, page 391).

   a) Express each equation in logarithmic form.

   b) Suggest why someone might prefer to have the equations in logarithmic form rather than in exponential form.

9. Write each equation in exponential form. Explain what each equation represents when it is written in exponential form.

   a) the equation $s = 10.7 \log A + 8.1$ in *Example 2* that represents the number of plant species in a square patch of lawn with area A square metres

   b) the equation $s = -18 \log t + 84$ in exercise 6 that represents the percent of words remembered after t minutes

Exercises 8 and 9 are significant because they emphasize the relationship between logarithmic and exponential functions. Since these are inverse functions, for any property of exponential functions there is a corresponding property of logarithmic functions.

10. **Thinking/Inquiry/Problem Solving** Choose one equation in exercise 9. It is written in logarithmic form using a base-10 logarithm. Change the equation so it is written in logarithmic form using a base-2 logarithm.

11. Consider $f(x) = \log_a x$, $a > 0$, $a \neq 1$. Here are the laws of exponents expressed using function notation:

   i) $f(x_1 + x_2) = f(x_1)f(x_2)$ ii) $f(x_1 - x_2) = \dfrac{f(x_1)}{f(x_2)}$

   iii) $f(nx_1) = \left(f(x_1)\right)^n$ iv) $f(-x_1) = \dfrac{1}{f(x_1)}$

   a) Write similar expressions that express the laws of logarithms using function notation.

   b) Prove that your expressions are correct.

12. Mathematicians have always been interested in determining larger and larger prime numbers. The largest known prime number, as of September, 2001, was found by a computer in 1999. The number is $2^{6\,972\,593} - 1$.

   a) How many digits are there in this number?

   b) What is the first digit of this number?

   c) What is the last digit?

Self-Check 6.5 – 6.7

1. Express each function as an equivalent exponential function with base 5.

   a) $y = 2^x$ b) $y = 6^x$ c) $y = 0.5^x$

2. Express the function $y = 10^x$ as an equivalent function with each base.

   a) 4 b) 0.5 c) 15

3. The amount of an investment, A dollars, after n years can be modelled by $A = 1500(1.035)^n$.

   a) What is the principal?

   b) What is the interest rate?

   c) Rewrite the equation for the amount using base 2.

   d) How long does it take for the investment to double? Explain.

4. How many times as intense is an earthquake with magnitude 7 than one with each magnitude below?

   a) 8 b) 6 c) 5 d) 9 e) 4

5. Use the Richter scale on page 407. How many times as intense as the 1999 Turkey earthquake was the 1938 Indonesia earthquake?

6. Use the decibel scale on page 411. How many times as loud as average street traffic is a car horn?

7. Use the pH scale on page 412. How many times as acidic as sea water is distilled water?

PERFORMANCE ASSESSMENT

8. A thermocouple is used to measure very high temperatures. A thermocouple is placed on the element of an electric range. The temperature of the element, T degrees Celsius, can be modelled by the equation $T = 150 \log 4t$, where t is the time in seconds after the element is turned on.

   a) Graph the function.

   b) What is the temperature after 30 s?

   c) Sketch a graph of the inverse of the function T.

   d) Determine the time when the temperature reaches 375°C.

   e) Pose a problem about this scenario. Solve the problem you posed.

Review Exercises

Mathematics Toolkit

Exponential and Logarithmic Tools

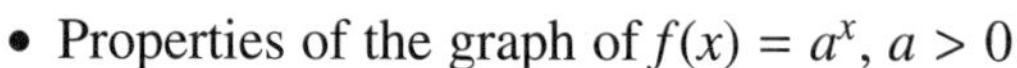

- Properties of the graph of $f(x) = a^x$, $a > 0$:
  - The vertical intercept is 1.
  - There is no horizontal intercept.
  - The domain is the set of real numbers; $D = \Re$.
  - The range is the set of positive real numbers; $R = \{y \mid y \in \Re, y > 0\}$.
  - When $a > 1$, the function increases.
  - When $0 < a < 1$, the function decreases.
  - When $a = 1$, the function is constant.
  - The graph is concave up except when $a = 1$.
  - The x-axis is a horizontal asymptote.

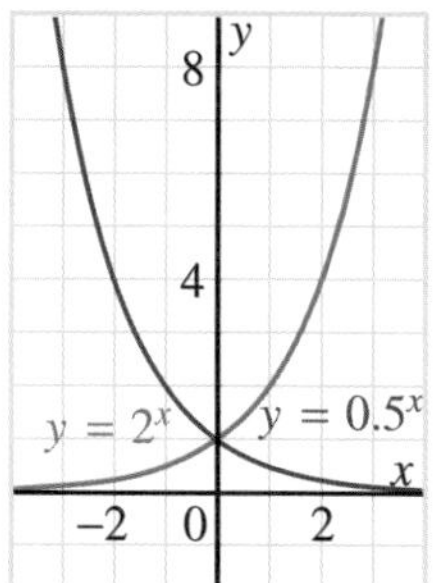

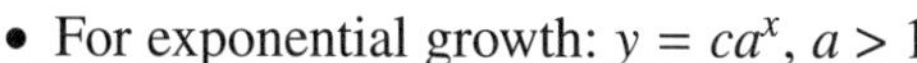

- For exponential growth: $y = ca^x$, $a > 1$
- For exponential decay: $y = ca^x$, $0 < a < 1$
- A logarithm is an exponent: $\log_a x = y$ means $a^y = x$, where $a > 0$, $a \neq 1$, and $x > 0$

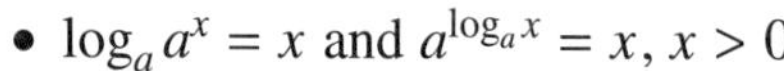

- $\log_a a^x = x$ and $a^{\log_a x} = x$, $x > 0$
- The function $y = \log_a x$ is the inverse of the function $y = a^x$.
- Properties of the graph of $f^{-1}(x) = \log_a x$, $a > 0$, $a \neq 1$:
  - There is no vertical intercept.
  - The horizontal intercept is 1.
  - The domain is the set of positive real numbers; $D = \{x \mid x \in \Re, x > 0\}$.
  - The range is the set of real numbers; $R = \Re$.
  - When $a > 1$, the function increases for $x > 0$.
  - When $0 < a < 1$, the function decreases for $x > 0$.
  - The graph is concave down for $a > 1$, and concave up for $0 < a < 1$.
  - The y-axis is a vertical asymptote.
- Laws of logarithms, where $a > 0$, $a \neq 1$:
  - For powers: x and n are real numbers and $x > 0$; $\log_a x^n = n \log_a x$
  - For multiplication: x and y are positive real numbers; $\log_a xy = \log_a x + \log_a y$
  - For division: x and y are positive real numbers; $\log_a \left(\frac{x}{y}\right) = \log_a x - \log_a y$
  - For roots: x is a positive real number, n is a natural number; $\log_a \sqrt[n]{x} = \frac{1}{n} \log_a x$

- Logarithmic scales:
  - The Richter scale: each increase in 1 unit of magnitude of an earthquake represents a 10-fold increase in intensity.
  - The decibel scale: each increase of 10 dB represents a 10-fold increase in the loudness of a sound.
  - The pH scale: for pH less than 7, each 1-unit decrease in pH represents a 10-fold increase in acidity; for pH greater than 7, each 1-unit increase in pH represents a 10-fold increase in alkalinity.

6.1 **1.** Sketch graphs of these functions on the same grid.

a) $y = 4^x$ **b)** $y = \left(\frac{1}{4}\right)^x$ **c)** $y = 2(4)^x - 2$

2. For each function in exercise 1, answer these questions.

i) What are the x- and y-intercepts?
ii) What are the domain and range?
iii) Is the function increasing or decreasing? Explain.
iv) Is the graph concave up or concave down? Explain.
v) Does the graph have an asymptote? Explain.

3. Consider the general exponential function $y = ca^x + b$.

a) i) On the same screen, graph $y = 2^x$, $y = 5^x$, $y = \left(\frac{1}{5}\right)^x$, and $y = \left(\frac{1}{2}\right)^x$.
ii) Explain the effect on the graph of $y = a^x$ when a varies.

b) i) On the same screen, graph $y = 2(2)^x$, $y = 5(2)^x$, $y = \frac{1}{5}(2)^x$, and $y = \frac{1}{2}(2)^x$.
ii) Explain the effect on the graph of $y = ca^x$ when c varies.

c) i) On the same screen, graph $y = 5(2)^x + 2$, $y = 5(2)^x + 10$, $y = 5(2)^x - 2$, and $y = 5(2)^x - 10$.
ii) Explain the effect on the graph of $y = ca^x + b$ when b varies.

4. Suppose a principal of P dollars is invested at 3.75% compounded annually. After n years, the amount is \$5000. This situation is modelled by the equation $P = 5000(1.0375)^{-n}$, where P is the present value.

a) How much should be invested today to have \$5000 after 10 years?

b) Suppose \$3000 are invested today. How long will it take until the amount is \$5000?

5. Use exercise 4 as a guide. Pose your own problem about the present value of an investment. Solve the problem you posed.

Review Exercises

6. a) On the same screen, graph $y = 2^x$ and $y = 2x$. For large values of x, which function has the greater rate of increase? Explain.

b) On the same screen, graph $y = 2^x$ and $y = x^{\frac{1}{2}}$. For large values of x, which function has the greater rate of increase? Explain.

6.2 **7.** Write in exponential form.

a) $\log_2 32 = 5$ **b)** $\log_3 3 = 1$ **c)** $\log_{10} 1 = 0$

d) $\log_4 \left(\frac{1}{16}\right) = -2$ **e)** $\log_5 0.008 = -3$ **f)** $\log_8 64 = 2$

8. Write in logarithmic form.

a) $2^{10} = 1024$ **b)** $10^2 = 100$ **c)** $10^{-2} = 0.01$

d) $25^{\frac{1}{2}} = 5$ **e)** $16^{\frac{3}{2}} = 64$ **f)** $1296^{0.25} = 6$

9. Evaluate each logarithm.

a) $\log 1$ **b)** $\log 10\,000$ **c)** $\log_3 729$

d) $\log_9 \left(\frac{1}{9}\right)$ **e)** $\log_4 0.0625$ **f)** $\log_2 0.125$

10. Simplify each expression.

a) $\log 10^4$ **b)** $\log_4 4^5$ **c)** $10^{\log 1000}$ **d)** $2^{\log_2 4}$

6.3 **11.** Solve each equation to 4 decimal places. Check the solution.

a) $10^x = 15$ **b)** $9^x = 30$ **c)** $8^x = 3$

d) $5^x = 100$ **e)** $3^x = 2$ **f)** $2^x = 3$

12. Express:

a) 26 as a power of 10 **b)** 5 as a power of 10

c) 75 as a power of 4 **d)** 3 as a power of 4

e) 100 as a power of 7 **f)** 6 as a power of 7

13. The number of mutual funds available in Canada, M, is modelled by the equation $M = 460(1.19)^n$, where n is the number of years since 1989.

a) When will the number of mutual funds reach 10 000?

b) How many years will it take for the number of mutual funds to triple?

6.4 **14.** Simplify each expression.

a) $\log 50 + \log 20$ **b)** $\log_2 64 + \log_2 0.5$

c) $\log_4 2 + \log_4 8$ **d)** $\log_5 62.5 + \log_5 2$

15. Simplify each expression.

a) $\log_3 108 - \log_3 4$ b) $\log 50 - \log 500$

c) $\log_6 108 - \log_6 3$ d) $\log_2 54 - \log_2 108$

16. Solve each equation to 4 decimal places.

a) $327 = 210(1.1)^x$ b) $5(1.03)^x = 8$ c) $42 = 50(0.9)^x$

d) $3 = 5(0.7)^x$ e) $5000(1.0375)^x = 6575$ f) $0.5 = 4(0.65)^x$

17. Solve for x. Give the answers to 4 decimal places where necessary.

a) $2 \log x = \log 25 + \log 4$ b) $3 \log x = \log 54 - \log 2$

c) $\log x^2 - \log 3x = \log 4$ d) $\log x + \log (x - 2) = 1$

18. Radioactive tritium has a half-life of 12 years. A sample of this material has a mass of 1000 g. An equation that models the mass, m grams, remaining after t years is $m = 1000(0.9439)^t$.

a) How much radioactive tritium remains after 100 years?

b) How long does it take until only 100 g of the radioactive tritium remain?

c) Pose a problem about radioactive tritium. Solve the problem you posed.

6.5 19. Write each function as an equivalent exponential function with base 4.

a) $y = 5^x$ b) $y = 10^x$ c) $y = 2.5^x$

20. Express the function $y = 2^x$ as an equivalent function with each base.

a) 3 b) 5 c) 7.5

21. Carbon-14 is radioactive. It is used to estimate the age of ancient specimens. All living matter contains traces of carbon-14, which decays. The age of an ancient specimen can be determined by measuring the radioactivity of the carbon-14 it contains, and comparing it with that of modern living matter. The percent, P, of carbon-14 that remains after n years is modelled by $P = 100(0.999\,88)^n$.

a) The Dead Sea Scrolls are about 2000 years old. What percent of carbon-14, relative to modern living matter, should be expected from a sample taken from the Scrolls?

b) Charcoal found at Stonehenge, England contains about 62.0% of carbon-14 relative to modern living matter. Determine the approximate age of the charcoal.

c) Write the equation for the percent of carbon-14, using base 0.5.

d) What is the half-life of carbon-14? Explain.

Review Exercises

6.6 **22.** In 1970, there was an earthquake in Peru that measured 7.7 on the Richter scale. Thirty years later, there was an earthquake in Papua New Guinea that measured 8.0 on the Richter scale.

a) **i)** Which earthquake was more intense? Explain.
ii) How do the intensities of the earthquakes compare? Explain.

b) What would be the magnitude of an earthquake that was twice as intense as the Peru earthquake?

c) What would be the magnitude of an earthquake that was one-half as intense as the Papua New Guinea earthquake?

23. The loudness of a referee's whistle is 130 dB. The loudness of normal conversation is 60 dB.

a) How many times as loud as normal conversation is a referee's whistle?

b) What is the measure in decibels of a sound that is 50 times as loud as normal conversation?

c) What is the loudness in decibels of a sound that is $\frac{1}{50}$ as loud as a referee's whistle?

24. Vinegar has a pH of 2.8. Tomato juice has a pH of 4.0.

a) **i)** Which is more acidic, vinegar or tomato juice? Explain.
ii) How do the acidities compare? Explain.

b) Determine the pH of something that is twice as acidic as tomato juice.

c) Determine the pH of something that is one-half as acidic as vinegar.

6.7 **25.** Write the equation of the inverse of each exponential function.

a) $y = 9^x$ **b)** $f(x) = 0.1^x$ **c)** $g(x) = \left(\frac{3}{4}\right)^x$

26. Write the equation of the inverse of each logarithmic function.

a) $y = \log x$ **b)** $y = \log_5 x$ **c)** $y = \log_8 x$

27. Sketch graphs of these functions on the same grid.

a) $y = \log x$ **b)** $y = \log_3 x$ **c)** $y = \log_6 x$

28. For each function in exercise 27, answer these questions.

i) What are the x- and y-intercepts?
ii) What are the domain and range?
iii) Is the function increasing or decreasing? Explain.
iv) Is the graph concave up or concave down? Explain.
v) Does the graph have an asymptote? Explain.

Self-Test

1. Evaluate without using a calculator.

   a) $\log_5 1$ b) $\log_4 16^2$ c) $\log_2 4 + \log_2 50 - \log_2 25$

2. Express as a single logarithm.

   a) $\log x - 4\log y$ b) $6\log_3 x - \frac{1}{2}\log_3 y + \log_3 2x$

3. **Knowledge/Understanding** Solve each equation.

   a) $7^{x-1} = 49^{4-2x}$

   b) $\log x = \frac{3}{2}\log 9 + \log 2$

   c) $\log_7(x+1) + \log_7(x-5) = 1$

4. Solve each equation to 4 decimal places.

   a) $4^x = 24$ b) $3^{2x} = 8$ c) $5^{x-2} = 150$

5. Express the function $y = 5^x$ as an equivalent function with base 3.

6. **Communication** Describe what you would do to the graph of $y = 4^x$ to obtain the graph of $y = -2(4)^x + 2$.

7. a) On the same grid, sketch the graph of each function.

   i) $y = \log_3 x$ ii) $y = 3^x$

   b) For each function in part a:

   i) State its intercepts.

   ii) State its domain and range.

   iii) State the equation of its asymptote.

8. **Application** Suppose money can be invested at 3.75% compounded semi-annually. Determine how long it takes for a principal of \$100 000 to triple in value.

9. The decibel (dB) is used to measure the loudness of a sound. The equation $D = 10\log I$ models the decibel level of a sound whose intensity is I watts per square metre (W/m^2). The decibel levels of a subway train and normal conversation are 115 dB and 60 dB, respectively. How many times as intense as normal conversation is the noise of a subway train?

10. **Thinking/Inquiry/Problem Solving**

    a) Solve the equation $(2^x)^2 - 4(2^x) + 3 = 0$.

    b) Graph the function $y = 2^{2x} - 4(2^x) + 3$.

    c) How is the graph in part b related to your answer in part a?

Self-Test

11. The first artificial satellites were put into orbit in the late 1950s. In 1960, the cumulative mass of satellites in orbit was 120 t. Since then, the cumulative mass has been growing at about 12% annually.

    a) Write an equation that expresses the cumulative mass, c tonnes, as a function of n, the number of years since 1960.

    b) Graph c as a function of n.

    c) Estimate the cumulative mass in 1969, the year when a person first walked on the moon.

    d) Estimate when the cumulative mass became 4000 t.

    e) Pose your own problem about the cumulative mass of satellites in orbit. Solve the problem you posed.

12. Determine each logarithm to 5 decimal places. Check the result.

    a) $\log_3 14$ b) $\log_2 42$ c) $\log_5 4250$

13. Prove the law of logarithms for roots: $\log_a \sqrt[n]{x} = \frac{1}{n}\log_a x \ (a > 0,\ a \neq 1)$

PERFORMANCE ASSESSMENT

14. The number of caribou, N, in a forest in northern Ontario is modelled by the equation $N = 2000 - 1000(3)^{-t}$, where t is the time in years.

    a) How many caribou were in the forest when their number was first recorded?

    b) How many caribou were in the forest 4 years later?

    c) Describe how the graphs of $N = 3^t$ and $N = 2000 - 1000(3)^{-t}$ are related.

    d) Sketch the graph of number against time.

    e) Label the asymptote on the graph. What does this asymptote represent?

    f) How long will it take for the number of caribou to reach 1600?

    g) Solve the equation $N = 2000 - 1000(3)^{-t}$ for t.

15. Consider $f(x) = a^x$, where $a > 0$, $a \neq 1$. Show that each equation is correct.

    a) $f(x_1 + x_2) = f(x_1)f(x_2)$ b) $f(x_1 - x_2) = \frac{f(x_1)}{f(x_2)}$

    c) $f(nx_1) = (f(x_1))^n$ d) $f(-x_1) = \frac{1}{f(x_1)}$

Cumulative Review (Chapters 1–6)

1. Determine the equation of the tangent to the graph of $y = 1 - 2x^3$ at the point on the graph where $x = -1$.

2. For the graph of each polynomial function, determine the intercepts, critical points, points of inflection, intervals of increase, decrease, and concavity; then sketch the graph of the function.

   a) $f(x) = x^3 + 5x^2 + 7x + 3$
   b) $f(x) = x^4 - 3x^2 + 2$

3. Differentiate each function.

   a) $y = 6x^3(5 - 3x)$
   b) $y = (1 - 5x)(3x + 2)$
   c) $y = \cos x$
   d) $y = \dfrac{x - 1}{x^2 - x}$
   e) $y = \dfrac{1}{3x^2 - 4}$
   f) $y = \dfrac{6x^3 - 4x^2 - 1}{x}$
   g) $y = \sin x$
   h) $y = \left(\dfrac{3x^2 - 4}{4 - x}\right)^6$

4. Determine $\dfrac{d^2y}{dx^2}$ for the function $y = 5x^2(1 - x^4)$. Why might we need to know the second derivative? Explain.

5. Determine the coordinates of the point(s) on the graph of $y = \dfrac{4x^3 - 8x}{2x - 3}$ where the tangent is horizontal.

6. Determine $\dfrac{dy}{dx}$ for each relation.

   a) $14x^2 + 7y^2 = 56$
   b) $3x^2 - 5xy + y^2 = 4$

7. Determine the equations of the tangents to the graph of $x^2 - xy + y^2 = 1$ at the points on the graph where $x = 1$.

8. The displacement of a particle travelling in a straight line after t seconds is given by $s(t) = t^3 - 9t^2 + 15t + 9$, where s is measured in metres.

   a) When is the particle at rest?
   b) When is the particle moving toward the origin? Justify your answer.
   c) When is the particle slowing down? Justify your answer.
   d) Determine the total distance travelled during the first 6 s of motion.

9. Consider $f(x) = x^2 + 1$ and $g(x) = \sqrt{2 - x}$. Determine:

   a) $g(f(x))$
   b) the domain of $g(f(x))$
   c) $g(f(-2))$
   d) $f(g(x))$

10. Evaluate each limit. Give a geometrical interpretation of each result.

a) $\lim\limits_{x \to -4} \dfrac{x^2 - 16}{x + 4}$ **b)** $\lim\limits_{x \to 5} \dfrac{x^2 - 4x - 5}{x - 5}$ **c)** $\lim\limits_{x \to \infty} \dfrac{6x^2 - 2x - 4}{4 - x^2}$

11. For each rational function below:

i) Determine the intercepts of its graph.

ii) Determine the equations of the vertical, horizontal, and oblique asymptotes if they exist.

iii) Determine $\lim\limits_{x \to \infty} f(x)$.

iv) Sketch the graph of $y = f(x)$.

a) $f(x) = \dfrac{2x^2 - 8}{x^2 - 1}$ **b)** $f(x) = \dfrac{4x^2}{2x + 1}$ **c)** $f(x) = \dfrac{2x + 1}{x^2 + 1}$

12. Draw a possible graph for each rational function.

a) $\lim\limits_{x \to -1^+} f(x) = -\infty$, $\lim\limits_{x \to -1^-} f(x) = \infty$, $f(-1)$ does not exist

b) $\lim\limits_{x \to \infty} f(x) = 1$, $\lim\limits_{x \to -\infty} f(x) = 1$, $\lim\limits_{x \to 2} f(x) = \infty$, $f(4) = f(6) = 0$

c) $y = x + 2$ is an oblique asymptote, $\lim\limits_{x \to 0^+} f(x) = \infty$, $\lim\limits_{x \to 0^-} f(x) = -\infty$, there are no intercepts

13. Evaluate without using a calculator.

a) $\log 500 - \log 5$

b) $\log_{12} 15 + \log_{12} 20 - \log_{12} 25$

c) $\frac{2}{5} \log_4 32 - \log_4 8$

14. Solve each equation. Give the solution to 4 decimal places where necessary.

a) $9^x = 5$ **b)** $3^{x-2} = 8$

c) $\log_3 (x + 3) = \log_3 8 - \log_3 2$ **d)** $\log_3 (x + 2) + \log_3 6 = 3$

e) $5(4)^{1-x} = 400$ **f)** $\log (x + 6) + \log (x - 6) = 2$

15. Sketch the graphs of each pair of functions on the same grid. Label the intercepts and asymptotes.

a) $y = 3^x$ and $y = 4(3)^x + 1$ **b)** $y = \log x$ and $y = 10^x$

16. A colony of bacteria contains 5000 bacteria at 2 P.M. and 97 000 bacteria at 4:30 P.M. Assume the population is increasing exponentially. Determine the population at 6 P.M.

17. Use one of the logarithmic scales on pages 407, 411, or 412. Use the information in that scale to pose a problem. Solve the problem you posed. Write a clear explanation of your problem and its solution.

Differentiating Exponential and Logarithmic Functions 7

Curriculum Expectations

By the end of this chapter, you will:

- Pose and solve problems related to models of exponential functions drawn from a variety of applications, and communicate the solutions with clarity and justification.
- Determine the limit of an exponential function.
- Identify e as $\lim_{n\to\infty}\left(1 + \frac{1}{n}\right)^n$ and approximate the limit, using informal methods.
- Define $\ln x$ as the inverse function of e^x.
- Determine the derivatives of the exponential functions a^x and e^x and the logarithmic functions $\log_a x$ and $\ln x$.
- Determine the derivatives of combinations of the basic polynomial, rational, exponential, and logarithmic functions, using the rules for sums, differences, products, quotients, and compositions of functions.
- Determine the equation of the tangent to the graph of an exponential or a logarithmic function.
- Determine, from the equation of an exponential function, the key features of the graph of the function, using the techniques of differential calculus, and sketch the graph by hand.
- Sketch the graphs of the first and second derivative functions, given the graph of the original function.

Necessary Skills

1. New: Proportionality

Suppose a square has sides x units long. Its area, A, is a function of x:

$A = x^2$

Suppose a circle has diameter x units. Its area, A, is also a function of x:

$A = \pi\left(\frac{x}{2}\right)^2$

$A = \frac{\pi}{4}x^2$

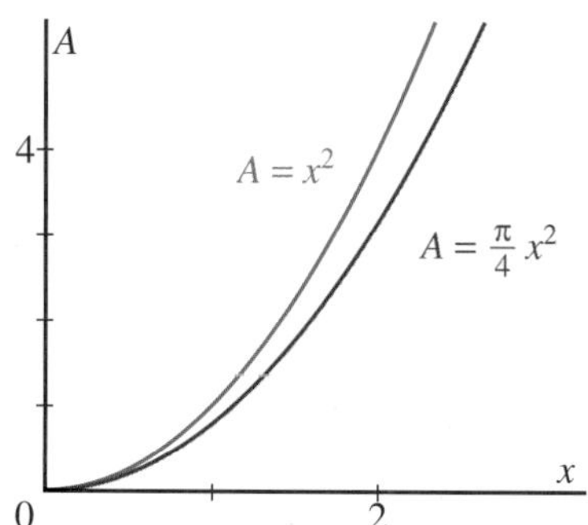

The area of the circle is a constant times the area of the square. The graph of the area of the circle is a vertical compression of the graph of the area of the square.

We say that the area of the circle is proportional to the area of the square. The constant of proportionality is $\frac{\pi}{4}$.

When two functions of the same variable have a constant ratio, each function is *proportional* to the other. The ratio is called the *constant of proportionality*. The graph of each function is a vertical expansion or compression of the graph of the other function.

Exercises

1. Use the above example.

a) Draw a graph to show the area of the circle plotted against the area of the square. Describe the graph.

b) The area of the square is also proportional to the area of the circle. What is the constant of proportionality?

2. Four functions are listed below. Describe how one of these functions is proportional to another. State the constant of proportionality.

circumference of a circle with diameter x: $C = \pi x$

surface area of a sphere with diameter x: $A = \pi x^2$

volume of a sphere with diameter x: $V = \frac{\pi}{6}x^3$

volume of a cube with edge length x: $V = x^3$

2. Review: The Chain Rule

When you differentiate exponential and logarithmic functions in this chapter, you will use the differentiation tools you learned earlier. You will most often use the chain rule.

Chain Rule

$$\frac{d}{dx}[f(g(x))] = f'(g(x)) \times g'(x)$$

Example 1

Differentiate $y = (2x^2 - 1)^4$.

Solution

$$y = (2x^2 - 1)^4$$
$$\frac{dy}{dx} = 4(2x^2 - 1)^3 \times 4x$$
$$= 16x(2x^2 - 1)^3$$

Example 2

Differentiate each function.

a) $y = \dfrac{1}{(1 - x^3)^2}$

b) $y = x(5x + 1)^2$

Solution

a) Use the power and chain rules.

$$y = (1 - x^3)^{-2}$$
$$\frac{dy}{dx} = -2(1 - x^3)^{-3} \times (-3x^2)$$
$$= 6x^2(1 - x^3)^{-3}$$
$$= \frac{6x^2}{(1 - x^3)^3}$$

b) Use the product and chain rules.

$$y = x(5x + 1)^2$$
$$\frac{dy}{dx} = (1)(5x + 1)^2 + (x)[2(5x + 1)(5)]$$
$$= (5x + 1)^2 + 10x(5x + 1)$$
$$= (5x + 1)(5x + 1 + 10x)$$
$$= (5x + 1)(15x + 1)$$

Exercises

1. Differentiate each function.

a) $y = (3x^3 - 1)^2$

b) $y = (1 + 2x)^5$

c) $y = x(4x - 3)^2$

d) $y = x(x^2 + 1)^2$

e) $y = \dfrac{1}{(2x^2 + 1)^2}$

f) $y = \dfrac{x}{(1 - 3x)^2}$

3. Review: Changing the Base of an Exponential Function

Recall that we can express any positive number as a power of any other positive number. This means that we can also express any exponential function as an equivalent exponential function with a different base.

Example 1

Express 62 as a power of 5.

Solution

Let $62 = 5^x$.
To solve for x, take the base-10 logarithm of each side.

$$\log 62 = x \log 5$$

$$x = \frac{\log 62}{\log 5}$$

$$\doteq 2.564\ 33$$

Therefore, $62 \doteq 5^{2.564\ 33}$

Example 2

Write the function $y = 2.5^x$ as an equivalent exponential function with base 1.5.

Solution

Express 2.5 as a power of 1.5.
Let $2.5 = 1.5^k$.
Take the base-10 logarithm of each side.

$$\log 2.5 = k \log 1.5$$

$$k = \frac{\log 2.5}{\log 1.5}$$

$$\doteq 2.259\ 85$$

Therefore, $2.5 \doteq 1.5^{2.259\ 85}$, and $y = 2.5^x$ is approximated as $y = 1.5^{2.259\ 85x}$.

Exercises

1. Express:

   a) 65 as a power of 4 b) 100 as a power of 3

2. Express each function as an equivalent exponential function with the indicated base.

   a) $y = 1.2^x$ Change to base 2.

   b) $y = 2(1.05)^x$ Change to base 3.

   c) $y = 4.25(1.75)^x$ Change to base 10.

4. Review: Compounding Periods for Compound Interest

Interest is frequently compounded more than once per year. Recall what each variable represents when we use the formula $A = P(1 + i)^n$:

A is the amount.
P is the principal.
i is the interest rate for each compounding period, expressed as a decimal.
n is the number of compounding periods.

Example

Calculate the amount when $5000 are invested for 8 years at 6.25% compounded semi-annually.

Solution

There are 2 compounding periods each year.
The interest rate for each compounding period is $\frac{6.25\%}{2} = 3.125\%$.
In 8 years, there are 8×2, or 16 compounding periods.
Substitute $P = 5000$, $i = 0.031\ 25$, and $n = 16$ in the formula.

$$\begin{aligned} A &= P(1 + i)^n \\ &= 5000(1.031\ 25)^{16} \\ &\doteq 8180.76 \end{aligned}$$

The amount after 8 years is $8180.76.

Exercises

1. Calculate each amount.

   a) $1000 invested for 7 years at 5.5% compounded semi-annually

   b) $2500 invested for 5 years at 6% compounded monthly

   c) $3000 invested for 9 years at 7% compounded quarterly

7.1 The Function $y = e^x$

We saw, in Chapter 6, how exponential functions arise in situations that involve multiplicative growth and decay. The derivatives of exponential functions provide information about the instantaneous rates of growth or decay in these situations.

Investigation

The Derivatives of $y = 2^x$ and $y = 3^x$

Work with a partner. One person uses the function $y = 2^x$. The other person uses the function $y = 3^x$.

1. Make a table of values with the headings shown below. Use integer values of x from –2 to 2 in the first column. Use the equation of the function to complete the second column.

x	y	slope at x	$\frac{\text{slope at } x}{y}$

2. A graph of the function is shown on the next page. Tangents are drawn to the graph at points A, B, C, D, and E, where $x = -2, -1, 0, 1,$ and 2, respectively. Another point A_1, B_1, C_1, D_1, and E_1, is marked on each tangent.
   a) Calculate the slope of each tangent rounded to 3 decimal places. Record the slopes in the third column of the table.
   b) Divide each slope by the y-coordinate of the point where the tangent meets the graph of the function. Round the quotient to 2 decimal places. Record the quotients in the fourth column.

3. a) The slopes of the tangents form a pattern. Describe the pattern.
   b) Predict the slopes of the tangents at the points where $x = 3, x = 4$, and $x = 10$.
   c) Suppose you know the x-coordinate of a point on the graph of the function. How can you calculate the slope of the tangent at that point?

4. a) On the same grid, graph the function and its derivative.
   b) Compare the two graphs. Explain how the graph of the derivative provides information about how the values of y are changing on the graph of the function, as x increases.
   c) Write the equation of the derivative function.

The coordinates of the points on the graphs below were calculated using the derivatives of the functions. You will confirm some of these coordinates in a later exercise.

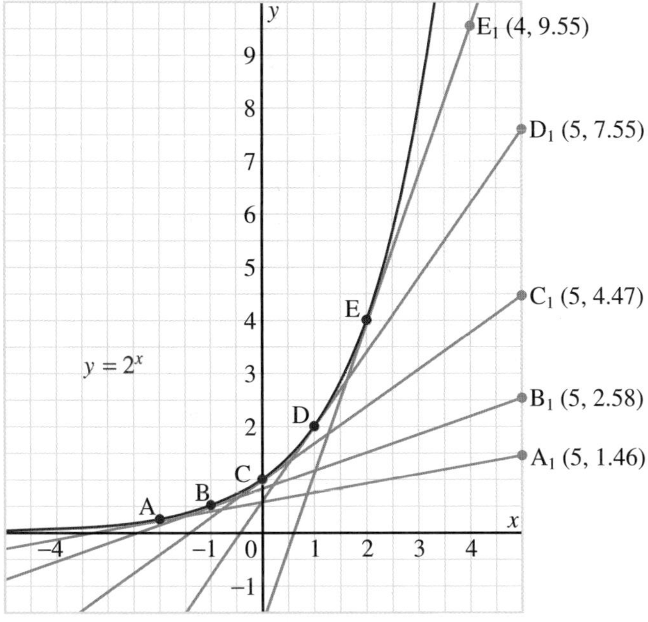

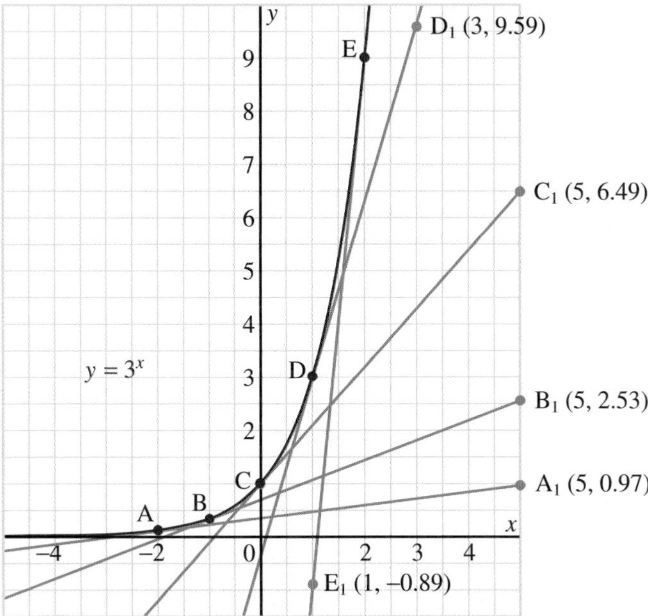

In the *Investigation*, you should have conjectured that:

- The derivative of the function $y = 2^x$ is approximately $y = 0.69 \times 2^x$.
- The derivative of the function $y = 3^x$ is approximately $y = 1.10 \times 3^x$.

These results are significant because they indicate that the derivative of an exponential function is proportional to the function. The reason will be seen below, and in Sections 7.3 and 7.4.

Take Note

The Derivatives of $y = 2^x$ and $y = 3^x$

$\frac{d}{dx}(2^x) \doteq 0.69 \times 2^x$

$\frac{d}{dx}(3^x) \doteq 1.10 \times 3^x$

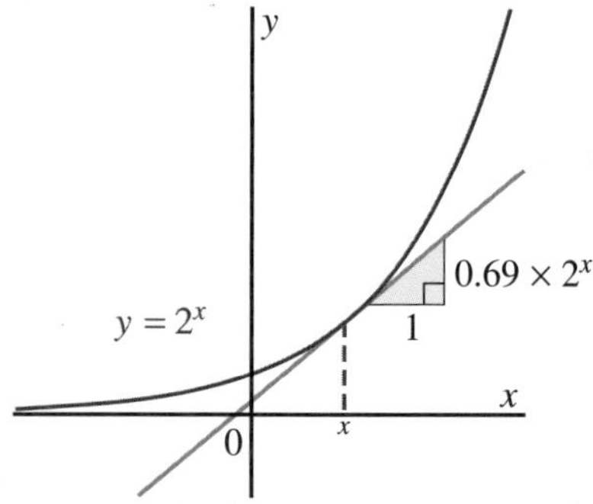

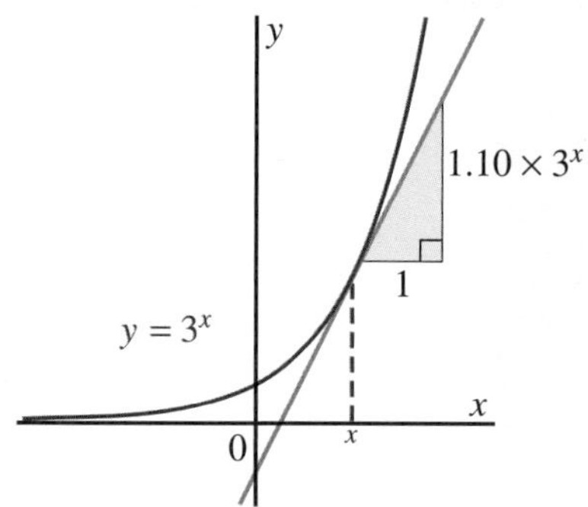

Visualize a point moving from left to right along either graph. The slope of the tangent is proportional to the value of the function at that point.

The graph of $y' = 0.69 \times 2^x$ is a vertical compression of the graph of $y = 2^x$.

The graph of $y' = 1.10 \times 3^x$ is a vertical expansion of the graph of $y = 3^x$.

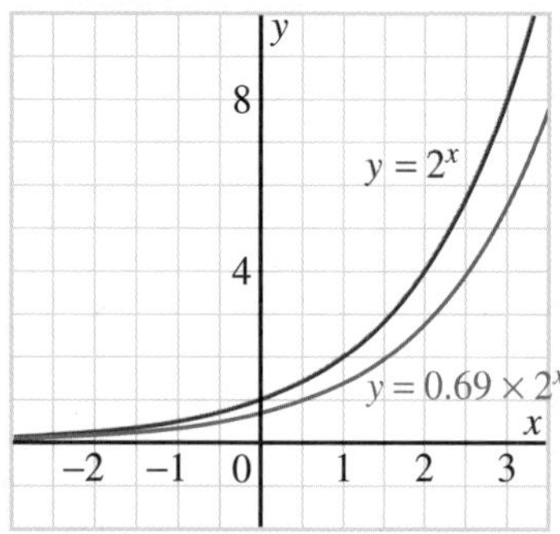

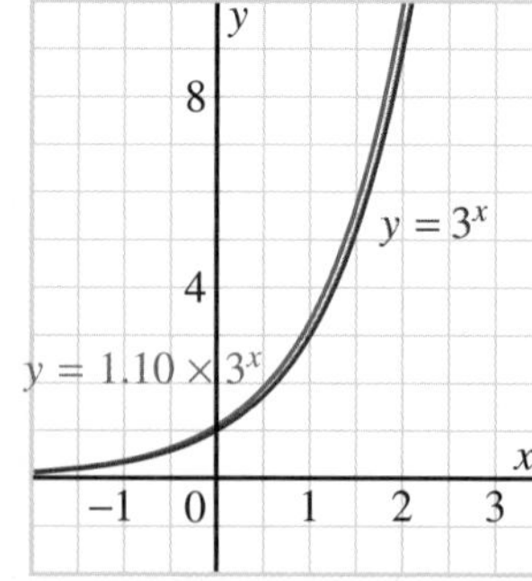

The program *Graphs! Graphs! Graphs!* can be used to graph the function $f(x) = a^x$ and its derivative on the same grid. You can use a slider to vary a, then observe the effect on the graphs. A few examples are shown at the right. Starting at the bottom, visualize how the graph of the function (shown in red) and its derivative (shown in blue) change as a increases. In each case, the graph of the derivative appears to be a vertical expansion or compression of the graph of the function. This provides more evidence that the derivative of an exponential function is proportional to the function.

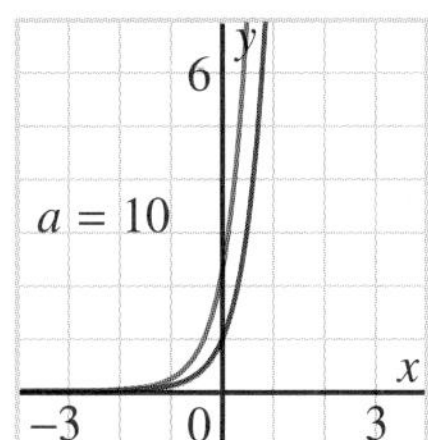

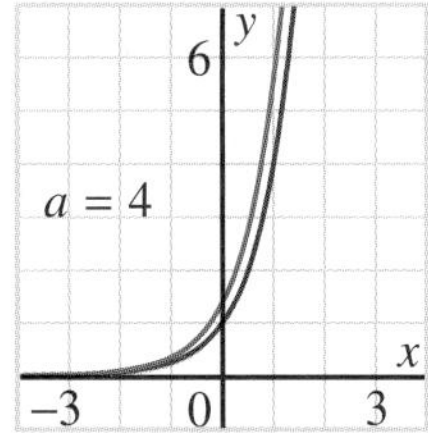

Both the examples from the *Investigation*, and the examples shown at the right suggest:

- $\frac{d}{dx}(a^x) = ka^x$, where k is a constant that depends on the value of a
- When $a \doteq 2.7$, the graph of f' appears to coincide with the graph of f. We will see on page 440 that a more accurate value is $a \doteq 2.718\ 28$. For this value of a, $\frac{d}{dx}(a^x) = a^x$
- When $a < 2.718\ 28$, the graph of f' is a vertical compression of the graph of f. For these values of a, $\frac{d}{dx}(a^x) < a^x$
- When $a > 2.718\ 28$, the graph of f' is a vertical expansion of the graph of f. For these values of a, $\frac{d}{dx}(a^x) > a^x$

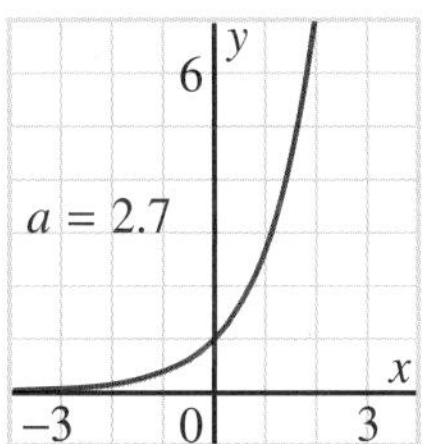

To confirm these results, we apply the first-principles definition of the derivative to the function $f(x) = a^x$, $a > 0, a \neq 1$.

$$f'(x) = \lim_{h \to 0} \frac{f(x+h) - f(x)}{h}$$

Replace $f(x)$ with a^x.

$$f'(x) = \lim_{h \to 0} \frac{a^{x+h} - a^x}{h}$$

$$= \lim_{h \to 0} \frac{a^x(a^h - 1)}{h}$$

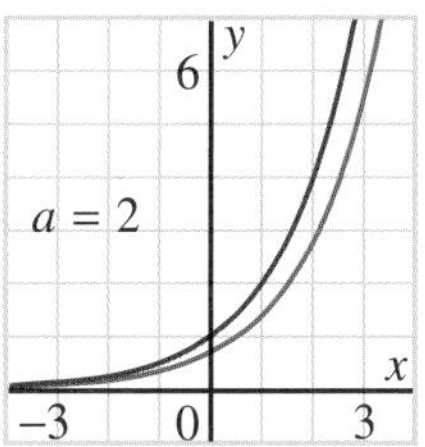

As h gets smaller and smaller, the value of a^x is not affected. Therefore:

$$f'(x) = a^x \lim_{h \to 0} \frac{a^h - 1}{h}$$

That is, $\frac{d}{dx}(a^x) = \left(\lim_{h \to 0} \frac{a^h - 1}{h}\right) \times a^x$ ... ①

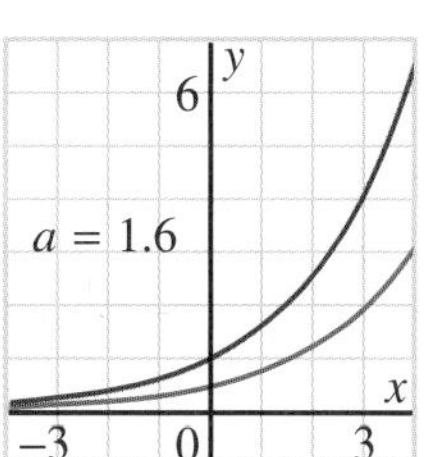

This result provides further evidence that the derivative of an exponential function is proportional to the function. The expression, $\lim_{h \to 0} \frac{a^h - 1}{h}$, corresponds to the constant, k, described on page 439. However, we cannot be certain that this limit exists. Its value, if it does exist, depends on the value of a. We will determine the value of the limit in Section 7.4. In the meantime, we can use equation ① to determine the derivative of an exponential function, but a separate calculation is required for every base.

Example 1

Determine $\frac{d}{dx}(5^x)$.

Solution

Use $\frac{d}{dx}(a^x) = \left(\lim_{h \to 0} \frac{a^h - 1}{h}\right) \times a^x$.

Substitute $a = 5$.

$$\frac{d}{dx}(5^x) = \left(\lim_{h \to 0} \frac{5^h - 1}{h}\right) \times 5^x$$

To estimate the limit, substitute a very small number for h, such as 0.001, into the expression $\frac{5^h - 1}{h}$.

When $h = 0.001$, $\frac{5^h - 1}{h} = \frac{5^{0.001} - 1}{0.001}$

$\doteq 1.6107$

Therefore, $\frac{d}{dx}(5^x) \doteq 1.6107 \times 5^x$

Introducing the Function $y = e^x$

Each of the 5 graphs on page 439 shows an exponential function and its derivative. These appear to coincide on the middle graph. Therefore, there should be a value of a near 2.7 such that equation ① becomes $\frac{d}{dx}(a^x) = a^x$. For this value of a, $\lim_{h \to 0} \frac{a^h - 1}{h} = 1$. To determine this value of a, we substitute numbers close to 2.7 for a, and a very small number for h, such as $h = 0.000\,001$. Some results are shown in the table. To 5 decimal places, the number a, for which $\frac{d}{dx}(a^x) = a^x$, is 2.718 28. We represent this number with the letter e.

a	$\frac{a^{0.000\,001} - 1}{0.000\,001}$
2.7	0.993 252 266
2.71	0.996 949 132
2.718	0.999 896 816
2.718 2	0.999 970 396
2.718 28	0.999 999 827

Using this value of a, equation ① becomes $\frac{d}{dx}(e^x) = e^x$. This result is significant because it provides an example of a function that equals its derivative. This function is $y = e^x$.

Take Note

The Derivative of the Function y = e^x

We define e to be the number such that $\lim_{h \to 0} \frac{e^h - 1}{h} = 1$.

To 5 decimal places, $e = 2.718\ 28$

The derivative of $y = e^x$ is the function itself.

$\frac{d}{dx}(e^x) = e^x$

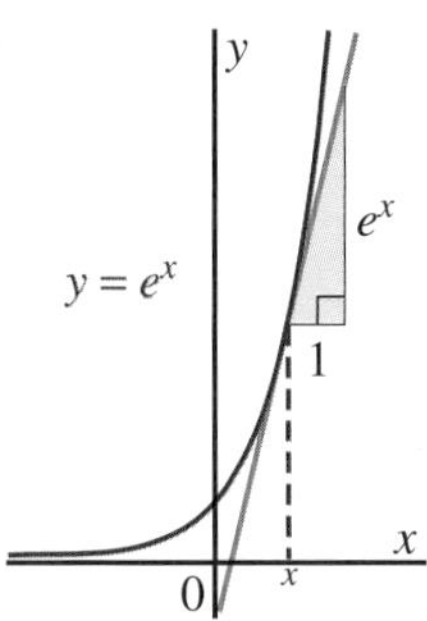

Visualize a point moving from left to right along the graph of $y = e^x$. The slope of the tangent is the same as the y-coordinate of the point.

The derivative of the function $y = e^x$ is equal to the function, and their graphs coincide.

Example 2

Differentiate each function.

a) $y = 3e^x$ **b)** $y = e^{3x}$ **c)** $y = e^{x+3}$ **d)** $y = xe^x$

Solution

a) $y = 3e^x$

Use the constant multiple rule.

$\frac{dy}{dx} = 3e^x$

b) $y = e^{3x}$

Use the chain rule.

$$\begin{aligned}\frac{dy}{dx} &= e^{3x} \times 3 \\ &= 3e^{3x}\end{aligned}$$

c) $y = e^{x+3}$

Use the chain rule.

$$\begin{aligned}\frac{dy}{dx} &= e^{x+3} \times 1 \\ &= e^{x+3}\end{aligned}$$

d) $y = xe^x$

Use the product rule.

$$\begin{aligned}\frac{dy}{dx} &= (1)(e^x) + (x)(e^x) \\ &= e^x(1 + x)\end{aligned}$$

In *Example 2b* and *c*, we used the chain rule to differentiate functions of the form $y = e^{f(x)}$.

Take Note

The Derivative of the Function $y = e^{f(x)}$

To differentiate $y = e^{f(x)}$, where $f(x)$ is differentiable, use the chain rule.

$$\frac{d}{dx}(e^{f(x)}) = e^{f(x)} \times f'(x)$$

7.1 Exercises

A

1. Use the equation $\frac{d}{dx}(a^x) = \left(\lim_{h \to 0} \frac{a^h - 1}{h}\right) \times a^x$ to confirm these results:

   a) $\frac{d}{dx}(2^x) \doteq 0.69 \times 2^x$ b) $\frac{d}{dx}(3^x) \doteq 1.10 \times 3^x$

2. Differentiate each function.

   a) $y = 2e^x$ b) $y = e^{2x}$ c) $y = e^{x+2}$ d) $y = e^{x^2}$ e) $y = e^{\frac{1}{x}}$ f) $y = e^{-x}$

B

3. You know that $\frac{d}{dx}(x^n) = nx^{n-1}$. Explain why we cannot use this rule to determine $\frac{d}{dx}(a^x)$.

Exercise 3 is significant because it highlights the difference between power functions and exponential functions. Although both x^n and a^x are powers, different differentiation rules are required to differentiate $y = x^n$ and $y = a^x$.

4. The graphs of $f(x) = 2^x$, $f(x) = 3^x$, and their derivatives are shown on page 438. For each graph, describe what happens to the slopes of the tangents to the graph of f as x increases. Explain how the graph of the derivative function shows what is happening to the slopes of these tangents.

5. **Communication** Decide whether this statement is true or false: "There is only one function that equals its derivative." Explain your decision.

6. Differentiate each function.

   a) $y = 2xe^x$ b) $y = xe^{2x}$ c) $y = x^2e^{-2x}$

   d) $y = x^2(e^x + 1)$ e) $y = \frac{x^2}{1 - e^x}$ f) $y = \frac{1 + e^{2x}}{1 - e^{2x}}$

7. **Knowledge/Understanding** Differentiate each function.

   a) $y = x^2e^x$ b) $y = x(1 + e^x)$ c) $y = e^x(1 + x)$

   d) $y = e^x(1 + e^{2x})$ e) $y = \frac{1 + e^x}{e^x}$ f) $y = \frac{1 - e^{-x}}{x^2}$

8. Differentiate each function.

a) $e^x + e^y = 1$ b) $e^{xy} = 1$ c) $e^{xy} = x$

9. Consider the exponential function $y = e^x$. Use the derivative, $y' = e^x$, to justify your answer to each question about the graph of $y = e^x$.

a) Are there any critical points?

b) Are there any points of inflection?

c) What are the intervals of increase and decrease?

d) Is the graph concave up or concave down?

10. **Application** Consider the graph of $y = e^{-x^2}$ at the right. Use calculus to determine each feature listed.

a) the coordinates of any maximum or minimum points

b) the coordinates of any points of inflection

c) the intervals of increase and decrease

d) the intervals of concavity

e) the equation of the tangent to the graph at the point $(1, \frac{1}{e})$

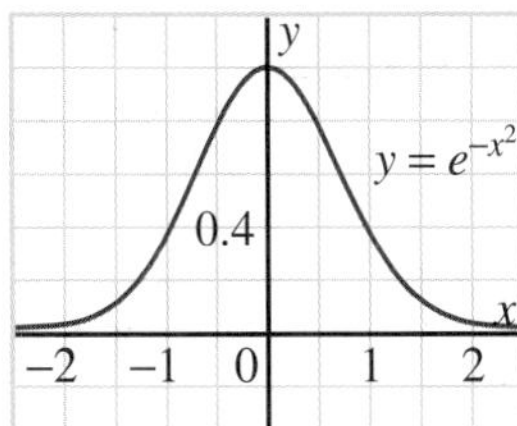

11. For the function in exercise 10:

a) Sketch the graph of the first derivative function.

b) Sketch the graph of the second derivative function.

12. **Thinking/Inquiry/Problem Solving** Use the graph of $y = e^{-x^2}$ in exercise 10. Determine the area of the largest rectangle that has one side on the x-axis and two vertices on the graph of $y = e^{-x^2}$.

13. a) Sketch the graph of the function $y = e^x$. Mark the points $(0, 1)$ and $(1, e)$ on the graph. Draw the tangent at each point.

b) Determine the equation of each tangent in part a.

14. Analyse the key features of the graph of each function, then sketch the graph.

a) $y = -e^x$ b) $y = e^{2x}$

15. Analyse the key features of the graph of each function, then sketch the graph.

a) $y = 2e^x$ b) $y = e^{-\frac{x}{2}}$

Key Features

Domain

Intercepts

Critical points

Intervals of increase/decrease

Points of inflection

Intervals of concavity

Asymptotes

Point discontinuities

16. Differentiate each function.

a) $y = e^{\sin x}$ b) $y = \sin e^x$ c) $y = e^x \sin x$

d) $y = e^{\cos 2x}$ e) $y = \sin e^{2x}$ f) $y = e^{2x} \cos 3x$

7.2 Natural Logarithms

In Section 7.1, we defined e as the number such that $\frac{d}{dx}(e^x) = e^x$. This provides a function whose derivative equals itself. That is, the derivative of $f(x) = e^x$ is $f'(x) = e^x$. This is more than an intellectual curiosity. It is a very important result, and we can do a lot with it. Its significance will become clear in Section 7.3, where we will use base e in applied problems.

When we use base e in numerical calculations, it is simpler to use logarithms to base e instead of base-10 logarithms. Base-e logarithms are called *natural logarithms*, and you can determine them using the [LN] key on a calculator. The expression $\ln x$ is read "the natural logarithm of x," or "lawn x." The expression $\ln x$ is the same as $\log_e x$.

Example 1

Use a calculator to determine each logarithm to 4 decimal places. Explain the meaning of each result.

a) $\ln 74.45$ **b)** $\ln 1.82$ **c)** $\ln 0.000\ 25$

Solution

a) Press: [LN] 74.45 [)] [ENTER] to display 4.310127759
$\ln 74.45 \doteq 4.3101$
This means that $e^{4.3101} \doteq 74.45$.

b) $\ln 1.82 \doteq 0.5988$
This means that $e^{0.5988} \doteq 1.82$.

c) $\ln 0.000\ 25 \doteq -8.2940$
This means that $e^{-8.2940} \doteq 0.000\ 25$.

Take Note

Definition of ln x

The expression $\ln x$ is an exponent. It is the exponent that e has when x is written as a power of e.

$\ln x = y$ means $e^y = x$, where $x > 0$.

To remember these relationships, think of them in these ways:

$$\ln x = y \qquad\qquad x = e^y$$

e raised to this number ... equals this number.

This number ... is the natural logarithm of this number.

Note: To input e^x, press 2nd LN. To determine the value of *e*, input 2nd LN 1) ENTER.

Special cases occur when we determine the natural logarithms of powers of *e*. For example:

$\ln e = 1$

$\ln e^2 = 2$

$\ln e^3 = 3$

```
ln(e^(1))
                1
ln(e^(2))
                2
ln(e^(3))
                3
```

Natural logarithms satisfy the laws of logarithms because these laws were developed for any base in Chapter 6. Therefore, we can do the same things with natural logarithms that we did with base-10 logarithms. For example, we can use natural logarithms to solve exponential equations.

Example 2

Use natural logarithms to solve the equation $6^x = 53$ to 4 decimal places.

Solution

$6^x = 53$

Take the natural logarithm of each side.

$\ln 6^x = \ln 53$

Use the law of logarithms for powers.

$x \ln 6 = \ln 53$

$$x = \frac{\ln 53}{\ln 6}$$

$$\doteq 2.2159$$

Compare the above solution with the solution of *Example 1*, page 387. All the steps are the same. The only difference is that the logarithms have different bases, but the results are the same.

Example 3

Solve each equation to 4 decimal places.

a) $\ln x = 3$ **b)** $e^x = 4$ **c)** $7 = 2e^{1.5x}$

Solution

a) $\ln x = 3$ means $e^3 = x$.

Hence, $x = e^3$

Press: [2nd] [LN] 3 [)] [ENTER]

to display 20.08553692

$x \doteq 20.0855$

b) $e^x = 4$

Take the natural logarithm of each side.

$x \ln e = \ln 4$ Recall that $\ln e = 1$.

$x = \ln 4$

$\doteq 1.3863$

c) $7 = 2e^{1.5x}$

Take the natural logarithm of each side.

$\ln 7 = \ln 2 + \ln e^{1.5x}$

$\ln 7 = \ln 2 + 1.5x$

$$x = \frac{\ln 7 - \ln 2}{1.5}$$

$\doteq 0.8352$

7.2 Exercises

A

1. **Knowledge/Understanding** Use a calculator to determine each logarithm to 4 decimal places. Explain the meaning of each result.

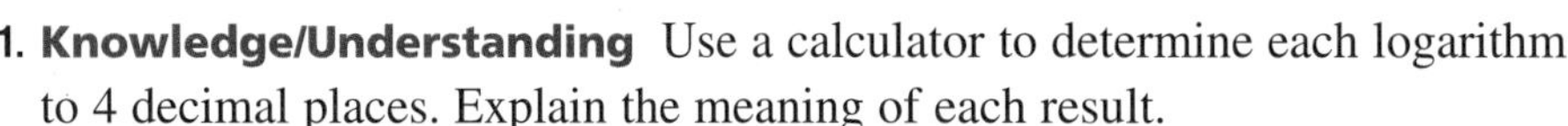

a) $\ln 465$ **b)** $\ln 10$ **c)** $\ln 2.8$ **d)** $\ln 2.6$ **e)** $\ln 0.6$ **f)** $\ln 0.000\,33$

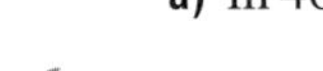

2. Refer to the logarithms in exercise 1. Explain your answer to each question.
   a) Why are the first four logarithms positive and the last two negative?
   b) Why is the logarithm in part c slightly greater than 1?
   c) Why is the logarithm in part d slightly less than 1?

3. What is the value of each logarithm? Explain.
   a) $\ln 1$ **b)** $\ln e$

B

4. Use natural logarithms to solve each equation to 4 decimal places.

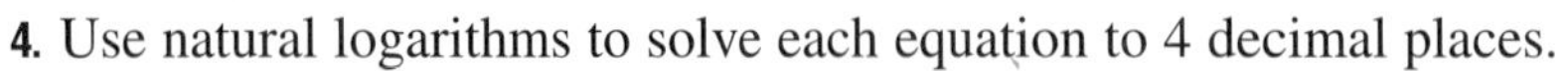

a) $2^x = 9$ **b)** $5^x = 18.6$ **c)** $3^x = 2$ **d)** $6^x = 3.25$

5. Solve each equation to 4 decimal places where necessary.

a) $\ln x = 4$ b) $\ln x = 1$ c) $\ln x = 0$ d) $\ln x = -1$

6. Solve each equation to 4 decimal places where necessary.

a) $e^x = 1$ b) $e^x = 2$ c) $e^x = e$ d) $e^x = 0.5$

e) $e^{2x} = 5$ f) $e^{0.5x} = 3$ g) $2e^{0.6x} = 3$ h) $4 = 5e^{1.2x}$

i) $3e^x = 4$ j) $2 = 3e^{1.2x}$ k) $5 = 2e^{0.25x}$ l) $6 = 1.2e^{0.02x}$

7. **Application** The calculator screen shows the results of two calculations. Each calculation includes the number 2.376, and the result is also 2.376. Is this a coincidence, or is there a reason why the result is the same as the number we started with? Explain.

Exercises 7, 8, and 9 are significant because they confirm that determining a natural logarithm and determining a power of e are inverse operations. When these operations are applied in succession they leave a number unchanged. Therefore, the following equations are true for all values of x for which the expressions are meaningful.

$\ln e^x = x$ $\quad e^{\ln x} = x, x > 0$

These properties are a direct consequence of the definition of $\ln x$.

8. Simplify each expression.

a) $\ln e^{3.5}$ b) $\ln e^{0.5}$ c) $\ln e^{-1}$ d) $\ln e^{-4.25}$

9. Simplify each expression.

a) $e^{\ln 3.2}$ b) $e^{\ln 1}$ c) $e^{\ln 0.5}$ d) $e^{\ln e}$

10. Write each expression as a single natural logarithm.

a) $\ln 4 + \ln 3$ b) $\ln 6 + \ln 2$ c) $\ln 12 + \ln 1$

d) $\ln 30 - \ln 10$ e) $\ln 30 - \ln 6$ f) $\ln 30 - \ln 2$

11. **Communication** If you use a calculator to determine ln 2 and ln 0.5, you will find that:

$\ln 2 \doteq 0.693\ 147\ 180\ 6$

$\ln 0.5 \doteq -0.693\ 147\ 180\ 6$

Explain, in two different ways, why these logarithms are opposites.

12. **Thinking/Inquiry/Problem Solving**

a) Use a calculator to determine the logarithms in each list.

i) log 5, log 50, log 500, log 5000

ii) ln 5, ln 50, ln 500, ln 5000

b) In the results of part a, there is an obvious pattern in the first list, but not an obvious pattern in the second. Find a pattern in the second list. Describe the pattern.

Self-Check 7.1, 7.2

1. Differentiate each function.

a) $y = 4e^x$ **b)** $y = e^{4x}$ **c)** $y = e^{x+4}$

d) $y = e^{x^4}$ **e)** $y = e^{\frac{4}{x}}$ **f)** $y = e^{-4x}$

2. Differentiate each function.

a) $y = 4xe^x$ **b)** $y = xe^{4x}$ **c)** $y = x^4e^{-4x}$

d) $y = x^4(e^x + 4)$ **e)** $y = \dfrac{x^4}{4 - e^x}$ **f)** $y = \dfrac{4 + e^{4x}}{4 - e^{4x}}$

3. Differentiate each function.

a) $y = x^2e^{3x}$ **b)** $y = 2x(3 + e^x)$ **c)** $y = e^{2x}(1 + x^2)$

d) $y = e^{3x}(1 + e^x)$ **e)** $y = \dfrac{1 + 2e^{4x}}{e^{2x}}$ **f)** $y = \dfrac{2 - e^{-2x}}{x^3}$

4. Use a calculator to determine each logarithm to 4 decimal places. Explain the meaning of each result.

a) $\ln 325$ **b)** $\ln 12$ **c)** $\ln 1.2$ **d)** $\ln 0.05$

5. Use natural logarithms to solve each equation to 4 decimal places.

a) $3^x = 25$ **b)** $4^x = 150$ **c)** $5^x = 3$

6. Solve each equation to 4 decimal places.

a) $\ln x = 5$ **b)** $e^x = 3$ **c)** $5 = 3e^{2x}$

7. Simplify each expression.

a) $\ln e^{x^2}$ **b)** $\ln e^{\frac{1}{x}}$ **c)** $\ln e^{1-x}$

8. The key features of the graph of a function are: domain; intercepts; critical points; intervals of increase, decrease, and concavity; points of inflection; asymptotes; and point discontinuities. Analyse the key features of the graph of $y = 2e^{2x}$, then sketch the graph.

PERFORMANCE ASSESSMENT

9. For the function at the right:

a) Sketch the graph of the first derivative function.

b) Sketch the graph of the second derivative function.

c) Determine the coordinates of any critical points.

d) Determine the equation of the tangent to the graph at the point $(1, \frac{e}{2})$.

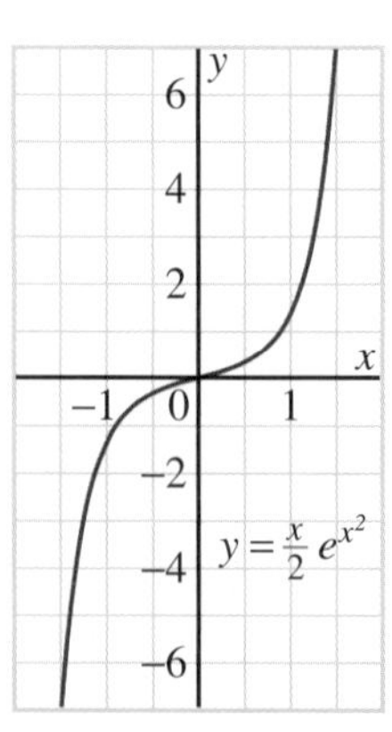

7.3 The Nature of Exponential Growth

At the beginning of the year 2000, the world population was approximately 6.0 billion, and was growing at an annual rate of about 1.5%. The population, P billion, can be modelled by the equation $P = 6.0(1.015)^t$, where t represents the time in years since 2000. When we use this equation, we assume that the growth rate is constant. We can express the growth rate as a percent or as a number.

Consider the equation $P = 6.0(1.015)^t$.
Suppose we change the base in this equation to e.
Let $1.015 = e^x$.
Solve for x. Take the natural logarithm of each side.

$\ln 1.015 = \ln e^x$

$\ln 1.015 = x$

$x \doteq 0.014\,888\,612\,5$

Therefore, $1.015 \doteq e^{0.014\,89}$

This means that the population equation can be written in these forms:

$$P = 6.0(1.015)^t \qquad \ldots ①$$

$$P = 6.0\,e^{0.014\,89t} \qquad \ldots ②$$

Equation ② has two significant features:

- The base is e. This means that we can differentiate the equation and solve problems involving the instantaneous growth rate.
- The coefficient in the exponent, 0.014 89, is slightly less than the decimal part of the base in equation ①.

Using Base *e*

Differentiate equation ②.

$$P = 6.0\,e^{0.014\,89t}$$

$$\frac{dP}{dt} = 6.0\,e^{0.014\,89t} \times 0.014\,89$$

$$\frac{dP}{dt} = 0.089\,34\,e^{0.014\,89t} \qquad \ldots ③$$

This result is significant because it is a constant times the population function. The derivative of the population function is proportional to the function. This is reasonable because the population is growing exponentially. When a quantity grows exponentially, the numerical growth rate is proportional to the size of the quantity. This is a fundamental property of exponential growth. For example, if the population were twice as great, the numerical growth rate would also be twice as great.

We can use equation ③ to determine the numerical instantaneous growth rate of the world population at any time. For example, if we substitute $t = 0$, we obtain the numerical instantaneous growth rate at the beginning of 2000.

When $t = 0$, $\frac{dP}{dt} = 0.089\,34\,e^0$
$= 0.089\,34$

At the beginning of 2000, the population was growing at the rate of about 0.089 34 billion, or 89.34 million, per year. This situation is illustrated below. The graph is not drawn to scale.

Point A represents the world population on January 1, 2000.

Point B represents the population on January 1, 2001 (calculated using either equation ① or equation ②).

Line AC is the tangent to the graph at A, and has slope 0.089 34. This is the instantaneous growth rate of the world population on January 1, 2000. Secant AB has slope 0.09. This is the numerical average growth rate during 2000.

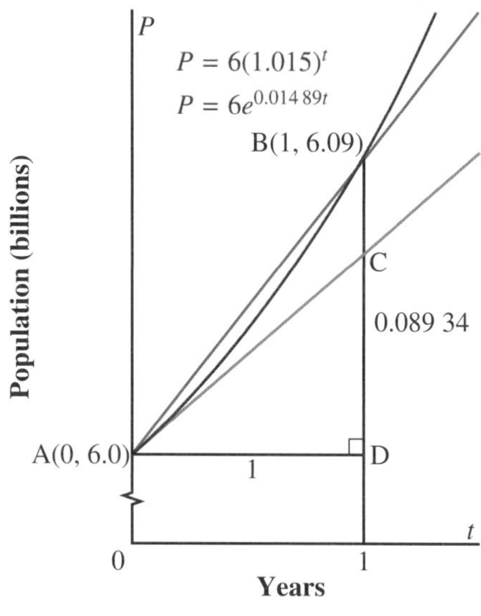

Interpreting the Exponent

The decimal part of the base in equation ① is 0.015. This represents an average growth rate of 1.5%. Since the population at the beginning of 2000 was 6 billion, the numerical average growth rate in 2000 is 1.5% of 6 billion, which is 0.09 billion, or 90 million. This is the slope of secant AB on the diagram.

The coefficient in the exponent in equation ② is 0.014 89. This number illustrates that the population growth is increasing continuously throughout the year, and not all at once at the end of the year. It represents the instantaneous growth rate of 1.489%. The numerical instantaneous growth rate at the beginning of 2000 is 1.489% of 6 billion, which is 0.089 34 billion, or 89.34 million. This is the slope of tangent AC on the diagram. The instantaneous rate of 1.489% is slightly less than the average rate of 1.5%.

Equations ① and ② represent the same function. Equation ① contains the *average* rate of increase (a percent expressed in decimal form) as the base of the power. Equation ② contains the *instantaneous* rate of increase (a percent expressed in decimal form) as a coefficient in the exponent. In many situations, such as the one above, these rates of growth are approximately equal. However, equation ② has the advantage that it can be easily differentiated.

Take Note

Equations of Exponential Functions

The equation of any exponential function can be expressed in two forms. In each case, y_0 is the value of y when $x = 0$.

Basic form $y = y_0(1 + r)^x$

- r represents the average rate of change (a percent expressed in decimal form).
- In growth situations, $r > 0$; in decay situations, $-1 < r < 0$

Base *e* form $y = y_0 e^{kx}$

- k represents the instantaneous rate of change (a percent expressed in decimal form).
- In growth situations, $k > 0$; in decay situations, $k < 0$

We can differentiate the equation in base-*e* form more easily than the equation in basic form. The derivative, $y' = ky_0 e^{kx}$, is k times the function; that is, $y' = ky$.

Example 1

At the beginning of the year 2000, the population of Indonesia was 221 million. It was growing at approximately 1.5% per annum.

a) Write an equation for the population, P million, after t years, using an exponential function with base e.

b) Assume the growth rate is constant.

i) Determine the predicted population at the beginning of 2012.

ii) Determine when the population will reach 250 million.

c) Predict the numerical instantaneous growth rate in the population at the beginning of 2005.

Solution

a) The population can be modelled by $P = 221(1.015)^t$.
From page 449, $1.015 \doteq e^{0.01489}$

So, an equation for the approximate population, in millions, is $P = 221e^{0.01489t}$, where t is the time in years after 2000.

b) **i)** Substitute $t = 12$.

$$P = 221\,e^{0.01489 \times 12}$$
$$\doteq 264.236$$

The predicted population at the beginning of 2012 is approximately 264 million.

ii) Substitute $P = 250$.

$$250 = 221e^{0.01489t}$$

Solve for t by taking the natural logarithm of each side.

$$\ln 250 = \ln 221 + \ln e^{0.01489t}$$

$$\ln 250 = \ln 221 + 0.01489t$$

$$t = \frac{\ln 250 - \ln 221}{0.01489}$$

$$= \frac{\ln\left(\frac{250}{221}\right)}{0.01489}$$

$$\doteq 8.28061$$

The population will reach 250 million during the 8th year after 2000; that is, in 2008.

c) Differentiate the function $P = 221e^{0.01489t}$.

$$\frac{dP}{dt} = 221e^{0.01489t} \times 0.01489$$
$$\doteq 3.291e^{0.01489t}$$

This is the numerical instantaneous growth rate at any time t.

Substitute $t = 5$.

$$\frac{dP}{dt} = 3.291e^{0.01489 \times 5}$$
$$\doteq 3.545$$

At the beginning of 2005, the population is increasing at approximately 3.5 million per year.

Example 2

Uganda has one of the world's fastest growing populations. At the beginning of 1995, its population was 19.2 million. At the beginning of 2000, it was 24.1 million.

a) Write an equation to represent the population of Uganda, P million, in terms of the time in years, t, since 1995.

b) Estimate the percent instantaneous growth rate of Uganda's population.

Solution

a) Assume the growth rate is exponential.

Let $P = 19.2e^{kt}$, where P is the population in millions, and k is to be determined.

Since the population in 2000 was 24.1 million, substitute $P = 24.1$ and $t = 5$.

$$24.1 = 19.2e^{5k}$$

Solve for k by taking the natural logarithm of each side.

$$\ln 24.1 = \ln 19.2 + \ln e^{5k}$$

$$\ln 24.1 = \ln 19.2 + 5k$$

$$k = \frac{\ln 24.1 - \ln 19.2}{5}$$

$$= \frac{\ln\left(\frac{24.1}{19.2}\right)}{5}$$

$$\doteq 0.045\,46$$

An equation to represent the approximate population is $P = 19.2e^{0.045t}$, where t is the time in years since 1995.

b) From the equation in part a, the coefficient of t in the exponent is 0.045. So, the instantaneous growth rate is approximately 4.5% per year.

In *Example 2*, the equation $P = 19.2e^{0.045t}$ represents the population of Uganda t years after 1995. The coefficient on the right side is the 1995 population. If we replace this coefficient with the year 2000 population of 24.1 million, the equation becomes $P = 24.1e^{0.045t}$. This represents the population of Uganda t years after 2000.

So far in this section, we have been dealing with exponential growth. *Example 3* deals with exponential decay.

Example 3

Water is brought to a boil then removed from the heat. The temperature of the water, T degrees Celsius, is modelled as a function of time, t minutes, by the equation $T = 80e^{-0.057t} + 20$.

a) Determine the temperature after 15 min.

b) Determine the rate at which the temperature is decreasing after 15 min.

c) Determine how long it takes for the temperature to reach 30°C.

d) Sketch graphs of the function and the derivative function.

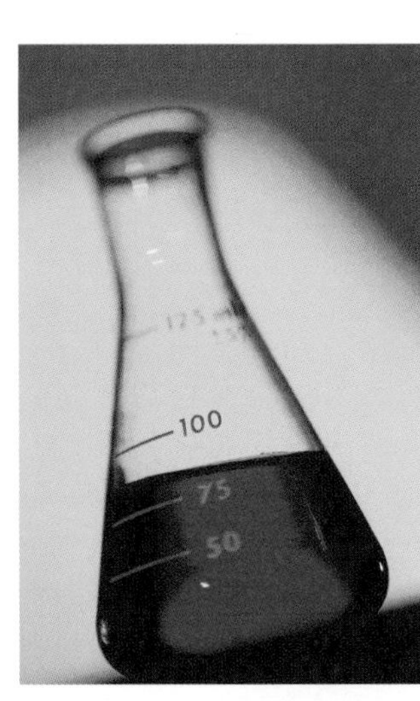

Solution

a) $T = 80e^{-0.057t} + 20$

Substitute $t = 15$.

$$T = 80e^{-0.057 \times 15} + 20$$
$$\doteq 54.02$$

The temperature after 15 min is about 54°C.

b) $T = 80e^{-0.057t} + 20$

Differentiate the function.

$$\frac{dT}{dt} = 80e^{-0.057t} \times (-0.057)$$
$$= -4.56e^{-0.057t}$$

Substitute $t = 15$.

$$\frac{dT}{dt} = -4.56e^{-0.057 \times 15}$$
$$\doteq -1.939$$

After 15 min, the temperature is decreasing at about 1.9°C/min.

c) $T = 80e^{-0.057t} + 20$

Substitute $T = 30$, then solve for t.

$$30 = 80e^{-0.057t} + 20$$
$$80\,e^{-0.057t} = 10$$
$$e^{-0.057t} = \frac{1}{8}$$

Take the natural logarithm of each side.

$$\ln e^{-0.057t} = \ln\left(\frac{1}{8}\right)$$
$$-0.057t = \ln\left(\frac{1}{8}\right)$$
$$t = \frac{\ln\left(\frac{1}{8}\right)}{-0.057}$$
$$\doteq 36.481$$

It takes about 36 min for the temperature to reach 30°C.

d) From the equation of the function $T = 80e^{-0.057t} + 20$:

When $t = 0$, $T = 80e^{0} + 20 = 100$

When $t = 60$, $T = 80e^{-0.057 \times 60} + 20 \doteq 22.6$

From part a, when $t = 15$, $T \doteq 54$

From part c, when $t \doteq 36$, $T = 30$

On a grid, plot the points (0, 100), (15, 54), (36, 30), and (60, 22.6). Join the points with a smooth curve for the graph of the function T.

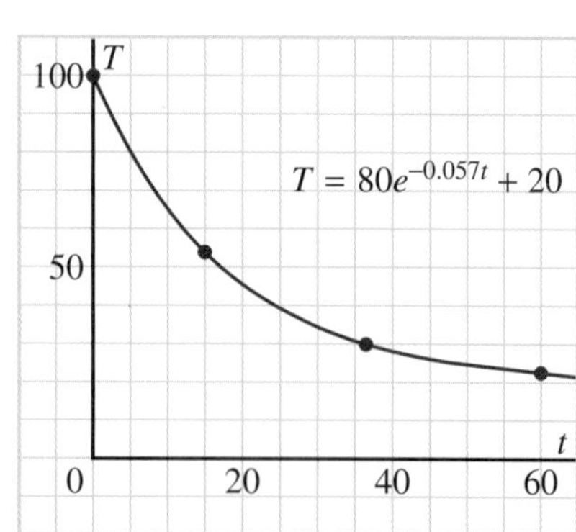

From the equation of the derivative function $T' = -4.56e^{-0.057t}$:

When $t = 0$, $T' = -4.56e^{0} = -4.56$

When $t = 60$, $T' = -4.56e^{-0.057 \times 60} \doteq -0.15$

From part b, when $t = 15$, $T' \doteq -1.9$

From the graph of the function T, as t increases, the slopes of the tangents increase.

On a grid, plot the points $(0, -4.56)$, $(15, -1.9)$, and $(60, -0.15)$. Join the points with a smooth curve for the graph of the derivative function T'.

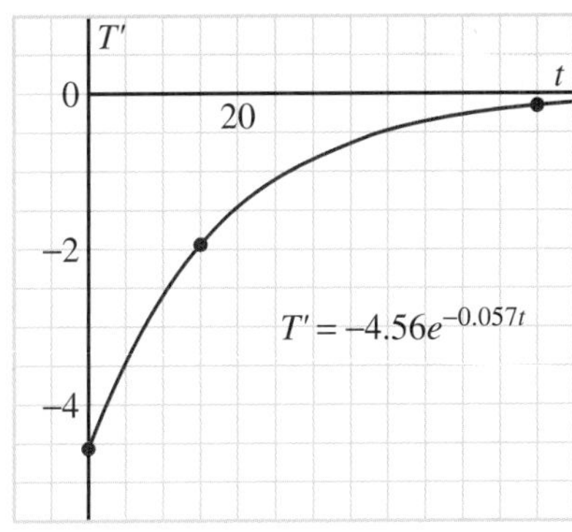

Four Definitions of *e*

The number e can be defined in several ways. Here are four possible definitions:

Definition 1

e is the number such that

$$\lim_{h \to 0} \frac{e^h - 1}{h} = 1$$

Definition 2

e is the number such that

$$\frac{d}{dx}(e^x) = e^x$$

Definition 3

$$e = \lim_{n \to \infty} \left(1 + \frac{1}{n}\right)^n$$

Definition 4

$$e = \lim_{m \to 0} (1 + m)^{\frac{1}{m}}$$

Mathematicians need to be certain that these definitions define the same number. In Section 7.1, we deduced $\frac{d}{dx}(e^x) = e^x$ from $\lim_{h \to 0} \frac{e^h - 1}{h} = 1$. So, Definitions 1 and 2 are equivalent. To show that Definitions 3 and 4 are equivalent, we replace n in Definition 3 with $\frac{1}{m}$. To show that Definitions 3 and 4 are equivalent to Definitions 1 and 2, we would need to show that $\lim_{h \to 0} \frac{e^h - 1}{h} = 1$ can be deduced from $e = \lim_{n \to \infty} \left(1 + \frac{1}{n}\right)^n$. This proof is beyond the scope of this course.

In exercise 15, you will consider different compounding periods that illustrate the approximation of $\lim_{n \to \infty} \left(1 + \frac{1}{n}\right)^n$.

7.3 Exercises

1. Each equation represents the population, P million, of a country t years after 2000. State the population at the beginning of 2000 and the percent instantaneous growth rate for each country.

   a) Mexico $P = 103e^{0.0178t}$

   b) United Kingdom $P = 59.4e^{0.0034t}$

   c) Bangladesh $P = 127e^{-0.0020t}$

2. The equation, $\frac{dP}{dt} = 0.089\,34e^{0.014\,89t}$, represents the numerical instantaneous growth rate of the world population at the beginning of the tth year after 2000. Calculate the numerical instantaneous growth rate at the beginning of each year.

   a) 2001 b) 2002 c) 2005

3. Refer to the graphs in *Example 3*, pages 454, 455.

   a) Explain why each graph has a horizontal asymptote.

   b) Explain why the horizontal asymptotes are different.

4. At the beginning of 2000, the population of Saudi Arabia was 18.7 million. It was growing at approximately 3.5% per annum.

   a) Write an equation for the population, P million, after t years, using an exponential function with base e.

   b) Assume the growth rate is constant.

      i) Determine the predicted population at the beginning of 2010.

      ii) Determine when the population will reach 50 million.

   c) Predict the numerical instantaneous growth rate in the population at the beginning of 2006.

5. Repeat exercise 4 for each country below, given its population at the beginning of 2000 and the growth rate at that time.

   a) United States 276 million 0.82%

   b) Argentina 37.6 million 1.7%

   c) South Africa 42.8 million –0.96%

6. **Communication** In *Example 1*, instead of using $P = 221e^{0.014\,89t}$ to represent Indonesia's population, we could have used $P = 221(1.015)^t$.

   a) Complete parts b and c of *Example 1* using this equation.

   b) Explain the advantages and disadvantages of the two equations.

7. India and China are the world's two most populous countries. At the beginning of 2000, their populations were 1026 million and 1264 million, respectively. India's population was growing at about 1.65% per annum, while China's was growing more slowly at about 0.90% per annum. Assume the growth rates remain constant. Determine the year when India's population will become greater than China's.

✓ 8. **Knowledge/Understanding** The populations of three countries, at the beginning of 1995 and 2000, are listed below. Choose one country. Assume the population growth is exponential.

a) Write an equation to model the population, P million, of the country as a function of t, the time in years since 1995.

b) Estimate the percent instantaneous growth rate of the population.

Country	1995 population (millions)	2000 population (millions)
Brazil	162	176
Nepal	21.9	25.4
Singapore	2.96	3.80

✓ 9. Choose one country from the list in exercise 8. Pose your own problem about population growth. Solve your problem. Write a clear explanation of your problem and its solution.

10. **Application** An important application of exponential decay is to estimate the age of ancient specimens using a method known as *carbon dating*. All living matter contains traces of carbon-14, which is a radioactive isotope of carbon. Carbon-14 decays with a half-life of about 5730 years. After t years, the percent, P, of carbon-14 remaining compared with modern living matter is modelled by $P = 100e^{-0.000\,12t}$.

a) In 1995, scientists discovered some pine logs that had been floating undisturbed in a lake in Algonquin Park. Tests showed that one log had fallen into the lake about 500 years earlier. Determine the percent of carbon-14 remaining in a sample taken from this log.

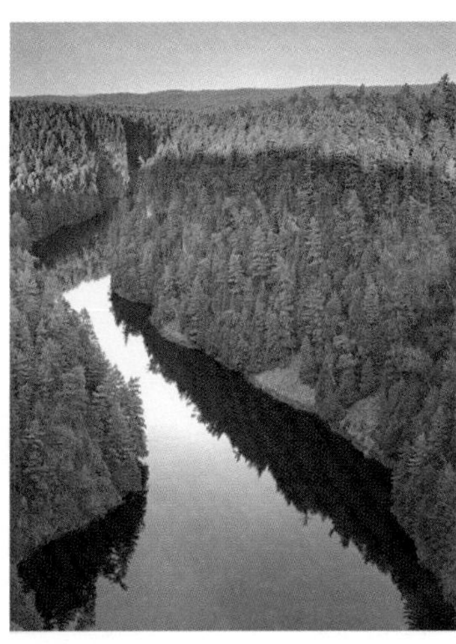

b) In 1993, miners at Last Chance Creek, Yukon, unearthed the remains of an ancient horse. Tests showed that the carbon-14 in the remains was about 4.83% relative to modern living matter. Estimate the age of the horse.

c) Solve the equation $P = 100e^{-0.000\,12t}$ for t.

d) Use the result in part c. Determine the approximate age of each specimen, given the percent of its carbon-14 content relative to modern living matter.

i) Shroud of Turin 92.0%

ii) specimen from the end of the last ice age 24.0%

iii) paintings in Lascaux Cave, France 15.3%

iv) paintings in Chauvet Cave, France 1.5%

✓ **11.** Refer to exercise 10. Show that the equation $P = 100e^{-0.000\,12t}$ is equivalent to the equation $P = 100(0.5)^{\frac{t}{5730}}$.

✓ **12.** Atmospheric pressure is an exponential function of altitude above sea level. The pressure at sea level is 130 kPa, and the pressure at 10 000 m is 27.6 kPa.

a) Determine an equation that expresses the pressure, P kilopascals, as a function of altitude, h metres.

b) Pose your own problem about atmospheric pressure. Solve the problem you posed. Write a clear explanation of your problem and its solution.

✓ **13.** A container of juice is removed from the refrigerator and left on the counter. The temperature of the juice, T degrees Celsius, is modelled as a function of time, t minutes, by the equation $T = 20 - 17e^{-0.045t}$.

a) Determine the temperature of the juice after 30 min.

b) Determine the rate at which the temperature is increasing after 30 min.

c) Determine how long it takes for the temperature to reach 18°C.

d) Sketch graphs of the temperature function and its derivative.

✓ **14.** The concentration of a certain prescription drug in the bloodstream after a single dose can be modelled by the equation $c = 100te^{-0.5t}$, where c represents the concentration of the drug, and t represents the time in hours since the drug was taken.

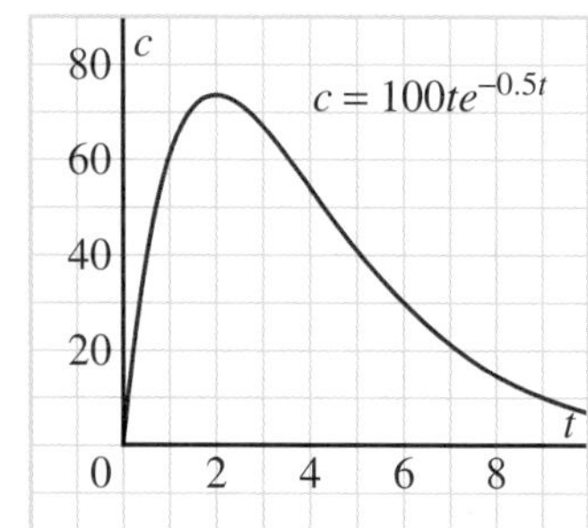

a) Determine $\frac{dc}{dt}$.

b) Determine the maximum concentration of the drug in the bloodstream, and when this occurred.

✓ **15. Thinking/Inquiry/Problem Solving** Here is a fanciful compound interest situation that has an important connection with the number e. Suppose you invest \$1 for one year at an interest rate of 100%.

Exercise 15 is significant because it illustrates Definition 3 on page 455:

$$e = \lim_{n\to\infty}\left(1 + \frac{1}{n}\right)^n.$$

a) Determine, as accurately as possible, the amount you would have for each compounding period.

i) annually ii) semi-annually iii) monthly

iv) daily v) every minute vi) every second

b) Explain how the results of part a illustrate that $e = \lim_{n\to\infty}\left(1 + \frac{1}{n}\right)^n$.

✓ **16.** The coefficient in the exponent of the equation $P = 6.0e^{0.014\,89t}$ is the instantaneous growth rate of the world population as a percent. We showed this on pages 449, 450.

a) Show this in another way, as follows. Determine the population at the end of one year, assuming an annual growth rate of 1.489% for each compounding period.

i) annually **ii)** semi-annually

iii) monthly **iv)** daily

b) Explain why your results in part a show that the coefficient, 0.014 89, represents the instantaneous growth rate.

17. The following growth and decay situations occurred in Chapter 6. Write each equation using base *e*.

a) Compound interest, $A = 500(1.065)^n$

b) Caffeine consumption, $P = 100(0.87)^n$

c) Hosts on the Internet, $h = 7.4(1.59)^n$

d) Chemical reaction, $t = 8.72(0.93)^T$

e) Canadian mutual funds, $M = 460(1.19)^n$

f) Debit card transactions, $D = 328(1.384)^n$

g) Chernobyl radiation, $P = 100(0.918)^t$

18. In exercise 17, the equations are given in basic form, $y = y_0(1 + r)^x$. Your answers were in base-*e* form, $y = y_0 e^{kx}$.

a) Make a table of values of r and k.

b) Graph k against r.

c) Determine an equation that relates k and r.

d) In many situations, the rates of growth represented by r and k are approximately equal. Explain how the graph in part b and the equation in part c support this statement.

C

19. An approximate rule to estimate the time for an investment to double is to divide 70 by the instantaneous interest rate. For example, an investment at 8% will double in approximately 9 years. Use an exponential function with base *e* to explain why this rule works.

20. The time of death for a recently deceased person can be determined from the body temperature. At 10 P.M., the temperature of the deceased was 27°C. One hour later, the temperature was 25°C. Determine the probable time of death. Assume the room temperature was constant at 20°C and normal body temperature is 37°C.

7.4 Differentiating $y = a^x$

In Section 7.2, we stated that we can do a lot with $\frac{d}{dx}(e^x) = e^x$. We used this result in Section 7.3 to differentiate the function $P = 6.0(1.015)^t$, which models the world population. The key was to change the base of this function to e, then differentiate the resulting equation.

This strategy can be applied to any exponential function $y = a^x$.

Change the function to base e.

Let $a = e^k$. ... ①

Solve for k by taking the natural logarithm of each side.

$$\ln a = \ln e^k$$
$$\ln a = k$$

Substitute for k in equation ① to express a as a power with base e.

$$a = e^{\ln a} \quad \text{... ②}$$

Substitute for a in $y = a^x$ to express this function with base e.

$$y = (e^{\ln a})^x$$
$$y = e^{(\ln a)x}$$

Differentiate using the chain rule.

$$\frac{dy}{dx} = e^{(\ln a)x} \times \ln a$$

Use equation ②.

$$\frac{dy}{dx} = a^x \times \ln a$$

Compare this expression with that obtained earlier:

$$\frac{d}{dx}(a^x) = \left(\lim_{h \to 0} \frac{a^h - 1}{h}\right) \times a^x$$

From the two expressions, $\lim_{h \to 0} \frac{a^h - 1}{h} = \ln a$

In Section 7.1, we determined the limit expression $\left(\lim_{h \to 0} \frac{a^h - 1}{h}\right) \times a^x$ numerically, with a separate calculation for each base. We now know that $\lim_{h \to 0} \frac{a^h - 1}{h} = \ln a$. Instead of using the limit expression, we simply determine $\ln a$.

Take Note

The Derivative of the Function $y = a^x$

The derivative of $y = a^x$ is proportional to the function. The constant of proportionality is $\ln a$.

$$\frac{d}{dx}(a^x) = \ln a \times a^x$$

Example

Differentiate each function.

a) $y = 5^x$ b) $y = 3^{1-2x}$ c) $y = \frac{x^2}{2^x}$

Solution

a) $y = 5^x$

$\frac{dy}{dx} = \ln 5 \times 5^x$

b) $y = 3^{1-2x}$

Use the chain rule.

$\frac{dy}{dx} = \ln 3 \times 3^{1-2x} \times (-2)$

$= -2\ln 3 \times 3^{1-2x}$

c) $y = \frac{x^2}{2^x}$

Use the quotient rule.

$\frac{dy}{dx} = \frac{(2x)(2^x) - (x^2)(\ln 2 \times 2^x)}{2^{2x}}$

$= \frac{2^x(2x - x^2 \ln 2)}{2^{2x}}$

$= \frac{2x - x^2 \ln 2}{2^x}$

In the *Example*, we can approximate the derivatives by substituting decimal approximations for the logarithms. The derivatives are:

a) $\frac{d}{dx}(5^x) \doteq 1.609 \times 5^x$

b) $\frac{d}{dx}(3^{1-2x}) \doteq -2.197 \times 3^{1-2x}$

c) $\frac{d}{dx}\left(\frac{x^2}{2^x}\right) \doteq \frac{2x - x^2(0.693)}{2^x}$

In part b of the *Example*, we differentiated a function of the form $y = a^{f(x)}$ using the chain rule.

Take Note

The Derivative of the Function $y = a^{f(x)}$

To differentiate $y = a^{f(x)}$, where $f(x)$ is differentiable, use the chain rule.

$\frac{d}{dx}(a^{f(x)}) = \ln a \times a^{f(x)} \times f'(x)$

7.4 Exercises

1. Use the formula $\frac{d}{dx}(a^x) = a^x \times \ln a$ to verify the following results that were determined in Section 7.1.

   a) $\frac{d}{dx}(2^x) \doteq 0.69 \times 2^x$ b) $\frac{d}{dx}(3^x) \doteq 1.10 \times 3^x$

B

2. Differentiate each function.

   a) $y = 4^x$ b) $y = 6^x$ c) $y = 10^x$

   d) $y = 3^{2x+1}$ e) $y = 2^{1-3x}$ f) $y = 5^{x^2}$

3. Differentiate each function.

   a) $y = x \times 7^x$ b) $y = (2x - 1) \times 3^x$ c) $y = x^2 \times 4^{-x}$

   d) $y = x^3 \times 2^{1-2x}$ e) $y = (x - 1)^2 \times 6^x$ f) $y = (1 - x)^2 \times 2^{2x}$

4. Differentiate.

   a) $y = \frac{x}{7^x}$ b) $y = \frac{2x - 1}{3^x}$ c) $y = \frac{x^2}{4^x}$

   d) $y = \frac{x^3}{2^{2x-1}}$ e) $y = \frac{(x - 1)^2}{6^x}$ f) $y = \frac{(1 - x)^2}{2^{2x}}$

5. Compare the function in exercise 3c with the function in exercise 4c.

   a) Explain why these functions are equal.

   b) Show that the derivatives of these functions are equal.

6. Repeat exercise 5 for the functions in exercises 3d and 4d.

7. **Knowledge/Understanding** Differentiate each function.

   a) $y = 7^x$ b) $y = 4^{-x}$ c) $y = 2^{2-3x}$

   d) $y = (3x - 1) \times 2^x$ e) $y = (x^2 + 1)^2 \times 2^{-x}$ f) $y = \frac{5^{1-x}}{(2x - 1)^2}$

8. Consider the function $y = 2^{2x}$. Differentiate this function in four different ways, as follows. Show that all the results are equal.

   a) Use the rule for differentiating $y = a^{f(x)}$.

   b) Write the function as $y = (2^x)(2^x)$, then use the product rule.

   c) Write the function as $y = (2^x)^2$, then use the power rule.

   d) Write the function as $y = 4^x$, then use the rule for differentiating $y = a^x$.

9. Explain why the derivative of a^x is $\ln a \times a^x$, and not a^x. Give as complete an explanation as you can.

✓ **10. Application** Consider the graph of $y = a^x$, where $a > 0$, $a \neq 1$. Use the derivative, $y' = \ln a \times a^x$, to justify your answer to each question.

a) Are there any critical points?

b) Are there any points of inflection?

c) What are the intervals of increase and decrease?

d) Is the graph concave up or concave down?

✓ **11. Thinking/Inquiry/Problem Solving** The graph, below left, shows a function $f(x) = a^x$ and its derivative. These graphs pass through the points (0, 1) and (0, 2), respectively.

a) Determine the value of a and, hence, the equation of the function.

b) Determine the equation of the function for a similar situation in which the graph of the derivative passes through (0, 3).

Exercises 11 and 12 are significant because they involve a key feature of exponential functions. The derivative of an exponential function is proportional to the function. That is, the graph of one is a vertical expansion or compression of the graph of the other.

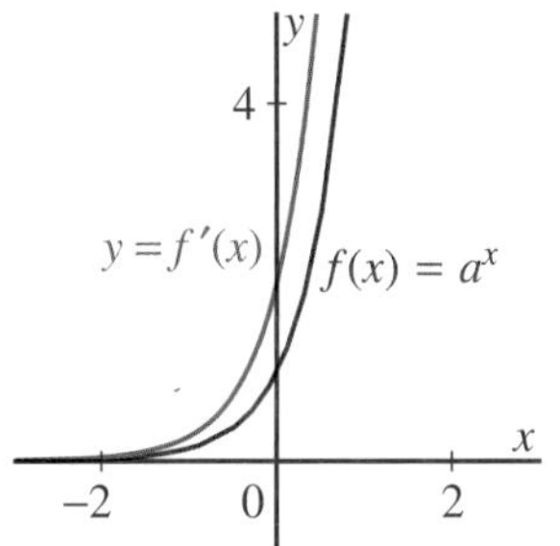

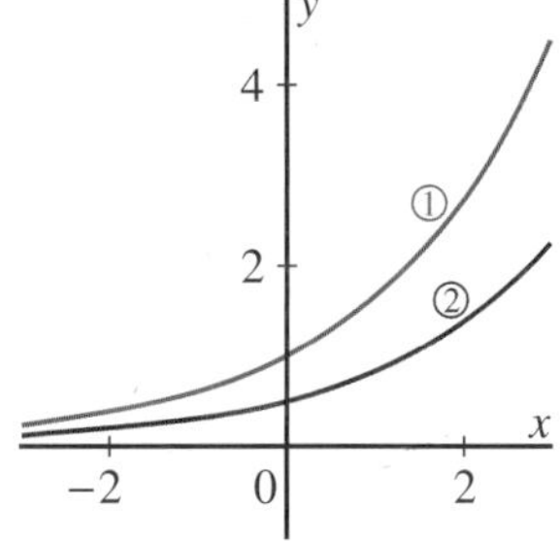

12. The graphs of an exponential function $y = ca^x$ and its derivative are shown above right. The y-intercept of one graph is 1. The y-intercept of the other graph is less than 1. In each situation below, state as much as possible about the values of a and c.

a) Graph ① is the function and graph ② is the derivative.

b) Graph ② is the function and graph ① is the derivative.

✓ **13. Communication** We have used the function $P = 6.0(1.015)^t$ to model the world population. We changed to base e, then differentiated the result. Use the formula $\frac{d}{dx}(a^x) = \ln a \times a^x$ to differentiate the function directly. Compare your result with the result on page 449. Explain any similarities or differences.

14. The graph at the right shows the function $y = 4^x$ and the tangent at the point C(1, 4) on the graph.

a) Determine the equation of the tangent.

b) Determine the x- and y-intercepts of the tangent.

c) Determine the coordinates of the point D on the graph of $y = 4^x$ such that the tangent at D passes through the origin.

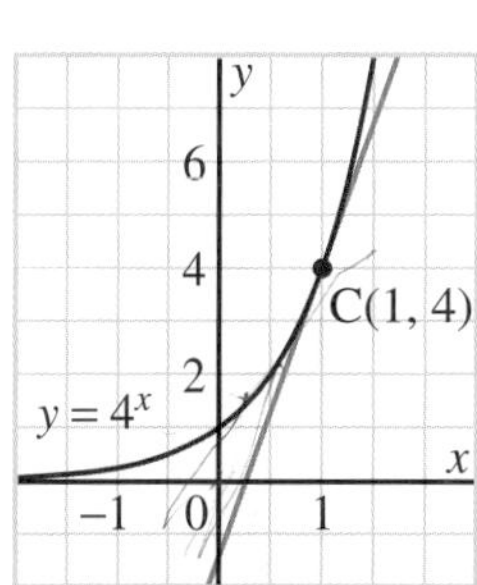

7.5 Differentiating Logarithmic Functions

The property $\frac{d}{dx}(e^x) = e^x$ is very useful. In Section 7.3, we used this property to solve problems involving exponential growth and decay. In Section 7.4, we used the property to differentiate $y = a^x$. In this section, we will use it to differentiate the logarithmic function $y = \ln x$, which is the inverse of $y = e^x$. We will also use it to differentiate $y = \log_a x$.

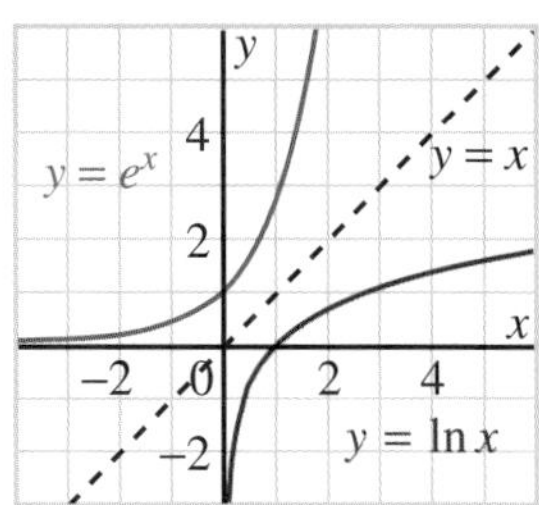

The Derivative of y = ln x

To determine the derivative of $y = \ln x$, we write the equation in exponential form:

$$e^y = x$$

This is an implicit equation, and we can differentiate each side with respect to x.

$$e^y \frac{dy}{dx} = 1$$

$$\frac{dy}{dx} = \frac{1}{e^y}$$

$$\frac{dy}{dx} = \frac{1}{x}$$

The derivative of $y = \ln x$ is $y' = \frac{1}{x}$.

Take Note

The Derivative of y = ln x

$$\frac{d}{dx}(\ln x) = \frac{1}{x},\ x > 0$$

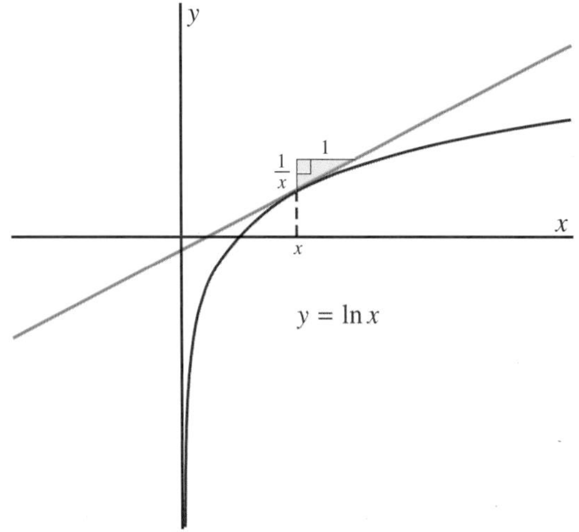

Visualize a point moving from left to right along the graph of $y = \ln x$. The slope of the tangent is $\frac{1}{x}$.

Example 1

Differentiate each function.

a) $y = \ln x^3$

b) $y = \ln(3 - x)$

Solution

a) $y = \ln x^3$

Use the chain rule.

$$\frac{dy}{dx} = \frac{1}{x^3}(3x^2)$$
$$= \frac{3}{x}$$

b) $y = \ln(3 - x)$

Use the chain rule.

$$\frac{dy}{dx} = \frac{1}{3 - x}(-1)$$
$$= \frac{1}{x - 3}$$

In *Example 1*, we differentiated functions of the form $y = \ln f(x)$.

Take Note

The Derivative of the Function $y = \ln f(x)$

To differentiate $y = \ln f(x)$, where $f(x)$ is differentiable, use the chain rule.

$$\frac{d}{dx}(\ln f(x)) = \frac{1}{f(x)} \times f'(x)$$

Example 2

Differentiate each function.

a) $y = x \ln 2x$

b) $y = \dfrac{x}{\ln x^2}$

Solution

a) $y = x \ln 2x$

Use the product rule and the chain rule.

$$\frac{dy}{dx} = (1)(\ln 2x) + (x)\left(\frac{1}{2x} \times 2\right)$$
$$= \ln 2x + 1$$

b) $y = \dfrac{x}{\ln x^2}$

Use the quotient rule and the chain rule.

$$\frac{dy}{dx} = \frac{(1)(\ln x^2) - (x)\left(\frac{1}{x^2} \times 2x\right)}{(\ln x^2)^2}$$
$$= \frac{\ln x^2 - 2}{(\ln x^2)^2}$$

The Derivative of $y = \log_a x$

$y = \log_a x$

To determine the derivative, we write the equation in exponential form.

$a^y = x$

Differentiate each side with respect to x.

$$\frac{d}{dx}(a^y) = \frac{d}{dx}(x)$$

$$\ln a \times a^y \times \frac{dy}{dx} = 1$$

$$\frac{dy}{dx} = \frac{1}{a^y \ln a}$$

$$= \frac{1}{x \ln a}$$

The derivative of $y = \log_a x$ is $y' = \frac{1}{x \ln a}$.

Take Note

The Derivative of y = log$_a$ x

$$\frac{d}{dx}(\log_a x) = \frac{1}{x \ln a},\ x > 0$$

THE BIG PICTURE

Filling the Gap

Recall that the derivative of $f(x) = x^n$ is $f'(x) = nx^{n-1}$. All possible integers can appear in the exponent of the function. Since the exponent in the derivative is 1 less than the exponent in the function, we might think that all possible integers can appear in the exponent of the derivative as well. However, let's look at this more closely.

Here is a list of some power functions and their derivatives.

Function	Derivative
⋮	⋮
x^3	$3x^2$
x^2	$2x^1$
x^1	$1x^0$
x^0	$0x^{-1}$
x^{-1}	$-1x^{-2}$
x^{-2}	$-2x^{-3}$
⋮	⋮

Each exponent in the derivative column is 1 less than the corresponding exponent in the function column. On the shaded line, the derivative equals 0. Therefore, x^{-1}, or $\frac{1}{x}$, is not represented in the derivative column. There is no function of the form $f(x) = x^n$ whose derivative is $\frac{1}{x}$. The function $f(x) = \ln x$ fills this gap, because its derivative is $\frac{1}{x}$.

7.5 Exercises

1. Differentiate each function.

   a) $y = \ln x$ b) $y = 2 + \ln x$ c) $y = 2 \ln x$

   d) $y = \ln 2x$ e) $y = \ln (x + 2)$ f) $y = \ln x^2$

2. Differentiate each function.

   a) $y = \ln x^3$ b) $y = \ln (1 - x)$ c) $y = \ln (2x + 1)$

   d) $y = \ln (1 + x)$ e) $y = \ln (x^2 + x + 1)$ f) $y = \ln (x^2 + 2x + 3)$

B

3. Refer to the graph of $y = \ln x$ on page 464. At first glance, it might appear that the graph has a horizontal asymptote. Explain why it does not.

4. Differentiate each function.

   a) $y = x^2 - \ln x$ b) $y = 2x \ln x$ c) $y = x \ln x^2$

   d) $y = \ln (\ln x)$ e) $y = \dfrac{x}{\ln x}$ f) $y = \dfrac{\ln x}{x}$

5. Differentiate each function.

   a) $y = x \ln (x + 1)$ b) $y = x^2 \ln (1 - 2x)$ c) $y = \dfrac{1 + \ln x}{1 - \ln x}$

   d) $y = \dfrac{\ln x^2}{x}$ e) $y = \dfrac{x}{\ln x^2}$ f) $y = \ln \left(\dfrac{x + 1}{x - 1}\right)$

6. **Knowledge/Understanding** Differentiate each function.

   a) $y = x - \ln x$ b) $y = x \ln x$ c) $y = \ln \left(\frac{1}{x}\right)$

   d) $y = x + \ln x^2$ e) $y = \ln e^x$ f) $y = \dfrac{1}{\ln x}$

7. Consider the function $y = \ln (x^2 - 1)$. Differentiate this function in two different ways, as follows. Show that the derivatives are equal.

   a) Use the rule for differentiating $y = \ln f(x)$.

   b) Write the function as $y = \ln ((x - 1)(x + 1))$, then apply the law of logarithms for multiplication before differentiating.

8. Consider the graph of the logarithmic function $y = \ln x$. Use the derivative, $y' = \frac{1}{x}$, to justify your answer to each question.

   a) Are there any critical points?

   b) Are there any points of inflection?

   c) What are the intervals of increase and decrease?

   d) Is the graph concave up or concave down?

9. Application The graphs of three functions are given. Determine the coordinates of all maximum and minimum points on the graph of each function.

a) 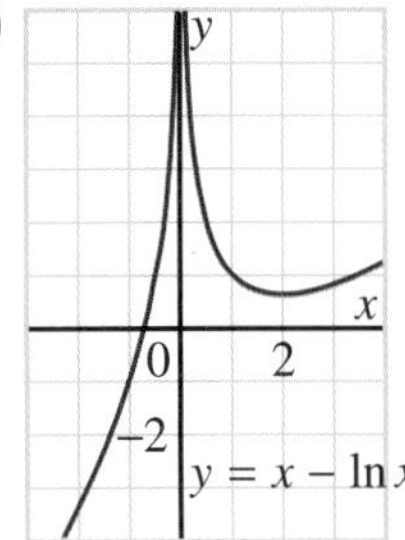

b) 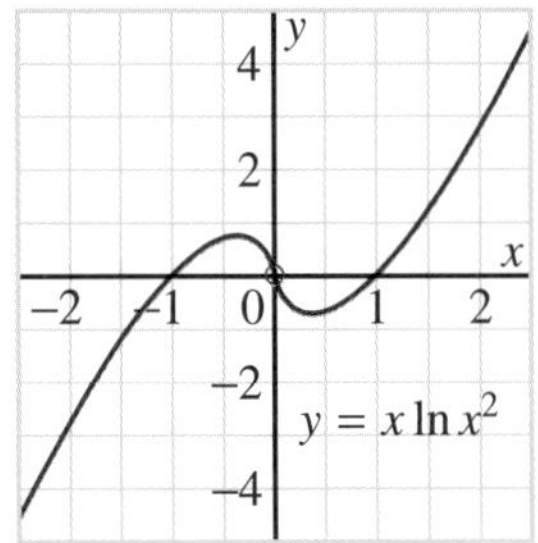

c) 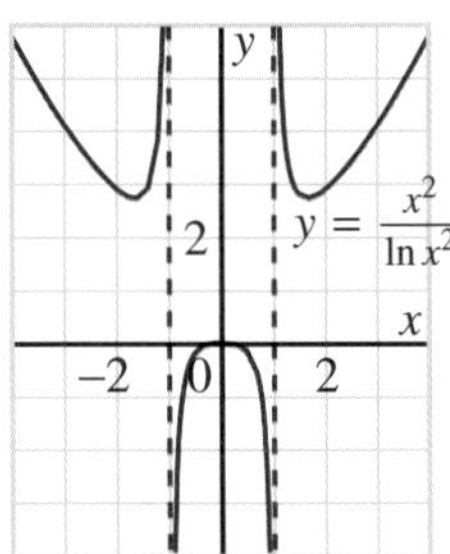

10. Refer to the graph in exercise 9a. Determine the equation of the tangent to the graph at the point (1, 1).

11. Refer to the graph in exercise 9b. Determine the equation of the tangent to the graph at the point (–1, 0).

12. Refer to the graph in exercise 9c. Determine the equation of the tangent to the graph at the point $\left(2, \frac{4}{\ln 4}\right)$.

13. **Communication**

a) Differentiate each function.

i) $y = \ln x$ ii) $y = \ln 2x$ iii) $y = \ln 3x$ iv) $y = \ln 4x$

b) Explain why the derivatives of all the functions in part a are the same.

14. **Thinking/Inquiry/Problem Solving** In Section 7.1, exercise 13, you determined the equations of the tangents at the points (0, 1) and (1, e) on the graph of $y = e^x$, at the right. The equations of these tangents are $y = x + 1$ and $y = ex$, respectively. Use these results to determine the equations of the tangents at the points (1, 0) and (e, 1) on the graph of $y = \ln x$.

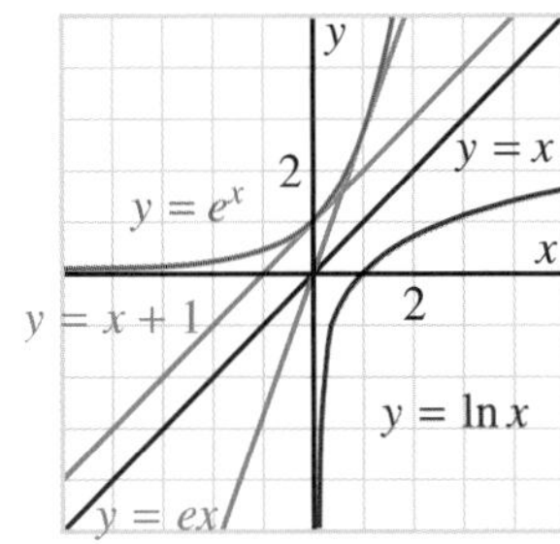

15. Explain why all the tangents to the graph of $y = \ln x$ have positive slopes.

16. Determine the equation of the tangent to the graph of $y = \ln x$ that has slope e. Show the result on a graph.

17. a) Determine the equation of the tangent to the graph of $y = \ln x$ that has slope k.

b) For what values of k does the tangent in part a have a y-intercept that is positive? negative?

c) For what values of k does the tangent in part a have an x-intercept that is positive? negative?

18. Differentiate each function.

a) $y = \log_2 x$ **b)** $y = \log x$ **c)** $y = \log_3 2x$

d) $y = \log x^2$ **e)** $y = \log_4 (2 - x)$ **f)** $y = (\ln x)(\log x)$

19. Differentiate each function.

a) $x = \ln y$ **b)** $x + \ln y = 1$ **c)** $x \ln y = 1$

d) $\ln xy = 1$ **e)** $\ln x + \ln y = 1$ **f)** $\ln (x + y) = 1$

20. The graph of the function $y = x^x$ is shown at the right.

a) Explain why neither the rule $\frac{d}{dx}(x^n) = nx^{n-1}$ nor the rule $\frac{d}{dx}(a^x) = \ln a \times a^x$ can be used to differentiate this function.

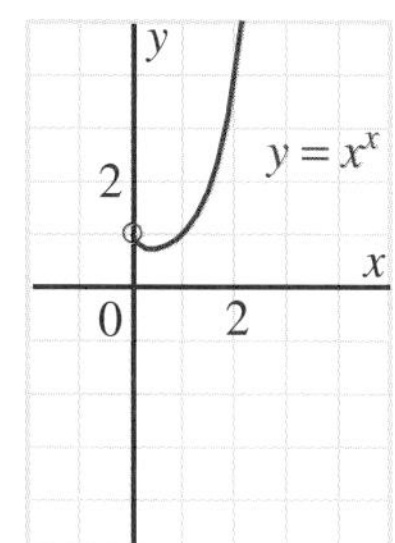

b) One method that can be applied to the function $y = x^x$ is to take the natural logarithm of each side. Do this, then differentiate each side to determine the derivative of this function.

c) Determine the coordinates of the minimum point of this function.

The method you used in exercise 20 is called *logarithmic differentiation*. By taking the natural logarithm of each side, we can differentiate functions of the form $f(x)^{g(x)}$. We can also use logarithmic differentiation to differentiate complicated products and quotients that would be cumbersome to differentiate directly using the product and quotient rules.

21. Differentiate each function.

a) $y = \ln (\sin x)$ **b)** $y = \sin (\ln x)$ **c)** $y = \ln (\sin 2x)$

d) $y = \sin (\ln 2x)$ **e)** $y = (\ln x)(\sin x)$ **f)** $y = \frac{\ln x}{\sin x}$

C

22. Two intersecting curves are perpendicular if the tangents at their point of intersection are perpendicular. The graphs of the functions $y = -\frac{1}{2}x^2 + 3$ and $y = \ln x$ are shown.

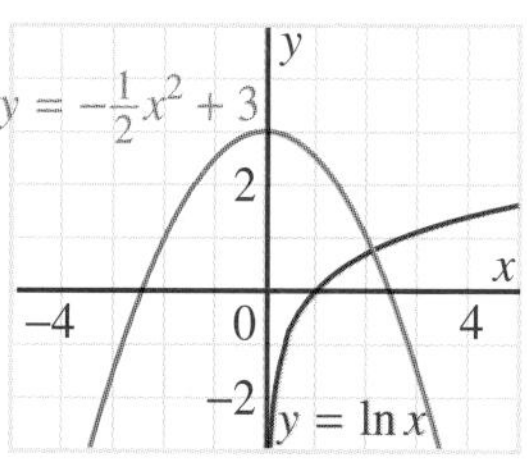

a) Show that the graphs of these functions are perpendicular.

b) Show that the 3 in the equation $y = -\frac{1}{2}x^2 + 3$ can be replaced with any other number and the graphs of the two functions will still be perpendicular.

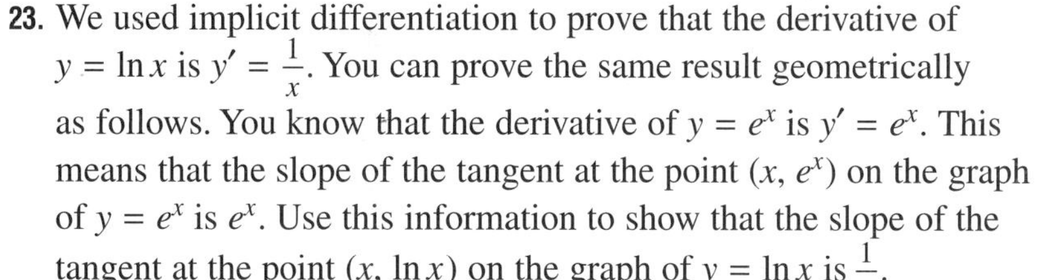
23. We used implicit differentiation to prove that the derivative of $y = \ln x$ is $y' = \frac{1}{x}$. You can prove the same result geometrically as follows. You know that the derivative of $y = e^x$ is $y' = e^x$. This means that the slope of the tangent at the point (x, e^x) on the graph of $y = e^x$ is e^x. Use this information to show that the slope of the tangent at the point $(x, \ln x)$ on the graph of $y = \ln x$ is $\frac{1}{x}$.

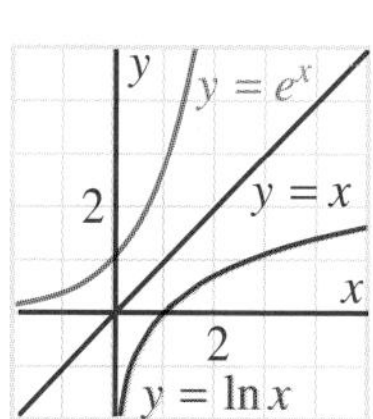

Self-Check 7.3 – 7.5

1. At the beginning of the year 2000, the population of Iceland was approximately 272 000. It was growing at approximately 0.57% per annum.
   a) Write an equation for the population, P, after t years, using an exponential equation with base e.
   b) Assume the growth rate is constant.
      i) Determine the predicted population at the beginning of 2005.
      ii) Determine when the population will reach 500 000.
   c) Predict the numerical instantaneous growth rate in the population at the beginning of 2008.

2. When an organism dies, the carbon-14 it contains continues to decay with a half-life of about 5730 years. After t years, the percent, P, of the carbon-14 remaining compared with modern living matter is $P = 100e^{-0.000\ 12t}$.
   a) Determine the percent of carbon-14 that remains in a sample of wood from a tree that died 1000 years ago.
   b) An ancient animal contains 31.3% carbon-14 relative to modern living matter. Estimate when the animal died.

3. Differentiate each function.
   a) $y = 6^x$
   b) $y = 4^{3+2x}$
   c) $y = 3^{x^2-2}$

4. Differentiate each function.
   a) $y = x^2 \times 5^x$
   b) $y = (x^2 + 1) \times 4^x$
   c) $y = (x - x^2) \times 7^x$

5. Differentiate each function.
   a) $y = \frac{3x}{4^x}$
   b) $y = \frac{2x^3 + 3}{2^x}$
   c) $y = \frac{2^{2x}}{2x^2 - x^3}$

6. Differentiate each function.
   a) $y = x^2 \ln 3x$
   b) $y = \frac{2x}{\ln x^3}$
   c) $y = 3 \ln x + x$
   d) $y = x^3 - \ln x^2$
   e) $y = \frac{2 + x\ln x}{2 - x\ln x}$
   f) $y = \frac{3}{\ln x^2}$

PERFORMANCE ASSESSMENT

7. The graph of $y = \frac{-x^2}{\ln x^2}$ is shown at the right.
   a) Determine the coordinates of all maximum and minimum points on the graph.
   b) Determine the equation of the tangent to the graph at the point $(3, \frac{-9}{\ln 9})$.

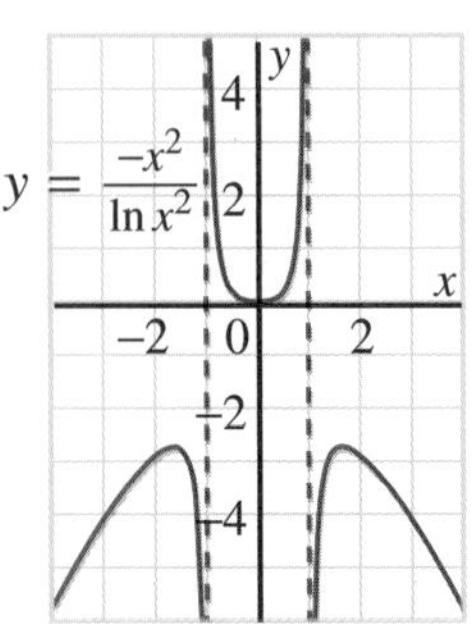

Mathematics Toolkit

Exponential and Logarithmic Tools

Exponents and Exponential Functions

- The number e is such that $\lim_{h \to 0} \frac{e^h - 1}{h} = 1$.

  To 5 decimal places, $e = 2.718\ 28$
- Two other definitions of e are:

  $e = \lim_{n \to \infty}\left(1 + \frac{1}{n}\right)^n$ and $e = \lim_{m \to 0}(1 + m)^{\frac{1}{m}}$
- The basic form of an exponential function is $y = y_0(1 + r)^x$, where: y_0 is the value of y when $x = 0$; r is the average rate of change as a percent expressed in decimal form; $r > 0$ for growth; and $-1 < r < 0$ for decay.
- The base e form of an exponential function is $y = y_0e^{kx}$, where: y_0 is the value of y when $x = 0$; k is the instantaneous rate of change as a percent expressed in decimal form; $k > 0$ for growth; and $k < 0$ for decay.

Logarithms and Logarithmic Functions

- Base-e logarithms are natural logarithms. The natural logarithm of x is denoted $\ln x$.
- The expression $\ln x$ is an exponent.

  $\ln x = y$ means $e^y = x$, where $x > 0$.
  $\ln e^x = x$ and $e^{\ln x} = x, x > 0$

Derivatives

- $\frac{d}{dx}(e^x) = e^x$
- $\frac{d}{dx}(e^{f(x)}) = e^{f(x)} \times f'(x)$
- $\frac{d}{dx}(y_0e^{kx}) = ky_0e^{kx}$
- $\frac{d}{dx}(a^x) = \ln a \times a^x$
- $\frac{d}{dx}(a^{f(x)}) = \ln a \times a^{f(x)} \times f'(x)$
- $\frac{d}{dx}(\ln x) = \frac{1}{x}, x > 0$
- $\frac{d}{dx}(\ln f(x)) = \frac{1}{f(x)} \times f'(x)$
- $\frac{d}{dx}(\log_a x) = \frac{1}{x \ln a}, x > 0$

Review Exercises

7.1 1. Differentiate each function.

a) $y = 10e^x$ b) $y = e^{10x}$ c) $y = e^{x-10}$

d) $y = e^{x^{10}}$ e) $y = e^{\frac{10}{x}}$ f) $y = e^{-10x}$

2. Differentiate each function.

a) $y = 10xe^x$ b) $y = xe^{10x}$ c) $y = x^{10}e^{-10x}$

d) $y = x^{10}(e^x - 10)$ e) $y = \dfrac{x^{10}}{10 + e^x}$ f) $y = \dfrac{10 - e^{10x}}{10 + e^{10x}}$

3. Differentiate each function.

a) $y = x^4e^{2x}$ b) $y = x^2(2x + e^x)$ c) $y = e^x(x + 2x^3)$

d) $y = e^{2x}(x - e^x)$ e) $y = \dfrac{3e^{2x} + x^2}{e^x}$ f) $y = \dfrac{x - e^{-x}}{x^2}$

4. The graphs of two functions are given. Write the equation of the derivative of each function, then sketch the graph of the derivative.

a) 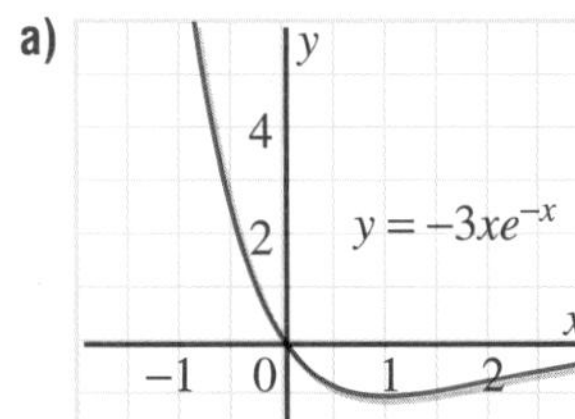

b) 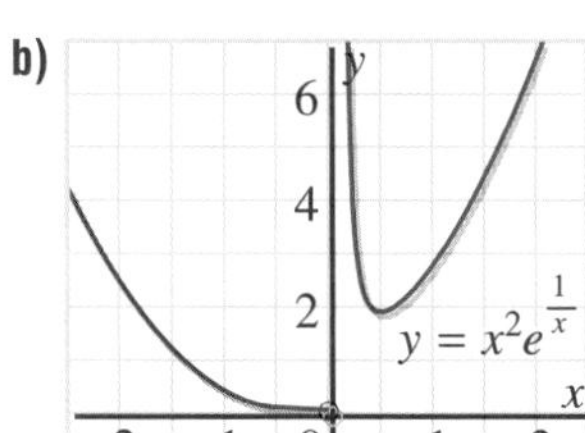

5. For each function in exercise 4, sketch the graph of the second derivative function.

6. For each function in exercise 4, determine the coordinates of the maximum point, the minimum point, and the point of inflection, if they exist.

7. Refer to the graph in exercise 4a. Determine the equation of the tangent to the graph of $y = -3xe^{-x}$ at the point on the graph where $x = 2$.

8. Refer to the graph in exercise 4b. Determine the equation of the tangent to the graph of $y = x^2e^{\frac{1}{x}}$ at the point on the graph where $x = -2$.

9. Analyse the key features of the graph of $y = 3e^{3x}$, then sketch the graph. Refer to page 443, exercise 14, for a list of the key features.

7.2 10. Use natural logarithms to solve each equation to 4 decimal places.

a) $12^x = 7$ b) $8^x = 72$ c) $5^x = 727$

11. Solve each equation to 4 decimal places.

a) $\ln x = 5$ b) $\ln x = 0.5$ c) $\ln x = 50$

12. Solve each equation to 4 decimal places.

a) $e^x = 7$ b) $1.5e^{2x} = 9$ c) $130 = 7e^{3x}$

13. Simplify each expression.

a) $\ln e^{1.2}$ b) $\ln e^{-7}$ c) $e^{\ln 4}$ d) $e^{\ln 0.8}$

14. Write each expression as a single logarithm.

a) $\ln 5 + \ln 10$ b) $\ln 5 - \ln 10$ c) $\ln 36 - \ln 3$

7.3 15. At the beginning of the year 2000, the population of Russia was 146 million. It was decreasing at the rate of 0.5% per annum.

a) Write an equation for the population, P million, after t years, using an exponential equation with base e.

b) Assume the rate is constant.

i) Determine the predicted population at the beginning of the year 2012.

ii) Determine when the population will reach 145 million.

c) Predict the numerical instantaneous decrease rate in the population at the beginning of the year 2015.

16. Liberia has one of the world's fastest growing populations. At the beginning of 1995, its population was 2.50 million. At the beginning of the year 2000, its population was 2.94 million.

a) Write an equation to represent the population, P million, of Liberia in terms of t, the time in years since 1995.

b) Estimate the percent instantaneous growth rate of the population of Liberia.

c) Pose then solve your own problem about Liberia's population.

17. The altitude of an airplane can be determined by measuring the air pressure. In the stratosphere (between 12 000 m and 30 000 m), the pressure, P kilopascals, is modelled by the equation $P = 130e^{-0.000\,155h}$, where h is the altitude in metres.

a) What is the pressure at an altitude of 15 000 m?

b) What is the altitude when the pressure is 2.69 kPa?

c) What is the percent instantaneous rate of change of pressure? Explain.

7.4 18. Differentiate each function.

a) $y = 3^x$ b) $y = 2^{x+3}$ c) $y = 4^{3+x^2}$

d) $y = x \times 6^x$ e) $y = (x^3 + x^2) \times 2^x$ f) $y = (2x - 3)^2 \times 5^x$

19. Differentiate each function.

a) $y = \dfrac{x^2}{3^x}$ b) $y = \dfrac{3^x}{x^2}$ c) $y = \dfrac{x^2 - 2x}{5^x}$

d) $y = \dfrac{x^3}{2^{2x-1}}$ e) $y = \dfrac{2^{2x+1}}{x^2}$ f) $y = \dfrac{x^3 - 3x^2}{2^{x^2}}$

20. The graph of $y = -2^x$ is shown below left. A tangent to the graph is drawn at the point B(1, –2).

a) Determine the equation of the tangent, and its x- and y-intercepts.

b) Determine the coordinates of a point, C, on the graph of $y = -2^x$ such that the tangent at C passes through the origin.

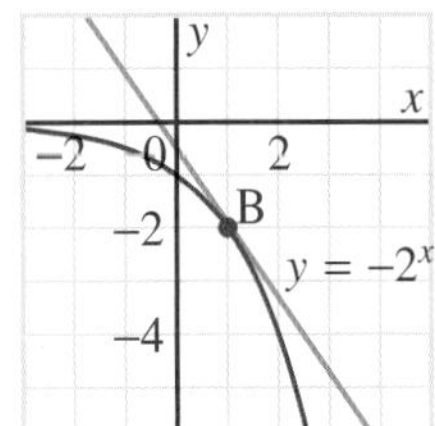

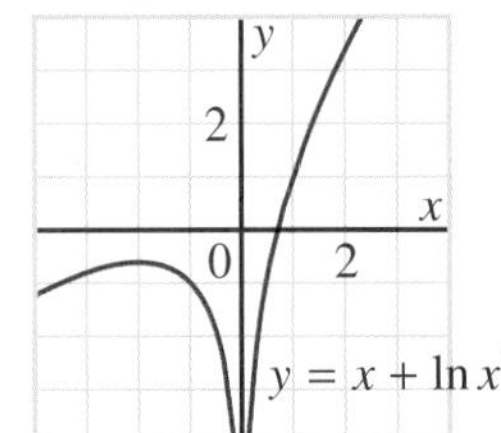

7.5 21. The graph of $y = x + \ln x^2$ is shown above right.

a) Determine the coordinates of any maximum or minimum points on the graph.

b) Determine the equation of the tangent to the graph at the point (1, 1).

22. Differentiate each function.

a) $y = \ln x$ b) $y = \ln x^3$ c) $y = \ln 2x^2$

d) $y = \ln (2 - x)$ e) $y = \ln (x^2 + 1)$ f) $y = \ln (2x^2 + x)$

23. Differentiate each function.

a) $y = x \ln 3x$ b) $y = -x^2 \ln 0.5x$ c) $y = \dfrac{x}{\ln 4x}$

d) $y = \dfrac{x^2}{3 \ln x}$ e) $y = \dfrac{5 \ln x}{x^2}$ f) $y = \dfrac{3}{\ln 2x}$

24. Differentiate each function.

a) $y = 2 \ln x + x^2$ b) $y = \ln\left(\dfrac{x}{x+1}\right)$ c) $y = \ln\left(\dfrac{x-1}{x^2}\right)$

d) $y = \ln x^2 - \ln x$ e) $y = 2 \ln e^x$ f) $y = x^2 \ln (3x + 2)$

25. Differentiate each function.

a) $y = \log_3 x$ b) $y = \log x^2$ c) $y = \log_4 3x$

26. Differentiate each function.

a) $x^2 = \ln y$ b) $x - \ln y = 3$ c) $\ln x^2 - \ln y = 2$

Self-Test

1. Simplify each expression without the aid of a calculator.

   a) $\ln e^{-3}$ b) $e^{\ln 5}$ c) $\ln 4 - \ln 8$ d) $e^{-3\ln 2 + \ln 3}$

2. **Communication** A teacher explained to her students that $e^{\ln 2} = 2$. One student said, "Therefore, the value of $e^{-\ln 2}$ must be equal to –2." Is this correct reasoning? Justify your response.

3. Solve each equation to 3 decimal places where necessary.

   a) $4e^{3x} = 125$ b) $3\ln x = 12$ c) $e^{x} - e^{-2x} = 0$

4. **Knowledge/Understanding** Differentiate each function.

   a) $y = 4^{2x-3}$ b) $y = e^{-2x}$ c) $y = \ln x^3$ d) $y = \log(3x-1)^4$

   e) $y = x^3e^{4x}$ f) $y = \dfrac{3-2e^{6x}}{e^{3x}}$ g) $y = \dfrac{\ln x^2}{e^{2x}}$ h) $y = \ln\left(\dfrac{1-x^2}{1+x^2}\right)$

5. Determine the coordinates of the points of inflection on the graph of $y = x^2\ln x$.

6. **Application** According to Newton's Law of Cooling, the temperature, T degrees Celsius, of a heated object placed in a room with air temperature 20°C is modelled by the equation $T = 20 + Be^{-kt}$, where t is the time in hours the object takes to cool, and B is a constant. A cup of coffee has an initial temperature of 90°C. It cools down to 30°C in 20 min when the room temperature is 20°C. How long did it take for the cup of coffee to cool to 50°C?

7. The graph of $y = 2x^2e^{-x}$ is shown below left. Determine the equation of the tangent to the graph at the point $(1, \frac{2}{e})$.

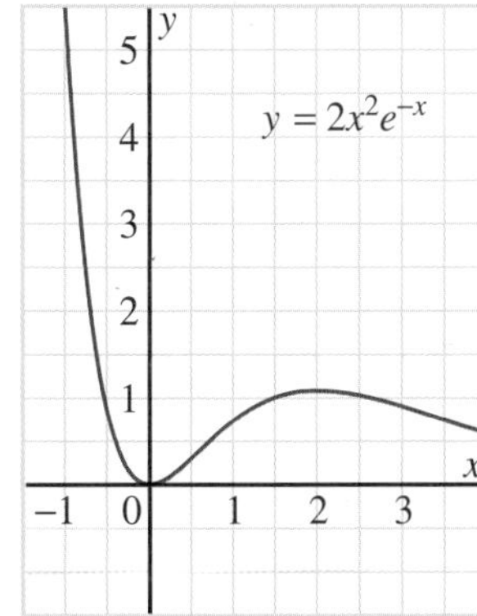

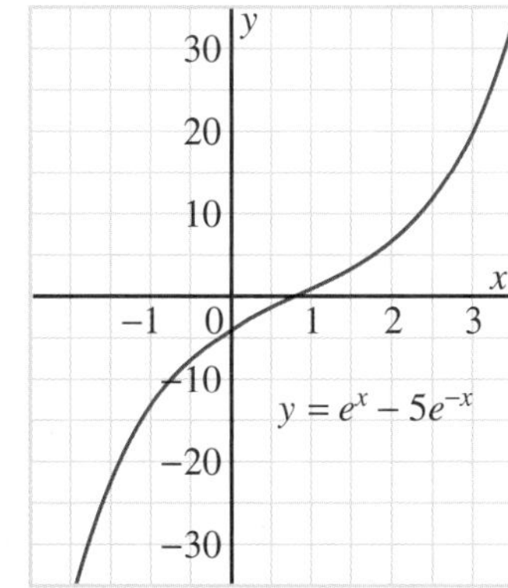

8. **Thinking/Inquiry/Problem Solving** The graph of $y = e^{x} - 5e^{-x}$ is shown above right. Determine the coordinates of all the points on the graph where the tangent is perpendicular to the line $x + 6y + 8 = 0$.

Self-Test

9. At the beginning of the year 2000, the population of Canada was about 31 million. It was growing at approximately 0.4% per annum.
   a) Write an equation for the population, P million, after t years, using an exponential equation with base e.
   b) Assume the growth rate is constant.
      i) Determine the predicted population at the beginning of the year 2006.
      ii) Determine when the population will reach 35 million.
   c) Predict the numerical instantaneous growth rate in the population at the beginning of the year 2009.

PERFORMANCE ASSESSMENT

10. Point P is any point on the graph of $y = e^x$ at the right. The tangent at P intersects the x-axis at A. Point Q is on the tangent such that right $\triangle$QPM has leg PM that is 1 unit long. Line QM is extended to meet the x-axis at N.

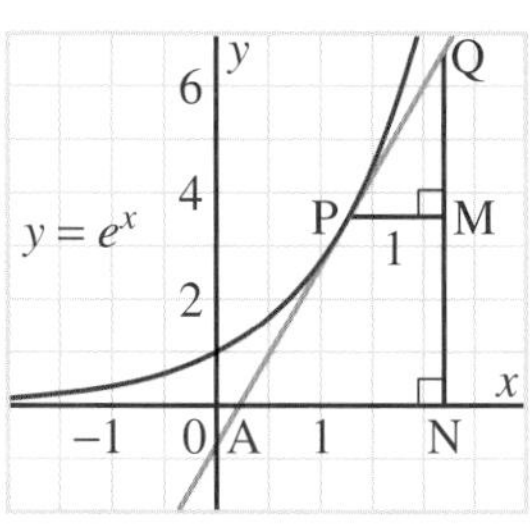

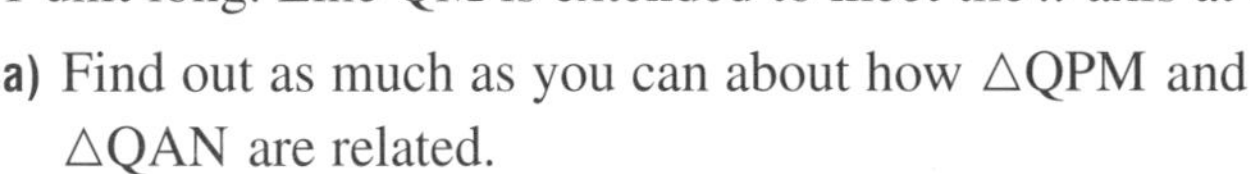

   a) Find out as much as you can about how $\triangle$QPM and $\triangle$QAN are related.
   b) What special case occurs when P is the point $(1, e)$?

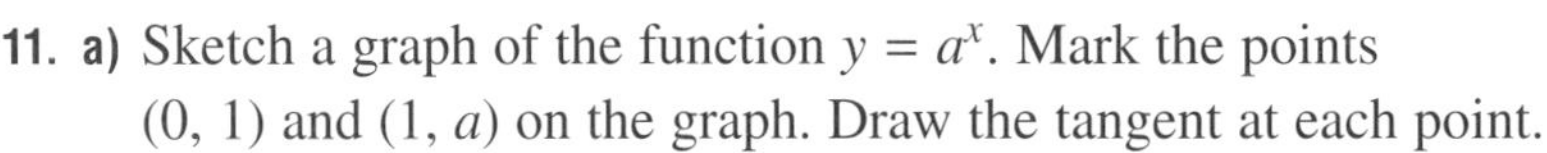

11. a) Sketch a graph of the function $y = a^x$. Mark the points $(0, 1)$ and $(1, a)$ on the graph. Draw the tangent at each point.
    b) Determine the equation of each tangent in part a.

12. Consider these two lists of functions.

List 1

$y = \ln x$, $y = \ln x^2$, $y = \ln x^3$, $y = \ln x^4$

List 2

$y = \ln\left(\frac{1}{x}\right)$, $y = \ln\left(\frac{1}{x^2}\right)$, $y = \ln\left(\frac{1}{x^3}\right)$, $y = \ln\left(\frac{1}{x^4}\right)$

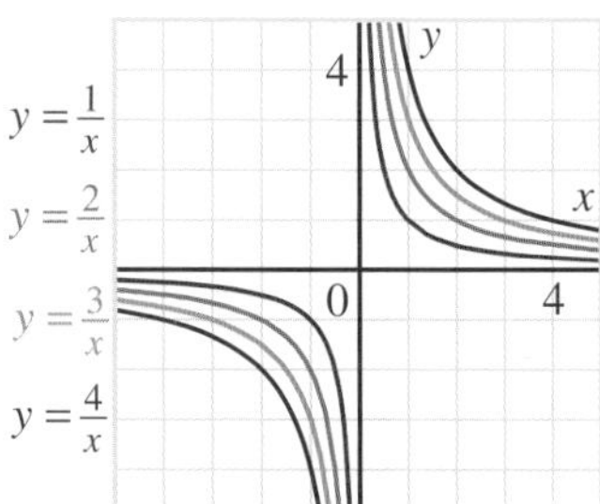

The graphs of the derivatives of the functions in List 1 are shown at the right.

   a) Differentiate each function in List 1.
   b) Account for the pattern in the graphs of the derivatives.
   c) Differentiate each function in List 2.
   d) Sketch a similar graph of the derivatives of the functions in List 2.

Mathematical Modelling

Quality Control

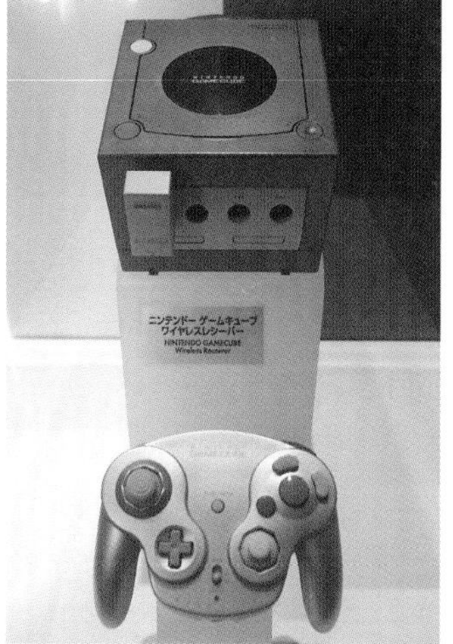

Suppose you run a plant that manufactures the latest game machine. The current cost to make a machine is \$100 (for parts and labour). However, 12% of the machines fail during the warranty period. A failed machine is returned to you for replacement. You estimate the cost of each failure (replacement, handling, and damage to the reputation of the product) to be \$200.

The average cost per machine, C dollars, is:

$$\begin{aligned} C &= \$100 + 12\% \text{ of } \$200 \\ &= \$100 + 0.12 \times \$200 \\ &= \$124 \end{aligned}$$

Formulating a Hypothesis

You wish to reduce the number of machines that fail. The only way to do this is to take more care in making the machines. This means increasing the labour cost. An increase in labour cost should improve the quality of the machines and reduce the failure rate. This causes the average cost to drop below \$124 per machine. Improvement in quality reaches a limit, beyond which the increased spending on labour only causes the average cost per machine to increase. Let x dollars represent the additional amount spent on labour per machine. Then the graph of average cost, C, against the increase in labour cost, x, should look similar to the graph at the right.

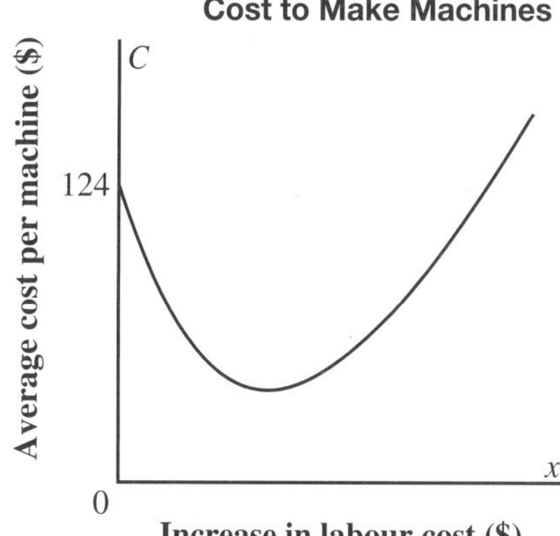

We make the following hypothesis:
There should be a minimum point on the graph that indicates the additional amount to spend on labour that minimizes the average cost per machine.

Developing a Mathematical Model

To determine the additional amount to spend on labour, we need to express C as a function of x.

An equation for the percent of the machines that fail

We estimate that each additional dollar spent on labour will reduce the percent of machines that fail by 10%.

This argument is valid only for integer values of x but it is reasonable to extend it to non-integer values.

Let p represent the percent of machines that fail.

If no additional money is spent on labour, $x = 0$ and $p = 12\%$, or 0.12. Hence, $p(0) = 0.12$

When an additional \$1 is spent on labour, $x = 1$, and p is reduced by 10%. That is the same as multiplying p by 0.90, so $p(1) = 0.12(0.90)$.

When an additional \$2 is spent on labour, $x = 2$, and p is reduced by 10% again. That is the same as multiplying p by 0.90 again, so $p(2) = 0.12(0.90)^2$.

This pattern continues. The equation is $p(x) = 0.12(0.90)^x$.

An equation for the overall cost per machine

Let C_M dollars represent the cost to manufacture each machine. Then,

$$C_M = 100 + x$$

Let C_R dollars represent the average replacement cost per machine. A certain percent, p, of the machines will fail, and each failure incurs an additional cost of \$200. Therefore, the average replacement cost per machine is:

$$C_R = 200p$$
$$C_R = 200(0.12)(0.90)^x$$
$$C_R = 24(0.90)^x$$

The overall cost, C dollars, per machine is:

$$C = C_M + C_R$$
$$C = 100 + x + 24(0.90)^x$$

This is the function we want. The graph of this function is shown at the right. A minimum for C seems to be approximately $x = 9$.

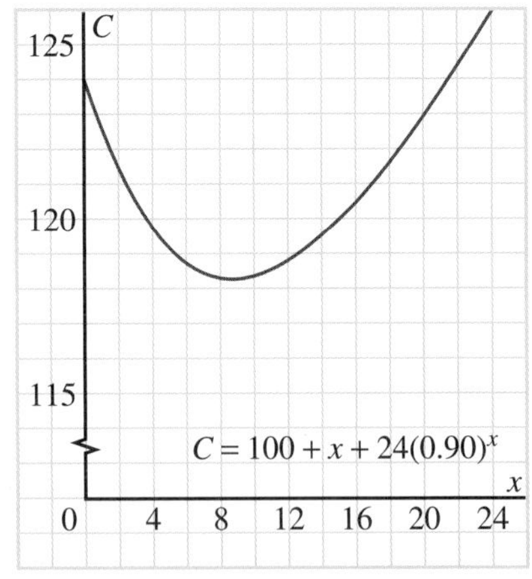

Key Features of the Model

1. Use the equation for C to calculate the overall cost per machine when $x = 0$. Explain the result.

2. Calculate the overall cost per machine for each additional amount spent on labour.

   a) \$2 b) \$5 c) \$10
   d) \$25 e) \$50 f) \$100

3. What happens to the graph of C as x becomes larger and larger? Explain your answer using the equation $C = 100 + x + 24(0.9)^x$ and also in terms of the application.

4. a) Differentiate the function $C = 100 + x + 24(0.9)^x$ to determine the derivative of the average cost, C', with respect to the additional amount spent on labour.

   b) Evaluate C' for each additional amount spent on labour in exercise 2.

   c) Explain the results of part b in terms of the graph of C.

5. Describe the graph of C.

Making Inferences from the Model

6. a) Determine the critical value(s) of C by equating its derivative to 0, then solving for x.

   b) In part a, you should have found only one critical value. Explain how you know that this is an absolute minimum value.

7. In exercise 6, you should have found that the minimum overall cost per machine is \$118.30, and that this occurs when an additional \$8.80 is spent per machine on labour. According to the model, what percent of the machines will fail in this case?

8. Suppose you wish to reduce the percent of machines that fail to 1%.

   a) How much additional money should be spent on labour?

   b) What is the average cost per machine?

9. Predict how the graph on page 478 will change in each situation.

   a) The basic cost to make a machine decreases from \$100 to \$50.

   b) The cost to replace a failed machine increases from \$200 to \$300.

Related Problems

10. Suppose you manufacture an espresso machine. The current cost to make a machine is \$100. However, 20% of the machines fail during the warranty period. A failed machine is returned, and you estimate the cost of each failure to be \$100. You estimate that each additional dollar spent on labour (per machine) will reduce the percent of machines that fail by 8%.

a) Find an expression for p, the percent of machines that fail, in terms of the additional amount, x dollars, spent on labour per machine.

b) Express the overall cost per machine, C dollars, as a function of x.

c) Calculate the value of x that minimizes C.

d) Suppose the current failure probability is 10% rather than 20%. What is the optimal value of x now? Explain.

11. Suppose you stand x metres away from a large garbage pail and try to throw beanbags into it. The percent, p, of beanbags that fall in is a decreasing function of x.

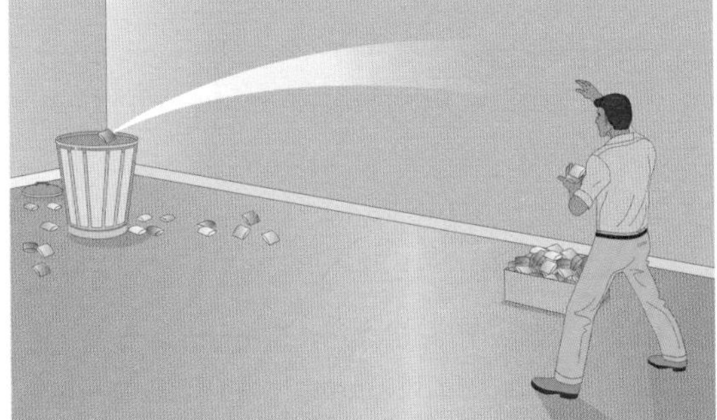

a) Sketch what you think the graph of p against x will look like for $0 < x < 20$.

b) Suppose you throw 100 beanbags at the pail from any distance x you choose. You get \$$x$ for each beanbag that falls into the pail. How can you determine x to maximize your payoff?

c) Consider the function $p(x) = 0.99^{x^2}$. Does this function have the same key features as your graph in part a? Explain.

d) Use the function in part c to obtain an algebraic solution to the maximum payoff problem, and provide the optimal value of x.

12. The quality control model developed above raises a fundamental question in economics: How good should you make a product that you are selling? Explain why this is a fundamental question. Include some equations and graphs in your account.

Applications of Differential Calculus 8

Curriculum Expectations

By the end of this chapter, you will:

- Determine the derivatives of combinations of the basic polynomial, rational, and exponential functions, using the rules for sums, differences, products, quotients, and compositions of functions.
- Solve optimization problems involving polynomial and rational functions.
- Solve related-rates problems involving polynomial and rational functions.
- Determine, from the equation of a simple combination of polynomial, rational, or exponential functions, the key features of the graph of the combination of functions, using the techniques of differential calculus, and sketch the graph by hand.
- Sketch the graphs of the first and second derivative functions, given the graph of the original function.

Necessary Skills

1. Review: Implicit Differentiation

Recall that, when we differentiated $x^2 + y^2 = 9$ implicitly, we took the derivative of y with respect to x to obtain: $2x + 2yy' = 0$. In Leibniz notation, this derivative is $2x\frac{dx}{dx} + 2y\frac{dy}{dx} = 0$. Since $\frac{dx}{dx} = 1$, it was not written in the first analysis.

In Sections 8.3 and 8.4, we will solve problems that involve quantities that change with respect to time. We will use the chain rule and Leibniz notation to differentiate the equations with respect to time, t.

Example

Determine the derivative with respect to t, then solve each equation for the indicated quantity.

a) $x^2 + y^2 = z^2$; $\frac{dy}{dt}$ **b)** $A = \frac{1}{2}bh$; $\frac{dh}{dt}$ **c)** $V = \frac{1}{3}\pi r^2 h$; $\frac{dr}{dt}$

Solution

a) $x^2 + y^2 = z^2$

Differentiate each side implicitly with respect to t.

$$2x\frac{dx}{dt} + 2y\frac{dy}{dt} = 2z\frac{dz}{dt}$$

Divide each side by 2.

$$x\frac{dx}{dt} + y\frac{dy}{dt} = z\frac{dz}{dt}$$

Solve for $\frac{dy}{dt}$.

$$y\frac{dy}{dt} = z\frac{dz}{dt} - x\frac{dx}{dt}$$

Divide each term by y, $y \neq 0$.

$$\frac{dy}{dt} = \frac{z}{y}\frac{dz}{dt} - \frac{x}{y}\frac{dx}{dt}$$

b) $A = \frac{1}{2}bh$

Differentiate each side implicitly with respect to t.

$$\frac{dA}{dt} = \frac{1}{2}\left(b\frac{dh}{dt} + \frac{db}{dt}h\right)$$

Multiply each side by 2.

$$2\frac{dA}{dt} = b\frac{dh}{dt} + \frac{db}{dt}h$$

Solve for $\frac{dh}{dt}$.

$$b\frac{dh}{dt} = 2\frac{dA}{dt} - \frac{db}{dt}h$$

Divide each term by b, $b \neq 0$.

$$\frac{dh}{dt} = \frac{2}{b}\frac{dA}{dt} - \frac{db}{dt}\frac{h}{b}$$

c) $V = \frac{1}{3}\pi r^2 h$

Differentiate each side with respect to t.

$$\frac{dV}{dt} = \frac{1}{3}\pi\left(2rh\frac{dr}{dt} + r^2\frac{dh}{dt}\right)$$

Multiply each side by $\frac{3}{\pi}$.

$$\frac{3}{\pi}\frac{dV}{dt} = 2rh\frac{dr}{dt} + r^2\frac{dh}{dt}$$

Solve for $\frac{dr}{dt}$.

$$2rh\frac{dr}{dt} = \frac{3}{\pi}\frac{dV}{dt} - r^2\frac{dh}{dt}$$

Divide each term by $2rh$, $r \neq 0$, $h \neq 0$.

$$\frac{dr}{dt} = \frac{3}{2\pi rh}\frac{dV}{dt} - \frac{r^2}{2rh}\frac{dh}{dt}$$

$$= \frac{3}{2\pi rh}\frac{dV}{dt} - \frac{r}{2h}\frac{dh}{dt}$$

Exercises

1. Determine the derivative with respect to t, then solve each equation for the indicated quantity.

a) $x^2 - y^2 = z^2$; $\frac{dx}{dt}$ **b)** $A = \pi r^2 h$; $\frac{dh}{dt}$ **c)** $V = \frac{4}{3}\pi r^3$; $\frac{dr}{dt}$

2. Determine the derivative with respect to t, then substitute to determine the value of the indicated quantity.

a) For $A = lw$, determine $\frac{dl}{dt}$ when $A = 48$, $\frac{dA}{dt} = -3$, $l = 8$, and $\frac{dw}{dt} = -3$.

b) For $x^2 + y^2 = 36$, determine $\frac{dx}{dt}$ when $\frac{dy}{dt} = 5$ and $x = -4$.

c) For $S = 4\pi r^2$, determine $\frac{dr}{dt}$ when $r = 9$ and $\frac{dS}{dt} = 7$.

2. Review: Similar Triangles

In grade 10, you learned that two triangles are *similar* if they have the same shape but not necessarily the same size. Recall the following properties of similar triangles.

In similar triangles, corresponding angles are equal and the ratios of corresponding sides are equal.

Given $\triangle ABC$ is similar to $\triangle PQR$,
then, $\angle ABC = \angle PQR$
$\angle BAC = \angle QPR$
$\angle ACB = \angle PRQ$
and $\frac{AB}{PQ} = \frac{BC}{QR} = \frac{AC}{PR}$

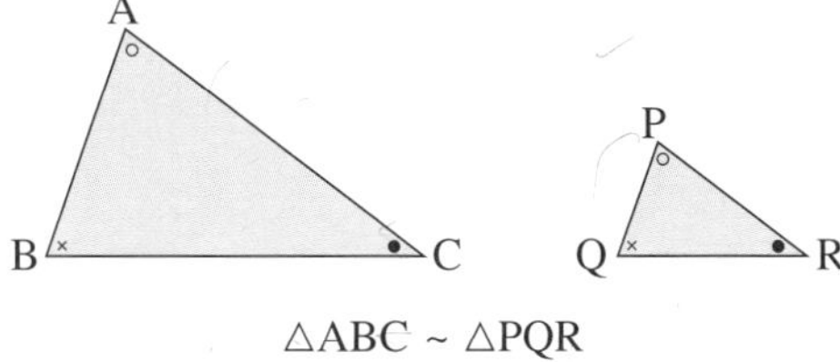

$\triangle ABC \sim \triangle PQR$

Recall that it is sufficient to show that two pairs of corresponding angles are equal to determine that two triangles are similar.

Example 1

In right $\triangle KMN$ and right $\triangle XYZ$, $\angle N = \angle Z = 90^\circ$, $\angle M = \angle Y$, NM = 5.74 cm, KN = 8.40 cm, and ZY = 4.10 cm

a) Explain why $\triangle KMN \sim \triangle XYZ$.

b) Determine the lengths of XZ, XY, and KM, to 2 decimal places where necessary.

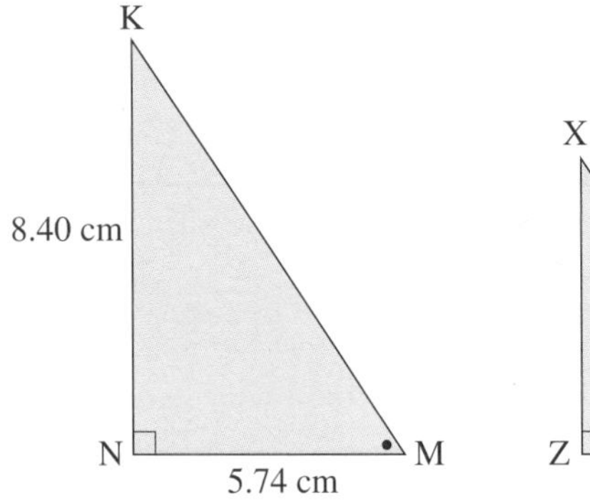

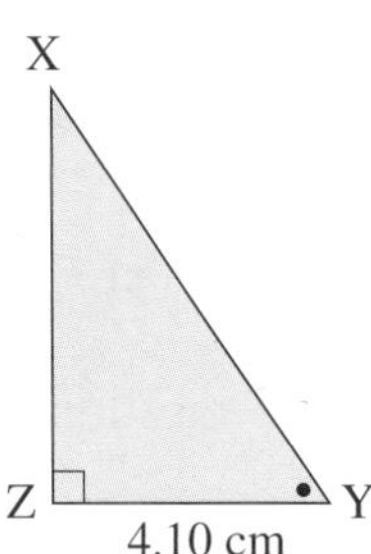

Solution

a) $\angle M = \angle Y$ and $\angle N = \angle Z = 90^\circ$

Since two pairs of corresponding angles are equal, then $\triangle KMN \sim \triangle XYZ$

b) Since $\triangle KMN \sim \triangle XYZ$, the ratios of corresponding sides are equal.

$$\frac{MN}{YZ} = \frac{KN}{XZ} = \frac{KM}{XY}$$

Substitute the given lengths.

$$\frac{5.74}{4.10} = \frac{8.40}{XZ} = \frac{KM}{XY}$$

To calculate XZ, use the first two fractions above.

$$\frac{5.74}{4.10} = \frac{8.40}{XZ}$$

$$XZ = \frac{8.40 \times 4.10}{5.74}$$

$$= 6.00$$

XZ is 6.00 cm.

To calculate XY, use the Pythagorean Theorem in $\triangle XYZ$.

$$XY^2 = ZY^2 + XZ^2$$

$$XY^2 = 4.1^2 + 6.0^2$$

$$XY = \sqrt{4.1^2 + 6.0^2}$$

$$\doteq 7.267$$

XY is approximately 7.27 cm.

To calculate KM, use the first and third fractions above.

$$\frac{5.74}{4.10} = \frac{KM}{7.267}$$

$$KM = \frac{5.74 \times 7.267}{4.10}$$

$$\doteq 10.174$$

KM is approximately 10.17 cm.

Discuss

What other method could we use to calculate KM? Use this method to check.

Some similar triangles are found in composite figures, as illustrated in *Example 2*.

Example 2

For the diagram at the right:

a) State which two triangles are similar. Give reasons.

b) Determine the lengths represented by x and y.

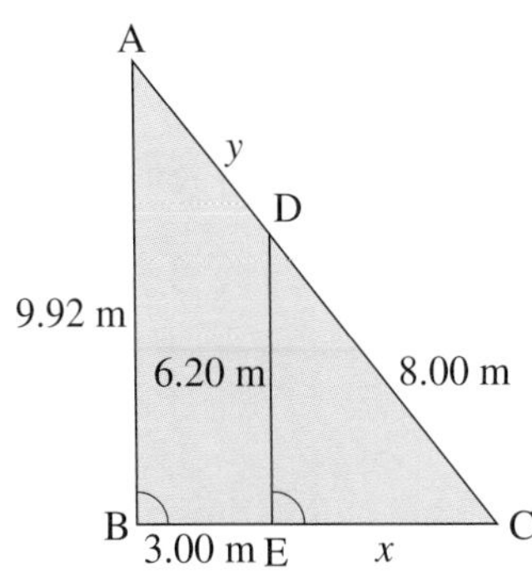

Solution

a) $\angle ABC = \angle DEC$

Since $\angle C$ is common to both triangles, $\angle ACB = \angle DCE$.

Since two pairs of corresponding angles are equal, then $\triangle ABC \sim \triangle DEC$.

b) Since $\triangle ABC \sim \triangle DEC$, the ratios of corresponding sides are equal.

$$\frac{AB}{DE} = \frac{BC}{EC} = \frac{AC}{DC}$$

Substitute the given lengths and variables.

$$\frac{9.92}{6.20} = \frac{3.00 + x}{x} = \frac{y + 8.00}{8.00}$$

To determine x, use the first two fractions above.

$$\frac{9.92}{6.20} = \frac{3.00 + x}{x}$$
$$1.6 = \frac{3.00 + x}{x}$$
$$1.6x = 3.00 + x$$
$$0.6x = 3.00$$
$$x = 5$$

The length x is 5.00 m.

To determine y, use the first and third fractions above.

$$\frac{9.92}{6.20} = \frac{y + 8.00}{8.00}$$
$$1.6 = \frac{y + 8.00}{8.00}$$
$$12.8 = y + 8.00$$
$$y = 4.8$$

The length y is 4.80 m.

Discuss

Why could we not use the Pythagorean Theorem in this solution?

Exercises

1. Why is it sufficient to show that two pairs of corresponding angles are equal to deduce that two triangles are similar?

2. For each pair of triangles below:

i) Explain why the triangles are similar.

ii) Determine the lengths indicated by x, y, and z. Where necessary, give the lengths to 1 decimal place.

a)

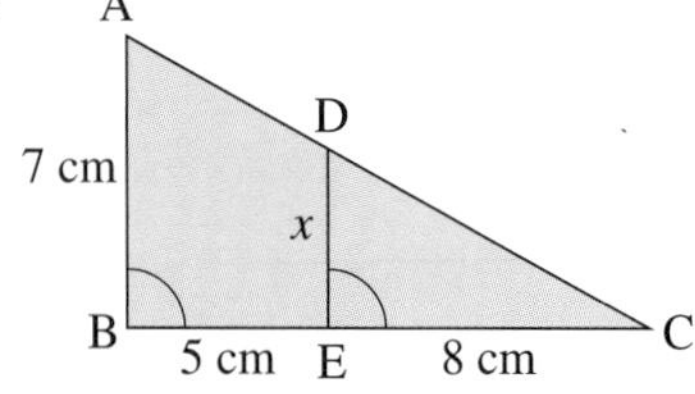

b)

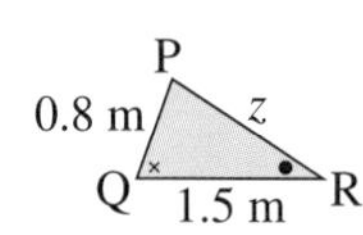

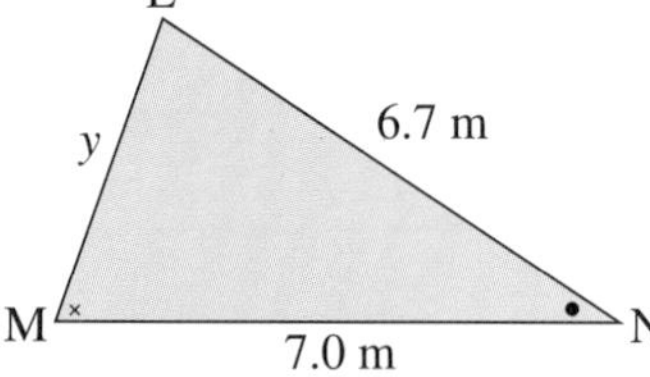

3. On a sunny day, a tree casts a shadow 3.5 m long. At the same time, a person who is 187 cm tall casts a shadow 2.2 m long. Estimate the height of the tree.

8.1 Graphing Combinations of Functions

In the previous chapters, we worked with polynomial, rational, and exponential functions. In this section, we will graph combinations of these functions.

Example 1

Analyse the key features of the graph of $f(x) = \frac{x}{e^x}$, then sketch the graph.

Solution

$f(x) = \frac{x}{e^x}$

Intercepts

The y-intercept is $f(0) = \frac{0}{e^0} = 0$.

Solve $f(x) = 0$ to determine the x-intercepts.

$$\frac{x}{e^x} = 0$$
$$x = 0$$

The x-intercept is 0.

Domain

Since the denominator, e^x, is always positive, and the numerator can be any real number, the domain is the set of real numbers.

$D = \Re$

Critical points

Determine the derivative, $f'(x)$.

$f(x) = \frac{x}{e^x}$

Differentiate. Use the quotient rule.

$$\begin{aligned} f'(x) &= \frac{(1)e^x - x(e^x)}{e^{2x}} \\ &= \frac{e^x(1-x)}{e^{2x}} \\ &= \frac{1-x}{e^x} \end{aligned}$$

For critical values, solve $f'(x) = 0$.

$$\frac{1-x}{e^x} = 0$$
$$1 - x = 0$$
$$x = 1$$

So, there is a critical point at $x = 1$. Determine the corresponding y-value.

$$f(1) = \frac{1}{e^1}$$
$$= \frac{1}{e}$$

The critical point is $\left(1, \frac{1}{e}\right)$, or approximately (1, 0.37).

Intervals of increase and decrease

Consider the sign of $f'(x)$ in each interval.

$$f'(x) = \frac{1-x}{e^x}$$

Since $e^x > 0$ for all values of x, consider only the sign of the numerator, $1 - x$.
When $x < 1$, say $x = 0.9$, $f'(0.9)$ is positive.
When $x > 1$, say $x = 1.1$, $f'(1.1)$ is negative.

Use a number-line analysis for the first derivative.

Direction of f	↗	→	↘
x		1	
Sign of f'	+	0	–

The graph of $f(x) = \frac{x}{e^x}$ is increasing when $x < 1$, and decreasing when $x > 1$. Therefore, the critical point $\left(1, \frac{1}{e}\right)$ is a maximum.

Points of inflection

Consider the second derivative, $f''(x)$.

$$f'(x) = \frac{1-x}{e^x}$$

Differentiate. Use the quotient rule.

$$f''(x) = \frac{(-1)e^x - (1-x)e^x}{e^{2x}}$$
$$= \frac{e^x(-1 - 1 + x)}{e^{2x}}$$
$$= \frac{x-2}{e^x}$$

For a point of inflection, solve $f''(x) = 0$.

$$\frac{x-2}{e^x} = 0$$
$$x - 2 = 0$$
$$x = 2$$

There is a possible point of inflection when $x = 2$.
Check the concavity to determine whether this point of inflection exists.

Intervals of concavity

Consider the sign of $f''(x) = \frac{x-2}{e^x}$ in each interval.

Since $e^x > 0$ for all values of x, consider only the sign of the numerator, $x - 2$. When $x < 2$, say $x = 1.9$, $f''(1.9)$ is negative, and the graph is concave down. When $x > 2$, say $x = 2.1$, $f''(2.1)$ is positive, and the graph is concave up. Use a number-line analysis for the second derivative.

Shape of f	∩	Point of inflection	∪
x		2	
Sign of f''	–	0	+

The graph of $y = \frac{x}{e^x}$ is concave down when $x < 2$ and concave up when $x > 2$. So, the graph has a point of inflection at $x = 2$.

Determine the y-coordinate when $x = 2$.

$f(2) = \frac{2}{e^2}$

The point of inflection has coordinates $\left(2, \frac{2}{e^2}\right)$, or approximately (2, 0.27).

Asymptotes and point discontinuities

There are vertical asymptotes or point discontinuities at values of x for which $f(x) = \frac{x}{e^x}$ is undefined. Since e^x is always positive, $f(x)$ is always defined. So, there are no vertical asymptotes or point discontinuities.

For a horizontal asymptote, determine $\lim\limits_{x \to \infty} \frac{x}{e^x}$ and $\lim\limits_{x \to -\infty} \frac{x}{e^x}$.

As $x \to \infty$, $e^x \to \infty$, and $\frac{x}{e^x} \to 0$, since the denominator increases faster than the numerator.

So, $\lim\limits_{x \to \infty} \frac{x}{e^x} = 0$, and there is a horizontal asymptote at $y = 0$ for $x \to \infty$.

As $x \to -\infty$, $e^x \to 0$, and $\frac{x}{e^x} \to -\infty$

So, $\lim\limits_{x \to -\infty} \frac{x}{e^x} = -\infty$, and there is no horizontal asymptote for $x \to -\infty$.

Use the information to sketch the graph of $f(x) = \frac{x}{e^x}$.

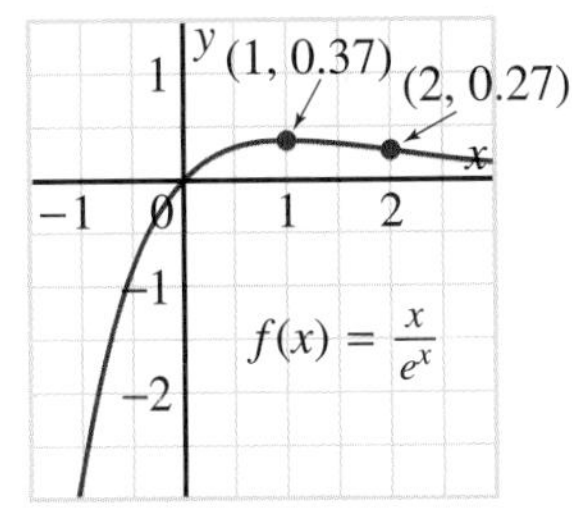

In *Example 1*, the graph of $f(x) = \frac{x}{e^x}$ has a horizontal asymptote only for $x \to \infty$. There is no asymptote for $x \to -\infty$. This is because $\lim\limits_{x \to \infty} \frac{x}{e^x} = 0$, but the limit does not exist as $x \to -\infty$.

Example 2

Analyse the key features of the graph of $f(x) = xe^{\frac{1}{x}}$, then sketch the graph.

Solution

$f(x) = xe^{\frac{1}{x}}$

Intercepts

The y-intercept is $f(0) = 0e^{\frac{1}{0}}$.

But $e^{\frac{1}{0}}$ is not defined, so there is no y-intercept.

Solve $f(x) = 0$ to determine the x-intercepts.

$xe^{\frac{1}{x}} = 0$

Either $x = 0$ or $e^{\frac{1}{x}} = 0$

But $e^{\frac{1}{x}} \neq 0$

And when $x = 0$, $e^{\frac{1}{0}}$ is not defined. So, there is no x-intercept.

Domain

$e^{\frac{1}{x}}$ is defined for all non-zero values of x.

$D = \{x \mid x \in \Re, x \neq 0\}$

Critical points

Determine the derivative, $f'(x)$.

$f(x) = xe^{\frac{1}{x}}, \ x \neq 0$

Differentiate. Use the product rule.

$$f'(x) = (1)e^{\frac{1}{x}} + x\left(e^{\frac{1}{x}} \times \frac{-1}{x^2}\right)$$
$$= e^{\frac{1}{x}} - \frac{e^{\frac{1}{x}}}{x}$$
$$= \frac{xe^{\frac{1}{x}} - e^{\frac{1}{x}}}{x}$$
$$= \frac{e^{\frac{1}{x}}(x-1)}{x}$$

For critical values, solve $f'(x) = 0$.

$$\frac{e^{\frac{1}{x}}(x-1)}{x} = 0$$
$$e^{\frac{1}{x}}(x-1) = 0$$

Either $e^{\frac{1}{x}} = 0$ or $x - 1 = 0$

But $e^{\frac{1}{x}} \neq 0$, so $x - 1 = 0$

$x = 1$

There is a critical point at $x = 1$. Determine the corresponding y-value.

$$f(1) = (1)e^1$$
$$= e$$

The critical point is $(1, e)$, or approximately $(1, 2.7)$.

Intervals of increase and decrease

Consider the sign of $f'(x)$ in each interval.

$$f'(x) = \frac{e^{\frac{1}{x}}(x - 1)}{x}, \; x \neq 0$$

Since $e^{\frac{1}{x}}$ is always positive, consider only the sign of $\frac{x-1}{x}$.
When $x < 0$, say $x = -0.1$, $f'(x)$ is positive.
When $0 < x < 1$, say $x = 0.5$, $f'(x)$ is negative.
When $x > 1$, say $x = 1.1$, $f'(x)$ is positive.
Use a number-line analysis for the first derivative.

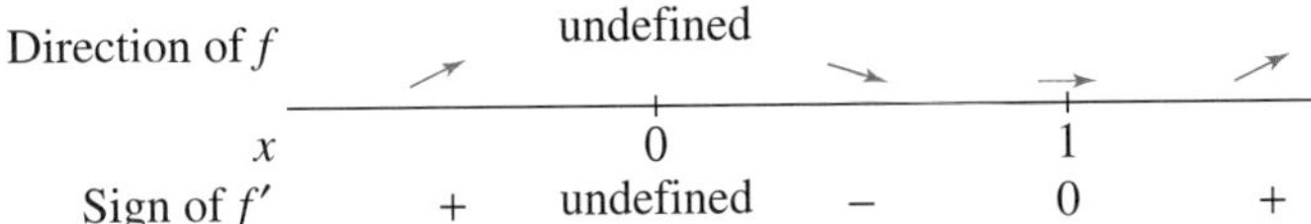

The graph of $f(x) = xe^{\frac{1}{x}}$ is increasing for $x < 0$ and $x > 1$, and decreasing for $0 < x < 1$. Therefore, the critical point $(1, e)$ is a minimum point.

Points of inflection

Consider the second derivative, $f''(x)$.

Use $f'(x)$ in this form: $f'(x) = e^{\frac{1}{x}} - e^{\frac{1}{x}}(x^{-1})$
Differentiate.

$$f''(x) = -\frac{1}{x^2}e^{\frac{1}{x}} - \left[e^{\frac{1}{x}}(-x^{-2}) + (x^{-1})e^{\frac{1}{x}}\left(-\frac{1}{x^2}\right)\right]$$
$$= -\frac{e^{\frac{1}{x}}}{x^2} + \frac{e^{\frac{1}{x}}}{x^2} + \frac{e^{\frac{1}{x}}}{x^3}$$
$$= \frac{e^{\frac{1}{x}}}{x^3}$$

For a point of inflection, solve $f''(x) = 0$.

$$\frac{e^{\frac{1}{x}}}{x^3} = 0$$

But $e^{\frac{1}{x}} \neq 0$, so there are no points of inflection.

Intervals of concavity

Consider the sign of $f''(x) = \frac{e^{\frac{1}{x}}}{x^3}$, $x \neq 0$, in each interval.

Since $e^{\frac{1}{x}} > 0$ for all values of x, consider only the sign of the denominator, x^3.
When $x < 0$, $f''(x)$ is negative, and the graph is concave down.
When $x > 0$, $f''(x)$ is positive, and the graph is concave up.
Use a number-line analysis for the second derivative.

Shape of f	$\frown$	undefined	$\smile$
x		0	
Sign of f''	$-$	undefined	$+$

The graph of $f(x) = xe^{\frac{1}{x}}$ is concave down when $x < 0$ and concave up when $x > 0$.

Asymptotes and point discontinuities

There are vertical asymptotes or point discontinuities at values of x for which $f(x) = xe^{\frac{1}{x}}$ is undefined.

$xe^{\frac{1}{x}}$ is undefined when $x = 0$, so there is a vertical asymptote or point discontinuity here.

Consider $\lim\limits_{x \to 0^+} xe^{\frac{1}{x}}$.

When x is very small and positive, $\frac{1}{x}$ is very large and positive, so $e^{\frac{1}{x}} \to \infty$.

$\lim\limits_{x \to 0^+} xe^{\frac{1}{x}} = \infty$

There is a vertical asymptote $x = 0$.

Consider $\lim\limits_{x \to 0^-} xe^{\frac{1}{x}}$.

When x is negative and has a very small absolute value, $\frac{1}{x}$ is negative and has a very large absolute value, so $e^{\frac{1}{x}} \to 0$.

$\lim\limits_{x \to 0^-} xe^{\frac{1}{x}} = 0$

Since $f(0)$ does not exist, there is a point discontinuity at $x = 0$.

For an oblique asymptote, determine $\lim\limits_{x \to \pm\infty} xe^{\frac{1}{x}}$.

As $x \to \pm\infty$, $e^{\frac{1}{x}} \to 1$, and $xe^{\frac{1}{x}} \to x$

There is an oblique asymptote $y = x$.

Use the information to sketch the graph of $f(x) = xe^{\frac{1}{x}}$.

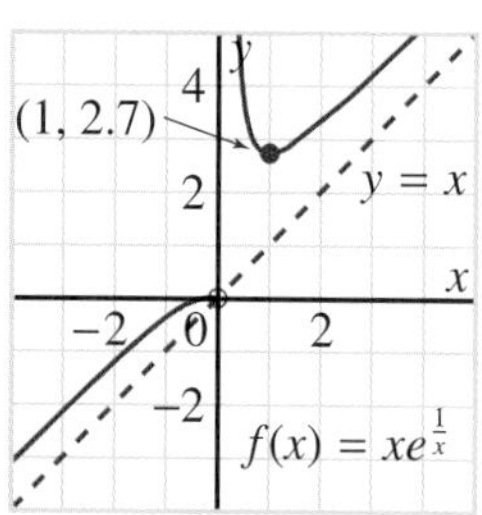

8.1 Exercises

1. For each function below:
   i) Determine the domain.
   ii) Determine the x- and y-intercepts of its graph.

   a) $f(x) = e^x - x$ b) $f(x) = \dfrac{e^x}{x - 1}$ c) $f(x) = x^2 + \dfrac{1}{x}$

B

2. Refer to the functions in exercise 1.

a) Which function(s) have an infinite discontinuity? Explain.

b) Do any functions have a point discontinuity? Explain.

3. Determine the equation of any vertical, horizontal, or oblique asymptote for the graph of each function.

a) $f(x) = \dfrac{x}{1 - e^x}$ **b)** $f(x) = x^2 + e^x$ **c)** $f(x) = \dfrac{e^x}{x}$

4. Which functions in exercise 3 are discontinuous? Explain.

5. Determine the critical values for each function.

a) $f(x) = e^x(x + 3)$ **b)** $f(x) = \dfrac{-e^{x+1}}{x}$ **c)** $f(x) = x^2e^x$

6. Determine the maximum and minimum values for each function.

a) $f(x) = x^3e^x$ **b)** $f(x) = x^2 + e^x$ **c)** $f(x) = \dfrac{1}{x} + e^x$

7. Determine the coordinates of the points of inflection for the graph of each function.

a) $f(x) = e^x - x^2$ **b)** $f(x) = e^x - e^{-x}$ **c)** $f(x) = \dfrac{x^2 + x}{e^x}$

8. Determine the domain and the equations of the asymptotes for each function.

a) $f(x) = \dfrac{x + 2}{e^x}$ **b)** $f(x) = \dfrac{x}{e^x + x}$ **c)** $f(x) = \dfrac{x^2 - 1}{e^x}$

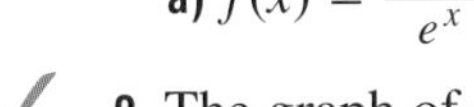

9. The graph of $y = xe^x$ is shown at the right. Use calculus to determine each feature listed.

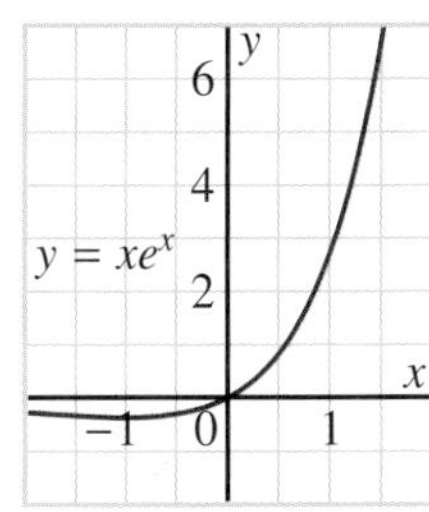

a) the coordinates of any maximum or minimum points

b) the coordinates of any points of inflection

c) the intervals of increase and decrease

d) the intervals of concavity

10. The graph of $y = e^{\frac{1}{x^2}}$ is shown at the right.

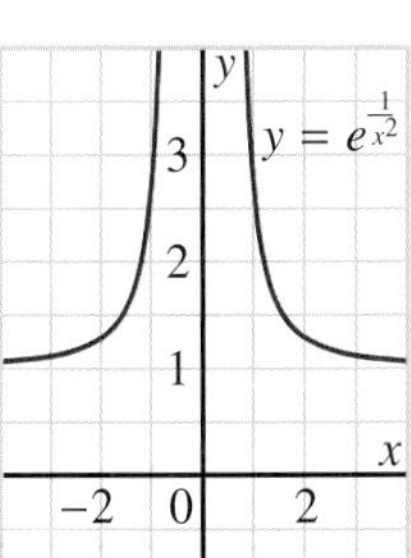

a) Write the equation of the derivative of this function, then sketch the graph of the derivative.

b) Use the sketch in part a to sketch the graph of the second derivative function.

11. Refer to the graph of $y = e^{\frac{1}{x^2}}$ in exercise 10. Show that this function has no maximum or minimum points, and no points of inflection.

12. Analyse the key features of the graph of the function $f(x) = \frac{1}{x^2} + e^x$, then sketch the graph.

13. **Knowledge/Understanding** Analyse the key features of the graph of the function $f(x) = \frac{x+1}{e^x - 1}$, then sketch the graph.

14. Analyse the key features of the graph of the function $f(x) = x^3e^{-x}$, then sketch the graph.

15. **Application** For each function $y = f(x)$, sketch graphs of $f'(x)$ and $f''(x)$.

a) 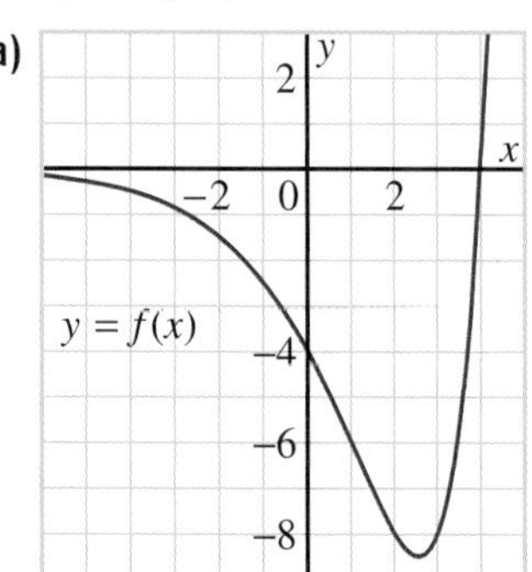

b) 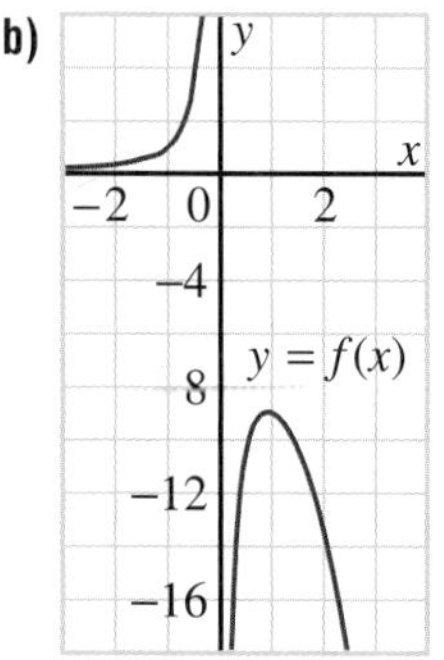

Key Features of the Graph of a Function

Domain
Intercepts
Critical points
Intervals of increase/decrease
Points of inflection
Intervals of concavity
Asymptotes
Point discontinuities

16. **Thinking/Inquiry/Problem Solving** Two functions are graphed below.

Graph 1

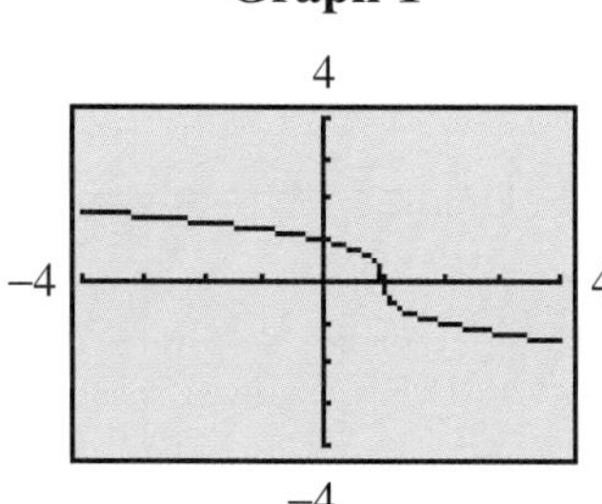

Graph 2

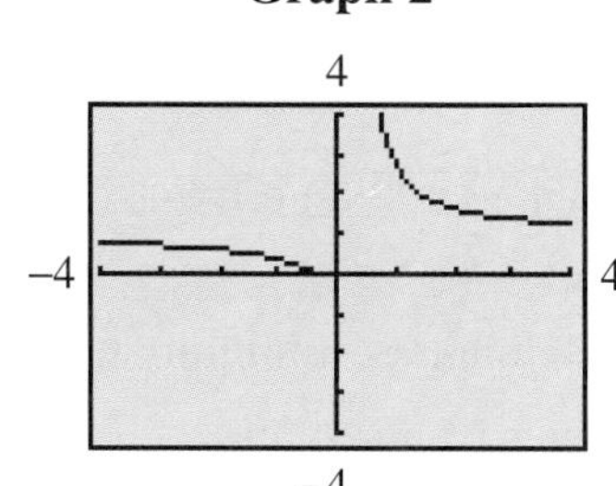

a) For each function, sketch graphs of the first and second derivative functions.

b) Use your graphs in part a to classify the following statements as either true or false. Justify your answers.

i) If $f(x)$ is continuous, then $f'(x)$ is continuous.

ii) If $f(x)$ is discontinuous, then $f'(x)$ is discontinuous.

iii) If $f(x)$ has a vertical tangent at $x = a$, then $f'(x)$ is undefined at $x = a$.

iv) If $f(x) > 0$, then $f'(x) > 0$

v) If $f'(x) < 0$, then $f''(x) < 0$

Exercise 16 is significant because the results illustrate the relationship between the continuity of a function and its derivatives.

17. **Communication** Explain how to draw the graph of a continuous function, $y = f(x)$, that satisfies the following conditions, then draw the graph: $f(3) = 5$, $f'(3) = 0$, $f''(x) < 0$ for $x < 3$ and $f''(x) > 0$ for $x > 3$

C

18. a) Sketch the graph of each function.
    i) $f(x) = x \sin x$ ii) $f(x) = x \cos x$
    b) Describe the symmetry illustrated by each graph.
    c) Determine the value(s) of x such that $y = x$ is a tangent to $y = f(x)$.
    d) Determine the value(s) of x such that $y = -x$ is a tangent to $y = f(x)$.

19. The graph of the function $f(x) = \sin \frac{1}{x}$ is shown below for different scales on the x-axis. The function is undefined when $x = 0$. From the first screen, it appears that the function has a point discontinuity at $x = 0$. However, from the second and third screens, the function becomes more erratic and oscillates between the values 1 and -1 more frequently as x gets closer and closer to 0. This discontinuity is called *oscillating discontinuity*.

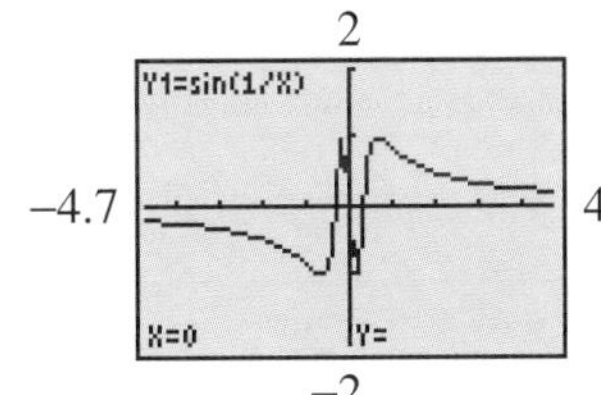

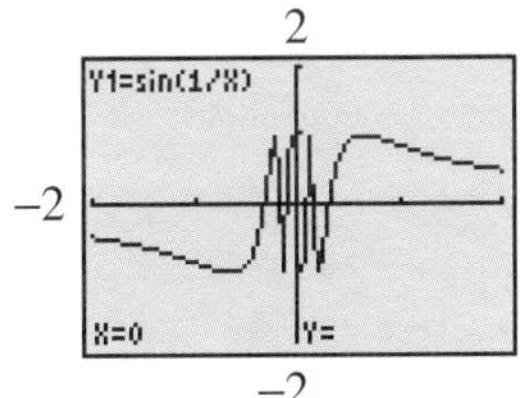

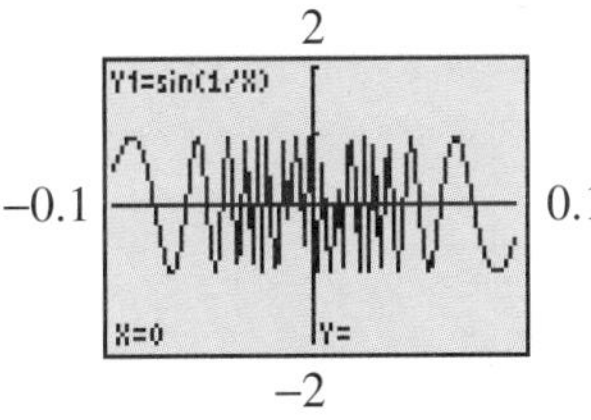

a) Explain why the values of the function oscillate between 1 and -1 more frequently as $x \to 0$.
b) Determine where each function has oscillating discontinuity. Explain.
   i) $f(x) = \cos \frac{1}{x-4}$ ii) $f(x) = \frac{1}{\sin \frac{1}{3x+8}}$
   iii) $f(x) = \cos\left(2 + \frac{1}{x}\right)$ iv) $f(x) = \tan \frac{1}{x+2}$

8.2 Optimization

In Chapter 1, you were introduced to calculus and its importance as a tool to describe how related quantities change. You now know how to determine the maximum and minimum values of many different types of functions. As seen in earlier chapters, functions can model physical situations. You can now use calculus as a tool to solve realistic problems in which a quantity is to be maximized or minimized. These types of problems are called *optimization* problems. In these problems, we need to determine the equation of the function to which we then apply calculus techniques. The power of calculus lies in the fact that the same methods apply, regardless of the type of function.

To illustrate the method, consider this problem. Suppose we have 64 m of fencing to enclose a rectangular plot. We can do this in different ways, and obtain different areas.

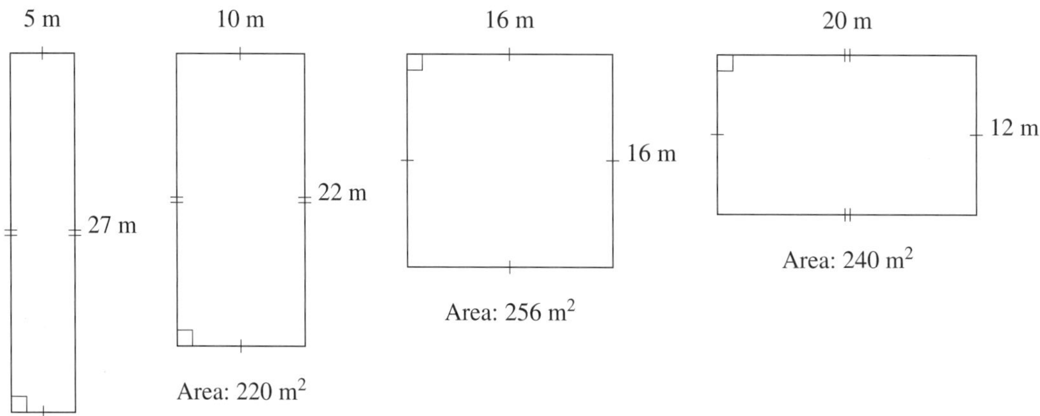

The rectangle with the greatest area seems to be the square, but we cannot be certain. To show that a square plot has the maximum area, we need a convincing argument similar to the one below. The steps shown can be applied to solve many similar problems.

Step 1. Identify the quantity to be maximized or minimized.

This quantity is the area of the rectangular plot.

Step 2. Write an equation for this quantity.

Draw a diagram. Let A represent the area. Let l and w represent the length and width, respectively. Then,

$$A = lw \qquad \ldots ①$$

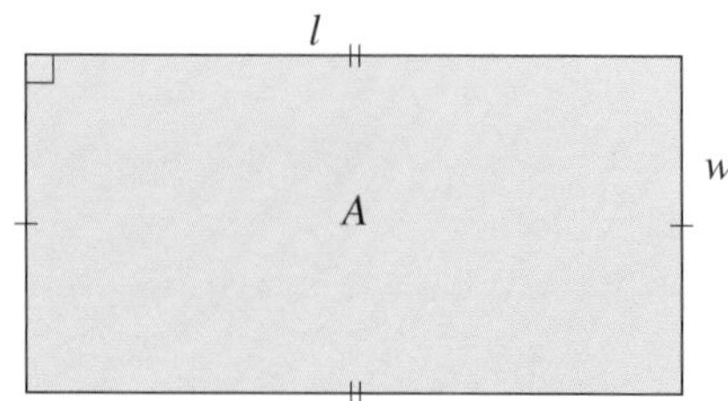

Step 3. Identify any constraints on the variables, and write appropriate equations.

The constraint is that the perimeter must be 64 m. Therefore,

$$2l + 2w = 64 \qquad \ldots ②$$

Step 4. Write the equation of a function that models the quantity to be maximized or minimized.

Use equation ② to write equation ① with only one variable on the right side. To do this, solve equation ② for either variable, say l.

$$2l + 2w = 64$$
$$l + w = 32$$
$$l = 32 - w \qquad \ldots ③$$

Substitute this expression for l in equation ①.

$$A = lw$$
$$A = (32 - w)w$$
$$A = 32w - w^2, \text{ or } A(w) = 32w - w^2$$

Step 5. Graph the function.

From equation ③, the domain is $0 \le w \le 32$. Since $A(w) = 32w - w^2$ is a quadratic function, its graph is a parabola. The parabola opens down and has a maximum value.

For the vertical intercept, determine $A(0) = 0$.

For the horizontal intercept, solve $A(w) = 0$.

$$32w - w^2 = 0$$
$$w(32 - w) = 0$$

Either $w = 0$ or $w = 32$

Differentiate to determine the critical value.

$$A(w) = 32w - w^2$$
$$A'(w) = 32 - 2w$$

Solve $A'(w) = 0$.

$$32 - 2w = 0$$
$$w = 16$$

The critical value is $w = 16$.

Determine the corresponding value of A.

$$A(16) = 32(16) - 16^2$$
$$= 256$$

The critical point is (16, 256).

Plot the intercepts and critical points. Sketch the graph of A.

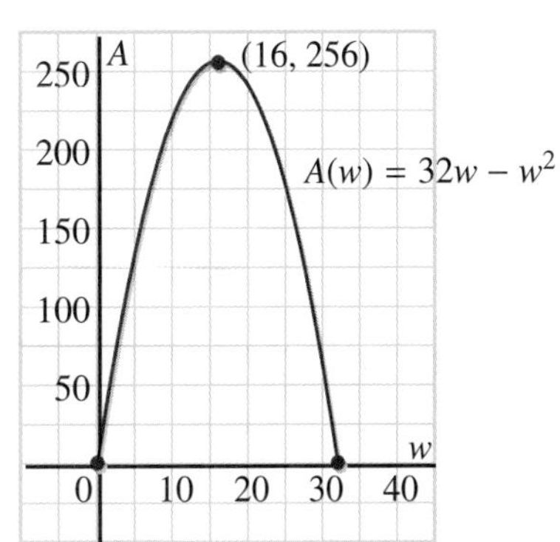

Step 6. Identify maximum or minimum values.

From the graph and the equation, we know that (16, 256) is a maximum point.
Check the value of A at each endpoint of the domain.
When $w = 0$, $A(0) = 0$
When $w = 32$, $A(32) = 0$
This confirms that (16, 256) is a maximum point.

Discuss

How could you use calculus to verify that the critical point (16, 256) is a maximum point?

Step 7. State the result.

Determine the length when the width is 16 m.
Substitute $w = 16$ in equation ③.

$$\begin{aligned} l &= 32 - 16 \\ &= 16 \end{aligned}$$

Hence, the length and width are equal when the rectangle has maximum area.

Therefore, the rectangle with the greatest area is a square.

The steps in the solution of the above problem can be used to solve all optimization problems. These steps are described below.

Take Note

Strategy for Solving Optimization Problems

1. **Identify the quantity to be maximized or minimized.**
   Read the problem carefully. Understand the problem. Identify the quantity to be maximized or minimized.

2. **Write an equation for this quantity.**
   Draw a diagram, and label appropriate parts. Use variables to represent the relevant quantities. Write an equation to calculate the quantity to be maximized or minimized.

3. **Identify any constraints on the variables, and write appropriate equations.**
   Your equation may have more than one variable on the right side. If so, you need one constraint for each extra variable. Write equation(s) to represent the constraint(s).

4. **Write the equation of a function to model the quantity to be maximized or minimized.**
   Use the constraint(s) in Step 3 to write the equation in Step 2 with only one variable on the right side (the independent variable). The result is the equation of a function.

5. **Graph the function.**
Identify the domain (the values of the independent variable that make sense in the problem). Graph the function over this domain, using calculus techniques.

6. **Identify the maximum or minimum values.**
Investigate the critical values and the endpoints of the domain to decide which value solves the problem.

7. **State the result.**
State the solution to the problem. Decide if it is reasonable.

We shall use these steps in the examples that follow.

We can use calculus to solve optimization problems that involve measurement in 3 dimensions.

Example 1

A box with no lid is constructed from a piece of cardboard that measures 16 cm by 30 cm. Congruent squares are cut from the corners of the cardboard and the resulting sides are bent upward to form a box.

a) What size should the squares be to obtain a box with the maximum volume?

b) What is the maximum volume?

Solution

a) *Step 1.* The quantity to be maximized is the volume of a box.

Step 2. Let x represent the side length of each cut-out square.
Let l and w represent the length and width of the box, respectively.
Then, the height of the box is x.

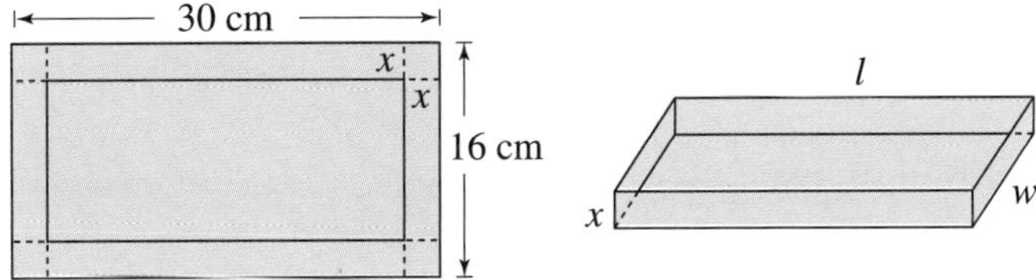

Let V represent the volume of the box.
Then $V = lwx$

Step 3. From the diagram, $l = 30 - 2x$ and $w = 16 - 2x$
The constraints on the dimensions are:
$l = 30 - 2x,\ 0 \le x \le 15$
$w = 16 - 2x,\ 0 \le x \le 8$

Step 4. Substitute for l and w in $V = lwx$.
$V = (30 - 2x)(16 - 2x)x$
$V = 4x^3 - 92x^2 + 480x$
The model function is $V(x) = 4x^3 - 92x^2 + 480x$.

Step 5. Since the side length of the cut-out square cannot exceed one-half the width of the cardboard, the domain is $0 \le x \le 8$.

To graph $V(x) = 4x^3 - 92x^2 + 480x$:
For the V-intercept, determine $V(0) = 0$.
For the x-intercept, solve $V(x) = 0$. Use $V(x)$ in factored form.
$(30 - 2x)(16 - 2x)x = 0$
$30 - 2x = 0$ or $16 - 2x = 0$ or $x = 0$
$x = 15$ $\quad x = 8$
The x-intercepts are 0, 8, and 15.
The intercept $x = 15$ is outside the domain.

Differentiate to determine the critical values.
$V(x) = 4x^3 - 92x^2 + 480x$
$V'(x) = 12x^2 - 184x + 480$

For the critical values, solve $V'(x) = 0$.
$12x^2 - 184x + 480 = 0$
Divide each side by 4.
$3x^2 - 46x + 120 = 0$
$(3x - 10)(x - 12) = 0$

Either $3x - 10 = 0$ or $x - 12 = 0$
$x = \frac{10}{3}$ $\quad x = 12$
The critical values are $x = \frac{10}{3}$ and $x = 12$.
The critical value $x = 12$ is outside the domain.
Determine the value of V when $x = \frac{10}{3}$.

$$V\left(\frac{10}{3}\right) = 4\left(\frac{10}{3}\right)^3 - 92\left(\frac{10}{3}\right)^2 + 480\left(\frac{10}{3}\right)$$
$$\doteq 725.9$$

The critical point is approximately (3.3, 726). To determine whether the critical point is a maximum or a minimum point, consider the second derivative.
$V'(x) = 12x^2 - 184x + 480$
$V''(x) = 24x - 184$

Determine the sign of the second derivative when $x = \frac{10}{3}$.

$V''\left(\frac{10}{3}\right) = 24\left(\frac{10}{3}\right) - 184$

$= -104$

Since the second derivative is negative, the graph of V is concave down in this interval, so the function has a maximum value at $x = \frac{10}{3}$. Sketch the graph of V in the given domain.

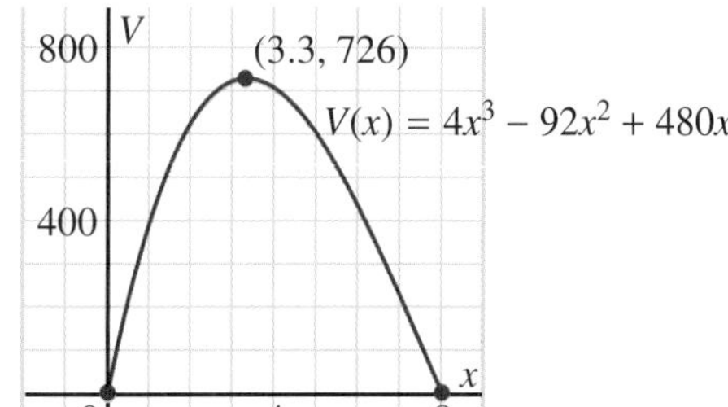

Step 6. Check the value of V at each endpoint.
When $x = 0$, $V(0) = 0$
When $x = 8$, $V(8) = 0$, since $x = 8$ is an intercept.
This confirms that $\left(\frac{10}{3}, 726\right)$ is a maximum point.

Step 7. For maximum volume, the cut-out squares should have side length approximately 3.3 cm.

b) From Step 6 in part a, the maximum volume is the V-coordinate of the maximum point. The maximum volume is approximately 726 cm^3.

Example 2

A cylindrical can is to be constructed to have a volume of 1000 cm^3.

a) Determine the dimensions of the can to minimize the cost of the material for the can.

b) What is the minimum surface area of the can?

Solution

a) *Step 1*. The quantity to be minimized is the surface area of the can.

Step 2. Let r and h represent the radius and height of the can, respectively. Let S represent the surface area of the can. Use the formula for the surface area of a cylinder.

$S = 2\pi r^2 + 2\pi rh \qquad \ldots ①$

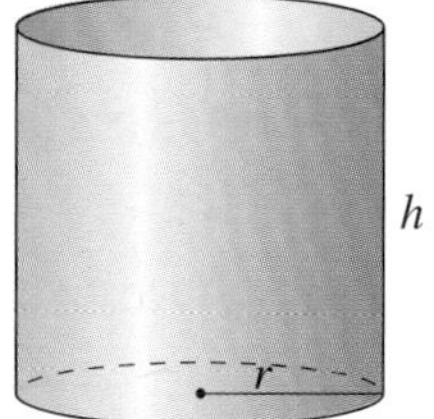

Step 3. The constraint is that the volume must be 1000 cm^3.
Use the formula for the volume of a cylinder: $V = \pi r^2 h$

Substitute $V = 1000$.

$1000 = \pi r^2 h \qquad \ldots ②$

Step 4. Solve equation ② for h.

$h = \frac{1000}{\pi r^2} \qquad \ldots ③$

Substitute for h in equation ①.

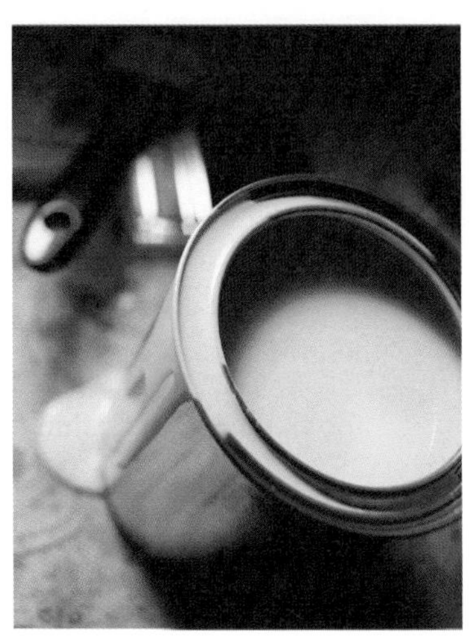

$S = 2\pi r^2 + 2\pi r\left(\frac{1000}{\pi r^2}\right)$

$S = 2\pi r^2 + \frac{2000}{r}$

The model function is $S(r) = 2\pi r^2 + \frac{2000}{r}$.

Step 5. Since the denominator in the model function cannot be zero, the domain is $r > 0$.

To graph $S(r) = 2\pi r^2 + \frac{2000}{r}$:

For the S-intercept, determine $S(0)$.

$S(0)$ is not defined, so there is no S-intercept.

For the r-intercept, solve $S(r) = 0$.

$2\pi r^2 + \frac{2000}{r} = 0$

Since $r > 0$, $2\pi r^2 + \frac{2000}{r} \neq 0$, so there is no r-intercept.

Differentiate to determine the critical values.

$S(r) = 2\pi r^2 + \frac{2000}{r}$

$S'(r) = 4\pi r - \frac{2000}{r^2}$

For the critical values, solve $S'(r) = 0$.

$$4\pi r - \frac{2000}{r^2} = 0$$

$$4\pi r = \frac{2000}{r^2}$$

$$r^3 = \frac{2000}{4\pi}$$

$$r = \sqrt[3]{\frac{2000}{4\pi}}$$

$$= \left(\frac{500}{\pi}\right)^{\frac{1}{3}}$$

$$\doteq 5.42$$

The critical value is $\left(\frac{500}{\pi}\right)^{\frac{1}{3}}$, or approximately 5.42.

Determine the value of S when $r = \left(\frac{500}{\pi}\right)^{\frac{1}{3}}$.

$$S\left(\left(\frac{500}{\pi}\right)^{\frac{1}{3}}\right) = 2\pi\left(\frac{500}{\pi}\right)^{\frac{2}{3}} + 2000\left(\frac{500}{\pi}\right)^{-\frac{1}{3}}$$

$$\doteq 553.6$$

The critical point is approximately (5.42, 554).

To determine whether the critical point is a maximum or a minimum point, consider the second derivative.

$S'(r) = 4\pi r - \frac{2000}{r^2}$

$S''(r) = 4\pi + \frac{4000}{r^3}$

For all values of r in the domain, $S''(r)$ is positive.
Since the second derivative is positive, the graph of S is concave up in this interval; so, the function has a minimum value at $x \doteq 5.42$.
Sketch the graph of S in the given domain.

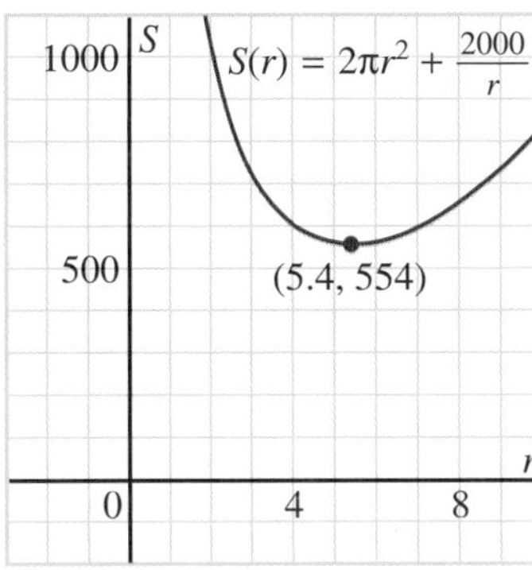

Step 6. Check the value of S at each endpoint.
When $r \to 0$, $S(r) \to \infty$
When $r \to \infty$, $S(r) \to \infty$
So, there are no endpoints to consider.
The critical value $r \doteq 5.42$ produces the minimum value of the function.

Step 7. Determine the height of the can for minimum surface area.

Substitute $r = \left(\frac{500}{\pi}\right)^{\frac{1}{3}}$ in equation ③: $h = \frac{1000}{\pi r^2}$

$$h = \frac{1000}{\pi\left(\frac{500}{\pi}\right)^{\frac{2}{3}}}$$
$$\doteq 10.84$$

For the cost of the can to be a minimum, its radius is approximately 5.4 cm and its height is approximately 10.8 cm.

Discuss

In Step 4, could we have completed the solution by solving and substituting for *r* instead of for *h*? Explain.

b) From Step 5 in part a, the minimum surface area of the can is the S-coordinate of the minimum point. The minimum surface area is approximately 554 cm^2.

We can use calculus to solve number problems.

Example 3

Determine the positive number such that the sum of the square of the number and the reciprocal of the number has a minimum value.

Solution

Step 1. The quantity to be minimized is the sum of the square of a positive number and the reciprocal of the number.

Step 2. Let x represent the number. Let S represent the sum.
Then, x^2 is the square of the number and $\frac{1}{x}$ is the reciprocal.
Then, $S = x^2 + \frac{1}{x}$

Step 3. Since there is only one variable on the right side of the equation, there are no constraints to consider.

Step 4. The model function is $S(x) = x^2 + \frac{1}{x}$.

Step 5. Since $x \neq 0$, and x is positive, the domain $D = \{x \mid x \in \Re, x > 0\}$.

To graph $S(x) = x^2 + \frac{1}{x}$, $x \neq 0$:
For the S-intercept, determine $S(0)$.
$S(0)$ is not defined, so there is no S-intercept.
For the x-intercept, solve $S(x) = 0$.

$$
\begin{aligned}
x^2 + \frac{1}{x} &= 0 \\
x^3 + 1 &= 0 \\
x^3 &= -1 \\
x &= \sqrt[3]{-1} \\
x &= -1
\end{aligned}
$$

The x-intercept is –1, but this is outside the domain.

Differentiate to determine the critical value(s).

$$
\begin{aligned}
S(x) &= x^2 + \frac{1}{x} \\
S'(x) &= 2x - \frac{1}{x^2}
\end{aligned}
$$

For the critical values, solve $S'(x) = 0$.

$$
\begin{aligned}
2x - \frac{1}{x^2} &= 0 \\
2x^3 - 1 &= 0 \\
2x^3 &= 1 \\
x^3 &= 0.5 \\
x &= \sqrt[3]{0.5} \\
&\doteq 0.79
\end{aligned}
$$

The critical value is $x = 0.5^{\frac{1}{3}}$, or $x \doteq 0.79$.

Determine the corresponding value of S.

$$S = 0.5^{\frac{2}{3}} + \frac{1}{0.5^{\frac{1}{3}}}$$
$$= \frac{0.5 + 1}{0.5^{\frac{1}{3}}}$$
$$= 1.5(0.5)^{-\frac{1}{3}}$$
$$\doteq 1.89$$

The critical point is $\left(0.5^{\frac{1}{3}},\ 1.5(0.5)^{-\frac{1}{3}}\right)$, or approximately $(0.79, 1.89)$.

To determine whether the critical point is a maximum or a minimum point, consider the second derivative.

$$S'(x) = 2x - \frac{1}{x^2}$$
$$S''(x) = 2 + \frac{2}{x^3}$$

Determine the sign of the second derivative when $x = 0.5^{\frac{1}{3}}$.

$S''\left(0.5^{\frac{1}{3}}\right) = 2 + \frac{2}{0.5}$, which is positive

Since the second derivative is positive, the graph of S is concave up in this interval, so the function has a minimum value at $x = 0.5^{\frac{1}{3}}$. Sketch the graph of S in the given domain.

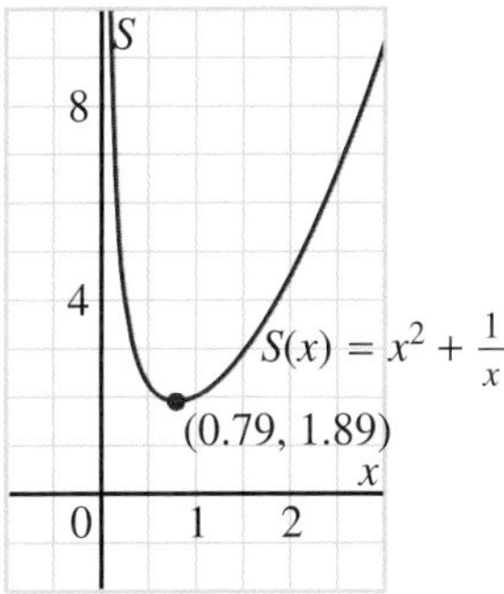

Step 6. When $x \to 0,\ S(x) \to \infty$

When $x \to \infty,\ S(x) \to \infty$

So, there are no endpoints to consider.

The critical value $x = 0.5^{\frac{1}{3}}$ produces the minimum value of the function.

Step 7. The sum of the square of a positive number and the reciprocal of the number is a minimum when the number is $0.5^{\frac{1}{3}}$, or approximately 0.79.

8.2 Exercises

1. State a formula for each measurement. Describe any variables used.
   a) the surface area of a sphere
   b) the area of a rectangle
   c) the surface area of a cylinder
   d) the volume of a cylinder
   e) the area of a triangle
   f) the volume of a rectangular prism
   g) the surface area of a rectangular prism
   h) the volume of a cone

2. Write the constraint equation for each measurement. Describe any variable used.
   a) The perimeter of a rectangle is 30 m.
   b) The area of a triangle is 21 cm^2.
   c) The volume of a cylinder is 500 cm^3.
   d) The surface area of a square-based prism is 800 cm^2.
   e) The volume of a square-based prism is 1200 cm^3.

3. **Communication** A farmer has 100 m of fencing and wishes to make a rectangular pen, with maximum area, to house her cows. The land borders on a river; therefore, no fence is required on that side. Explain how to determine the equation, $A(x) = 100x - 2x^2$, which is used to solve this problem. Describe what each variable represents. State the constraints and explain how you determined them.

4. The number of bacteria, N, in a culture at time, t hours, is modelled by the equation $N = 1000\left(50 + te^{-\frac{t}{10}}\right)$. Determine the maximum number of bacteria in the culture between $t = 0$ and $t = 40$ h.

5. A box with no lid has a square base and a total surface area of 3600 cm^2. You are to determine the dimensions of the box with the maximum volume.
   a) What quantity is to be maximized?
   b) What is the constraint equation?
   c) Express the quantity in part a as a function of a single variable.
   d) Solve the problem.
   e) Determine the maximum volume.

6. a) What changes need to be made in the solution of exercise 5 if the box has a lid? Explain.
   b) Assume the box in exercise 5 has a lid. Determine the maximum volume of the box.

7. **Knowledge/Understanding** The sum of two numbers is 25. The sum of their squares is a minimum. Determine the numbers.

8. a) Determine a positive number such that the sum of the number and the square of its reciprocal is a minimum.
   b) Compare the answer to part a with the minimum sum in *Example 3*. What do you notice?
   c) How could you complete part a using the result of *Example 3*?

9. A box with no lid is to be made by cutting congruent squares from the corners of a 40-cm by 60-cm sheet of plywood.
   a) Determine the side length of the cut-out squares so the volume of the box is a maximum.
   b) What is the maximum volume?

10. A lifeguard wishes to enclose a rectangular swimming area at a beach with 350 m of rope. One side of the rectangle is a straight sandy shore.
    a) What are the dimensions of the rectangle that produce the maximum swimming area?
    b) What is the maximum swimming area?

11. A farmer wants to enclose and subdivide a rectangular field into 3 congruent plots of land. He has 4000 m of fencing. The area to be enclosed is a maximum. Determine the dimensions of one plot of land.

12. **Application** The perimeter of an isosceles triangle is to be 48 cm. Determine the lengths of the sides of the triangle that produce a maximum area.

13. **Thinking/Inquiry/Problem Solving** In $\triangle ABC$, $\angle B = 90^\circ$ and the length of AC is 12 cm.

    a) Show that the area of $\triangle ABC$ is a maximum when AB = BC.

    b) Will the result of part a be true if the length of AC is 8 cm, or 15 cm, or 20 cm? Explain.

    c) What conclusion can be made for any constant length l of AC? Explain.

14. A straight section of railroad track crosses two highways at points that are 400 m and 600 m, respectively, from an intersection. Determine the dimensions of the largest rectangular lot that can be laid out in the triangle formed by the railroad and highways.

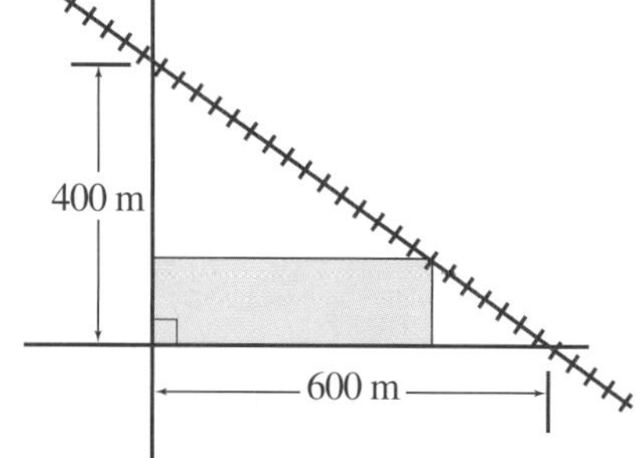

15. A piece of rope, 100 m long, is to be cut into two pieces. One piece is used to enclose a circular garden and the other piece is used to enclose a square garden. How should the rope be cut in each case?

    a) The sum of the enclosed areas is a maximum.

    b) The sum of the enclosed areas is a minimum.

Exercise 15 is significant because it illustrates that, for a fixed perimeter, the maximum area occurs when the figure is a circle.

16. The cost price of a particular model of car is \$15 000. When the dealer sells each car for \$25 000, she sells 24 cars per month. For each reduction of \$600 in the selling price, the dealer sells 2 more cars each month. The model function for the profit, P dollars, is $P = (25\,000 - 600x)(24 + 2x) - 15\,000(24 + 2x)$.

    a) Explain how the profit function was determined. Describe what x represents.

    b) Determine the selling price of a car for maximum monthly profit.

17. Determine the dimensions of a cylinder of maximum volume that can be inscribed in a sphere of radius 10 cm.

18. Determine the maximum volume of a cone that can be inscribed in a sphere of radius 10 cm. The volume, V, of a cone is given by $V = \frac{1}{3}\pi r^2 h$.

19. The strength of a rectangular wooden beam is proportional to the product of its width and the square of its depth. Determine the dimensions of the strongest beam that can be cut from a 30-cm diameter cylindrical log.

Self-Check 8.1, 8.2

1. For each function below:

   i) Determine the domain.

   ii) Determine the x- and y-intercepts of its graph.

   a) $f(x) = x + e^x$ b) $f(x) = x^2 - e^x$ c) $f(x) = \frac{1}{x} - e^x$

2. Determine the critical values of each function in exercise 1.

3. Consider the graph of $y = xe^{-x}$.

   a) Write the equation of the derivative function.

   b) Sketch the graph of the derivative function.

   c) Determine the coordinates of the maximum point and the point of inflection.

   d) Sketch the graph of the second derivative function.

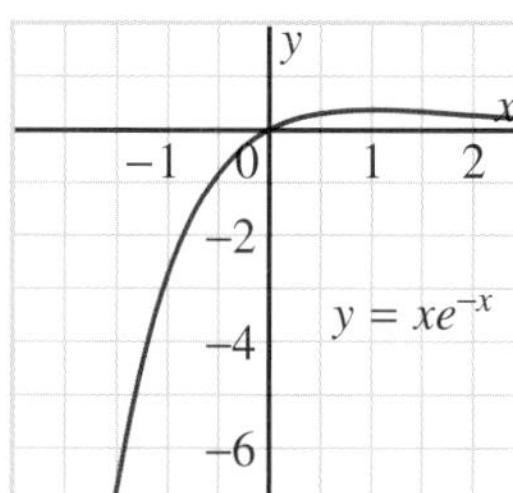

4. Analyse the key features of the graph of $f(x) = \frac{x^2 + x}{e^x}$, then sketch the graph.

5. Two numbers have a difference of 10. The result of adding their sum and their product is a minimum. Determine the numbers.

6. A 30-cm piece of wire is cut in two. One piece is bent to form a square. The other piece is bent to form a circle. What are the lengths of the two pieces in each case?

   a) The sum of the areas of the square and circle is a minimum.

   b) The sum of the areas of the square and circle is a maximum.

7. Determine the maximum volume of a box with surface area 3000 cm^2. The box has a square base and no top.

8. A manufacturer is to produce a soup can with a capacity of 450 mL. Determine the dimensions of the can that requires the minimum amount of material. Recall that 1 mL = 1 cm^3.

PERFORMANCE ASSESSMENT

9. A semicircular piece of wood has radius 3 m. A rectangle is to be cut from the semicircle. What are the dimensions of the rectangle with the maximum area?

8.3 Related Rates: Part I

In this section and the next, we will use calculus as a tool to solve problems involving two or more quantities that vary. If these quantities are related, then their rates of change will also be related.

For example, suppose we pour water into a cylindrical container to form a cylinder of water with volume V and height h. If we know the rate of change of V, we should be able to calculate the rate of change of h.

Suppose we pour water into a conical container to form a cone of water with volume V, height h, and radius r. If we know the rate of change of V, we should be able to calculate the rate of change of both h and r.

Investigation

Comparing Rates of Change of Depth

Water is poured into each container below at a constant rate of 5 cm^3/s. The containers are initially empty.

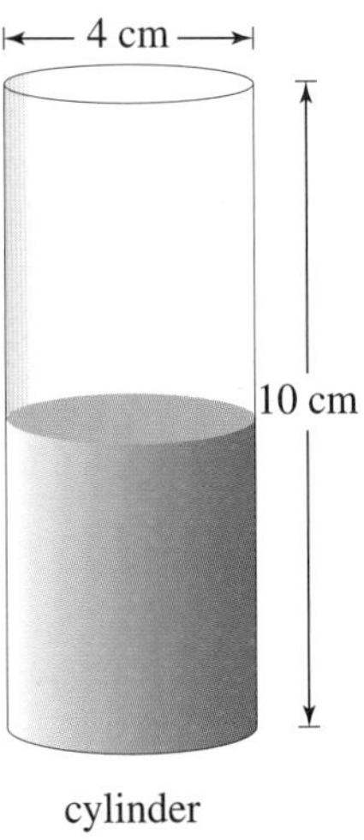

cylinder

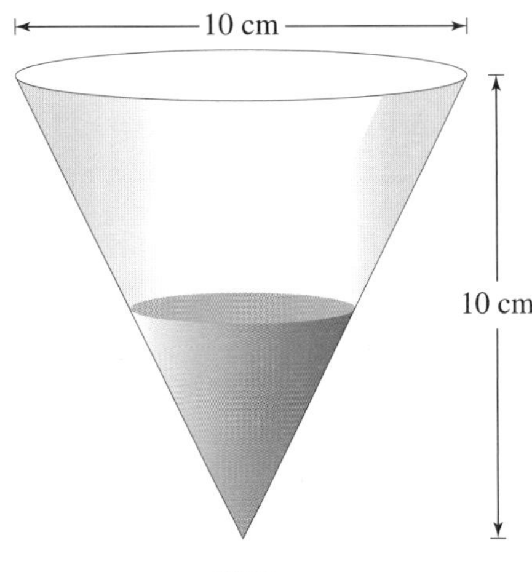

cone

1. **a)** After 1 s, the cylinder contains 5 cm^3 of water. Use the formula $V = \pi r^2 h$. Substitute $V = 5$ and $r = 2$. Solve for h. This is the height of water after 1 s.

   b) Repeat part a to determine the heights of water after 2 s, 3 s, 4 s, and 5 s. Record the times and heights in a table.

   c) **i)** Graph height, h, against time, t.

   ii) Describe the graph.

   iii) Use the graph of h against t to describe the derivative function $\frac{dh}{dt}$.

2. The volume of the cone is $V = \frac{1}{3}\pi r^2 h$. Consider a vertical section through the vertex of the cone.

At any time t, the radius of the water is r and its height is h. Use similar triangles.

a) Express r as a function of h.

b) Express V as a function of h.

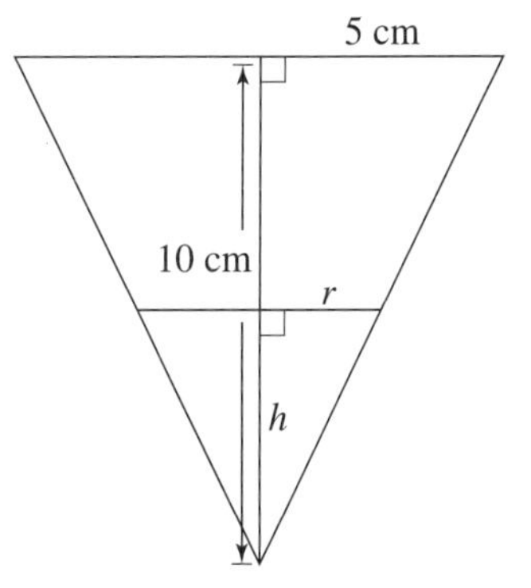

3. Repeat exercise 1 for the cone. Use the formula you obtained in exercise 2b to calculate h.

4. Compare the functions and their graphs in exercises 1 and 3. Why are the derivative functions different?

In the *Investigation*, you drew graphs of h against t, then described the functions $\frac{dh}{dt}$. We shall now determine each $\frac{dh}{dt}$ algebraically. Recall that the rate of change of the volume is 5 cm³/s.

Cylinder

As the water is poured in, it forms a cylinder of water with increasing height. The volume of this cylinder of water is $V = \pi r^2 h$.

The radius is constant. Substitute $r = 2$.

$V = \pi(2)^2 h$

$V = 4\pi h$

This equation relates the volume of the water cylinder to its height. We can use it to determine how the rates of change of volume and height are related.

Differentiate each side with respect to t.

$\frac{dV}{dt} = 4\pi\frac{dh}{dt}$

This equation relates the rate of change of volume, $\frac{dV}{dt}$, and the rate of change of height, $\frac{dh}{dt}$. If we know one of these we can calculate the other.

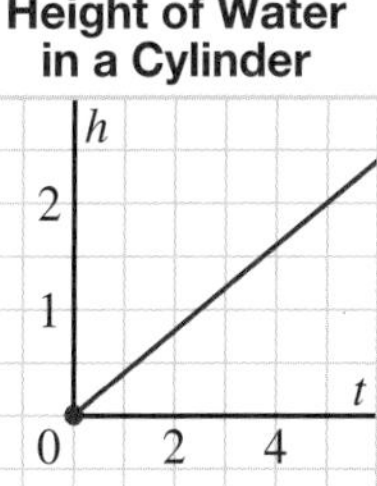

We know that $\frac{dV}{dt} = 5$ cm³/s. Substitute $\frac{dV}{dt} = 5$.

$5 = 4\pi\frac{dh}{dt}$

$\frac{dh}{dt} = \frac{5}{4\pi}$

$\frac{dh}{dt}$ is constant, as you found in the *Investigation*, exercise 1.

Cone

For the cone of water, $r = \frac{1}{2}h$, so the volume $V = \frac{1}{3}\pi\left(\frac{h}{2}\right)^2 h = \frac{1}{12}\pi h^3$

The volume of the cone is $V = \frac{1}{12}\pi h^3$.

Differentiate each side with respect to t.

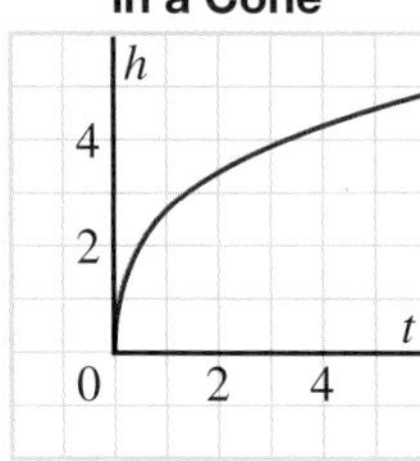

$$\frac{dV}{dt} = \frac{1}{12}\pi(3h^2)\frac{dh}{dt}$$

$$\frac{dV}{dt} = \frac{1}{4}\pi h^2 \frac{dh}{dt}$$

This equation relates the rate of change of volume, $\frac{dV}{dt}$, and the rate of change of height, $\frac{dh}{dt}$. If we know one of these, and also the height of the water cone, we can calculate the other.

Substitute $\frac{dV}{dt} = 5$.

$$5 = \frac{1}{4}\pi h^2 \frac{dh}{dt}$$

$$\frac{dh}{dt} = \frac{20}{\pi h^2}$$

$\frac{dh}{dt}$ is proportional to $\frac{1}{h^2}$. As h increases, $\frac{dh}{dt}$ decreases, as you found in the *Investigation*, exercise 3.

Strategy for Solving Related-Rates Problems

Step 1. Identify the quantity whose rate of change is to be determined.
Read the problem carefully. Understand the problem. Identify the quantity whose rate of change is known and the quantity whose rate of change is to be determined.

Step 2. Write an equation for this quantity.
Draw a diagram, and label appropriate parts. Use variables to represent the relevant quantities. Write an equation to calculate the quantity whose rate of change is required.

Step 3. Identify any constraints on the variables, and write appropriate equations.
If the equation has more than one variable on the right side, write equations to represent the constraints for these variable(s), if necessary.

Step 4. Write the equation of a function to model the quantity whose rate of change is required.
Use the constraints in Step 3 to write the equation of the model function.

Step 5. Differentiate the function.
Differentiate the model function with respect to time, t.

Step 6. Calculate the required rate of change.
Substitute the known rate of change and, if necessary, the corresponding value of the variable.

Step 7. State the result.
State the solution to the problem. Decide if it is reasonable.

Example 1

A cylindrical container has radius 2 cm and height 4 cm. Water drains from the base of the container at a constant rate of 6 cm^3/s. How fast does the depth of the water decrease?

Solution

Step 1. The rate of change of the volume of the water is known: -6 cm^3/s. The negative sign indicates that the volume of water in the cylinder is decreasing. The rate of change of the depth of the water is to be determined.

Step 2. At any time, t seconds:
Let r and h represent the radius and height of the surface of the water, respectively.
Let V represent the volume of the water in the cylinder.
$V = \pi r^2 h$ ... ①

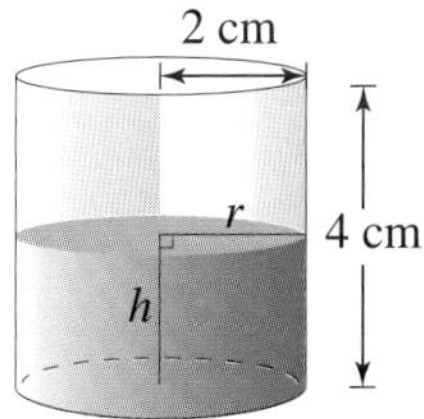

Step 3. The constraint is that the radius is constant: $r = 2$ cm

Step 4. Substitute $r = 2$ in equation ①.

$$V = \pi(2)^2 h$$
$$V = 4\pi h$$
$$h = \frac{V}{4\pi}$$

The model function is $h = \frac{V}{4\pi}$.

Step 5. Differentiate the function with respect to t, since $\frac{dh}{dt}$ is to be determined.

$$\frac{dh}{dt} = \frac{1}{4\pi}\frac{dV}{dt}$$

Step 6. Substitute $\frac{dV}{dt} = -6$.

$$\begin{aligned}\frac{dh}{dt} &= \frac{1}{4\pi}(-6)\\ &= \frac{-3}{2\pi}\\ &\doteq -0.48\end{aligned}$$

Step 7. The depth of water is decreasing at approximately 0.5 cm/s.

Example 2

A conical reservoir is filling with water at a constant rate of 3 m^3/min. The reservoir is 3 m deep and has a maximum diameter of 8 m. Determine the rate at which the depth of the water is increasing when the depth is 2 m.

Solution

Step 1. The rate of change of the volume of the water is known: 3 m^3/min
The rate of change of the depth of the water is to be determined.

Step 2. At any time, t seconds:
Let r and h represent the radius and height of the surface of the water, respectively.
Let V represent the volume of the water in the reservoir.

$$V = \frac{1}{3}\pi r^2 h \qquad \ldots ①$$

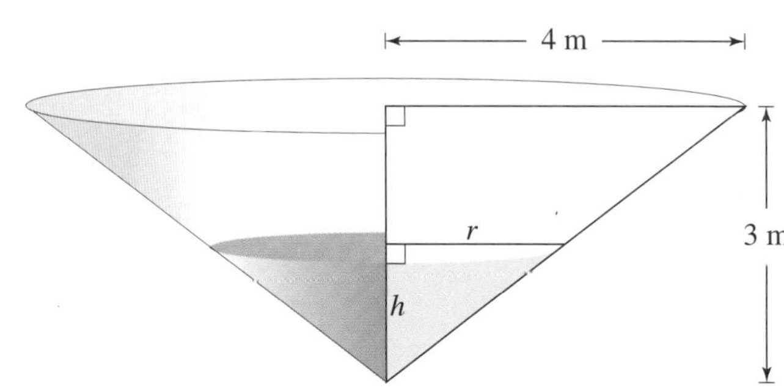

Step 3. Determine a relationship between r and h.
The right triangles in the diagram above are similar.
So, the ratios of corresponding sides are equal.

$$\frac{r}{h} = \frac{4}{3} \qquad \ldots ②$$

Since we want to determine $\frac{dh}{dt}$, solve equation ② for r.

$$r = \frac{4}{3}h$$

Step 4. Substitute $r = \frac{4}{3}h$ in $V = \frac{1}{3}\pi r^2 h$.

$$V = \frac{1}{3}\pi\left(\frac{4}{3}h\right)^2 h$$

$$V = \frac{16}{27}\pi h^3$$

The model function is $V = \frac{16}{27}\pi h^3$.

Step 5. Differentiate the function with respect to t.

$$\frac{dV}{dt} = \frac{16}{27}\pi(3h^2)\frac{dh}{dt}$$

$$\frac{dV}{dt} = \frac{16\pi h^2}{9}\frac{dh}{dt}$$

Solve for $\frac{dh}{dt}$.

$$\frac{dh}{dt} = \frac{9}{16\pi h^2}\frac{dV}{dt}$$

Step 6. Substitute $\frac{dV}{dt} = 3$ and $h = 2$.

$$\frac{dh}{dt} = \frac{9}{16\pi(2)^2}(3)$$
$$= \frac{27}{64\pi}$$
$$\doteq 0.13$$

Step 7. When the depth of the water is 2 m, the depth is increasing at approximately 0.13 m/s.

Discuss

In Step 4, we could have solved the model function for *h* before differentiating in Step 5. Why did we not do this?

Example 3

A square is expanding so that its area increases at 10 cm^2/min.

a) How fast is the side length increasing when the area is 52 cm^2?

b) How fast is the perimeter increasing when the side length is 8 cm?

Solution

a) *Step 1.* The rate of change of the area is known: 10 cm^2/min
The rate of change of the side length is to be determined.

Step 2. At time t minutes:
Let s and A represent the side length and area, respectively.

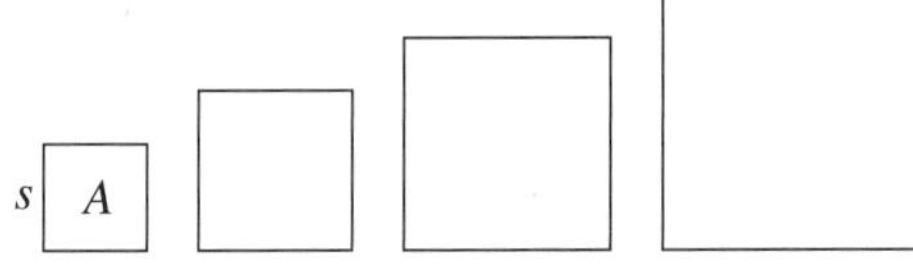

Then, $A = s^2$

Step 3. Since there is only 1 variable on the right side of the equation, there are no constraints to consider.

Step 4. The model function is $A = s^2$.

Step 5. Differentiate the function with respect to t.

$$A = s^2$$
$$\frac{dA}{dt} = 2s\frac{ds}{dt}$$
$$\frac{ds}{dt} = \frac{1}{2s}\frac{dA}{dt} \quad \ldots ①$$

Step 6. Determine the side length, s, when $A = 52\text{ cm}^2$.
That is, $s = \sqrt{52}$ cm

Substitute $s = \sqrt{52}$ and $\frac{dA}{dt} = 10$ in equation ①.

$$\begin{aligned}\frac{ds}{dt} &= \frac{1}{2\sqrt{52}}(10)\\ &= \frac{5}{\sqrt{52}}\\ &\doteq 0.69\end{aligned}$$

Step 7. When the area is 52 cm^2, the side length is increasing at approximately 0.7 cm/min.

b) *Step 1.* The rate of change of the area is known: $10\text{ cm}^2/\text{min}$
The rate of change of the perimeter is to be determined.

Step 2. Use the diagram in part a.
Let P represent the perimeter at time t minutes.
Then, $P = 4s$

Step 3. There are no constraints to consider.

Step 4. The model function is $P = 4s$.

Step 5. Differentiate the function with respect to t.

$$\frac{dP}{dt} = 4\frac{ds}{dt} \qquad \dots ②$$

Step 6. We do not know $\frac{ds}{dt}$, but we do know $\frac{dA}{dt}$.

Use equation ① in part a: $\frac{ds}{dt} = \frac{1}{2s}\frac{dA}{dt}$

Substitute $s = 8$ and $\frac{dA}{dt} = 10$.

$$\begin{aligned}\frac{ds}{dt} &= \frac{1}{2(8)}(10)\\ &= \frac{5}{8}\end{aligned}$$

Substitute $\frac{ds}{dt} = \frac{5}{8}$ in equation ②.

$$\begin{aligned}\frac{dP}{dt} &= 4\left(\frac{5}{8}\right)\\ &= 2.5\end{aligned}$$

Step 7. When the side length is 8 cm, the perimeter is increasing at 2.5 cm/min.

Discuss

How is the perimeter of a square related to its area? How could you use an equation that relates *P* and *A* to complete part b?

8.3 Exercises

1. Solve each equation for the indicated rate.

   a) $\frac{dA}{dt} = 8\pi r\frac{dr}{dt}$; for $\frac{dr}{dt}$

   b) $\frac{dA}{dt} = 2\pi r\frac{dr}{dt}$; for $\frac{dr}{dt}$

   c) $\frac{dV}{dt} = 4\pi r\frac{dr}{dt}$; for $\frac{dr}{dt}$

   d) $\frac{dV}{dt} = \frac{4}{9}\pi h^2\frac{dh}{dt}$; for $\frac{dh}{dt}$

   e) $\frac{dP}{dt} = 4\frac{ds}{dt}$; for $\frac{ds}{dt}$

   f) $\frac{dV}{dt} = \frac{48}{25}\pi h^2\frac{dh}{dt}$; for $\frac{dh}{dt}$

B

2. State an equation you could differentiate to determine each rate of change with respect to time.

   a) the rate of change of the perimeter of a square

   b) the rate of change of the volume of a spherical balloon

   c) the rate of change of the surface area of a cube

   d) the rate of change of the circumference of a circle

   e) the rate of change of the radius of a circle as its area decreases

3. As a spherical snowball melts, its volume decreases at a rate of 4 cm^3/min. What is the rate of decrease of the radius when the radius is 8 cm?

4. **Knowledge/Understanding** Water is draining from a tank at a rate of 1.5 m^3/min. The tank has the shape of a cone, with a base radius of 3 m at the top, and a height of 6 m. Determine the rate at which the depth of the water is decreasing when the depth is 3.2 m.

5. The side of a square is increasing at a rate of 1.4 m/min. How fast is the perimeter of a square increasing when its side length is 5 cm?

6. a) A spherical balloon is filling with air. The radius is increasing at a constant rate of 2 cm/s. What is the rate of increase of the volume of the balloon when the radius is 3 cm?

   b) What is the rate of increase of the surface area of the balloon when the radius is 3 cm?

7. **Communication** Explain the term *related-rates problem*. How can you distinguish related-rates problems from optimization problems?

8. The edge length of an ice cube is melting at a rate of 1 mm/min. What is the rate of change of the surface area of the ice cube when the edge length is 8 mm?

9. The area of a square is increasing at a rate of 8 cm^2/min.

a) Determine the rate of increase of the side length when the area is 32 cm^2.

b) Determine the rate of increase of the perimeter when the side length is 6 cm.

10. A ball is thrown into a pond and creates circular ripples that travel outward at 8 cm/s. Determine the rate of increase of the area enclosed by the ripples, after 3 s.

11. A rectangle is expanding. The ratio of its length to its width is always 3 : 5. Determine the rate of increase of the area of the rectangle when the perimeter is increasing at 12 cm/s and the perimeter is 32 cm.

12. **Application** The sides of an equilateral triangle decrease at the rate of 5 cm/s. Determine the rate of decrease of the area of the triangle when the area is 100 cm^2.

13. Top soil is poured to form a conical pile at a rate of 2.1 m^3/s. The ratio of the base diameter to the height is always 5 : 4. How quickly is the height of the pile growing when the pile is 6 m high?

14. A spherical hot-air balloon deflates at a rate of 9 m^3/min.

a) How fast is the radius decreasing when the radius is 4 m?

b) How fast is the radius decreasing when the volume is 60 m^3?

15. **Thinking/Inquiry/Problem Solving** At a certain instant, the area and radius of a circular oil spill on the ocean are increasing at the same rate. What is the radius, in kilometres, at that instant?

C

16. The cross section of a water trough is an isosceles triangle. The triangle has a height of 0.5 m and a base of 1.4 m. The trough is 4 m long. A rainstorm fills the trough with water so that its depth increases at a rate of 3 cm/min. How fast is the volume increasing at the instant when the depth is 5 cm?

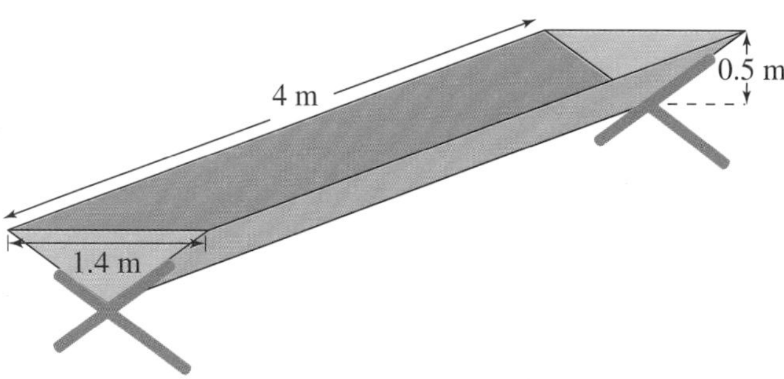

17. In a circular region of a forest, trees are cut down so that the area of the region decreases at a rate equal to its circumference. Show that the radius is decreasing at a constant rate.

18. Helium is pumped into a spherical balloon so that its volume increases at a rate equal to its surface area. Show that the radius of the balloon is increasing at a constant rate.

8.4 Related Rates: Part II

In Section 8.3, the related-rates problems involved areas, volumes, and perimeters. We used formulas to establish relationships between the different rates of change, such as $\frac{dA}{dt}$, $\frac{dr}{dt}$, $\frac{dV}{dt}$, $\frac{dh}{dt}$, and so on. In this section, we will work with quantities that are changing; however, the quantities are not related by measurement formulas. As before, we analyse the quantities that are changing and those that remain constant. We use the steps on pages 512, 513.

Example 1

A bird of prey is perched at the top of a tree that is 40 m high. A squirrel runs away from the base of the tree at a rate of 2 m/s. What is the rate of change of the distance between the bird and the squirrel when the squirrel is 30 m from the tree?

Solution

Step 1. The rate of change of the distance between the squirrel and the base of the tree is known: 2 m/s
The rate of change of the distance between the bird and the squirrel is to be determined.

Step 2. Let x represent the distance between the base of the tree and the squirrel.
Let z represent the corresponding distance between the bird and the squirrel.

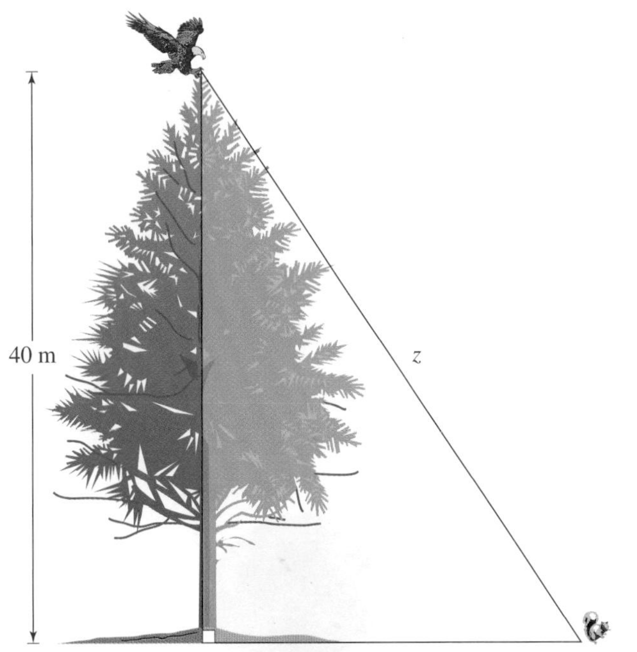

We want to determine $\frac{dz}{dt}$ when $x = 30$ m.

Use the Pythagorean Theorem to relate x and z:

$x^2 + 40^2 = z^2$ ... ①

Step 3. There are no constraints to consider.

Step 4. The model relation is $x^2 + 40^2 = z^2$.

Step 5. Differentiate with respect to t.

$$2x\frac{dx}{dt} = 2z\frac{dz}{dt}$$

$$\frac{dz}{dt} = \frac{x}{z}\frac{dx}{dt} \quad \text{... ②}$$

Step 6. We know that $x = 30$. Use equation ① to determine the corresponding value of z.

$$x^2 + 40^2 = z^2$$
$$30^2 + 40^2 = z^2$$
$$z = 50$$

Substitute $x = 30$, $z = 50$, and $\frac{dx}{dt} = 2$ in equation ②.

$$\frac{dz}{dt} = \frac{30}{50}(2)$$
$$= 1.2$$

Step 7. When the squirrel is 30 m from the tree, the distance between the bird and the squirrel is increasing at 1.2 m/s.

Example 2

A 5-m ladder rests in a vertical position against the side of a building. The base of the ladder begins to slip at a constant rate of 0.5 m/min. How fast is the top of the ladder sliding down the building after 4 min?

Solution

Step 1. The rate of change of the distance between the wall and the base of the ladder is known: 0.5 m/min

The rate of change of the distance between the top of the ladder and the base of the building is to be determined.

Step 2. Let x represent the distance between the wall and the base of the ladder. Let y represent the corresponding distance between the top of the ladder and the base of the building.

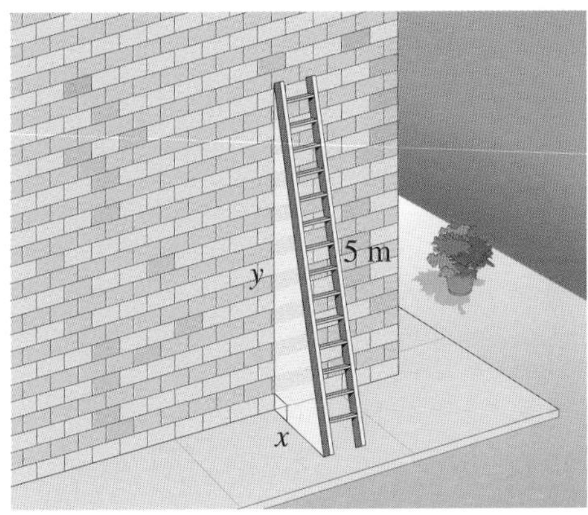

We want to determine $\frac{dy}{dt}$ after 4 min.

Use the Pythagorean Theorem to relate x and y:

$$x^2 + y^2 = 5^2 \qquad \ldots \text{①}$$

Step 3. There are no constraints to consider.

Step 4. The model relation is $x^2 + y^2 = 5^2$.

Step 5. Differentiate with respect to t.

$$2x\frac{dx}{dt} + 2y\frac{dy}{dt} = 0$$

$$\frac{dy}{dt} = -\frac{x}{y}\frac{dx}{dt} \qquad \ldots \text{②}$$

Step 6. We know that $\frac{dx}{dt} = 0.5$. We need to determine the values of x and y after 4 min. The base of the ladder is moving at 0.5 m/min.

So, after 4 min, it has moved 2 m, so $x = 2$.

Use equation ① to determine the corresponding value of y.

$$x^2 + y^2 = 5^2$$

$$2^2 + y^2 = 5^2$$

$$y^2 = 21$$

$$y = \sqrt{21}$$

Substitute $x = 2$, $y = \sqrt{21}$, and $\frac{dx}{dt} = 0.5$ in equation ②.

$$\frac{dy}{dt} = -\frac{2}{\sqrt{21}}(0.5)$$

$$\doteq -0.218$$

Discuss

Why is the rate negative? Explain.

Step 7. After 4 min, the top of the ladder is sliding down at a rate of approximately 0.2 m/min.

Example 3

A person is walking away from a streetlight at a rate of 1 m/s. The person is 1.8 m tall and the light is 4 m high. How fast is the length of the person's shadow increasing when she is 3 m from the base of the streetlight?

Solution

Step 1. The rate of change of the distance between the person and the streetlight is known: 1 m/s
The rate of change of the length of the person's shadow is to be determined.

Step 2. Let x represent the person's distance from the base of the light.
Let s represent the length of the person's shadow.

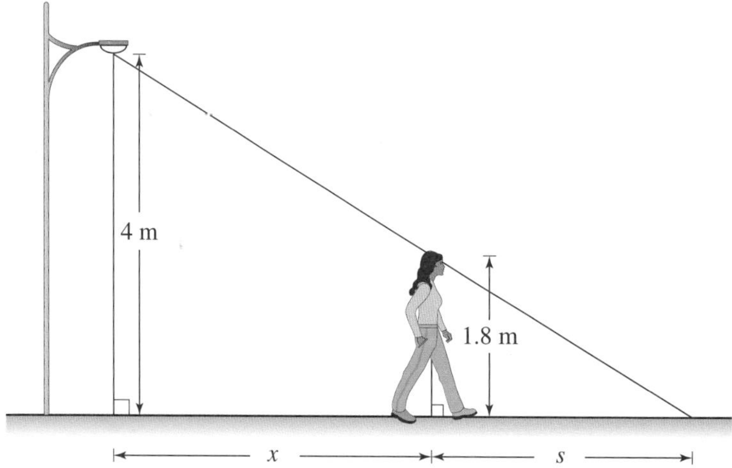

We want to determine $\frac{ds}{dt}$ when $x = 3$ m.
Use similar triangles to determine a relationship between x and s.

The right triangles in the diagram are similar; so the ratios of corresponding sides are equal.

$$\frac{4}{x+s} = \frac{1.8}{s}$$

$$4s = 1.8x + 1.8s$$

$$2.2s = 1.8x$$

Step 3. There are no constraints to consider.

Step 4. The model function is $2.2s = 1.8x$.

Step 5. Differentiate with respect to t.

$$2.2\frac{ds}{dt} = 1.8\frac{dx}{dt}$$

$$\frac{ds}{dt} = \frac{1.8}{2.2}\frac{dx}{dt} \qquad \dots ①$$

Notice that the speed of the shadow depends only on the speed of the person, $\frac{dx}{dt}$.

Step 6. Substitute $\frac{dx}{dt} = 1$ in equation ①.

$$\frac{ds}{dt} = \frac{1.8}{2.2}(1)$$
$$\doteq 0.82$$

Step 7. When the person is 3 m from the streetlight, the length of her shadow increases at approximately 0.8 m/s.

In the solution of *Example 3*, we did not use the fact that the person is 3 m from the base of the streetlight. This is because the rate of change of the length of the person's shadow is not dependent on how far away she is from the light. Since she is walking at a constant speed, the length of her shadow is changing at a constant speed.

8.4 Exercises

1. Solve each equation for the indicated rate.

a) $x\frac{dx}{dt} + y\frac{dy}{dt} = 0$; for $\frac{dy}{dt}$

b) $x\frac{dx}{dt} + y\frac{dy}{dt} = z\frac{dz}{dt}$; for $\frac{dy}{dt}$

c) $\frac{dP}{dt} = 2\left(\frac{dl}{dt} + \frac{dw}{dt}\right)$; for $\frac{dl}{dt}$

d) $\frac{dA}{dt} = \frac{1}{2}\left(b\frac{dh}{dt} + h\frac{db}{dt}\right)$; for $\frac{db}{dt}$

2. Match each derivative in exercise 1 with one of the following equations.

a) $A = \frac{bh}{2}$ **b)** $x^2 + y^2 = 15$ **c)** $P = 2(l + w)$ **d)** $x^2 + y^2 = z^2$

B

3. A person is 1.7 m tall. He walks at 1.3 m/s. The person walks away from a spotlight that is 8 m above the ground. Determine the rate at which the length of his shadow increases when he is 12 m from the base of the spotlight.

4. Knowledge/Understanding A 6-m ladder is propped against a wall. The base of the ladder slips away from the wall at 25 cm/s. How fast is the top of the ladder sliding down the wall when the base of the ladder is 3 m from the wall?

5. David and Andrew plan to meet at St. Mark High School to play basketball. David drives west at 65 km/h and Andrew drives south at 70 km/h. Both boys are approaching the school, which is located at the intersection of the two roads. At what rate is the distance between the two cars decreasing when David is 0.5 km from the school and Andrew is 0.4 km from the school?

6. **Communication** Compare exercises 4 and 5. How are the problems and their solutions similar? How are they different? Explain.

7. Two motorcycles approach an intersection, one from the south and the other from the west. When one motorcycle is 160 m south and the other is 120 m west of the intersection, the distance between them decreases at 25 m/s. The two motorcycles are travelling at the same speed. What is this speed?

8. A hot-air balloon rises vertically above a straight road at 0.4 m/s. A van travelling at 50 km/h passes under the balloon when the balloon is 34 m high. How quickly is the distance between the van and the balloon increasing 5 s later?

9. Two cars leave an intersection at the same time. One car travels north at 75 km/h and the other travels east at 70 km/h.

   a) What is the distance between the two cars after 10 min?

   b) How fast is the distance between the two cars increasing after 10 min?

10. **Application** The area of a triangle increases at 6 cm^2/min and the length of its base increases at 0.8 cm/min. Determine the rate of increase of the length of the altitude of the triangle when the area is 64 cm^2 and the altitude is 15 cm.

11. Right $\triangle ABC$ changes size, while still maintaining its right angle at A. The length of one leg, AB, of the triangle increases at 3 cm/s. The length of the other leg, AC, decreases at 2 cm/s. How fast is the length of the hypotenuse, BC, changing when AB = 8 cm and AC = 6 cm?

12. A cone is growing while still maintaining its conical shape. The radius increases at 2 cm/s and the height increases at 4 cm/s. At what rate is the volume increasing when the height is 5 cm and the radius is 3 cm?

13. **Thinking/Inquiry/Problem Solving** At 9:00 A.M., a tugboat is 100 km north of a cruise ship. The tugboat sails east at 40 km/h and the cruise ship sails south at 60 km/h. What is the rate of change of the distance between them at 12 noon?

14. A severe storm system is located at an altitude of 10 km. It moves horizontally at a speed of 28 km/h and will pass directly over the town of Manotick.

   a) What is the rate of change of the distance between the storm and Manotick when the storm is 25 km from Manotick?

   b) How long will it take the storm to reach Manotick?

15. A kite is 90 m above the ground. It is blown horizontally by the wind at a speed of 3 km/h. Determine the rate at which the string is unwinding at the instant when 200 m of string are unwound.

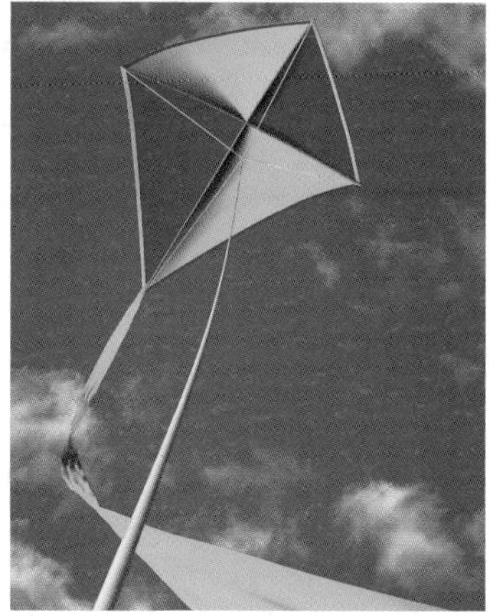

C

16. The focal length of a lens is the distance from the centre of the lens to the point where incoming parallel light rays converge on the other side of the lens. When an object is placed on one side of the lens, its image appears on the other side. The focal length, f, object distance, p, and image distance, q, are related by the lens equation $f = \frac{pq}{p + q}$.

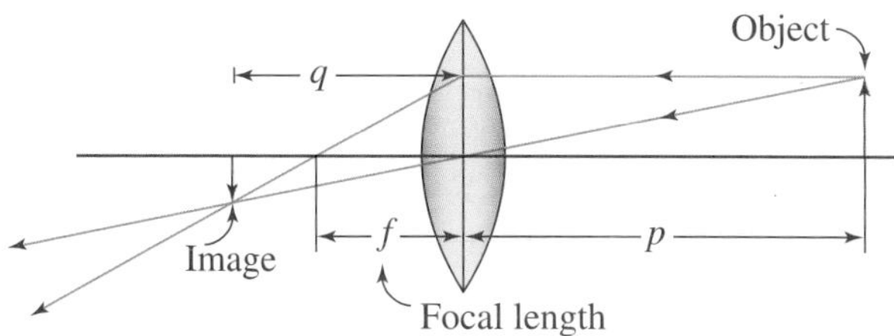

Suppose a lens has a focal length of 65 mm. The object moves away from the lens at 53 mm/s. Determine the rate at which the position of the image changes when the object is 120 mm from the lens.

17. To make a jump, a skier skis up a ramp that measures 7 m along the base and is 3 m high. The speed of the skier is 54 km/h. How quickly is the skier rising as she takes off from the ramp?

18. In $\triangle$XYZ, side XY increases at 3 cm/min and side XZ increases at 5 cm/min. The angle between these two sides, $\angle$X, increases at 2°/min. Determine the rate of change of the area of $\triangle$XYZ when XY = 35 cm, XZ = 50 cm, and $\angle$X = 45°. (Note: convert the angles to radians.)

19. A spotlight shines from the top of a 35-m pole. A ball is dropped from the same height at a point 20 m from the spotlight. The height of the ball is given by the equation $h(t) = 35 - 4.9t^2$, where h is measured in metres, and t is the time in seconds since the ball was dropped.

 a) In which direction does the shadow of the ball move along the ground?

 b) How fast is the shadow of the ball moving along the ground 1 s after the ball was dropped?

Self-Check 8.3, 8.4

1. Sand is poured at 9 m^3/min into a conical pile. The height of the pile is always two-fifths of the base diameter. How fast are the radius and height increasing when the sand pile is 7 m high?

2. When the circular base of a frying pan is heated, its radius expands at 0.05 mm/s. What is the rate of increase of the area of the circular base when the radius is 15 cm?

3. A square is expanding so that its area increases at a rate of 14 cm^2/min.
   a) Determine how fast the side length increases when the area is 80 cm^2.
   b) Determine how fast the perimeter increases when the side length is 12 cm.

4. a) A spherical balloon is filling with helium. The radius increases at a constant rate of 3.1 cm/s. What is the rate of increase in the volume of the balloon when the radius is 7 cm?
   b) What is the rate of increase of the surface area of the balloon when the radius is 7 cm?

5. A person, 1.5 m tall, walks away from a lamppost at 1.3 m/s. A spotlight on the post is 12 m above the ground.
   a) At what rate is the length of her shadow changing at that time?
   b) At what rate is the end of her shadow moving when she is 15 m from the base of the post?

6. The top of a 12-m ladder slides down a wall at 20 cm/s. How fast is the base of the ladder sliding away from the wall when the base of the ladder is 4 m from the wall?

7. Two cars approach an intersection. At a certain time, each car is 5 km from the intersection. One car travels west at 90 km/h and the other travels south at 80 km/h.
   a) What is the distance between the two cars at this time?
   b) How fast is the distance between the two cars decreasing after 2 min?

PERFORMANCE ASSESSMENT

8. Right $\triangle$DEF changes in size while still maintaining its right angle at E. The length of one leg, DE, decreases at 1.8 cm/s, while the length of the other leg, EF, increases at 3 cm/s. How fast is the length of the hypotenuse, DF, changing when DE = 12 cm and EF = 5 cm?

Review Exercises

Mathematics Toolkit

Problem Solving Tools

- Problems that involve the maximum or minimum value of a quantity are called *optimization* problems. See pages 498, 499 for the steps to solve these problems.
- Problems that involve relationships between quantities that change are called *related-rates* problems. See pages 512, 513 for the steps to solve these problems.

8.1 1. **a)** Determine the local maximum or minimum values of the function $f(x) = x - e^x$.

b) Determine any point(s) of inflection for the graph of this function.

2. Determine the equation(s) of any vertical or horizontal asymptotes for the graph of each function.

a) $f(x) = x - e^{-x}$ **b)** $f(x) = -3x + e^x$ **c)** $f(x) = \frac{e^x}{x+2}$

3. Determine the domain and the equations of the asymptotes for each function.

a) $f(x) = \frac{x}{e^x - 1}$ **b)** $f(x) = \frac{1}{x^2} - e^x$ **c)** $f(x) = \frac{e^x + 1}{e^x - 1}$

4. Consider the graph of $f(x) = x^2e^x$ at the right.

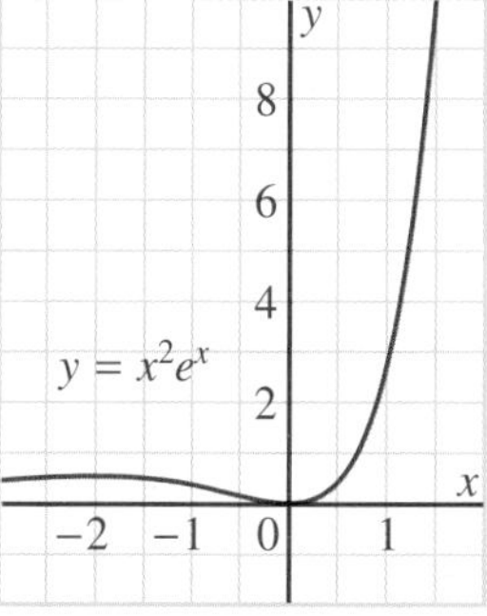

a) Determine the coordinates of the maximum point, the minimum point, and the point of inflection.

b) Write the equation of the derivative f', then sketch its graph.

c) Sketch the graph of the second derivative, f''.

5. Analyse the key features of the graph of each function, then sketch the graph.

a) $f(x) = \frac{e^x}{x^2}$

b) $f(x) = e^x + e^{-x} - 4$

c) $f(x) = x + e^{-x}$

8.2 6. The sum of two numbers is 56. The sum of their squares is a minimum. Determine the numbers.

7. A container with a lid is to be made by cutting congruent squares from the corners of a 60-cm by 80-cm sheet of wood. The container is to have a maximum volume. Determine the side length of the squares to be cut.

8. A farmer wants to enclose and subdivide a rectangular field into 4 congruent plots of land. She has 5800 m of fencing. The area to be enclosed is a maximum. Determine the dimensions of one plot of land.

9. The perimeter of an isosceles triangle is 90 cm. Determine the lengths of the sides of the triangle that produce a maximum area.

8.3 10. A spherical meteorite enters Earth's atmosphere and disintegrates. Its volume decreases at 5 m^3/min.

a) What is the rate of decrease of the radius when the radius is 10 m?

b) What is the rate of change of the surface area at this time?

11. Water drains from a cylindrical tank at 4.5 m^3/min. The tank has radius 4.6 m, and height 9 m. Determine the rate at which the depth of the water is decreasing when the water is 6.2 m deep.

12. The area of a square increases at 16 cm^2/min.

a) Determine the rate of increase of the side length when the area is 70 cm^2.

b) Determine the rate of increase of the perimeter when the side length is 13 cm.

13. Crushed stone is poured into a conical pile at 3.6 m^3/s. The base diameter of the pile is 14 m and the height is 14 m. The ratio of height to base diameter is constant. What is the rate of change of the height of the pile when the pile is 12 m high?

8.4 14. A person, 1.9 m tall, walks away from a spotlight at 1.6 m/s. The spotlight is 15 m above the ground. Determine the rate at which the person's shadow is growing when she is 14 m from the base of the spotlight.

15. The base of a 13-m ladder slides away from a wall at 36 cm/s. How fast is the top of the ladder sliding down the wall when the base of the ladder is 5 m from the wall?

16. Julia and Cassandra plan to meet at the movie theatre. Julia drives east at 58 km/h and Cassandra drives north at 65 km/h. The two friends are approaching the theatre, which is located at the intersection of the two roads. At what rate is the distance between the two cars decreasing when Julia is 2.7 km from the intersection and Cassandra is 0.4 km from the intersection?

17. The area of a triangle decreases at 4 cm^2/min and the length of its base increases at 1.2 cm/min. Determine the rate of change of the length of the altitude when the area is 90 cm^2 and the altitude is 13 cm.

Self-Test

1. Consider the function $f(x) = \frac{e^x}{x^2 - 4}$.

   a) Determine the local maximum and minimum values of the function.

   b) Determine the equations of the asymptotes.

2. **Knowledge/Understanding** Analyse the key features of the graph of the function $f(x) = \frac{e^x}{2x + 1}$, then sketch the graph.

3. A farmer has 60 m of fencing to enclose two animal pens with a common side. One pen is rectangular; the other is square. Determine the dimensions of the two pens so that the total area is a maximum.

4. A box with no lid is to have a surface area of 4000 cm^2. The width of the base is one-third the length of the base. Determine the dimensions of the box that will have the maximum volume. Determine the maximum volume.

5. **Communication** Explain how to determine the maximum area of an isosceles triangle for which the two equal sides have measure 8 cm.

6. a) A spherical balloon is filling with air. The radius is increasing at a constant rate of 3.2 cm/s. What is the rate of increase of the volume of the balloon when the radius is 7.7 cm?

   b) What is the rate of increase of the surface area of the balloon at this time?

7. How fast is the perimeter of a square increasing when its side length is 21 cm and is increasing at 3.8 m/min?

8. **Application** A rectangle is expanding so that the ratio of its length to width is 7 : 4. Determine the rate of increase of the area of the rectangle when the perimeter is increasing at 16 cm/s and the perimeter is 56 cm.

9. A 12-m ladder is leaning against a wall. The ladder begins to slide. The base of the ladder slides at 3.5 cm/s. Determine the rate of descent of the top of the ladder in each case.

   a) The base of the ladder is 2 m from the wall.

   b) The top of the ladder is 9 m above the ground.

   c) The angle between the top of the ladder and the wall is 40°.

10. A jogger is 1.87 m tall. He jogs away from a streetlight that is at the top of a cement post 8.5 m tall. The jogger runs at 90 m/min. How quickly is the length of his shadow increasing when he is 10 m from the base of the post?

Self-Test

11. Thinking/Inquiry/Problem Solving At 9 A.M., ship A is 100 nautical miles due south of ship B. Ship A travels east at 20 knots and ship B travels west at 16 knots. Determine the rate of change of the distance between the two ships at 11:30 A.M. (One knot is one nautical mile per hour.)

12. One car travels north at 95 km/h and approaches an intersection. At a certain time, the car is 9 km from the intersection. At the same time, a second car leaves the intersection and travels west at 80 km/h.

a) What is the distance between the two cars after 5 min?

b) How fast is the distance between the two cars changing at this time?

PERFORMANCE ASSESSMENT

13. Consider the graphs of the functions $f(x) = e^x + x$ and $g(x) = e^x - x$.

a) Determine which graph represents each function.

b) Show that (0, 1) is a minimum point on the graph of one function. Then show that the other function has no minimum point.

c) Show that neither graph has a point of inflection.

d) Show that the graphs of both functions are concave up for all values of x.

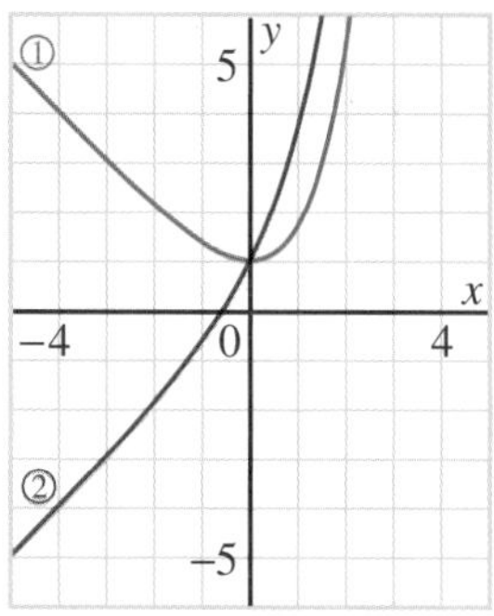

14. Coffee drains from a conical filter into a cylindrical coffee pot at 2 cm^3/s. The dimensions of the filter and the pot are shown on the diagram at the right. At a certain time, the coffee in the filter is 13 cm deep.

a) How fast is the depth of coffee in the filter changing at that time?

b) How fast is the depth of coffee in the pot changing at that time?

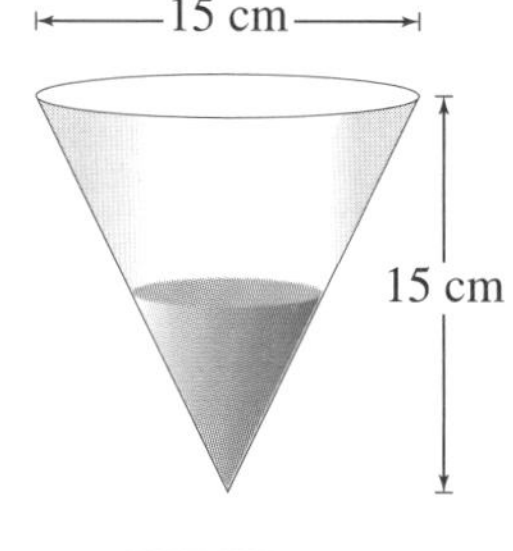

15. A stained glass window is to have the shape of a rectangle capped by a semicircle. The perimeter of the window is 10 m. Determine the width of the window that will admit the maximum amount of light.

Mathematical Modelling

Optimal Consumption Time

A bird spends time foraging for food, then, having found a source of berries, it begins to eat. At the beginning, the berries are easy to find and the bird gains food energy at a high rate. As time passes, the berry patch is depleted, and the rate of energy gain decreases. At some point, the bird should leave the patch and search for a new one. We want to determine how long a bird should stay in a patch before moving on; that is, the optimal consumption time.

Formulating a Hypothesis

If the bird stays too long in the berry patch, it will spend time looking for berries, and will consume them at a low rate of energy gain. If the bird leaves the patch near the beginning of its consumption, then most of its time is spent foraging (with no energy gain). We make the following hypothesis:

There should be an optimum time for the bird to stay at a berry patch.

Developing a Mathematical Model

Let t minutes represent the time the bird is foraging for food, then eating it. Then, for $0 \le t \le s$, the bird is foraging, and for $t > s$, it is eating. Let E calories represent the energy obtained in time t. Then, for $0 \le t \le s$, $E = 0$, and for $t > s$, E increases with t. However, the rate of energy gain decreases as t increases, so, the slope of the E-graph decreases and the graph is concave down. A possible graph is drawn at the right. We have assumed a foraging time of $s = 10$ min.

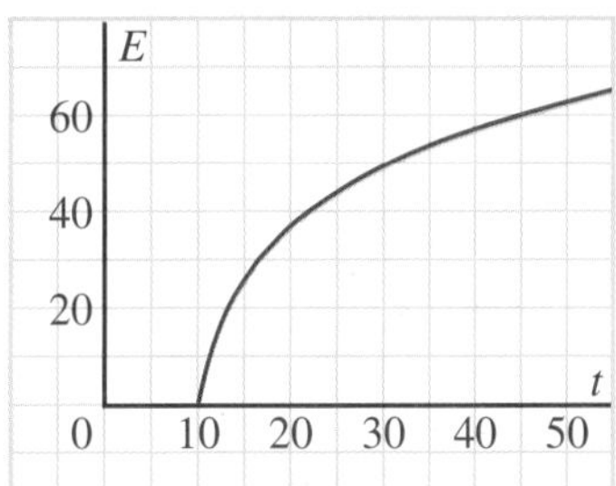

We want to determine the departure time, t minutes, that maximizes the average rate of energy intake, R calories per minute, over a complete cycle (foraging plus eating):

$$R = \frac{\text{energy}}{\text{time}} = \frac{E}{t}$$

We shall consider two methods—graphical and algebraic.

Analysing the Model

Graphical Method

We want to find the point on the graph with the highest possible value of $\frac{E}{t}$. The E-value of a point is its height above the t-axis. The t-value is its distance to the right of the E-axis. The quotient, $\frac{E}{t}$, is the slope of the line drawn from the origin to the point (t, E) on the graph. For example, the point A(13, 20) has an R value of $\frac{20}{13} \doteq 1.54$. That is, the average rate of energy intake, R, at $t = 13$ min, is 1.54 calories/min.

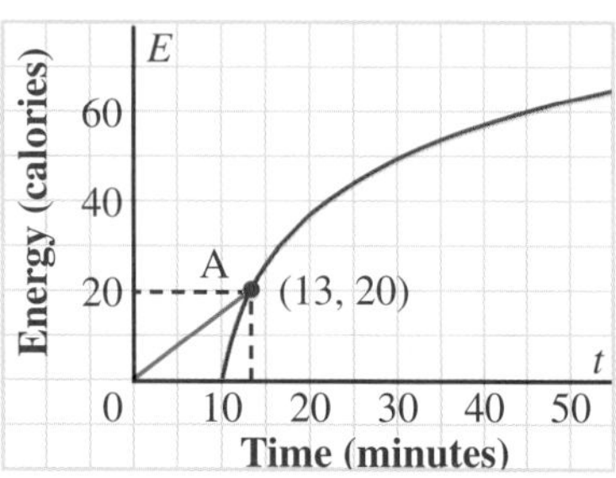

At the point $t = 13$, $E \doteq 20$, and R is the slope of the line segment drawn from the origin.

The point on the graph that has the greatest value of $\frac{E}{t}$ is the point at which the line drawn from the origin to the graph is a tangent to the graph. This tangent has been drawn at the right. Any line with greater slope will not intersect the graph. The optimal cycle time (foraging plus eating) is approximately 19 min with 35 calories extracted from the patch. The maximum average rate of energy intake is $R = \frac{35}{19} \doteq 1.84$. The optimal cycle time is 19 min for an average rate of energy intake of 1.84 calories/min. So, the optimal consumption time is 19 min – 10 min = 9 min.

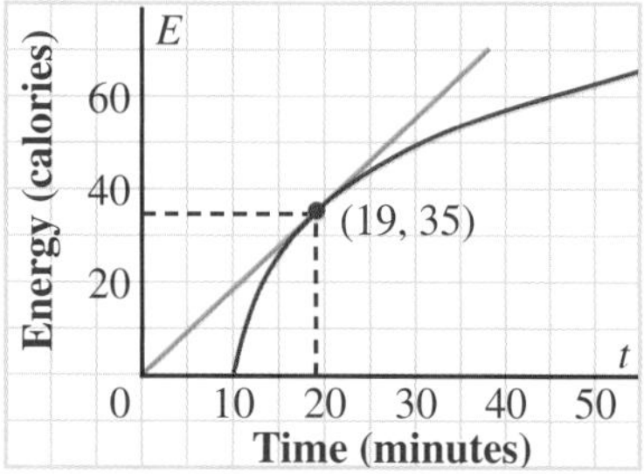

Algebraic Method

We shall develop an equation for the energy graph. We expect the rate of energy intake to decrease as the patch becomes depleted. Each calorie takes a little longer to extract than the previous calories. There are two assumptions to make about this extraction time: it increases additively or it increases multiplicatively. We will consider the multiplicative case here and leave the additive case to exercise 3.

We assume that once the bird has arrived at the patch, the first calorie takes 0.1 min to extract, and each subsequent calorie takes 5% longer to obtain. Then, the first calorie takes 0.1 min, the second takes 0.1(1.05), the third takes $0.1(1.05)^2$, and so on. The time required to obtain E calories is:

$$\begin{aligned} t &= 10 + \left[0.1 + 0.1(1.05) + 0.1(1.05)^2 + 0.1(1.05)^3 + \ldots + 0.1(1.05)^{E-1}\right] \\ &= 10 + 0.1[1 + (1.05) + (1.05)^2 + (1.05)^3 + \ldots + (1.05)^{E-1}] \end{aligned}$$

In the square brackets, there is a finite geometric series that can be summed:

$$t = 10 + 0.1\left[\frac{1.05^E - 1}{1.05 - 1}\right]$$
$$= 10 + \frac{0.1}{0.05}\left[1.05^E - 1\right]$$
$$= 10 + 2\left[1.05^E - 1\right]$$

Recall that the sum to n terms of the geometric series $a + ar + ar^2 + ... + ar^{n-1}$ is $S_n = \frac{a(r^n - 1)}{r - 1}$.

The equation is $t = 8 + 2(1.05)^E$.

This equation expresses t as a function of E. To get an equation for E in terms of t, we solve the equation for E.

Solve $t = 8 + 2(1.05)^E$ for E.

$$t - 8 = 2(1.05)^E$$
$$\frac{t-8}{2} = 1.05^E$$
$$0.5t - 4 = 1.05^E$$

Take the natural logarithm of each side.

$$\ln(0.5t - 4) = E \ln 1.05$$
$$E = \frac{\ln(0.5t - 4)}{\ln 1.05}$$

We want to maximize $\frac{E}{t}$, so divide each side by t.

$$\frac{E}{t} = \frac{\ln(0.5t - 4)}{t \ln 1.05} = R$$
$$R = \frac{\ln(0.5t - 4)}{t \ln 1.05}$$

Differentiate with respect to t. Remove $\frac{1}{\ln 1.05}$ as a constant.

$$\frac{dR}{dt} = \frac{1}{\ln 1.05}\frac{d}{dt}\left(\frac{\ln(0.5t - 4)}{t}\right)$$
$$= \frac{1}{\ln 1.05} \times \frac{\frac{0.5t}{0.5t - 4} - \ln(0.5t - 4)}{t^2}$$
$$= \frac{0.5t - (0.5t - 4)\ln(0.5t - 4)}{t^2(0.5t - 4)\ln 1.05}$$

Most equations cannot be solved analytically. An equation that combines exponential terms and polynomials can almost never be solved analytically. Guess and check is a possible method for solving such equations.

For maximum R, the numerator is 0.
Let the numerator be represented by $f(t)$. That is, $f(t) = 0.5t - (0.5t - 4)\ln(0.5t - 4)$
Then, for maximum R, $f(t) = 0$. We cannot solve this equation analytically, but we can use guess and check.

From page 532, R is a maximum when $t \doteq 19$.
Substitute $t = 19$ in $f(t)$.

$$f(19) = 0.5(19) - (0.5(19) - 4)\ln(0.5(19) - 4)$$
$$\doteq 0.12$$

Check values of t close to 19.

$$f(19.1) = 0.5(19.1) - (0.5(19.1) - 4)\ln(0.5(19.1) - 4)$$
$$\doteq 0.038$$
$$f(19.2) = 0.5(19.2) - (0.5(19.2) - 4)\ln(0.5(19.2) - 4)$$
$$\doteq -0.05$$

The value of t for which $f(t)$ is closest to 0 is approximately 19.15. On average, a bird should spend approximately 19 min foraging and then eating before moving to the next patch.

Key Features of the Model

We developed both a graphical and an algebraic model for this problem. In each case, we produced a relationship between the cycle time and the energy gain per cycle. From the graph, we estimated the optimal cycle time. Algebraically, we made a multiplicative assumption about how the amount of time required to gain each new calorie would increase. We used this assumption to determine the equation of the graph.

Graphically, the slope of the line drawn to the graph from the origin represents the average rate of energy gain; so, this is what we needed to maximize. This occurred where this line was a tangent to the graph.

Algebraically, we determined t as a function of E, then rearranged the equation to get E as a function of t. We divided by t, then differentiated with respect to t.

Making Inferences from the Models

1. Suppose the foraging time is $s = 20$ min, and once the bird has arrived at the patch, the first calorie takes 0.2 min to extract. Each subsequent calorie takes 4% longer to obtain.

   a) Determine an equation for the cycle time, t minutes, in terms of the energy, E calories, extracted from a patch. The graph of this relationship is shown at the right. Use the graph to check your equation at any two points.

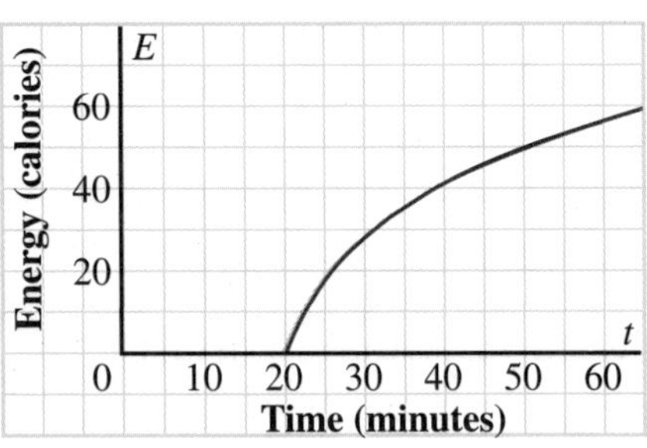

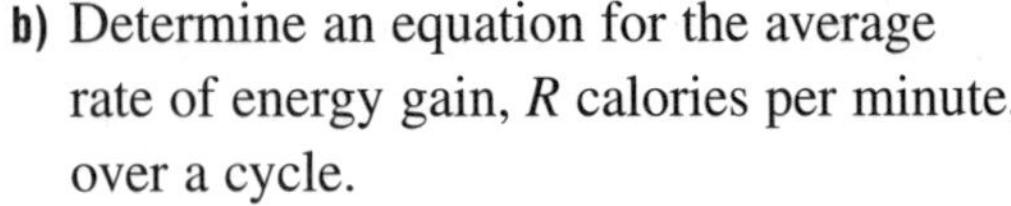

   b) Determine an equation for the average rate of energy gain, R calories per minute, over a cycle.

c) Explain how R can be represented on the graph on page 534. Show how to use the graph to obtain a geometric solution to the problem of maximizing R.

d) Use the result of part c to approximate a solution to $\frac{dR}{dt} = 0$. Determine the optimal time a bird should spend eating.

2. Use the information in exercise 1, but the foraging time is now 10 min instead of 20 min.

a) Use a copy of the graph in exercise 1. Make a construction that demonstrates the optimal cycle time.

b) Use the graph in part a. Explain how changes in the foraging time affect the optimal cycle time.

3. Consider an additive model. Suppose the foraging time is $s = 10$ min, and once the bird has arrived at the patch, the first calorie takes 0.1 min to extract. Each subsequent calorie takes 0.02 min longer to obtain than the previous one.

a) Determine an equation for the cycle time, t minutes, in terms of the energy, E calories, extracted from a patch. The graph of this relationship is shown at the right. Use the graph to check your equation at any two points.

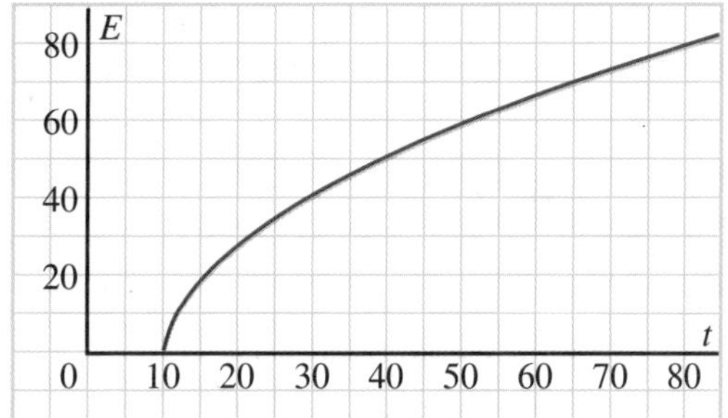

b) Determine an equation for the average rate of energy gain, R calories per minute, over a cycle. Interpret this gain with a construction on the graph.

c) Use the graph to determine approximate values of t and E that make R a maximum.

d) Equate the derivative of R to 0. Solve to determine the exact values of t and E that make R a maximum. Check that these agree with your answer in part c.

4. Write a report to describe what you have learned about optimal cycles for foraging and consumption. Include graphs and equations in your report.

Cumulative Review (Chapters 1–8)

1. Use the first-principles definition to determine the derivative of each function.

   a) $f(x) = -4x^2$ b) $f(x) = \frac{1}{4x - 1}$ c) $f(x) = \sqrt{3 - 2x}$

2. Consider the function $y = 2x - \frac{4}{x^2}$.

   a) Determine the average rate of change of the function between $x = 1$ and $x = 4$.

   b) Determine the instantaneous rate of change of y with respect to x when $x = 2$.

   c) Describe the graphical meaning of each answer in parts a and b.

3. For each derivative function given below, sketch a possible graph of $y = f(x)$, then sketch the graph of $y = f''(x)$.

   a) 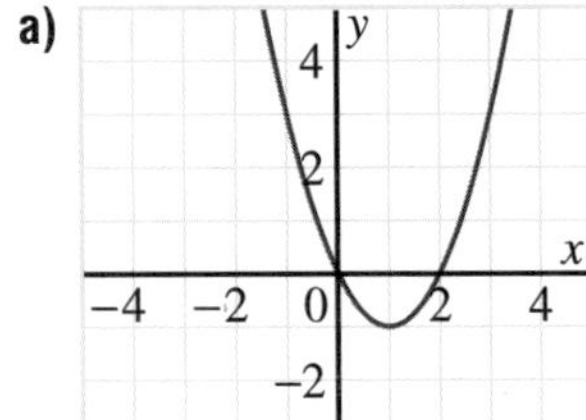

   b) 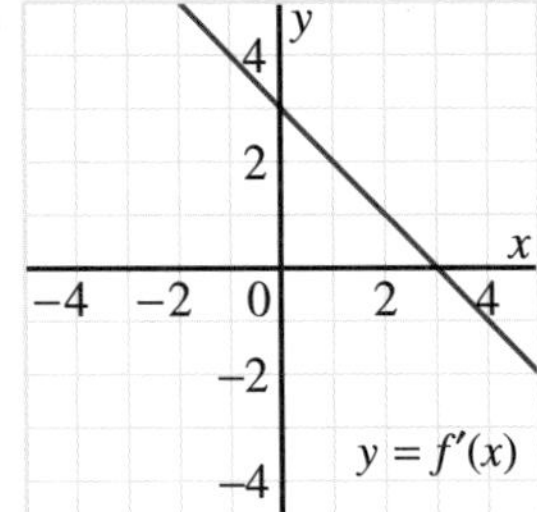

   c) 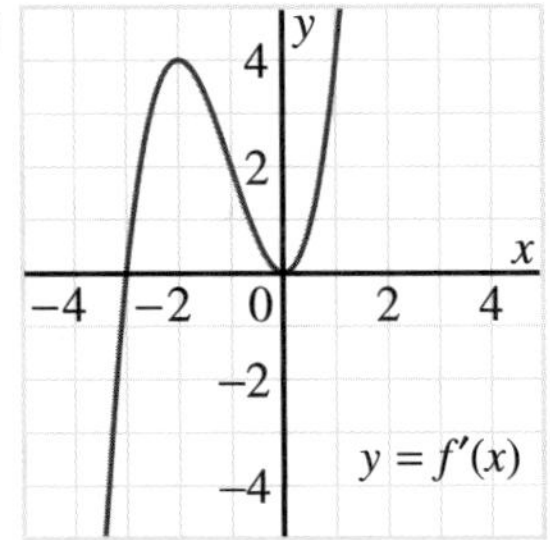

4. Differentiate each function.

   a) $y = 4x^2 - 3x + 7$ b) $y = \frac{1}{2x^2}$ c) $y = \frac{3}{(x-4)^3}$

   d) $y = \sqrt{6x - 5}$ e) $y = \sin 2x$ f) $y = \sin(\cos x)$

   g) $y = 2^{4-x}$ h) $y = e^{x^2-2}$ i) $y = \ln(2x + 1)$

   j) $y = e^3$ k) $y = \cos^3(4x + 3)$ l) $y = \frac{8x^3 - 4x^2 - 1}{x}$

5. Determine $\frac{dy}{dx}$ for each function.

   a) $y = \sqrt{4x^2 - 1}(x^2 + 1)^4$ b) $y = \frac{1 - 4x^3}{3x + 5}$ c) $y = \left(\frac{4 - 3x}{7x - 2}\right)^5$

   d) $y = \ln\left(\frac{1-x}{1+x}\right)$ e) $y = x^3e^{3x}$ f) $y = \frac{\ln 2x}{x^2}$

6. Determine $\frac{dy}{dx}$ for each relation.

   a) $3x^2 - 2xy - y^2 = 6$ b) $x \sin y - \pi^2 = xy^2$ c) $\cos y^2 = \sin 2x$

7. Determine the equation of the tangent to the graph of each relation at the indicated point.

   a) $y = 3x^2 - 1$; $(1, 2)$ b) $5x^2 - 9y^2 = 36$; $(-3, 1)$ c) $y = x \ln x$; (e, e)

8. Solve each inequality.

   a) $x(x-1)^2(x+3) > 0$

   b) $x^3 - 7x - 6 \leq 0$

9. When the polynomial $x^5 - kx^3 + 2x^2 - 3k$ is divided by $x + 1$, the remainder is -1. Determine the value of k.

10. a) At which points on the graph of $y = \frac{2x^2}{x-2}$ is the slope of the tangent 0?

    b) What is the geometrical significance of your answer to part a?

11. Tangents are drawn from the point (0, 8) to the graph of $y = -2x^2$. Determine the equations of these tangents.

12. The displacement, s metres, of a particle travelling in a straight line, after t seconds, is given by $s(t) = 2t^3 - 9t^2 + 12t + 2$.

    a) Determine the velocity and acceleration of the particle at any time, t.

    b) When is the particle at rest?

    c) Determine the average velocity during the first 3 s of motion.

    d) Determine the total distance travelled during the first 6 s of motion.

13. Consider the functions $f(x) = 3x^2 - 2x$ and $g(x) = 4 - x$.

    a) Determine $f(g(x))$.

    b) Determine $f'(g(x))$.

    c) Determine $f'(g(x))$ using a method different from that used in part b.

14. Evaluate each limit, then give a geometrical interpretation of each result.

    a) $\lim_{x \to -3} \frac{x^3 + 27}{x + 3}$

    b) $\lim_{x \to 6} \frac{12 + 4x - x^2}{x - 6}$

    c) $\lim_{x \to \infty} \frac{3x^2 - x + 6}{2x^2 - 5}$

15. Determine the key features of the graph of each function, then sketch the graph.

    a) $y = x^4 - 13x^2 + 36$

    b) $y = -x^3 + 2x^2 + 7x + 4$

    c) $y = \frac{x^2 + 2x - 3}{x^2}$

    d) $y = \frac{2x^2 - 8}{x - 6}$

16. Determine the coordinates of the points of inflection for the graph of each function.

    a) $y = x^4 - 2x^2 + 1$

    b) $y = -3x^2 \ln x$

17. Sketch a possible graph for a function that satisfies each set of conditions.

    a) $f(-2) = f(2) = -3; f'(-2) = f'(2) = 0; f''(-2) > 0, f''(2) < 0$

    b) $f(1) = 2; f'(x) > 0$ for all x; $f''(x) > 0$ for $x < 1$; $f''(x) < 0$ for $x > 1$

18. Determine a possible equation for a rational function $y = f(x)$ with each set of properties.

a) vertical asymptotes $x = \pm 1$; horizontal asymptote $y = -3$; $f(2) = -1$

b) oblique asymptote $y = 2x$; vertical asymptote $x = 3$; y intercept: -1

19. Determine the exact value of each expression.

a) $64^{-\frac{2}{3}} - 4^{-1} + 8^0$ **b)** $\left(\frac{2}{3}\right)^{-2} + 81^{0.25}$ **c)** $\ln e^{-3}$

d) $\log 500 - \log 5$ **e)** $e^{-\ln 4}$ **f)** $2^{\log_2 5 + \log_2 3}$

20. Solve each equation.

a) $16^{4-x} = 128^{2x+1}$ **b)** $x^{-\frac{4}{3}} = 16$

c) $\log_2 x + \log_2 (x - 2) = 3$

21. Solve each equation to 2 decimal places.

a) $5(2)^{x-3} = 800$ **b)** $e^{3x-5} = 125$ **c)** $4 - \ln x = \ln 8$

22. A chief executive officer of a company receives a severance package of \$650 000. She invests it at 4.5% compounded semi-annually. How long will it take for the money to double?

23. Determine the key features of the graph of each function, then sketch the graph.

a) $y = x^2 e^{-4x}$ **b)** $y = e^{2x} + e^{x^2}$ **c)** $y = \dfrac{e^{2x}}{e^{x^2}}$

24. A particle moves along a straight line with velocity, v metres per second, where $v = 2t - \dfrac{1}{(t+2)^2}$. Determine the maximum acceleration of the particle.

25. Three sides of a rectangular enclosure are to be made from fencing that costs \$30 per metre. The fourth side is to be made from railings that cost \$6 per metre. The area of the enclosure is 250 m^2. Determine the minimum cost to make the enclosure.

26. An isosceles triangle has a perimeter of 30 cm. Determine the dimensions of the triangle for the area of the triangle to be a maximum.

27. A manager wants to install a rectangular notice board, ABCD, with area 1600 cm^2, in the office meeting room. The notice board is to be subdivided by two thin strips of red tape AC and PQ, where PQ is parallel to AB, the shorter side of the rectangle. The length of red tape is to be a minimum. Determine the dimensions of the notice board.

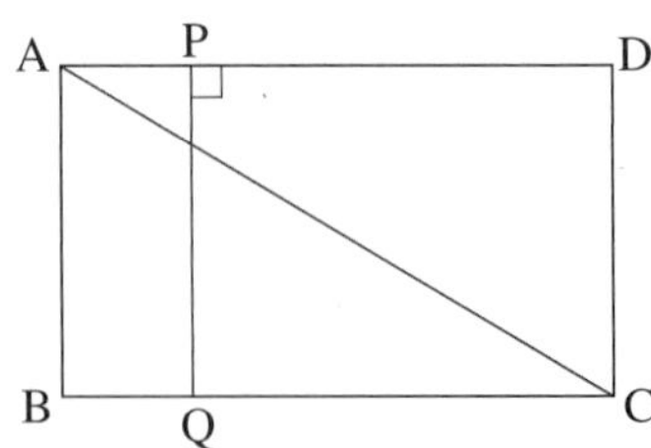

28. A cedar chest with a lid is to be constructed by cutting equal squares from the corners of a square sheet of wood with side length 3 m. Determine the dimensions of the chest so that it has a maximum volume.

29. A rectangular swimming pool is being filled with water. The pool has length 15 m, width 8 m, and depth 1.6 m.

a) How fast is the depth of water increasing when the pool is filled at 8 m^3/min?

b) At what rate is the pool being filled when the depth of the water is increasing at 0.1 m/min?

30. A block of ice is a cube with edge length 20 cm. It melts so that each dimension decreases at a rate of 0.2 cm/s. At what rate is the volume changing when the edge length is 5 cm?

31. A rectangle is expanding so that its width is always one-third of its length. The width is increasing at 0.3 cm/s. Determine the rate of increase of the area of the rectangle when the perimeter is 60 cm.

32. A spherical soap bubble expands so that its volume increases at a constant rate of 8 mm^3/s. How fast is the radius increasing when the surface area is 400 mm^2?

33. The sides of an equilateral triangle are increasing at a rate of 0.4 cm/s. At what rate is the area increasing at the instant when the sides are 10 cm long?

34. A water trough 3 m long has a cross section that is an equilateral triangle with sides 80 cm. Water is poured into the trough at the rate of 10 cm^3/s. How fast is the depth of the water changing when the depth of the water is 35 cm?

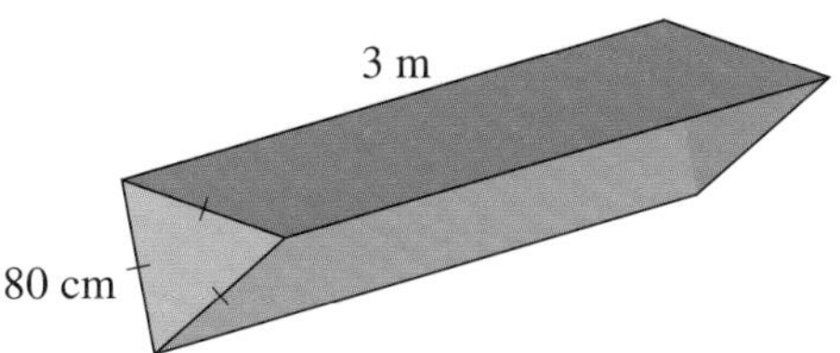

35. A toy duck is thrown into a swimming pool. It creates circular ripples that travel outward at 7.6 cm/s. Determine the rate of increase of the area enclosed by the ripples after 4 s.

36. A hot-air balloon rises vertically above a straight road at 1.3 m/s. A bicyclist travelling at 30 km/h passes under the balloon just when the balloon is 39 m high. What is the rate of change of the distance between the bicyclist and the balloon 6 s later?

Answers

Chapter 1 Analysing Change

Necessary Skills

1 Review: Slope of a Line Segment

Exercises, page 3

1. a) $\frac{1}{2}$; line segment NM goes up to the right
b) Undefined; line segment RS is vertical
c) $\frac{1}{4}$; line segment DC goes up to the right

2. a) $k = 87$ **b)** $k = -73$ **c)** $k = 7$

2 Review: Equations of a Line

Exercises, page 5

1. a) i) $y = 2x + 1$ **II)** $2x - y + 1 = 0$
b) I) $y = -4x + 7$ **ii)** $4x + y - 7 = 0$
c) i) $y = \frac{3}{4}x + \frac{1}{2}$ **ii)** $3x - 4y + 2 = 0$

2. a) $y = -3x + 11;\ m = -3$ **b)** $y = \frac{2}{5}x + 3;\ m = \frac{2}{5}$
c) $y = \frac{1}{6}x + \frac{1}{3};\ m = \frac{1}{6}$

3. $3x + y - 15 = 0$

3 Review: Function Concepts

Exercises, page 7

1. a) This relation is a function.

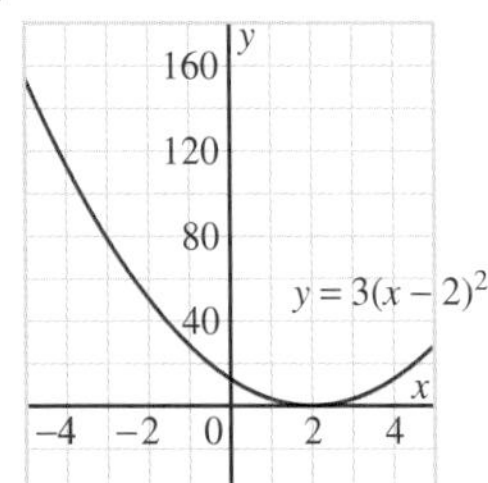

b) This relation is not a function.

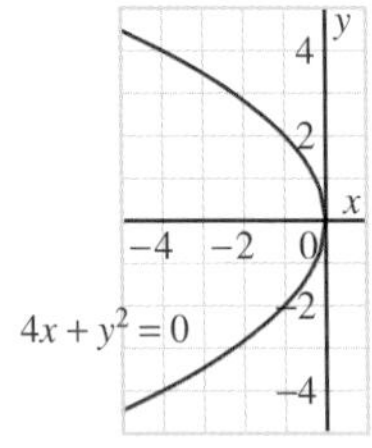

c) This relation is a function.

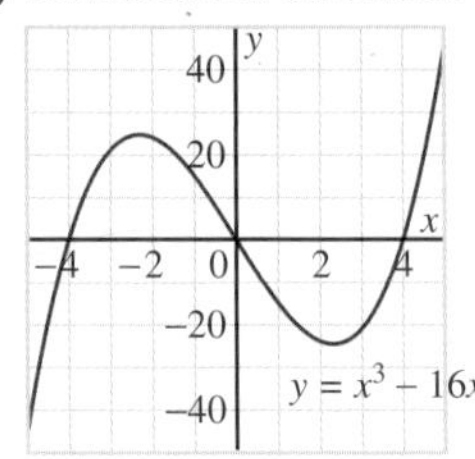

2. a) $D = \Re;\ R = \{y \mid y \in \Re, y \geq 0\}$
b) $D = \{x \mid x \in \Re, x \leq 0\};\ R = \Re$
c) $D = \Re;\ R = \Re$

1.1 Exercises, page 13

1. a) 3°C **b)** 10°C **c)** 17°C

2. a)

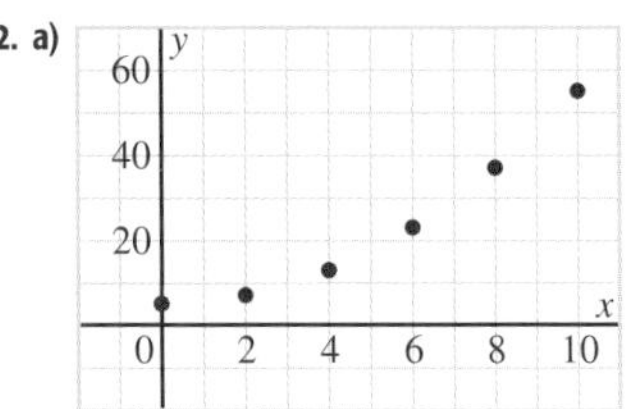

b) i) 1 **ii)** 2 **iii)** 5 **iv)** 8

3. a) 2; 6; 18; 54; 162; 486
b) –0.02; –0.028; –0.034; –0.04; 0.044; –0.05

4. a) –1 **b)** 2 **c)** 0

6. a) 3.33°C/s **b)** 1.49°C/s **c)** 0.68°C/s **d)** 0.31°C/s

7. a) i) 0.72°C/s **ii)** 0.67°C/s

8. Yes

9. a) More water would take longer to reach the boiling point.

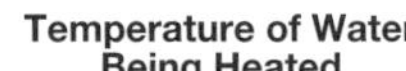

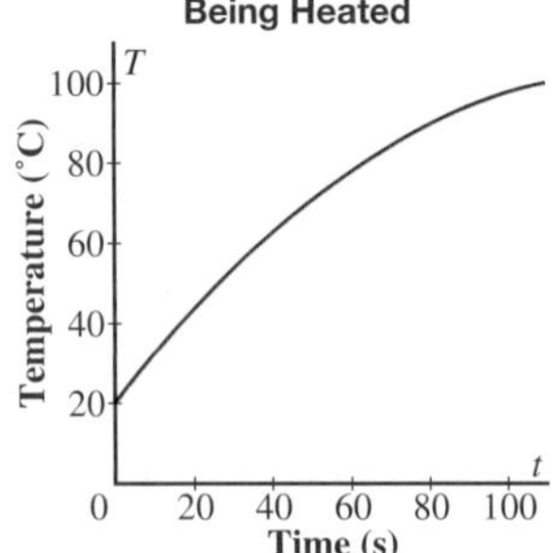

b) Less water would take less time to reach the boiling point.

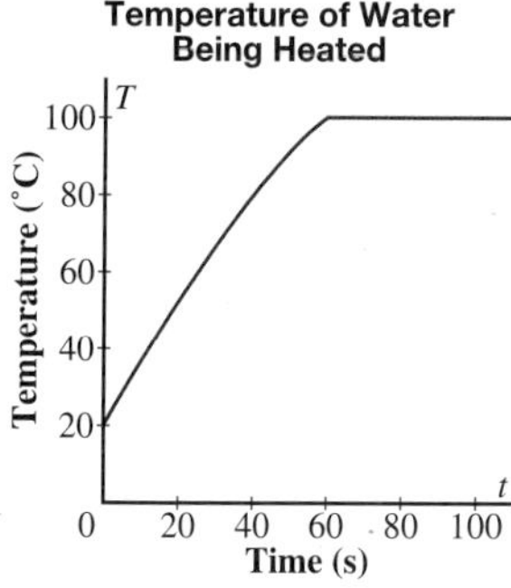

c) Warmer water will take less time to reach the boiling point.

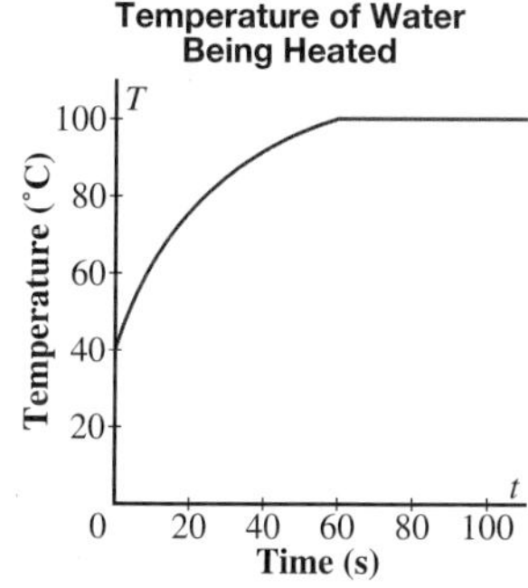

d) Colder water will take longer to reach the boiling point.

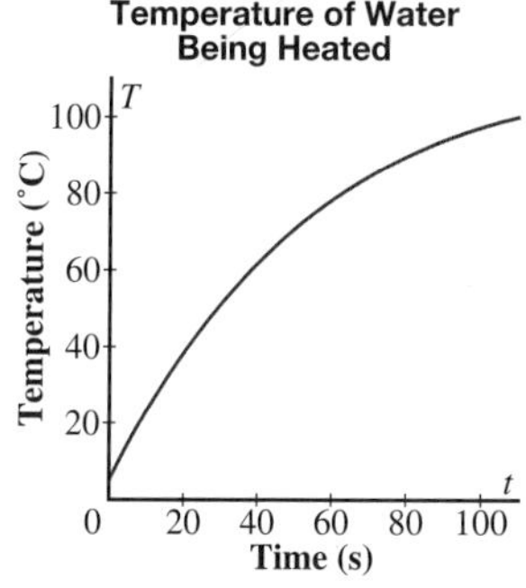

11. a) 5°C/h; 7°C/h **b)** 12 noon

12. a), b) **Pop Bottle Water Level Height**

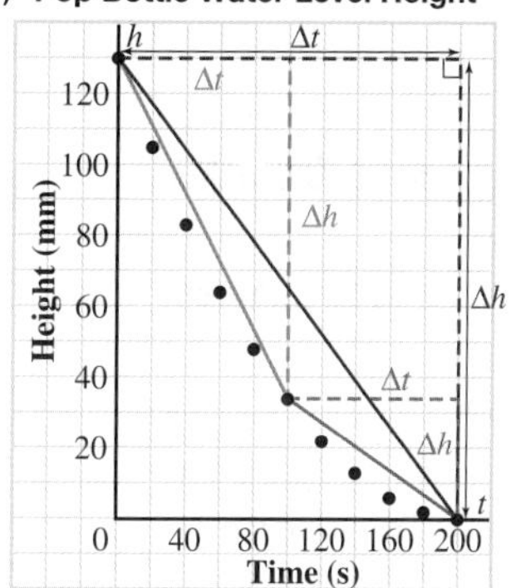

b) i) −0.96 mm/s **ii)** −0.34 mm/s **iii)** −0.65 mm/s

c) Estimates may vary. −0.875 mm/s

14. a) **Pop Bottle Water Level Height**

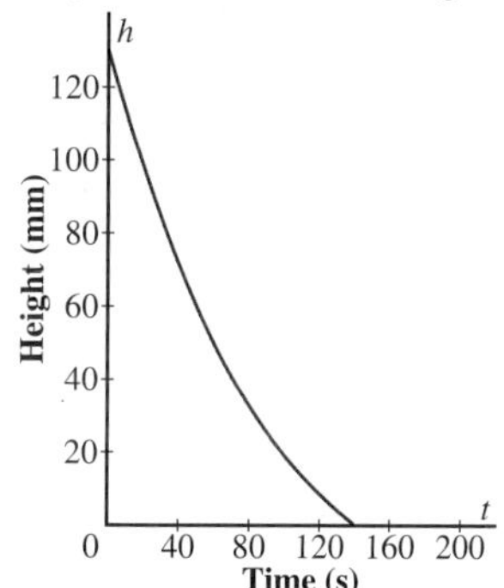

b) **Pop Bottle Water Level Height**

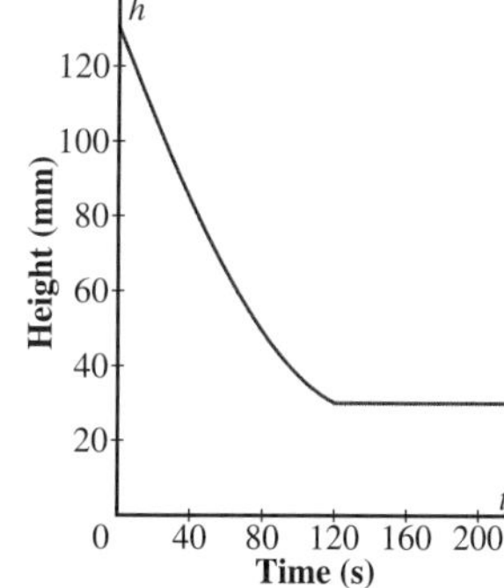

c) **Pop Bottle Water Level Height**

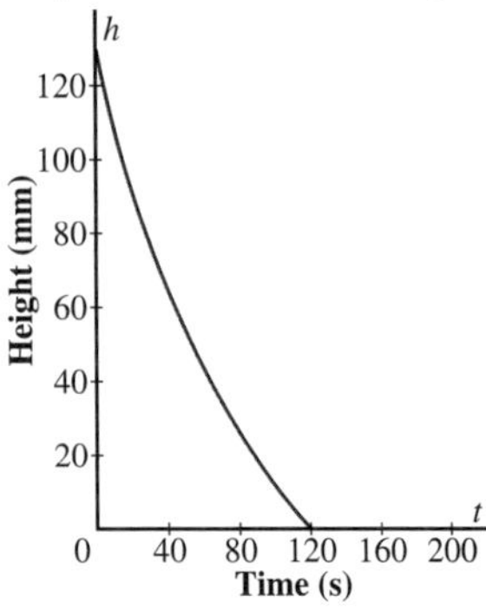

15. a) Answers may vary. –0.675 kPa/mL
b) –0.5 kPa/mL

16. a) **Growth Curve for North American Girls**

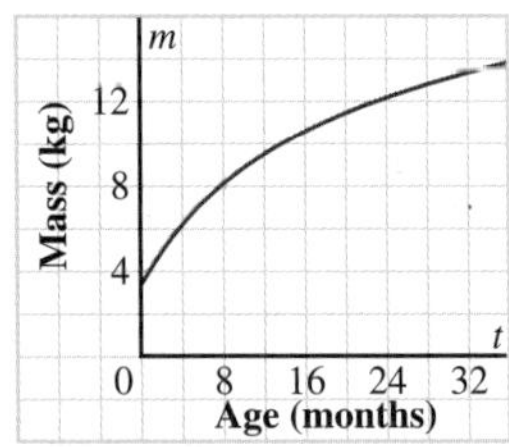

b) The graph is a smooth increasing curve that is concave down.
c) As a girl ages, her mass increases quickly in the first year, then increases at a slower rate for the next two years.
d) The slopes of the tangents in the first year are greater than the slopes in the second and third years.

1.2 Exercises, page 22

1. a) 3 **b)** 4 **c)** 5
d) –2 **e)** –1 **f)** 0

2. a) i) $\frac{1}{2}$ **ii)** $\frac{1}{2}$ **iii)** $\frac{1}{2}$
b) All the average rates of change are $\frac{1}{2}$ because the function is linear and the graph has slope 0.5.

4. a) 1.59°C/s **b)** 0.55°C/s **c)** 0.31°C/s **d)** 4.76°C/s

5. a) –0.8125 mm/s
b) i) –1.30 mm/s **ii)** –0.91 mm/s
iii) –0.52 mm/s **iv)** –0.13 mm/s

6. a) –0.35 kPa/mL
b) i) –7.87 kPa/mL **ii)** –0.88 kPa/mL
iii) –0.32 kPa/mL **iv)** –0.16 kPa/mL

7. a) $-2.8\ \text{W/m}^2/\text{m}$; $-0.4\ \text{W/m}^2/\text{m}$
b) $-2.7\ \text{W/m}^2/\text{m}$; $-0.35\ \text{W/m}^2/\text{m}$

8. a) 10 m/s; 19.5 m/s **b)** 9.85 m/s; 19.65 m/s

9. Estimates may vary.
a) i) 40 **ii)** 10 **iii)** 2.5
b) i) 0.25 **ii)** 0.5 **iii)** 1.0
c) i) –0.06 **ii)** –1.0 **iii)** –15.4

11. Estimates may vary.
a) 0.316; 0.632; 0.948

12. Estimates may vary.
a) 6.754; 6.754; 6.754
b) The instantaneous rates of change are equal.

13. Estimates may vary.
a) 0.69; 1.39 **b)** 1.10; 3.30

14. Estimates may vary.
a) 1.0; 0.87 **b)** –0.005; –0.50

15. a) CP: $\frac{\Delta y}{\Delta x} = 1.22$

h	P	Second point	Average rate of change
0.01	$(0.5, 2(0.5)^3)$	$(0.49, 2(0.49)^3)$	1.470 2
0.001	$(0.5, 2(0.5)^3)$	$(0.499, 2(0.499)^3)$	1.497 002
0.0001	$(0.5, 2(0.5)^3)$	$(0.4999, 2(0.4999)^3)$	1.499 700 02

b) CQ: $\frac{\Delta y}{\Delta x} = 1.52$

h	First point	Second point	Average rate of change
0.01	$(0.49, 2(0.49)^3)$	$(0.51, 2(0.51)^3)$	1.500 2
0.001	$(0.499, 2(0.499)^3)$	$(0.501, 2(0.501)^3)$	1.500 002
0.0001	$(0.4999, 2(0.4999)^3)$	$(0.5001, 2(0.5001)^3)$	1.500 000 02

16. a) i) 6 **ii)** 6 **iii)** 6 **iv)** 6
b) The instantaneous rate of change of y with respect to x at $x = 3$ is 6.

18. a) $2x_1$ **b)** $2x_1$

20.

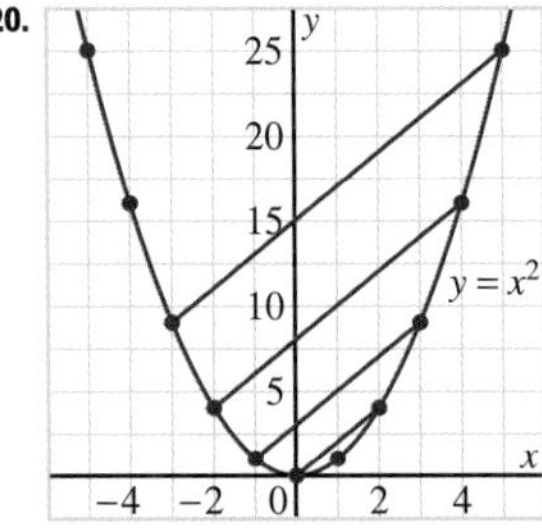

a) All average rates of change are 2.
b) (1, 2); (1, 5); (1, 10); (1, 17)
d) 2

1.3 Exercises, page 31

2. a) **Derivative of Pop Bottle Water Level Height**

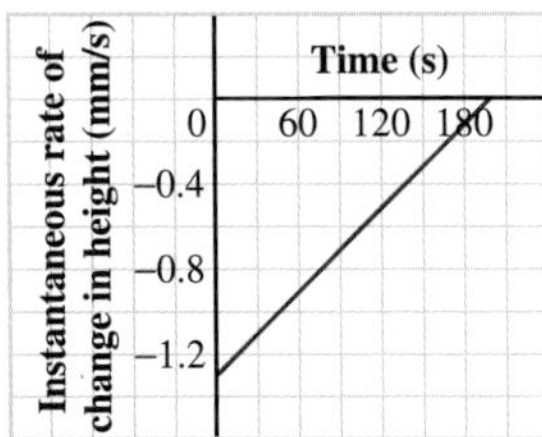

b) The derivative function appears to be linear.

3. a) Ball Height

b) i) 8.04 m/s **ii)** 3.15 m/s
iii) −1.74 m/s **iv)** −6.63 m/s

c) The derivative function appears to be linear.

Derivative of Ball Height

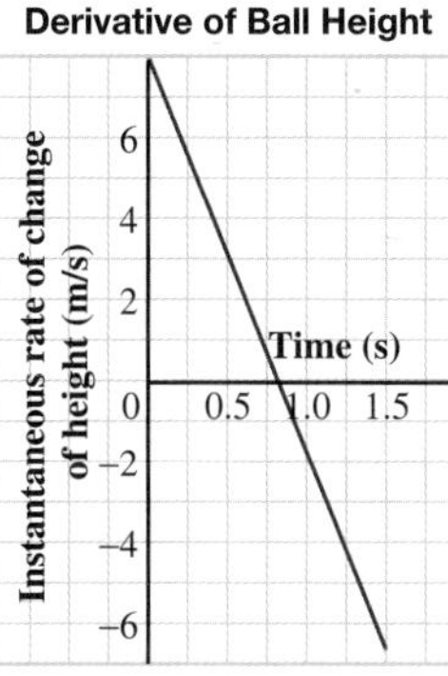

4. a) $y = 0.0065x - 1.30$

5. a) $y = -9.78x + 8.04$

6. Estimates may vary.

a) i) 2000/h **ii)** 15 700/h **iii)** 132 000/h
b) i) 5940/h **ii)** 47 500/h

c) Derivative of Bacteria Growth

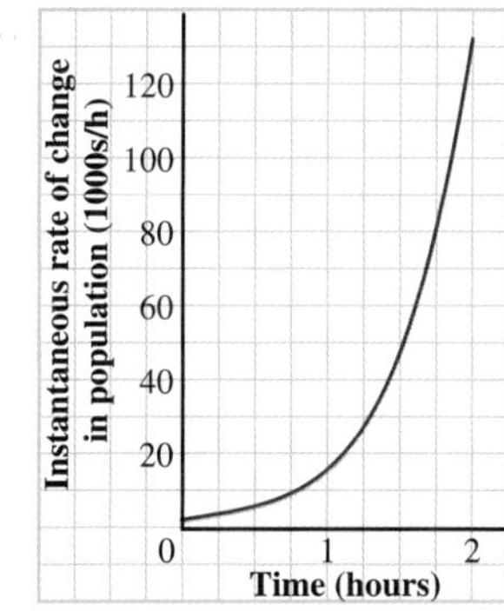

7. Estimates may vary.

a) i) −14%/h **ii)** −2.6%/h **iii)** −0.5%/h
b) i) −6.0%/h **ii)** −1.1%/h

c) Derivative of Caffeine Consumption

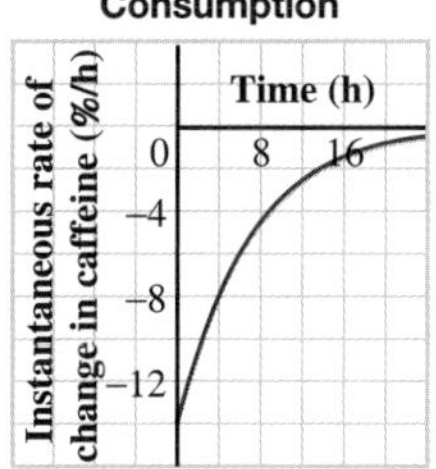

8. a) Derivative of Helium Pressure

10. a) Height of a Tomato Plant

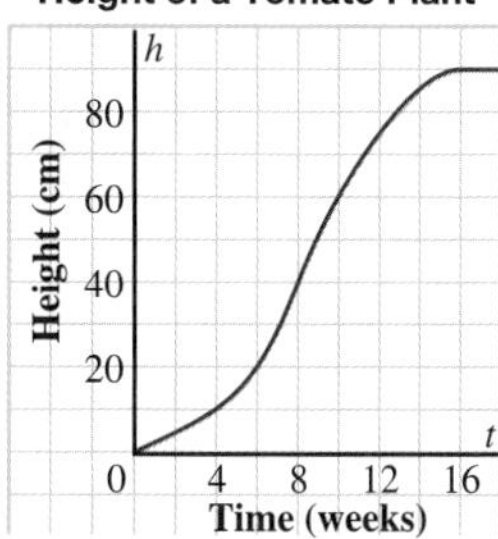

b)

Time (weeks)	Instantaneous growth rate (cm/week)
1	2.5
3	2.5
5	5
7	10
9	10
11	7.5
13	5
15	2.5
17	0

c) Derivative of Tomato Plant Height

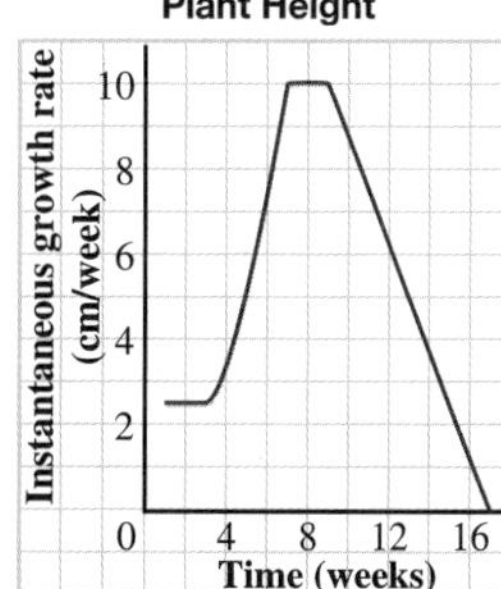

d) The tomato plant is growing fastest during weeks 7–9.

11. Graphs may vary.

a) **Height of a Tomato Plant**

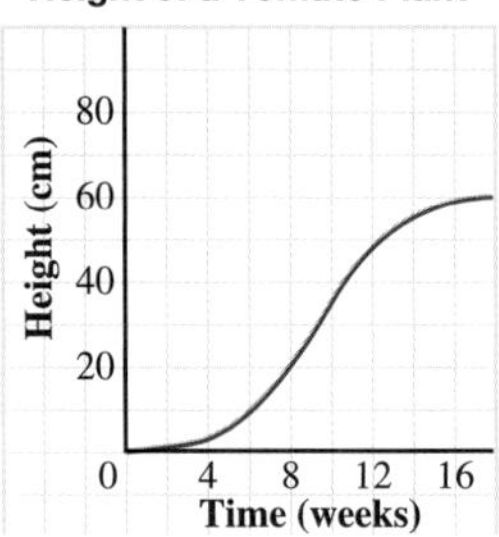

Derivative of Tomato Plant Height

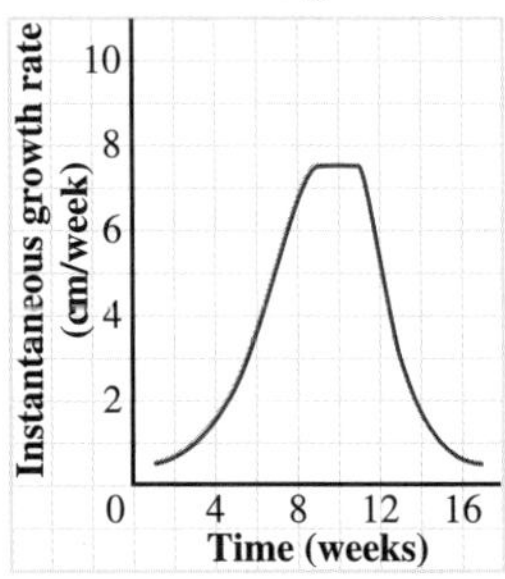

b) **Height of a Tomato Plant**

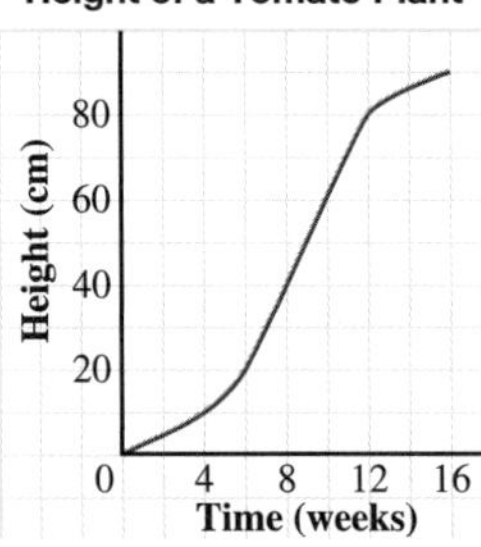

Derivative of Tomato Plant Height

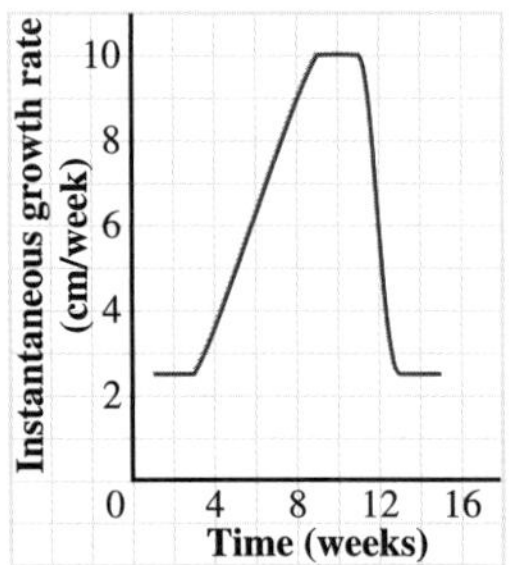

c) **Height of a Tomato Plant**

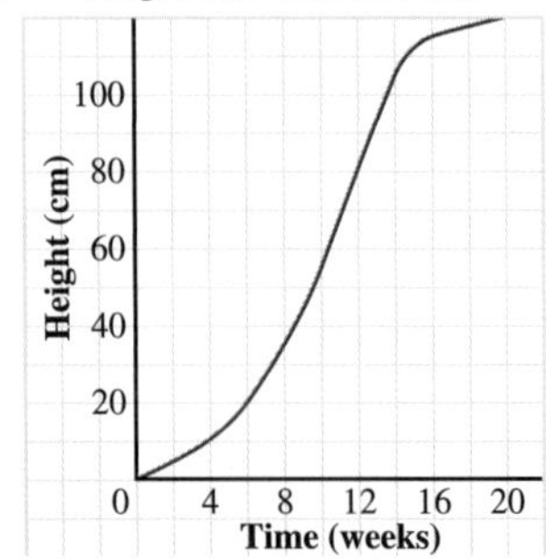

Derivative of Tomato Plant Height

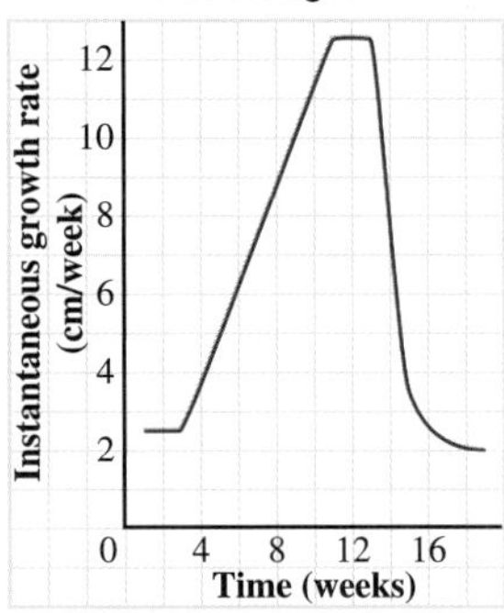

12. a) **Speed of Bungee Jumper**

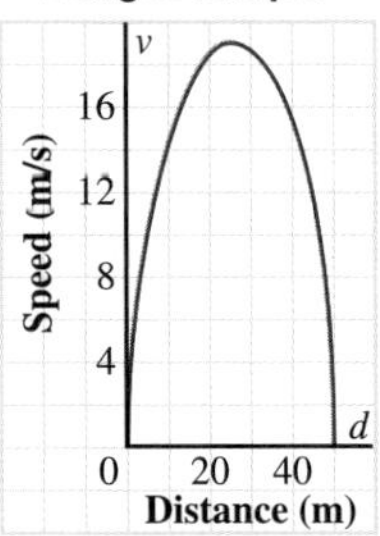

c) Approximately 25 m

d) Approximately 25 m

e)

Distance (m)	Rate of change of speed with distance (m/s/m)
2.5	2.00
7.5	0.82
12.5	0.56
17.5	0.30
22.5	0.12
27.5	–0.06
32.5	–0.22
37.5	–0.44
42.5	–0.76
47.5	–2.32

f) **Derivative of Bungee Jumper Speed**

Rate of change of speed with distance (m/s/m)
2.0
1.0
0
−1.0
−2.0
Distance (m)
20 40

g) The speed is increasing from the beginning of the jump to a distance of approximately 25 m.

h) Approximately 25 m

13. a) **Hertzberg Papers**

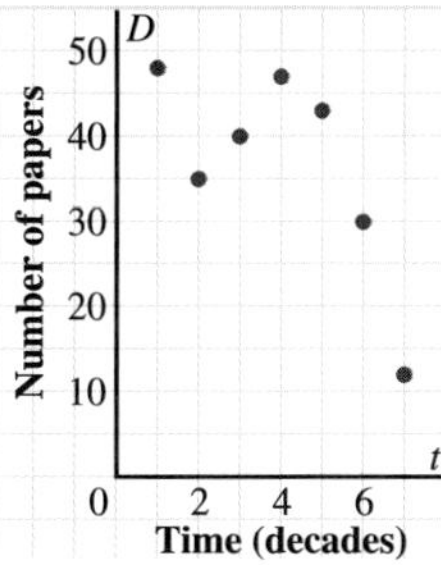

b)

Year	Total number of papers
1925	0
1935	48
1945	83
1955	123
1965	170
1975	213
1985	243
1995	255

Hertzberg Papers

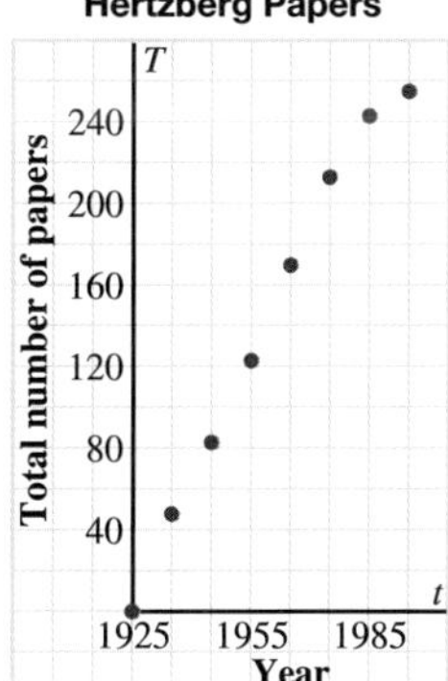

Self-Check 1.1–1.3, page 37

1. a) **Ottawa Airfares**

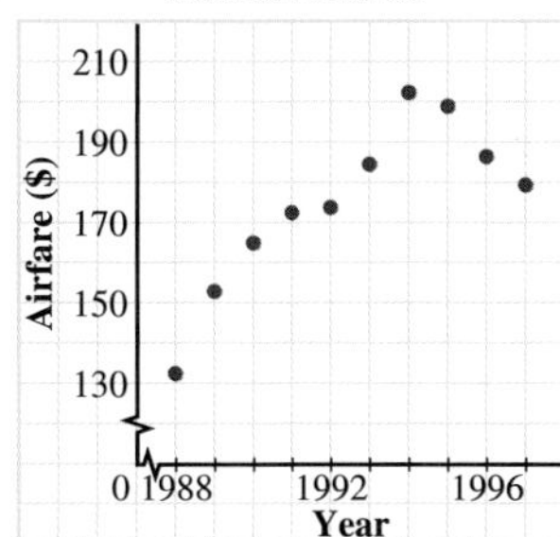

b)

Year	Average rate of change of price ($/year)
1988–1989	20.40
1989–1990	12.00
1990–1991	7.70
1991–1992	1.30
1992–1993	10.70
1993–1994	17.80
1994–1995	−3.40
1995–1996	−12.50
1996–1997	−7.00

c) The fares increase from 1988 to 1994. No, the increase is not constant.

2. a)

E

4800

3200

1600

0 2 4 6 8

d

$E = \frac{6000}{d^2}$

b) i) −5926 newtons per coulomb/m
ii) −1440 newtons per coulomb/m
iii) −617 newtons per coulomb/m
iv) −276 newtons per coulomb/m
c) −768 newtons per coulomb/m

3. a) 2 **b)** −4 **c)** 4

4. a) Estimates may vary. The estimates below were calculated using a 0.0001-m interval.

Distance (m)	Instantaneous rate of change of electric field (newtons/coulomb/m)
0.75	−28 439
1.25	−6 143
1.75	−2 239
2.25	−1 053
2.75	−577
3.25	−350
3.75	−228
4.25	−156

b) **Derivative of Electric Field**

Instantaneous rate of change of electric field (newtons/coulomb/m)

Distance (m)

0 2 4

−2000

−4000

−6000

1.4 Exercises, page 43

1. a) $y' = 3$ **b)** $y' = -0.5$ **c)** $y' = -1$
d) $y' = 1$ **e)** $y' = 2x$ **f)** $y' = 3x^2$

2. a) The graph of f is a linear function with a positive slope.
b) The graph of f is a linear function with a negative slope.

3. a) $y' = -2$

b)

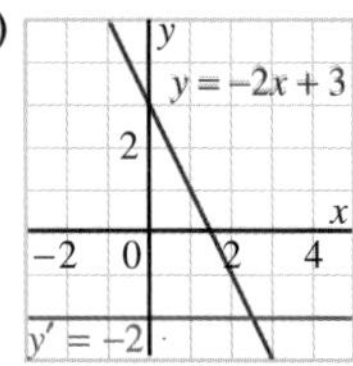

c) Answers may vary. $y = -2x + 5$

4. a) $y' = 0.5$; answers may vary. $y = 0.5x + 1$

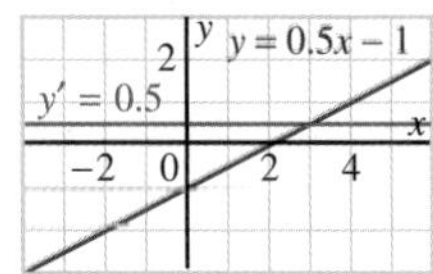

b) $y' = -1$; answers may vary. $y = -8 - x$

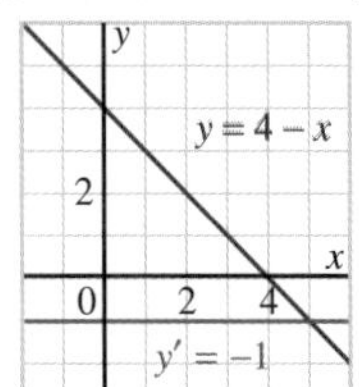

c) $y' = 0$; answers may vary. $y = 7$

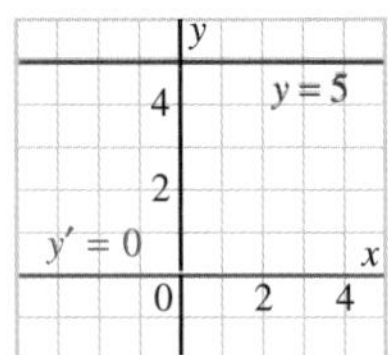

5. a) i) 8 **ii)** −10 **iii)** 1 **iv)** 0

b) i) When $x = 4$, the slope of the tangent is 8.
ii) When $x = -5$, the slope of the tangent is −10.
iii) When $x = 0.5$, the slope of the tangent is 1.
iv) When $x = 0$, the slope of the tangent is 0.

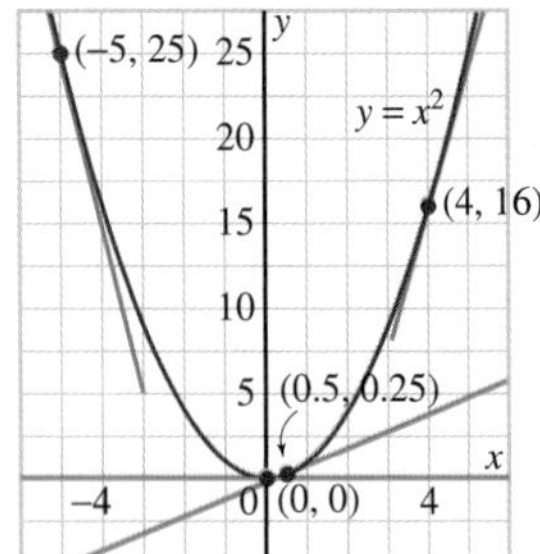

6. a) i) 48 **ii)** 75 **iii)** 0.75 **iv)** 0

When $x = 4$, the slope of the tangent is 48.
When $x = -5$, the slope of the tangent is 75.

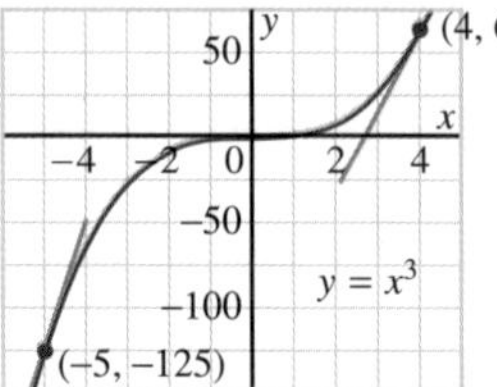

When $x = 0.5$, the slope of the tangent is 0.75.
When $x = 0$, the slope of the tangent is 0.

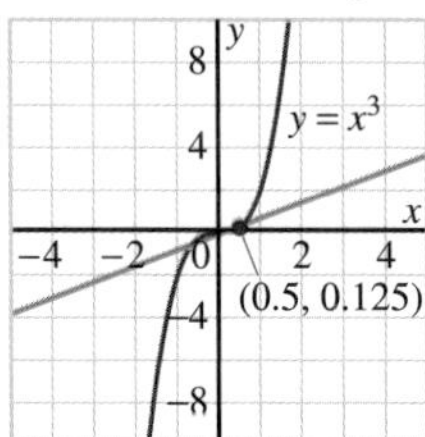

b) i), ii), iii), iv) 2

Since the function is linear, the tangent at any point coincides with the graph of the function.

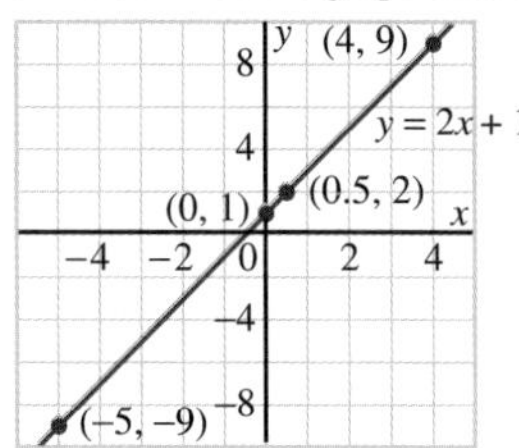

7. The derivative at $x = -4$ is the opposite of the derivative at $x = 4$.

8. The derivative at $x = -4$ is equal to the derivative at $x = 4$.

9. a) i) $y = 2x - 1$ **ii)** $y = 4x - 4$ **iii)** $y = 6x - 9$

b)

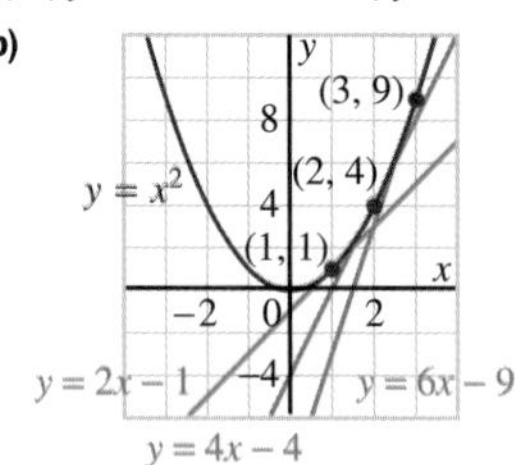

10. a) i) $y = 3x - 2$ **ii)** $y = 12x + 16$ **iii)** $y = 0$

b)

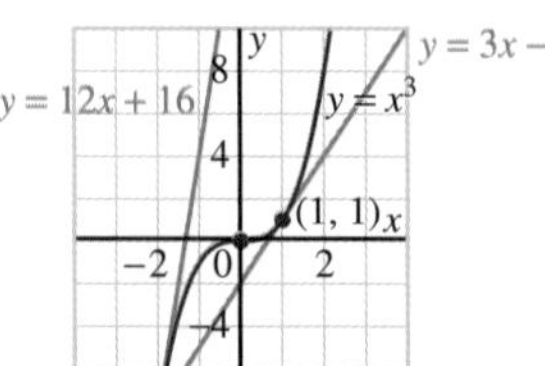

11. $y = 3x - \frac{9}{4}$; $y = -3x - \frac{9}{4}$

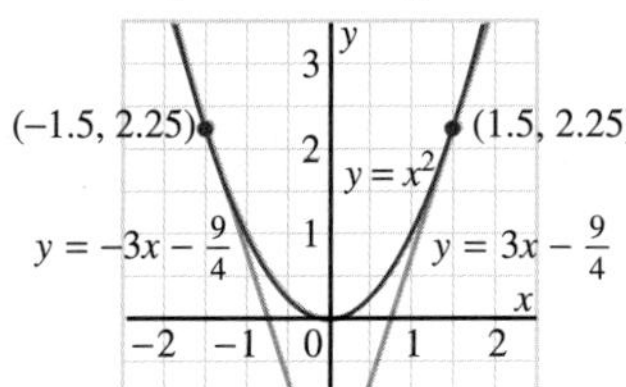

12. $y = \frac{3}{4}x - \frac{1}{4}$; $y = \frac{3}{4}x + \frac{1}{4}$

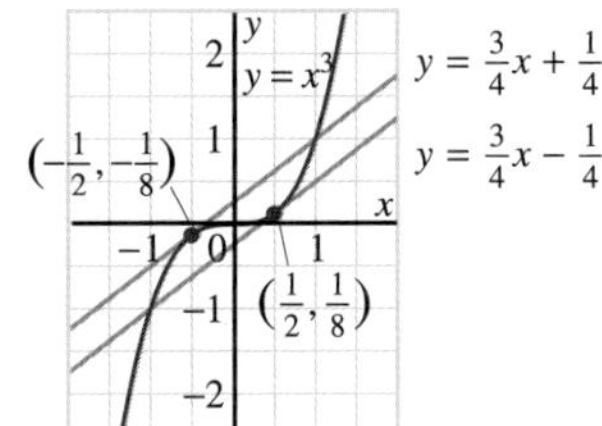

13. a), b), c) $y' = 2x$

14. a) $y' = 2x$ **b)** $y' = 2x$ **c)** $y' = 2x$
d) $y' = 3x^2$ **e)** $y' = 3x^2$ **f)** $y' = 3x^2$

16. a) $y' = 2x$ **b)** $y' = 4x$ **c)** $y' = x$

17. a) $y' = 8x$ **b)** $y' = 0.5x$ **c)** $y' = -4x$ **d)** $y' = 6x^2$

18. a) At $\left(-2, -\frac{1}{2}\right)$, $m = -\frac{1}{4}$; at $\left(\frac{1}{2}, 2\right)$, $m = -4$;
at $\left(-\frac{1}{2}, -2\right)$, $m = -4$

19. a) $\frac{1}{2}$; $\frac{1}{4}$; $\frac{1}{6}$

20. a) 1; $\frac{1}{2}$; 0; $-\frac{1}{\sqrt{2}}$

21. a) 0; $-\frac{1}{\sqrt{2}}$; $-\frac{1}{2}$; $\frac{1}{\sqrt{2}}$

23. a) -108; -32; -4; 0; 4; 32; 108
b) $y' = 4x^3$ **c)** $y' = 5x^4$

24. a)

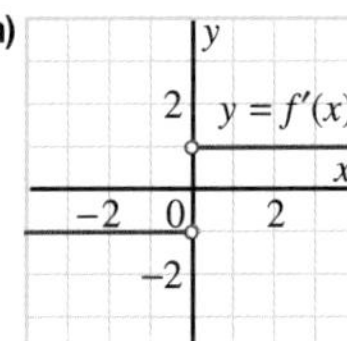

b)

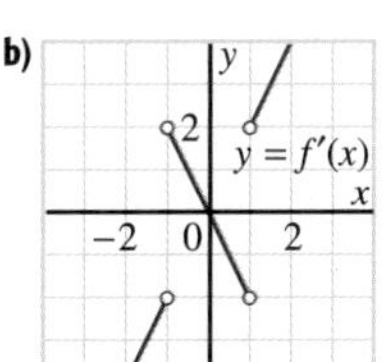

c)

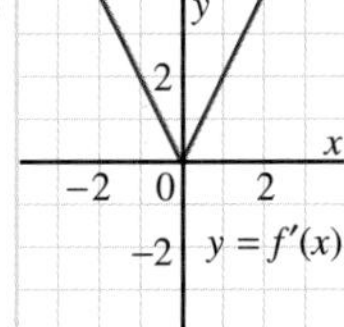

25. a) $y = mx - \frac{m^2}{4}$ **b)** $y = (2x_1)x + (y_1 - 2x_1^2)$

26. a) $y = mx \pm 2\left(\sqrt{\frac{m}{3}}\right)^3$ **b)** $y = (3x_1^2)x + (y_1 - 3x_1^3)$

1.5 Exercises, page 51

1. a)

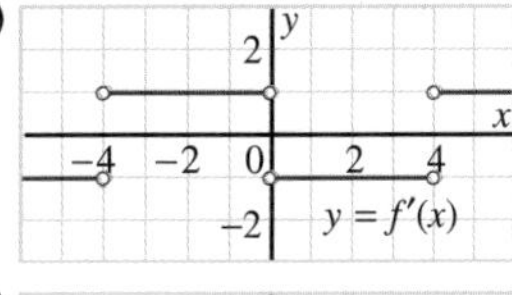

b)

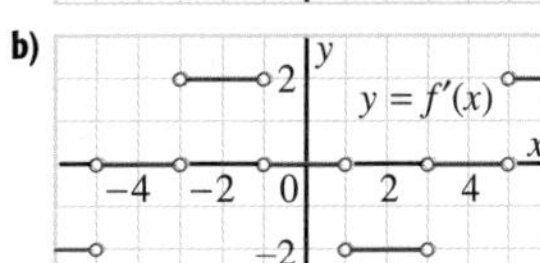

2. a)

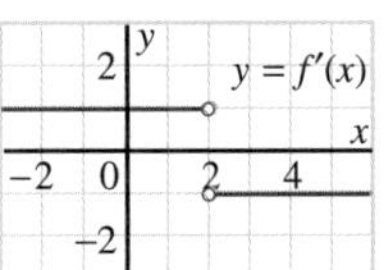

b)

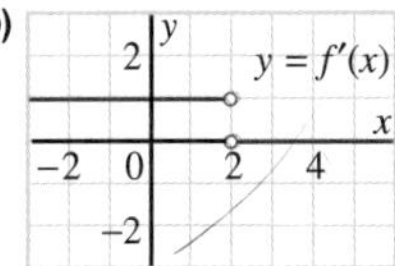

c) y = f′(x)

3. Diagrams may vary. No

4. Diagrams may vary.
a) Yes **b)** Yes **c)** Yes

5. a)

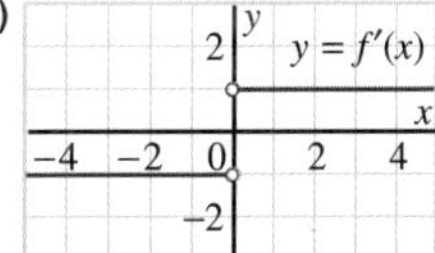

6. a) i) 3 **ii)** 5 **iii)** −4 **iv)** −3
c)

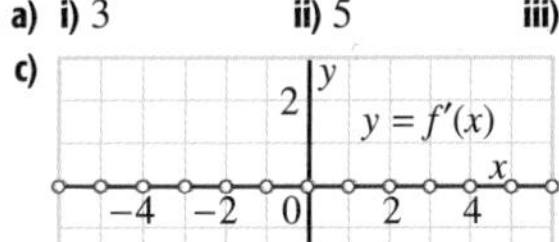

7. a) i) 0.75 **ii)** 0 **iii)** 0 **iv)** 0.85
c)

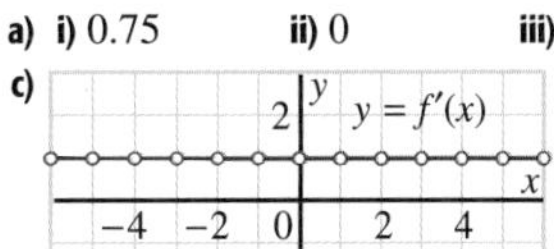

8. b)

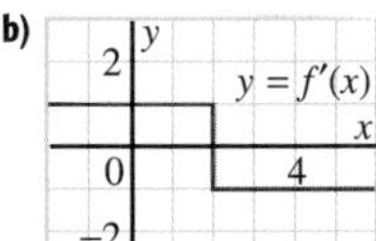

9. a) iii **b)** i **c)** ii

11. b)

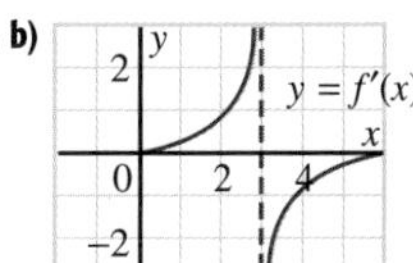

12. $y' = \frac{1}{2\sqrt{x}}$

1.6 Exercises, page 57

1. a) ② is f; ① is f'.
b) ③ is f'; ④ is f.
c) ⑤ is f'; ⑥ is f.

3. a) ① is f; ② is f'.
b) ③ is f; ④ is f'.
c) ⑤ is f; ⑥ is f'.

4. a)
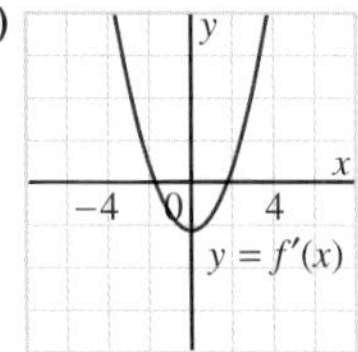

b)
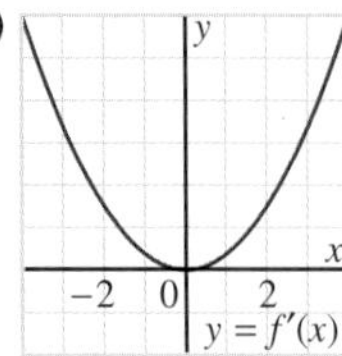

c)
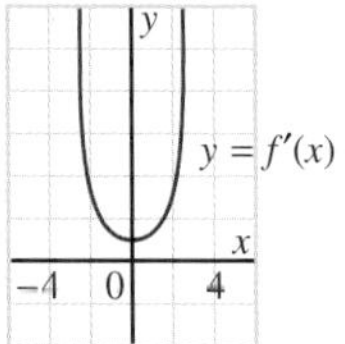

d)
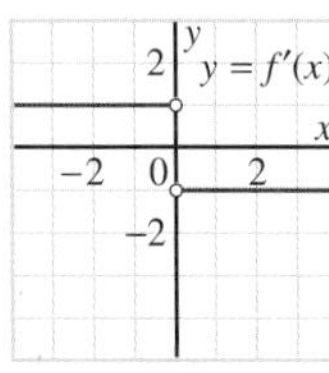

e)
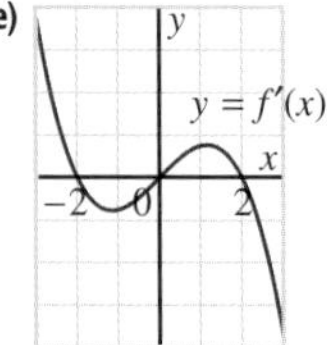

f)
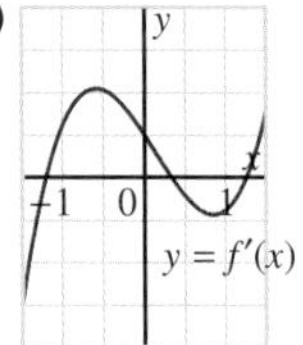

5. a) For $x > 0$
i) The graph of the function rises to the right.
ii) The graph of the derivative has $y > 0$.
b) For $x < 0$
i) The graph of the function falls to the right.
ii) The graph of the derivative has $y < 0$.
c) For $x = 0$
i) The graph of the function has a horizontal tangent.
ii) The graph of the derivative changes from negative y-values to positive y-values.

6. a) For $x \in \Re,\ x \neq 0$
i) The graph of the function always rises to the right.
ii) The graph of the derivative has $y > 0$.
b) There are no x-values.
i) The graph of the function never falls to the right.
ii) The graph of the derivative has no negative y-values.
c) For $x = 0$
i) The graph of the function has a horizontal tangent.
ii) The graph of the derivative has $y = 0$.

7. a) ① is f'; ② is f.
b) ③ is f; ④ is f'.
c) ⑤ is f; ⑥ is f'.

8. a) Graph 2 **b)** Graph 3 **c)** Graph 1

9. a) The graphs of the given functions have a horizontal tangent at $x = 0$.
b) The graphs of the given functions are symmetrical about the y-axis.

10. a)
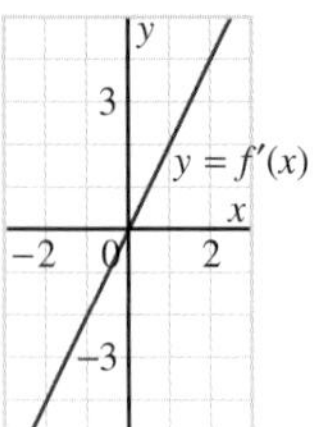

b)
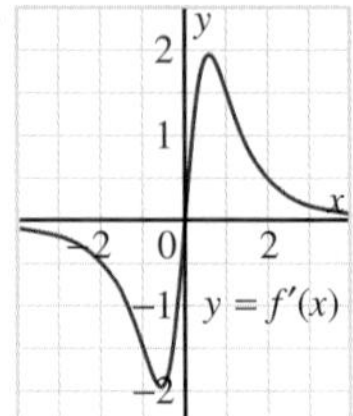

c)
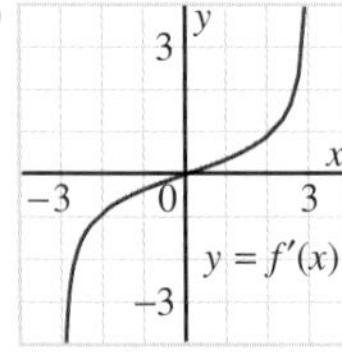

11. Graphs may vary.

a)

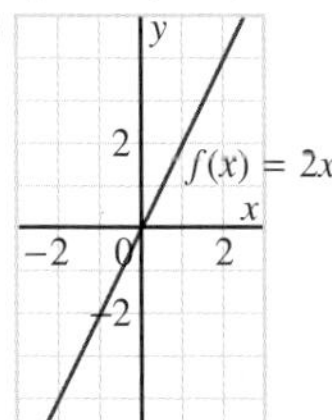

b)
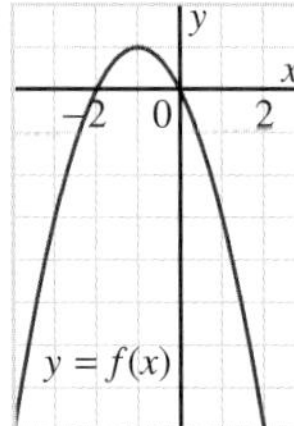

c)
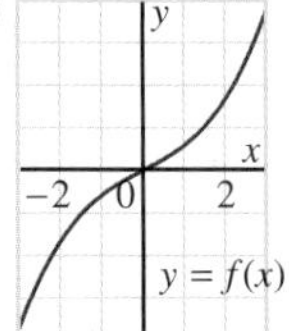

12. a) The graphs of the original functions are always increasing.
b) The graphs of the original functions are unchanged by a 180° rotation about the origin.

13. a)
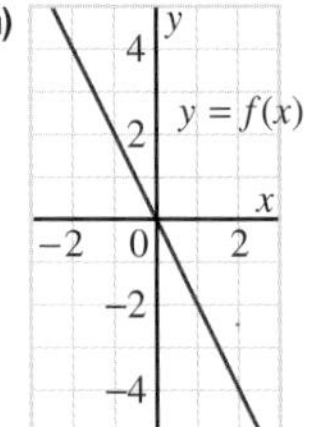

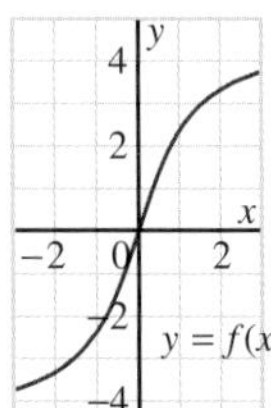

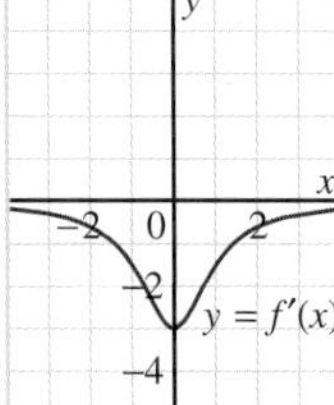

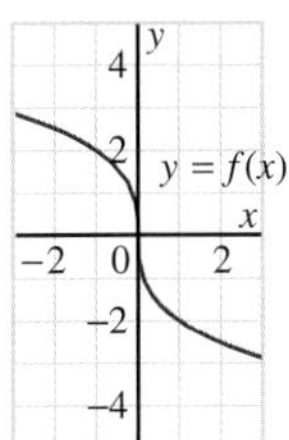

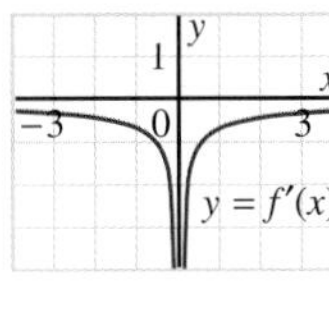

b) The graph of each derivative is the image of the previous derivative graph after a reflection in the x-axis.

14. a)

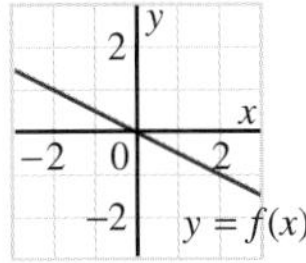

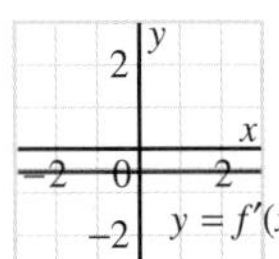

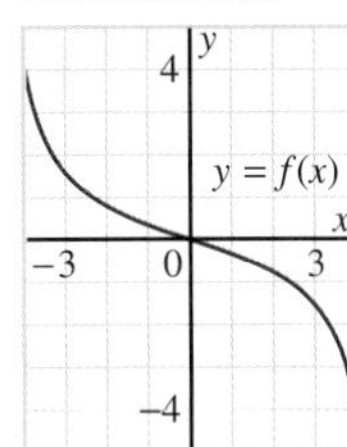

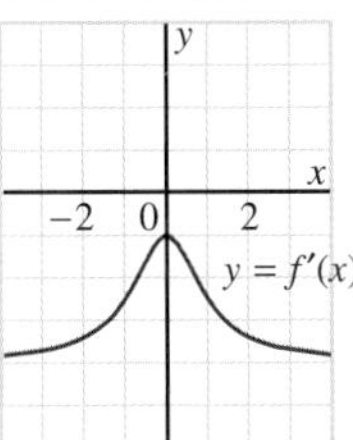

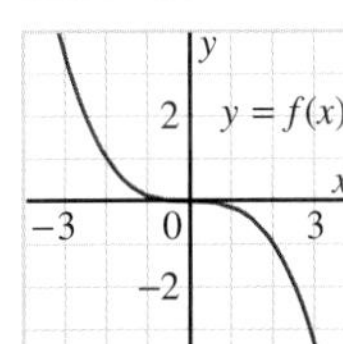

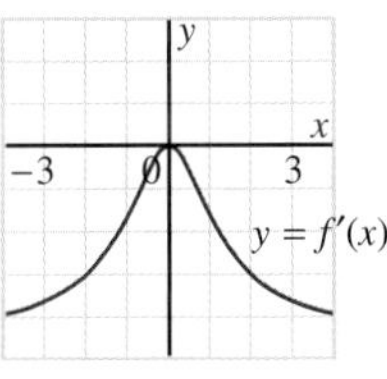

15. a)

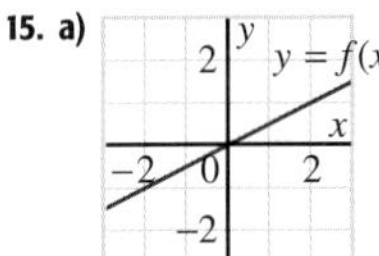

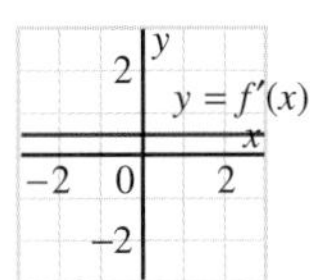

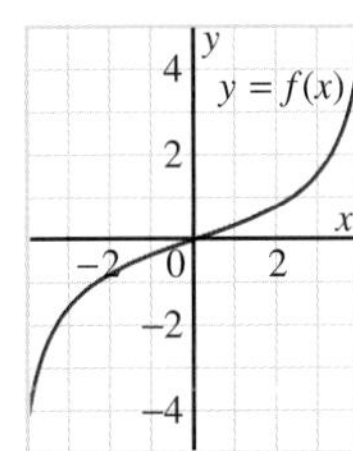

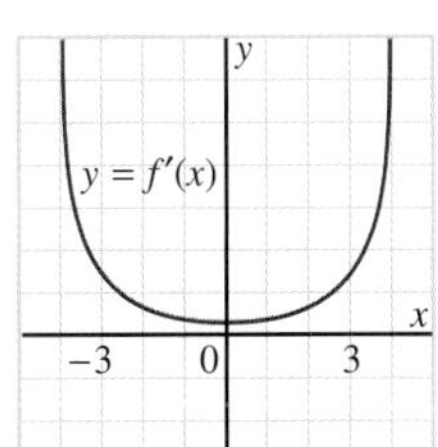

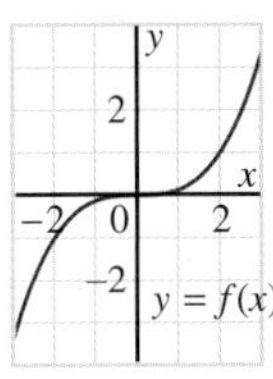

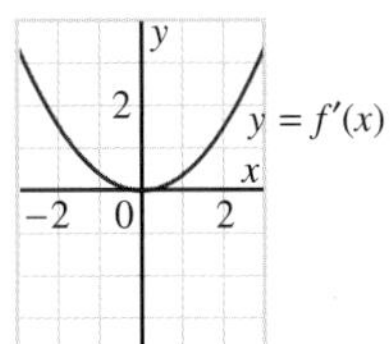

16.

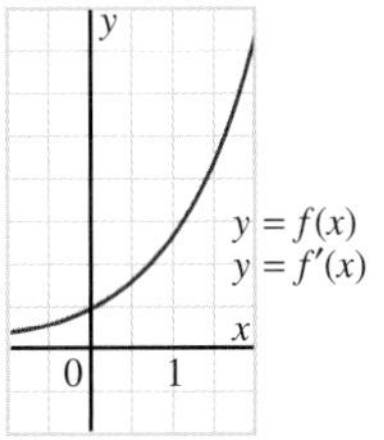

Self-Check 1.4–1.6, page 61

1. a) $y' = 1$ **b)** $y' = 2$ **c)** $y' = 0.5$
d) $y' = 2x$ **e)** $y' = -6x^2$ **f)** $y' = 15x^2$

2. Answers may vary. Sample answers follow:
a) $y = x - 3$ **b)** $y = 2x + 4$ **c)** $y = 0.5x$
d) $y = x^2 - 1$ **e)** $y = -2x^3 + 5$ **f)** $y = 5x^3 - 2$

3. a) i) $y' = 6x$ **ii)** $y' = x$ **iii)** $y' = -8x$ **iv)** $y' = -6x$
b) Vertical stretch rule

4. a)

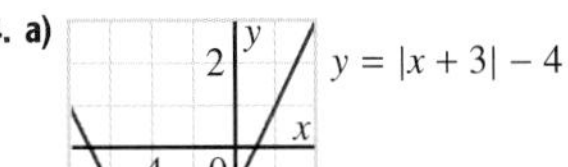

b)

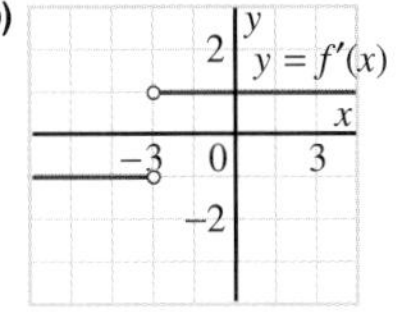

c) At $x = -3$

5. ② is f; ① is f'.

6.

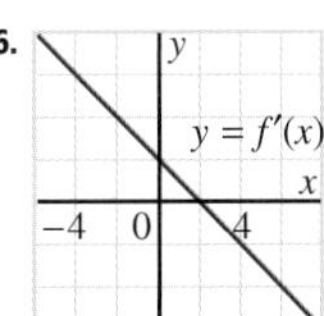

7. a) −18.75 **b)** $y = -18.75x + 31.25$
c) $y = -18.75x - 31.25$

Chapter 1 Review Exercises, page 63

1. a) 1; 3; 5; 7

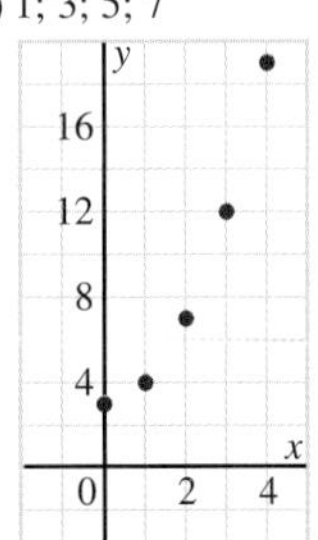

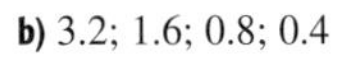
b) 3.2; 1.6; 0.8; 0.4

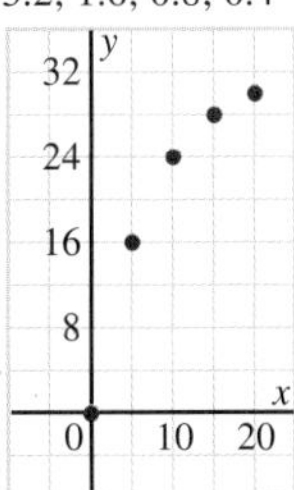

c) −0.1; −0.2; −0.3; −0.4

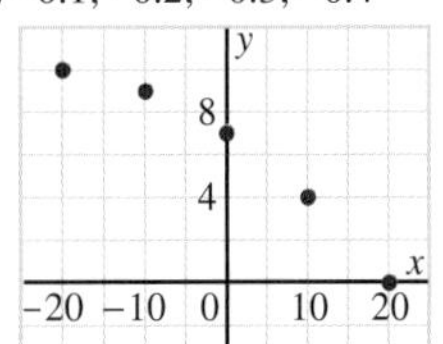

2. a) 0.67 ppt/m **b)** 1.4 ppt/m **c)** 0.2 ppt/m

3. a)

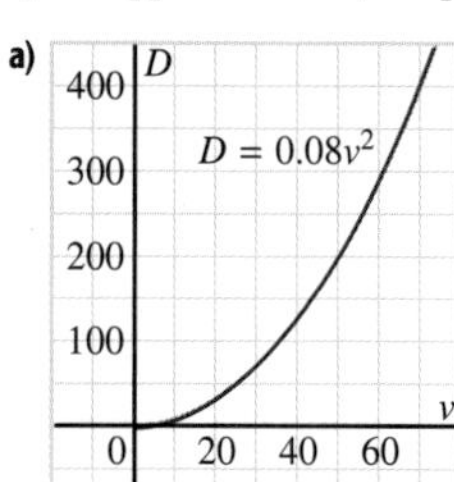

b) 5.6 kN/m/s **c)** Estimates may vary. 5.6 kN/m/s

4. a) 7 **b)** 0.21 **c)** −1.2 **d)** 0

5. Estimates may vary.
a) 4; 4 **b)** −4; 6 **c)** −12; −27 **d)** 3; 3

6. Estimates may vary. 5

7. a)

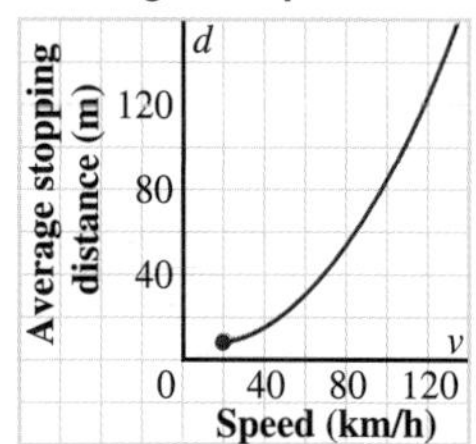

b) i) 0.16 m/km/h **ii)** 0.56 m/km/h **iii)** 0.96 m/km/h
iv) 1.36 m/km/h **v)** 1.76 m/km/h

c) The graph of the derivative function appears to be linear.

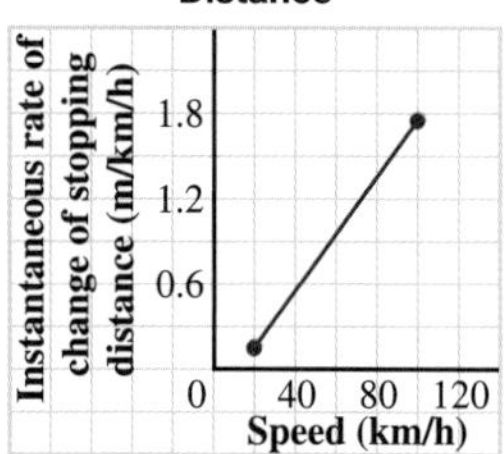

8. a)

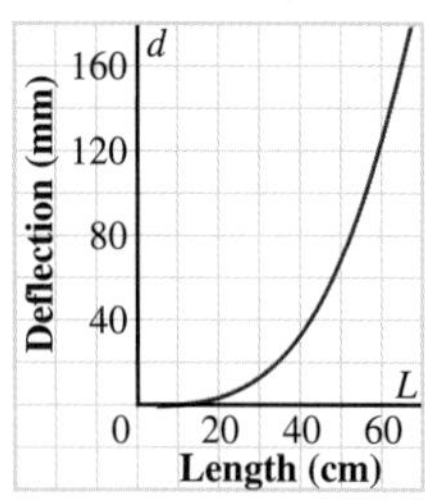

b)

Length (cm)	Estimated instantaneous rate of change of deflection (mm/cm)
7.5	0.2
12.5	0.2
17.5	0.2
22.5	0.8
27.5	1.2
32.5	1.6
37.5	2.6
42.5	3.0
47.5	3.8
52.5	5.0
57.5	5.8
62.5	7.0

c)

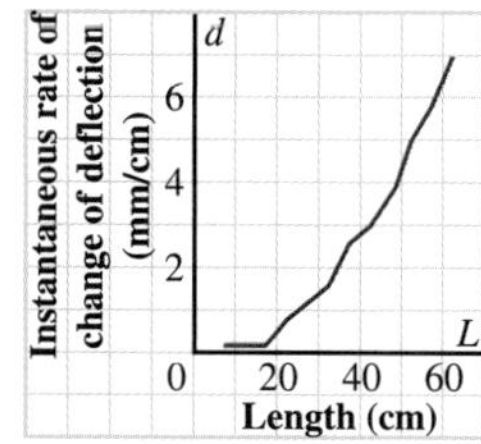

d) The rate of change of deflection is greatest between lengths of 60 cm and 65 cm.

9. a) $y' = 0$ **b)** $y' = 12x^2$ **c)** $y' = -2x$ **d)** $y' = 7$

10. a) 1 **b)** −3 **c)** 0.5 **d)** −2

11. $y = -8x + 7$

12. a) $y' = \frac{1}{2\sqrt{x}}$ **b)** $y' = \frac{1}{x^2}$
c) $y' = 2\cos x$ **d)** $y' = -\sin x$

13. a) $y = \frac{3}{4}x + 3$ **b)** $y = -\frac{1}{8}x + 1$ **c)** $y = \frac{1}{4}x - 3$

14. a) 1 **b)** $\pm\sqrt{\frac{2}{3}}$ **c)** No values
d) $\frac{1}{4}$ **e)** $\pm\sqrt{2}$ **f)** All real values

15. c)

16.

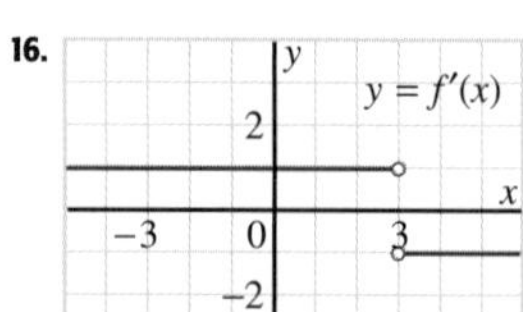

17. a) ii **b)** iii **c)** i

18. Answers may vary.
$y = 3x,\ y = 3x + 1,\ y = 3x - 1,\ y = 3x - 2$

19. Answers may vary.
$y = x^2$, $y = x^2 + 1$, $y = x^2 - 1$, $y = x^2 - 2$

20. a)

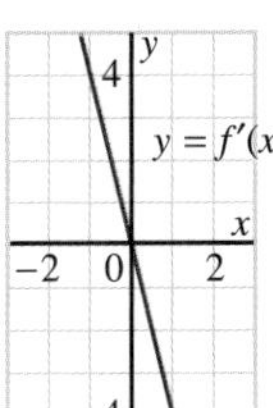

b)

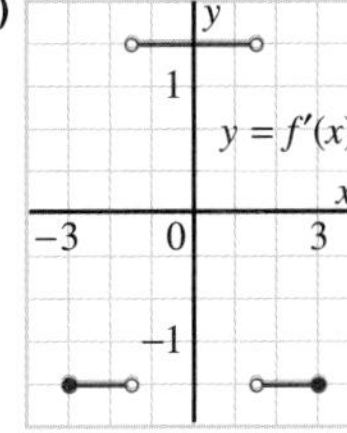

c)

21. Graphs may vary.

a)

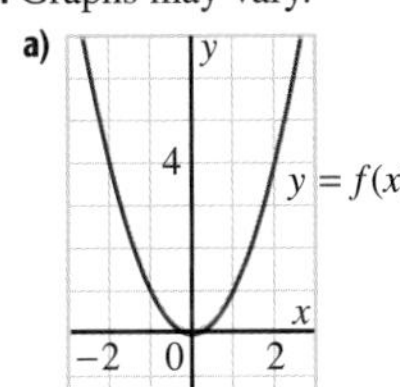

b)

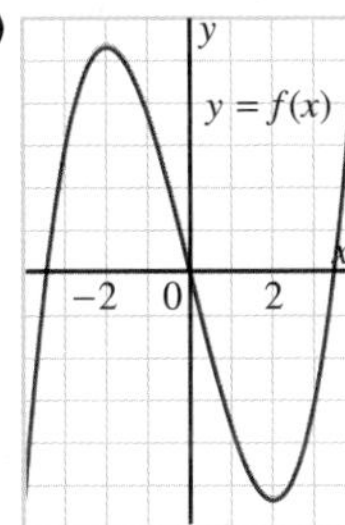

c)

Chapter 1 Self-Test, page 67

1. a) −0.76, −0.96, −0.88, −0.84, −0.46, −0.22, −0.12

2. a) Estimates may vary. 2
b) $y' = 2x$ **c)** $y' = 2$

3.

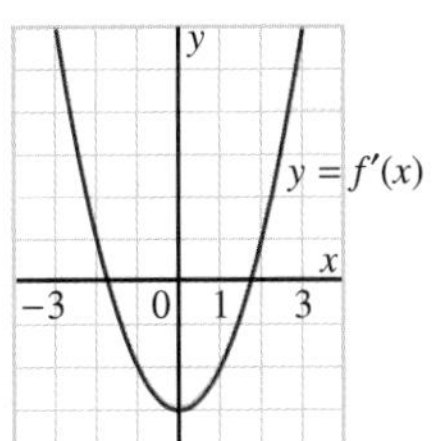

4. Graphs may vary.

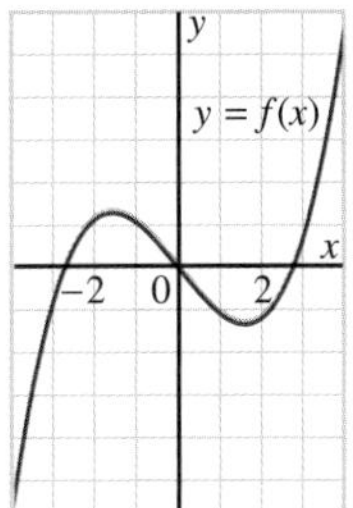

5. a) i) 0.32 **ii)** 1 **iii)** 3.16 **iv)** 10
b) The function $y = \sqrt{x}$ has a vertical tangent at $x = 0$.

6. a) 72.2 W/K **b)** Estimates may vary. 72.9 W/K

7. $y = -4x + 6$

8. a)

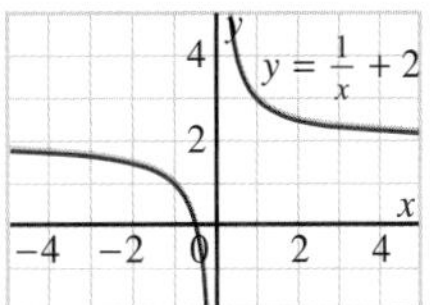

b) For values of x greater than 5, the y-values get closer and closer to 2, but never reach 2. For values of x less than −5, the y-values get closer and closer to 2 but never reach 2.

c)

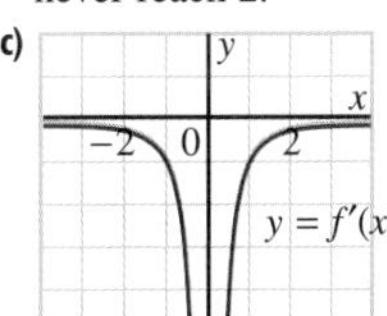

9. c) $y = 0.8x + 5.6$ **d)** (−7, 0)

Chapter 2 Calculating Derivatives from First Principles

Necessary Skills

1 New: Powers of Binomials

Exercises, page 72

1. a) $a^3 - 3a^2b + 3ab^2 - b^3$
b) $a^4 - 4a^3b + 6a^2b^2 - 4ab^3 + b^4$
c) $8x^3 + 12x^2 + 6x + 1$
d) $1 + 9x + 27x^2 + 27x^3$
e) $1 - 8x + 24x^2 - 32x^3 + 16x^4$
f) $9a^2 + 12ab + 4b^2$
g) $27a^3 + 54a^2b + 36ab^2 + 8b^3$
h) $81a^4 + 216a^3b + 216a^2b^2 + 96ab^3 + 16b^4$

2. b) $a^5 + 5a^4b + 10a^3b^2 + 10a^2b^3 + 5ab^4 + b^5$

2 New: Factoring Differences of Powers

Exercises, page 73

1. a) $(x - 1)(x^2 + x + 1)$ **b)** $(x - 2)(x^2 + 2x + 4)$
c) $(x - 3)(x^2 + 3x + 9)$ **d)** $(x - 4)(x^2 + 4x + 16)$
e) $(x - 1)(x^3 + x^2 + x + 1)$ **f)** $(x - 2)(x^3 + 2x^2 + 4x + 8)$

g) $(x-1)(x^4+x^3+x^2+x+1)$
h) $(x-2)(x^4+2x^3+4x^2+8x+16)$

2. a) $b(2a+b)$ b) $b(3a^2+3ab+b^2)$
c) $b(2a+b)(2a^2+2ab+b^2)$

2.1 Exercises, page 78

1. a) $y'=1$ b) $y'=3$ c) $y'=3$ d) $y'=m$

2. a) $y'=0$ b) $y'=0$

5. a) $y'=2x$, this is equal to the derivative of $y=x^2$.

6. $f'(x)=4x-3$

7. a) $y'=2x-5$ b) $y'=2x+1$
c) $f'(x)=8x+2$ d) $f'(x)=2x-1$
e) $g'(x)=4x+1$ f) $g'(x)=3+4x$

8. a) $y'=10x$; the derivative is 5 times the derivative of $y=x^2$.
b) $y'=-2x$
c) The derivative of $y=cf(x)$ is $y=cf'(x)$.

9. a), b) $f'(x)=2x+2$

11. $y'=3x^2$

12. $y'=4x^3$

13. $y'=5x^4$; $y'=6x^5$

15. a) $y'=2x-1$
b)

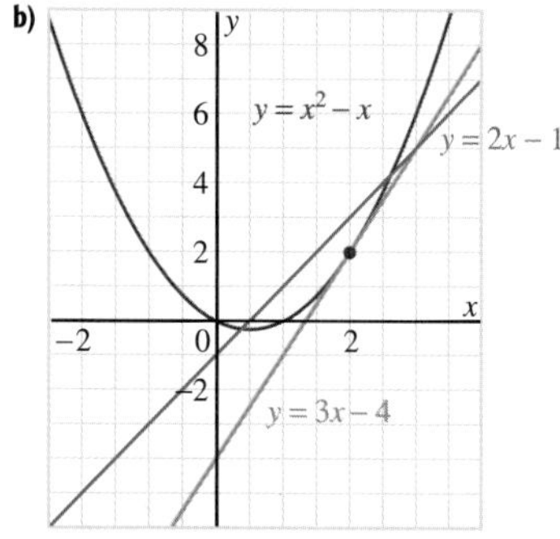

c) $y=3x-4$

16. a) $y'=-2x$
b)

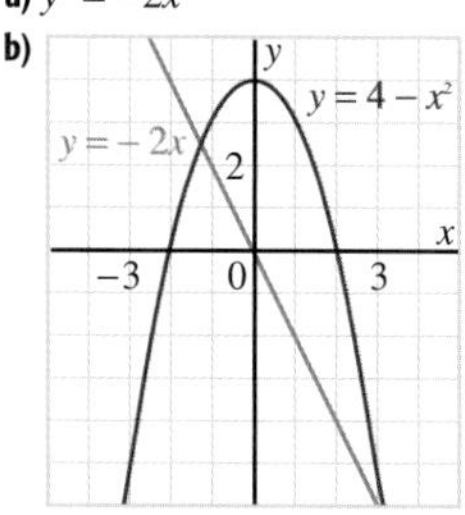

c) -6; this is the slope of the tangent to the graph of $y=4-x^2$ at the point on the graph where $x=3$.

17. Diagrams may vary.
a) False b) False c) False

2.2 Exercises, page 85

1. a) $y'=5x^4$ b) $y'=10x^9$ c) $f'(x)=17x^{16}$

2. a) $9x^8$ b) $20x^{19}$ c) 1

4. a) $y'=3x^2$; $y'=3x^2$; $y'=3x^2$; $y'=3x^2$
b) $y'=4x^3$; $y'=4x^3$; $y'=4x^3$; $y'=4x^3$
c) $y'=3x^2$; $y'=3(x-1)^2$; $y'=3(x-2)^2$; $y'=3(x-3)^2$

5. a) $y=\frac{3}{4}x-\frac{1}{4}$ b) $(-1,-1)$

7. a) i) Graph A is the function. Graph B is the derivative.
ii) n is even.
b) i) Graph B is the function. Graph A is the derivative.
ii) n is odd.

8. a) Yes b) Yes

9. a) Yes b) Yes

10. a) Graphs may vary.
b) n is odd:
$y'=nx^{n-1}$ for $x\ge 0$ (where $x=0$ exists)
$y'=-nx^{n-1}$ for $x<0$
n is even:
$y'=nx^{n-1}$ for all x-values

2.3 Exercises, page 91

1. a) $y'=8x$ b) $y'=-10x^4$
c) $y'=3x^2+1$ d) $y'=4x^3$

2. a) $y'=20x^4-6x^2$ b) $y'=14x^6+12x^3-5$
c) $f'(x)=8x^3-14x-1$ d) $f'(x)=-3+30x^4$

3. a) $y'=6x-2$ b) $y'=24x^3-6x^2+1$
c) $f'(x)=40x^7+6x$ d) $f'(x)=-4+27x^2$

4. $\frac{dA}{dr}=2\pi r$; the instantaneous rate of change of the area with respect to the radius equals the circumference.

5. $\frac{dV}{dr}=4\pi r^2$; the instantaneous rate of change of the volume with respect to the radius equals the surface area.

6. a) $y'=2x-5$ b) $y'=10-12x$
c) $y'=12x+5$ d) $y'=8x+12$

7. a) $y'=1+\cos x$ b) $y'=6x^2-3\sin x$
c) $f'(x)=\cos x-\sin x$ d) $f'(x)=5\cos x+2\sin x$

9. a) The points have the common property that their x-coordinates are 2.
b) $y'=2ax$
c) The points of intersection of the graphs of a function and its derivative satisfy both equations.

10. a) Yes. The points of intersection for $y=ax^3$ are the origin and the points with x-coordinate 3. The points of intersection for $y=ax^4$ are the origin and the points with x-coordinate 4.
b) Yes. The points of intersection are the origin and the points with x-coordinate n.

11. −1.3 mm/s; −0.52 mm/s; −0.13 mm/s

12. a) $d'=9.8t$ b) 9.8 m/s; 19.6 m/s

c) **Derivative of Distance Fallen**

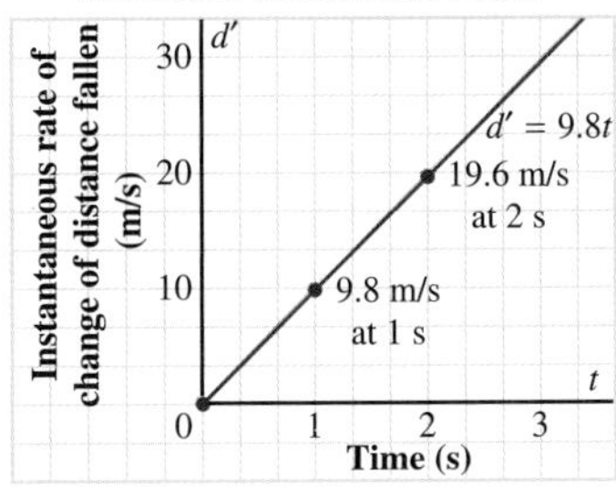

13. a) $h' = -9.78t + 8.09$
d) 8.09 m/s

14. a), b) $y' = 2(x - 3)$

15. a) Yes, $y' = 3(x + 5)^2$
b) No, $y' = 8x - 4$
c) No, $y' = 6x^5 - 6x^2$

16. a)

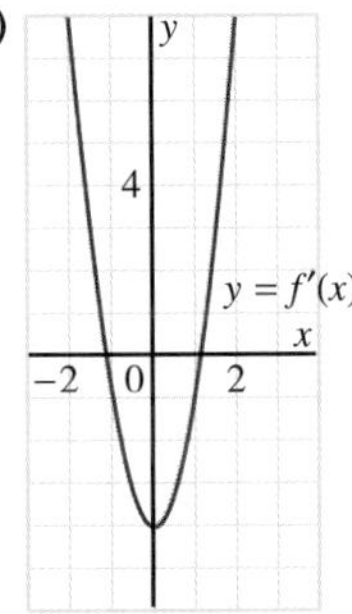

b) $f'(x) = 3x^2 - 4$

17. a) $y = 4x$ **b)** $y = 4x - \frac{32}{27}$

18. a) $f'(x) = 4x^3 + 4x$; $g'(x) = 6x^5 + 12x^3 + 6x$
b) $h(x) = 2x$; $k(x) = 2x$
c) The functions h and k are the derivative of x^2, which is a term in the equations for f and g.
d) $y' = n(f(x))^{n-1} \times f'(x)$

Self-Check 2.1–2.3, page 95

1. a) $y' = 1$ **b)** $y' = 2$ **c)** $y' = 3$ **d)** $y' = -3$

2. a) $y' = 2x + 2$ **b)** $y' = 2x - 1$
c) $f'(x) = 4x - 3$ **d)** $g'(x) = -2x - 1$

3. a) $y' = 3x^2$ **b)** $y' = 7x^6$
c) $f'(x) = 5x^4$ **d)** $f'(x) = 4(x - 2)^3$

4. a) $8x^7$ **b)** $12x^{11}$
c) $10x^9$ **d)** $7(x - 3)^6$

5. a) $y' = 2x$; $y' = 2x$; $y' = 2x$; $y' = 2x$
b) $y' = 4x^3$; $y' = 4x^3$; $y' = 4x^3$; $y' = 4x^3$
c) $y' = 5x^4$; $y' = 5(x - 1)^4$; $y' = 5(x - 2)^4$; $y' = 5(x - 3)^4$

6. $y = -32x - 48$

7. (1, 1)

8. a) $y' = 9x^2 - 2$ **b)** $y' = -8x^3 + 6x$
c) $y' = 4x + 7$ **d)** $y' = 18x + 30$

9. a)

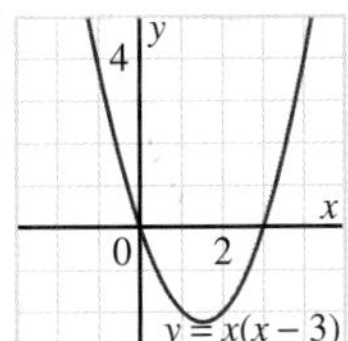

b)

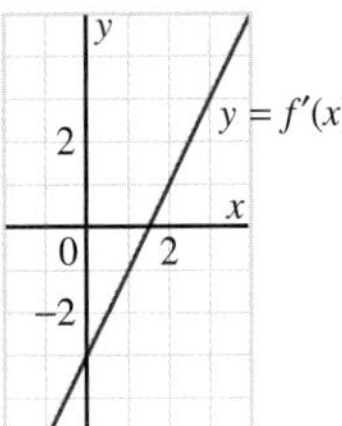

10. a) $y' = -2x + 2$

b)

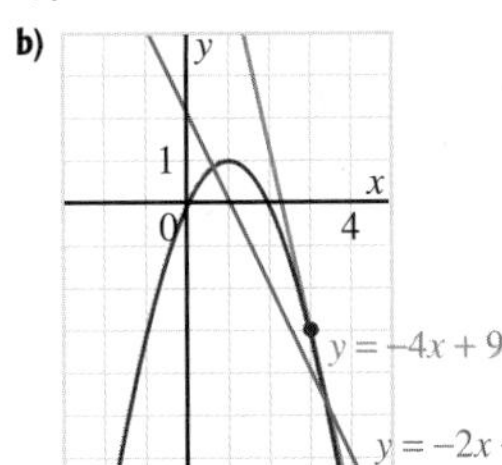

c) $y = -4x + 9$

2.4 Exercises, page 100

1. a) $f'(x) = \frac{4}{(1 - 2x)^2}$ **b)** $f'(x) = \frac{6}{(1 - 2x)^2}$
c) $f'(x) = \frac{-1}{(1 - 2x)^2}$ **d)** $f'(x) = \frac{5}{(1 - 2x)^2}$

2. a) i) -16 **ii)** -1
iii) $-\frac{1}{4}$ **iv)** $-\frac{1}{16}$

b)

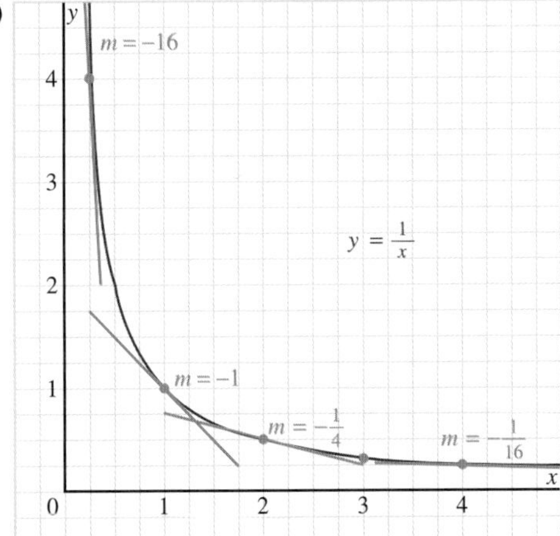

3. a) $f'(x) = \frac{-1}{(x - 1)^2}$ **b)** $f'(x) = \frac{-1}{(x - 1)^2}$

4. a) When two different functions have the same derivative, their graphs are parallel.
b) The graph of $y = \frac{1}{x - 1}$ is the image of the graph of $y = \frac{x}{x - 1}$ after a vertical translation of -1 unit.

5. 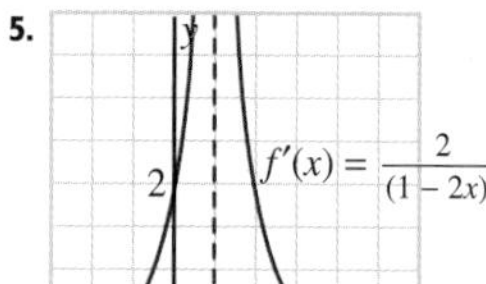

6. a) $y' = 12x^3 - \frac{1}{x^2}$ b) $y' = 4x + \frac{3}{x^2}$
c) $f'(x) = 4x + 3 - \frac{4}{x^2}$ d) $f'(x) = -\frac{2}{x^2} - 6 + 9x^2$

7. a) $y' = \frac{3}{(x+3)^2}$ b) $y' = \frac{3}{(x+3)^2}$
c) $y' = \frac{6}{(x+3)^2}$

8. The sum of the functions in parts a and b is equal to the function in part c. The sum of the derivative functions in parts a and b is equal to the derivative function in part c.

9. $y = 2x - 3$

10. a) $P' = -\frac{79\,500}{V^2}$; approximately –0.5 kPa/mL

b)

Volume (mL)	Instantaneous rate of change of pressure (kPa/mL)
100	–7.95
200	–1.9875
300	$-0.88\overline{3}$
400	–0.496 875
500	–0.318
600	$-0.220\,8\overline{3}$
700	–0.162 244 9
800	–0.124 218 75

11. $y = \frac{1}{9}x - \frac{8}{9}$

12. a) $\frac{-2}{(x-1)^2}$ b) $\frac{-1}{(2x-1)^2}$ c) $\frac{-4x}{(x^2-1)^2}$

13. No

14. Yes

15. a) $y = -x + 2$ b) $y = -\frac{1}{16}x + \frac{1}{2}$
c) $y = -16x + 8$

16. a) i) $y = -\frac{1}{9}x + \frac{2}{3}$; $y = -\frac{1}{9}x - \frac{2}{3}$
ii) Not possible iii) Not possible

17. a) $y' = \frac{-x^2+1}{(x^2+1)^2}$ b) $y' = \frac{x^2+2x}{(x+1)^2}$
c) $y' = \frac{-2x-1}{(x^2+x)^2}$

18. a) 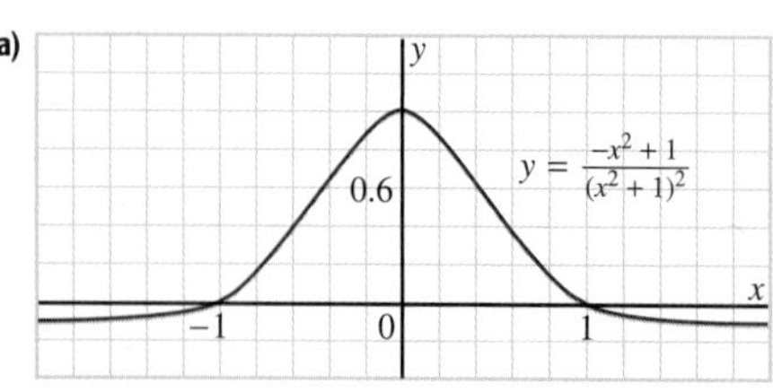

b) 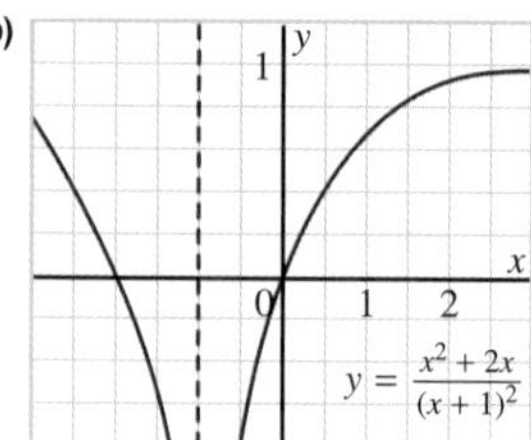

c) 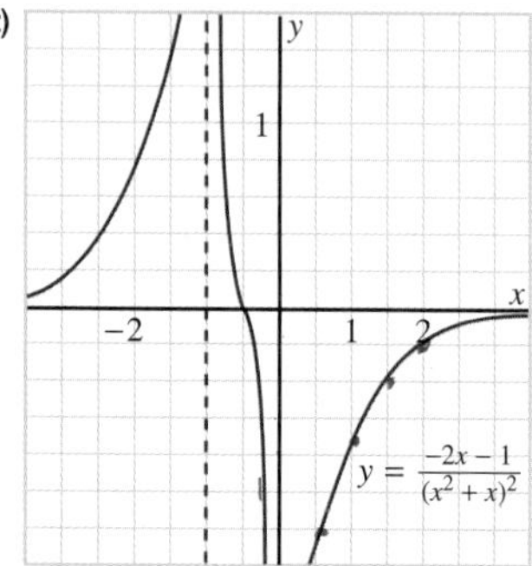

19. a) $y = -4x + 9$ and $y = -4x + 1$
b) Yes

20. $f'(x) = \frac{1}{x^2}$

21. a) $y = mx \pm \frac{2}{x}$, $m < 0$ b) $y = \frac{-x}{x_1^2} + \left(y_1 + \frac{1}{x_1}\right)$

22. a) $y' = \frac{-2}{x^3}$ b) $y' = \frac{-3}{x^4}$ c) $y' = \frac{-n}{x^{n+1}}$

23. $f'(x) = \frac{-2}{(x+1)^3}$

2.5 Exercises, page 105

1. a) $f'(x) = \frac{1}{\sqrt{x}}$ b) $f'(x) = \frac{3}{2\sqrt{x}}$
c) $f'(x) = 1 + \frac{1}{2\sqrt{x}}$ d) $f'(x) = 2x - \frac{1}{2\sqrt{x}}$

2. a) $f'(x) = \frac{4}{\sqrt{2x+1}}$ b) $f'(x) = 1 - \frac{1}{\sqrt{2x+1}}$

3. a) $y' = \frac{5}{2\sqrt{x}}$ b) $y' = 2x + \frac{1}{\sqrt{x}}$
c) $f'(x) = \frac{1}{2\sqrt{x}} - \frac{1}{x^2}$

4. a) $f'(x) = \frac{1}{2\sqrt{x+1}}$ b) $f'(x) = \frac{-1}{\sqrt{1-2x}}$

5. a) f is defined when $x = 0$ because $\sqrt{0}$ is in the numerator. f' is not defined when $x = 0$ because $\sqrt{0}$ is in the denominator.

b) 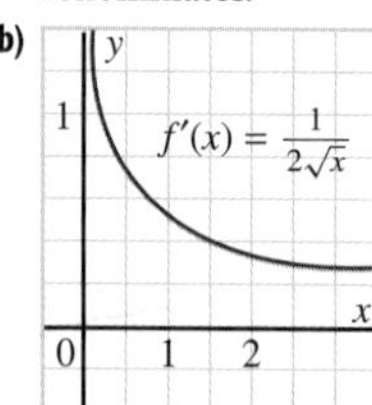

6. No

7. a) $y = \frac{1}{2}x + \frac{1}{2}$ b) $y = \frac{1}{4}x + 1$ c) $y = \frac{1}{6}x + \frac{3}{2}$

Self-Check 2.4, 2.5, page 106

1. a) $f'(x) = \frac{-4}{(1+2x)^2}$ **b)** $f'(x) = \frac{-1}{(1+2x)^2}$
c) $f'(x) = \frac{4}{(1+2x)^2}$

2. a) $f'(x) = \frac{2}{(1+2x)^2}$ **b)** $f'(x) = \frac{-2}{(1+2x)^2}$
c) $f'(x) = \frac{-1}{3(1+2x)^2}$

3.

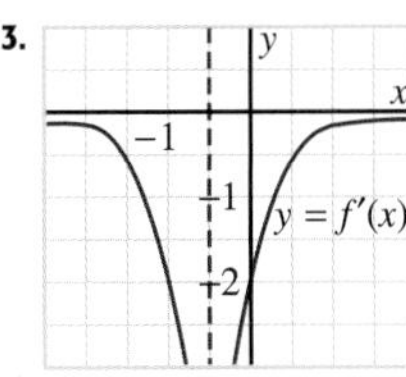

4.

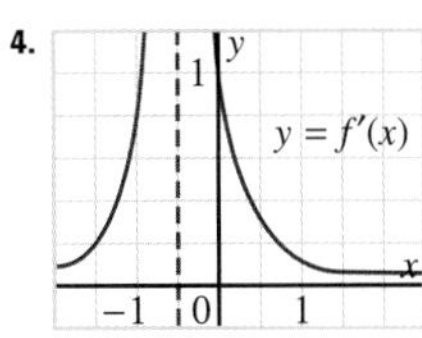

5. a) $y' = 4x + \frac{1}{x^2}$ **b)** $y' = 3x^2 - 2x - \frac{1}{x^2}$
c) $y' = -8x - 1 + \frac{3}{x^2}$

6. a) $y' = \frac{-1}{(x+1)^2}$ **b)** $y' = \frac{1}{(x+2)^2}$
c) $y' = \frac{-3}{(x+3)^2}$

7. a) $y' = \frac{-1}{\sqrt{x}}$ **b)** $y' = \frac{1}{2\sqrt{x}} - 9x^2$
c) $y' = \frac{2}{\sqrt{x}} - 4x^3 + 4x$

8. a) $y = -x - 2$
b)

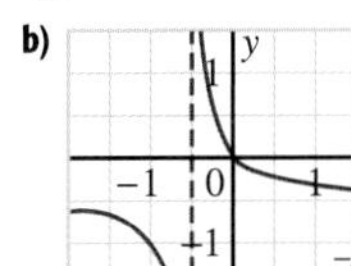

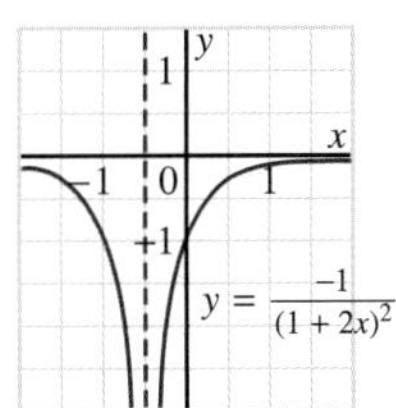

Chapter 2 Review Exercises, page 108

1. a) $y' = 4$ **b)** $y' = 4x + 5$ **c)** $y' = 56x - 7$ **d)** $y' = 0$

2. a) $y' = 2x + 1$
b)

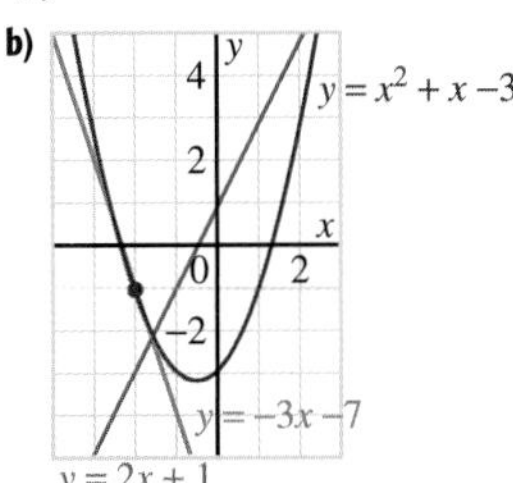

c) $y = -3x - 7$

3. $y = 4x - 7$

4. a) $y' = 6x^5$ **b)** $y' = 15x^{14}$ **c)** $y' = 30x^{29}$

5. $y = 4x - 3$

6. a) $y = -6x - 5$ **b)** No

7. a) $y' = 14x + 3$ **b)** $y' = -3x^2 - 5$
c) $y' = 88x + 115$
d) $y' = 36x^8 + 42x^6 - 30x^5 + 3$

9. a) $y' = 24x^2 - 72x + 54$ **b)** $y' = \frac{2}{5}x^3 - \frac{1}{5}$
c) $y' = 7\cos x - 2\sin x$ **d)** $y' = \sin x + 2\cos x$

10. a) $y' = 4(x+2)^3$
b) $y' = 4x^3 + 24x^2 + 48x + 32$

11. a)

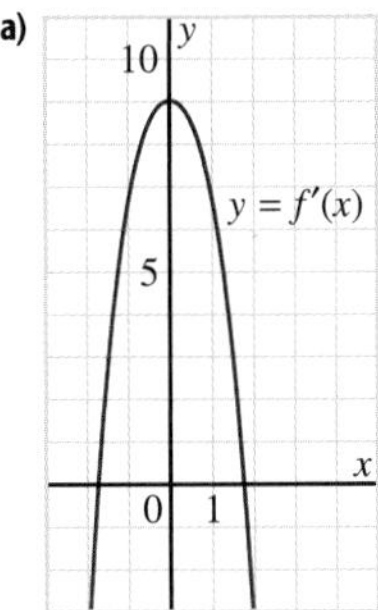

b) $f'(x) = 9 - 3x^2$

12. a) $y = -7x - 4$ **b)** $y = -7x + \frac{392}{27}$

13. a) $y' = \frac{-3}{(x+2)^2}$ **b)** $y' = \frac{4x^2 - 4x}{(2x-1)^2}$

14.

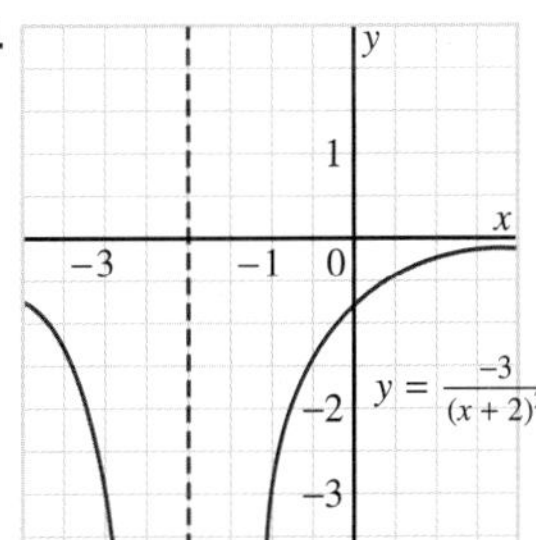

15. a) $y' = 4 + \frac{5}{x^2}$ **b)** $y' = 4x - \frac{1}{4x^2}$
c) $y' = 4 + \frac{15}{x^2}$ **d)** $y' = -\frac{3}{2x^2} + \frac{7}{6x^2}$

16.

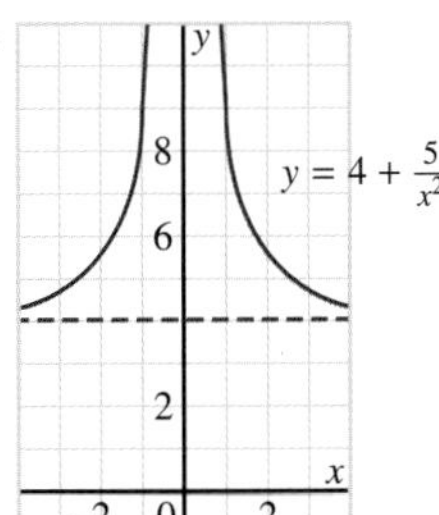

17. $y = -3x - 15$

18. a) $y = 2x + 1;\ y = 2x - 3$ **b)** Yes

19. a) Not possible
b) $y = -\frac{1}{4}x - \sqrt{2};\ y = -\frac{1}{4}x + \sqrt{2}$

20. a) $y' = \frac{-2}{\sqrt{x}}$ **b)** $y' = 3 - \frac{1}{2\sqrt{x}}$
c) $y' = 4x - \frac{5}{2\sqrt{x}}$ **d)** $y' = \frac{-1}{2\sqrt{x}} + \frac{3}{x^2}$

21. a) $y' = \frac{-2}{\sqrt{3-x}}$ **b)** $y' = \frac{4}{\sqrt{8x+2}}$

22. $y = x + 1$

23. a) $(0, 1)$ **b)** No

24. a) $y = 2x - \frac{3}{4}$ **b)** No

Chapter 2 Self-Test, page 111

1. a) $y' = 4x - 1$ **b)** $y' = \frac{-1}{(1-x)^2}$

2. a) $y' = -10x + 6$ **b)** $y' = -\frac{1}{x^2} - 5 + 3x^2$

c) $y' = 4x^3 - \frac{3}{2\sqrt{x}} + 6x^2$ **d)** $y' = -\frac{2}{x^2} - 2x^3$

3. a) $y = 7x - 4$ **b)** $y = \frac{-1}{16}x + \frac{9}{16}$

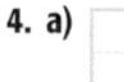

4. a)

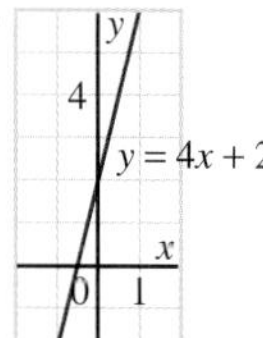

b)

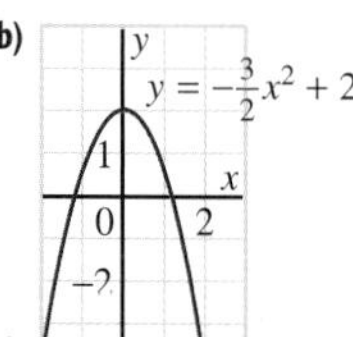

5. a) $\frac{dh}{dt} = -9.8t + 30$; the instantaneous rate of change of the ball's height with respect to time

b) 20.2 m/s; the ball is 25.1 m above the ground after 1 s.

6. $y' = 4x$; explanations may vary.

7. $y = 8x - 25$; $y = -8x - 9$

8. a) $(1, -2)$; $\left(\frac{1}{3}, -\frac{10}{27}\right)$ **b)** Yes

9. a) i) $\frac{1}{2}$ **ii)** $\frac{1}{2\sqrt{2}}$ **iii)** $\frac{1}{4}$ **iv)** 1

b) i) At $x = 1$, the slope of the tangent is $\frac{1}{2}$.

ii) At $x = 2$, the slope of the tangent is $\frac{1}{2\sqrt{2}}$.

iii) At $x = 4$, the slope of the tangent is $\frac{1}{4}$.

iv) At $x = 0.25$, the slope of the tangent is 1.

10. a) $y = x + \frac{1}{4}$ **b)** $y = 2x + \frac{1}{8}$ **c)** $y = 3x + \frac{1}{12}$

11. a) $f'(x) = \lim_{h \to 0} \frac{f(x+h) - f(x)}{h}$

b) i) $f'(x) = 4x$ **ii)** $f'(x) = 4x - 3$

iii) $f'(x) = \frac{-2}{(2x+1)^2}$

12. a)

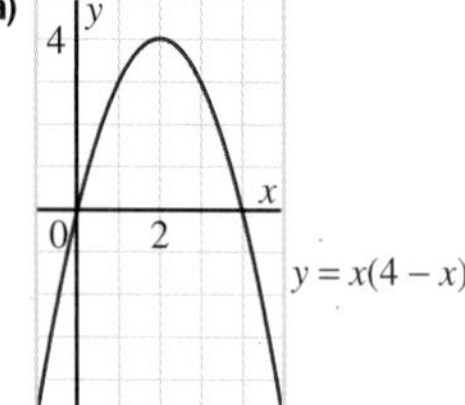

b) $y = 4x$; $y = -4x + 16$ **c)** 32 square units

d) The tangent at (0, 0) is a median in the triangle described in part c.

13. a) $y' = 5x^4$ **b)** $y' = 6x^5$

14. b) i) $f'(x) = 2x$ **ii)** $f'(x) = 8x + 1$ **iii)** $f'(x) = 6 - 2x$

Mathematical Modelling: Optimal Cooking Time

Exercises, page 115

Answers determined from graphs may vary.

1. b) $T = 10 + 4t$

2. a) 16 min

b)

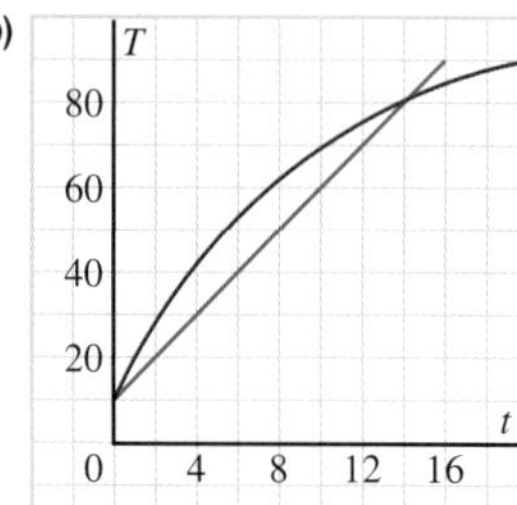

c) Approximately 13 min

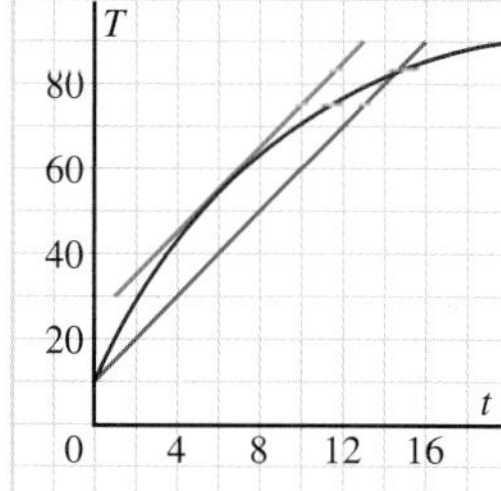

3. a)

b) Minimum cooking time is 12 min.

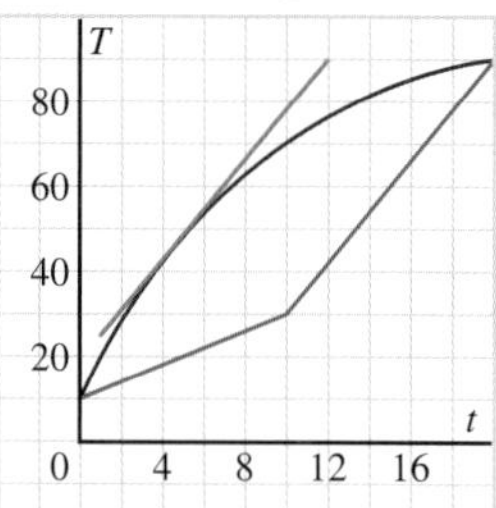

4. a) $2.80

b) Time is 14.5 min at approximately 82°C. Minimum cost is \$1.85.

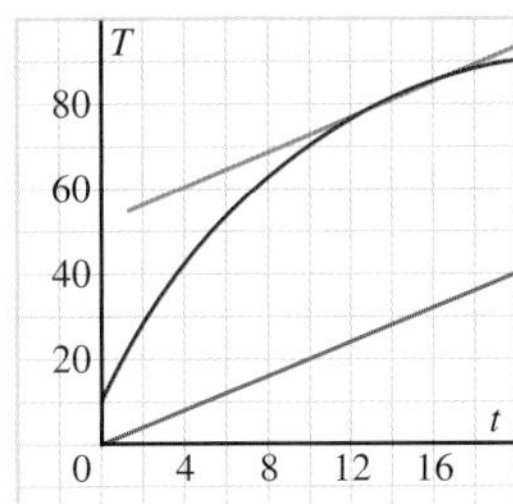

5. a)

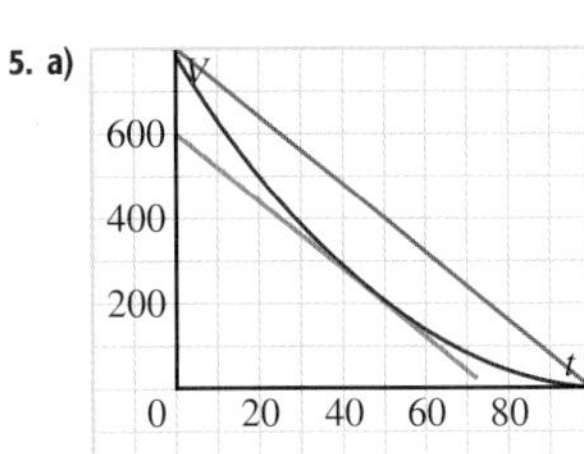

b) 75 min

c) $V = 0.08(t - 100)^2$

d) $\frac{dV}{dt} = 0.16(t - 100)$

Chapter 3 Polynomial Functions

Necessary Skills

1 Review: Quadratic Functions

Exercises, page 120

1. a) i) $(1, 3)$ **ii)** 3 **iii)** $x = 1$

iv)

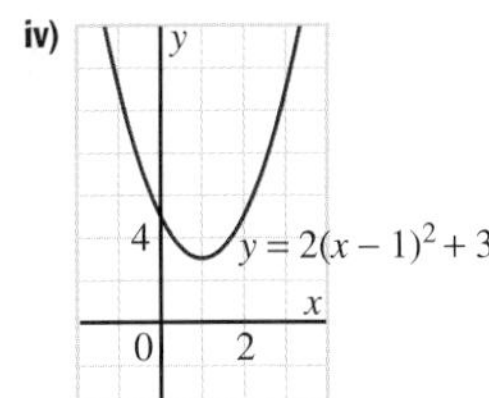

v) $D = \Re$; $R = \{y \mid y \in \Re,\ y \geq 3\}$

b) i) $(-4, -5)$ **ii)** -5 **iii)** $x = -4$

iv)

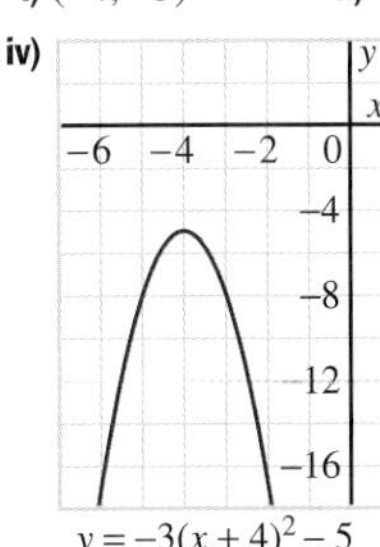

v) $D = \Re$; $R = \{y \mid y \in \Re,\ y \leq -5\}$

c) i) $(-1, -64)$ **ii)** -64 **iii)** $x = -1$

iv)

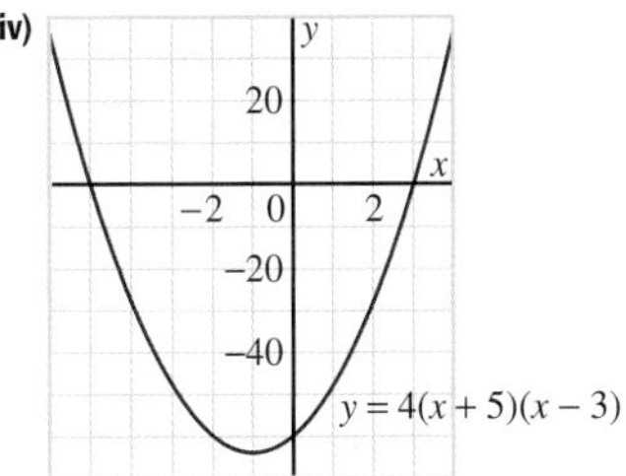

v) $D = \Re$; $R = \{y \mid y \in \Re,\ y \geq -64\}$

2. a) i) The function is increasing for $x > 1$ and decreasing for $x < 1$.

b) i) The function is decreasing for $x > -4$ and increasing for $x < -4$.

c) i) The function is increasing for $x > -1$ and decreasing for $x < -1$.

3. a) $y = -(x - 2)^2 + 4$; $D = \Re$; $R = \{y \mid y \in \Re,\ y \leq 4\}$

b) $y = 2(x + 3)^2 + 2$; $D = \Re$; $R = \{y \mid y \in \Re,\ y \geq 2\}$

2 New: Solving Quadratic Inequalities

Exercises, page 124

1. a) i) $-2 < x < 3$ **ii)** $x \leq -2$ or $x \geq 3$

b) i) $x \leq -1$ or $x \geq 3$ **ii)** $-1 < x < 3$

2. a) $x < -5$ or $x > 2$ **b)** $0 < x < 5$

c) $n \leq -4$ or $n \geq 1$ **d)** $-1 \leq x \leq 2$

3. a) $x < -3$ or $x > -2$

–5 –4 –3 –2 –1 0 1

b) $m < -2$ or $m > 4$

–4 –2 0 2 4

c) $y < -6$ or $y > 3$

–6 –4 –2 0 2 4

d) $-1 < x < 3$

–3 –2 –1 0 1 2 3 4

e) $1 \leq x \leq 9$

–2 0 2 4 6 8 10

f) $-2.5 < x < 3$

–4 –2 0 2 4

4. a) $x \neq 2,\ x \in \Re$ **b)** $x \leq -0.5$ or $x \geq 0.5$

c) $x \leq -0.8$ or $x \geq 2.1$ **d)** $-\frac{5}{2} < x < \frac{4}{3}$

e) $x \in \Re$ **f)** No solution

g) $x = \frac{2}{3}$ **h)** $-3.1 \leq x \leq 0.6$

5. a) $x < -4.4$ or $x > -1.6$ **b)** $-0.6 < x < 2.6$

c) $x < -6.9$ or $x > -5.1$

3 New: Dividing a Polynomial by a Binomial

Exercises, page 127

1. a) $2x^2 - x + 5 = (x + 3)(2x - 7) + 26;\ x \neq -3$

b) $x^3 - x - 10 = (x + 4)(x^2 - 4x + 15) - 70;\ x \neq -4$

c) $-10x^3 + x - 8 + 21x^2 = (x - 2)(-10x^2 + x + 3) - 2;\ x \neq 2$

d) $4x^3 - 10x^2 + 6x - 18 = (2x - 5)(2x^2 + 3) - 3;\ x \neq \frac{5}{2}$

2. a) $6x^3 + 31x^2 + 3x - 10 = (2x - 1)(3x + 2)(x + 5)$
b) $2x^3 + 3x^2 - 8x + 3 = (2x - 1)(x + 3)(x - 1)$
c) $2x^3 - x^2 - 8x + 4 = (2x - 1)(x + 2)(x - 2)$
d) $4x^3 + 4x^2 - x - 1 = (2x - 1)(2x + 1)(x + 1)$

3. $5x^3 - 8x^2 + 11x + 5$

4. $6x^4 - 8x^3 - 3x^2 + 7x + 4$

5. a) $2m^3 - 3m^2 - 8m - 3 = (2m + 1)(m - 3)(m + 1)$
b) $-a^4 + 3a^2 - a + 1 = (a + 2)(-a^3 + 2a^2 - a + 1) - 1$
c) $c^3 - 39c + 70 = (c - 2)(c + 7)(c - 5)$
d) $2x^3 - x^2 - 13x - 6 = (2x + 1)(x + 2)(x - 3)$
e) $10x^3 - 21x^2 + 6 - x = (5x - 3)(2x + 1)(x - 2)$
f) $5x^4 + x^2 - 3 + 4x = (x - 1)(5x^3 + 5x^2 + 6x + 10) + 7$

4 New: Factoring by Grouping

Exercises, page 128

1. a) $(x - 3)(x^2 + 2)$ **b)** $(x + 2)(x^2 - 5)$
c) $(x + 1)(x^2 - 6)$ **d)** $(x - 1)(-x^2 - 2)$
e) $(x + 4)(x^2 + 1)$ **f)** $(x - 8)(x^2 - 3)$

2. a) $(2x + 1)(x^2 - 2)$ **b)** $(2x - 1)(x^2 + 3)$
c) $(2x - 1)(-x^2 - 4)$ **d)** $(2x - 3)(x^2 + 2)$
e) $(3x + 2)(x^2 - 3)$
f) $(4x + 1)(x - 2)(x + 2)$

5 Review: Absolute Value

Exercises, page 130

1. a) 8 **b)** 3.9 **c)** $\frac{4}{3}$ **d)** 23

2. a) −6 −4 −2 0 2 4 6
b) −6 −4 −2 0 2 4 6
c) −6 −4 −2 0 2 4 6
d) −6 −4 −2 0 2 4 6

3. a) −6 −4 −2 0 2 4 6
b) −4 −2 0 2 4
c) −6 −4 −2 0 2 4 6
d) −6 −4 −2 0 2 4 6

3.1 Exercises, page 138

1. a) This is a polynomial function of degree 3.
b) This is a polynomial function of degree 2.
c) This is a polynomial function of degree 1.
d) This is not a polynomial function because it is a square-root function.
e) This is not a polynomial function because it has the form of a rational expression.
f) This is a polynomial function of degree 3.
g) This is a polynomial function of degree 2.
h) This is not a polynomial function because it contains a square-root function.

2. Graphs b and c could represent cubic functions.

3. Graphs a and d could represent quartic functions.

4. Graphs a and d represent functions with negative leading coefficients.

5. a) 1 **b)** 2 **c)** 4 **d)** 6

6. a) $g(x) = x^4 - 10x^2 - 5x + 5$
b) $k(x) = 5x^4 - 14x^3$
c) $f(x) = x^3 - 3x^2 - 5x + 16$

7. Graphs may vary. Examples follow.

a)
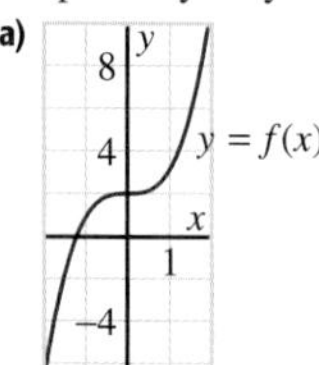

b)
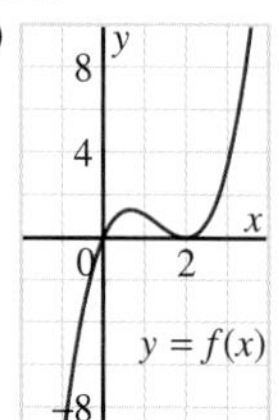

c)
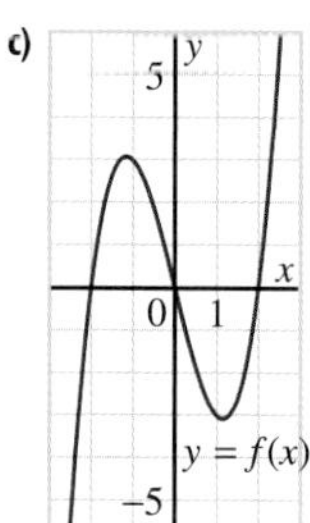

8. Graphs may vary. Examples follow.

a)
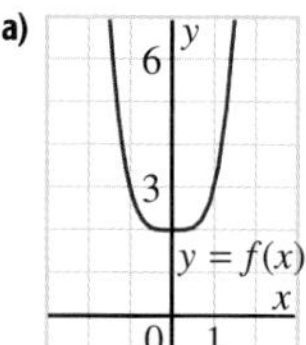

b)
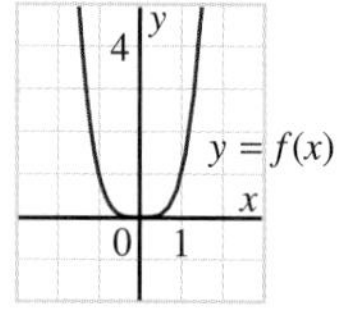

c)
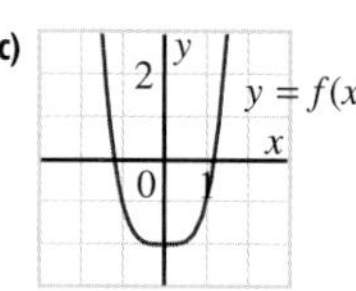

d)
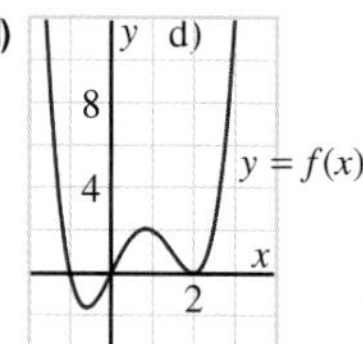

e)
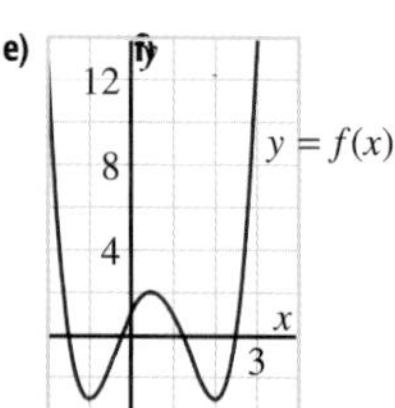

9. Graphs may vary. Examples follow.

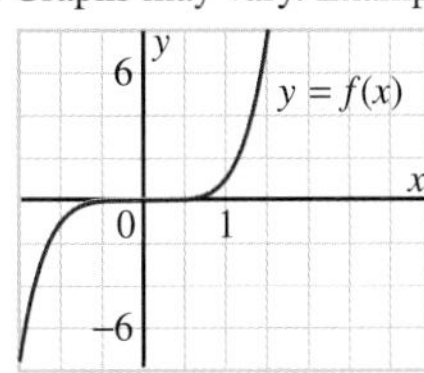

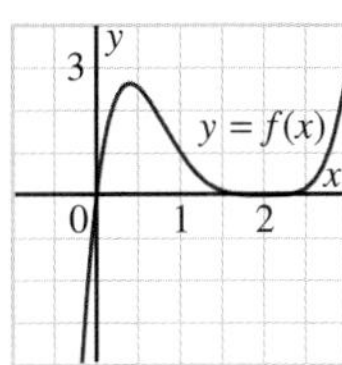

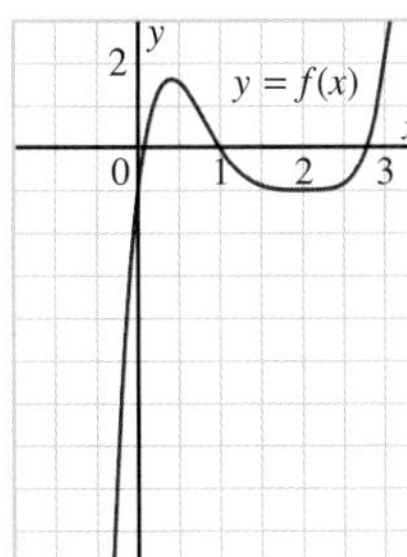

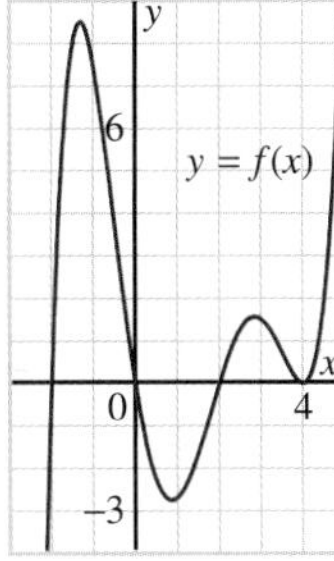

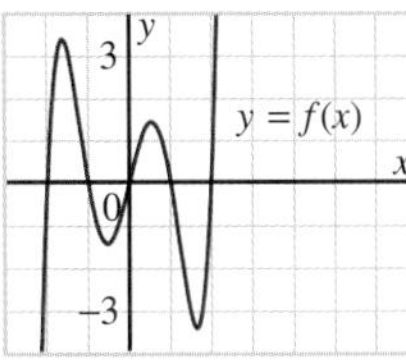

11. Graphs a, c, d, and f could be the graphs of polynomial functions.

12. a) Neither
b) Neither
c) Point symmetry
d) Line symmetry. When the graph is reflected in any line perpendicular to itself, the graph coincides with itself.
e) Line symmetry
f) Line symmetry

13. No

14. a) Point symmetry
b) Point symmetry
c) Line symmetry

15. a)

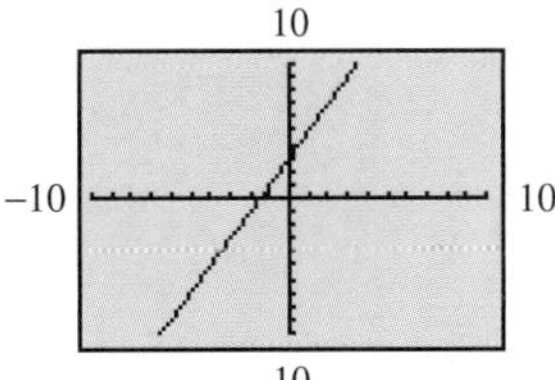

b), c)

x	y	First differences	Second differences
−4	−5		
		2	
−3	−3		0
		2	
−2	−1		0
		2	
−1	1		0
		2	
0	3		0
		2	
1	5		0
		2	
2	7		0
		2	
3	9		0
		2	
4	11		

d) i) The first differences are constant.
ii) Yes
e) i) All the second differences are 0.
ii) Yes
f) The first differences indicate that as x increases, the rate of change of y with respect to x is constant.
g) Linear functions have degree 1, which is why the first differences are constant.

16. a)

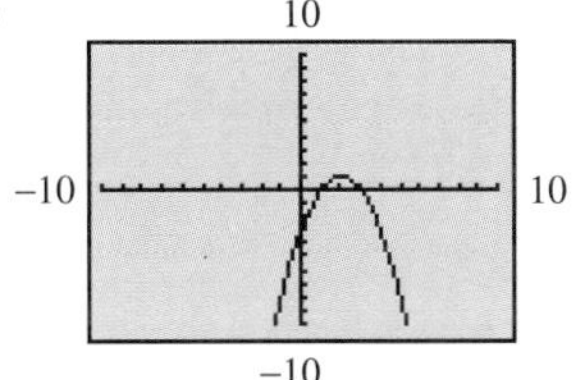

b)

x	y	First differences	Second differences
−4	−35		
		11	
−3	−24		−2
		9	
−2	−15		−2
		7	
−1	−8		−2
		5	
0	−3		−2
		3	
1	0		−2
		1	
2	1		−2
		−1	
3	0		−2
		−3	
4	−3		

c) i) The first differences decrease by the same number.
ii) Yes
d) i) The second differences are constant.
ii) Yes
e) The first differences indicate that the rate of change of y with respect to x is decreasing.
f) The second differences indicate that the rate of change of the rate of change of y with respect to x is constant.
g) Quadratic functions have degree 2, which is why the second differences are constant.

17. a)

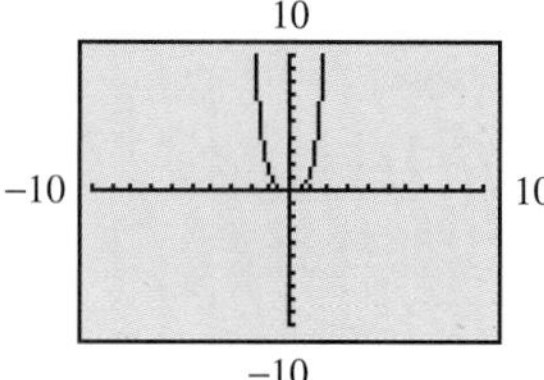

b)

x	y	1st dif.	2nd dif.	3rd dif.	4th dif.
−4	256				
		−175			
−3	81		110		
		−65		−60	
−2	16		50		24
		−15		−36	
−1	1		14		24
		−1		−12	
0	0		2		24
		1		12	
1	1		14		24
		15		36	
2	16		50		24
		65		60	
3	81		110		
		175			
4	256				

c) i) The first differences are increasing as x increases, so the slopes of the tangents are increasing.

ii) The second differences are positive, so the graph of the function is concave up.

d) The third differences are increasing. The fourth differences are constant.

e) Quartic functions have degree 4, which is why the fourth differences are constant.

19. Yes

20. a) i) All the graphs of these polynomial functions are transformations of $y = x^3$.

ii) $y = x^3 + 2$ is a vertical translation 2 units up.

iii) $y = (x - 2)^3$ is a horizontal translation 2 units right.

iv) $y = 2x^3$ is a vertical stretch by a factor of 2.

v) $y = -2(x + 1)^3 - 3$ is a reflection in the x-axis, vertical stretch by a factor of 2, horizontal translation 1 unit left, and a vertical translation 3 units down.

21. a) The graph of $y = -2(x - 5)^4 + 7$ is the image of the graph of $y = x^4$ after a reflection in the x-axis, a vertical stretch by a factor of 2, a horizontal translation 5 units right, and a vertical translation 7 units up.

b)

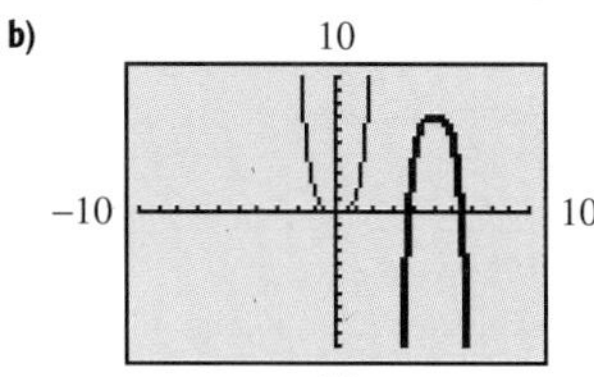

23. a) i)

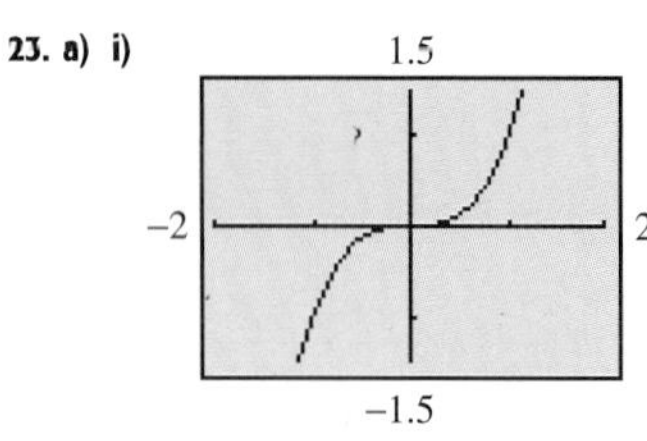

ii)

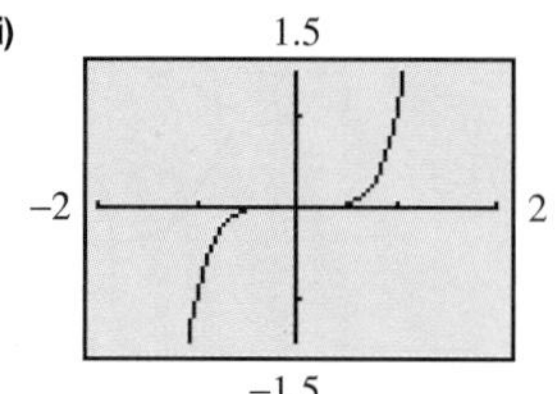

iii)

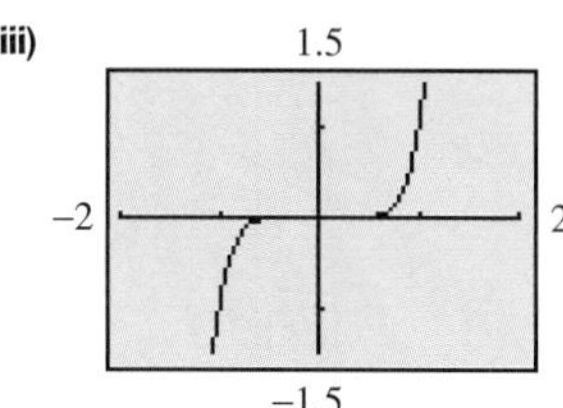

b) If n were a much greater odd number than those in part a, the graph of $y = x^n$ would have steeper branches and the branches would be farther apart.

c) As n increases through odd values of n, the graph of $y = x^n$ has steeper branches that are farther apart.

3.2 Exercises, page 148

1. a) The greatest number of zeros is 3 since it is a cubic function.

b) The greatest number of zeros is 2 since it is a quadratic function.

c) The greatest number of zeros is 3 since it is a cubic function.

d) The greatest number of zeros is 4 since it is a quartic function.

e) The greatest number of zeros is 5 since it is a quintic function.

f) The greatest number of zeros is 4 since it is a quartic function.

3. The function in part c, $f(x) = (x + 2)^2(x - 2)$, has the x-axis as a tangent at $x = -2$. The function does not change sign at $x = -2$ since this zero has order 2.

4. a) $y = a(x + 1)(x - 4)$ Functions may vary.
$y = 2(x + 1)(x - 4)$; $y = -3(x + 1)(x - 4)$

b) $y = \frac{3}{2}(x + 1)(x - 4)$

c)

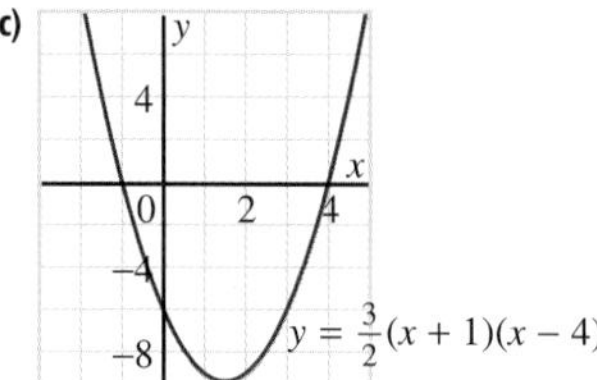

5. a)

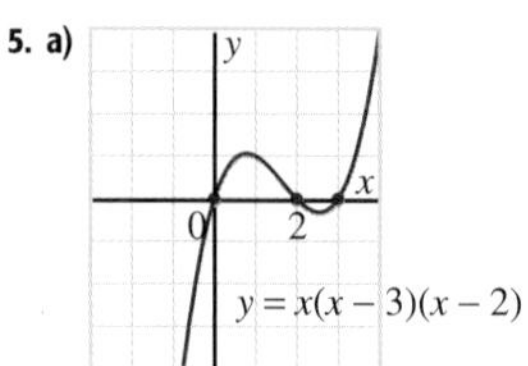

b)

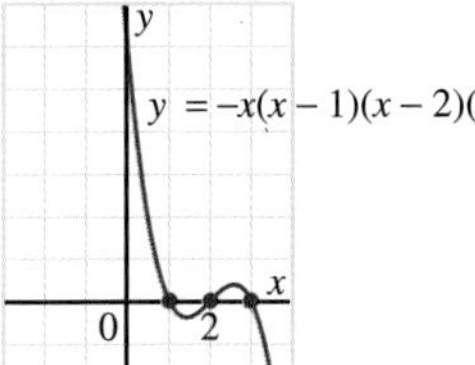

c)

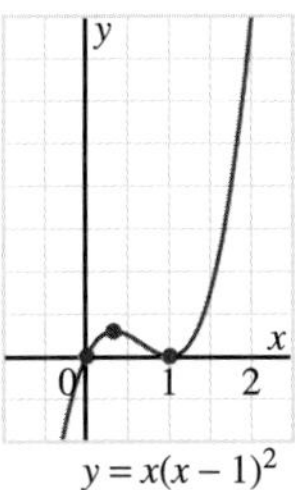

d)

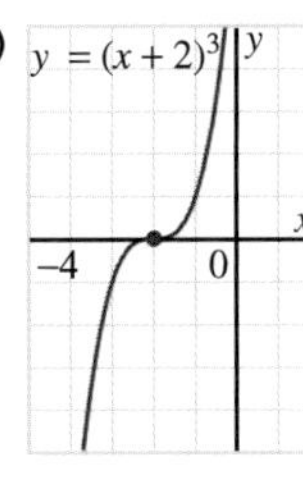

6. No equation is unique. Equations may vary.

a) $y = (x + 2)(x - 1)(x - 4)$

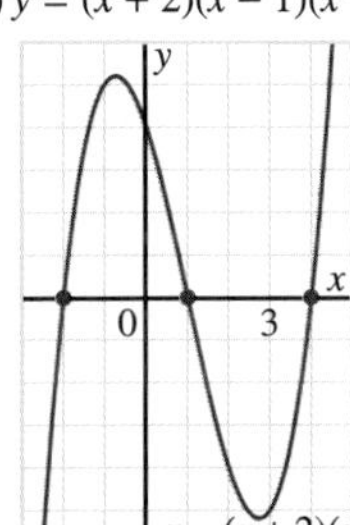

b) $y = (x - 3)^3$

c) $y = (x + 2)(x - 2)^2$

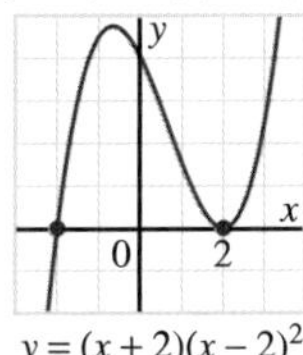

7. a)

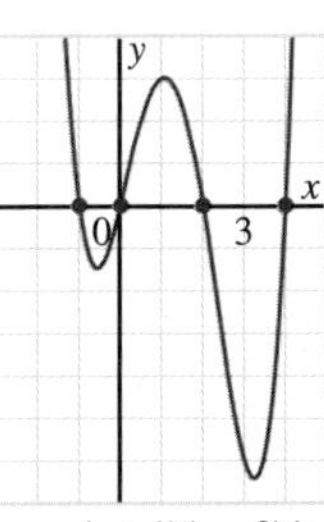

b)

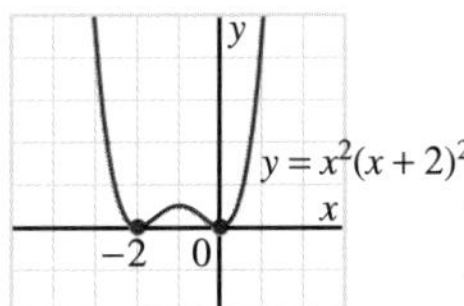

c)

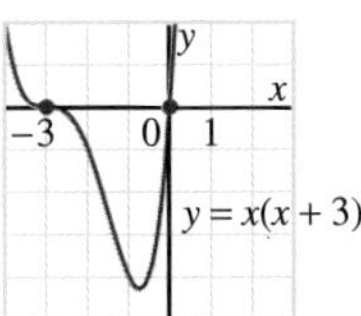

d)

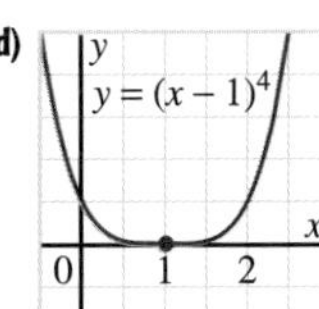

8. Answers may vary. Sample answers follow.

a) $y = (x + 3)(x + 2)(x - 1)(x - 4)$
$y = 2(x + 3)(x + 2)(x - 1)(x - 4)$
$y = -3(x + 3)(x + 2)(x - 1)(x - 4)$

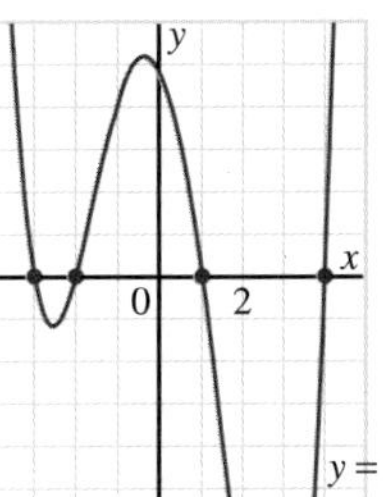

b) $y = (x + 1)^2(x - 2)^2$;
$y = 2(x + 1)^2(x - 2)^2$;
$y = -3(x + 1)^2(x - 2)^2$

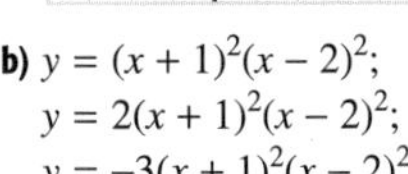

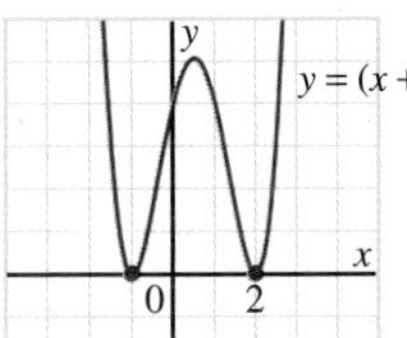

c) $y = x(x + 3)(x - 3)^2$;
$y = 2x(x + 3)(x - 3)^2$;
$y = -3x(x + 3)(x - 3)^2$

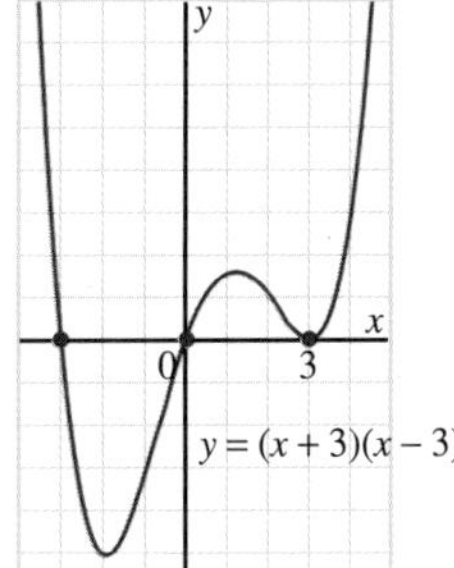

d) $y = (x - 3)^4$;
$y = 2(x - 3)^4$;
$y = -3(x - 3)^4$

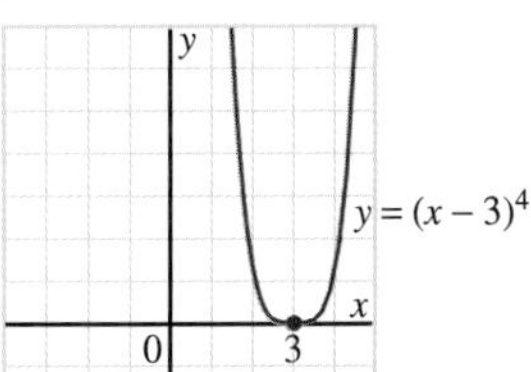

9. a) True **b)** False
c) True **d)** False

10. a) $y = 2(x + 1)(x - 3)$ **b)** $y = -\frac{1}{2}(x + 4)(x + 2)$

11. a)

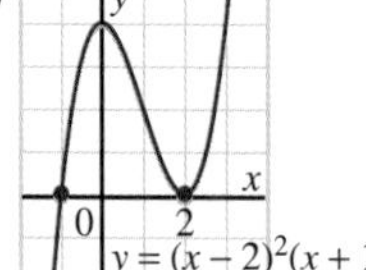

$y = (x - 2)^2(x + 1)$

b)

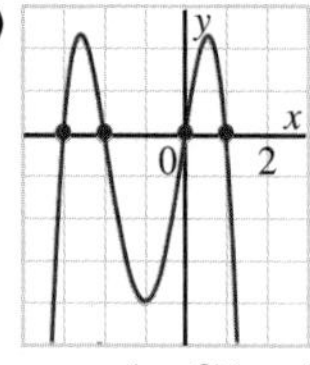

$y = -x(x + 3)(x - 1)(x + 2)$

c)

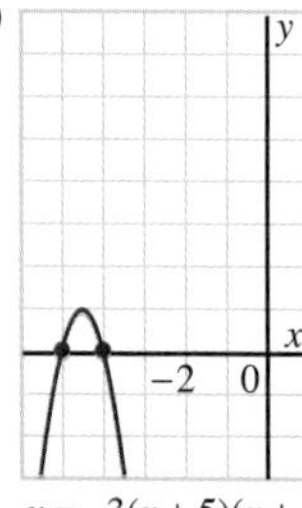

$y = -3(x + 5)(x + 4)$

d)

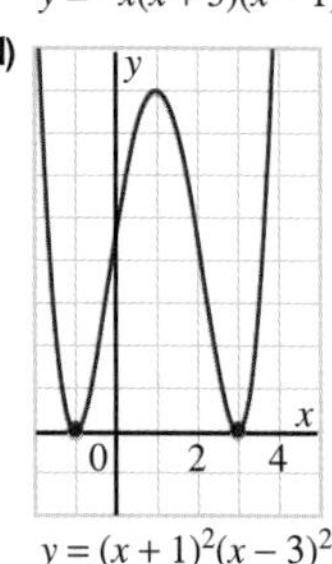

$y = (x + 1)^2(x - 3)^2$

12. a) The functions in parts b and c have no zeros of order 2 as none of their factors is raised to the power 2.
b) The function in part a has one zero of order 2 as the binomial $(x - 2)$ is raised to the power 2.
c) The function in part d has two zeros of order 2 as each of the two binomials $(x + 1)$ and $(x - 3)$ is raised to the power 2.

13. a) $y = -2(x + 3)(x + 1)(x - 2)$
b)

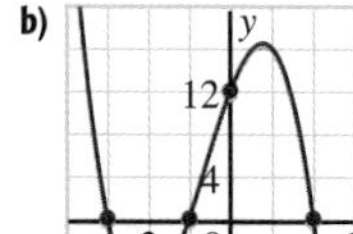

$y = -2(x + 3)(x + 1)(x - 2)$

14. a) $y = 3(x - 2)^2$

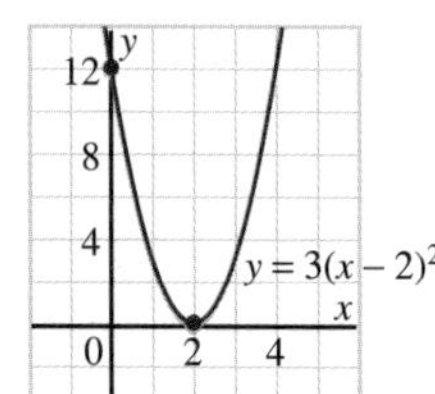

b) $y = 3(x + 2)(x - 1)(x - 4)$

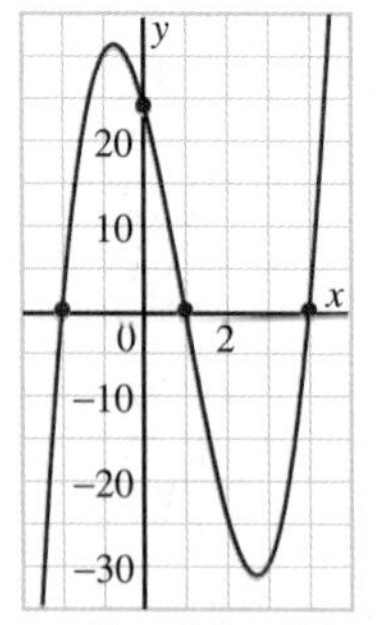

$y = 3(x + 2)(x - 1)(x - 4)$

c) $y = -2(x + 2)(x - 2)^2$

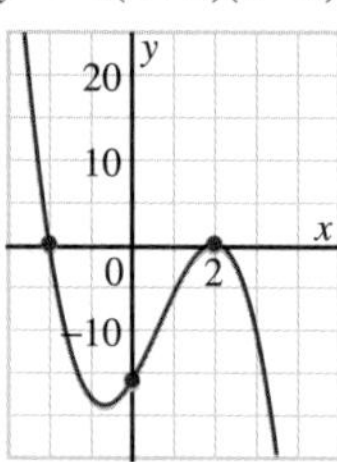

$y = -2(x + 2)(x - 2)^2$

d) $y = -3x(x - 2)(x - 4)$

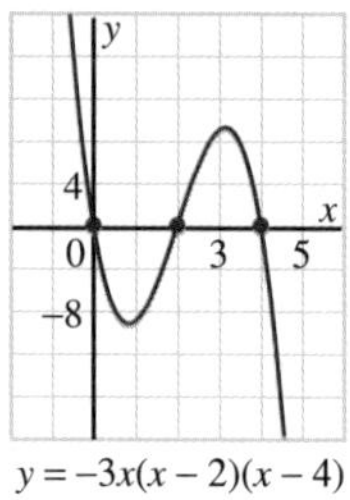

$y = -3x(x - 2)(x - 4)$

15. a) 2, 8 **b)** 0, 3, 4 **c)** –2, 0, 7

16. a) $y = -2(x - 1)(x - 2)(x - 3)$
b) $y = \frac{1}{3}(x + 3)(x - 2)^2$

17. a) $y = (x + 1)^2(x - 2)$
b) $y = -(x + 2)^2(x - 1)(x - 2)$
c) $y = -(x + 3)(x + 2)(x - 2)$
d) $y = \frac{1}{2}(x + 3)^2(x - 1)^2$

18. a) The function changes sign at both zeros shown on the screen, so neither zero has order 2. Since the function is cubic, a third zero exists.
b) The third zero is positive.
c) No

19. If a zero has an odd order, then the function changes sign at the zero. If a zero has an even order, then the function does not change sign at the zero.

20. a) When $k = 0$, $f(x) = x^3 - 2x^2$, which has a common factor of x^2. Therefore, $f(x) = 0$ has a double root of 0.
c) $k = \frac{32}{27}$

Self-Check NS, 3.1, 3.2, page 152

1. a) $x \le -3$ or $x \ge \frac{1}{2}$ **b)** $-\frac{2}{5} < x < 3$

2. a) Approximately $-0.4 < x < 2.7$
b) No solution

3. a) Quotient: $3x + 8$; remainder: 11; $x \ne 2$
b) Quotient: $x^2 - 8x + 16$; remainder: –28; $x \ne -2$
c) Quotient: $2x^2 + 3$; remainder: 0; $x \ne \frac{5}{2}$
d) Quotient: $x^4 + x^3 + x^2 + x + 1$; remainder: 0; $x \ne 1$

4. a) ii **b)** iii **c)** i

5. The graph in part ii has point symmetry about the origin.

6. a)

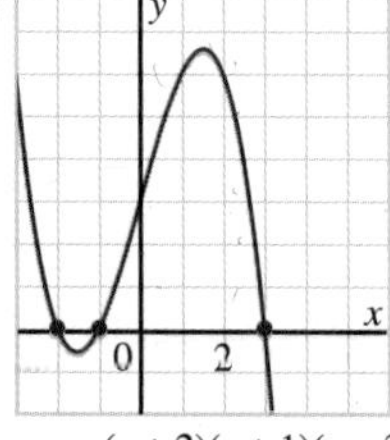

$y = -(x + 2)(x + 1)(x - 3)$

b)

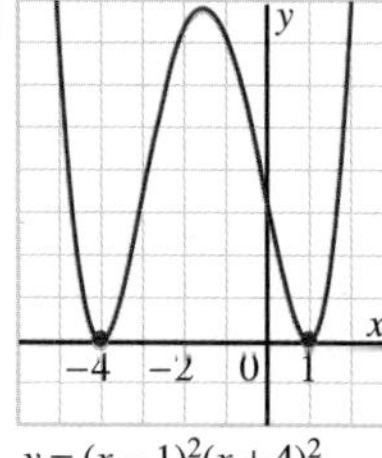

$y = (x - 1)^2(x + 4)^2$

7. a) 0 (of order 2), ±4 **b)** ±2, 3, 6

8. a) $y = -\frac{3}{8}(x + 1)(x - 2)(x - 4)$
b) Answers may vary. $y = (x + 3)(x + 2)(x - 2)^2$
c) The equation in part a is unique. The equation in part b is not unique.

3.3 Exercises, page 162

1.

	Local minimum points	Local maximum points
a)	(1, –4)	(–1, 4)
b)	(–2, –16), (2, –16)	(0, 0)
c)	(0, 3)	(–1, 4), (1, 4)

2. a) i) $x = \pm 1$
ii) $x < -1$ and $x > 1$ $(|x| > 1)$
iii) $-1 < x < 1$ $(|x| < 1)$

b) i) $x = \pm 2, 0$
ii) $-2 < x < 0$ and $x > 2$
iii) $x < -2$ and $0 < x < 2$

c) i) $x = \pm 1, 0$
ii) $x < -1$ and $0 < x < 1$
iii) $-1 < x < 0$ and $x > 1$

3. a) $x = -2$ **b)** $x = -2, 3$ **c)** $x = -3, -1, 2$

4. a) i) Local maximum at $x = -2$
ii) $x < -2$
iii) $x > -2$
iv) Graphs may vary.

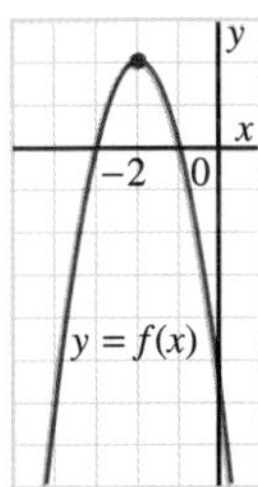

b) i) Local maximum at $x = -2$; local minimum at $x = 3$
ii) $x < -2$ and $x > 3$
iii) $-2 < x < 3$
iv) Graphs may vary.

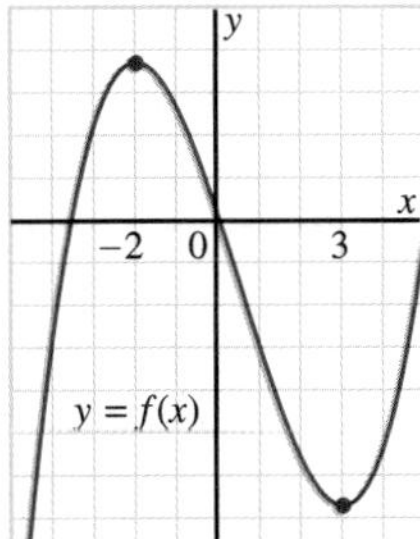

c) i) Local maximum at $x = -3, 2$; local minimum at $x = -1$
ii) $x < -3$ and $-1 < x < 2$
iii) $-3 < x < -1$ and $x > 2$
iv) Graphs may vary.

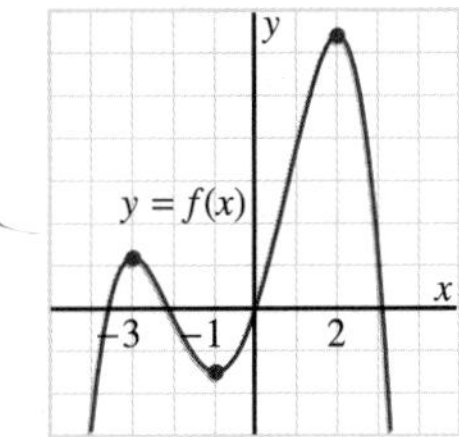

5. a) Increase: $x < -2$ and $x > 4$; decrease: $-2 < x < 4$
b) Increase: $-1 < x < 0$ and $x > 1$; decrease: $x < -1$ and $0 < x < 1$

6. a) i) $x = -2, 4$
ii) The critical values $x = -2$ and $x = 4$ are local maximum and minimum points, respectively.
b) i) $x = \pm 1, 0$
ii) The critical values $x = \pm 1$ are absolute minimum points.

7. a) $x = 0, -\frac{3}{2}$ **b)** $x = -1, -\frac{3}{5}$

8. a) $x = -\frac{4}{3}, -1, -2$ **b)** $x = 0, -\frac{1}{6}, -7, \frac{1}{2}$

9. a) $y' = 9x^2 + 4$ **b)** $y' = 4x^3 - 6x^2 + 1$
c) $y' = -2x + 3$ **d)** $y' = -8x^3 + 10x$
e) $y' = -12x^2 - 2x + 1$ **f)** $y' = 10x^4 + 9x^2 + 1$

10. The degree of the derivative function is 1 less than the degree of the original function. Both the original and derivative function are polynomials.

12. a) $x < \frac{2}{3}$ **b)** $-3 < x < 0$ and $x > 1$

13. a) $(\sqrt{2}, -4\sqrt{2})$ and $(-\sqrt{2}, 4\sqrt{2})$
b) $(\sqrt{3}, -6\sqrt{3})$ and $(-\sqrt{3}, 6\sqrt{3})$
c) (0, 0) and $\left(\frac{3}{2}, -\frac{27}{16}\right)$
d) (0, 0) and $\left(-\frac{3}{2}, -\frac{27}{16}\right)$
e) (0, 0) and approximately (–2.25, –8.54)
f) (0, 0) and (3, –27)

14. a) x-intercepts: $\pm\sqrt{3}, 0$; y-intercept: 0
b) (1, 6) and (–1, –6)
c) (1, 6) is a local maximum point.
(–1, –6) is a local minimum point.
d)

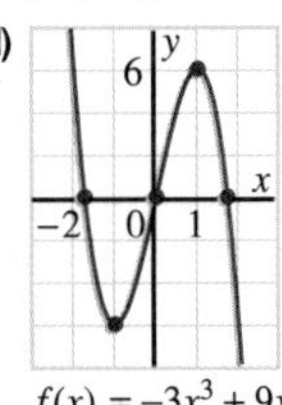

$f(x) = -3x^3 + 9x$

e) Increase: $-1 < x < 1$; decrease: $x < -1$ and $x > 1$

15. a) x-intercepts: $0, \frac{3}{2}$; y-intercept: 0
b) (0, 0) and (1, 2)
c) (0, 0) is a local minimum point.
(1, 2) is a local maximum point.

d)

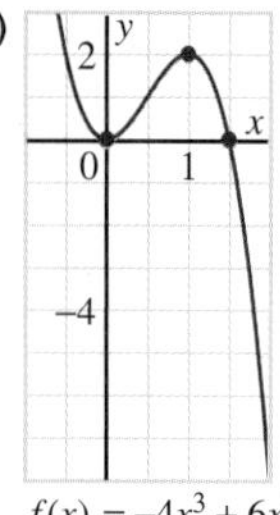

$f(x) = -4x^3 + 6x^2$

e) Increase: $0 < x < 1$; decrease: $x < 0$ and $x > 1$

16. a) x-intercepts: $\pm\sqrt{8}$, 0; y-intercept: 0

b) (0, 0), (2, −16) and (−2, −16)

c) (0, 0) is a local maximum point.
(2, −16) and (−2, −16) are local minimum points.

d)

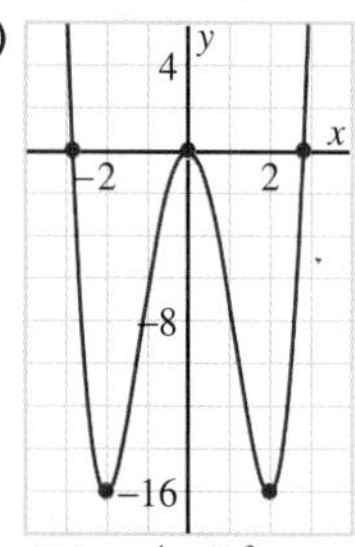

$f(x) = x^4 - 8x^2$

e) Increase: $-2 < x < 0$ and $x > 2$;
decrease: $x < -2$ and $0 < x < 2$

17. Graphs may vary.

a)

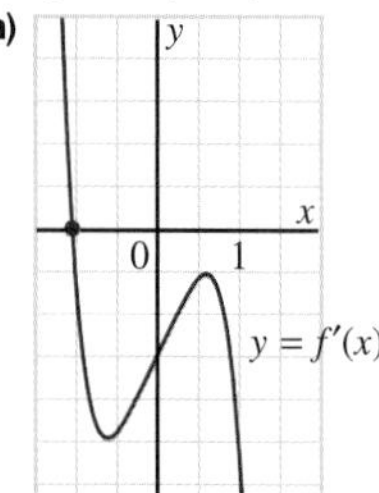

b)

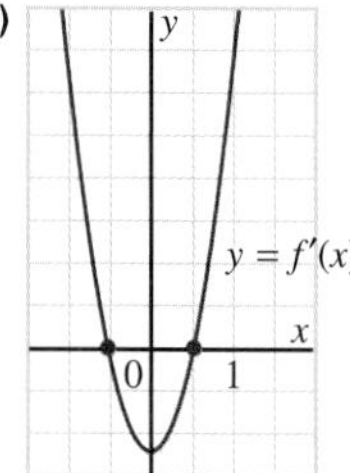

c)

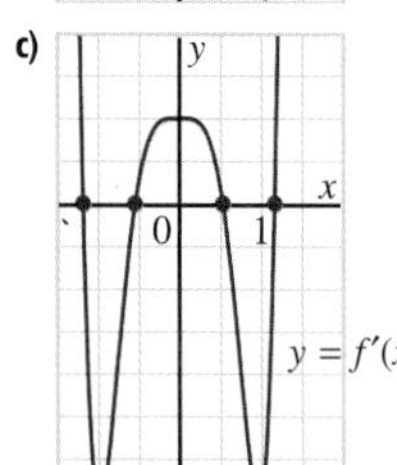

18. a) $m = -12$ b) −39

19. $m = -3$, $n = -36$

22. No, the correct x-coordinates are $\pm\sqrt{\frac{25}{6}}$.

24. a) x-intercepts: $\pm\sqrt{15}$, 0; y-intercept: 0

b) (0, 0), (3, 162) and (−3, −162)

c) (0, 0) is a point of inflection.
(3, 162) is a local maximum point.
(−3, −162) is a local minimum point.

d)

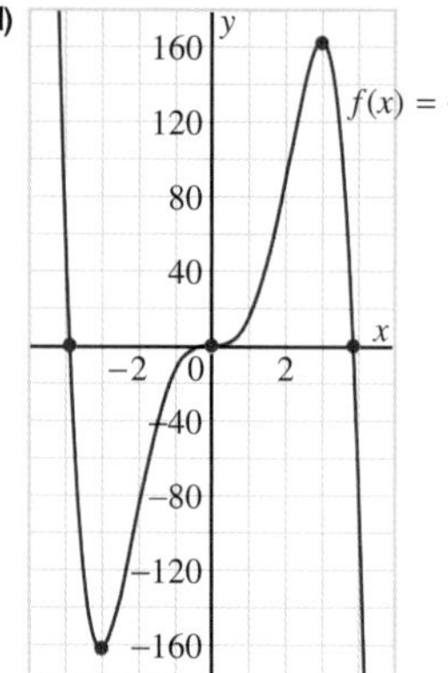

e) Increase: $-3 < x < 0$ and $0 < x < 3$;
decrease: $x < -3$ and $x > 3$

25. b) Yes, any cubic function of the form $f(x) = x^3 - bx$, $b > 0$, has critical points that lie on the graph of $y = 2x^3$.

26. b) Yes, any quartic function of the form $f(x) = x^4 + bx^3$ has critical points that lie on the graph of $y = -\frac{x^4}{3}$.

27. Graphs may vary. A sample graph follows.

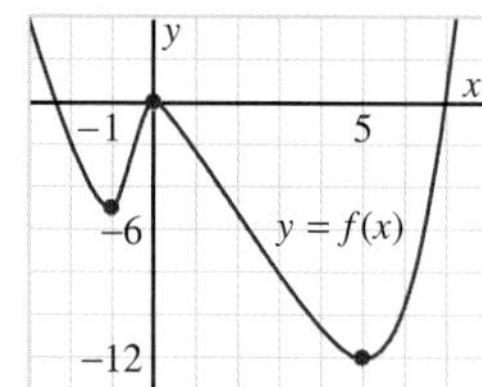

28. No

29. Approximately (−1.52, −10.19)

3.4 Exercises, page 174

1. a) −3 b) 7 c) −4 d) 3

2. a) $y'' = -6x$ b) $y'' = 4$
c) $y'' = 12x + 6$ d) $y'' = 36x^2 + 10$
e) $y'' = -12x^2 + 6x - 6$ f) $y'' = -40x^3 + 48x^2$

3. a) 0, $-\frac{3}{2}$ b) $\frac{5}{4}$, −2, $\frac{1}{3}$
c) 2, 3 d) ±2

4. a) $f''(x) = 18x$ b) $g''(x) = 60x^2 - 12x$
c) $y'' = -4$ d) $h''(x) = 12x + 8$
e) $y'' = -96x - 32$ f) $p''(x) = 48x + 24$

5. a) Graph ② is f and graph ① is f''.
b) Graph ① is f and graph ② is f''.
c) Graph ② is f and graph ① is f''.

7. Estimates may vary.

a) i) Local maximum: (1.25, 9)
ii) Increase: $x < 1.25$; decrease: $x > 1.25$
iii) No points of inflection
iv) Concave down: $x \in \Re$

b) i) No local maximum or minimum points
ii) Increase: $x < 2$ and $x > 2$
iii) Point of inflection: (2, −3)
iv) Concave down: $x < 2$; concave up: $x > 2$

c) i) Local maximum: (−1.8, 8.2) and (1.8, 8.2); local minimum: (0, 2)
ii) Increase: $x < -1.8$ and $0 < x < 1.8$; decrease: $-1.8 < x < 0$ and $x > 1.8$
iii) Point of inflection: (−1.2, 5) and (1.2, 5)
iv) Concave down: $|x| > 1.2$; concave up: $|x| < 1.2$

8. a) i) Degree of f: 2; degree of f'': 0
ii) Negative
b) i) Degree of f: 3; degree of f'': 1
ii) Positive
c) i) Degree of f: 4; degree of f'': 2
ii) Negative

9. a) Graph ② is f, graph ① is f', and graph ③ is f''.
b) Graph ① is f, graph ③ is f', and graph ② is f''.
c) Graph ③ is f, graph ① is f', and graph ② is f''.

10. a) $0, \frac{1}{2}$ b) −2, 0 c) $\sqrt[3]{-\frac{3}{2}}$

11. a) x-intercepts: 0, ±2; y-intercept: 0
b), c) Local minimum: approximately (−1.15, −3.08); local maximum: approximately (1.15, 3.08)
d) (0, 0)
e)
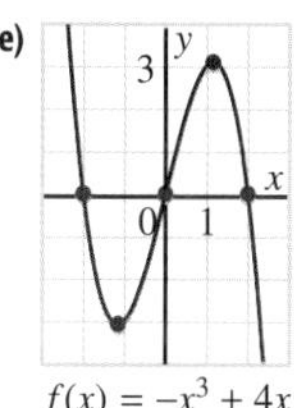
$f(x) = -x^3 + 4x$

f) Increase: $|x| < 1.15$; decrease: $|x| > 1.15$
g) Concave up: $x < 0$; concave down: $x > 0$

12. a) x-intercepts: $0, \pm\frac{1}{\sqrt{2}}$; y-intercept: 0
b), c) Local maximum points: (±0.5, 0.125); local minimum point: (0, 0)
d) Approximately (±0.29, 0.07)
e)
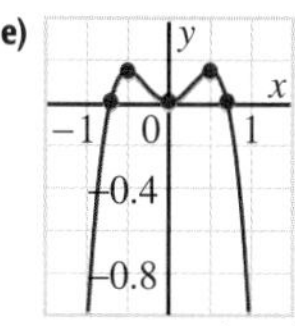
$f(x) = -2x^4 + x^2$

f) Increase: $x < -\frac{1}{2}$ and $0 < x < \frac{1}{2}$; decrease: $-\frac{1}{2} < x < 0$ and $x > \frac{1}{2}$
g) Concave up: $|x| < 0.29$; concave down: $|x| > 0.29$

13. a) x-intercepts: ±3, −0.4; y-intercept: −18
b), c) Local minimum: approximately (1.60, −64.4); local maximum: approximately (−1.87, 40.4)
d) Approximately (−0.13, −11.98)
e)
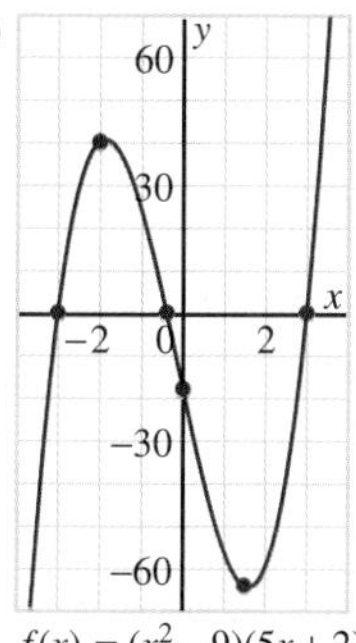
$f(x) = (x^2 - 9)(5x + 2)$

f) Increase: $x < -1.87$ and $x > 1.60$; decrease: $-1.87 < x < 1.60$
g) Concave up: $x > -0.13$; concave down: $x < -0.13$

14. a) x-intercepts: $0, \pm\sqrt{8}$; y-intercept: 0
b), c) Local minimum: (0, 0); local maximums: (±2, 32)
d) Approximately (±1.15, 17.78)
e)
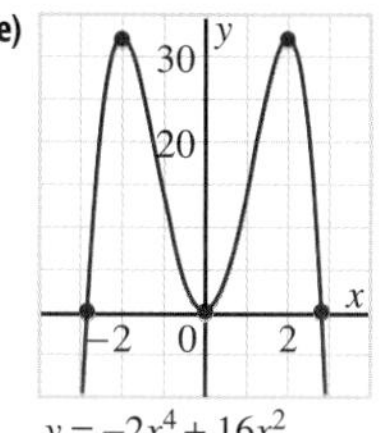
$y = -2x^4 + 16x^2$

f) Increase: $x < -2$ and $0 < x < 2$; decrease: $-2 < x < 0$ and $x > 2$
g) Concave up: $|x| < 1.15$; concave down: $|x| > 1.15$

15. Graphs may vary.
a) i) 0 times
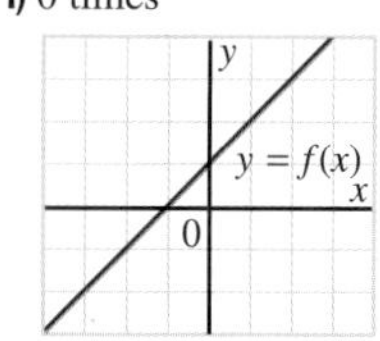

ii) Once
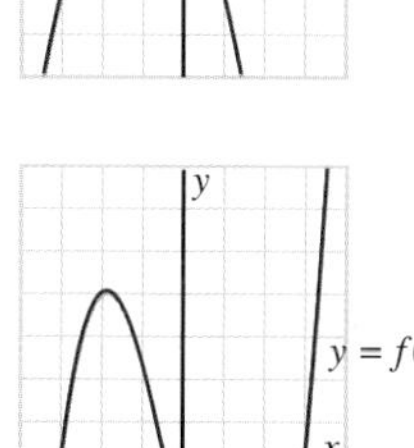

iii) 0 times or twice
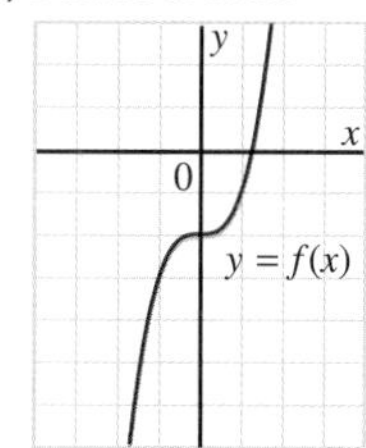

iv) Once or 3 times

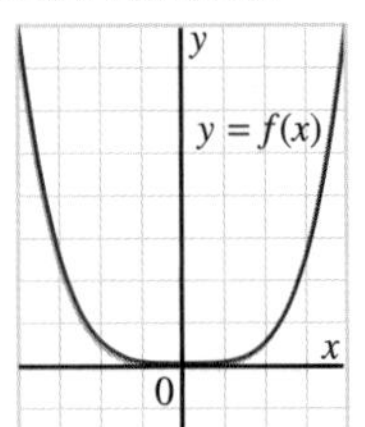

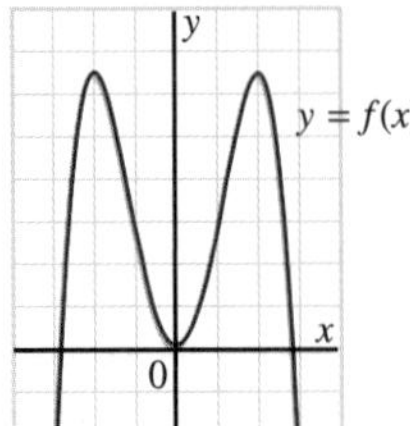

b) $(n - 1)$ times

16. a) True **b)** False **c)** True
d) False **e)** True

17. a) $(7, -2)$ and $(3, 5)$
b) Increase: $x < 3$ and $x > 7$; decrease: $3 < x < 7$
c) $(5, 2)$
d) Graphs may vary.

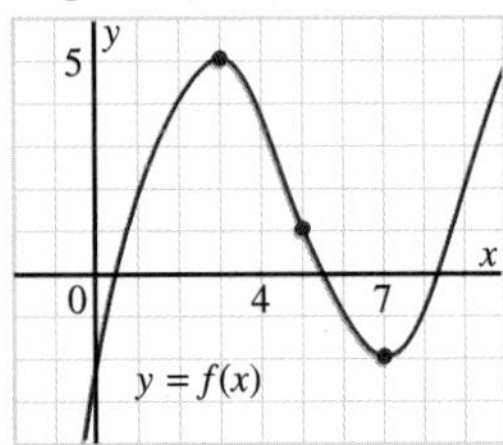

e) Concave up: $x > 5$; concave down: $x < 5$

18. Graphs may vary.

a)

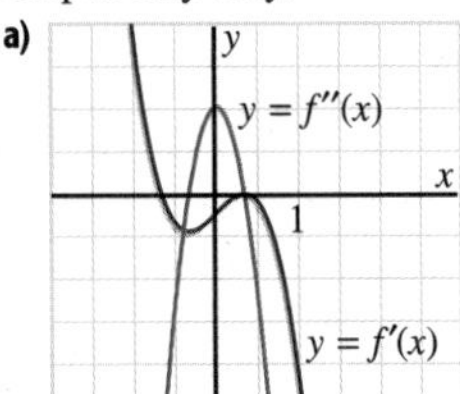

b)

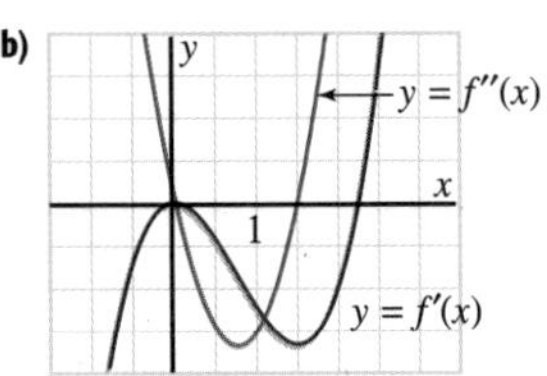

c)

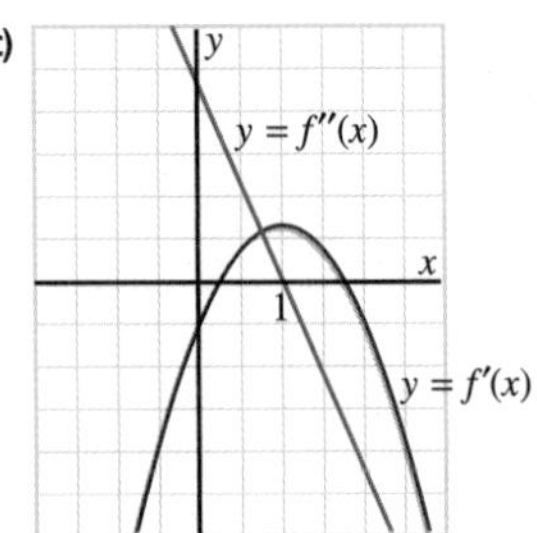

19. a) $a = 6,\ b = -5$ **b)** $y = -11x - 13$

20. a) $f''(c)$ is positive. **b)** $f''(c)$ is negative.
c) If $f'(c) = 0$ and $f''(c) > 0$, then a local minimum occurs at $x = c$.
If $f'(c) = 0$ and $f''(c) < 0$, then a local maximum occurs at $x = c$.

21. a) Local minimum **b)** Neither **c)** Neither
d) Local maximum **e)** Neither **f)** Local minimum

22. a) Increase: $|x| < 1$; decrease: $|x| > 1$
b) Local minimum: $(-1, -2)$; point of inflection: $(0, 0)$; local maximum: $(1, 2)$
c) Approximately $(-0.71, -1.24)$, $(0, 0)$, and $(0.71, 1.24)$
d) Concave up: $x < -0.71$ and $0 < x < 0.71$; concave down: $-0.71 < x < 0$ and $x > 0.71$
e)

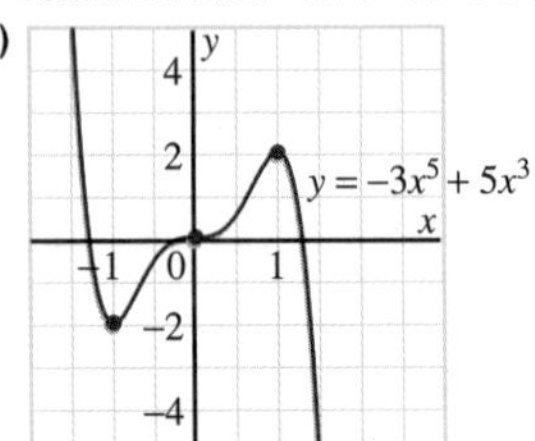

23. a) Approximately $(166.\overline{6}, -2\ 310\ 815)$
b) Approximately $(83.\overline{3}, -1\ 153\ 407)$
c) Answers may vary.
$-200 \le x \le 400,\ -3\ 000\ 000 \le y \le 3\ 000\ 000$

24. $(0, 47)$

3.5 Exercises, page 184

1. a) -11 **b)** 0 **c)** 165
d) -3 **e)** 4 **f)** -35

2. $x - 2$

3. a) 97 **b)** 11 **c)** 42

4. 0

5. a) $x + 2$ **b)** $x + 2;\ x - 1$
c) $x + 1;\ x - 1;\ x - 2$

6. a) $2x^3 - 5x - 6$ **c)** $x^4 - 7x^2 + 12$

7. $\pm1, \pm2, \pm3, \pm4, \pm6, \pm12$

8. a) $k = -3$ **b)** $k = -2$ **c)** $k = 5$

9. b) $2x - 1$

10. b) $x - 1;\ x - 4$

11. a) $(x - 1)(x - 2)(x + 1)$ **b)** $(x + 1)(x + 3)(x - 2)$
c) $(x - 1)(x + 2)(x + 3)$

12. a) $(x - 1)(x + 2)(x + 4)$ **b)** $(x + 1)(x + 3)(x + 5)$
c) $(x - 2)(x + 2)(x + 3)$

13. a) $(x - 2)^2(x + 5)$ **b)** $(x - 2)(x - 5)(x + 2)$

14. a) $(x + 1)(x^2 - x + 1)$ **b)** $(x - 1)(x^2 + x + 1)$

15. a) $(x + 1)(x - 2)(2x + 1)$ **b)** $(x - 1)(x + 2)(2x + 3)$
c) $(x - 3)(x + 4)(3x + 1)$

16. a) $(x - 1)(x + 2)(3x - 1)$ **b)** $(x + 2)(x - 3)(x - 4)$
c) $(x - 1)(x + 3)(5x + 1)$

17. $(x + 1)(x + 3)(x - 2)(x - 4)$

18. a) $b = -2$ **b)** -1

19. a) $a = 2$ **b)** 6

20. a) $b = -5$ **b)** -23

21. a) $k = -1$ **b)** $k = 24$ **c)** $k = -2$

22. $a = 4$

23. $b = 8$

24. b) $x = \frac{1}{2}$

25. a) 11 **b)** −4

26. $x \doteq 1.38,\ 3.62$

27. a) Yes **b)** $(2x - 3)(x + 2)(3x + 2)$

28. $k = -\frac{3}{4},\ p = 1$

Self-Check 3.3–3.5, page 187

1.

	Local maximum points	Local minimum points	Points of inflection
a)	$(\pm\sqrt{1.5}, 4.5)$	$(0, 0)$	$(\pm\frac{1}{\sqrt{2}}, 2.5)$
b)	$(1, -6.6)$	$(0, -7)$; $(2, -10.2)$	$(1.69, -8.86)$; $(0.71, -6.76)$
c)	$(0, 7)$	$(0.63, 4.74)$; $(-2, -45)$	$(-1.25, -23.27)$; $(0.33, -5.81)$

2. a) i) x-intercepts: $0, \frac{3}{2}$; y-intercept: 0
ii) $(0, 0)$ and $(1, 1)$
iii) Local maximum point: $(1, 1)$; local minimum point: $(0, 0)$
iv) $\left(\frac{1}{2}, \frac{1}{2}\right)$
v)

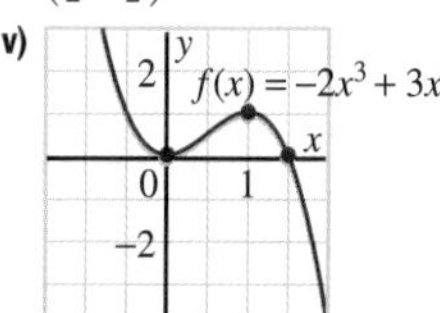

vi) Increase: $0 < x < -1$; decrease: $x < 0$ and $x > 1$
vii) Concave up: $x < \frac{1}{2}$; concave down: $x > \frac{1}{2}$

b) i) x-intercepts: $0, \pm 2.45$; y-intercept: 0
ii) $(0, 0)$ and $(\pm 1.73, -9)$
iii) Local maximum point: $(0, 0)$; local minimum point: $(\pm 1.73, -9)$
iv) $(\pm 1, -5)$
v)

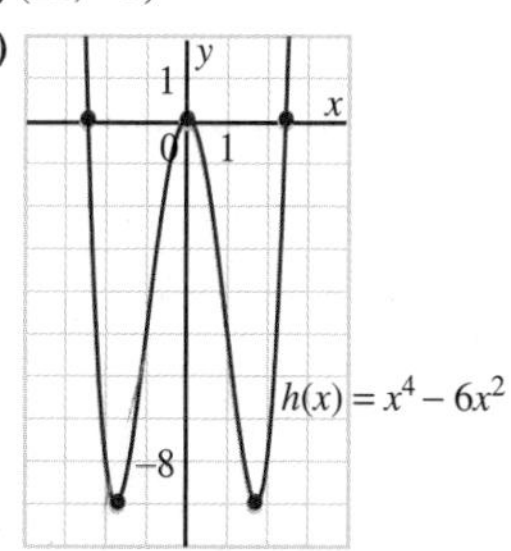

vi) Increase: $-1.73 < x < 0$ and $x > 1.73$; decrease: $x < -1.73$ and $0 < x < 1.73$
vii) Concave up: $|x| > 1$; concave down: $|x| < 1$

3. a) 3 **b)** 14 **c)** 43
d) 11 **e)** 18 **f)** 19

4. a) $(x - 1)(x - 4)(x + 2)$ **b)** $(x + 1)(x + 5)(x - 2)$
c) $(x - 2)^2(x - 4)$ **d)** $(x + 2)^2(x - 3)$
e) $(x - 1)(x + 3)(x - 4)$ **f)** $(x - 2)(x + 1)(x + 4)$

5. a) $b = -2$ **b)** −1

6. a) Increase: $-4 < x < 0$ and $x > 2$; decrease: $x < -4$ and $0 < x < 2$
b) $x = -4,\ 0,\ 2$
c) Concave up: $x < -2.5$ and $x > 1$; concave down: $-2.5 < x < 1$
d) Graphs may vary.

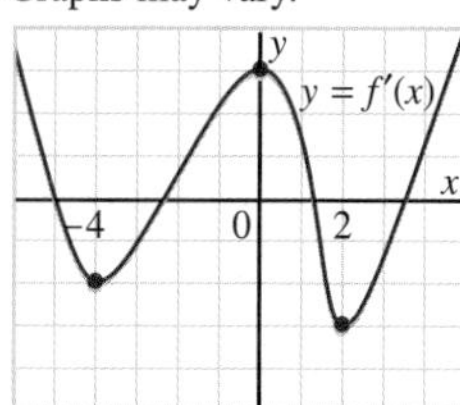

3.6 Exercises, page 194

1. a) $x = 4,\ -3$ **b)** $x = 5,\ -\frac{1}{2},\ \frac{2}{3}$
c) $x = 3,\ -1,\ 4,\ -5$ **d)** $x = 0,\ -2,\ \frac{5}{4}$

2. Estimates may vary.
a) $x \doteq 1.7$ **b)** $x \doteq 2.3$
c) $x \doteq -2.7,\ -1,\ 0.5,\ 3.3$ **d)** $x \doteq -2.9,\ -1.4,\ 0,\ 1.4,\ 2.9$

3. a) None of the graphs touches the x-axis, so no equation has a double root.
b) The equations in parts a and b have complex roots.

4. Answers may vary.
a) $x < 1.7$ **b)** $x < 2.3$
c) $x \le -2.7$; $-1 \le x \le 0.5$; $x \ge 3.3$
d) $x \le -2.9$; $-1.4 \le x \le 0$; $1.4 \le x \le 2.9$

5. Answers may vary.
a) $y = x^2 - x - 6$ **b)** $y = x^3 - 2x^2 - 3x$
c) $y = x^3 - 5x^2 - 26x + 120$

6. a) $x = 0,\ 5,\ -4$ **b)** $x = -3,\ \pm 1,\ 4$
c) $-4 < x < 0$ and $x > 5$ **d)** $-3 \le x \le -1$ and $1 \le x \le 4$

7. a) $x = 0,\ -7,\ -3$ **b)** $x = \pm 2,\ 3$
c) $x = -3,\ 1.5,\ 2$ **d)** $x = \pm 3,\ -1,\ 2$

8. a) $x < -7$ and $-3 < x < 0$ **b)** $-2 < x < 2$ and $x > 3$
c) $-3 \le x \le 1.5$ and $x \ge 2$ **d)** $-3 \le x \le -1$ and $2 \le x \le 3$

9. a) $x \doteq 1.85$
b) $x \doteq -1.53,\ -0.35,\ 1.88$
c) $x < 1.85$
d) $-1.53 \le x \le -0.35$ and $x \ge 1.88$

10. a) $x = -1 \pm \sqrt{6}$ **b)** $x = -\frac{1}{2} \pm \frac{i\sqrt{23}}{2}$
c) $x = -1 \pm i$ **d)** $x = -3,\ 2$

11. a) $x = 2$ and $x = \pm i$
b) $x = -4$ and $x = 2 \pm i2\sqrt{3}$
c) $x = 1$ and $x = -\frac{1}{2} \pm \frac{i\sqrt{3}}{2}$
d) $x = -1,\ 2$ (of order 2)

12. a) $x = 0$ (of order 2) and $x = \pm i$
b) $x = \pm\sqrt{2}$ and $x = \pm i\sqrt{2}$
c) $x = \pm i,\ \pm 2i$
d) $x = 1,\ -2,\ -\frac{1}{2} \pm i\frac{\sqrt{3}}{2}$

13. a) 0 or 2 complex roots
b) 0 or 2 complex roots
c) 0, 2, or 4 complex roots

14. a) $x = -2,\ 1,\ 4$ **b)** $x = -5,\ -1,\ 2$
c) $x = 2$ (of order 2), 4 **d)** $x = -2,\ \frac{-1 \pm \sqrt{5}}{2},\ 3$
e) $x = 2$ (of order 2) **f)** $x = 2$ and $x = 1 \pm i\sqrt{3}$

15. a) $x \doteq -3.5,\ -0.7,\ 4.2$ **b)** $x \doteq 2.3$
c) $x \doteq \pm 3.7,\ \pm 1.2$ **d)** $x \doteq -2.3,\ 3.2$

16. a) $x > 1.6$ **b)** $x \le 3.5$
c) $-3.3 < x < -0.6$ and $1.3 < x < 2.7$
d) $x \le -2.1$ and $x \ge 3.2$ **e)** $-0.5 < x < 1$ and $x > 3$
f) $x < -1$ and $0.\overline{3} < x < 2$

17. a) $x \ge -2.6$ **b)** $x \le -\frac{2}{3}$ and $-\frac{1}{2} \le x \le 2$
c) $-2.5 \le x \le -2$ and $x \ge 1$
d) $x < -3.1$ and $0.7 < x < 2.4$
e) $-0.6 < x < 1.6$ and $x > 4$
f) $x \le -2$ and $-1.3 \le x \le 0.8$

19. a) $x = -3,\ \pm 2$ **b)** $x = -2$ (of order 2), 3
c) $x = -4$ and $x = \frac{1}{2} \pm \frac{i\sqrt{3}}{2}$ **d)** $x = \pm 2,\ \pm 1$

20. a) $k = \pm 15$ **b)** $k = 1$

21. $m = 20;\ x = \frac{9 \pm \sqrt{57}}{6}$

22. a) $m = 9$ **b)** $m = \pm 20$

23. a) i) $x^3 - 12x + 10 = 0$
ii) $x^3 - 12x + 20 = 0$ and $x^3 - 12x - 20 = 0$
b) i) $|k| < 16$ **ii)** $k = \pm 16$ **iii)** $|k| > 16$

24. a) Answers may vary.
i) $(x - \sqrt{3})(x + \sqrt{3}) = 0$
ii) $(x - 2 - \sqrt{3})(x - 2 + \sqrt{3}) = 0$
iii) $(x - 4 + \sqrt{3})(x - 4 - \sqrt{3}) = 0$
b) i) $y = a(x - \sqrt{3})(x + \sqrt{3})$
ii) $y = a(x - 2 - \sqrt{3})(x - 2 + \sqrt{3})$
iii) $y = a(x - 4 + \sqrt{3})(x - 4 - \sqrt{3})$

25. a) $n = 1$ **b)** $n = \pm 24$

26. (0, 0) and (6, −108)

28. (2, −21)

29. a) Answers may vary.
i) $(x - i\sqrt{3})(x + i\sqrt{3}) = 0$
ii) $(x - i - \sqrt{3})(x + i - \sqrt{3}) = 0$
iii) $(x - 1 - i\sqrt{3})(x - 1 + i\sqrt{3}) = 0$
b) i) $y = a(x - i\sqrt{3})(x + i\sqrt{3})$
ii) $y = a(x - i - \sqrt{3})(x + i - \sqrt{3})$
iii) $y = a(x - 1 - i\sqrt{3})(x - 1 + i\sqrt{3})$

30. a) Answers may vary.
i) $(x - i)(x + i)(x - 1) = 0$
ii) $(x - 3)(x + 5)(x - 2i)(x + 2i) = 0$
iii) $(x - 2)(x - 3 - i\sqrt{5})(x - 3 + i\sqrt{5}) = 0$
b) i) $y = a(x - i)(x + i)(x - 1)$
ii) $y = a(x - 3)(x + 5)(x - 2i)(x + 2i)$
iii) $y = a(x - 2)(x - 3 - i\sqrt{5})(x - 3 + i\sqrt{5})$

31. $4x^2 + 16x - 105 - 0$

3.7 Exercises, page 207

1. a) $x = -\frac{3}{2},\ 5,\ -1$ **b)** $x = 0,\ -2,\ \frac{1}{3},\ -1$

2. a) $x = \frac{1}{2},\ -2,\ -6$ **b)** $x = \frac{3}{4},\ \pm 1,\ -\frac{5}{3}$

3. a) Local maximum: (2, −1) **b)** Local minimum: (−5, 3)
c) Local minimum: (−1, 8) **d)** Local maximum: (2.5, 10)

4. a) (−4, 3) **b)** None **c)** (−1.5, 2) **d)** (3, 12)

5. a) Local maximum when $x = -\frac{1}{3}$; local minimum when $x = 1.5$
b) Local maximum when $x = 1$; local minimums when $x = -1,\ \frac{3}{2}$

6. Local maximum value approximately 3.71 when $x \doteq -0.63$; local minimum value approximately −13.71 when $x \doteq 2.63$

7. a) Local minimum: (2, −16); local maximum: (−1, 11)
b) Local maximum: approximately (−1.34, 6.90); local minimum: approximately (1.34, 1.10)

8. a) $x = \pm 2,\ -\frac{2}{3}$ **b)** $x = -2,\ -1,\ 0,\ 4$

9. a) x-intercepts: ±1, 4; y-intercept: 4
b) Local maximum: (−0.12, 4.06); local minimum: (2.79, −8.21)
c) Approximately (1.33, −2.07)
d)

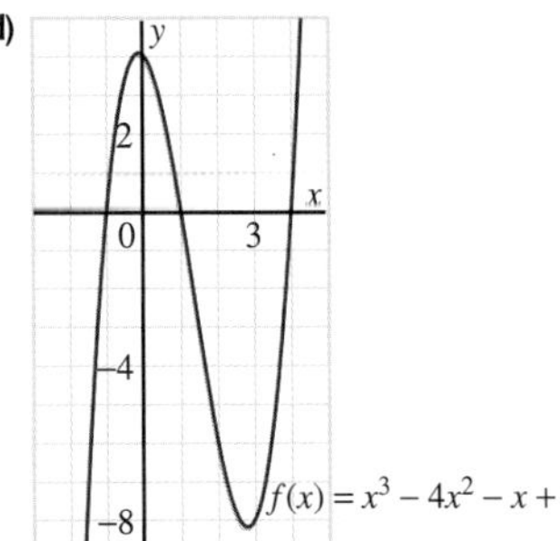

e) Increase: $x < -0.12$ and $x > 2.79$; decrease: $-0.12 < x < 2.79$
f) Concave up: $x > 1.33$; concave down: $x < 1.33$

10. a) x-intercepts: $x = -1,\ 0.5,\ 2$; y-intercept: −2
b) Local maximum: (1.37, 2.60); local minimum: (−0.37, −2.60)
c) (0.5, 0)
d)

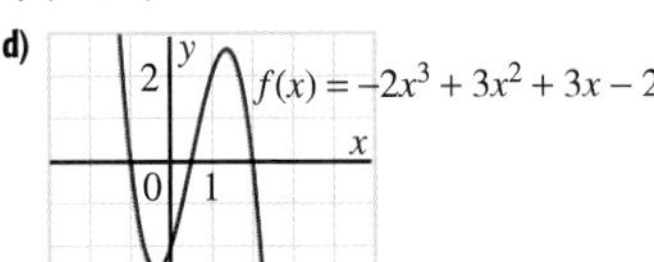

e) Increase: $-0.37 < x < 1.37$; decrease: $x < -0.37$ and $x > 1.37$
f) Concave up: $x < 0.5$; concave down: $x > 0.5$

11. a) x-intercepts: −2, 1; y-intercept: −16
b) Local maximum: (−2, 0); local minimum: (0, −16)
c) (−1, −8)
d)

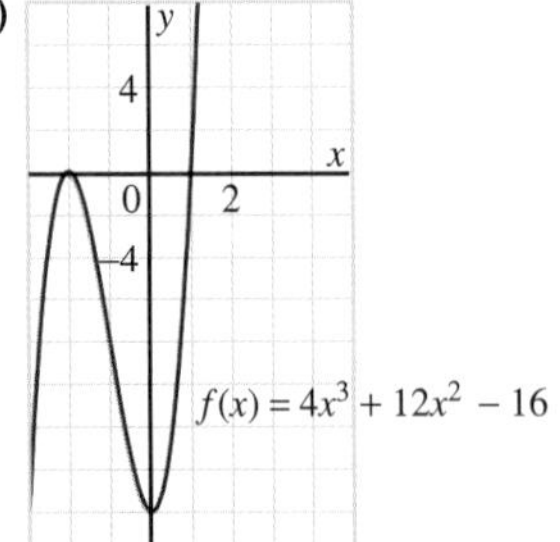

e) Increase: $x < -2$ and $x > 0$; decrease: $-2 < x < 0$
f) Concave up: $x > -1$; concave down: $x < -1$

12. a) x-intercepts: -2, 1, 3; y-intercept: -18
b) Local maximum: (1.75, 4.39);
local minimums: $(-1, -32)$ and (3, 0)
c) (0.07, -16.56) and (2.43, 2.06)
d)

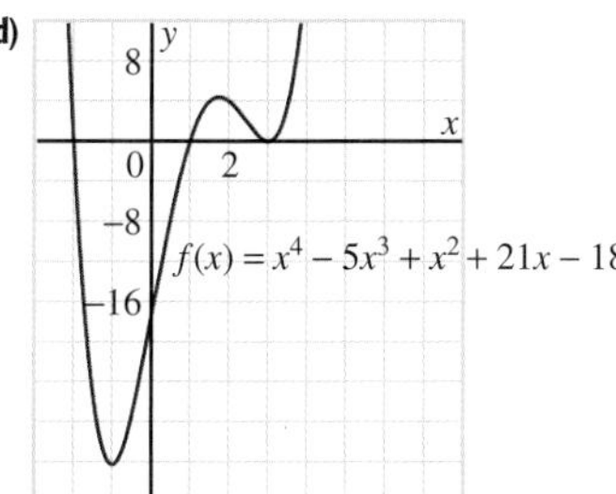

e) Increase: $-1 < x < 1.75$ and $x > 3$;
decrease: $x < -1$ and $1.75 < x < 3$
f) Concave up: $x < 0.07$ and $x > 2.43$;
concave down: $0.07 < x < 2.43$

13. a) 3 **b)** 2
c) Local minimum: (0, 3); no maximum points;
no points of inflection

14. a) i) Local and absolute minimum: $(4, -201)$;
absolute maximum: $(-2, 123)$
ii) Local and absolute minimums: $(\pm 2, -9)$; local
maximum: (0, 7); absolute maximum: $(\pm 3, 16)$
b) i) Local minimum: $(4, -201)$; local maximum: $(-3, 142)$
ii) Local and absolute minimums: $(\pm 2, -9)$; local
maximum: (0, 7)

16. a) $f'(x) = 3x^2 - 3$
b) Answers may vary.
$f'(x) = 12x^3 - 48x^2 + 12x + 72$

17.

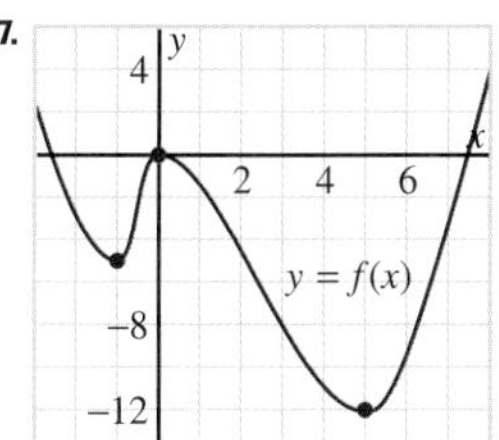

18. a) $m = 6,\ n = -2$
b) Concave up: $x > -1$; concave down: $x < -1$;
local maximum: $(-2.22, 12.35)$;
local minimum: $(0.22, -2.35)$

19. a) $m = -12$ **b)** $(6, -39)$

20. $m = -3,\ n = -36$

3.8 Exercises, page 216

1.

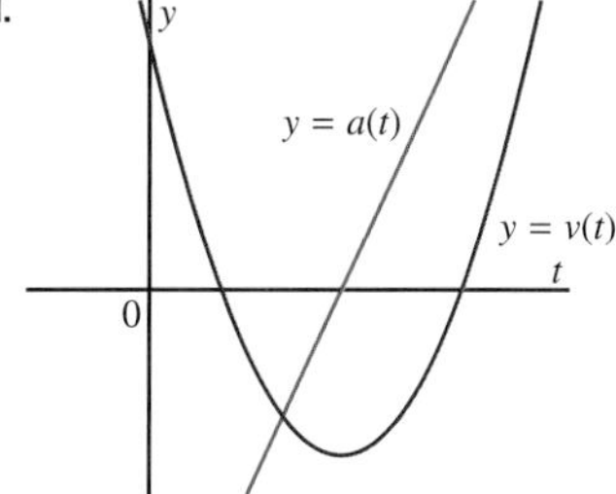

2. a) $y = v(t)$ is increasing; $y = a(t)$ is positive.
b) $y = v(t)$ is decreasing; $y = a(t)$ is negative.
c) $y = v(t)$ is increasing; $y = a(t)$ is positive.

3. a) D to E **b)** A to B **c)** C to D **d)** B to C

4. a) i) Positive **ii)** Negative **iii)** Decreasing
b) i) Positive **ii)** Positive **iii)** Increasing

5. a) $s'(t) = -15t^2 + 16t$; $v(3) = -87$ m/s
b) $s'(t) = 12t^3 - 6t^2 + 5$; $v(3) = 275$ m/s
c) $s'(t) = 4t - 7$; $v(3) = 5$ m/s
d) $s'(t) = -11$; $v(3) = -11$ m/s

6. a) $s''(t) = 6t - 2$; $a(2) = 10$ m/s^2
b) $s''(t) = -30t + 8$; $a(2) = -52$ m/s^2

7. a) $t = 1$ s: moving away; $t = 5$ s: moving away
b) $t = 1$ s: moving away; $t = 5$ s: moving toward
c) $t = 1$ s: moving away; $t = 5$ s: moving away

8. a) $t = 0$ s, 2 s
b) Moving left for $0 < t < 2$; moving right for $t > 2$
c)

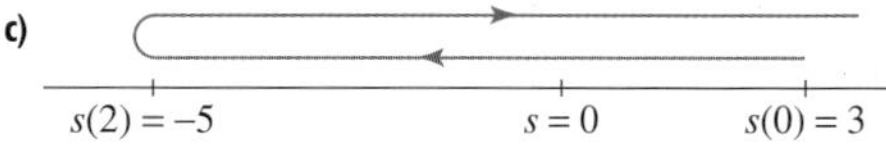

d) 656 m

9. a) 6.4 m/s **b)** 4 s

10. a) 0 m/s **b)** 12 m/s^2
c) Moving right for $0 < t < 1$ and $t > 3$;
moving left for $1 < t < 3$
d)

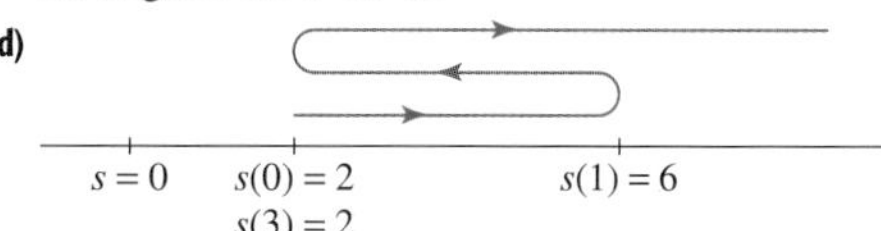

e) 28 m

11. a) 2.2 m/s; -17.4 m/s **b)** Approximately 4.5 s
c) Approximately 1.2 s **d)** Approximately 52.3 m

12. a) Faster at C **b)** Accelerating at A and C
c) Stopped from D to E
d) At F, the motorcycle has returned to its starting position.

13. a) 6500 **b)** 7669; 8825
c) $P'(x) = -0.5x + 235$ **d)** In approximately 15 years
e) In 20 years

14. a) 36 h **b)** 1 680 000 microbes/mL/h

15. a) \$60 000/box/\$; \$40 000/box/\$
b) The company's revenue will increase by \$60 000
at \$10/box and \$40 000 at \$15/box.
c) $\$0 \le x \le \$25$ **d)** \$25/box
e) The price in part d is the R-coordinate of the maximum
point on the graph of the revenue function.

16. **a)** -2.4 m/s^2 **b)** 127.5 m

17. **a)** 122.5 m **b)** $v(t) = 9.8t$
c) 58.8 m/s **d)** 88.5 m/s

Self-Check 3.6–3.8, page 220

1. **a)** $x = -1, 2, 4$ **b)** $x = -2, 1, 3$
2. **a)** $x < -0.5$ and $1 < x < 1.5$
   b) Solutions are approximate.
   $x < -2.3$ and $-2 < x < 0$ and $x > 1.3$
3. $m = -1, 3$
4. Answers may vary. $x^2 - 9 = 0$
5. Answers may vary.
   a) $(x + 1)(x + i)(x - i) = 0$
   b) $(x - 2 - \sqrt{3})(x - 2 + \sqrt{3})(x - 1) = 0$
6. **a)** x-intercepts: approximately -1.55, 0.22, and -1
   y-intercept: -1
   b) Local minimum: $(-0.26, -1.36)$; local maximum: $(-1.30, 0.34)$
   c) $(-0.78, -0.51)$
   d)

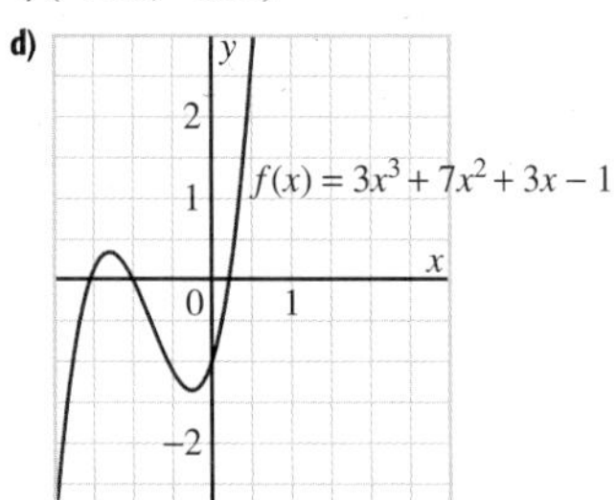

   e) Increase: $x < -1.30$ and $x > -0.26$;
   decrease: $-1.30 < x < -0.26$
   f) Concave up: $x > -0.78$; concave down: $x < -0.78$
7. **a)** -11 m/s; -32 m/s^2 **b)** 159 m/s; 160 m/s^2
8. **a)** The velocity is negative because the car is moving in a negative direction.
   b) The acceleration is positive because the velocity is increasing.
   c) The velocity is increasing because it is negative but approaching 0.
9. **a)** At rest: 1 s and 5 s; positive direction: $0 \le t < 1$ and $t > 5$
   b) -12 m/s; -5 m
   c)

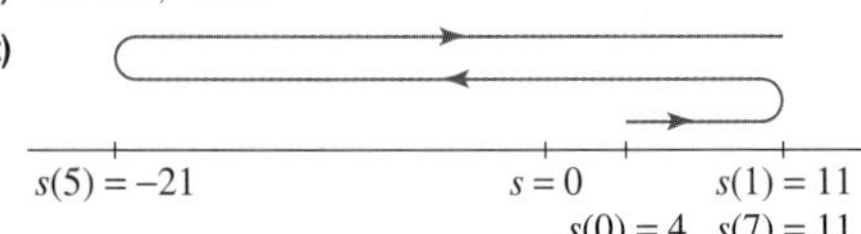

   d) 71 m

Chapter 3 Review Exercises, page 223

1. **a)** $x < -4$ and $x > -3$
   −6 −4 −2 0 2
   b) $-9 < x < 3$
   −10 −8 −6 −4 −2 0 2 4 6
   c) $-2.5 \le x \le -2$
   −5 −4 −3 −2 −1 0 1
   d) $x \le -0.5$ and $x \ge 0.4$
   −0.5 0 0.5
2.

	Quotient	Remainder	Restrictions
a)	$x + 3$	-1	$x \neq \frac{1}{2}$
b)	$3x^2 - 4x + 3$	-5	$x \neq -1$
c)	$3x^2 + 3x + 2$	-1	$x \neq 1$
d)	$3x^2 - 4x + 4$	-15	$x \neq -\frac{3}{2}$

3. $3x^4 + 4x^3 + 6x^2 - 7x - 22$
4. **a)** $f(x) = x^3 - 3x^2 + 1$ **b)** $g(x) = -2x^3 + 4x^2 + 7x - 3$
   c) $h(x) = -x^4 - 4x^3 + 10$ **d)** $k(x) = 2x^4 - 3x^2 - 21$
5. **a) ii)**

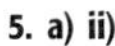

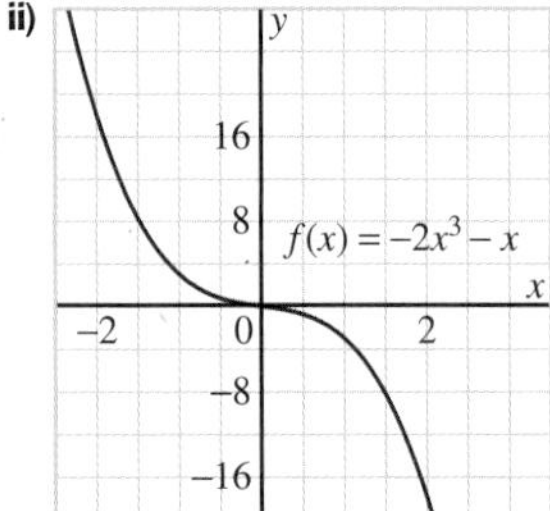

   b) ii)

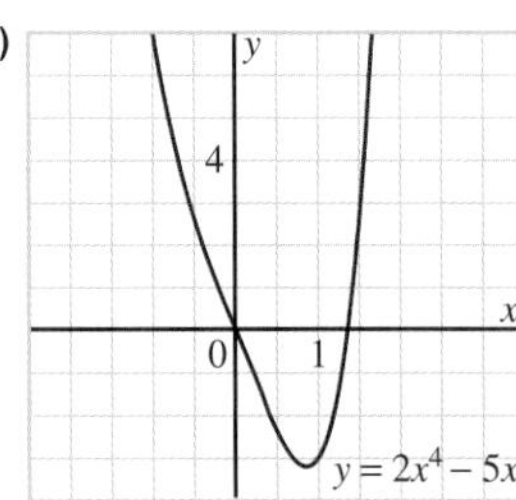

   c) ii)

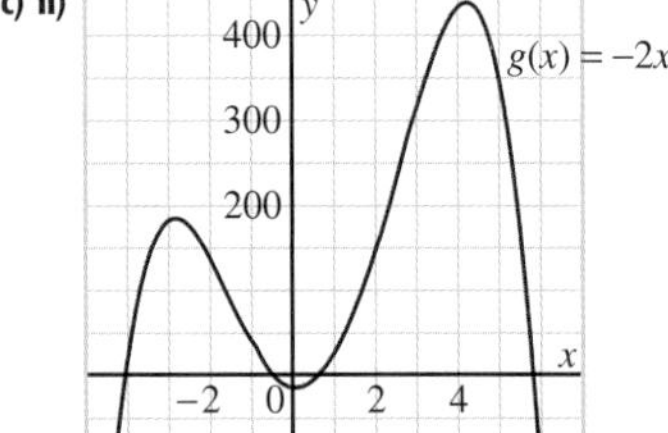

6. **i)** $y = f(x)$, $y = h(x)$, $y = k(x)$
   ii) $y = g(x)$ **iii)** $y = f(x)$, $y = g(x)$
   iv) $y = h(x)$, $y = k(x)$
   v) Point symmetry: $y = f(x)$, $y = h(x)$, $y = k(x)$
7. **a)** $x = 3, -\frac{5}{2}, -2$ (of order 2)
   b) $x = 0$ (of order 2), $-\frac{4}{3}, -5, 1$
8. **i)** b **ii)** d
   iii) Answers may vary.
   a) $y = (x - 3)(x + 2)(x - 1)$ **b)** $y = x^2(x - 1)$
   c) $y = (x + 4)(x - 5)(x + 1)$ **d)** $y = (x + 3)^3$

9. a) $x = -3,\ 4$

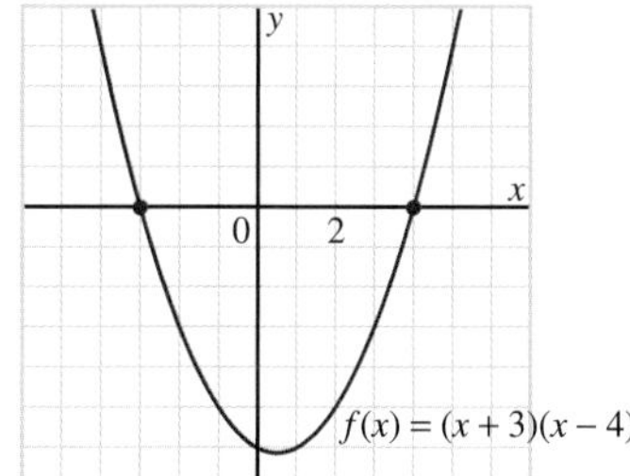

b) $x = -3$ (of order 2), 4

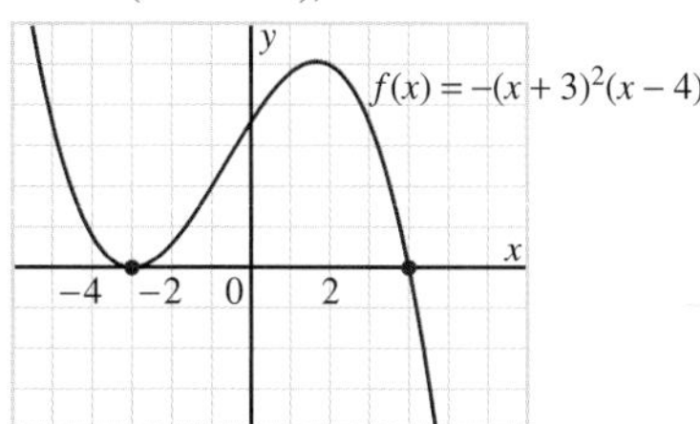

c) $x = 1,\ -2,\ 3$

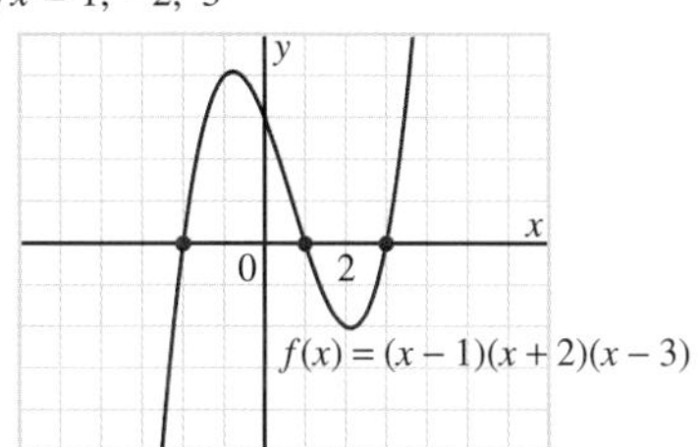

10. a) $y = 1.5(x + 3)(x - 4)^2$

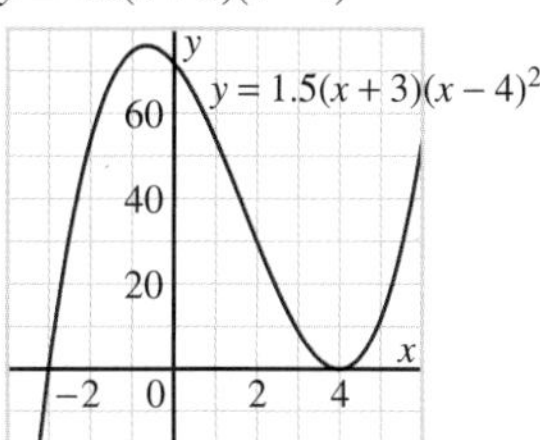

b) $y = -\frac{1}{3}x^2(x + 2)(x - 1)$

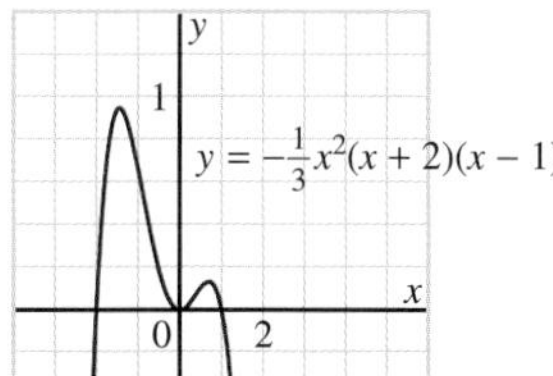

12. Graphs may vary.

a) i)

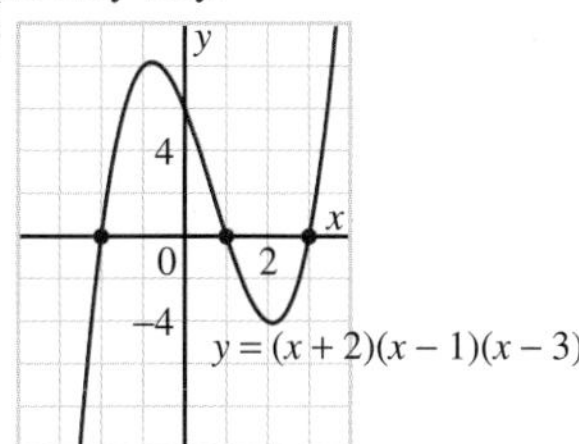

ii) $y = a(x + 2)(x - 1)(x - 3)$

b) i)

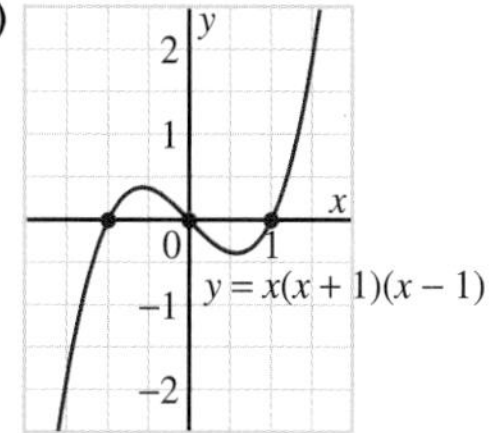

ii) $y = ax(x + 1)(x - 1)$

c) i)

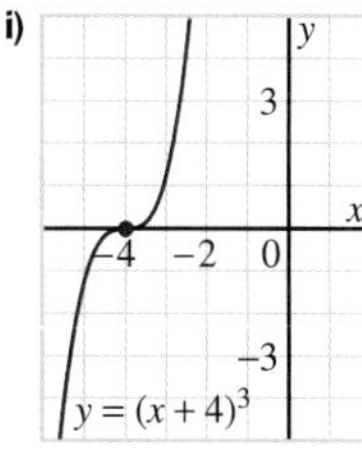

ii) $y = a(x + 4)^3$

13. a)

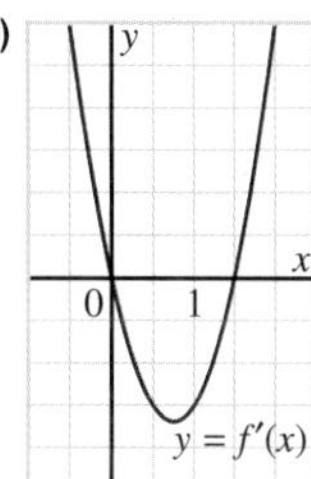

b)

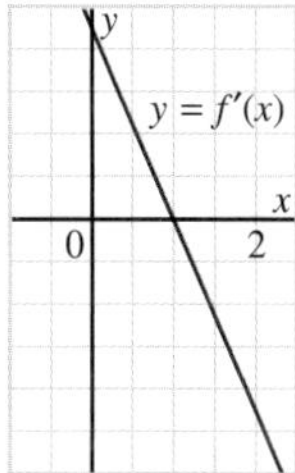

14. a) i) Local maximum: (0, 0); local minimum: (1.5, −6.8)

ii) Increase: $x < 0$ and $x > 1.5$; decrease: $0 < x < 1.5$

b) i) Local maximum: (1, 5)

ii) Increase: $x < 1$; decrease: $x > 1$

15. a) $x = 0,\ 2.25$ **b)** $x = 0,\ 2$

c) $x = 0$ **d)** $x = \frac{2}{3},\ 2$

16. a) Local minimum: (2.25, −3.54)

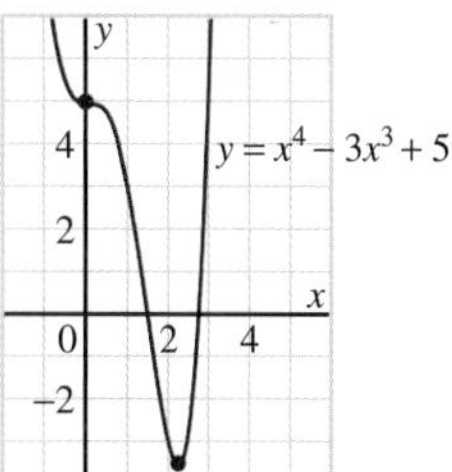

b) Local minimum: (0, −11); local maximum: (2, 1)

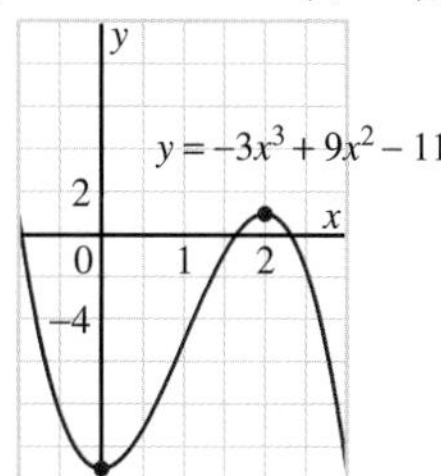

c) Local maximum: $(0, -9)$

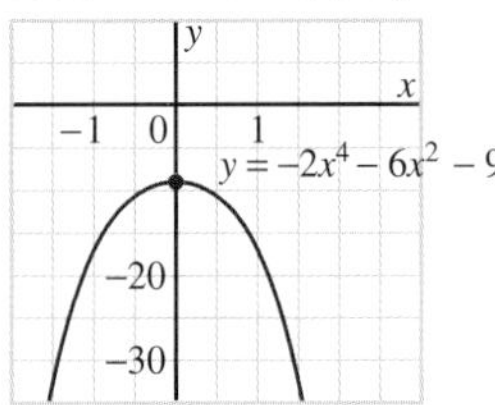

d) Local minimum: $(2, 0)$; local maximum: $(0.67, 1.19)$

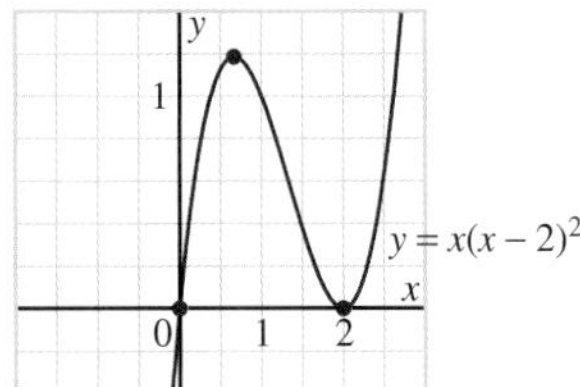

17. a) i) x-intercepts: $-0.75, 0$; y-intercept: 0

ii) $(0, 0)$ and $(-0.5, -0.25)$

iii) Local maximum: $(0, 0)$; local minimum: $(-0.5, -0.25)$

iv)

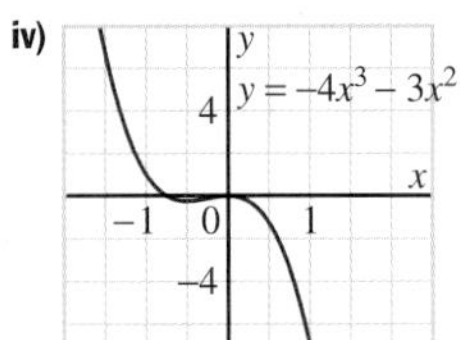

v) Increase: $-0.5 < x < 0$; decrease: $x < -0.5$ and $x > 0$

b) i) x-intercepts: $0, \pm 1.41$; y-intercept: 0

ii) $(0, 0)$ and $(\pm 1, -1)$

iii) Local maximum: $(0, 0)$; local minimum: $(\pm 1, -1)$

iv)

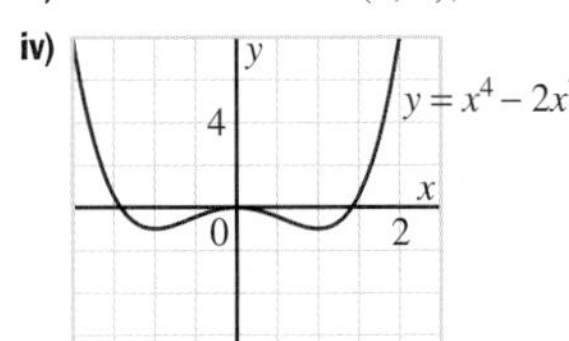

v) Increase: $-1 < x < 0$ and $x > 1$; decrease: $x < -1$ and $0 < x < 1$

18. a)

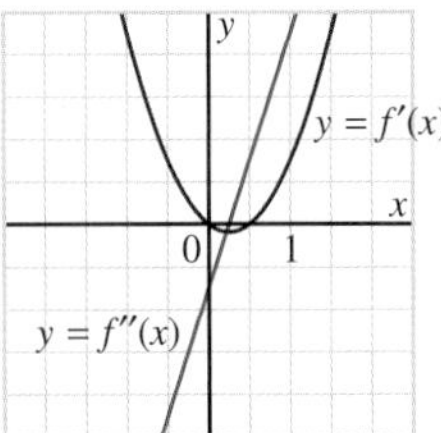

b)

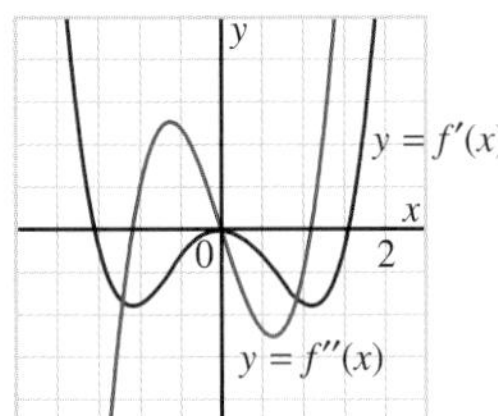

c)

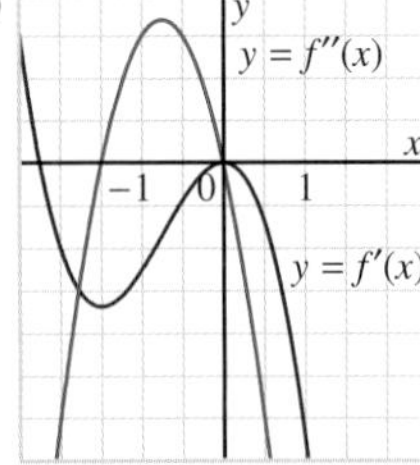

19. a) $f''(x) = 12x^2 - 12x + 32$ **b)** $f''(x) = 54x - 30$
c) $f''(x) = 108x^2 + 72x - 10$ **d)** $f''(x) = -72x - 4$

20. a) No point of inflection **b)** $(0.56, -6.42)$
c) $(0.12, -5.46)$ and $(-0.78, -0.05)$
d) $(-0.06, 0.72)$

21. a) x-intercepts: $0, 0.8$; y-intercept: 0
b) $(0, 0)$ and $(0.6, 0.22)$
c) Local maximum: $(0.6, 0.22)$
d) $(0, 0)$ and $(0.4, 0.13)$
e)

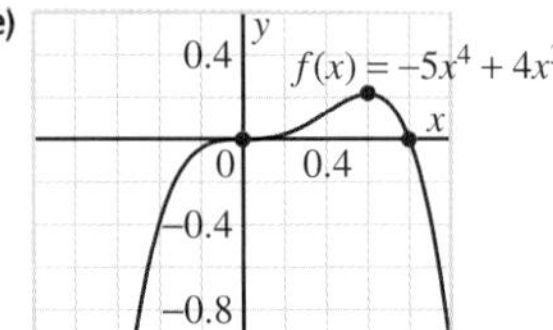

f) Increase: $x < 0$ and $0 < x < 0.6$; decrease: $x > 0.6$
g) Concave up: $0 < x < 0.4$; concave down: $x < 0$ and $x > 0.4$

22. a) x-intercepts: $-2, 1$; y-intercept: 6
b) $(1, 0)$ and $(-1, 12)$
c) Local maximum: $(-1, 12)$; local minimum: $(1, 0)$
d) $(0, 6)$
e)

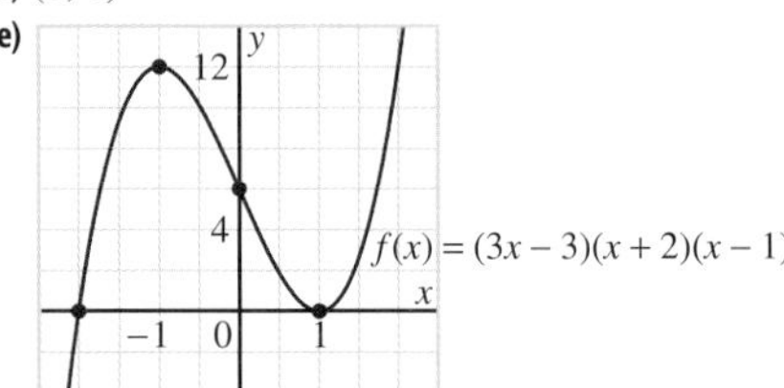

f) Increase: $x < -1$ and $x > 1$; decrease: $-1 < x < 1$
g) Concave up: $x > 0$; concave down: $x < 0$

23.

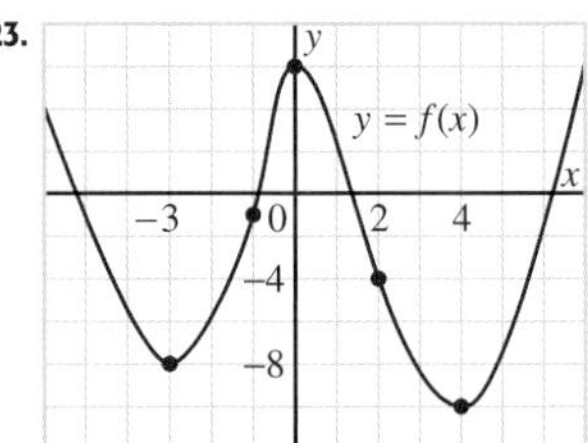

24. a) i) 4 **ii)** 0 **iii)** 0 **iv)** 10
b) A zero remainder means that the binomial divisor is a factor of the polynomial.

25. $x - 1$ is a factor of the polynomial in part a.

26. a) $(x + 1)(x - 3)(x - 2)$ **b)** $(x + 1)(x + 5)(x - 4)$
c) $(x + 1)(x + 5)(x - 2)$

27. a) $(x - 1)(2x - 1)(3x - 4)$
b) $(x + 1)(x - 1)(x + 4)(2x - 3)$
c) $-(x + 1)(x - 2)(x + 3)(2x - 1)$

28. a) $m = 8$ **b)** $m = -1$

29. a) $x = -1,\ 0.5,\ 3$ **b)** $x = \pm 3,\ 1,\ 2$

30. a) $x < -1$ and $0.5 < x < 3$
b) $x < -3$, $1 < x < 2$, and $x > 3$

31. a) $x = 3$ and $x = -\frac{3}{2} \pm \frac{i3\sqrt{3}}{2}$

b) $x = \frac{-3 \pm \sqrt{41}}{2}$
c) $x = 1$ and $x = -\frac{1}{2} \pm \frac{i\sqrt{3}}{2}$
d) $x = -2$ and $x = 1 \pm i\sqrt{3}$
e) $x = 0$ and $x = \pm\frac{i}{\sqrt{2}}$
f) $x = \pm 2$ and $x = \pm 2i$

32. a) $k = \pm 3\sqrt{3.5}$ **b)** $k = \pm 6\sqrt{2}$

33. a) $k = \pm 4$ **b)** $k = 12$
c) $k = 4$ **d)** $k = \pm 8$
e) $k = 48$ **f)** $k = \pm 48$

34. a) $m = \pm 6$ **b)** $m = 49$

35. a) $x < 1$ **b)** $x \leq -2.43$ and $x \geq 1.84$
c) $x \leq -\sqrt{3}$ and $0 \leq x \leq \sqrt{3}$

36. Equations may vary.
a) $(x + 2)(x - 4)(5x - 2)(x + 1) = 0$
b) $(x + 1)(x - 1)(x - 2 + \sqrt{3})(x - 2 - \sqrt{3}) = 0$
c) $(x + 2)(x + 3 - 4i)(x + 3 + 4i) = 0$
No equation is unique.

37. a) Local maximum: $(1, 4)$; local minimums: $(-1, -28)$ and $(2, -1)$
b) No local maximum or minimum points

38. a) $(2, 3)$
b) $(0.63, 4.15)$ and $(2.37, 9.34)$

39. a) x-intercepts: ± 2, ± 3; y-intercept: 72
b) Local maximum: $(0, 72)$;
local minimums: $(\pm 2.55, -12.5)$
c) $(\pm 1.47, 25.06)$
d)

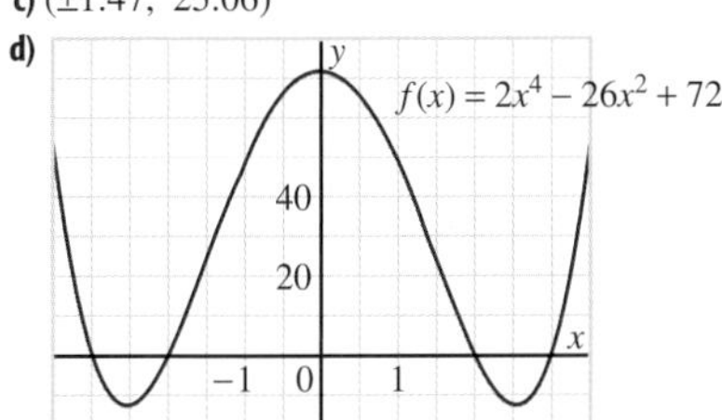

e) Increase: $-2.55 < x < 0$ and $x > 2.55$;
decrease: $x < -2.55$ and $0 < x < 2.55$
f) Concave up: $|x| > 1.47$; concave down: $|x| < 1.47$

40. a) 28 m **b)** -2 m/s
c) 28.2 m **d)** -23.7 m/s

41. a) 36 m/s **b)** 84 m/s^2
c) 2 s and 0 s **d)** At all times
e) $s(0) = 0$ $s(2) = 16$
f) 475 m

Chapter 3 Self-Test, page 229

1. a) i) Negative **ii)** 3 **iii)** $x = 1, \pm 2$
iv) -4 **v)** $y = -(x - 1)(x - 2)(x + 2)$
b) i) Positive **ii)** 4
iii) $x = -3, -1, 2$ (of order 2)
iv) 12 **v)** $y = (x + 3)(x + 1)(x - 2)^2$

2. a)

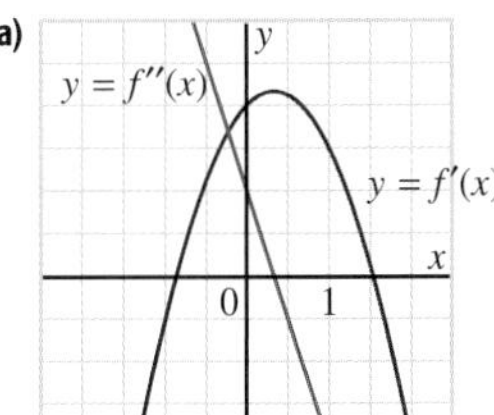

b)

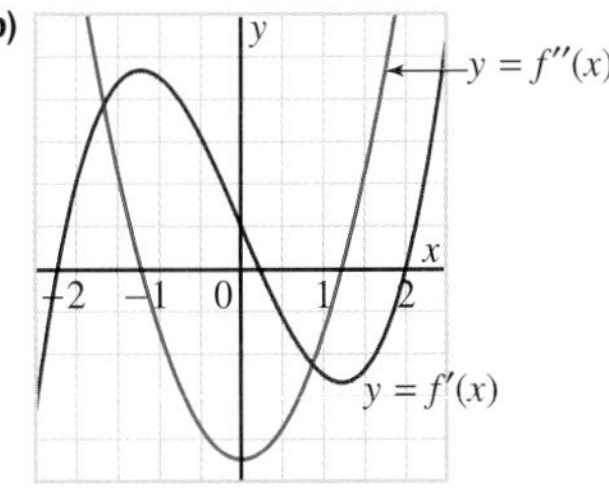

3. a) x-intercepts: $0, -1.5$; y-intercept: 0
b) Local maximum: $(-1, 1)$; local minimum: $(0, 0)$
c) $(-0.5, 0.5)$
d)

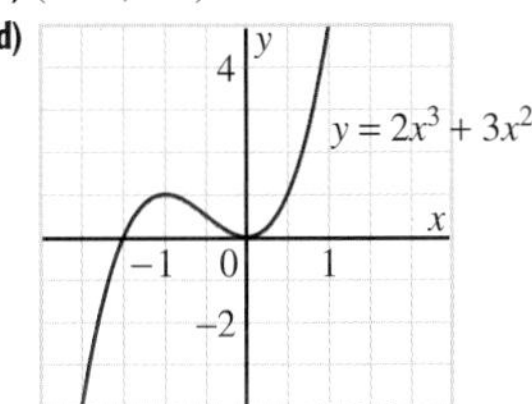

e) Increase: $x < -1$ and $x > 0$; decrease: $-1 < x < 0$
f) Concave up: $x > -0.5$; concave down: $x < -0.5$

4. $m = -10$

6. a) $x = -\frac{1}{3}$, 2 and $x = -\frac{3}{2} \pm \frac{i\sqrt{11}}{2}$
b) $x < -\frac{1}{3}$ and $x > 2$

7. a) $s(t) = 30 + 55t - 4.9t^2$ **b)** 11.7 s
c) -60.1 m/s **d)** 184.3 m, 5.6 s

8. $y = -2(x + 1)(3x - 2)(x - 2 + i)(x - 2 - i)$

9. $(0, 5)$

Cumulative Review (Chapters 1–3), page 231

1.

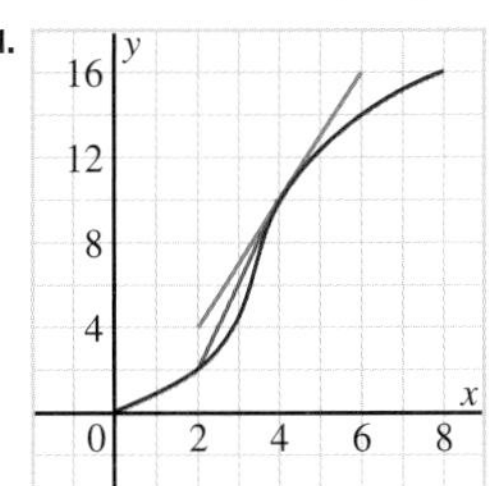

a) 4
b) Estimates may vary.

2. a)

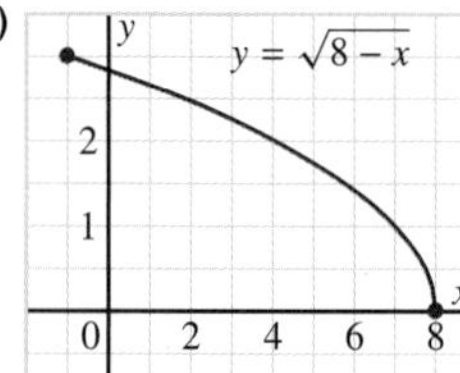

b) Approximately –0.22 **c)** Estimates may vary. –0.25

d) The average rate of change in part b is the slope of the secant joining $x = 1$ and $x = 4$.
The instantaneous rate of change in part c is the slope of the tangent at $x = 4$.

3. a)

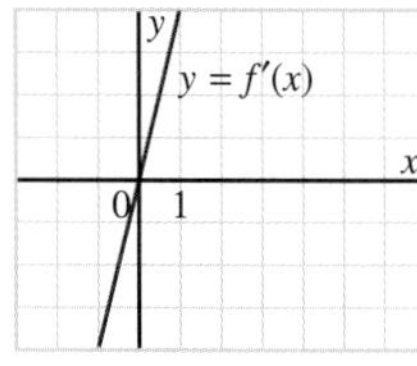

b)

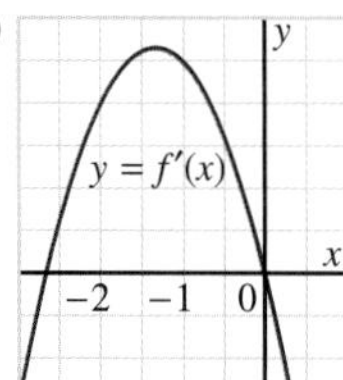

c)

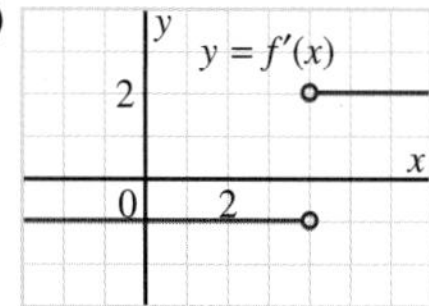

4. $y = -3x - 6$

5. a) $f'(x) = 8x$ **b)** $f'(x) = \dfrac{-1}{(3x - 1)^2}$
c) $f'(x) = \dfrac{-2}{\sqrt{1 - 4x}}$

6. $\left(\frac{2}{3}, \frac{20}{27}\right)$ and $\left(-\frac{1}{2}, -\frac{3}{4}\right)$

7. $y' = 12x^2$

8. Graphs may vary.

a)

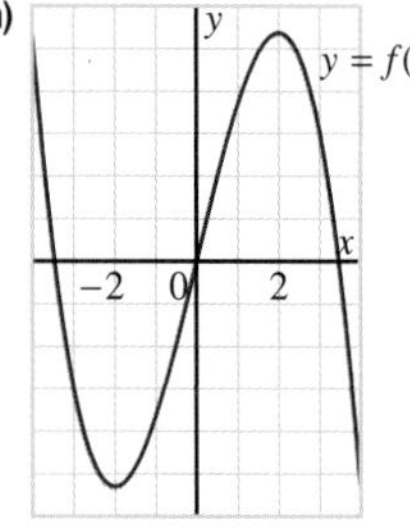

b)

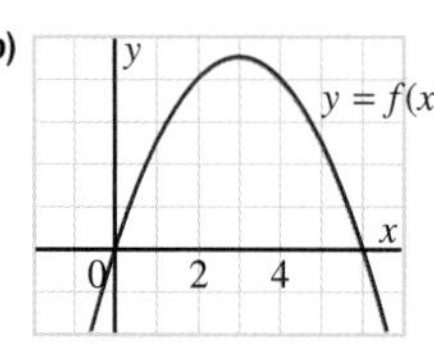

c)

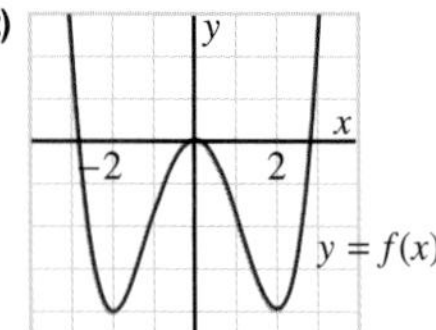

9. a) $\dfrac{dh}{dt} = -9.8t + 49$; the speed of the rocket
b) 123.5 m **c)** 34.3 m/s

10. $1 < x < 3$

11. Equations may vary. $y = (x - 3)(x - 2 + \sqrt{3}i)(x - 2 - \sqrt{3}i)$

12. $k = \frac{1}{2}$

13. a) x-intercepts: ±3; y-intercept: –27;
local maximum: (–3, 0);
local minimum: (1, –32);
point of inflection: (–1, –16);
increase: $x < -3$ and $x > 1$; decrease: $-3 < x < 1$;
concave up: $x > -1$; concave down: $x < -1$

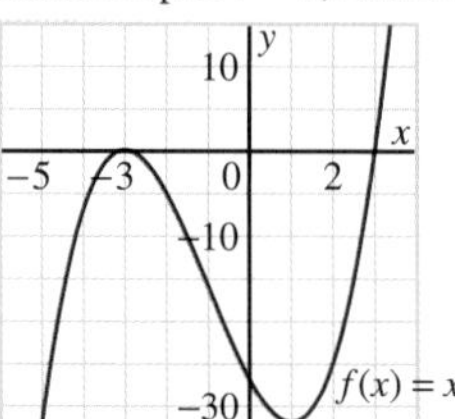

b) x-intercepts: ±1.41, ±4; y-intercept: –32;
local maximums: (±3, 49);
local minimum: (0, –32);
points of inflection: (±1.73, 13);
increase: $x < -3$ and $0 < x < 3$;
decrease: $-3 < x < 0$ and $x > 3$;
concave up: $-1.73 < x < 1.73$;
concave down: $x < -1.73$ and $x > 1.73$

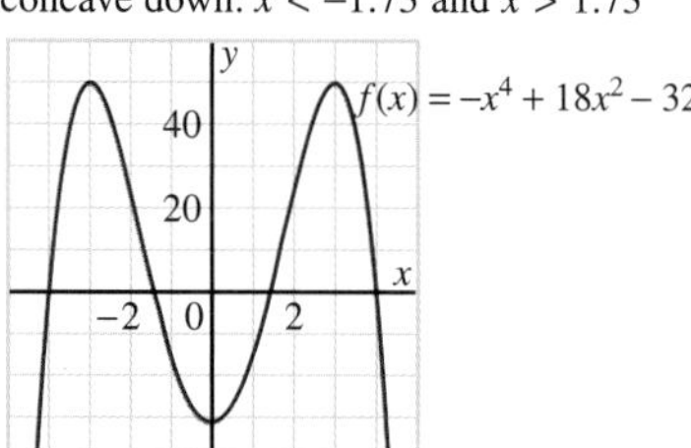

14. a) $v(t) = 3t^2 - 10t + 3$; $a(t) = 6t - 10$
b) $t = 0.\overline{3}$ s, 3 s **c)** $t > 1.\overline{6}$ s
d) Approximately 33.96 m **e)** –1 m/s

Chapter 4 Calculating Derivatives Using Differentiation Rules

Necessary Skills

1 Review: Algebraic Operations

Exercises, page 235

1. a) $6x^2 + 12x + 3$ **b)** $-12x^2 + 22x + 3$
c) $8x(2x^2 + 1)$ **d)** $3x^2(4x^3 - 5)$
e) $2x^2(x - 1)(5x - 3)$ **f)** $4x(2 - x)(1 - x)$

2. a) $(x + 1)(-2x^2 + 2x + 1)$ **b)** $2x(3 - 2x)(3 - 4x)$
c) $-x(4x^2 - 15x + 12)$ **d)** $-12x^3 - 33x^2 + 16x + 7$
e) $-24x^3 + 15x^2 - 14x + 2$ **f)** $2x(2x - 3)(4x - 3)$

3. a) $\dfrac{x^3 - 6}{2x^3}$ **b)** $\dfrac{-x^2 + 4x + 2}{(2 + x^2)^2}$
c) $\dfrac{x^2 - 2x - 3}{(x - 1)^2}$ **d)** $\dfrac{-2(x^2 + 3x - 3)}{(x^2 - 2x)^2}$

2 Review: Inverse of a Function

Exercises, page 237

1. a) $y = -x + 3$; yes, the inverse is a function because there is only one value of y for each value of x.

b) $y = 2x - 10$; yes, the inverse is a function because there is only one value of y for each value of x.

c) $y = \pm\sqrt{x + 4}$; no, the inverse is not a function because there is more than one value of y for each value of $x > -4$.

d) $y = \frac{1-x}{x}$; yes, the inverse is a function because there is only one value of y for each value of x.

e) $y = \frac{x+1}{x-1}$; yes, the inverse is a function because there is only one value of y for each value of x.

f) $y = \pm\sqrt{\frac{3-x}{x}}$; no, the inverse is not a function because there is more than one value of y for each value of $x < 0 < 3$.

2. a) i)

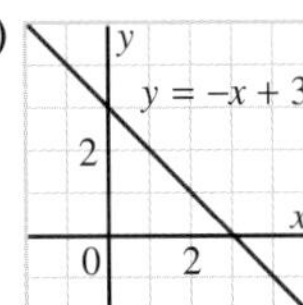

ii) The inverse is a function since its graph passes the vertical line test.

b) i)

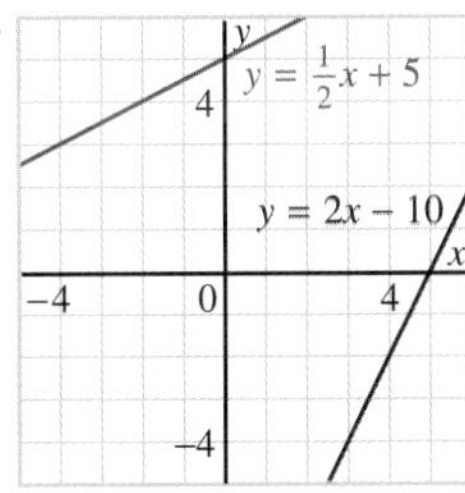

ii) The inverse is a function since its graph passes the vertical line test.

c) i)

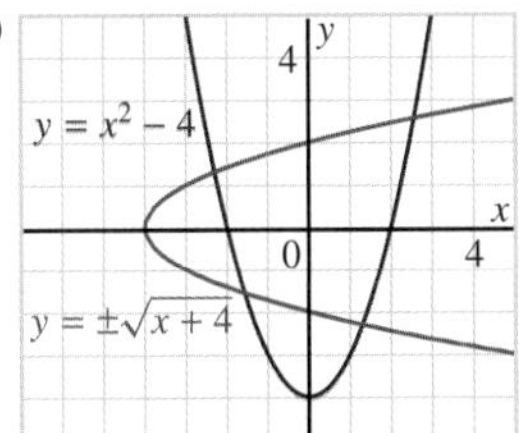

ii) The inverse is not a function since its graph fails the vertical line test.

3. a) $f^{-1}(x) = -\frac{1}{3}x + \frac{2}{3}$; f^{-1} is a function since there is only one value of f^{-1} for each x-value.

b) $f^{-1}(x) = \frac{4}{x-2}$; f^{-1} is a function since there is only one value of f^{-1} for each permissible x-value.

c) $f^{-1}(x) = \frac{x}{2x-1}$; f^{-1} is a function since there is only one value of f^{-1} for each permissible x-value.

4.1 Exercises, page 242

1. a) $\frac{dy}{dx} = 3(4x + 1)$ **b)** $\frac{dy}{dx} = x(9x + 4)$

c) $\frac{dy}{dx} = 9x^2 + 1$ **d)** $\frac{dy}{dx} = 5(1 - 4x)$

e) $\frac{dy}{dx} = 3x(2 - x)$ **f)** $\frac{dy}{dx} = 2(9x^2 + 12x + 5)$

2. a) $\frac{dy}{dx} = 2x + 5$ **b)** $\frac{dy}{dx} = 12x + 1$

c) $\frac{dy}{dx} = -4(3x + 4)$ **d)** $\frac{dy}{dx} = 2(9x^2 + 3x + 2)$

e) $\frac{dy}{dx} = 60x - 61$ **f)** $\frac{dy}{dx} = 2(2x^3 - 3x^2 + 1)$

3. a) i) $\frac{dy}{dx} = 2x + 1$ **ii)** $\frac{dy}{dx} = 2x + 3$

iii) $\frac{dy}{dx} = 2x + 5$ **iv)** $\frac{dy}{dx} = 2x + 7$

b) $\frac{dy}{dx} = 2x + 2k + 1$

5. a) $\frac{dy}{dx} = 10x(4x^2 + 3)$ **b)** $\frac{dy}{dx} = 2x(1 - 2x^2)$

c) $\frac{dy}{dx} = 8x^3 - 3x^2 + 4x - 1$

d) $\frac{dy}{dx} = 30x^4 + 36x^3 + 39x^2 + 5$

e) $\frac{dy}{dx} = -48x^3 - 99x^2 - 46x + 15$

f) $\frac{dy}{dx} = 3x^2 - 1$

6. a) $\frac{dy}{dx} = 2(x^2 + 3x)(2x + 3)$

b) $\frac{dy}{dx} = 2(2x - x^2)(2 - 2x)$

c) $\frac{dy}{dx} = 2(5x^3 - 2x^2)(15x^2 - 4x)$

d) $\frac{dy}{dx} = 2(3x^2 + 2x - 1)(6x + 2)$

10. $y = -6x + 5$

11. $(0, 1)$, $\left(\pm\frac{1}{\sqrt{2}}, 0\right)$

12. $\frac{dr}{dm} = -m(m - k)$

13. a) i) $\frac{dy}{dx} = 5$ **ii)** $\frac{dy}{dx} = 10x$ **iii)** $\frac{dy}{dx} = -\frac{5}{x^2}$

14. a) $\frac{dy}{dx} = 2x$ **b)** $\frac{dy}{dx} = 3x^2$

c) $\frac{dy}{dx} = 4x^3$ **d)** $\frac{dy}{dx} = 5x^4$

15. a) $\frac{dy}{dx} = -2x$ **b)** $\frac{dy}{dx} = 3x^2$

c) $\frac{dy}{dx} = -4x^3$ **d)** $\frac{dy}{dx} = 5x^4$

16. a) $\frac{dy}{dx} = 3x^2 + 2x + 2$

b) $\frac{dy}{dx} = 6x^5 + 5x^4 + 8x^3 + 15x^2 + 6x + 6$

c) $\frac{dy}{dx} = 10x^9 + 9x^8 + 16x^7 + 35x^6 + 42x^5 + 50x^4 + 56x^3 + 60x^2 + 24x + 24$

17. a) $(fgh)' = f'gh + fg'h + fgh'$

c) $(fghk)' = f'ghk + fg'hk + fgh'k + fghk'$

18. a) $C(t) = 65t(20 - t)(8 - 0.5t)$

b) Approximately $0 \le t \le 5.89$ and $18.11 \le t \le 20$

19. $x = \pm 1$

20. a) $\frac{dy}{dx} = -\sin^2 x + \cos^2 x$

b) $\frac{dy}{dx} = x(x\cos x + 2\sin x)$

c) $\frac{dy}{dx} = x^2(x\cos x + 3\sin x)$

d) $\frac{dy}{dx} = 2\sin x\cos x$

e) $\frac{dy}{dx} = 4(x\cos x + \sin x)$

f) $\frac{dy}{dx} = -2(\sin^2 x - \cos^2 x)$

21. a) $\frac{dy}{dx} = \frac{3x}{2\sqrt{x}}$ **b)** $\frac{dy}{dx} = \frac{3x+1}{2\sqrt{x}}$ **c)** $\frac{dy}{dx} = \frac{5x^2}{2\sqrt{x}}$

23. (0, −64), (64, 64)

4.2 Exercises, page 247

1. a) $\frac{dy}{dx} = \frac{1}{(x+1)^2}$ **b)** $\frac{dy}{dx} = \frac{x(x+2)}{(x+1)^2}$

c) $\frac{dy}{dx} = \frac{x^2(2x+3)}{(x+1)^2}$ **d)** $\frac{dy}{dx} = \frac{-1}{(x+1)^2}$

e) $\frac{dy}{dx} = \frac{-x(x+2)}{(x+1)^2}$ **f)** $\frac{dy}{dx} = \frac{-x^2(2x+3)}{(x+1)^2}$

2. a) $\frac{dy}{dx} = \frac{1}{x^2}$ **b)** $\frac{dy}{dx} = \frac{-x+2}{x^3}$

c) $\frac{dy}{dx} = \frac{-2x+3}{x^4}$ **d)** $\frac{dy}{dx} = -\frac{1}{x^2}$

e) $\frac{dy}{dx} = \frac{x-2}{x^3}$ **f)** $\frac{dy}{dx} = \frac{2x-3}{x^4}$

3. a) $\frac{dy}{dx} = \frac{1}{(x+1)^2}$ **b)** $\frac{dy}{dx} = \frac{3}{(x+3)^2}$

c) $\frac{dy}{dx} = \frac{1}{(3x+1)^2}$ **d)** $\frac{dy}{dx} = \frac{1}{(x+3)^2}$

e) $\frac{dy}{dx} = \frac{-1}{(x-2)^2}$ **f)** $\frac{dy}{dx} = \frac{2}{(5x+2)^2}$

4. a) $\frac{dy}{dx} = \frac{3}{(2x+1)^2}$ **b)** $\frac{dy}{dx} = \frac{2}{(3x+2)^2}$

c) $\frac{dy}{dx} = \frac{-1}{(3x+1)^2}$ **d)** $\frac{dy}{dx} = \frac{-5}{(x-3)^2}$

e) $\frac{dy}{dx} = \frac{7}{(3x+2)^2}$ **f)** $\frac{dy}{dx} = \frac{-11}{(5x-6)^2}$

5. a) i) $\frac{dy}{dx} = \frac{1}{(x+1)^2}$ **ii)** $\frac{dy}{dx} = \frac{1}{(x+2)^2}$

iii) $\frac{dy}{dx} = \frac{1}{(x+3)^2}$ **iv)** $\frac{dy}{dx} = \frac{1}{(x+4)^2}$

b) $\frac{dy}{dx} = \frac{1}{(x+k+1)^2}$

7. a) $\frac{dy}{dx} = \frac{-2x}{(x^2+2)^2}$ **b)** $\frac{dy}{dx} = \frac{2x}{(1-x^2)^2}$

c) $\frac{dy}{dx} = \frac{1}{(x-1)^2}$ **d)** $\frac{dy}{dx} = \frac{-2x-1}{(x^2+x)^2}$

e) $\frac{dy}{dx} = \frac{-1+2x}{(x-x^2)^2}$ **f)** $\frac{dy}{dx} = \frac{-2(x+1)}{(x^2+2x+3)^2}$

8. a) $\frac{dy}{dx} = \frac{-x^2+2}{(x^2+2)^2}$ **b)** $\frac{dy}{dx} = \frac{4x}{(1-x^2)^2}$

c) $\frac{dy}{dx} = \frac{2x-x^2}{(x-1)^2}$ **d)** $\frac{dy}{dx} = \frac{3}{(x+1)^2}$

e) $\frac{dy}{dx} = \frac{2x-x^2}{(x-1)^2}$ **f)** $\frac{dy}{dx} = \frac{4x(x+3)}{(x^2+2x+3)^2}$

9. a) $\frac{dy}{dx} = \frac{-2-3x}{x^3(1+x)^2}$ **b)** $\frac{dy}{dx} = \frac{-15}{(5-x)^2}$

c) $\frac{dy}{dx} = \frac{2x}{(1-2x^2)^2}$

12. a) i) $\frac{dy}{dx} = \frac{-1}{x^2}$ **ii)** $\frac{dy}{dx} = \frac{-2}{x^3}$ **iii)** $\frac{dy}{dx} = \frac{-3}{x^4}$

iv) $\frac{dy}{dx} = \frac{-4}{x^5}$ **v)** $\frac{dy}{dx} = \frac{-5}{x^6}$

b) i) $\frac{dy}{dx} = \frac{-1}{x^2}$ **ii)** $\frac{dy}{dx} = \frac{-2}{x^3}$ **iii)** $\frac{dy}{dx} = \frac{-3}{x^4}$

iv) $\frac{dy}{dx} = \frac{-4}{x^5}$ **v)** $\frac{dy}{dx} = \frac{-5}{x^6}$

c) i) $\frac{dy}{dx} = \frac{-4}{x^3}$ **ii)** $\frac{dy}{dx} = \frac{-3}{2x^4}$ **iii)** $\frac{dy}{dx} = \frac{20}{x^5}$

iv) $\frac{dy}{dx} = \frac{9}{x^7}$

13. $\frac{dy}{dx} = -\frac{f'(x)}{[f(x)]^2}$

14. $y = \frac{1}{4}x + \frac{1}{4}$

15. (0, 0) and (2, 4)

16. a) $y = \frac{-x}{(a-1)^2} + \frac{a^2}{(a-1)^2}$ **b)** $x = a^2$

18. a) $\frac{dy}{dx} = \frac{3(\sin x - x\cos x)}{\sin^2 x}$ **b)** $\frac{dy}{dx} = \frac{-x\sin x - \cos x}{3x^2}$

c) $\frac{dy}{dx} = \frac{x\cos x - \sin x}{x^2}$ **d)** $\frac{dy}{dx} = \frac{1}{\cos^2 x}$

e) $\frac{dy}{dx} = \frac{-1}{\sin^2 x}$ **f)** $\frac{dy}{dx} = \frac{\sin x}{\cos^2 x}$

19. a) $\frac{dy}{dx} = \frac{-3}{2\sqrt{x}(2\sqrt{x}-1)^2}$ **b)** $\frac{dy}{dx} = \frac{-x-2}{2\sqrt{x}(x-2)^2}$

c) $\frac{dy}{dx} = \frac{6\sqrt{x}\cos x - 2x\cos x + \sin x}{2\sqrt{x}(3-\sqrt{x})^2}$

4.3 Exercises, page 254

1. a) $\frac{dy}{dx} = 12(2x+2)^5$ **b)** $\frac{dy}{dx} = 20(2x-5)^9$

c) $\frac{dy}{dx} = -15(9-3x)^4$ **d)** $\frac{dy}{dx} = 6(5+3x)$

2. a) $\frac{dy}{dx} = 6(2x+1)^2$ **b)** $\frac{dy}{dx} = -35(1-5x)^6$

c) $\frac{dy}{dx} = -12(5-3x)^3$ **d)** $\frac{dy}{dx} = 35(7x-5)^4$

3. a) $\frac{dy}{dx} = -8(4x+1)^{-3}$ **b)** $\frac{dy}{dx} = 12(3-2x)^{-7}$

c) $\frac{dy}{dx} = -12(3x+5)^{-5}$ **d)** $\frac{dy}{dx} = -15(2+5x)^{-4}$

4. a) $\frac{dy}{dx} = -24(4x+2)^{-7}$ **b)** $\frac{dy}{dx} = 8(1-2x)^{-5}$

c) $\frac{dy}{dx} = 15(-2-3x)^{-6}$ **d)** $\frac{dy}{dx} = -18(6x-5)^{-4}$

5. a) $\frac{dy}{dx} = 6x(x^2+1)^2$ **b)** $\frac{dy}{dx} = -20x(2x^2+3)^4$

c) $\frac{dy}{dx} = 2(6x+5)(3x^2+5x-1)$

d) $\frac{dy}{dx} = -8(x-1)(x^2-2x+5)^3$

6. a) $\frac{dy}{dx} = -4x(1-x^2)^{-3}$

b) $\frac{dy}{dx} = 5(1+6x)(3-x-3x^2)^{-6}$

c) $\frac{dy}{dx} = -42x(2-3x^2)^{-8}$

d) $\frac{dy}{dx} = -4(4x+3)(2x^2+3x+1)^{-5}$

7. a) $\frac{dy}{dx} = \frac{-8x}{(1+x^2)^5}$ **b)** $\frac{dy}{dx} = \frac{-18x^2}{(2-3x^3)^3}$

c) $\frac{dy}{dx} = \frac{-3(2x+1)}{(x^2+x+1)^4}$ **d)** $\frac{dy}{dx} = \frac{-8x(1+x^2)}{(2x^2+x^4)^3}$

8. a) $\frac{dy}{dx} = \frac{-3x^2}{(x-1)^4}$ **b)** $\frac{dy}{dx} = \frac{x-2}{x^3}$

c) $\frac{dy}{dx} = \frac{5(x^2+4)(-4+x^2)^4}{x^6}$ **d)** $\frac{dy}{dx} = \frac{-4x^7(x^3-18)}{(x^3+9)^5}$

9. a) $\frac{dy}{dx} = (x-1)(x+3)^2(5x+3)$

b) $\frac{dy}{dx} = 4(2x^2+3)^2(1-x)^3(5x^2-3x+3)$

c) $\frac{dy}{dx} = 2(2x-x^2)^3(x+5)(-5x^2-14x+20)$

d) $\frac{dy}{dx} = -6x(x^2-2)^2(x^2+3)^2(2x^2+1)$

10. a) $\frac{dy}{dx} = \frac{-(x+1)(x-1)}{2x^2}$

b) $\frac{dy}{dx} = \frac{2(x^2+1)^2(2x^2-1)}{x^3}$

c) $\frac{dy}{dx} = \frac{-3(x+2)^2(x-2)}{(3x+2)^3}$

d) $\frac{dy}{dx} = \frac{(1-x)(x+1)}{x^2}$

e) $\frac{dy}{dx} = \frac{-3(1-x^2)^2(x^2+1)}{x^4}$

f) $\frac{dy}{dx} = \frac{(-4x^2-13x-8)(1-x-x^2)^2}{(x+2)^3}$

12. a) $\frac{dy}{dx} = 5(4x-3)(2x^2-3x+6)^4$

b) $\frac{dy}{dx} = \frac{-18x}{(2-3x^2)^4}$

c) $\frac{dy}{dx} = 2x(x^4-x^2)(3x^2-1)^2(21x^4-19x^2+2)$

d) $\frac{dy}{dx} = \frac{(x^2-x)^4(24x^2+31x-20)}{(3x+4)^3}$

13. a) $\frac{dy}{dx} = -\sin x(\sin^2 x - 2\cos^2 x)$

b) $\frac{dy}{dx} = \cos x(\cos^2 x - 2\sin^2 x)$

c) $\frac{dy}{dx} = 2\sin x\cos x(\cos^2 x - \sin^2 x)$

d) $\frac{dy}{dx} = \cos x(\cos^3 x - 3\sin^2 x\cos x)$

14. a) 64 cm^3 **b)** 25 s **c)** -4.8 cm^3/s

d) i) At $t = 5$ s **ii)** -6.4 cm^3/s

e) 4.8 cm^3/s

15. a) $\frac{dy}{dx} = \frac{-3f'(x)}{[f(x)]^4}$

16. a) $\frac{dy}{dx} = \frac{-nf'(x)}{[f(x)]^{n+1}}$

17. $y = 56x - 96$

18. $y = \frac{3}{8}x - \frac{1}{2}$

19. (2.5, 2.25) and (1.5, -0.25)

20. a) 2 m

b) -2 m/s^2; the velocity of the object is decreasing.

21. b) $\frac{dy}{dx} = \frac{-x^3-2x^2+9x+12}{x^5}$

22. (1.5, -0.25)

23. a) $f'(x) = \frac{4x(x^2-1)^2(x^2+2)}{(x^2+1)^2}$

b)

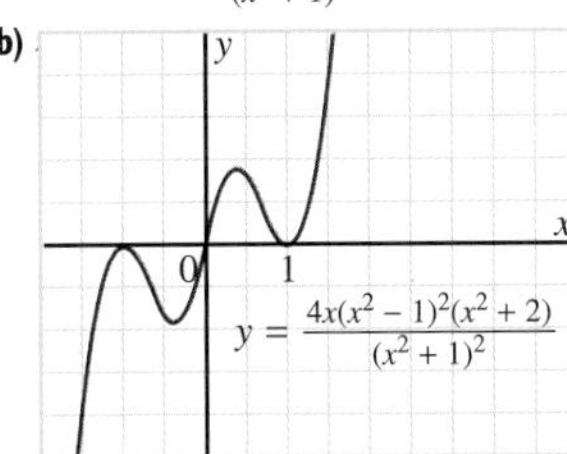

Self-Check 4.1–4.3, page 257

1. a) $\frac{dy}{dx} = 3x^2 + 3$ **b)** $\frac{dy}{dx} = 8x^3 - 6x^2$

c) $\frac{dy}{dx} = 50x^4 + 60x^3 - 45x^2$ **d)** $\frac{dy}{dx} = -3x^2 - 4x + 1$

2. a) $\frac{dy}{dx} = \frac{x^2+6x}{(x+3)^2}$ **b)** $\frac{dy}{dx} = \frac{-4}{(2+x)^2}$

c) $\frac{dy}{dx} = \frac{-3}{(x-1)^2}$ **d)** $\frac{dy}{dx} = \frac{-4}{(x-2)^2}$

3. a) $\frac{dy}{dx} = \frac{-6x-1}{(3x^2+x)^2}$ **b)** $\frac{dy}{dx} = \frac{4x^4+12x^2}{(x^2+1)^2}$

c) $\frac{dy}{dx} = 6x^2 - 4x + 3$ **d)** $\frac{dy}{dx} = -9x^2 - 10x + 3$

4. a) $\frac{dy}{dx} = -\sin^2 x + \cos x + \cos^2 x$

b) $\frac{dy}{dx} = x^2(x\cos x + 3\sin x + 6)$

c) $\frac{dy}{dx} = \frac{1+\sin x - x\cos x}{(1+\sin x)^2}$ **d)** $\frac{dy}{dx} = \frac{-1}{2\sin^2 x}$

5. a) $\frac{dy}{dx} = 18x(4+3x^2)^2$

b) $\frac{dy}{dx} = -4(1-4x)(x-2x^2)^3$

c) $\frac{dy}{dx} = -5(21x^2+1)(7x^3+x)^{-6}$

d) $\frac{dy}{dx} = 2(1-12x)(1+x-6x^2)^{-3}$

6. $y = -\frac{1}{2}x + \frac{7}{2}$

7. (5, 5) and (-3, 3)

4.4 Exercises, page 264

1. a) 1 **b)** 7 **c)** 11 **d)** 10

2. $g(f(x)) = 4x - 1$; $f(g(x)) = 4x + 2$

3. a) 1 **b)** -3 **c)** 3 **d)** -7

4. $f \circ g(x) = 2 - 9x^2$; $g \circ f(x) = -6 + 3x^2$

5. a) $f(g(x)) = -x^2 + 6x - 8$; $g(f(x)) = x^2 + 2$

b) $f(g(x)) = -2x + 64x^3$; $g(f(x)) = -2x + 4x^3$

c) $f(g(x)) = 8x^4 - 8x^3 + 20x^2 - 9x + 12$;
$g(f(x)) = -8x^4 + 8x^3 - 16x^2 + 7x - 8$

6. $C = 2\pi r$

7. $V = \frac{1}{8}\pi d^3$

8. a) $P = 4\sqrt{A}$ **b)** $A = \frac{P^2}{16}$

9. a) $V = \left(\frac{A}{6}\right)^{\frac{3}{2}}$ **b)** $A = 6V^{\frac{2}{3}}$

10. a) $f(g(x)) = 3x - 10$ **b)** $g(f(x)) = 3x - 2$

c) $f(h(x)) = \frac{2x+5}{x+1}$ **d)** $h(f(x)) = \frac{1}{3x+3}$

e) $g(h(x)) = \frac{-4x-3}{x+1}$ **f)** $h(g(x)) = \frac{1}{x-3}$

11. a) $x \in \Re$ **b)** $x \in \Re$

c) $\{x \mid x \in \Re, x \neq -1\}$ **d)** $\{x \mid x \in \Re, x \neq -1\}$

e) $\{x \mid x \in \Re, x \neq -1\}$ **f)** $\{x \mid x \in \Re, x \neq 3\}$

12. a) $f \circ g = \frac{8-2x}{5-x}$ **b)** $g \circ f = \frac{2}{3+x}$

c) $f \circ h = \frac{2x-3}{x-1}$ **d)** $h \circ f = \frac{1}{1-x}$

e) $g \circ h = \frac{2x-2}{5x-6}$ **f)** $h \circ g = \frac{5-x}{-3+x}$

13. a) $\{x \mid x \in \Re,\ x \neq 5\}$ **b)** $\{x \mid x \in \Re,\ x \neq -3\}$
c) $\{x \mid x \in \Re,\ x \neq 1\}$ **d)** $\{x \mid x \in \Re,\ x \neq 1\}$
e) $\{x \mid x \in \Re,\ x \neq 1, \frac{6}{5}\}$ **f)** $\{x \mid x \in \Re,\ x \neq 3, 5\}$

14. a) $f(f(x)) = x + 10$ **b)** $f(f(x)) = 4x + 3$
c) $f(f(x)) = x^4 + 2x^3 + 2x^2 + x$
d) $f(f(x)) = \sqrt{\sqrt{x+1} + 1}$

15. a) $f^{-1}(x) = \frac{x-3}{2}$

16. a) $f^{-1}(x) = \frac{1-x}{2}$ **b)** $f^{-1}(x) = \pm\sqrt{x^2 - 1}$ **c)** $f^{-1}(x) = \frac{1-5x}{x}$

18. a) i) $D_f = \Re;\ D_g = \Re$
ii) $R_f = \Re;\ R_g = \Re$
iii) $f \circ g = 10x + 7$
iv) $D = \Re$ **v)** $R = \Re$
b) i) $D_f = \Re;\ D_g = \Re$
ii) $R_f = \{y \mid y \in \Re,\ y \geq 2\};\ R_g = \Re$
iii) $f \circ g = 9x^2 + 30x + 27$
iv) $D = \Re$ **v)** $R = \{y \mid y \in \Re,\ y \geq 2\}$
c) i) $D_f = \Re;\ D_g = \{x \mid x \in \Re,\ x \geq 0\}$
ii) $R_f = \{y \mid y \in \Re,\ y \leq 4\};\ R_g = \{y \mid y \in \Re,\ y \geq 0\}$
iii) $f \circ g = 4 - x$
iv) $D = \{x \mid x \in \Re,\ x \geq 0\}$ **v)** $R = \{y \mid y \in \Re,\ y \leq 4\}$
d) i) $D_f = \{x \mid x \in \Re,\ x \neq 0\};\ D_g = \Re$
ii) $R_f = \{y \mid y \in \Re,\ y > 0\};\ R_g = \{y \mid y \in \Re,\ y \leq 5\}$
iii) $f \circ g = \frac{1}{(5 - x^2)^2}$
iv) $D = \{x \mid x \in \Re,\ x \neq \pm\sqrt{5}\}$ **v)** $R = \{y \mid y \in \Re,\ y > 0\}$

20. a) $f(x) = x^2;\ g(x) = x + 1$ **b)** $f(x) = -x^2;\ g(x) = x + 1$
c) $f(x) = \frac{1}{x};\ g(x) = x + 1$ **d)** $f(x) = -\frac{1}{x};\ g(x) = x + 1$
e) $f(x) = \sqrt{x};\ g(x) = x + 1$ **f)** $f(x) = -2\sqrt{x};\ g(x) = x + 1$

21. a) $f(x) = x^4;\ g(x) = x^2 + 3$ **b)** $f(x) = \sqrt{x};\ g(x) = 2x - 1$
c) $f(x) = \frac{1}{x};\ g(x) = 3x + 2$ **d)** $f(x) = -x^3;\ g(x) = 4x - 5$
e) $f(x) = x^2 + 3x - 4;\ g(x) = 2x + 3$
f) $f(x) = -2\sqrt{x};\ g(x) = x - 1$

22. a) $f(x) = x^2 - 2x + 3;\ g(x) = x + 1$
b) $f(x) = \sqrt{x};\ g(x) = 4x - 5$
c) $f(x) = \frac{1}{x};\ g(x) = 2x + 3$
d) $f(x) = x^2 - 4x + 4;\ g(x) = x^2$

23. Answers may vary.
a) $f(x) = \sqrt{x};\ g(x) = 1 - x^2$ or $f(x) = \sqrt{1 - x};\ g(x) = x^2$
b) $f(x) = x^2;\ g(x) = 3x^2 + 4$ or $f(x) = (x + 4)^2;\ g(x) = 3x^2$
c) $f(x) = \frac{1}{x};\ g(x) = 2x - 4$ or $f(x) = \frac{1}{x-4};\ g(x) = 2x$

24. $k(t) = 38.416t^2 - 384.16t + 960.4$

25. Answers may vary. $f(x) = x;\ g(x) = x + 1$

4.5 Exercises, page 271

1. a) $\frac{dy}{dx} = 8(4x + 3)$ **b)** $\frac{dy}{dx} = 10x(x^2 - 3)^4$
c) $\frac{dy}{dx} = 6x^2(x^3 + 1)$ **d)** $\frac{dy}{dx} = 12(3x - 2)^3$
e) $\frac{dy}{dx} = 10x(x^2 - 2)^4$
f) $\frac{dy}{dx} = 10(x - 1)(x^2 - 2x - 3)^4$

2. a) $\frac{dy}{dx} = -9(2 - 3x)^2$ **b)** $\frac{dy}{dx} = -8x(1 - 2x^2)$
c) $\frac{dy}{dx} = -48x^2(3 - 4x^3)^3$ **d)** $\frac{dy}{dx} = 4x(5 + x^2)$
e) $\frac{dy}{dx} = 3(-6x - 1)(-3x^2 - x)^2$ **f)** $\frac{dy}{dx} = 80x^3(5 + 10x^4)$

3. a) $\frac{dy}{dx} = -4(1 + 2x)^{-3}$ **b)** $\frac{dy}{dx} = (2x - 3)(-x^2 + 3x)^{-2}$
c) $\frac{dy}{dx} = 3(x - 2)^{-4}$
d) $\frac{dy}{dx} = 3(2x - 3x^2)(-x^2 + x^3)^{-4}$
e) $\frac{dy}{dx} = -4(1 + 6x)(2 + x + 3x^2)^{-5}$
f) $\frac{dy}{dx} = (6x + 6x^2)(3x^2 + 2x^3)^{-2}$

4. a) $\frac{dy}{dx} = \frac{-2}{(2x-1)^2}$ **b)** $\frac{dy}{dx} = \frac{12}{(2-4x)^4}$
c) $\frac{dy}{dx} = \frac{2x+2}{(x^2-2x)^2}$ **d)** $\frac{dy}{dx} = \frac{-6x^2}{(x^3+3)^3}$
e) $\frac{dy}{dx} = \frac{2+12x}{(2x+6x^2)^2}$ **f)** $\frac{dy}{dx} = \frac{4(2x-1)}{(x^2-x-1)^5}$

5. a) $\frac{dy}{dx} = \frac{1}{2\sqrt{1+x}}$ **b)** $\frac{dy}{dx} = \frac{2x-1}{2\sqrt{x^2-x}}$
c) $\frac{dy}{dx} = -\frac{3x^2-1}{2\sqrt{x^3-x}}$ **d)** $\frac{dy}{dx} = \frac{-1+2x}{2\sqrt{2-x+x^2}}$
e) $\frac{dy}{dx} = \frac{2x}{\sqrt{3-2x^2}}$ **f)** $\frac{dy}{dx} = \frac{1+2x^3}{\sqrt{2x+x^4}}$

6. a) $\frac{dy}{dx} = \frac{-6(x+1)^2}{(x-1)^4}$ **b)** $\frac{dy}{dx} = \frac{2(x^2+1)(x^2+2x-1)}{(x+1)^3}$
c) $\frac{dy}{dx} = \frac{3(1-2x)^2(4x^2-4x-2)}{(1+2x^2)^4}$ **d)** $\frac{dy}{dx} = \frac{2(x^2+x)(-x^4-2x^3+x^2)}{(x+x^3)^3}$
e) $\frac{dy}{dx} = \frac{4(1+x+x^2)^3(4x+2)}{(1-x-x^2)^5}$ **f)** $\frac{dy}{dx} = \frac{3(2-x^2)^2(x^4-6x^2-4x)}{(2+x^3)^4}$

7. a) $\frac{dy}{dx} = 5(4x - 3)(2x^2 - 3x)^4$ **b)** $\frac{dy}{dx} = -\frac{(1-2x+3x^2)}{(x-x^2+x^3)^2}$
c) $\frac{dy}{dx} = \frac{21(4x-3)^2}{(4-3x)^4}$ **d)** $\frac{dy}{dx} = \frac{-8x}{(x^2-3)^5}$

8. a) $\frac{dy}{dx} = 2\cos 2x$ **b)** $\frac{dy}{dx} = -3\sin 3x$
c) $\frac{dy}{dx} = 2x\cos x^2$ **d)** $\frac{dy}{dx} = -3x^2 \sin x^3$
e) $\frac{dy}{dx} = \frac{-1}{x^2}\cos\frac{1}{x}$ **f)** $\frac{dy}{dx} = \frac{2}{x^3}\sin\frac{1}{x^2}$

10. $y = -2x + 10$

11. $y = 0$ and $y = 1$

12. $\frac{dT}{dt} = 0.05$°C/s

13. $\frac{dF}{dt} = 5.4$°F/min

14. a) 5 **b)** $2y'\sqrt{5x + 1}$ **c)** $y' = \frac{5}{2\sqrt{5x+1}}$

15. a) $y' = \frac{-1}{(x-1)^2};\ y'' = \frac{2}{(x-1)^3}$
b) $y''' = \frac{-6}{(x-1)^4};\ y'''' = \frac{24}{(x-1)^5}$
c) $y^{(n)} = \frac{(-1)^n n!}{(x-1)^{n+1}}$

16. a) $y = 3.5x - 1.5$
b) Approximately (–2.32, –9.63)

17. (–1, 3) and (5, –21)

4.6 Exercises, page 279

1. a) Explicit **b)** Implicit **c)** Explicit
d) Implicit **e)** Implicit **f)** Implicit

2. a) $\frac{dy}{dx} = -\frac{3}{5}$ **b)** $\frac{dy}{dx} = \frac{2}{7}$ **c)** $\frac{dy}{dx} = \frac{2}{3}$

3. a) $y = -\frac{3}{5}x + \frac{4}{5};\ \frac{dy}{dx} = -\frac{3}{5}$ **b)** $y = \frac{2}{7}x - 2;\ \frac{dy}{dx} = \frac{2}{7}$
c) $y = \frac{2}{3}x + \frac{4}{3};\ \frac{dy}{dx} = \frac{2}{3}$

4. a) $\frac{dy}{dx} = -\frac{2}{3}$ **b)** $\frac{dy}{dx} = -\frac{2}{3}$

5. a) $\frac{dy}{dx} = -\frac{y}{x}$ **b)** $\frac{dy}{dx} = -\frac{2y}{x}$ **c)** $\frac{dy}{dx} = -\frac{y}{2x}$
d) $\frac{dy}{dx} = \frac{1-y}{x}$ **e)** $\frac{dy}{dx} = -\frac{y}{x}$ **f)** $\frac{dy}{dx} = \frac{1-2xy^2}{2x^2y}$

6. a) $\frac{dy}{dx} = \frac{5x}{y}$ **b)** $\frac{dy}{dx} = -2x$ **c)** $\frac{dy}{dx} = -\frac{1}{2y}$
d) $\frac{dy}{dx} = -3x^2$ **e)** $\frac{dy}{dx} = \frac{x}{2y}$ **f)** $\frac{dy}{dx} = -\frac{x^2}{y^2}$

7. a) $\frac{dy}{dx} = \frac{1}{3}x^{-\frac{2}{3}}$ **b)** $\frac{dy}{dx} = \frac{3}{4}x^{-\frac{1}{4}}$ **c)** $\frac{dy}{dx} = -\frac{2}{3}x^{-\frac{5}{3}}$

9. a) $\frac{dy}{dx} = \frac{3}{2}(3x-2)^{\frac{-1}{2}}$ **b)** $\frac{dy}{dx} = -\frac{1}{2}(2x+1)(x^2+x)^{-\frac{3}{2}}$
c) $\frac{dy}{dx} = -3x(4-x^2)^{\frac{1}{2}}$
d) $\frac{dy}{dx} = \frac{1}{3}(2x+1)(x^2+x-2)^{\frac{-2}{3}}$
e) $\frac{dy}{dx} = x^2(-5-x^3)^{-\frac{4}{3}}$
f) $\frac{dy}{dx} = -\frac{3}{2}(6-8x^3)(6x-2x^4)^{-\frac{5}{2}}$

10. a) $\frac{dy}{dx} = -\frac{x}{2y}$ **b)** $\frac{dy}{dx} = -\frac{5x}{y}$ **c)** $\frac{dy}{dx} = -\frac{x}{16y}$

11. a) $\frac{dy}{dx} = \frac{x}{y}$ **b)** $\frac{dy}{dx} = \frac{2x}{3y}$
c) $\frac{dy}{dx} = \frac{4x}{3y}$ **d)** $\frac{dy}{dx} = \frac{9x}{4y}$

12. a) i) $\frac{dy}{dx} = \frac{1}{2y};\ y \neq 0$ **ii)** $\frac{dy}{dx} = \frac{1}{2y};\ y \neq 0$
iii) $\frac{dy}{dx} = \frac{1}{2(y-2)};\ y \neq 2$

b) i)

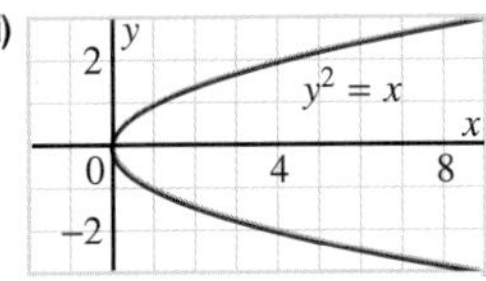

ii)

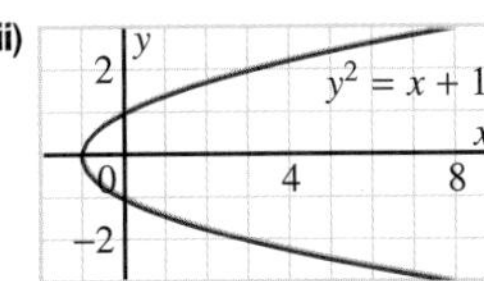

iii)

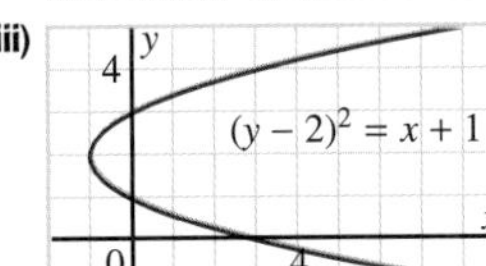

13. a) i) $\frac{dy}{dx} = -\frac{x}{y}$ **ii)** $\frac{dy}{dx} = -\frac{x}{y}$ **iii)** $\frac{dy}{dx} = -\frac{x}{y}$
b) All the values of $\frac{dy}{dx}$ in part a are the same.

14. a) $\frac{dy}{dx} = -\frac{3x}{4y}$ **b)** $y = -\frac{1}{2}x + 2$

15. a) $\frac{dy}{dx} = -\frac{16x}{9y}$ **b)** $y = \frac{4x}{3\sqrt{3}} + \frac{8}{\sqrt{3}}$

16. a) $\frac{dy}{dx} = \frac{x}{4y}$ **b)** $y = -\frac{x}{2\sqrt{5}} - \frac{4}{\sqrt{5}}$

19. $\frac{dy}{dx} = \frac{f'(x)}{2\sqrt{f(x)}}$

20. b) $y = \frac{7}{4}x - \frac{1}{2}$

22. a) $\frac{dy}{dx} = \frac{7}{8}x^{-\frac{1}{8}}$ **b)** $\frac{dy}{dx} = 3(4x+3)^{-\frac{1}{4}}$
c) $\frac{dy}{dx} = \left(\frac{-x}{y}\right)^{-\frac{2}{3}}$

23. $\frac{dy}{dx} = \frac{x^3 - y}{x - y^3}$

24. a) $\frac{dy}{dx} = \frac{x(1-y^2)}{y(x^2-1)}$ **b)** $\frac{dy}{dx} = \frac{y^2 - 2xy - 2x}{x^2 - 2xy + 2y}$
c) $\frac{dy}{dx} = \frac{y-2x}{2y-x}$

25. a) $\frac{d^2y}{dx^2} = \frac{2y^2 + 2xy}{(x+2y)^3}$ **b)** $\frac{d^2y}{dx^2} = \frac{6x^2 - 6xy + 6y^2}{(x-2y)^3}$

Self-Check 4.4–4.6, page 282

1. a) $f(g(x)) = 15x + 9;\ g(f(x)) = 15x - 1$
b) $f(g(x)) = x;\ g(f(x)) = x$

2. a) $f(g(x))$: $D = \Re$; $g(f(x))$: $D = \Re$
b) $f(g(x))$: $D = \{x \mid x \in \Re,\ x \neq 0\}$;
$g(f(x))$: $D = \{x \mid x \in \Re,\ x \neq 3\}$

3. a) $D_f = \Re$; $R_f = \{y \mid y \in \Re,\ 0 \leq y \leq 15\}$;
$D_g = \Re$; $R_g = \{y \mid y \in \Re,\ y \geq -1\}$
b) $f(g(x)) = \frac{15}{x^4 - 2x^2 + 2}$
c) $D = \Re$
d) $R = \{y \mid y \in \Re,\ 0 \leq y \leq 15\}$

4. Answers may vary.
a) $f(x) = \sqrt{x};\ g(x) = \frac{3x+1}{2}$
b) $f(x) = x^{-2};\ g(x) = 3x + 2$
c) $f(x) = x^2 + 1;\ g(x) = x^2 + 1$

5. a) $f^{-1}(x) = \frac{3-x}{2x}$ **b)** $f^{-1}(x) = \frac{2x+1}{x-1}$

6. a) $\frac{dy}{dx} = 12x(x^2+2)^5$
b) $\frac{dy}{dx} = (2x-1)(x+1)^2(10x+1)$
c) $\frac{dy}{dx} = -12(3x+5)^{-5}$

7. a) $\frac{dy}{dx} = \frac{-4(2x+3)}{(x^2+3x)^5}$ **b)** $\frac{dy}{dx} = -\frac{(2x+1)^2}{9x^4}$
c) $\frac{dy}{dx} = \frac{-12(1-3x^2)^3(3x^2+4x+1)}{(3x+2)^5}$

8. $y = -\frac{16}{5}x - \frac{7}{5}$ and $y = \frac{16}{5}x + \frac{7}{5}$

9. a) $\frac{dy}{dx} = -1$
b) The graph is 2 parallel lines with slope −1.

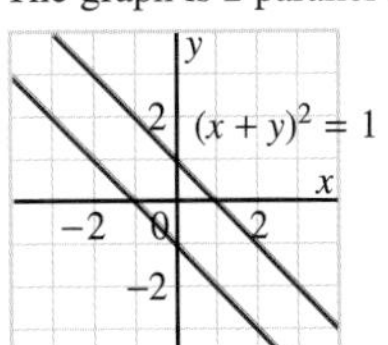

10. b) $y = 3x + 6$

11. a) $\frac{dy}{dx} = \frac{-x}{3y}$ **b)** $\frac{dy}{dx} = \frac{4x}{5y}$
c) $\frac{dy}{dx} = \frac{y(y-2x)}{x(x-2y)}$ **d)** $\frac{dy}{dx} = \frac{3}{4}x^{-\frac{1}{4}}$
e) $\frac{dy}{dx} = \frac{7-4x}{2(7x-2x^2)^{\frac{1}{2}}}$ **f)** $\frac{dy}{dx} = \frac{-y}{2y + x - 3y^2}$

Chapter 4 Review Exercises, page 283

1. a) $\frac{dy}{dx} = 2(1 - 6x)$ **b)** $\frac{dy}{dx} = -6x(2 + x)$
c) $\frac{dy}{dx} = -45x^2 + 12x - 5$ **d)** $\frac{dy}{dx} = x^2(5x^2 - 20x + 21)$
e) $\frac{dy}{dx} = -2(8x^3 - 3x^2 + 2)$
f) $\frac{dy}{dx} = -2(10x^4 + 4x^3 - 12x - 3)$

2. a) $\frac{d^2y}{dx^2} = -12$ **b)** $\frac{d^2y}{dx^2} = -12 - 12x$
c) $\frac{d^2y}{dx^2} = -90x + 12$ **d)** $\frac{d^2y}{dx^2} = 20x^3 - 60x^2 + 42x$
e) $\frac{d^2y}{dx^2} = -48x^2 + 12x$ **f)** $\frac{d^2y}{dx^2} = -80x^3 - 24x^2 + 24$

3. $y = 13x + 9$

4. a) $P = (250 + 10t)(200 + 7t)$ **b)** 3750 apples/year

5. a) $\frac{dy}{dx} = \sin^2 x - \cos^2 x$ **b)** $\frac{dy}{dx} = -x^2(x \sin x - 3\cos x)$
c) $\frac{dy}{dx} = \frac{5x^2}{2\sqrt{x}}$ **d)** $\frac{dy}{dx} = \frac{3x + 2}{2\sqrt{x + 1}}$

6. a) $\frac{dy}{dx} = \frac{x(x - 6)}{(x - 3)^2}$ **b)** $\frac{dy}{dx} = \frac{-x^2(x^2 + 6)}{(x^2 + 2)^2}$
c) $\frac{dy}{dx} = \frac{-x^2 - 2x - 5}{(x + 1)^2}$ **d)** $\frac{dy}{dx} = \frac{x^4 - 8x^3 + 2x - 4}{(x^3 - 1)^2}$
e) $\frac{dy}{dx} = \frac{-2x^2 - 12x}{(x^2 - x - 3)^2}$ **f)** $\frac{dy}{dx} = \frac{2x^4 + 2x^3 + 12x^2 - 6}{(2x^2 + x + 2)^2}$

7. a) $\frac{dy}{dx} = \frac{-1}{x^2}$ **b)** $\frac{dy}{dx} = \frac{-2x}{(x^2 + 2)^2}$
c) $\frac{dy}{dx} = \frac{2x}{(5 - x^2)^2}$ **d)** $\frac{dy}{dx} = \frac{-(3x^2 + 2)}{(x^3 + 2x)^2}$
e) $\frac{dy}{dx} = \frac{1 - 6x^2}{(x - 2x^3)^2}$ **f)** $\frac{dy}{dx} = \frac{9x^2 + 2}{(3x^3 + 2x)^2}$

8. a) $\frac{dy}{dx} = \frac{-1}{2\sqrt{x^3}}$ **b)** $\frac{dy}{dx} = \frac{-\cos x}{\sin^2 x}$
c) $\frac{dy}{dx} = \frac{1}{\sin^2 x}$ **d)** $\frac{dy}{dx} = \frac{1}{2\sqrt{x}}$
e) $\frac{dy}{dx} = \frac{x - 1}{2\sqrt{x}(x + 1)^2}$ **f)** $\frac{dy}{dx} = \frac{5(\cos x + x \sin x)}{\cos^2 x}$

9. $y = \frac{3}{4}x + \frac{1}{2}$

10. $y = -3x + 5$

11. a) $\frac{dy}{dx} = 12(3x - 5)^3$ **b)** $\frac{dy}{dx} = -21(2 - 7x)^2$
c) $\frac{dy}{dx} = -5(2x + 8)(x^2 + 8x)^4$ **d)** $\frac{dy}{dx} = -8(3 + 4x)^{-3}$
e) $\frac{dy}{dx} = -3(2x + 1)(x^2 + x)^{-4}$ **f)** $\frac{dy}{dx} = -(3x^2 - 5)(x^3 - 5x)^{-2}$

12. a) $\frac{dy}{dx} = \frac{-15}{(1 + 3x)^6}$ **b)** $\frac{dy}{dx} = \frac{3(2x - 3x^2)}{(x^2 - x^3)^4}$
c) $\frac{dy}{dx} = \frac{-2(-2 + 6x)}{(5 - 2x + 3x^2)^3}$

13. a) $\frac{dy}{dx} = \frac{6x^5}{(x + 1)^7}$ **b)** $\frac{dy}{dx} = \frac{32(x - 3)^3}{(x + 5)^5}$
c) $\frac{dy}{dx} = \frac{8x}{(1 - 3x)^3}$
d) $\frac{dy}{dx} = \frac{3(x - 2x^2)^2(-2x^2 - 4x + 1)}{(x + 1)^4}$
e) $\frac{dy}{dx} = \frac{2(5x^2 - x^3)(-2x^3 + 2x^2 + 10x)}{(x + 1)^3}$
f) $\frac{dy}{dx} = \frac{3(x^3 + 1)^2(-2x^3 - 3x^2 + 1)}{(x^3 - x)^4}$

14. a) $\frac{dy}{dx} = \frac{x - 1}{(x + 1)^3}$ **b)** $\frac{dy}{dx} = \frac{(2 + 5x)(5x - 2)}{x^2}$
c) $\frac{dy}{dx} = \frac{-2(x + 1)(-2x^2 - 3x + 3)}{(3 + x^2)^4}$

15. a) $\frac{dy}{dx} = 2\sin x \cos x(\sin^2 x - \cos^2 x)$
b) $\frac{dy}{dx} = \frac{-2\sin x}{\cos^3 x}$
c) $\frac{dy}{dx} = \frac{\sin^2 x(3\cos^2 x + \sin^2 x)}{\cos^2 x}$

16. a) $f(g(x)) = x^2 - 1$; $g(f(x)) = x^2 + 6x + 5$
b) $f(g(x)) = \frac{-1}{2 - x^3}$; $g(f(x)) = \frac{(x + 1)^3 + 1}{(x + 1)^3}$
c) $f(g(x)) = \sqrt{2x + 2}$; $g(f(x)) = 2\sqrt{x - 1} + 3$
d) $f(g(x)) = \frac{1}{1 - \sqrt{1 + x}}$; $g(f(x)) = \sqrt{\frac{2 - x}{1 - x}}$

17. a) $D = \{x \mid x \in \Re,\ x > -1\}$ **b)** $D = \{x \mid x \in \Re,\ x < \frac{5}{3}\}$
c) $D = \{x \mid x \in \Re,\ x \geq -2\}$ **d)** $D = \{x \mid x \in \Re,\ x \leq 2\}$
e) $D = \{x \mid x \in \Re,\ x \geq -2\}$ **f)** $D = \{x \mid x \in \Re,\ x > -1\}$

18. a) $V = \frac{1}{6}\pi d^3$ **b)** $A = \pi d^2$
c) $V = \frac{1}{6}\left(\frac{A^3}{\pi}\right)^{\frac{1}{2}}$ **d)** $A = (4\pi)^{\frac{1}{3}}(3V)^{\frac{2}{3}}$

19. a) i) $D_f = \Re$; $D_g = \{x \mid x \in \Re,\ x \geq -1\}$
ii) $R_f = \{y \mid y \in \Re,\ y \geq 1\}$; $R_g = \{y \mid y \in \Re,\ y \geq 0\}$
iii) $f(g(x)) = x + 2$
iv) $D = \{x \mid x \in \Re,\ x \geq -1\}$
v) $R = \{y \mid y \in \Re,\ y \geq 1\}$
b) i) $D_f = \{x \mid x \in \Re,\ x \neq 0\}$; $D_g = \Re$
ii) $R_f = \{y \mid y \in \Re,\ y < 0\}$; $R_g = \{y \mid y \in \Re,\ y \leq 1\}$
iii) $f(g(x)) = \frac{-1}{(1 - x^2)^2}$
iv) $D = \{x \mid x \in \Re,\ x \neq \pm 1\}$
v) $R = \{y \mid y \in \Re,\ y < 0\}$

20. a) $f(f(x)) = -x^4 + 4x^3 - 6x^2 + 4x$
b) $f(f(x)) = \sqrt{2 + \sqrt{2 + x}}$
c) $f(f(x)) = \frac{(x^2 - 3)^2}{1 - 3(x^2 - 3)^2}$

21. a) $f(x) = \frac{1}{x^2}$; $g(x) = x + 1$
b) $f(x) = -\frac{1}{x}$; $g(x) = x - 2$
c) $f(x) = x^2 + 2x$; $g(x) = x + 3$
d) $f(x) = \sqrt{x}$; $g(x) = 2x + 5$
e) $f(x) = x^2$; $g(x) = 4 - 3x$
f) $f(x) = x^2 + 2x + 1$; $g(x) = x^2$

22. Answers may vary.
a) $g(x) = x$ **b)** $g(x) = \frac{2 - x}{x}$ **c)** $g(x) = x$

23. Graph ① is $y = g(f(x))$.
Graph ② is $y = f(g(x))$.

24. $y = 6x + 7$

25. a) $\frac{dy}{dx} = 6(2x - 6)^2$ **b)** $\frac{dy}{dx} = -8x(x^2 - 5)^3$
c) $\frac{dy}{dx} = 5(-2 + 6x)(-2x + 3x^2)^4$
d) $\frac{dy}{dx} = 4x(-x^2 + 2)^{-3}$ **e)** $\frac{dy}{dx} = 6x(x^2 - 2)^{-4}$
f) $\frac{dy}{dx} = 4(1 - 6x^2)(x - 2x^3)^{-5}$

26. a) $\frac{dy}{dx} = \frac{-3(2x + 3)}{(x^2 + 3x)^4}$ **b)** $\frac{dy}{dx} = \frac{-6x^2}{(5 - x^3)^3}$
c) $\frac{dy}{dx} = \frac{-4(-2 + 4x^3)}{(-2x + x^4)^5}$

27. a) $\frac{dy}{dx} = \frac{-25(3 - x)^4}{(2 + x)^6}$ **b)** $\frac{dy}{dx} = \frac{3(3 - x^2)^2(x^2 - 4x + 3)}{(2 - x)^4}$
c) $\frac{dy}{dx} = \frac{4(3x - x^2)^3(x^4 - 6x^3 - 4x + 6)}{(2 + x^3)^5}$

28. a) $\frac{dy}{dx} = \frac{3}{2y}$ **b)** $\frac{dy}{dx} = \frac{1}{2 - 2y}$

c) $\frac{dy}{dx} = 2 - 2x$ **d)** $\frac{dy}{dx} = \frac{-1}{3y^2}$

e) $\frac{dy}{dx} = \frac{1 + 4x}{4y}$ **f)** $\frac{dy}{dx} = \frac{2x}{6y - 1}$

29. a) $\frac{dy}{dx} = \frac{x}{y}$ **b)** $y = -\frac{5}{4}x - \frac{9}{4}$

30. a) $\frac{dy}{dx} = \frac{-2x}{y}$ **b)** $y = -x + 3$ and $y = x - 3$

31. a) $\frac{dy}{dx} = \frac{4x}{3}(2x^2 + 1)^{-\frac{2}{3}}$ **b)** $\frac{dy}{dx} = \frac{-1}{2}(1 - 3x^2)(x - x^3)^{-\frac{3}{2}}$

c) $\frac{dy}{dx} = -\left(\frac{y}{x}\right)^{\frac{1}{3}}$

32. a) $\frac{dy}{dx} = \frac{x^2(1 - y^3)}{y^2(x^3 - 1)}$ **b)** $\frac{dy}{dx} = -1$

Chapter 4 Self-Test, page 287

1. a) $\frac{dy}{dx} = 2x(9x - 10)$ **b)** $\frac{dy}{dx} = \frac{-7x^2 + 16x + 9}{(x^2 - 5x + 7)^2}$

c) $\frac{dy}{dx} = 40x(5x^2 - 1)^3$ **d)** $\frac{dy}{dx} = \frac{-5}{2\sqrt{x}(\sqrt{(x - 5)^3})}$

e) $\frac{dy}{dx} = \frac{9x}{y}$ **f)** $\frac{dy}{dx} = \frac{1 - y^2}{2y(x + 3y)}$

2. a) $\frac{d^2y}{dx^2} = 4(9x - 5)$

b) $\frac{d^2y}{dx^2} = \frac{-2(-7x^3 + 24x^2 + 27x - 101)}{(x^2 - 5x + 7)^3}$

c) $\frac{d^2y}{dx^2} = 40(5x^2 - 1)^2(35x^2 - 1)$

3. $y = -\frac{3}{2}x + \frac{25}{2}$ and $y = \frac{3}{2}x + \frac{25}{2}$

4. $(-1, 1)$

5. $k > 0$

7. Answers may vary: $f(x) = x^2 + 1$; $g(x) = x$; these functions are not unique.

8. 4000 L/min

9. a) $\frac{dB}{dh} = \frac{-110}{h^3}$ **b)** Approximately −20.52

10. a) i) $\frac{dy}{dx} = x(x + 1) + (x + 1)(x + 2) + x(x + 2)$

ii) $\frac{dy}{dx} = (x + 1)(x + 2) + (x + 2)(x + 3) + (x + 1)(x + 3)$

iii) $\frac{dy}{dx} = (x + 2)(x + 3) + (x + 3)(x + 4) + (x + 2)(x + 4)$

iv) $\frac{dy}{dx} = (x + 3)(x + 4) + (x + 4)(x + 5) + (x + 3)(x + 5)$

b) $\frac{dy}{dx} = (x + k)(x + k + 1) + (x + k + 1)(x + k + 2) + (x + k)(x + k + 2)$

11. a) i) $f(f(x)) = \frac{2x + 3}{3x + 5}$ **ii)** $f(f(f(x))) = \frac{5x + 8}{8x + 13}$

iii) $f(f(f(f(x)))) = \frac{13x + 21}{21x + 34}$

12. b) Answers may vary. $g(x) = -2x - 3$

13. Answers may vary.

a) $f(x) = \sqrt{x}$; $g(x) = \frac{1}{x + 3}$; $h(x) = 2x$

14. (1.8, −3.2) and (−1.8, 3.2)

Mathematical Modelling: Optimal Profits

Exercises, page 291

1. a) $P = \frac{12\ 000(x - 15)}{x + 10} - (1000 + 80x)$

b) $\frac{dP}{dx} = \frac{300\ 000}{(x + 10)^2} - 80$

c) $x = 51$

2. a) $P = \frac{12\ 000(x - 15)}{x + 10} - (1000 + 100x)$

b) 45 workers **c)** Approximately \$125/day

3. a) Approximately 51 workers

b) Approximately \$100/day

4. Approximately 35 workers

5. a) $\frac{dP}{dx} = \frac{dR}{dx} - 80$

6. b) Approximately 31 **c)** \$80/min

d) $R = 2500 - 0.8(x - 50)^2$

Chapter 5 Rational Functions

Necessary Skills

1 Review: Rational Expressions

Exercises, page 295

1. a) $\frac{1}{2}$ **b)** 0 **c)** Undefined

d) Undefined **e)** $\frac{1}{8}$

2. The expression in part a is undefined at $x = 0$ because the denominator is 0.

3. a) 0 **b)** −5

c) ±1 **d)** All values are permissible.

e) −4, 6 **f)** 9

4. a) 0 **b)** 3 **c)** $\pm\frac{3}{2}$

d) No values **e)** ±1 **f)** 0, ±5

2 New: Equivalent Forms of Rational Expressions

Exercises, page 297

1. a) $1 + \frac{3}{x}$, $x \neq 0$ **b)** $x - 5 + \frac{4}{x}$, $x \neq 0$

c) $2x^2 + 5x - \frac{5}{2x}$, $x \neq 0$

2. a) $2x + 3 - \frac{4}{x + 1}$, $x \neq -1$ **b)** $3x + 8 + \frac{11}{x - 2}$, $x \neq 2$

c) $x^2 - 2x + 4 - \frac{3}{x - 3}$, $x \neq 3$ **d)** $x - 4 + \frac{14}{x + 4}$, $x \neq -4$

e) $3x^2 - \frac{3}{2}x + \frac{3}{4} - \frac{3}{4(2x + 1)}$, $x \neq -\frac{1}{2}$

f) $3x - 6 + \frac{21}{2x + 4}$, $x \neq -2$

3. a) $x + 3$, $x \neq 3$ **b)** $x - 3$, $x \neq -3$

c) $x + 3 + \frac{18}{x - 3}$, $x \neq 3$ **d)** $x - 3 + \frac{18}{x + 3}$, $x \neq -3$

4. When the remainder is 0, the divisor is a factor of the dividend.

5. a) $4x^3 + x^2 - 2x + 1 = (x^2 - 3)(4x + 1) + 10x + 4$

b) $4x + 1 + \frac{10x + 4}{x^2 - 3}$, $x \neq \pm\sqrt{3}$

c) The remainder is not always constant. The degree of the remainder is 1 less than the degree of the divisor.

3 Review: Graphs of Polynomial Functions

Exercises, page 301

1. Estimates may vary. Increase: $x < 0$ and $x > 2$; decrease: $0 < x < 2$; concave up: $x > 1$; concave down: $x < 1$

2. Estimates may vary.
$f'(x) = 0$ at $x = 0, \pm 2.2$;
$f'(x) > 0$ for $x < -2.2$ and $0 < x \le 2.2$;
$f'(x) < 0$ for $-2.2 < x < 0$ and $x > 2.2$;
$f''(x) = 0$ at $x = \pm 1$;
$f''(x) > 0$ for $|x| < 1$; $f''(x) < 0$ for $|x| > 1$

3. 

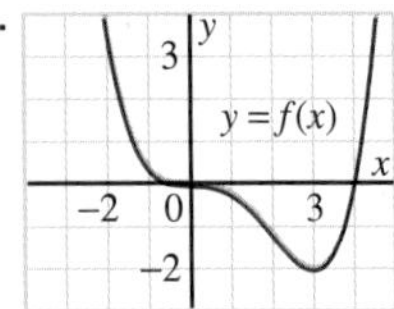

4. a) Intercepts: $x = -3, 1$; $y = -3$; critical point at $x = -1$; increase: $x > -1$; decrease: $x < -1$; no points of inflection; concave up for $x \in \Re$

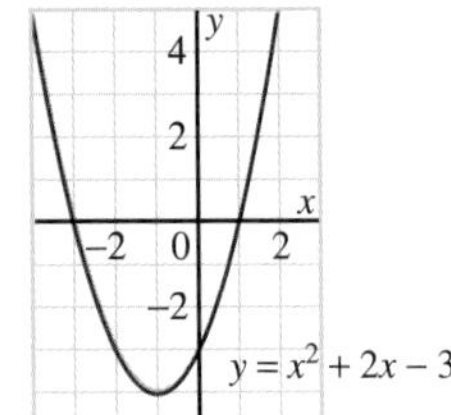

b) Intercepts: $x = -2$; $y = 20$; critical points at $x = 0, 2$; increase: $x < 0$ and $x > 2$; decrease: $0 < x < 2$; point of inflection at $x = 1$; concave up for $x > 1$; concave down for $x < 1$

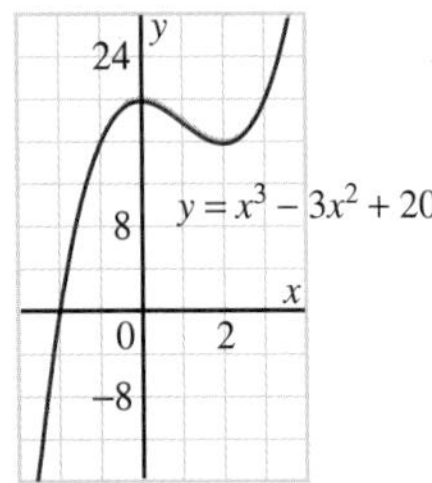

c) Intercepts: $x \doteq \pm 1.7$; $y = 9$; critical points at $x = 0, \pm 1.7$; increase: $-1.7 < x < 0$ and $x > 1.7$; decrease: $x < -1.7$ and $0 < x < 1.7$; points of inflection at $x = \pm 1$; concave up for $|x| > 1$; concave down for $|x| < 1$

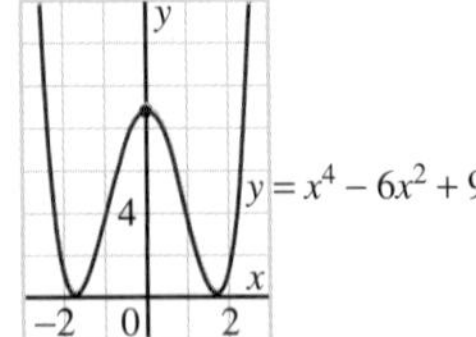

d) Intercepts: $x = 0$; $y = 0$; critical points at $x = 0, -2$; increase: $x > 0$; decrease: $x < 0$, $x \ne -2$; points of inflection at $x = -2, -\frac{2}{3}$; concave up for $x < -2$ and $x > -\frac{2}{3}$; concave down for $-2 < x < -\frac{2}{3}$

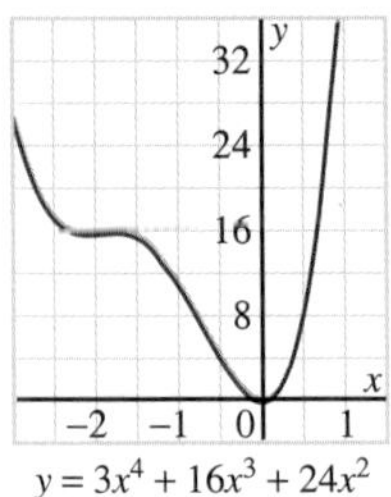

$y = 3x^4 + 16x^3 + 24x^2$

4 Review: Differentiation Tools

Exercises, page 303

1. a) $y' = \frac{-1}{x^2}$; $y'' = \frac{2}{x^3}$ **b)** $y' = \frac{-1}{(x+2)^2}$; $y'' = \frac{2}{(x+2)^3}$

c) $y' = \frac{-2x}{(x^2-4)^2}$; $y'' = \frac{2(3x^2+4)}{(x^2-4)^3}$

d) $y' = \frac{-2}{(x-3)^3}$; $y'' = \frac{6}{(x-3)^4}$ **e)** $y' = \frac{6}{(x+6)^2}$; $y'' = \frac{-12}{(x+6)^3}$

f) $y' = \frac{-7}{(x-5)^2}$; $y'' = \frac{-14}{(x-5)^3}$

2. a) $y' = \frac{10x}{(x^2+3)^2}$ **b)** $y' = \frac{x^4 - 75x^2}{(x^2-25)^2}$

c) $y' = \frac{x^3+2}{x^3}$ **d)** $y' = \frac{-x^2-12}{(x^2-x-12)^2}$

e) $y' = \frac{2x(-2x^2+1)}{(x^2+1)^4}$ **f)** $y' = \frac{-5x^3-5}{(x^3-5)^3}$

3. a) $y' = 1,\ x \ne 2$ **b)** $y' = \frac{-1}{(x+2)^2}, x \ne 2$

c) $y' = 2x,\ x \ne 0$ **d)** $y' = \frac{-1}{(x+3)^2},\ x \ne 0$

5.1 Exercises, page 308

1. Equations in parts a, b, c, d, and e represent rational functions; a polynomial function is a rational function with denominator 1.

2. All graphs could be graphs of rational functions. Graphs in parts b and d could be graphs of polynomial functions.

3. a) iii **b)** iv **c)** i **d)** ii

4. a) $x - 1$ is a factor of the denominator.
b) Graphs in parts a, b, and c
c) Graph in part d
d) $x - 1$ is a factor of the numerator.

5. a) ii **b)** iii **c)** iv **d)** i

6. a) i)

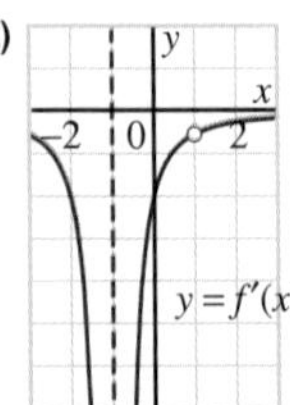

ii)

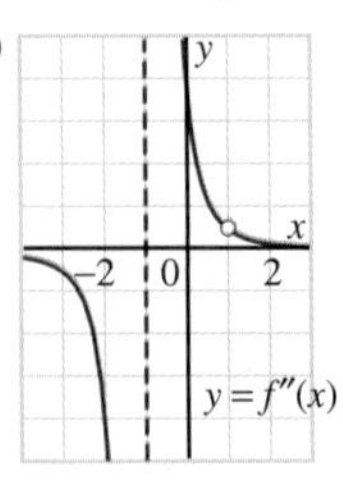

b) i)

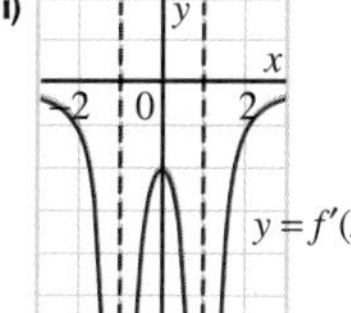

ii)

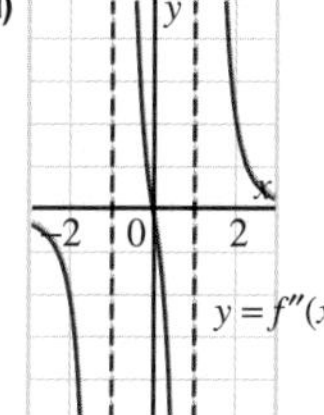

c) i)

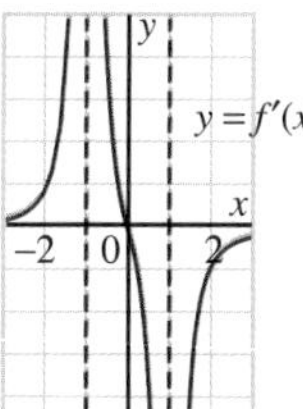

ii)

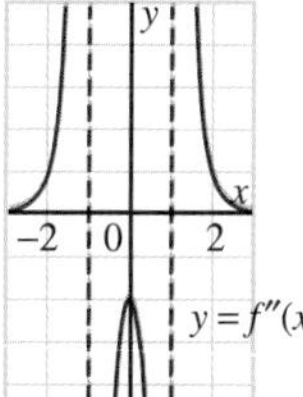

d) i)

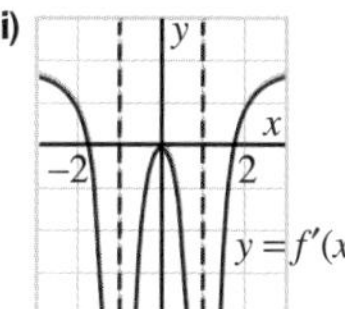

ii)

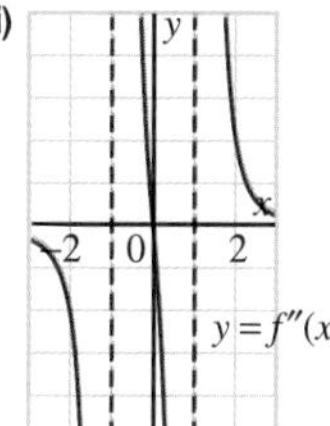

7. a) $x - 1$ and $x + 1$ are factors of the denominator.

b) Graphs in parts b, c, and d

c) Graph in part a

d) $x - 1$ is a factor of the numerator.

e) The graph in part a is not symmetrical. The graphs in parts b and d have point symmetry about the origin. The graph in part c has line symmetry about the y-axis.

8. a) i)

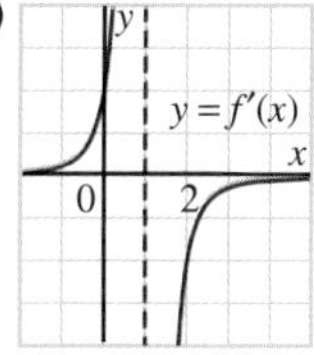

ii)

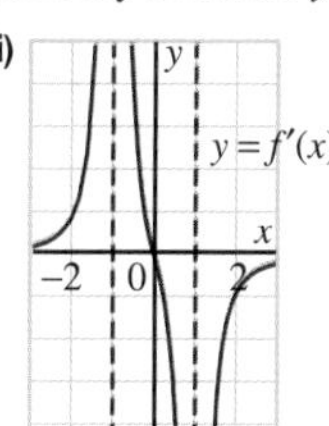

iii)

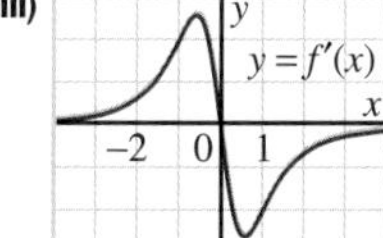

c) i) −1 **ii)** 4

d) iii) $\left(\frac{1}{\sqrt{3}}, 3\right)\left(-\frac{1}{\sqrt{3}}, 3\right)$

9. a) $y = \frac{1}{x - 2}$ **b)** $y = \frac{1}{x^2 - 1}$

c) $y = \frac{1}{\sqrt{x + 1}}$ **d)** $y = \frac{1}{x^2 - x - 6}$

e) $y = \frac{1}{2^x}$ **f)** $y = \frac{1}{x^3 + x}$

10. The reciprocal functions in exercise 9 a, b, d, and f are rational functions.

11. a) Yes **b)** No

12. a)

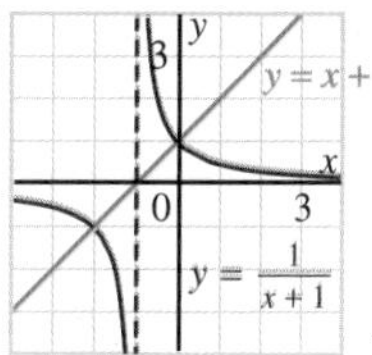

b)

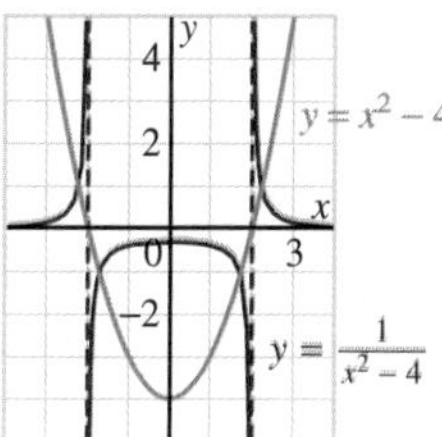

c)

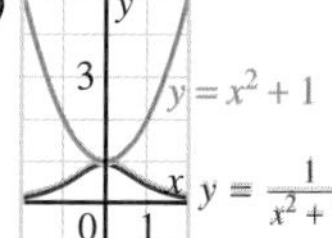

13. a) $y = -\frac{1}{4}x + \frac{3}{4}$ **b)** $y = -\frac{2}{9}x - \frac{1}{9}$ **c)** $y = \frac{-1}{2}x + 1$

14. a) Yes **b)** No

c) No **d)** $f(a) = 0$

15. No

16. Graph 1 is the function and graph 2 is the derivative.

17. Estimates for intervals may vary. Increase: $|x| > 1.8$; decrease: $-1.8 < x < -1$, $-1 < x < 0$, $0 < x < 1$, $1 < x < 1.8$; critical points: $x = \pm 1.8, 0$; point of inflection: $x = 0$; concave up: $-1 < x < 0$ and $x > 1$; concave down: $x < -1$ and $0 < x < 1$

18. Graphs may vary.

a)

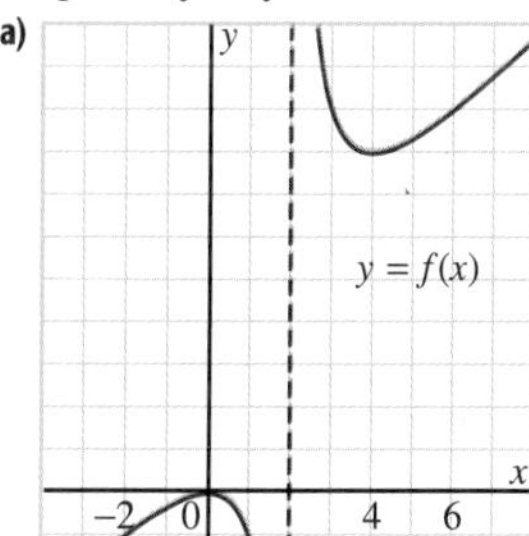

b)

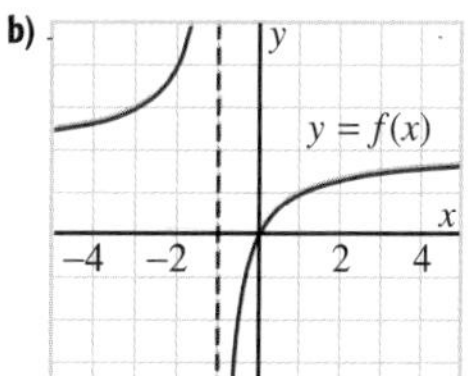

19. a)

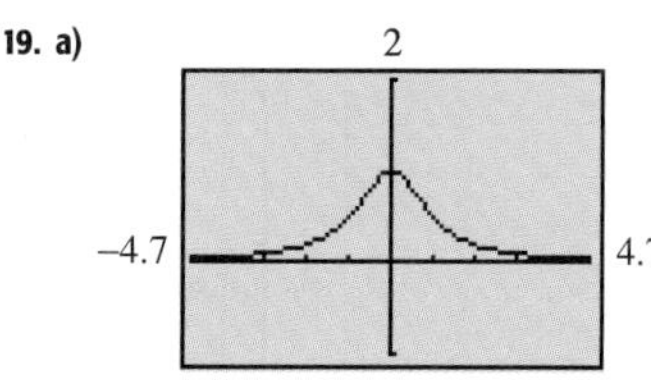

b)

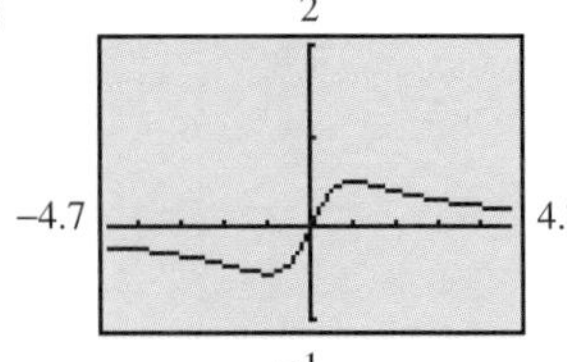

c)

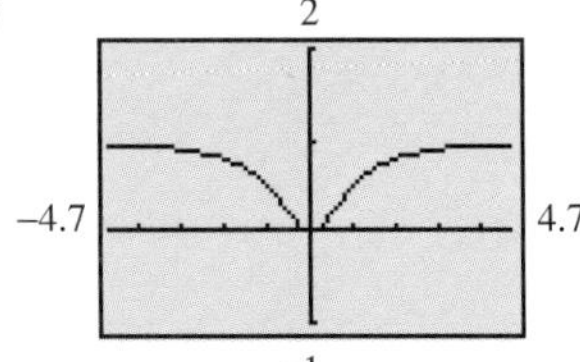

d)

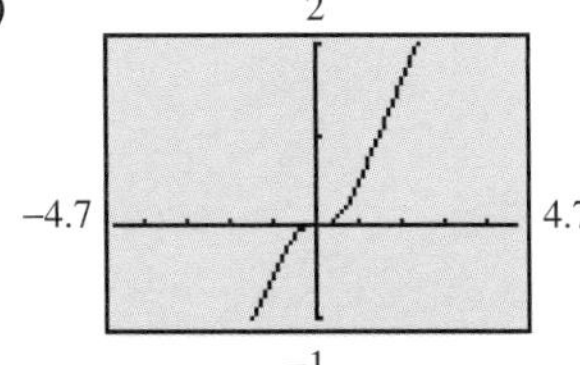

e)

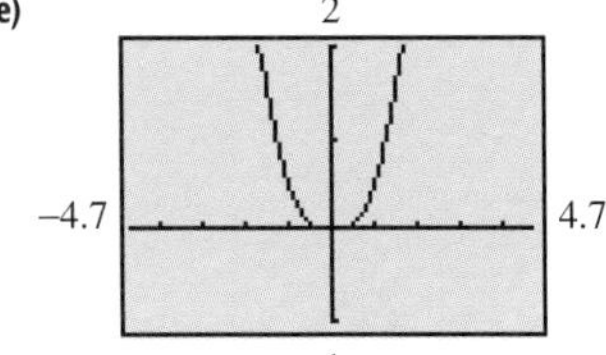

f)

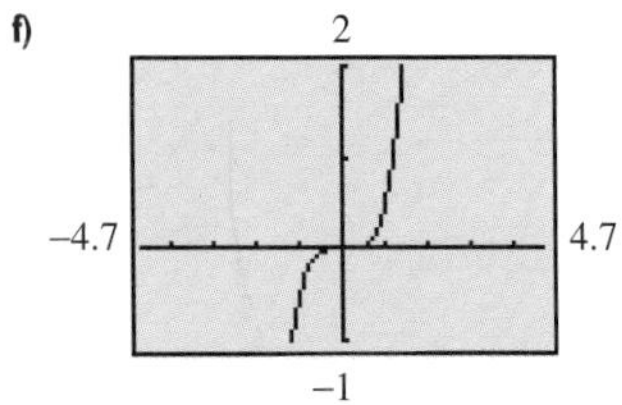

i) All the functions are continuous because each denominator, $x^2 + 1$, is defined for all real values of x.
ii) The graphs in parts a, c, and e have line symmetry. The graphs in parts b, d, and f have point symmetry.

20. a) $y = -1$ **b)** $y = -x$ **c)** $y = -x^2$
d) $y = -x^3$ **e)** $y = -x^4$ **f)** $y = -x^5$

21. Yes

5.2 Exercises, page 319

1. a) $\pm\infty$ **b)** ∞ **c)** 6

2. a) 2 **b)** Undefined **c)** $-\infty$ **d)** ∞

3. a) 0.5 **b)** Undefined **c)** ∞
d) $-\infty$ **e)** $\pm\infty$ **f)** Undefined
g) 0 **h)** 0

4. a) i) ∞ **ii)** $-\infty$ **iii)** $\pm\infty$
iv) $-\infty$ **v)** ∞ **vi)** $\pm\infty$
vii) 1 **viii)** 1
b) i) ∞ **ii)** $-\infty$ **iii)** $\pm\infty$
iv) ∞ **v)** $-\infty$ **vi)** $\pm\infty$
vii) ∞ **viii)** $-\infty$

5. a) 5.9, 5.99, 5.999, 6.1, 6.01, 6.001; 6
b) 10, 100, 1000, −10, −100, −1000; does not exist
c) 100, 10 000, 1 000 000, 100, 10 000, 1 000 000; ∞

6. a) 6 **b)** $\frac{1}{6}$ **c)** 4 **d)** $\frac{1}{4}$
e) −4 **f)** $-\frac{1}{4}$ **g)** $\frac{1}{2}$ **h)** $-\frac{5}{2}$

7. a) 1 **b)** $\pm\infty$ **c)** 0 **d)** $\pm\infty$
e) $\pm\infty$ **f)** $\pm\infty$ **g)** ∞ **h)** ∞

8. a) 0 **b)** 1 **c)** 3 **d)** $\frac{2}{7}$
e) ∞ **f)** ∞ **g)** 0 **h)** 0

10. a) As x gets very large, the function $\frac{3x}{x+2}$ gets closer and closer to 3.
b) $5998 \le x \le 6002$

11. a) As x gets closer to −2, the function $\frac{1}{(x+2)^2}$ gets very large.
b) $x = -1.99, \ -2.01$

12. a) As x gets closer to 2, the function $\frac{x^2-4}{x-2}$ gets closer and closer to 4.
b) $1.999 \le x \le 2.001, \ x \ne 2$

13. a) As x approaches 0 from the right side, the function $\frac{1}{x}$ gets very large.
b) $x = 10^{-10}$

14. a) ∞ **b)** ∞ **c)** $-\infty$ **d)** ∞
e) 0 **f)** $-\infty$ **g)** $-\frac{5}{3}$ **h)** $-\infty$

15. a) i) Does not exist **ii)** 6
b) The graph is the line $y = x + 3$ with a "hole" at (3, 6).

16. a) i) 6 **ii)** 6
b) The graph is the line $y = x + 3$.

17. a) $\frac{2}{3}$ **b)** $\frac{3}{2}$

18. Answers may vary. $f(x) = \frac{x^2-16}{x+4}$

19. Yes

Self-Check 5.1, 5.2, page 323

1. a)

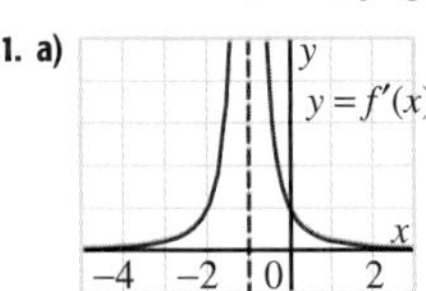

b)

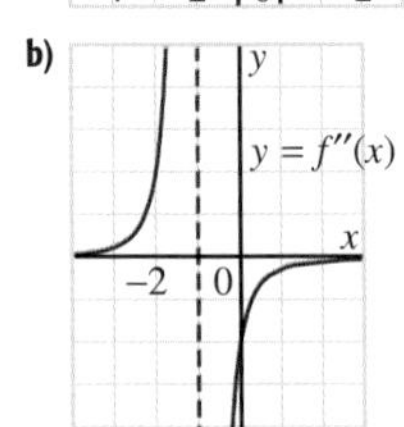

2. $y' = \frac{1}{(x+1)^2}$

3. Increase: $x < -1$ and $x > -1$; critical points: none; points of inflection: none; concave up: $x < -1$; concave down: $x > -1$

4. $y = x + 4$

5. a) -4 b) $\frac{1}{10}$ c) $\frac{1}{7}$ d) 7
e) Does not exist f) Does not exist
g) ∞ h) $-\infty$ i) $\frac{1}{6}$

6. a) 0 b) ∞ c) $-\frac{5}{2}$

7. a) Answers may vary. $f(x) = \frac{x^2 - 4}{x + 2}$
b) Answers may vary. $f(x) = \frac{1}{x - 1}$

8. a) i) ∞ ii) $-\infty$ iii) $\pm\infty$
b) None of the limits exists.

5.3 Exercises, page 330

1.

	x-intercept(s)	y-intercept
a)	none	$-\frac{1}{3}$
b)	0	0
c)	none	none
d)	none	-2
e)	0	0
f)	none	none
g)	-2	2
h)	none	3

2. a) iii b) i c) iv d) ii

3. a) x-intercept: none; y-intercept: -1; vertical asymptote: $x = 1$; horizontal asymptote: $y = 0$; point discontinuity: none
b) x-intercept: 0; y-intercept: 0; vertical asymptote: $x = 1$; horizontal asymptote: $y = 1$; point discontinuity: none
c) x-intercept: -2; y-intercept: 2; vertical asymptote: none; horizontal asymptote: none; point discontinuity at $x = 2$
d) x-intercept: 0; y-intercept: 0; vertical asymptotes: $x = \pm 2$; horizontal asymptote: $y = 2$; point discontinuity: none

4. There is a common factor of $x - 2$ in the numerator and the denominator.

5. a) Vertical asymptote: $x = -4$
b) Vertical asymptote: $x = 2$
c) Point discontinuity at $x = 0$
d) Point discontinuity at $x = 3$
e) Vertical asymptote: $x = -1$; point discontinuity at $x = 1$
f) Vertical asymptote: $x = 3$; point discontinuity at $x = -2$

6. a) $y = 0$ b) $y = 0$
c) $y = 1$ d) $y = 1.5$
e) $y = 0$ f) No horizontal asymptote

7. a) $x = -5$ b) $x = \frac{1}{3}$ c) $x = \pm 5$
d) None e) $x = -3,\ x = 4$ f) $x = -2$

8. $x = -1,\ x = 4$

9. There are no vertical asymptotes.

10. a) i) 7.5 g/L ii) 15 g/L iii) 18 g/L
b) 20 g/L
c) As the time increases, the concentration of salt in the tank gets closer and closer to 20 g/L.

11. a) Vertical asymptotes: $x = \pm 2$; horizontal asymptote: $y = 0$
b) Vertical asymptotes: $x = \pm 2$; horizontal asymptote: $y = 0$
c) Vertical asymptote: $x = -2$; horizontal asymptote: $y = 0$; point discontinuity at $x = 2$
d) Vertical asymptotes: $x = \pm 2$; horizontal asymptote: $y = 1$
e) Vertical asymptotes: $x = \pm 2$
f) Point discontinuity at $x = 4$

12. a) x-intercepts: ± 1; y-intercept: -1
b) Horizontal asymptote: $y = 1$
c)

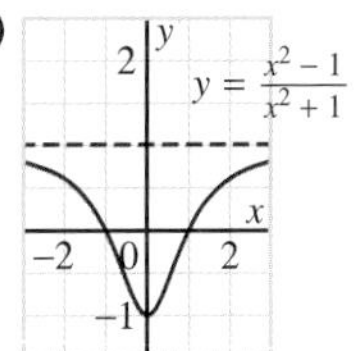

13. a) x-intercepts: ± 5; y-intercept: -1
b) Horizontal asymptote: $y = 1$
c)

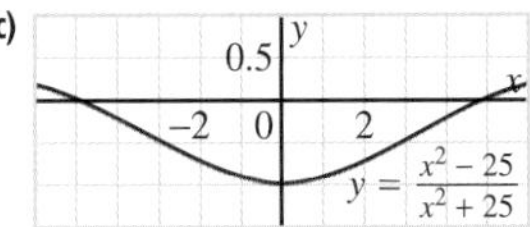

14. a) x-intercepts: $\pm a$; y-intercept: -1
b) Horizontal asymptote: $y = 1$

15. a) x-intercepts: none; none; none; y-intercepts: -1; -1; -1
b) Vertical asymptotes: $x = \pm 1$; $x = \pm 5$; $x = \pm a$; horizontal asymptotes: $y = 1$; $y = 1$; $y = 1$

16. Graphs may vary.
a)

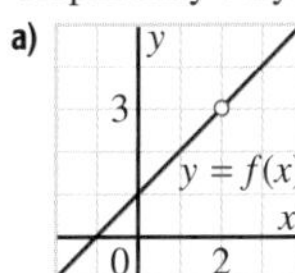

b)

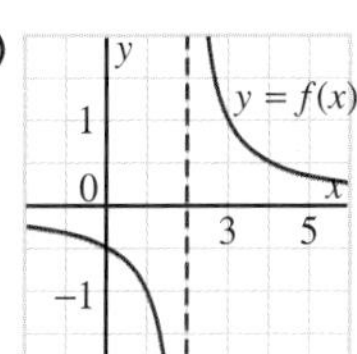

c)

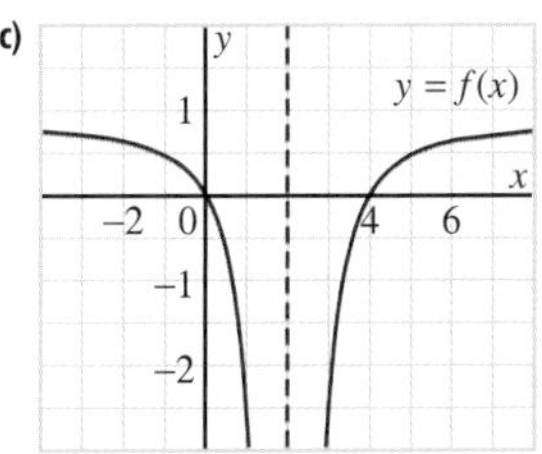

17. Limits in parts c and d could be true.

18. Limits in parts a and b must be true.

19. Limit in part d must be true.

20. a) i) $f(x) = \frac{x + 5}{x^2 - 25}$
ii) Vertical asymptote: $x = 5$; horizontal asymptote: $y = 0$
iii) $x = -5$
b) i) $f(x) = \frac{x}{x^2 - 3x}$
ii) Vertical asymptote: $x = 3$; horizontal asymptote: $y = 0$
iii) $x = 0$

c) i) $f(x) = \frac{x^2}{x^2+3}$
ii) Horizontal asymptote: $y = 1$
iii) No point discontinuity

d) i) $f(x) = \frac{x^3}{x^3-27}$
ii) Vertical asymptote: $x = 3$; horizontal asymptote: $y = 1$
iii) No point discontinuity

e) i) $f(x) = \frac{x^2}{x+7}$
ii) Vertical asymptote: $x = -7$
iii) No point discontinuity

f) i) $f(x) = \frac{x^4}{x^3-4x}$
ii) Vertical asymptote: $x = \pm 2$
iii) $x = 0$

21. a) $g(x) = \frac{x^2-25}{x+5}$; point discontinuity at $x = -5$
b) $g(x) = \frac{x^2-3x}{x}$; point discontinuity at $x = 0$
c) $g(x) = \frac{x^2+3}{x^2}$; vertical asymptote: $x = 0$; horizontal asymptote: $y = 1$
d) $g(x) = \frac{x^3-27}{x^3}$; vertical asymptote: $x = 0$; horizontal asymptote: $y = 1$
e) $g(x) = \frac{x+7}{x^2}$; vertical asymptote: $x = 0$; horizontal asymptote: $y = 0$
f) $g(x) = \frac{x^3-4x}{x^4}$; vertical asymptote: $x = 0$; horizontal asymptote: $y = 0$

22. a) Vertical asymptote: $x = -1$; horizontal asymptote: $y = 1$
b) $f'(x) = \frac{1}{(x+1)^2}$; vertical asymptote: $x = -1$; horizontal asymptote: $y = 0$
c) $f''(x) = \frac{-2}{(x+1)^3}$; vertical asymptote: $x = -1$; horizontal asymptote: $y = 0$
d) The vertical asymptote remains the same for f, f', and f''. Yes, this will always be true for any function of the form $f(x) = \frac{x}{x+k}$.

23. $\lim_{x \to a^+} f'(x) = -\infty$; $\lim_{x \to a^-} f'(x) = \infty$

5.4 Exercises, page 340

1. Estimates may vary.
a) $x = 1$; $y = -2$ **b)** $x = -1$; $y = x - 1$
c) $x = -3$; $y = 0$ **d)** $x = \pm 2$; $y = 3$

2. a) Horizontal asymptote
b) Horizontal asymptote
c) Oblique asymptote
d) No horizontal or oblique asymptote

3. a) $y = 0$ **b)** $y = 1$ **c)** $y = x - 5$

4. a) $y = x + 4$ **b)** $y = x + 3$
c) $y = 4x + 4$ **d)** $y = 4x + 2$
e) $y = x$ **f)** $y = x$

5. a) i) $y = x - 6$ **ii)** $y = 2x - 12$ **iii)** $y = 3x - 18$
b) i) $y = 4x - 24$ **ii)** $y = 10x - 60$ **iii)** $y = kx - 6k$

6. a) i) $y = x + 3$ **ii)** $y = 2x + 6$ **iii)** $y = 3x + 9$
b) i) $y = 4x + 12$ **ii)** $y = 10x + 30$ **iii)** $y = kx + 3k$

7. a) $y = 2.5$ **b)** $y = 0.5$ **c)** $y = 4$

8. a) The quotient of the coefficients of the highest-degree terms in $p(x)$ and $q(x)$
b) There is no oblique asymptote.

9. b) i) $y = x + 5$ **ii)** $y = x - 6$ **iii)** $y = x + a$

10. $y = \frac{1}{2}x + \frac{3}{2}$; $y = x - 3$; $y = \frac{5}{6}x + \frac{5}{3}$

11. a) i) $y = x - 10$ **ii)** $y = \frac{3}{2}x + 6$ **iii)** $y = x$
iv) $y = x$ **v)** $y = x$ **vi)** $y = x$
b) The last four functions in part a

12. a) i) $x = -10$ **ii)** $x = 4$ **iii)** $x = \pm 3$
iv) $x = \pm 3$ **v)** $x = \pm 3$ **vi)** $x = -2$

13. a) $f'(x) = \frac{x(x-2)}{(x-1)^2}$; $f'(x) = 0$ at $x = 0, 2$
b) $f(x)$: vertical asymptote: $x = 1$; oblique asymptote: $y = x + 1$; $f'(x)$: vertical asymptote: $x = 1$; horizontal asymptote: $y = 1$. Both graphs have the same vertical asymptote $x = 1$.

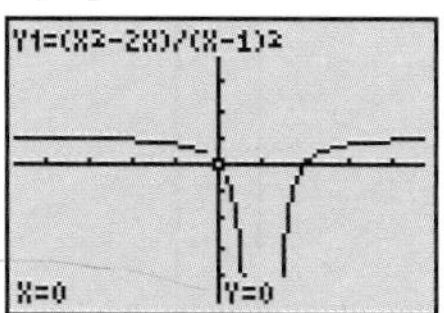

14. a) $S(n) = n$

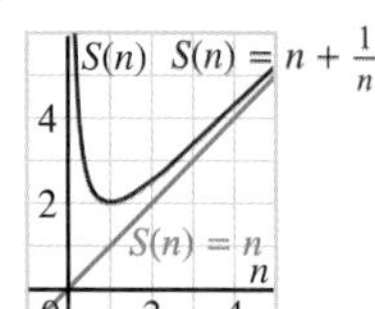

b) As $n \to \infty$, the value of $\frac{1}{n}$ gets closer and closer to 0. Therefore, the function resembles $S(n) = n$ as $n \to \infty$.
c) $S'(n) = 1 - \frac{1}{n^2}$; $n = 1$

5.5 Exercises, page 348

1. a) iii **b)** i **c)** iii, iv

2. Estimates may vary.
a) $f'(x) = 0$: $x = \pm 1$; $f'(x) > 0$: $|x| < 1$; $f'(x) < 0$: $|x| > 1$
$f''(x) = 0$: $x = \pm 2, 0$; $f''(x) > 0$: $-2 < x < 0$ and $x > 2$; $f''(x) < 0$: $x < -2$ and $0 < x < 2$
b) $f'(x) \neq 0$; $f'(x) < 0$: $x \in \Re$, $x \neq \pm 3$;
$f''(x) = 0$: $x = 0$; $f''(x) > 0$: $-3 < x < 0$ and $x > 3$;
$f''(x) < 0$: $x < -3$ and $0 < x < 3$

3. Graphs may vary.
a)

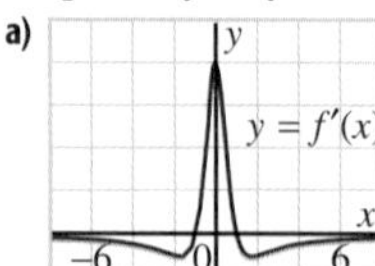

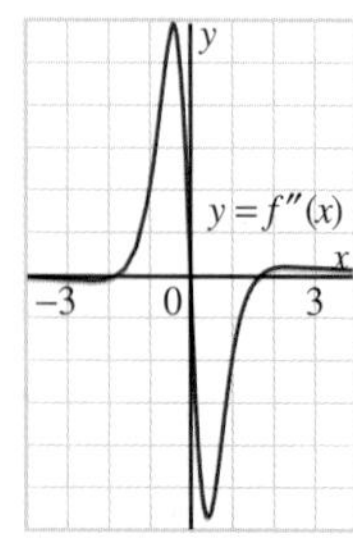

b) 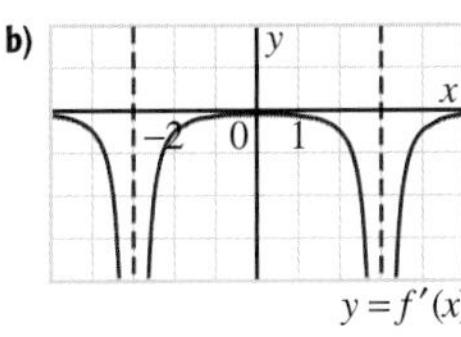

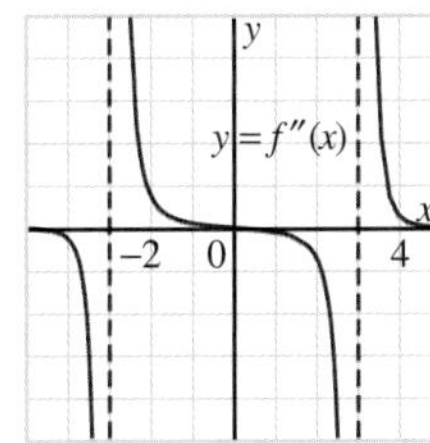

4. a) y-intercept: $-\frac{1}{5}$; domain: $\{x \mid x \in \Re,\ x \neq \pm 5\}$; decrease: $x \in \Re\ (x \neq \pm 5)$; concave up: $x > 5$; concave down: $x < 5$; vertical asymptote: $x = 5$; point discontinuity at $x = -5$; horizontal asymptote: $y = 0$

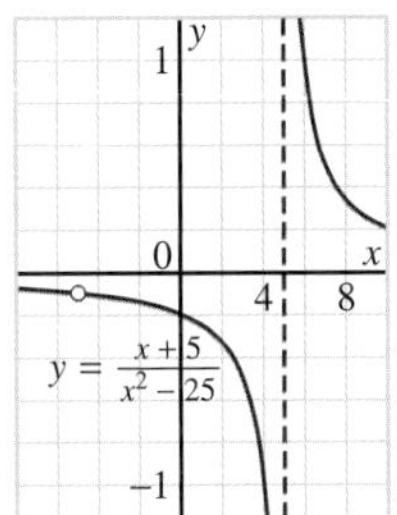

b) y-intercept: $\frac{1}{2}$; domain: $\{x \mid x \in \Re,\ x \neq \pm 2\}$; decrease: $x \in \Re\ (x \neq \pm 2)$; concave up: $x > -2$; concave down: $x < -2$; vertical asymptote: $x = -2$; point discontinuity at $x = 2$; horizontal asymptote: $y = 0$

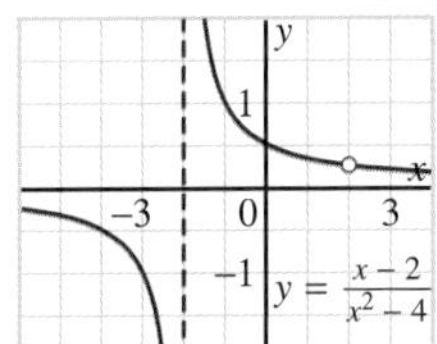

5. a) x-intercept: 0; y-intercept: 0; domain: $\{x \mid x \in \Re,\ x \neq \pm 2\}$; critical point at $x = 0$; increase: $x < 0\ (x \neq -2)$; decrease: $x > 0\ (x \neq 2)$; concave up: $|x| > 2$; concave down: $|x| < 2$; vertical asymptotes: $x = \pm 2$; horizontal asymptote: $y = 1$

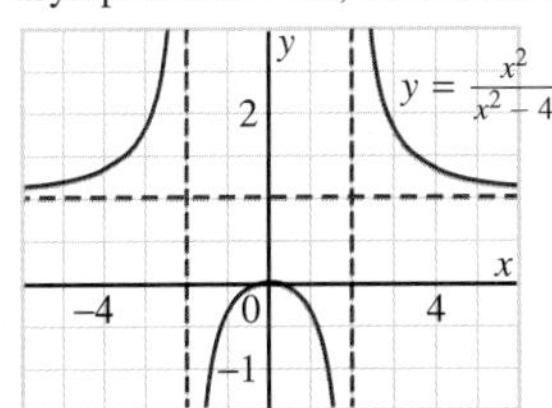

b) x-intercept: 0; y-intercept: 0; domain: $\{x \mid x \in \Re,\ x \neq \pm 1\}$; critical point at $x = 0$; increase: $x < 0\ (x \neq -1)$; decrease: $x > 0\ (x \neq 1)$; concave up: $|x| > 1$; concave down: $|x| < 1$; vertical asymptotes: $x = \pm 1$; horizontal asymptote: $y = 3$

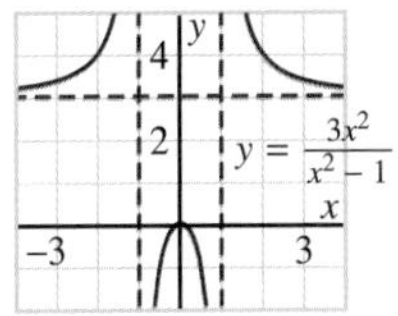

6. a) x-intercepts: ± 1; domain: $\{x \mid x \in \Re,\ x \neq 0\}$; increase: $x \in \Re\ (x \neq 0)$; concave up: $x < 0$; concave down: $x > 0$; vertical asymptote: $x = 0$; oblique asymptote: $y = x$

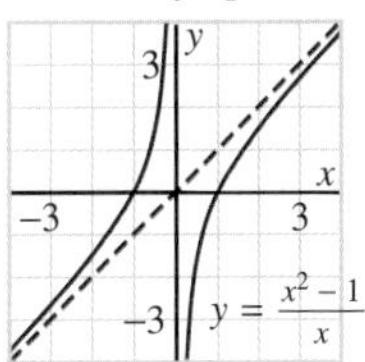

b) Domain: $\{x \mid x \in \Re,\ x \neq 0\}$; critical points at $x \doteq \pm 1.7$; increase: $|x| > 1.7$; decrease: $|x| < 1.7\ (x \neq 0)$; concave up: $x > 0$; concave down: $x < 0$; vertical asymptote: $x = 0$; oblique asymptote: $y = \frac{1}{2}x$

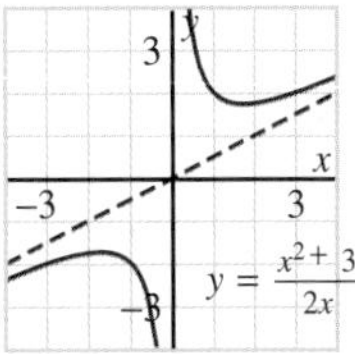

7. a) y-intercept: 2; domain: $x \in \Re$; critical point at $x = 0$; increase: $x < 0$; decrease: $x > 0$; points of inflection at $x \doteq \pm 1.2$; concave up: $|x| > 1.2$; concave down: $|x| < 1.2$; horizontal asymptote: $y = 0$

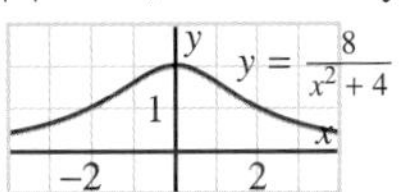

b) x-intercept: 0; y-intercept: 0; domain: $x \in \Re$; critical points at $x = \pm 2$; increase: $|x| < 2$; decrease: $|x| > 2$; points of inflection at $x = 0$ and $x \doteq \pm 3.5$; concave up: $-3.5 < x < 0$ and $x > 3.5$; concave down: $x < -3.5$ and $0 < x < 3.5$; horizontal asymptote: $y = 0$

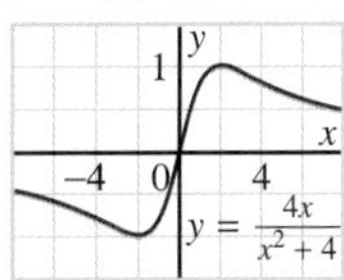

8. a) x-intercept: 5; y-intercept: -5; domain: $\{x \mid x \in \Re,\ x \neq -5\}$; increase: $x \in \Re\ (x \neq -5)$; point discontinuity at $x = -5$

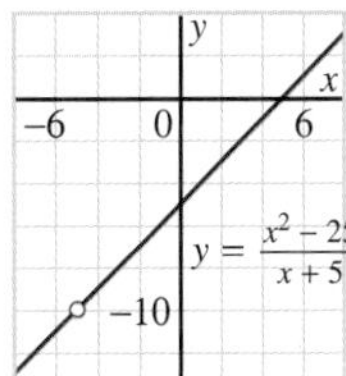

b) y-intercept: 1; domain: $\{x \mid x \in \Re,\ x \neq \pm 1\}$; critical point at $x = 0$; increase: $x > 0$ $(x \neq 1)$; decrease: $x < 0$ $(x \neq -1)$; concave up: $|x| < 1$; concave down: $|x| > 1$; vertical asymptotes: $x = \pm 1$; horizontal asymptote: $y = -1$

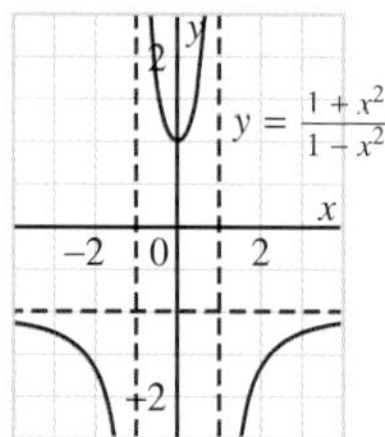

c) Domain: $\{x \mid x \in \Re,\ x \neq 0\}$; critical points at $x = \pm 2$; increase: $|x| > 2$; decrease: $|x| < 2$ $(x \neq 0)$; concave up: $x > 0$; concave down: $x < 0$; vertical asymptote: $x = 0$; oblique asymptote: $y = x$

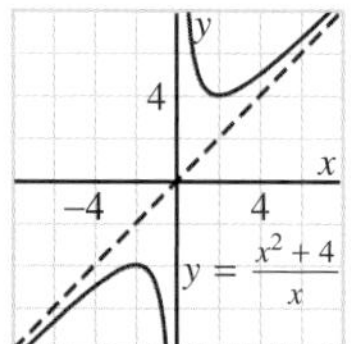

d) y-intercept: 3; domain: $x \in \Re$; critical point at $x = 0$; increase: $x < 0$; decrease: $x > 0$; points of inflection: $x \doteq \pm 0.6$; concave up: $|x| > 0.6$; concave down: $|x| < 0.6$; horizontal asymptote: $y = 0$

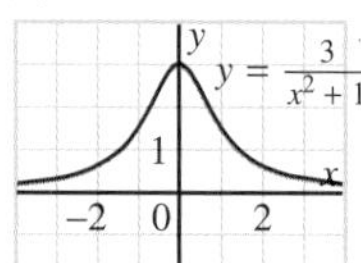

9. a) x-intercept: -2; y-intercept: 2; domain: $\{x \mid x \in \Re,\ x \neq 2\}$; increase: $x \in \Re$ $(x \neq 2)$; point discontinuity at $x = 2$

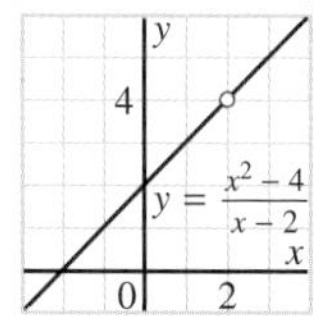

b) x-intercept: 0; y-intercept: 0; domain: $\{x \mid x \in \Re,\ x \neq \pm 1\}$; critical point at $x = 0$; increase: $x > 0$ $(x \neq 1)$; decrease: $x < 0$ $(x \neq -1)$; concave up: $|x| < 1$; concave down: $|x| > 1$; vertical asymptotes: $x = \pm 1$; horizontal asymptote: $y = -1$

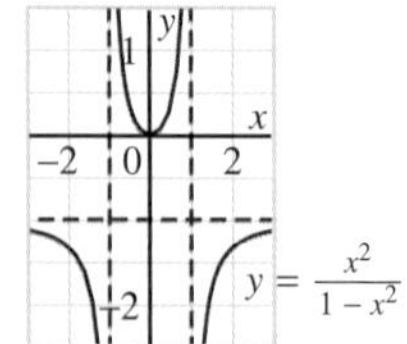

c) Domain: $\{x \mid x \in \Re,\ x \neq 0\}$; critical points at $x = \pm 1$; increase: $|x| > 1$; decrease: $|x| < 1$ $(x \neq 0)$; concave up: $x > 0$; concave down: $x < 0$; vertical asymptote: $x = 0$; oblique asymptote: $y = x$

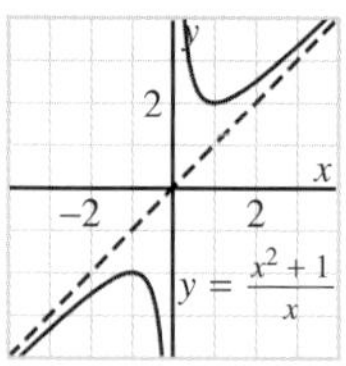

d) y-intercept: 1; domain: $x \in \Re$; critical point at $x = 0$; increase: $x < 0$; decrease: $x > 0$; points of inflection at $x \doteq \pm 0.8$; concave up: $|x| > 0.8$; concave down: $|x| < 0.8$; horizontal asymptote: $y = 0$

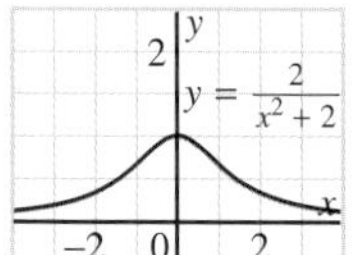

10. a) i) Increase: $x < 0$ and $x > 2$; decrease: $0 < x < 2$ $(x \neq 1)$
ii) $x = 0$ (local maximum) and $x = 2$ (local minimum)

b) Graphs may vary.

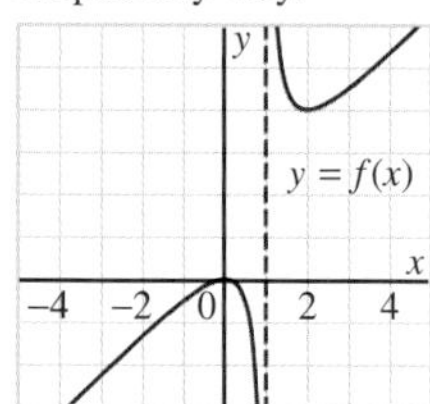

11. a) $f'(x) = \dfrac{x^2 - 2x}{(x - 1)^2}$

12. Graphs may vary.

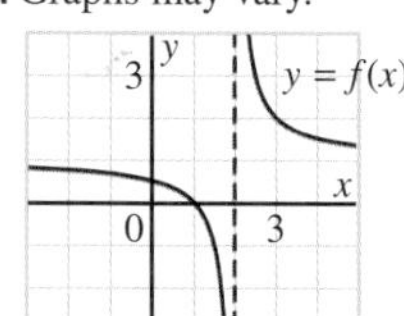

13. Graphs may vary.

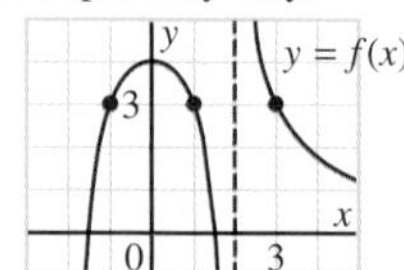

14. Graphs may vary.

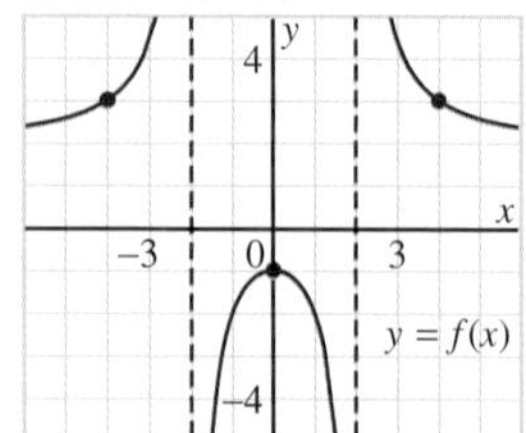

16. x-intercepts: approximately –4.4, 0.4; domain: $\{x \mid x \in \Re,\ x \neq 0\}$; critical point at $x = 1$; increase: $0 < x < 1$; decrease: $x < 0$ and $x > 1$; point of inflection at $x = 1.5$; concave up: $x > 1.5$; concave down: $0 < x < 1.5$ and $x < 0$; vertical asymptote: $x = 0$; horizontal asymptote: $y = 1$

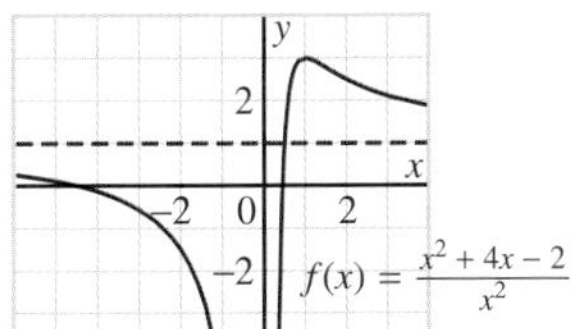

17. a) $k > 0$ **b)** $k < 0$ **c)** $k = 0$

18. a) $k > 1$ **b)** $k < 1$ **c)** $k = 1$

Self-Check 5.3–5.5, page 351

1. a) y-intercept: –0.5
b) x-intercept: –1; y-intercept: 1
c) x-intercept: –1; y-intercept: $-\frac{1}{6}$

2. a) Vertical asymptote: $x = 2$
b) Point discontinuity at $x = -9$
c) Vertical asymptote: $x = -1.5$; point discontinuity at $x = 0$

3. a) $y = 0$ **b)** $y = 1$ **c)** $y = \frac{5}{3}$

4. a) i) Horizontal asymptote **ii)** $y = 0$
b) i) Oblique asymptote **ii)** $y = x - 5$
c) i) Oblique asymptote **ii)** $y = x - 6$

5. a) Vertical asymptote: $x = 2$; oblique asymptote: $y = x + 2$
b) Vertical asymptote: $x = 1$; oblique asymptote: $y = x + 2$
c) Vertical asymptote: $x = 0$; oblique asymptote: $y = 3x + 2$

6. a) x-intercept: 0; y-intercept: 0; domain: $\{x \mid x \in \Re,\ x \neq 5\}$; decrease: $x \in \Re\ (x \neq 5)$; concave up: $x > 5$; concave down: $x < 5$; vertical asymptote: $x = 5$; horizontal asymptote: $y = 1$

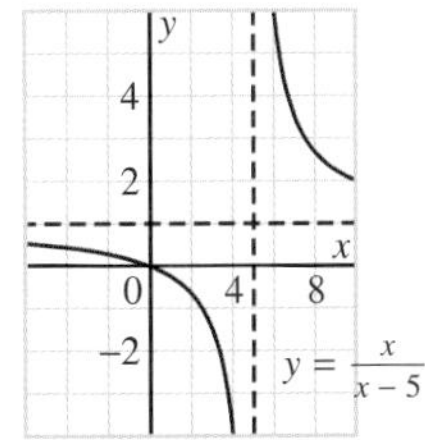

b) x-intercept: 0; y-intercept: 0; domain: $\{x \mid x \in \Re,\ x \neq \pm 5\}$; critical point at $x = 0$; increase: $x < 0\ (x \neq -5)$; decrease: $x > 0\ (x \neq 5)$; concave up: $|x| > 5$; concave down: $|x| < 5$; vertical asymptotes: $x = \pm 5$; horizontal asymptote: $y = 1$

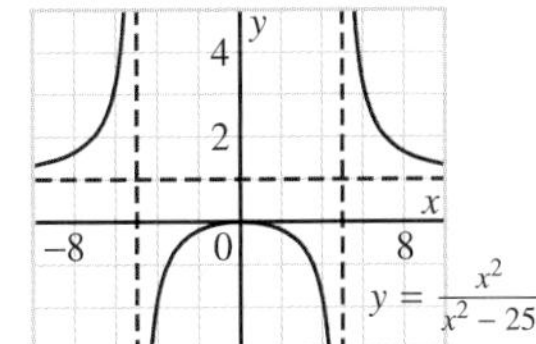

c) y-intercept: 0.2; domain: $\{x \mid x \in \Re,\ x \neq \pm 5\}$; decrease: $x \in \Re\ (x \neq \pm 5)$; concave up: $x > -5\ (x \neq 5)$; concave down: $x < -5$; vertical asymptote: $x = -5$; point discontinuity at $x = 5$; horizontal asymptote: $y = 0$

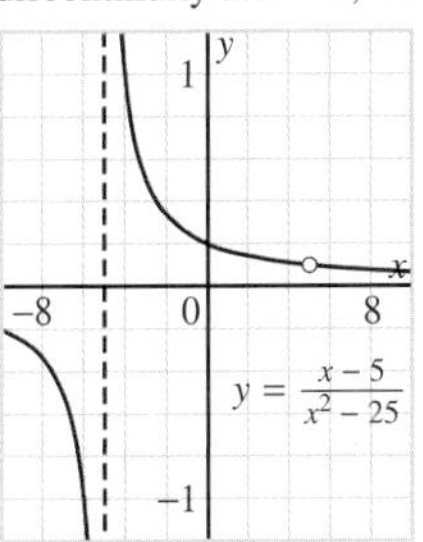

d) y-intercept: 2; domain: $x \in \Re$; critical point at $x = 0$; increase: $x < 0$; decrease: $x > 0$; points of inflection at $x \doteq \pm 1.3$; concave up: $|x| > 1.3$; concave down: $|x| < 1.3$; horizontal asymptote: $y = 0$

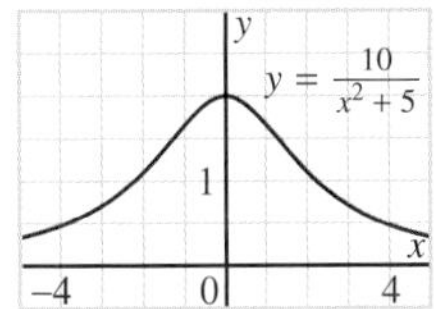

e) Domain: $\{x \mid x \in \Re,\ x \neq 0\}$; critical points at $x \doteq \pm 1.2$; increase: $|x| > 1.2$; decrease: $|x| < 1.2\ (x \neq 0)$; concave up: $x > 0$; concave down: $x < 0$; vertical asymptote: $x = 0$; oblique asymptote: $y = 2x$

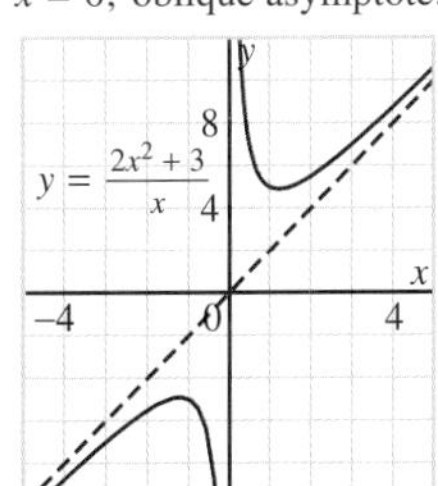

f) y-intercept: –2; domain: $\{x \mid x \in \Re,\ x \neq 1\}$; critical points at $x \doteq -0.7,\ 2.7$; increase: $x > 2.7$ and $x < -0.7$; decrease: $-0.7 < x < 2.7\ (x \neq 1)$; concave up: $x > 1$; concave down: $x < 1$; vertical asymptote: $x = 1$; oblique asymptote: $y = x + 1$

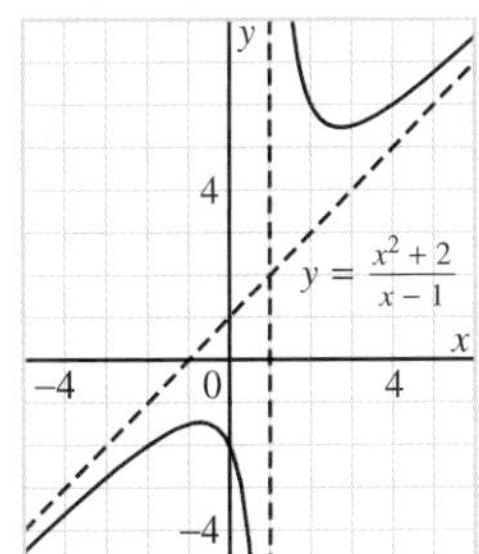

7. Graphs may vary.

a)

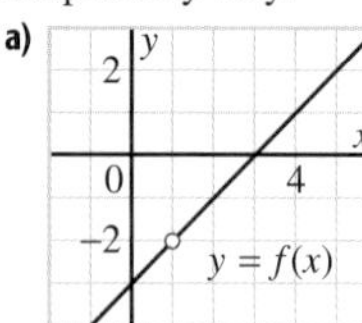

b)

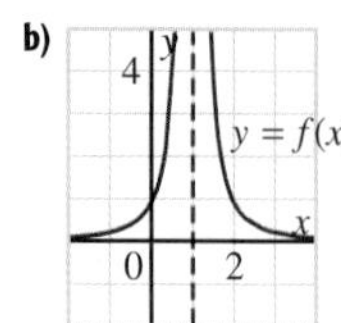

Chapter 5 Review Exercises, page 352

1. All except the equation in part e.

2. a) ii b) iii c) iv d) i

3. a) For $y = \frac{-1}{x+1}$: $y = \frac{1}{9}x - \frac{5}{9}$

b) For $y = \frac{1}{(x+1)^2}$: $y = -\frac{2}{27}x + \frac{7}{27}$

c) For $y = \frac{x^2-1}{x+1}$: $y = x - 1$

d) For $y = \frac{x}{x+1}$: $y = \frac{1}{9}x + \frac{4}{9}$

4. a) All the equations have a factor of $x + 1$ in the denominator.

b) Graphs in parts a, b, and d

c) Graph in part c

d) The equation has a factor of $x + 1$ in the numerator and denominator.

e) For part a: $y = 0$; for part b: $y = 0$; for part d: $y = 1$

5. a)

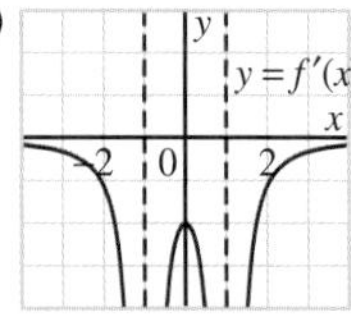

b)

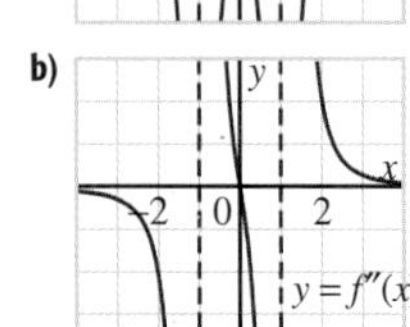

6. Graphs may vary.

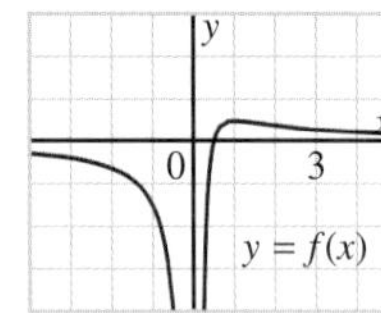

7. a) $\pm\infty$ b) Does not exist c) 0.5 d) 0.5 e) 1 f) 1

8. a) −8 b) $-\frac{1}{8}$ c) $-\frac{1}{2}$ d) 0 e) $\pm\infty$ f) $\pm\infty$ g) ∞ h) ∞

9. a) 1 b) 1 c) 0 d) 0 e) ∞ f) ∞ g) $\frac{3}{4}$ h) $-\frac{1}{2}$

10. a) i) 12 ii) 12 iii) 12

b) The graph is a straight line with a point discontinuity at (6, 12).

11. a) y-intercept: $\frac{1}{2}$

b) x-intercept: −3; y-intercept: 3

c) x-intercept: −2; y-intercept: −2

12. a) Vertical asymptote: $x = 1$

b) Vertical asymptote: $x = -3$

c) Point discontinuity at $x = 0$

d) Point discontinuity at $x = -2$

e) Vertical asymptote: $x = -2$; point discontinuity at $x = 2$

f) No vertical asymptote or point discontinuity

13. a) $y = 1$ b) $y = 6$ c) $y = 0$ d) $y = 0$ e) No horizontal asymptote f) No horizontal asymptote

14. The limits in parts c and d

15. a) Horizontal asymptote b) Oblique asymptote c) Oblique asymptote d) Horizontal asymptote

16. a) $y = 5$ b) $y = x - 6$ c) $y = x$ d) $y = 0$

17. a) $y = x$ b) $y = x + 1$ c) $y = 4x + 2$ d) $y = \frac{3}{5}x - \frac{3}{25}$

18. a) 5 b) $\frac{6}{5}$ c) 6

19. $y = ax + b$

20. Estimates may vary.

a) $f'(x) = 0$: $x = 0$; $f'(x) > 0$: $x > 0$; $f'(x) < 0$: $x < 0$; $f''(x) = 0$: $x = \pm 0.5$; $f''(x) > 0$: $|x| < 0.5$; $f''(x) < 0$: $|x| > 0.5$

b) $f'(x) = 0$: $x = 0$; $f'(x) > 0$: $x < 0$ $(x \neq -1)$; $f'(x) < 0$: $x > 0$ $(x \neq 1)$; $f''(x) \neq 0$; $f''(x) > 0$: $|x| > 1$; $f''(x) < 0$: $|x| < 1$

21. Estimates may vary.

a) i) Horizontal asymptote: $y = 2$ ii) (0, 0) iii) (±0.5, 0.5)

b) i) Horizontal asymptote: $y = 1$; vertical asymptotes: $x = \pm 1$ ii) (0, 0) iii) No points of inflection

22. a) Graphs may vary.

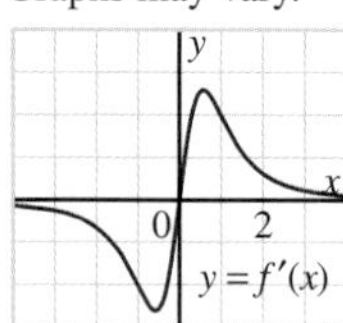

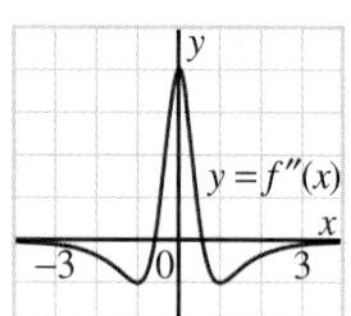

b) Graphs may vary.

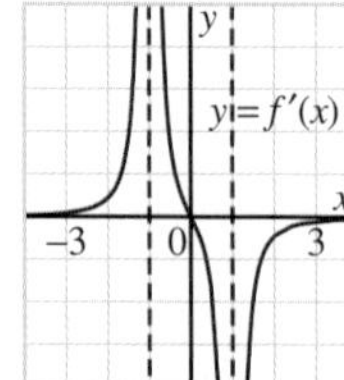

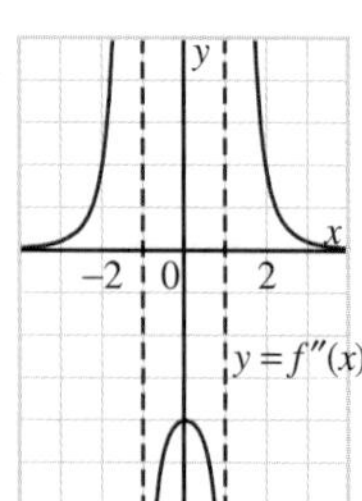

23. a) i) Increasing: $x < 0$ $(x \neq -1)$; decreasing: $x > 0$ $(x \neq 1)$

ii) $x = 0$ (local maximum)

iii) Graph may vary.

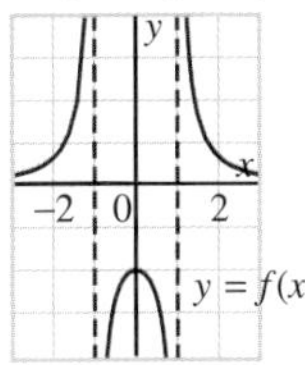

b) i) Decreasing: $x \in \Re$ $(x \neq -1)$

ii) No maximum or minimum

iii) Graph may vary.

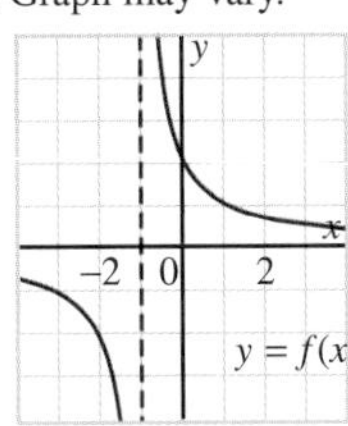

24. $f'(x) = \dfrac{-1}{(x+1)^2}$

25. a) y-intercept: $0.\overline{3}$; domain: $\{x \mid x \in \Re,\ x \neq -3\}$; decrease: $x \in \Re$ $(x \neq -3)$; concave up: $x > -3$; concave down: $x < -3$; vertical asymptote: $x = -3$; horizontal asymptote: $y = 0$

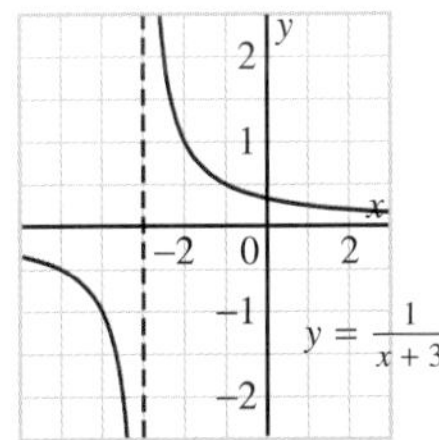

b) x-intercept: 0; y-intercept: 0; domain: $x \in \Re$; critical point at $x = 0$; increase: $x > 0$; decrease: $x < 0$; points of inflection at $x \doteq \pm 0.82$; concave up: $|x| < 0.82$; concave down: $|x| > 0.82$; horizontal asymptote: $y = 1$

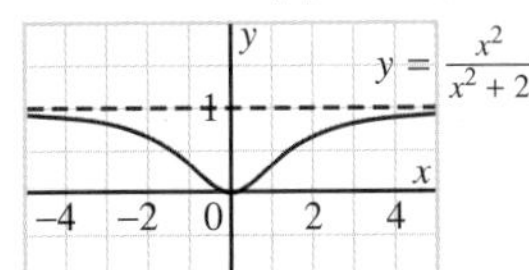

c) y-intercept: 0.5; domain: $\{x \mid x \in \Re,\ x \neq -2,\ 3\}$; decrease: $x \in \Re$ $(x \neq -2,\ 3)$; concave up: $x > -2$ $(x \neq 3)$; concave down: $x < -2$; vertical asymptote: $x = -2$; point discontinuity at $x = 3$; horizontal asymptote: $y = 0$

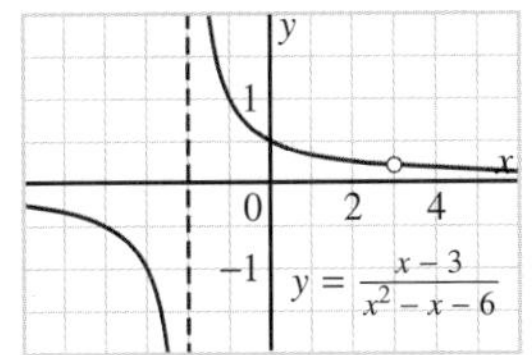

d) x-intercepts: $\pm\sqrt{2}$; y-intercept: -2; domain: $\{x \mid x \in \Re,\ x \neq -1\}$; increase: $x \in \Re$ $(x \neq -1)$; concave up: $x < -1$; concave down: $x > -1$; vertical asymptote: $x = 1$; oblique asymptote: $y = x - 1$

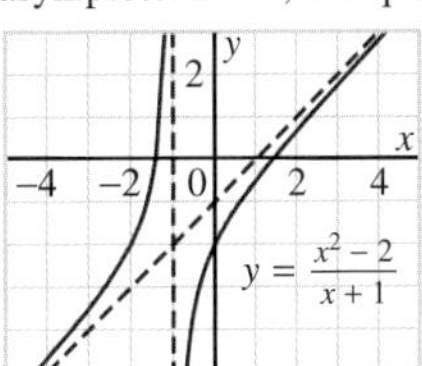

Chapter 5 Self-Test, page 357

1. a) 10 **b)** $-\frac{1}{3}$ **c)** $\pm\infty$

d) $\pm\infty$ **e)** 0 **f)** 1

g) $-\infty$ **h)** $-\frac{3}{5}$

2. Graphs may vary.

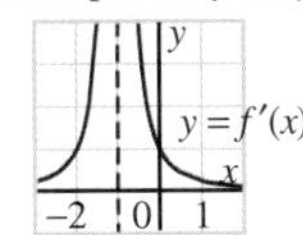

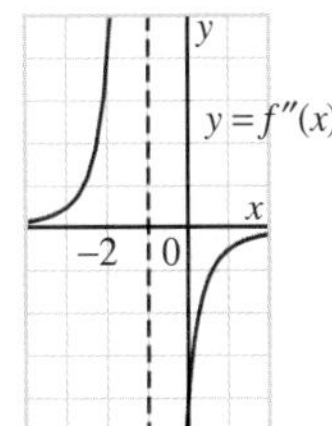

3. a) x-intercept: 2; y-intercept: -2; domain: $\{x \mid x \in \Re,\ x \neq -2\}$; increase: $x \in \Re$ $(x \neq -2)$; point discontinuity at $x = -2$

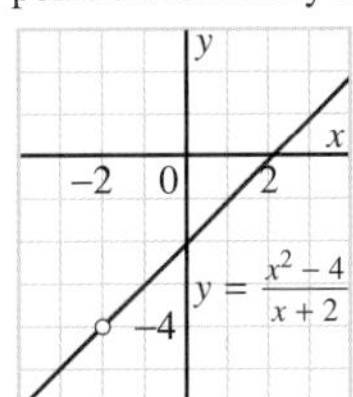

b) x-intercept: 0; y-intercept: 0; domain: $\{x \mid x \in \Re,\ x \neq \pm 3\}$; critical point at $x = 0$; increase: $x < 0$ $(x \neq -3)$; decrease: $x > 0$ $(x \neq 3)$; concave up: $|x| > 3$; concave down: $|x| < 3$; vertical asymptotes: $x = \pm 3$; horizontal asymptote: $y = 1$

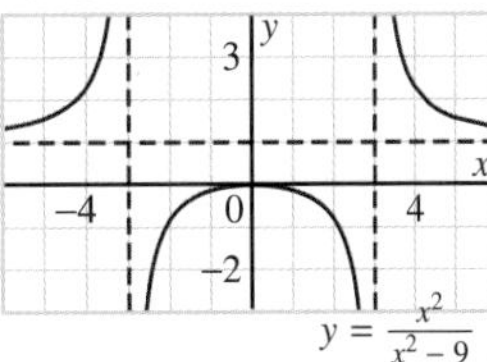

5. a) $y = n + 2$ **b)** $n = 0, 4$ **c)** 8 **d)** 0

6. Answers may vary. $y = \dfrac{2x^2 - 3}{x^2 - 1}$

7. a) i **b)** No **c)** Yes

8. a) i) A point discontinuity occurs at $x = 2$ with y approaching 4 near $x = 2$.

ii) A vertical asymptote occurs at $x = 2$ with y approaching infinity near $x = 2$.

iii) A vertical asymptote occurs at $x = 2$ with y approaching negative infinity from the left side of $x = 2$ and approaching positive infinity from the right side of $x = 2$.

b) Answers may vary.

i) $y = \frac{x^2 - 4}{x - 2}$ **ii)** $y = \frac{1}{(x - 2)^2}$ **iii)** $y = \frac{1}{x - 2}$

9. a) Vertical asymptote: $x = 2$; horizontal asymptote: $y = 1$

b) $f'(x) = \frac{-2(x - 1)}{(x - 2)^3}$

c) Vertical asymptote: $x = 2$; horizontal asymptote: $y = 0$; the vertical asymptote remains the same.

Mathematical Modelling: Predicting Populations

Exercises, page 362

1. a) $\Delta P = 0.1969P - 0.0225P^2$

b)

Year	Population (billions)
2005	6.449
2010	6.784
2015	7.085
2020	7.351
2025	7.584

2. a) Population of Canada

P

Population (millions)

30
20
10

1951 1971 1991

t

Year

b)

Year	$\frac{\Delta P}{P}$
1951 – '56	0.178
1956 – '61	0.134
1961 – '66	0.097
1966 – '71	0.078
1971 – '76	0.090
1976 – '81	0.059
1981 – '86	0.052
1986 – '91	0.073
1991 – '96	0.066
1996 – '01	0.055

c)

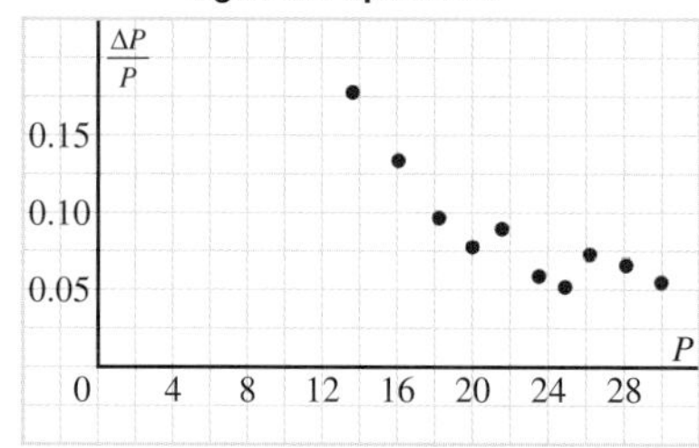

d) $\Delta P = 0.2346P - 0.0066P^2$

e) Approximately 32.4 million in 2006 and approximately 33.1 million in 2011.

Chapter 6 Exponential and Logarithmic Functions

Necessary Skills

1 Review: Exponent Laws for Integer Exponents

Exercises, page 365

1. a) 1 **b)** $\frac{1}{3}$ **c)** $\frac{1}{25}$ **d)** $\frac{4}{25}$ **e)** $\frac{8}{27}$
f) $\frac{4}{3}$ **g)** $\frac{16}{9}$ **h)** $\frac{16}{9}$ **i)** 8

2. a) m^{-4} **b)** b^4 **c)** $8a^{-4}$ **d)** x^2 **e)** $3x^{-2}$
f) a^4 **g)** c^6 **h)** $9x^{-4}$ **i)** a^6

2 Review: Exponent Laws for Rational Exponents

Exercises, page 367

1. a) 2 **b)** 2 **c)** 8 **d)** 8
e) $\frac{1}{3}$ **f)** $\frac{1}{9}$ **g)** $\frac{1}{125}$ **h)** $\frac{1}{32}$

2. a) 1.047 **b)** 1.050 **c)** 0.283 **d)** 0.967

3. a) $\left(\sqrt[100]{1.15}\right)^{328}$ **b)** $\left(\sqrt[100]{1.08}\right)^{633}$
c) $\frac{1}{\left(\sqrt[100]{1.56}\right)^{284}}$ **d)** $\frac{1}{\left(\sqrt[100]{2.45}\right)^{37}}$

4. a) $a^{4.5}$ **b)** $c^{0.2}$ **c)** $s^{-0.75}$ **d)** $x^{1.4}$ **e)** m^2
f) $a^{-1.6}$ **g)** x^3 **h)** $t^{-7.5}$ **i)** $\frac{1}{4}k^3$

3 Review: Transformations of Graphs

Exercises, page 369

1. a) $y = (x - 3)^3 + 1$ **b)** $y = \sqrt{x - 3} + 1$
c) $y = \frac{1}{x - 3} + 1$ **d)** $y = x - 2$
e) $y = 2^{x - 3} + 1$ **f)** $y = f(x - 3) + 1$

2. a) $y = 3\left(\frac{1}{2}x\right)^3$ **b)** $y = 3\sqrt{\frac{1}{2}x}$
c) $y = \frac{6}{x}$ **d)** $y = \frac{3}{2}x$
e) $y = 3(2)^{\frac{x}{2}}$ **f)** $y = 3f\left(\frac{1}{2}x\right)$

3. a) 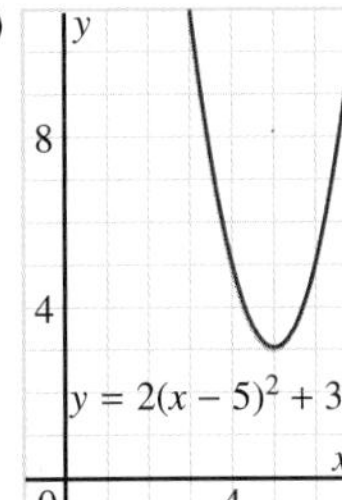

b) 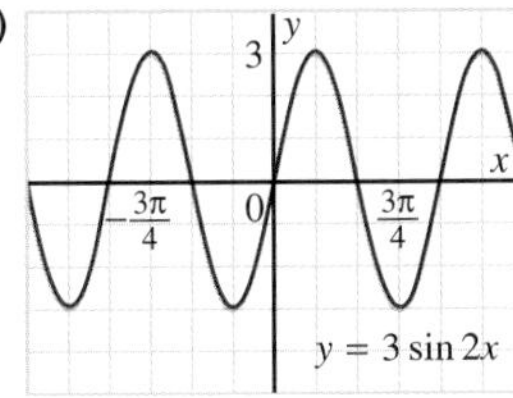

c) 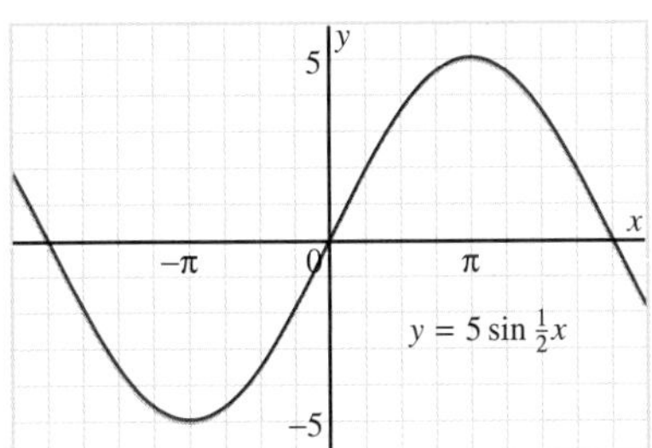

d) 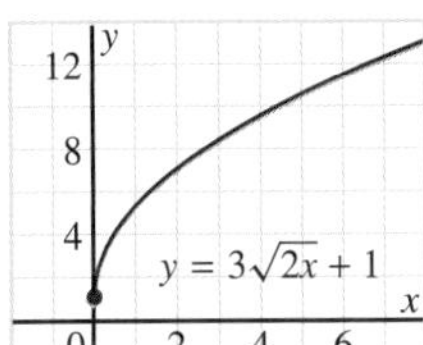

e) 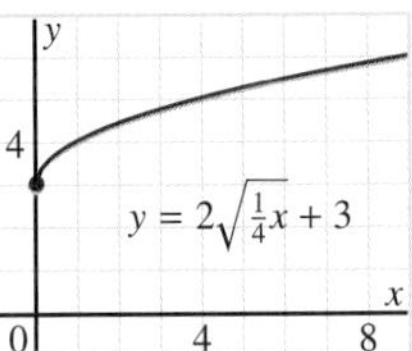

f) 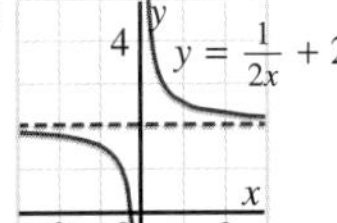

6.1 Exercises, page 376

1. a) iv b) iii c) i d) ii

2. a), b) 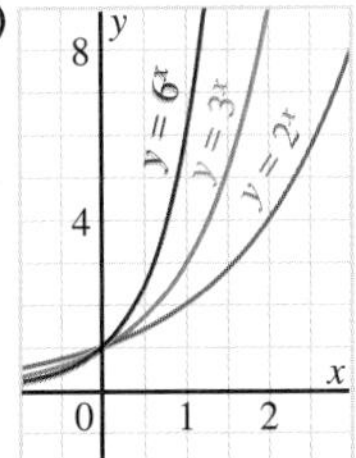

3. 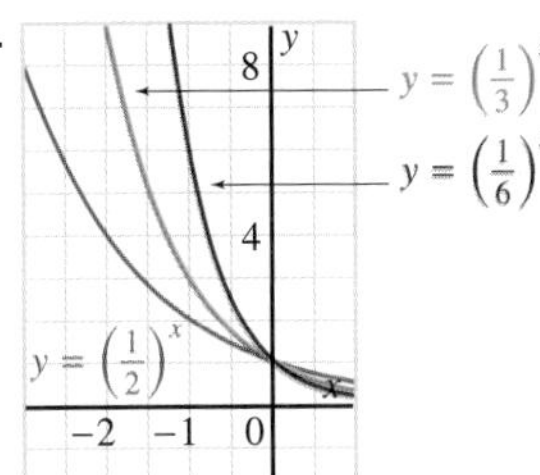

5. a) \$700 b) Approximately 11 years

c) i) The graph has a greater A-intercept and increases more quickly. The equation is a number greater than 500 multiplied by $(1.065)^n$.

ii) The graph has a lesser A-intercept and increases more slowly. The equation is a number less than 500 multiplied by $(1.065)^n$.

iii) The graph increases more quickly. The equation is 500 multiplied by a power with a base greater than 1.065.

iv) The graph increases more slowly. The equation is 500 multiplied by a power with a base less than 1.065.

6. a) Approximately 11 h

7. a) 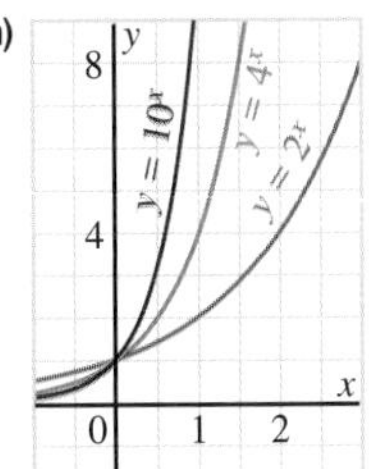

b) 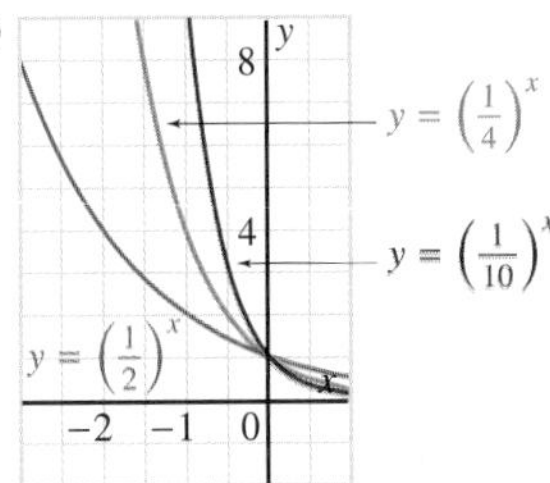

8. a) 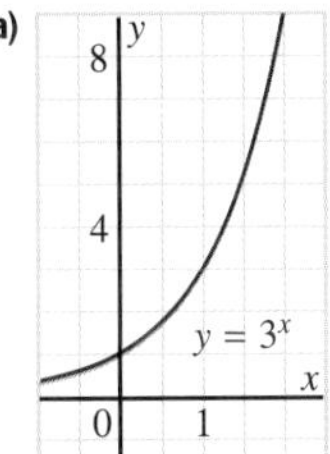

9. a) 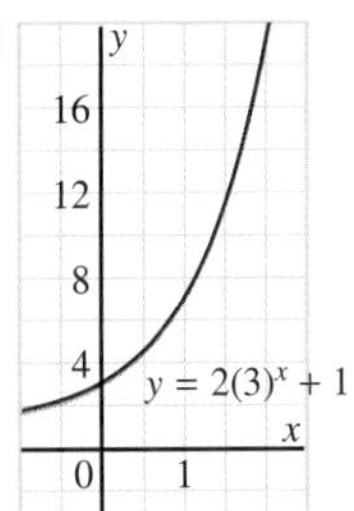

10. a) 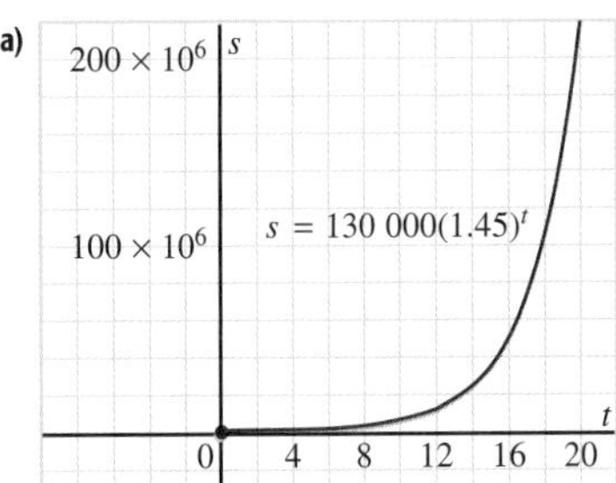

b) Assume we calculate at the end of 2005, and the number of users continues to be exponential. Approximately 104 368 000

c) In 1999 d) 45%

11. a) $y = 2^x$ **b)** $y = 2^x$ **c)** Yes

12. a) Approximately 36.4 million
b) 2008
c) The graph of P is the image after a vertical stretch by a factor of 24 of the graph of $y = 1.014^x$.
e) The population continues to grow according to the given multiplicative growth equation.

13. a) \$685.33 **b)** Approximately 11 years

14. a) Approximately 2.04 s **b)** Approximately –17°C

15. b) $P = 0.939^n$; $t = 8.72(1.075)^{-T}$

16. a) $P = 100(0.95)^n$
b)

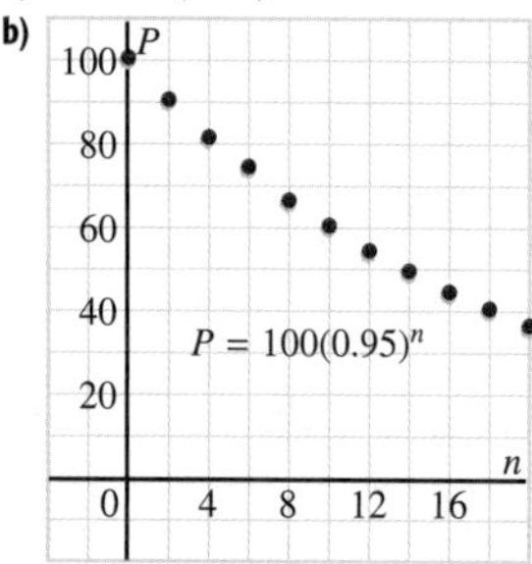

c) 13 or 14 panes

17. a) $P = 100(0.975)^d$
b)

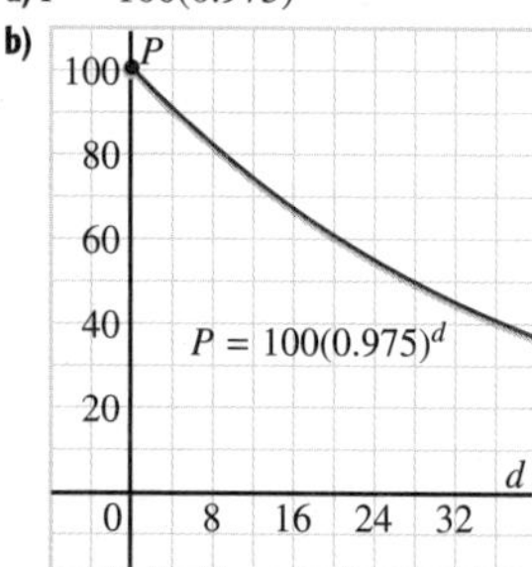

d) 77.6% **e)** Approximately 27 m

6.2 Exercises, page 383

1. a) 0.9191; $10^{0.9191} \doteq 8.3$ **b)** 2.8028; $10^{2.8028} \doteq 635$
c) 3.9931; $10^{3.9931} \doteq 9842$ **d)** –0.4318; $10^{-0.4318} \doteq 0.37$
e) –1.1024; $10^{-1.1024} \doteq 0.079$
f) –2.3468; $10^{-2.3468} \doteq 0.0045$

3. a) 0, 1, 2, 3 **b)** 0, 1, 2, 3 **c)** 0, 1, 2, 3

6. a) $2^3 = 8$ **b)** $3^4 = 81$ **c)** $10^4 = 10\ 000$
d) $10^{-2} = 0.01$ **e)** $5^3 = 125$ **f)** $16^{0.25} = 2$
g) $2^{-2} = \frac{1}{4}$ **h)** $10^y = x$ **i)** $a^y = x$

7. a) $\log_7 49 = 2$ **b)** $\log 1000 = 3$ **c)** $\log_2 1024 = 10$
d) $\log_5 625 = 4$ **e)** $\log_2 0.5 = -1$ **f)** $\log_{16} 4 = \frac{1}{2}$
g) $\log_{81} 27 = \frac{3}{4}$ **h)** $\log_{16} 2 = 0.25$ **i)** $\log_9\left(\frac{1}{3}\right) = -0.5$

8. a) 3 **b)** 2 **c)** 3 **d)** 5
e) 0 **f)** 0 **g)** 4 **h)** 2
i) –1 **j)** –1 **k)** –4 **l)** 1

10. a) 2 **b)** Answers may vary. $\log_7 7^2$

12. a) 3 **b)** 4 **c)** 3
d) –1 **e)** –2 **f)** x

13. a) 1000 **b)** 81 **c)** 216
d) $\frac{1}{2}$ **e)** $\frac{1}{25}$ **f)** x

14. a) i) 0.3010, 1.3010, 2.3010, 3.3010
ii) 0.4771, 1.4771, 2.4771, 3.4771
b) Each logarithm is 1 greater than the previous one since the value of x in $\log x$ increases by a factor of 10 and $\log 10 = 1$.

15. Answers may vary.

$\log 0.1$	$\log_5 0.2$	$\log_a a^{-1}$
$\log 0.01$	$\log_5 0.04$	$\log_a a^{-2}$
$\log 0.001$	$\log_5 0.008$	$\log_a a^{-3}$
$\log 0.0001$	$\log_5 0.0016$	$\log_a a^{-4}$

16. Answers are rounded.
a) i) 0.18 **ii)** 0.12 **iii)** 0.10 **iv)** 0.08
v) 0.07 **vi)** 0.06 **vii)** 0.05 **viii)** 0.05
b)

P
0.32
0.24
0.16
0.08
0
2
4
6
8
n
$P = \log\left(1 + \frac{1}{n}\right)$

17. 2.718 28

18. a) 10 **b)** –6 **c)** $\frac{5}{2}$

19. a) i) 3, $\frac{1}{3}$ **ii)** 2, $\frac{1}{2}$
b) $\log_a b = \frac{1}{\log_b a}$

6.3 Exercises, page 390

1. a) 0.699, 1.398, 2.097, 2.796, 3.495;
All the logarithms of powers of 5 are multiples of the logarithm of 5.
b) –0.301, –0.602, –0.903, –1.204, –1.505;
All the logarithms of powers of 0.5 are multiples of the logarithm of 0.5.

2. a) 0.301, 0.602, 0.903, 1.204, 1.505
b) 0.301, 0.602, 0.903, 1.204, 1.505

3. Answers may vary.

a)	b)
log 3	log 3
log 9	2 log 3
log 27	3 log 3
log 81	4 log 3
log 243	5 log 3

4. a) 3.4594 **b)** 2.5789 **c)** 0.8982

5. a) 0.6309, 1.2619, 1.8928, 2.5237
b) 0.6309, 0.3155, 0.2103, 0.1577
c) 1.5850, 0.7925, 0.5283, 0.3962

7. a) 3.3219 **b)** 4.1918 **c)** 2.5740
d) 5.8074 **e)** 0.9845 **f)** 2.4307

8. a) $2^{4.7004}$ **b)** $5^{2.3370}$ **c)** $4^{3.2379}$
d) $6^{3.3167}$ **e)** $10^{0.8451}$ **f)** $3^{0.6309}$

9. a) Approximately 11 years
   b) Approximately 11 years
   c) \$603.97; approximately 11 years
   d) Under exponential growth, the doubling time depends only on the base, 1.065, in the equation.

10. b)

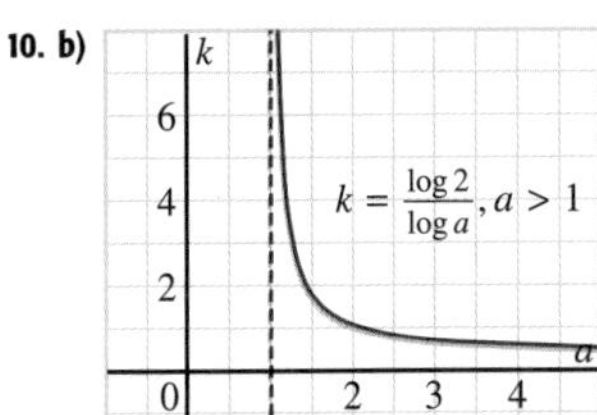

11. a) Approximately 2305 million
    b) Between 2002–2003

12. a) $L \doteq 17.83$ m; $T \doteq 0.05$ m
    b) 13 passes; approximately 0.04 m
    c) 33 passes; approximately 820 m

6.4 Exercises, page 396

1. a) 0.699, 1.699, 2.699, 3.699, 4.699
   b) –0.301, –1.301, –2.301, –3.301, –4.301

2. a) log 36, log 36, log 36, log 36, log 36
   b) log 2, log 2, log 2, log 2, log 2

3. Answers may vary.
   a) log 1 + log 12; log 2 + log 6; log 3 + log 4
   b) log 3 – log 1; log 6 – log 2; log 9 – log 3

4. a) 1 b) 2 c) 3 d) 4 e) 4
   f) 4 g) 1 h) 2 i) 1

5. a) 2 b) 4 c) 2

6. a) 2 b) 1 c) 3
   d) 3 e) 3 f) 4

7. Answers may vary. $\log_9 3 + \log_9 27$; $\log_4 8 + \log_4 2$

8. a) 1 b) 2 c) 2 d) 3

9. a) log 75 b) log 2 c) log 96 d) $\log x^{\frac{1}{2}}y^{\frac{3}{2}}$

10. a) 16.5490 b) 6.6617 c) 12.2192 d) 2.7757

11. a) 5 b) 2 c) 10 d) 10
    e) 12.5 f) 4 g) 20 h) 5

12. a) Approximately 5 h b) Approximately 5 h
    c) Approximately 76% d) Approximately 5 h
    e) Under exponential decay, the half-life depends only on the base, 0.87, in the equation.

13. b)

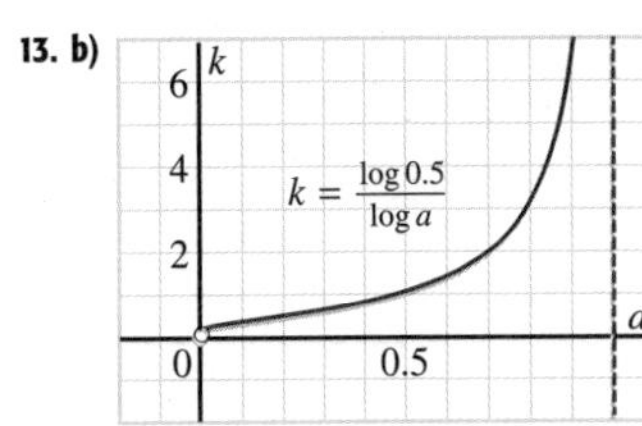

14. Answers may vary: the following answers are for iodine-131:
    a)

Time (days)	Percent remaining
0	100
8	50
16	25
24	12.5
32	6.25
40	3.125

Percent of iodine-131 remaining

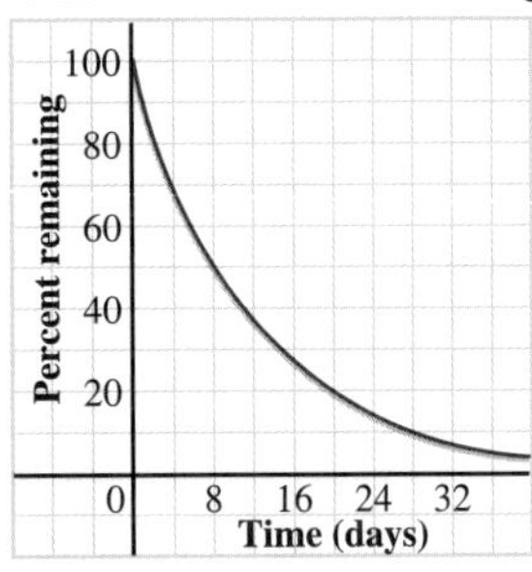

b) 3.125% c) 9.54×10^{-5}%

16. a)

Time (h)	Mass remaining (mg)
0	40
14.9	20
29.8	10
44.7	5
59.6	2.5
74.5	1.25

Mass of sodium-24 remaining

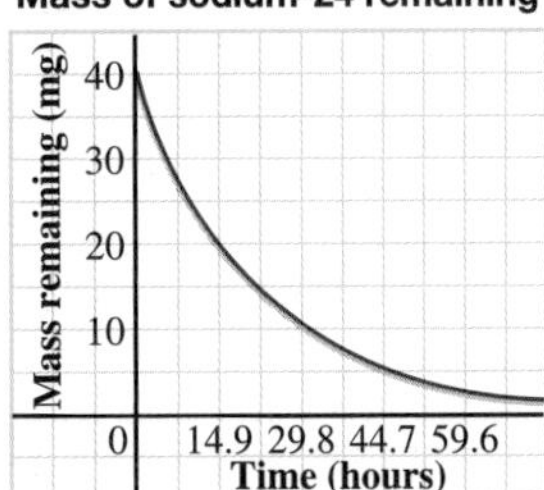

b) Approximately 4.29 mg c) Approximately 79.3 h

17. a)

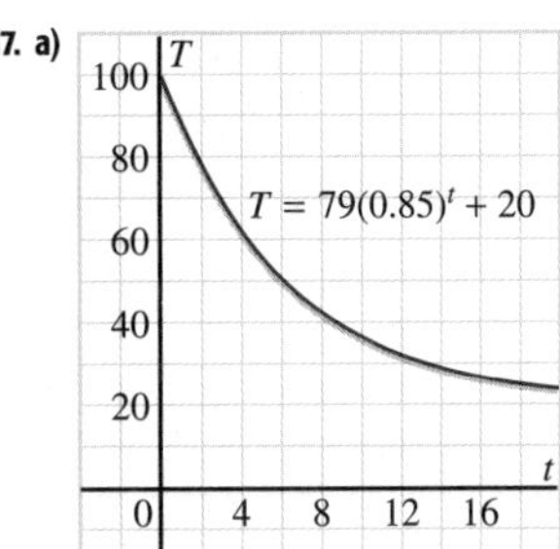

b) Approximately 55.1°C c) Approximately 8.5 min

18. a) 20 represents the surrounding temperature in degrees Celsius. 79 represents the initial difference in temperature of the hot chocolate in degrees Celsius; that is, 79°C = 99°C – 20°C. 0.85 represents the decay factor; the rate at which the hot chocolate is cooling.

b) The equation can be written as $T - 20 = 79(0.85)^t$, where the right side of the equation is an exponential function.

c) The graph of $y = 0.85^x$ is stretched vertically by a factor of 79 and translated 20 units up to result in the graph of $T = 79(0.85)^t + 20$.

20. a) $C = 10.6(1.07)^t$ **b)** 1995–1996

22. $A = 1000(1.062)^n$

23. a) 44 **b)** 2

24. a) 2021–2022

b) i) 2042–2043 **ii)** 2075–2076 **iii)** 2149–2150

Self-Check 6.1–6.4, page 400

1. a)

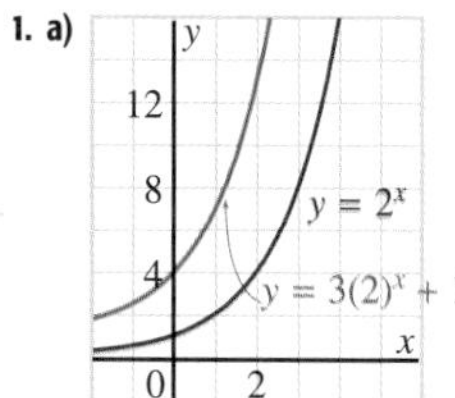

b)

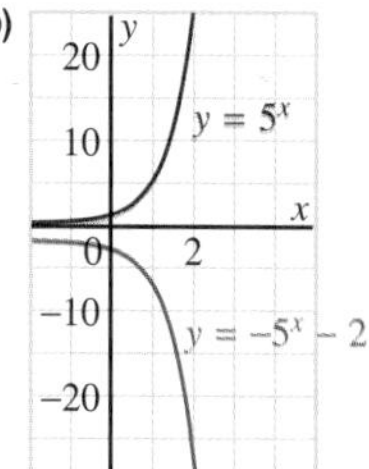

2. a) $3^3 = 27$ **b)** $3^{-2} = \frac{1}{9}$ **c)** $2^{-1} = 0.5$

3. a) $\log_6 1296 = 4$ **b)** $\log_6\left(\frac{1}{1296}\right) = -4$

c) $\log_{10} 1000 = 3$ **d)** $\log_{10} 0.001 = -3$

e) $\log_{27} 9 = \frac{2}{3}$ **f)** $\log_{25} 5 = 0.5$

4. a) 2 **b)** –2 **c)** 4 **d)** –3

5. a) 2.8074 **b)** 3.0959 **c)** 0.8617

6. a) $2^{5.0444}$ **b)** $7^{2.0499}$

7. a) 3 **b)** 3 **c)** log 1.5

8. a) 11.4698 **b)** 4 **c)** $\frac{6}{5}$

9. a) 5.2 million **b)** 2008–2009

c) The population continues to grow according to the given model.

d) Disease epidemics, war, and medical cures could change the growth pattern of the population.

e) The graph of $y = 1.022^x$ is stretched vertically by a factor of 2.76 to result in the graph of $P = 2.76(1.022)^t$.

6.5 Exercises, page 405

1. b) We know that the equations are equivalent for all values of n, not just for the 3 values in part a.

2. a) $y = 3^{1.4650x}$ **b)** $y = 3^{2.0959x}$ **c)** $y = 3^{0.3691x}$

3. a) $y = 2^{3.3219x}$ **b)** $y = 5^{1.4307}$ **c)** $y = 20^{0.7686x}$

4. a) $P = 8.48(2)^{0.0257t}$ **b)** Approximately 39 years

5. a) $h = 7.4(2)^{0.6690t}$ **b)** $P = 24.0(2)^{0.0201t}$

c) $M = 460(2)^{0.2510n}$

6. a) $P = 100(2)^{-0.2009n}$ **b)** $P = 100(2)^{-0.1234t}$

c) $T = 79(2)^{-0.2345t} + 20$

8. a) $c = 1000(2)^{\frac{n}{1.5}}$ **b)** $c = 1000(10)^{0.2007n}$

9. a)

b) The graph of $p = 100(0.98)^n$ is obtained from $y = 0.98^x$ with a vertical stretch by a factor of 100. The graph of $p = 100 - 50(0.98)^n$ is obtained from $y = 0.98^x$ with a reflection in the x-axis, a vertical stretch by a factor of 50, and a translation 100 units up.

c) Answers may vary. A disadvantage of using $p = 100 - 50(0.98)^n$ is that the model is only defined for $50 \le p < 100$.

6.6 Exercises, page 409

1. a) 10 **b)** 100 **c)** 1000

d) 10 000 **e)** 100 000 **f)** 1 000 000

2. 63 times

4. Answers are approximate.

a) i) 500 times **ii)** 3160 times **iii)** 6 times

b) Earthquakes occur along Canada's west coast because there is a fault line in the area.

6. Answers may vary.

a) Papua New Guinea was 2 times as intense as Peru.

b) Mexico was 4 times as intense as Turkey.

c) B.C. Coast was 8 times as intense as Mexico.

d) Alaska was 16 times as intense as Papua New Guinea.

7. Each number is a difference in measures on the Richter scale.

a) 2.3 **b)** 0.3 **c)** 1.3

8. a) 6 or 7 times as frequent

b) 36 to 49 times as frequent

c) 216 to 243 times as frequent

9. a) $E = E_0(31)^{M - M_0}$

b) i) 31 times **ii)** 961 times

c) 165 869 times

10. a) 100 000 times **b)** 1000 times **c)** 3.16×10^{11} times

11. a) 3 162 278 times **b)** 10 000 times

12. 316 times

13. 10 000 times

14. $L = L_0(10)^{\frac{d}{10}}$

15. a) 8 times **b)** 7943 times

16. a) i) 30 min **ii)** 1.875 min

b) $E = 8(0.5)^{\frac{s-90}{5}}$

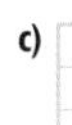

c)

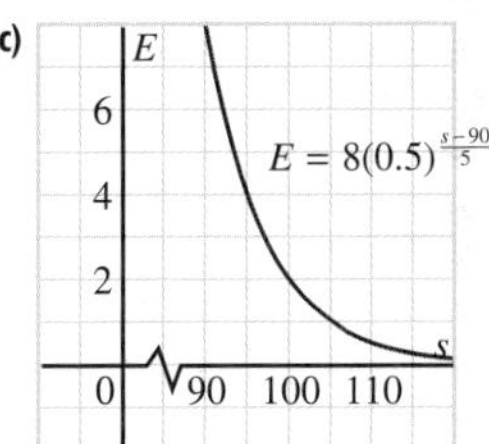

17. a) 15 849 times **b)** 2512 times

c) 10 000 times **d)** 16 times

18. a) pH 5.2 **b)** pH 5.8

19. a) 10 times

b) i) 32 times **ii)** 501 times

6.7 Exercises, page 418

1. a) $y = \log_{10} x$ **b)** $y = \log_2 x$

c) $f^{-1}(x) = \log_5 x$ **d)** $f^{-1}(x) = \log_{0.3} x$

e) $g^{-1}(x) = \log_{\frac{3}{2}} x$ **f)** $h^{-1}(x) = \log_{1.5} x$

2. a) $y = 10^x$ **b)** $y = 3^x$ **c)** $f^{-1}(x) = 7^x$

3. When $a = 1$, $y = a^x$ is $y = 1^x$, which is a horizontal line. The inverse of $y = 1$ is a vertical line, which is not a function.

4. a)

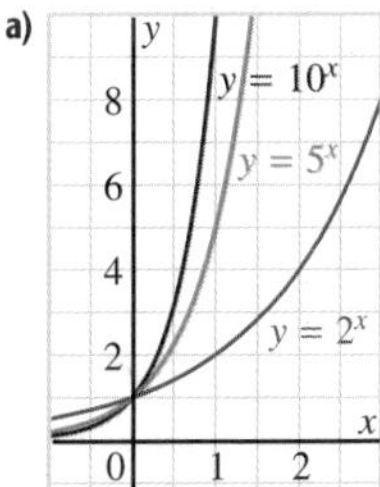

b)

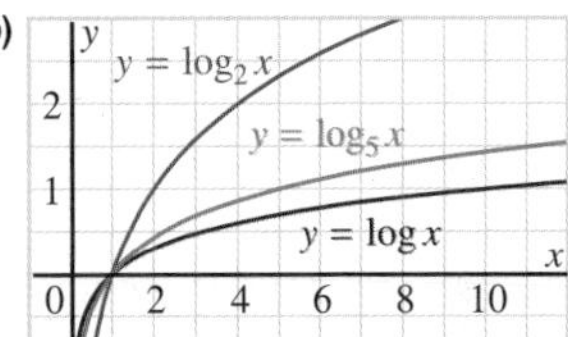

c) The graphs in part b are the inverses of the graphs in part a.

5. a)

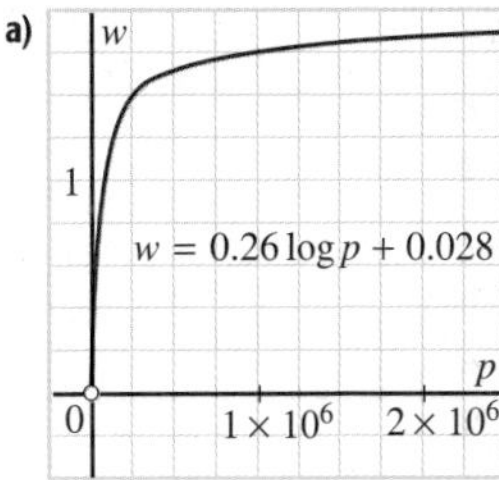

b) i) 0.69 m/s **ii)** 1.17 m/s

iii) 1.51 m/s **iv)** 1.67 m/s

c) 5476

6. a)

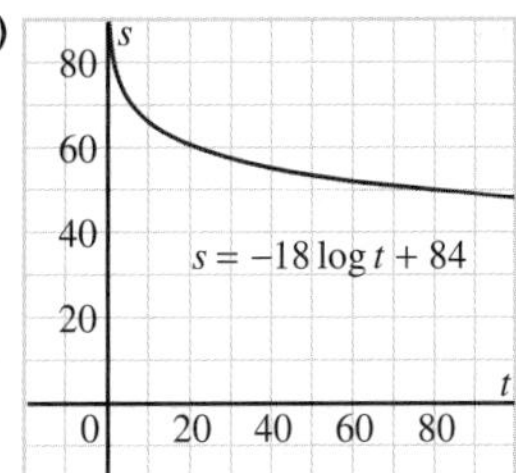

b) i) 57.4% **ii)** 52.0% **iii)** 27.1% **iv)** 11.9%

c) Approximately 77.4 min

8. a) Approximate equations:

$n = 12.63 \log\left(\frac{L}{2.00}\right)$; $n = -12.36 \log\left(\frac{T}{0.50}\right)$

b) If the equation is in logarithmic form, it is easier to determine the number of passes required to obtain a certain thickness.

9. a) $A = (10)^{\frac{s-8.1}{10.7}}$ **b)** $t = (10)^{\frac{84-s}{18}}$

10. a) $s = \frac{10.7 \log_2 A}{\log_2 10} + 8.1$; or $s = 3.22 \log_2 A + 8.1$

b) $s = \frac{-18 \log_2 t}{\log_2 10} + 84$; or $s = -5.42 \log_2 t + 84$

11. a) i) $f(x_1x_2) = f(x_1) + f(x_2)$ **ii)** $f\left(\frac{x_1}{x_2}\right) = f(x_1) - f(x_2)$

iii) $f(x_1{}^n) = nf(x_1)$ **iv)** $f\left(\frac{1}{x_1}\right) = -f(x_1)$

b) Answers may vary.

12. a) 2 098 960 **b)** 4 **c)** 1

Self-Check 6.5–6.7, page 421

1. a) $y = 5^{0.4307x}$ **b)** $y = 5^{1.1133x}$ **c)** $y = 5^{-0.4307x}$

2. a) $y = 4^{1.6610x}$ **b)** $y = 0.5^{-3.3219x}$ **c)** $y = 15^{0.8503x}$

3. a) \$1500 **b)** 3.5%

c) $A = 1500(2)^{0.0496n}$

d) Approximately 20 years

4. a) $\frac{1}{10}$ **b)** 10 **c)** 100 **d)** $\frac{1}{100}$ **e)** 1000

5. 10

6. 316

7. Approximately 16

8. a)

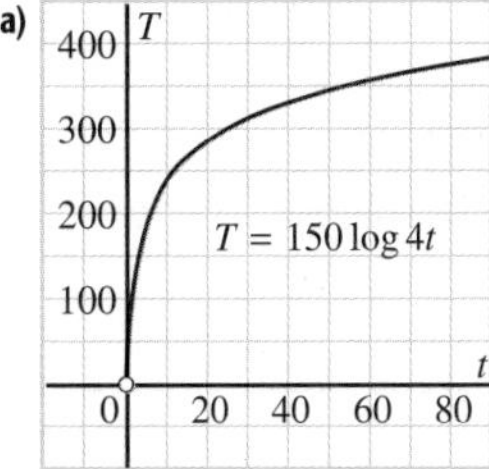

b) Approximately 312°C

c)

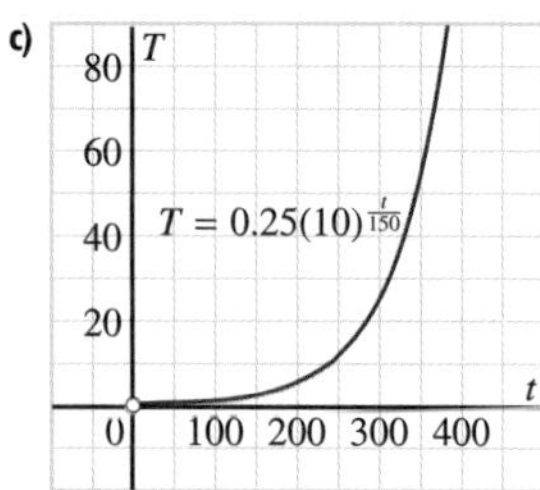

d) Approximately 79 s

Chapter 6 Review Exercises, page 423

1.

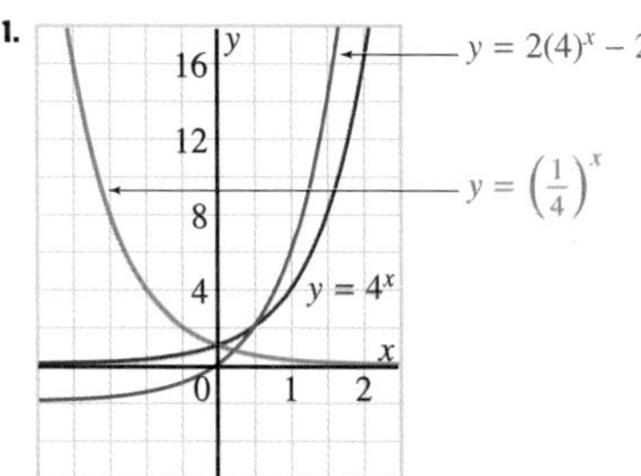

2. a) i) x-intercept: none; y-intercept: 1
 ii) $D = \Re$; $R = \{y \mid y \in \Re,\ y > 0\}$
 iii) The function is increasing.
 iv) The graph is concave up.
 v) The x-axis is a horizontal asymptote.
 b) i) x-intercept: none; y-intercept: 1
 ii) $D = \Re$; $R = \{y \mid y \in \Re,\ y > 0\}$
 iii) The function is decreasing.
 iv) The graph is concave up.
 v) The x-axis is a horizontal asymptote.
 c) i) x-intercept: 0; y-intercept: 0
 ii) $D = \Re$; $R = \{y \mid y \in \Re,\ y > -2\}$
 iii) The function is increasing.
 iv) The graph is concave up.
 v) The line $y = -2$ is a horizontal asymptote.

3. a) i)

b) i)

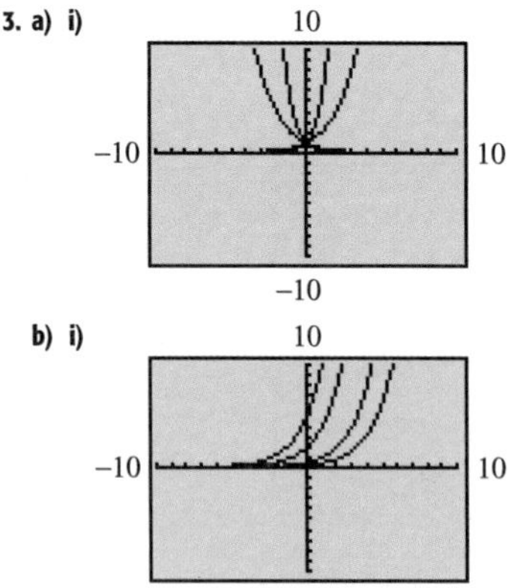

c) i)

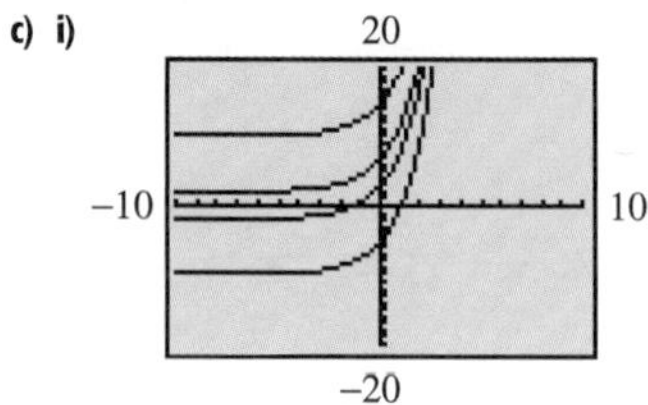

4. a) \$3460.10 b) 13.9 years

6. a) $y = 2^x$ b) $y = 2^x$

7. a) $2^5 = 32$ b) $3^1 = 3$ c) $10^0 = 1$
 d) $4^{-2} = \frac{1}{16}$ e) $5^{-3} = 0.008$ f) $8^2 = 64$

8. a) $\log_2 1024 = 10$ b) $\log 100 = 2$
 c) $\log 0.01 = -2$ d) $\log_{25} 5 = \frac{1}{2}$
 e) $\log_{16} 64 = \frac{3}{2}$ f) $\log_{1296} 6 = 0.25$

9. a) 0 b) 4 c) 6
 d) −1 e) −2 f) −3

10. a) 4 b) 5 c) 1000 d) 4

11. a) 1.1761 b) 1.5480 c) 0.5283
 d) 2.8614 e) 0.6309 f) 1.5850

12. a) $10^{1.4150}$ b) $10^{0.6990}$ c) $4^{3.1144}$
 d) $4^{0.7925}$ e) $7^{2.3666}$ f) $7^{0.9208}$

13. a) 2006–2007 b) 6.3 years

14. a) 3 b) 5 c) 2 d) 3

15. a) 3 b) −1 c) 2 d) −1

16. a) 4.6464 b) 15.9006 c) 1.6548
 d) 1.4322 e) 7.4384 f) 4.8271

17. a) 10 b) 3 c) 12 d) $1 + \sqrt{11}$

18. a) 3.11 g b) Approximately 40 years

19. a) $y = 4^{1.1610x}$ b) $y = 4^{1.6610x}$ c) $y = 4^{0.6610x}$

20. a) $y = 3^{0.6309x}$ b) $y = 5^{0.4307x}$ c) $y = 7.5^{0.3440x}$

21. a) 78.66%
 b) Approximately 3983 years old
 c) $P = 100(0.5)^{0.000\,17n}$
 d) Approximately 5776 years

22. a) i) The earthquake in Papua New Guinea
 ii) The Papua New Guinea earthquake was approximately twice as intense as the Peru earthquake.
 b) 8.0 c) 7.7

23. a) 10 000 000 b) 77 dB c) 113 dB

24. a) i) Vinegar
 ii) Vinegar is approximately 16 times as acidic as tomato juice.
 b) 3.7 c) 3.1

25. a) $y = \log_9 x$ b) $f^{-1}(x) = \log_{0.1} x$ c) $g^{-1}(x) = \log_{\frac{3}{4}} x$

26. a) $y = 10^x$ b) $y = 5^x$ c) $y = 8^x$

27. 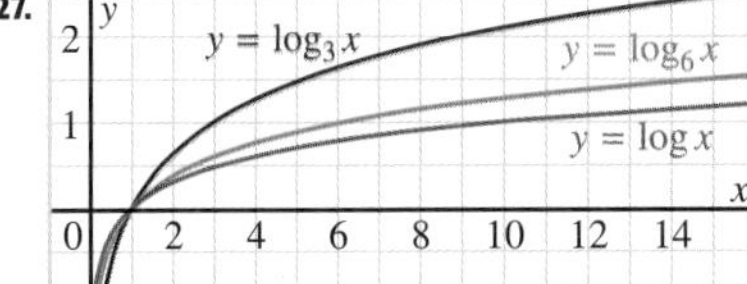

28. a), b), c)

i) x-intercept: 1; y-intercept: none

ii) $D = \{x \mid x \in \Re,\ x > 0\}$; $R = \Re$

iii) The function is increasing.

iv) The graph is concave down.

v) The y-axis is a vertical asymptote.

Chapter 6 Self-Test, page 427

1. a) 0 b) 4 c) 3

2. a) $\log\left(\frac{x}{y^4}\right)$ b) $\log_3\left(\frac{2x^7}{\sqrt{y}}\right)$

3. a) $\frac{9}{5}$ b) 54 c) 6

4. a) 2.2925 b) 0.9464 c) 5.1133

5. $y = 3^{1.4650x}$

6. Perform the following transformations: reflection in the x-axis; vertical expansion by a factor of 2; translation 2 units up.

7. a) 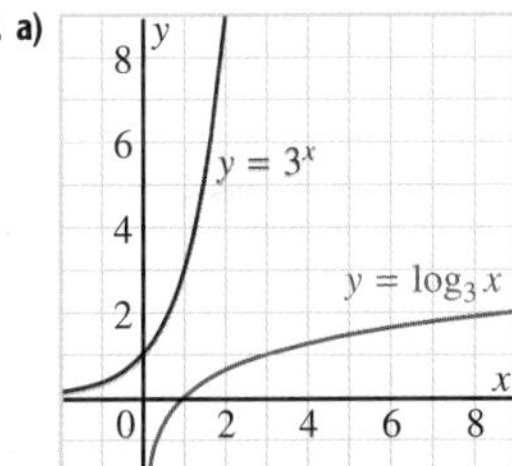

b) For $y = \log_3 x$:

i) x-intercept: 1; y-intercept: none

ii) Domain: $\{x \mid x \in \Re, x > 0\}$; range: $\Re$

iii) Asymptote: $x = 0$

For $y = 3^x$:

i) x-intercept: none; y-intercept: 1

ii) Domain: $\Re$; range: $\{y \mid y \in \Re,\ y > 0\}$

iii) Asymptote: $y = 0$

8. Approximately 30 years

9. 316 228

10. a) $x = 0$ and $x \doteq 1.585$

b) 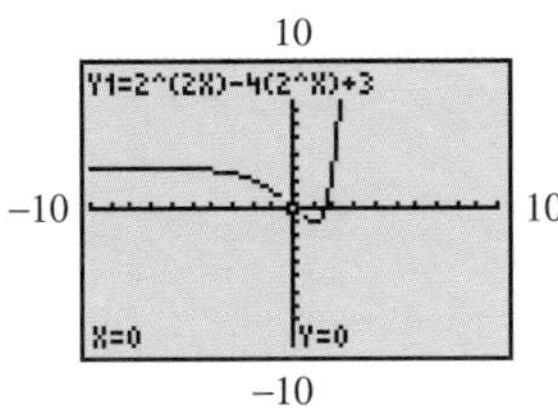

11. a) $c = 120(1.12)^n$

b) 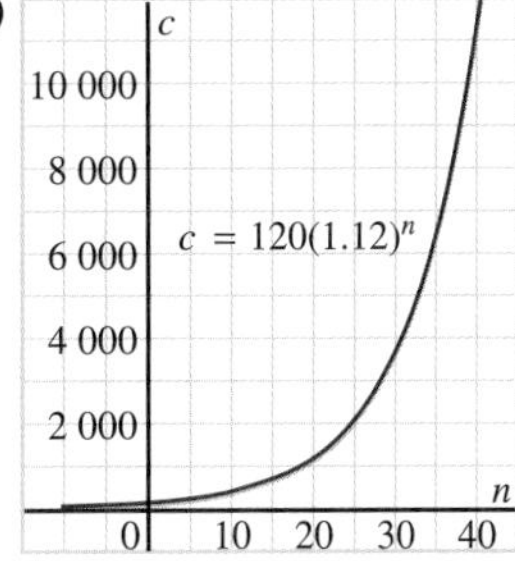

c) 333 t d) 1990–1991

12. a) 2.4022 b) 5.3923 c) 5.1911

14. a) 1000 b) Approximately 1988

d) 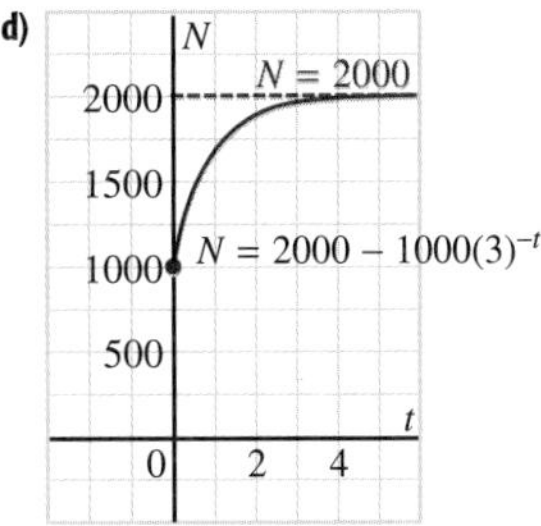

e) $y = 2000$

f) Approximately 10 months

g) $t = \frac{-\log(2 - 0.001N)}{\log 3}$

Cumulative Review (Chapters 1–6), page 429

1. $y = -6x - 3$

2. a) x-intercepts: -3, -1; y-intercept: 3; maximum point: $\left(-\frac{7}{3}, \frac{32}{27}\right)$; minimum point: $(-1, 0)$; point of inflection: $\left(-\frac{5}{3}, \frac{16}{27}\right)$; increase: $x < -\frac{7}{3}$ and $x > -1$; decrease: $-\frac{7}{3} < x < -1$; concave up: $x > -\frac{5}{3}$; concave down: $x < -\frac{5}{3}$

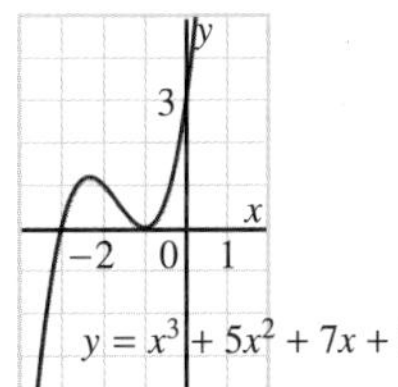

b) x-intercepts: ±1, approximately ±1.4; y-intercept: 2; maximum point: $(0, 2)$; minimum points: approximately $(\pm1.2, -0.25)$; points of inflection: approximately $(\pm0.7, 0.75)$; increase: $-1.2 < x < 0$ and $x > 1.2$; decrease: $x < -1.2$ and $0 < x < 1.2$; concave up: $|x| > 0.7$; concave down: $|x| < 0.7$

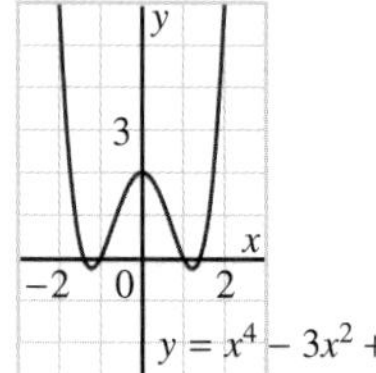

3. a) $y' = 18x^2(5 - 4x)$ **b)** $y' = -30x - 7$
c) $y' = -\sin x$ **d)** $y' = \frac{-1}{x^2}$, $x \neq 1$
e) $y' = \frac{-6x}{(3x^2 - 4)^2}$ **f)** $y' = 12x - 4 + \frac{1}{x^2}$
g) $y' = \cos x$ **h)** $y' = \frac{6(3x^2 - 4)^5(-3x^2 + 24x - 4)}{(4 - x)^7}$

4. $\frac{d^2y}{dx^2} = 10(1 - 15x^4)$: the second derivative might be needed to calculate the x-coordinates of points of inflection and the intervals of concavity for the graph of the function.

5. $(-0.71, -0.96)$, $(1.77, 14.85)$, and $(1.19, 4.48)$

6. a) $\frac{dy}{dx} = \frac{-2x}{y}$ **b)** $\frac{dy}{dx} = \frac{5y - 6x}{2y - 5x}$

7. $y = 2x - 2$ and $y = -x + 2$

8. a) 1 s and 5 s
b) When $1 < t < 3$ and $5 < t < 6.5$
c) When $0 < t < 1$ and $3 < t < 5$
d) 46 m

9. a) $g(f(x)) = \sqrt{1 - x^2}$ **b)** $\{x \mid x \in \Re,\ |x| \leq 1\}$
c) Undefined **d)** $f(g(x)) = 3 - x$

10. a) -8 **b)** 6 **c)** -6

11. a) i) x-intercepts: ±2; y-intercept: 8
ii) Vertical asymptotes: $x = \pm1$; horizontal asymptote: $y = 2$
iii) 2
iv)

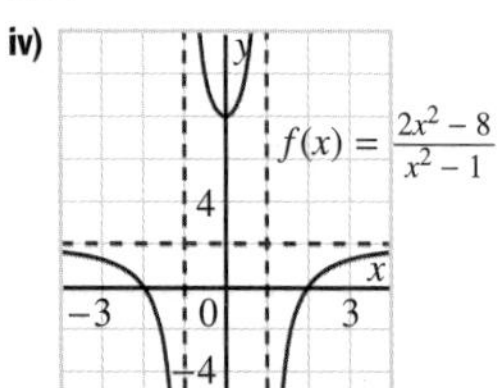

b) i) x-intercept: 0; y-intercept: 0
ii) Vertical asymptote: $x = -\frac{1}{2}$; oblique asymptote: $y = 2x - 1$
iii) ∞
iv)

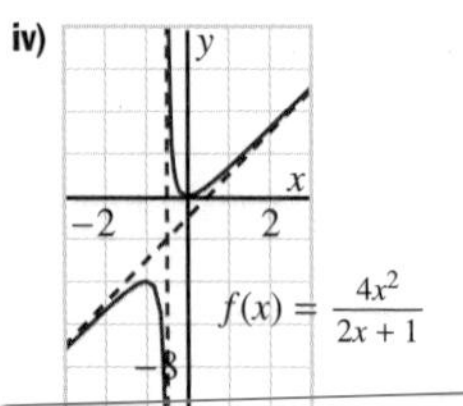

c) i) x-intercept: $-\frac{1}{2}$; y-intercept: 1
ii) Horizontal asymptote: $y = 0$
iii) 0
iv)

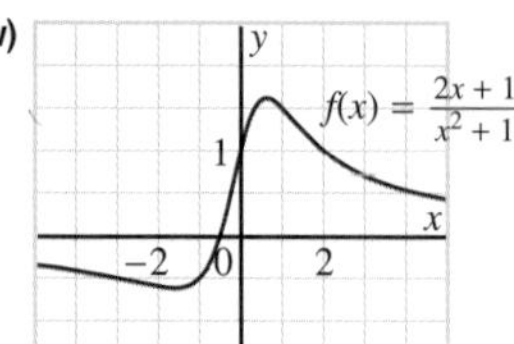

12. Graphs may vary.
a)

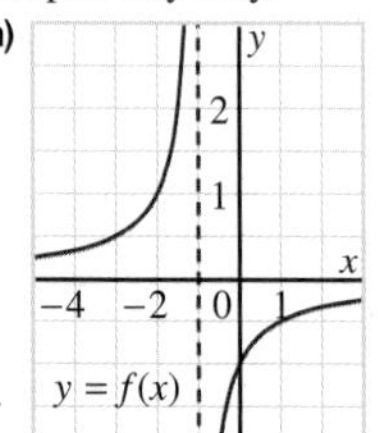

b)

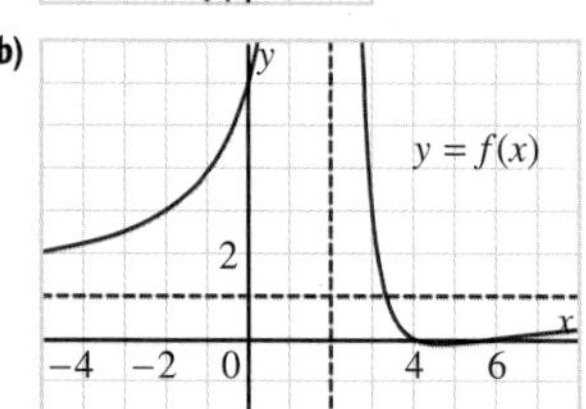

c)

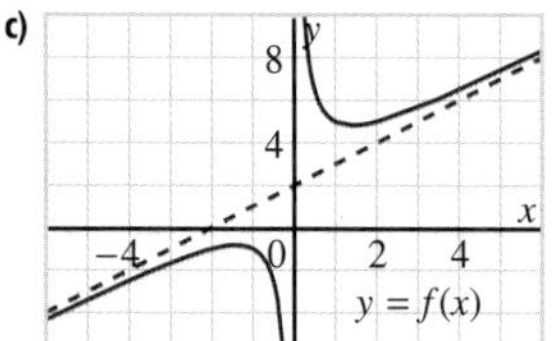

13. a) 2 **b)** 1 **c)** $-\frac{1}{2}$

14. a) 0.7325 **b)** 3.8928 **c)** 1
d) 2.5 **e)** -2.1610 **f)** 11.6619

15. a) For $y = 3^x$: y-intercept: 1; asymptote: $y = 0$
For $y = 4(3)^x + 1$: y-intercept: 5; asymptote: $y = 1$

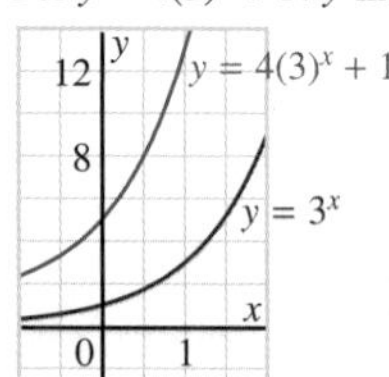

b) For $y = \log x$: x-intercept: 1; asymptote: $x = 0$
For $y = 10^x$: y-intercept: 1; asymptote: $y = 0$

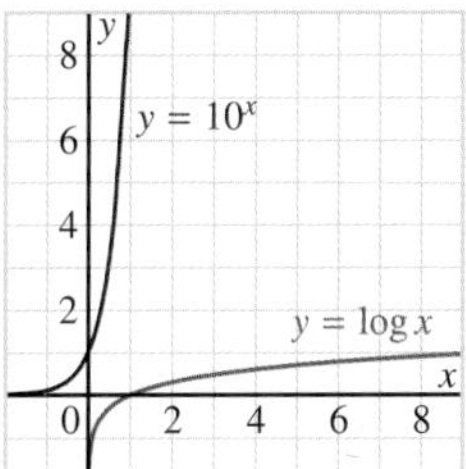

16. 574 715

Chapter 7 Differentiating Exponential and Logarithmic Functions

Necessary Skills

1 New: Proportionality

Exercises, page 432

1. a)

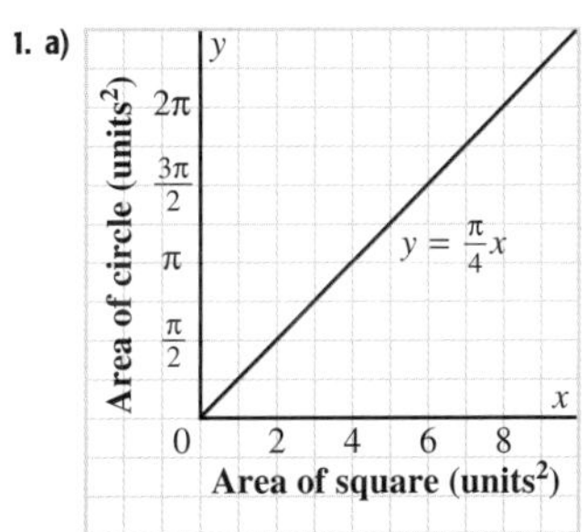

b) $\frac{4}{\pi}$

2. The volume of a sphere with diameter x is proportional to the volume of a cube with edge length x. The constant of proportionality is $\frac{\pi}{6}$.

2 Review: The Chain Rule

Exercises, page 433

1. a) $\frac{dy}{dx} = 18x^2(3x^3 - 1)$ **b)** $\frac{dy}{dx} = 10(1 + 2x)^4$

c) $\frac{dy}{dx} = (4x - 3)(12x - 3)$ **d)** $\frac{dy}{dx} = (x^2 + 1)(5x^2 + 1)$

e) $\frac{dy}{dx} = \frac{-8x}{(2x^2 + 1)^3}$ **f)** $\frac{dy}{dx} = \frac{3x + 1}{(1 - 3x)^3}$

3 Review: Changing the Base of an Exponential Function

Exercises, page 435

1. a) $4^{3.0112}$ **b)** $3^{4.1918}$

2. a) $y = 2^{0.2630x}$ **b)** $y = 2(3)^{0.0444x}$

c) $y = 4.25(10)^{0.2430x}$

4 Review: Compounding Periods for Compound Interest

Exercises, page 435

1. a) \$1461.99 **b)** \$3372.13 **c)** \$5602.22

7.1 Exercises, page 442

2. a) $\frac{dy}{dx} = 2e^x$ **b)** $\frac{dy}{dx} = 2e^{2x}$ **c)** $\frac{dy}{dx} = e^{x+2}$

d) $\frac{dy}{dx} = 2xe^{x^2}$ **e)** $\frac{dy}{dx} = \frac{-e^{\frac{1}{x}}}{x^2}$ **f)** $\frac{dy}{dx} = -e^{-x}$

4. The slopes of the tangents increase exponentially. The derivative is also an exponential function.

5. False

6. a) $\frac{dy}{dx} = 2e^x(x + 1)$ **b)** $\frac{dy}{dx} = e^{2x}(2x + 1)$

c) $\frac{dy}{dx} = -2xe^{-2x}(x - 1)$ **d)** $\frac{dy}{dx} = x(xe^x + 2e^x + 2)$

e) $\frac{dy}{dx} = \frac{x(2 - 2e^x + xe^x)}{(1 - e^x)^2}$ **f)** $\frac{dy}{dx} = \frac{4e^{2x}}{(1 - e^{2x})^2}$

7. a) $\frac{dy}{dx} = xe^x(x + 2)$ **b)** $\frac{dy}{dx} = 1 + xe^x + e^x$

c) $\frac{dy}{dx} = e^x(2 + x)$ **d)** $\frac{dy}{dx} = e^x(1 + 3e^{2x})$

e) $\frac{dy}{dx} = \frac{-1}{e^x}$ **f)** $\frac{dy}{dx} = \frac{xe^{-x} - 2 + 2e^{-x}}{x^3}$

8. a) $\frac{dy}{dx} = -e^{x-y}$ **b)** $\frac{dy}{dx} = \frac{-y}{x}$

c) $\frac{dy}{dx} = \frac{1 - ye^{xy}}{xe^{xy}}$

9. a) There are no critical points since $y' = e^x \neq 0$.
b) There are no points of inflection since $y'' = e^x \neq 0$.
c) The function is always increasing since $e^x > 0$.
d) The graph is concave up since $y'' > 0$.

10. a) $(0, 1)$ is a local maximum point.
b) $\left(\pm\frac{1}{\sqrt{2}}, \frac{1}{\sqrt{e}}\right)$ are points of inflection.
c) Increase: $x < 0$; decrease: $x > 0$
d) Concave up: $|x| > \frac{1}{\sqrt{2}}$; concave down: $|x| < \frac{1}{\sqrt{2}}$
e) $y = \frac{-2}{e}x + \frac{3}{e}$

11. a)

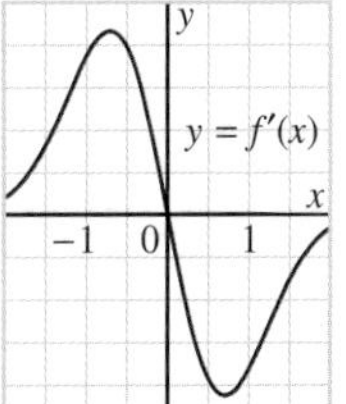

b)

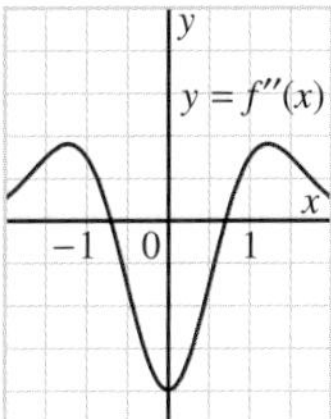

12. Approximately 0.86 units^2

13. a)

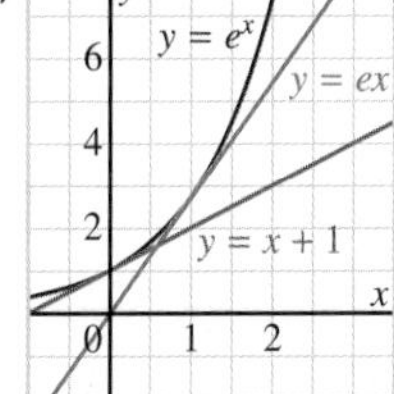

b) $y = x + 1$ for $(0, 1)$; $y = ex$ for $\left(1, \frac{1}{e}\right)$

14. a) y-intercept: -1; domain: $\Re$; decrease: $x \in \Re$; concave down: $x \in \Re$; horizontal asymptote: $y = 0$

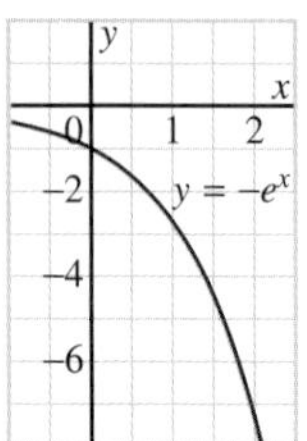

b) y-intercept: 1; domain: $\Re$; increase: $x \in \Re$; concave up: $x \in \Re$; horizontal asymptote: $y = 0$

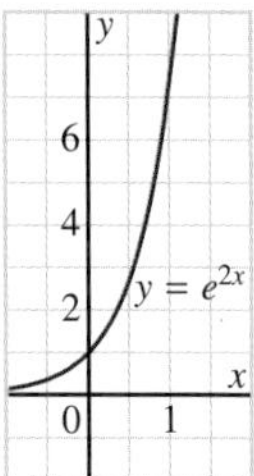

15. a) y-intercept: 2; domain: $\Re$; increase: $x \in \Re$; concave up: $x \in \Re$; horizontal asymptote: $y = 0$

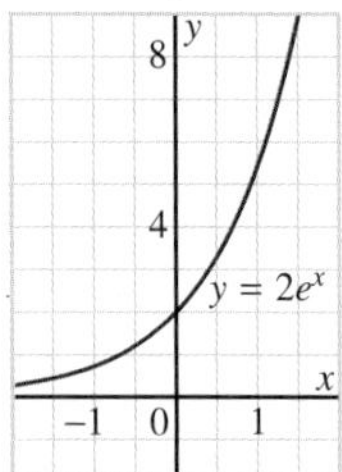

b) y-intercept: 1; domain: $\Re$; decrease: $x \in \Re$; concave up: $x \in \Re$; horizontal asymptote: $y = 0$

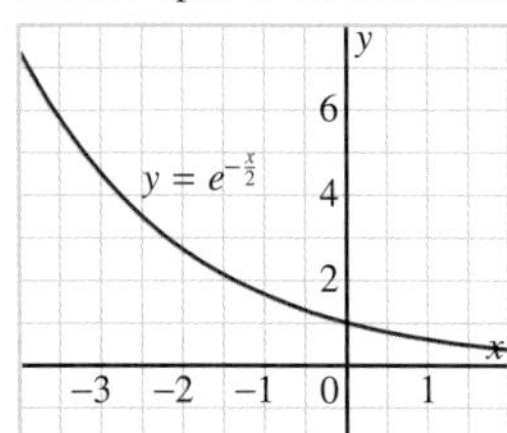

16. a) $\frac{dy}{dx} = e^{\sin x} \cos x$ **b)** $\frac{dy}{dx} = e^x \cos e^x$

c) $\frac{dy}{dx} = e^x(\cos x + \sin x)$ **d)** $\frac{dy}{dx} = -2e^{\cos 2x} \sin 2x$

e) $\frac{dy}{dx} = 2e^{2x} \cos e^{2x}$

f) $\frac{dy}{dx} = -e^{2x}(3 \sin 3x - 2 \cos 3x)$

7.2 Exercises, page 446

1. a) 6.1420; $e^{6.1420} \doteq 465$ **b)** 2.3026; $e^{2.3026} \doteq 10$

c) 1.0296; $e^{1.0296} \doteq 2.8$ **d)** 0.9555; $e^{0.9555} \doteq 2.6$

e) -0.5108; $e^{-0.5108} \doteq 0.6$

f) -8.0164; $e^{-8.0164} \doteq 0.000\ 33$

3. a) 0 **b)** 1

4. a) 3.1699 **b)** 1.8163 **c)** 0.6309 **d)** 0.6578

5. a) 54.5982 **b)** 2.7183 **c)** 1 **d)** 0.3679

6. a) 0 **b)** 0.6931 **c)** 1 **d)** -0.6931
e) 0.8047 **f)** 2.1972 **g)** 0.6758 **h)** -0.1860
i) 0.2877 **j)** -0.3379 **k)** 3.6652 **l)** 80.4719

8. a) 3.5 **b)** 0.5 **c)** -1 **d)** -4.25

9. a) 3.2 **b)** 1 **c)** 0.5 **d)** e

10. a) ln 12 **b)** ln 12 **c)** ln 12
d) ln 3 **e)** ln 5 **f)** ln 15

12. a) i) 0.6990, 1.6990, 2.6990, 3.6990
ii) 1.6094, 3.9120, 6.2146, 8.5172

Self-Check 7.1, 7.2, page 448

1. a) $\frac{dy}{dx} = 4e^x$ **b)** $\frac{dy}{dx} = 4e^{4x}$ **c)** $\frac{dy}{dx} = e^{x+4}$

d) $\frac{dy}{dx} = 4x^3e^{x^4}$ **e)** $\frac{dy}{dx} = \frac{-4e^{\frac{4}{x}}}{x^2}$ **f)** $\frac{dy}{dx} = -4e^{-4x}$

2. a) $\frac{dy}{dx} = 4e^x(x + 1)$ **b)** $\frac{dy}{dx} = e^{4x}(4x + 1)$

c) $\frac{dy}{dx} = -4x^3e^{-4x}(x - 1)$ **d)** $\frac{dy}{dx} = x^3(xe^x + 4e^x + 16)$

e) $\frac{dy}{dx} = \frac{x^3(16 - 4e^x + xe^x)}{(4 - e^x)^2}$ **f)** $\frac{dy}{dx} = \frac{32e^{4x}}{(4 - e^{4x})^2}$

3. a) $\frac{dy}{dx} = xe^{3x}(3x + 2)$ **b)** $\frac{dy}{dx} = 2(3 + xe^x + e^x)$

c) $\frac{dy}{dx} = 2e^{2x}(1 + x^2 + x)$ **d)** $\frac{dy}{dx} = e^{3x}(3 + 4e^x)$

e) $\frac{dy}{dx} = \frac{2(2e^{4x} - 1)}{e^{2x}}$ **f)** $\frac{dy}{dx} = \frac{2xe^{-2x} - 6 + 3e^{-2x}}{x^4}$

4. a) 5.7838; $e^{5.7838} \doteq 325$ **b)** 2.4849; $e^{2.4849} \doteq 12$
c) 0.1823; $e^{0.1823} \doteq 1.2$ **d)** -2.9957; $e^{-2.9957} \doteq 0.05$

5. a) 2.9299 **b)** 3.6144 **c)** 0.6826

6. a) 148.4132 **b)** 1.0986 **c)** 0.2554

7. a) x^2 **b)** $\frac{1}{x}$ **c)** $1 - x$

8. y-intercept: 2; domain: $\Re$; increase: $x \in \Re$; concave up: $x \in \Re$; horizontal asymptote: $y = 0$

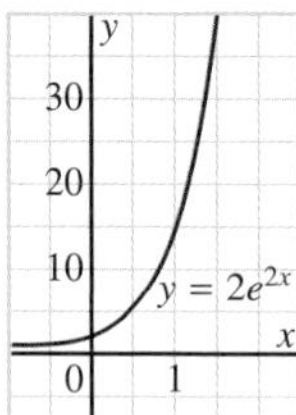

9. a) y = f′(x) **b)** y = f″(x)

c) No critical points **d)** $y = \frac{3}{2}ex - e$

7.3 Exercises, page 456

1. a) 103 million; 1.78% **b)** 59.4 million; 0.34%
c) 127 million; –0.2%

2. a) 90.68 million/year **b)** 92.04 million/year
c) 96.25 million/year

4. a) $P = 18.7e^{0.034\ 40t}$
b) i) 26.38 million
ii) 2028
c) Approximately 790 750/year

5. *United States*:
a) $P = 276e^{0.008\ 17t}$
b) i) 299.50 million
ii) Population is already greater than 50 million.
c) 2.37 million/year

Argentina:
a) $P = 37.6e^{0.016\ 86t}$
b) i) 44.51 million
ii) 2016
c) 0.70 million/year

South Africa:
a) $P = 42.8e^{-0.009\ 65t}$
b) i) 38.86 million
ii) Population will never reach 50 million at the current growth rate.
c) –0.39 million/year

6. a) 264.23 million; 2008; 3.54 million/year

7. 2028–2029

8. *Brazil*:
a) $P = 162e^{0.016\ 58t}$ **b)** 1.7%/year
Nepal:
a) $P = 21.9e^{0.029\ 65t}$ **b)** 3.0%/year
Singapore:
a) $P = 2.96e^{0.049\ 96t}$ **b)** 5.0%/year

10. a) 94.18%
b) Approximately 25 250 years old
c) $t \doteq -8333.3 \ln\left(\frac{P}{100}\right)$
d) i) 695 years **ii)** 11 890 years
iii) 15 644 years **iv)** 35 000 years

12. a) $P = 130e^{-0.000\ 15h}$

13. a) 15.6°C **b)** 0.20°C/min **c)** 47.6 min
d)

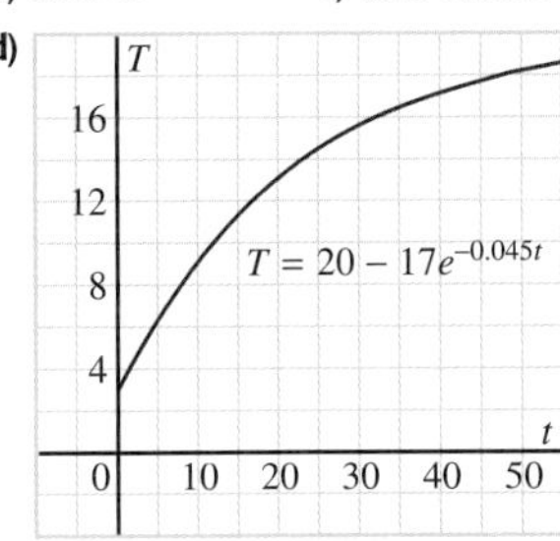

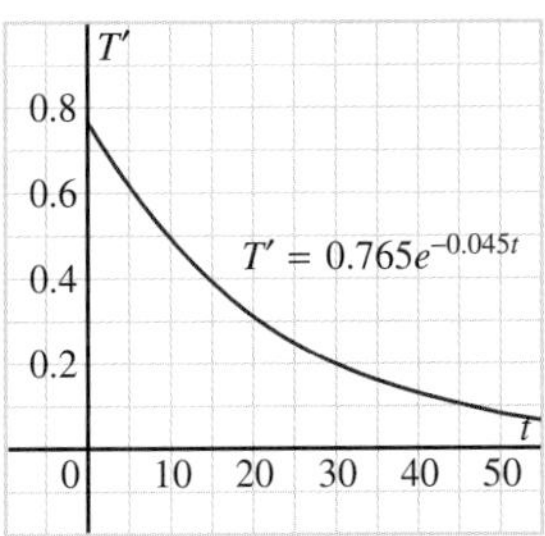

14. a) $\frac{dc}{dt} = -100e^{-0.5t}(0.5t - 1)$ **b)** 73.6% after 2 h

15. a) i) \$2 **ii)** \$2.25 **iii)** \$2.61
iv) \$2.71 **v)** \$2.72 **vi)** \$2.72

16. a) i) 6.089 34 billion **ii)** 6.089 67 billion
iii) 6.089 95 billion **iv)** 6.090 00 billion

17. a) $A = 500e^{0.062\ 97n}$ **b)** $P = 100e^{-0.139\ 26n}$
c) $h = 7.4e^{0.463\ 73n}$ **d)** $t = 8.72e^{-0.072\ 57T}$
e) $M = 460e^{0.173\ 95n}$ **f)** $D = 328e^{0.324\ 98n}$
g) $P = 100e^{-0.085\ 56t}$

18. a)

r	k
0.065	0.062 97
–0.13	–0.139 26
0.59	0.463 73
–0.07	–0.072 57
0.19	0.173 95
0.384	0.324 98
–0.082	–0.085 56

b)

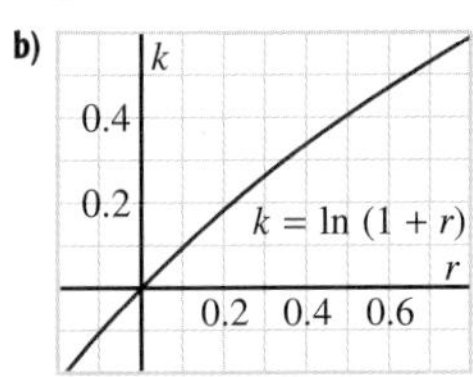

c) $k = \ln(1 + r)$

20. 7:22 P.M.

7.4 Exercises, page 462

2. a) $\frac{dy}{dx} = \ln 4 \times 4^x$ **b)** $\frac{dy}{dx} = \ln 6 \times 6^x$
c) $\frac{dy}{dx} = \ln 10 \times 10^x$ **d)** $\frac{dy}{dx} = 2\ln 3 \times 3^{2x+1}$
e) $\frac{dy}{dx} = -3\ln 2 \times 2^{1-3x}$ **f)** $\frac{dy}{dx} = 2x\ln 5 \times 5^{x^2}$

3. a) $\frac{dy}{dx} = 7^x(x\ln 7 + 1)$ **b)** $\frac{dy}{dx} = 3^x(2x\ln 3 - \ln 3 + 2)$
c) $\frac{dy}{dx} = -x4^{-x}(x\ln 4 - 2)$ **d)** $\frac{dy}{dx} = x^2 2^{1-2x}(3 - 2x\ln 2)$
e) $\frac{dy}{dx} = (x - 1)6^x(x\ln 6 - \ln 6 + 2)$
f) $\frac{dy}{dx} = 2^{2x}(1 - x)(2\ln 2 - 2x\ln 2 - 2)$

4. a) $\frac{dy}{dx} = \frac{1 - x\ln 7}{7^x}$ **b)** $\frac{dy}{dx} = \frac{2 - 2x\ln 3 + \ln 3}{3^x}$
c) $\frac{dy}{dx} = \frac{x(2 - x\ln 4)}{4^x}$ **d)** $\frac{dy}{dx} = \frac{x^2(3 - 2x\ln 2)}{2^{2x-1}}$
e) $\frac{dy}{dx} = \frac{(x-1)(2 - x\ln 6 + \ln 6)}{6^x}$ **f)** $\frac{dy}{dx} = \frac{-2(1-x)(1 + \ln 2 - x\ln 2)}{2^{2x}}$

5. a) $4^{-x} = \frac{1}{4^x}$

6. a) $2^{1-2x} = \frac{1}{2^{2x-1}}$

7. a) $\frac{dy}{dx} = \ln 7 \times 7^x$ **b)** $\frac{dy}{dx} = -\ln 4 \times 4^{-x}$

c) $\frac{dy}{dx} = -3\ln 2 \times 2^{2-3x}$ **d)** $\frac{dy}{dx} = 2^x(3x\ln 2 - \ln 2 + 3)$

e) $\frac{dy}{dx} = -(x^2+1)2^{-x}(x^2\ln 2 + \ln 2 - 4x)$

f) $\frac{dy}{dx} = \frac{-5^{1-x}(2x\ln 5 - \ln 5 + 4)}{(2x-1)^3}$

8. $\frac{dy}{dx} = 2\ln 2 \times 2^{2x}$

10. a) No **b)** No

c) If $0 < a < 1$, then the function is decreasing for all x values since $y' < 0$. If $a > 1$, then the function is increasing for all x values since $y' > 0$.

d) Since $y'' > 0$, then the function is always concave up.

11. a) $a = e^2$; $f(x) = e^{2x}$ **b)** $f(x) = e^{3x}$

12. a) $c = 1$; $1 < a < e$ **b)** $0 < c < 1$; $a = \sqrt[c]{e}$; $a > e$

13. $\frac{dP}{dt} = 6.0\ln 1.015 \times (1.015)^t$

14. a) $y = (4\ln 4)x - 4(\ln 4 - 1)$

b) x-intercept: approximately 0.28; y-intercept: approximately -1.55

c) $D\left(\frac{1}{\ln 4}, e\right)$

7.5 Exercises, page 467

1. a) $\frac{dy}{dx} = \frac{1}{x}$ **b)** $\frac{dy}{dx} = \frac{1}{x}$ **c)** $\frac{dy}{dx} = \frac{2}{x}$

d) $\frac{dy}{dx} = \frac{1}{x}$ **e)** $\frac{dy}{dx} = \frac{1}{x+2}$ **f)** $\frac{dy}{dx} = \frac{2}{x}$

2. a) $\frac{dy}{dx} = \frac{3}{x}$ **b)** $\frac{dy}{dx} = \frac{-1}{1-x}$ **c)** $\frac{dy}{dx} = \frac{2}{2x+1}$

d) $\frac{dy}{dx} = \frac{1}{1+x}$ **e)** $\frac{dy}{dx} = \frac{2x+1}{x^2+x+1}$ **f)** $\frac{dy}{dx} = \frac{2(x+1)}{x^2+2x+3}$

4. a) $\frac{dy}{dx} = 2x - \frac{1}{x}$ **b)** $\frac{dy}{dx} = 2 + 2\ln x$ **c)** $\frac{dy}{dx} = 2 + \ln x^2$

d) $\frac{dy}{dx} = \frac{1}{x\ln x}$ **e)** $\frac{dy}{dx} = \frac{\ln x - 1}{(\ln x)^2}$ **f)** $\frac{dy}{dx} = \frac{1 - \ln x}{x^2}$

5. a) $\frac{dy}{dx} = \frac{x}{x+1} + \ln(x+1)$

b) $\frac{dy}{dx} = \frac{-2x^2}{1-2x} + 2x\ln(1-2x)$

c) $\frac{dy}{dx} = \frac{2}{x(1-\ln x)^2}$ **d)** $\frac{dy}{dx} = \frac{2 - \ln x^2}{x^2}$

e) $\frac{dy}{dx} = \frac{\ln x^2 - 2}{(\ln x^2)^2}$ **f)** $\frac{dy}{dx} = \frac{-2}{x^2-1}$

6. a) $\frac{dy}{dx} = 1 - \frac{1}{x}$ **b)** $\frac{dy}{dx} = 1 + \ln x$ **c)** $\frac{dy}{dx} = \frac{-1}{x}$

d) $\frac{dy}{dx} = 1 + \frac{2}{x}$ **e)** $\frac{dy}{dx} = 1$ **f)** $\frac{dy}{dx} = \frac{-1}{x(\ln x)^2}$

7. $\frac{dy}{dx} = \frac{2x}{x^2-1}$

8. a) No **b)** No

c) Since $x > 0$ and $y' = \frac{1}{x}$, then the function is always increasing as $y' > 0$.

d) Since $x > 0$ and $y'' = \frac{-1}{x^2}$, then the graph is concave down as $y'' < 0$.

9. a) Minimum point: $(2,\ 2 - \ln 4)$

b) Maximum point: $\left(-\frac{1}{e}, \frac{2}{e}\right)$; minimum point: $\left(\frac{1}{e}, -\frac{2}{e}\right)$

c) Maximum point: $(0, 0)$; minimum points: $(\pm\sqrt{e}, e)$

10. $y = -x + 2$

11. $y = 2x + 2$

12. $y = \frac{4x(\ln 4 - 1)}{(\ln 4)^2} + \frac{8 - 4\ln 4}{(\ln 4)^2}$

13. a) i) $\frac{dy}{dx} = \frac{1}{x}$ **ii)** $\frac{dy}{dx} = \frac{1}{x}$ **iii)** $\frac{dy}{dx} = \frac{1}{x}$ **iv)** $\frac{dy}{dx} = \frac{1}{x}$

14. $y = x - 1$; $y = \frac{1}{e^x}$

16. $y = ex - 2$

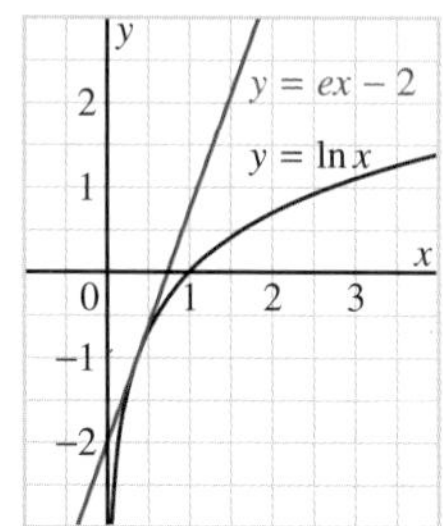

17. a) $y = kx - (\ln k + 1)$ **b)** $0 < k < \frac{1}{e}$; $k > \frac{1}{e}$

c) $k > \frac{1}{e}$; $0 < k < \frac{1}{e}$

18. a) $\frac{dy}{dx} = \frac{1}{x\ln 2}$ **b)** $\frac{dy}{dx} = \frac{1}{x\ln 10}$

c) $\frac{dy}{dx} = \frac{1}{x\ln 3}$ **d)** $\frac{dy}{dx} = \frac{2}{x\ln 10}$

e) $\frac{dy}{dx} = \frac{-1}{(2-x)\ln 4}$ **f)** $\frac{dy}{dx} = \frac{\ln x + \ln 10 \log x}{x\ln 10}$

19. a) $\frac{dy}{dx} = y$ **b)** $\frac{dy}{dx} = -y$

c) $\frac{dy}{dx} = \frac{-y\ln y}{x}$ **d)** $\frac{dy}{dx} = \frac{-y}{x}$

e) $\frac{dy}{dx} = \frac{-y}{x}$ **f)** $\frac{dy}{dx} = -1$

20. b) $\frac{dy}{dx} = x^x(1 + \ln x)$ **c)** $\left(\frac{1}{e}, \frac{1}{e^{e^{-1}}}\right)$

21. a) $\frac{dy}{dx} = \frac{\cos x}{\sin x}$ **b)** $\frac{dy}{dx} = \frac{\cos(\ln x)}{x}$

c) $\frac{dy}{dx} = \frac{2\cos 2x}{\sin 2x}$ **d)** $\frac{dy}{dx} = \frac{\cos(\ln 2x)}{x}$

e) $\frac{dy}{dx} = \cos x\ln x + \frac{\sin x}{x}$ **f)** $\frac{dy}{dx} = \frac{\sin x - x\cos x\ln x}{x\sin^2 x}$

Self-Check 7.3–7.5, page 470

1. a) $P = 272\ 000e^{0.005\ 68t}$

b) i) 279 836 **ii)** 2107

c) 1617/year

2. a) 88.69% **b)** 9680 years ago

3. a) $\frac{dy}{dx} = \ln 6 \times 6^x$ **b)** $\frac{dy}{dx} = 2\ln 4 \times 4^{3+2x}$

c) $\frac{dy}{dx} = 2x\ln 3 \times 3^{x^2-2}$

4. a) $\frac{dy}{dx} = x\ 5^x(x\ln 5 + 2)$ **b)** $\frac{dy}{dx} = 4^x(x^2\ln 4 + \ln 4 + 2x)$

c) $\frac{dy}{dx} = 7^x(x\ln 7 - x^2\ln 7 + 1 - 2x)$

5. a) $\frac{dy}{dx} = \frac{3(1 - x\ln 4)}{4^x}$ **b)** $\frac{dy}{dx} = \frac{6x^2 - 2x^3\ln 2 - 3\ln 2}{2^x}$

c) $\frac{dy}{dx} = \frac{x2^{2x}(4x\ln 2 - 2 - 2x^2\ln 2 - 4 + 3x)}{(2x^2 - x^3)^2}$

6. a) $\frac{dy}{dx} = x(1 + 2\ln 3x)$ **b)** $\frac{dy}{dx} = \frac{2(\ln x^3 - 3)}{(\ln x^3)^2}$

c) $\frac{dy}{dx} = \frac{3}{x} + 1$ **d)** $\frac{dy}{dx} = 3x^2 - \frac{2}{x}$

e) $\frac{dy}{dx} = \frac{4(1 + \ln x)}{(2 - x\ln x)^2}$ **f)** $\frac{dy}{dx} = \frac{-6}{x(\ln x^2)^2}$

7. a) Maximum points: $(\pm\sqrt{e}, -e)$; minimum point: (0, 0)

b) $y = \frac{(-6\ln 9 + 6)}{(\ln 9)^2}x + \frac{9\ln 9 - 18}{(\ln 9)^2}$

Chapter 7 Review Exercises, page 472

1. a) $\frac{dy}{dx} = 10e^x$ **b)** $\frac{dy}{dx} = 10e^{10x}$

c) $\frac{dy}{dx} = e^{x-10}$ **d)** $\frac{dy}{dx} = 10x^9e^{x^{10}}$

e) $\frac{dy}{dx} = -\frac{10e^{\frac{10}{x}}}{x^2}$ **f)** $\frac{dy}{dx} = -10e^{-10x}$

2. a) $\frac{dy}{dx} = 10e^x(x+1)$ **b)** $\frac{dy}{dx} = e^{10x}(10x+1)$

c) $\frac{dy}{dx} = -10x^9e^{-10x}(x-1)$ **d)** $\frac{dy}{dx} = x^9(xe^x + 10e^x - 100)$

e) $\frac{dy}{dx} = \frac{x^9(100 + 10e^x - xe^x)}{(10+e^x)^2}$ **f)** $\frac{dy}{dx} = \frac{-200e^{10x}}{(10+e^{10x})^2}$

3. a) $\frac{dy}{dx} = 2x^3e^{2x}(x+2)$ **b)** $\frac{dy}{dx} = x(6x + xe^x + 2e^x)$

c) $\frac{dy}{dx} = e^x(1 + x + 6x^2 + 2x^3)$ **d)** $\frac{dy}{dx} = e^{2x}(2x + 1 - 3e^x)$

e) $\frac{dy}{dx} = \frac{3e^{2x} + 2x - x^2}{e^x}$ **f)** $\frac{dy}{dx} = \frac{xe^{-x} + 2e^{-x} - x}{x^3}$

4. a) $\frac{dy}{dx} = (3x-3)e^{-x}$

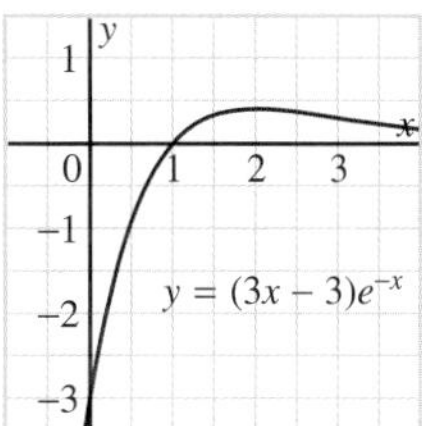

b) $\frac{dy}{dx} = (2x-1)e^{\frac{1}{x}}$

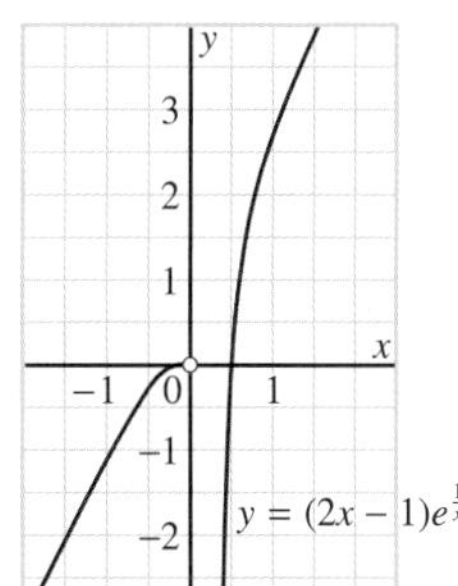

5. a)

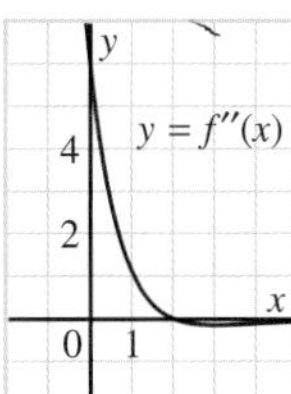

b)

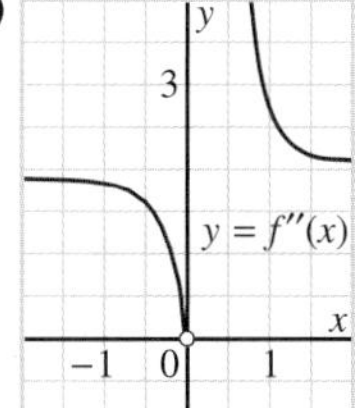

6. a) Minimum point: $\left(1, -\frac{3}{e}\right)$; point of inflection: $\left(2, -\frac{6}{e^2}\right)$

b) Minimum point: $\left(0.5, \frac{e^2}{4}\right)$

7. $y = \frac{3}{e^2}x - \frac{12}{e^2}$

8. $y = -\frac{5x}{\sqrt{e}} - \frac{6}{\sqrt{e}}$

9. y-intercept: 3; domain: $x \in \Re$; increase: $x \in \Re$; concave up: $x \in \Re$; horizontal asymptote: $y = 0$

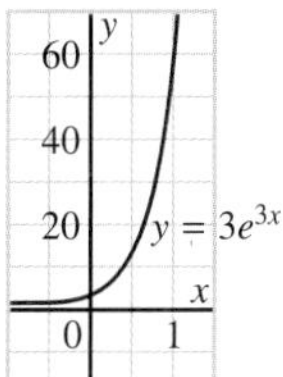

10. a) 0.7831 **b)** 2.0566 **c)** 4.0939

11. a) 148.4132 **b)** 1.6487 **c)** 5.1847×10^{21}

12. a) 1.9459 **b)** 0.8959 **c)** 0.9739

13. a) 1.2 **b)** −7 **c)** 4 **d)** 0.8

14. a) ln 50 **b)** $\ln\left(\frac{1}{2}\right)$ **c)** ln 12

15. a) $P = 146e^{-0.005\,01t}$

b) i) 137.5 million **ii)** 2001

c) Approximately 678 500/year

16. a) $P = 2.5e^{0.032\,42t}$ **b)** 3.24%

17. a) 12.71 kPa **b)** 25 019 m **c)** −0.0155%

18. a) $\frac{dy}{dx} = \ln 3 \times 3^x$ **b)** $\frac{dy}{dx} = \ln 2 \times 2^{x+3}$

c) $\frac{dy}{dx} = 2x\ln 4 \times 4^{3+x^2}$ **d)** $\frac{dy}{dx} = 6^x(x\ln 6 + 1)$

e) $\frac{dy}{dx} = x2^x(x^2\ln 2 + x\ln 2 + 3x + 2)$

f) $\frac{dy}{dx} = (2x-3)(5^x)[(2x-3)\ln 5 + 4]$

19. a) $\frac{dy}{dx} = \frac{x(2 - x\ln 3)}{3^x}$ **b)** $\frac{dy}{dx} = \frac{3^x(x\ln 3 - 2)}{x^3}$

c) $\frac{dy}{dx} = \frac{2x - 2 - x^2\ln 5 + 2x\ln 5}{5^x}$

d) $\frac{dy}{dx} = \frac{x^2(3 - 2x\ln 2)}{2^{2x-1}}$ **e)** $\frac{dy}{dx} = \frac{2^{2x+2}(x\ln 2 - 1)}{x^3}$

f) $\frac{dy}{dx} = \frac{x(3x - 6 - 2x^3\ln 2 + 6x^2\ln 2)}{2^{x^2}}$

20. a) $y = (-2\ln 2)x + 2\ln 2 - 2$; x-intercept: $1 - \frac{1}{\ln 2}$; y-intercept: $2\ln 2 - 2$

b) $\left(\frac{1}{\ln 2}, -e\right)$

21. a) Maximum point: $(-2, \ln 4 - 2)$

b) $y = 3x - 2$

22. a) $\frac{dy}{dx} = \frac{1}{x}$ **b)** $\frac{dy}{dx} = \frac{3}{x}$ **c)** $\frac{dy}{dx} = \frac{2}{x}$

d) $\frac{dy}{dx} = \frac{-1}{2-x}$ **e)** $\frac{dy}{dx} = \frac{2x}{x^2+1}$ **f)** $\frac{dy}{dx} = \frac{4x+1}{2x^2+x}$

23. a) $\frac{dy}{dx} = 1 + \ln 3x$ **b)** $\frac{dy}{dx} = -x(1 + 2\ln 0.5x)$

c) $\frac{dy}{dx} = \frac{\ln 4x - 1}{(\ln 4x)^2}$ **d)** $\frac{dy}{dx} = \frac{x(2\ln x - 1)}{3(\ln x)^2}$

e) $\frac{dy}{dx} = \frac{5(1 - 2\ln x)}{x^3}$ **f)** $\frac{dy}{dx} = \frac{-3}{x(\ln 2x)^2}$

24. a) $\frac{dy}{dx} = \frac{2}{x} + 2x$ **b)** $\frac{dy}{dx} = \frac{1}{x(x+1)}$

c) $\frac{dy}{dx} = \frac{-x+2}{x(x-1)}$ **d)** $\frac{dy}{dx} = \frac{1}{x}$

e) $\frac{dy}{dx} = 2$ **f)** $\frac{dy}{dx} = \frac{3x^2}{3x+2} + 2x\ln(3x+2)$

25. a) $\frac{dy}{dx} = \frac{1}{x\ln 3}$ **b)** $\frac{dy}{dx} = \frac{2}{x\ln 10}$ **c)** $\frac{dy}{dx} = \frac{1}{x\ln 4}$

26. a) $\frac{dy}{dx} = 2xy$ **b)** $\frac{dy}{dx} = y$ **c)** $\frac{dy}{dx} = \frac{2y}{x}$

Chapter 7 Self-Test, page 475

1. a) -3 **b)** 5 **c)** $\ln\left(\frac{1}{2}\right)$ **d)** $\frac{3}{8}$

3. a) 1.147 **b)** 54.598 **c)** 0

4. a) $\frac{dy}{dx} = 2\ln 4 \times 4^{2x-3}$ **b)** $\frac{dy}{dx} = -2e^{-2x}$
c) $\frac{dy}{dx} = \frac{3}{x}$ **d)** $\frac{dy}{dx} = \frac{12}{(3x-1)\ln 10}$
e) $\frac{dy}{dx} = x^2e^{4x}(4x+3)$ **f)** $\frac{dy}{dx} = \frac{-3(3+2e^{6x})}{e^{3x}}$
g) $\frac{dy}{dx} = \frac{2(1 - x\ln x^2)}{xe^{2x}}$ **h)** $\frac{dy}{dx} = \frac{-4x}{(1-x^2)(1+x^2)}$

5. $(e^{-1.5}, -1.5e^{-3})$

6. Approximately 9 min

7. $y = \frac{2}{e}x$

8. $(0, -4)$, $(\ln 5, 4)$

9. a) $P = 31e^{0.003\,99t}$
b) i) Approximately 31.8 million **ii)** 2030
c) Approximately 128 200/year

11. a) Graphs may vary.
b) $y = (\ln a)x + 1$ at $(0, 1)$;
$y = (a \ln a)x + a(1 - \ln a)$ at $(1, a)$

12. a) $\frac{dy}{dx} = \frac{1}{x}$; $\frac{dy}{dx} = \frac{2}{x}$; $\frac{dy}{dx} = \frac{3}{x}$; $\frac{dy}{dx} = \frac{4}{x}$
c) $\frac{dy}{dx} = \frac{-1}{x}$; $\frac{dy}{dx} = \frac{-2}{x}$; $\frac{dy}{dx} = \frac{-3}{x}$; $\frac{dy}{dx} = \frac{-4}{x}$
d)

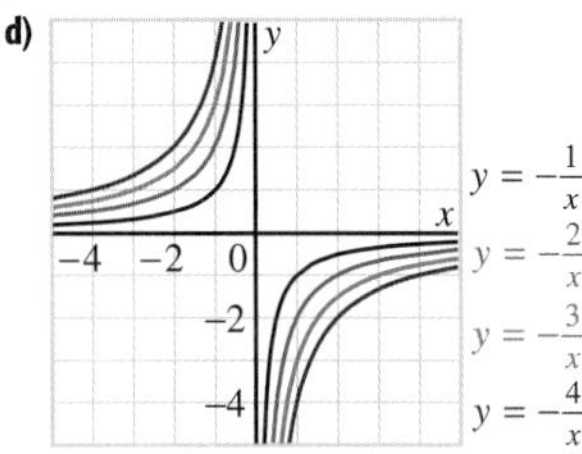

Mathematical Modelling: Quality Control

Exercises, page 479

1. \$124

2. a) \$121.44 **b)** \$119.17 **c)** \$118.37
d) \$126.72 **e)** \$150.12 **f)** \$200.00

4. a) $C'(x) = 1 + 24\ln 0.9 \times (0.9)^x$
b) -1.05; -0.49; 0.12; 0.82; 0.99; 1.00

6. a) $x \doteq 8.8$; $C = \$118.30$

7. Approximately 4.75%

8. a) \$23.58 **b)** \$125.58

9. a) The graph is translated 50 units down.
b) The C-intercept increases and the minimum point moves to the right.

10. a) $p = 0.20(0.92)^x$ **b)** $C = 100 + x + 20(0.92)^x$
c) \$6.13 **d)** 0

11. a) Graphs may vary. **d)** Approximately 7 m

Chapter 8 Applications of Differential Calculus

Necessary Skills

1 Review: Implicit Differentiation

Exercises, page 483

1. a) $\frac{dx}{dt} = \frac{y}{x}\frac{dy}{dt} + \frac{z}{x}\frac{dz}{dt}$ **b)** $\frac{dh}{dt} = \frac{1}{\pi r^2}\frac{dA}{dt} - \frac{2h}{r}\frac{dr}{dt}$
c) $\frac{dr}{dt} = \frac{1}{4\pi r^2}\frac{dV}{dt}$

2. a) $\frac{dl}{dt} = 3.5$ **b)** $\frac{dx}{dt} = \pm\frac{5\sqrt{5}}{2}$
c) $\frac{dr}{dt} = \frac{7}{72\pi}$

2 Review: Similar Triangles

Exercises, page 486

2. a) ii) $x \doteq 4.3$ cm
b) ii) $y \doteq 3.7$ cm, $z \doteq 1.4$ cm

3. 2.975 m

8.1 Exercises, page 492

1. a) i) $D = \Re$
ii) x-intercept: none; y-intercept: 1
b) i) $D = \{x \mid x \in \Re, x \neq 1\}$
ii) x-intercept: none; y-intercept: -1
c) i) $D = \{x \mid x \in \Re, x \neq 0\}$
ii) x-intercept: -1; y-intercept: none

2. a) $f(x) = \frac{e^x}{x-1}$ and $f(x) = x^2 + \frac{1}{x}$
b) None of the functions has a point discontinuity.

3. a) Horizontal asymptote: $y = 0$; oblique asymptote: $y = x$
b) No asymptotes
c) Vertical asymptote: $x = 0$; horizontal asymptote: $y = 0$

4. $f(x) = \frac{x}{1-e^x}$ and $f(x) = \frac{e^x}{x}$

5. a) $x = -4$ **b)** $x = 1$ **c)** $x = -2, 0$

6. Values are approximate.
a) Minimum value: -1.34
b) Minimum value: 0.83
c) Minimum value: 3.44

7. a) $\left(\ln 2,\ 2 - (\ln 2)^2\right)$
b) $(0, 0)$ **c)** $(0, 0)$ and $\left(3, \frac{12}{e^3}\right)$

8. a) $D = \Re$; horizontal asymptote: $y = 0$
b) $D = \{x \mid x \in \Re, x \neq -0.57\}$;
vertical asymptote: $x \doteq -0.57$;
horizontal asymptotes: $y = 1$ and $y = 0$
c) $D = \Re$; horizontal asymptote: $y = 0$

9. a) Minimum point: $\left(-1, -\frac{1}{e}\right)$ **b)** $\left(-2, -\frac{2}{e^2}\right)$
c) Increase: $x > -1$; decrease: $x < -1$
d) Concave up: $x > -2$; concave down: $x < -2$

10. a) $y' = \frac{-2e^{\frac{1}{x^2}}}{x^3}$ **b)**

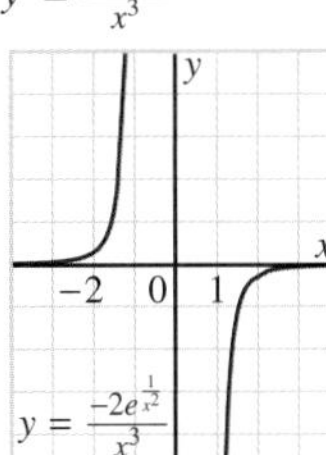

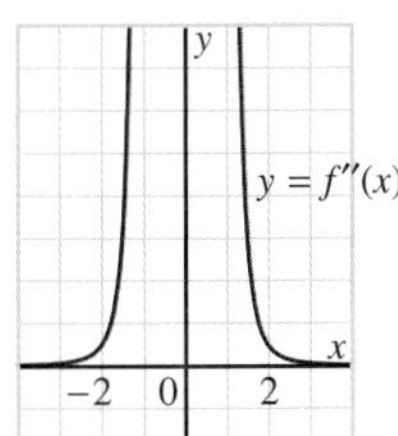

12. x-intercept: none; y-intercept: none; domain: $\{x \mid x \in \Re, x \neq 0\}$; minimum point: (0.93, 3.69); increase: $x < 0$ and $x > 0.93$; decrease: $0 < x < 0.93$; point of inflection: none; concave up: $x \in \Re$ $(x \neq 0)$; vertical asymptote: $x = 0$; horizontal asymptote: $y = 0$

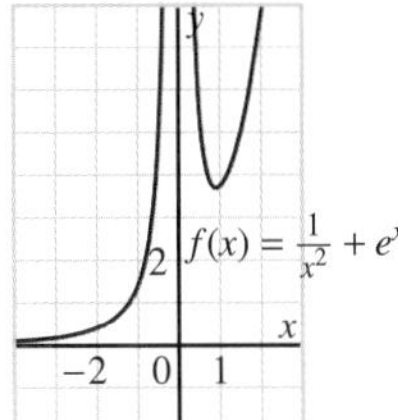

13. x-intercept: −1; y-intercept: none; domain: $\{x \mid x \in \Re, x \neq 0\}$; decrease: $x \in \Re$ $(x \neq 0)$; concave up: $x > 0$; concave down: $x < 0$; vertical asymptote: $x = 0$; horizontal asymptote: $y = 0$; oblique asymptote: $y = -x - 1$

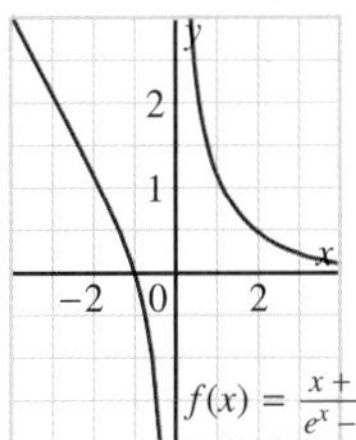

14. x-intercept: 0; y-intercept: 0; domain: $x \in \Re$; maximum point: $\left(3, \frac{27}{e^3}\right)$; increase: $x < 0$ and $0 < x < 3$; decrease: $x > 3$; points of inflection: (0, 0), (1.27, 0.58), (4.73, 0.93); concave up: $0 < x < 1.27$ and $x > 4.73$; concave down: $x < 0$ and $1.27 < x < 4.73$; horizontal asymptote: $y = 0$

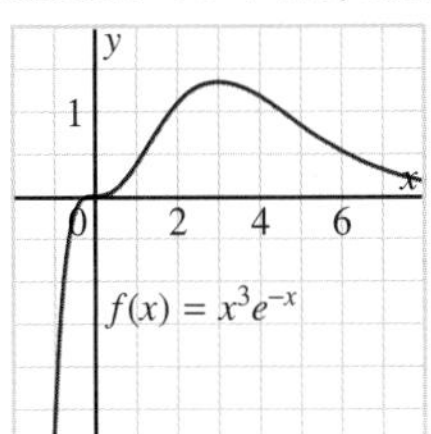

15. Graphs may vary.

a)

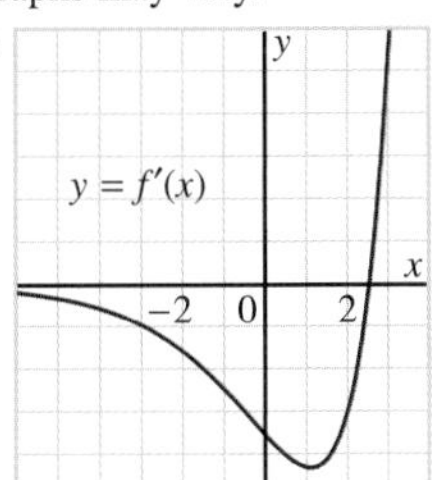

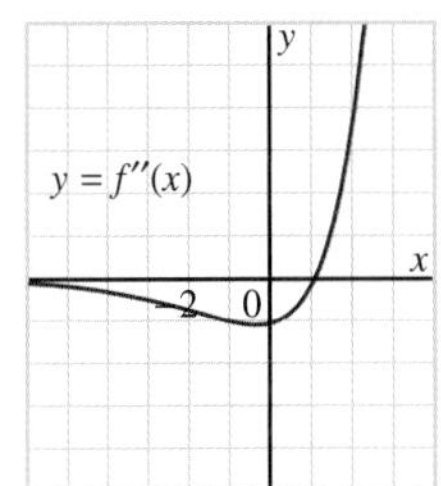

b)

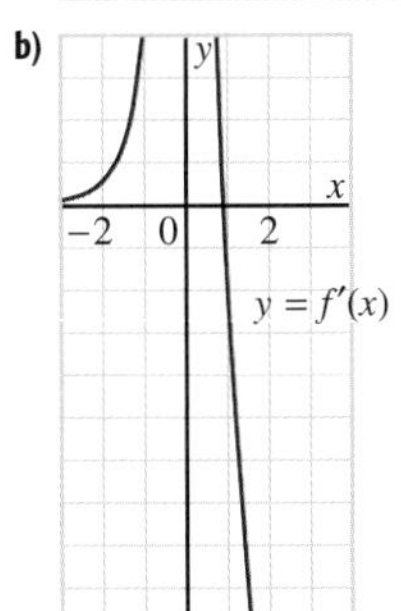

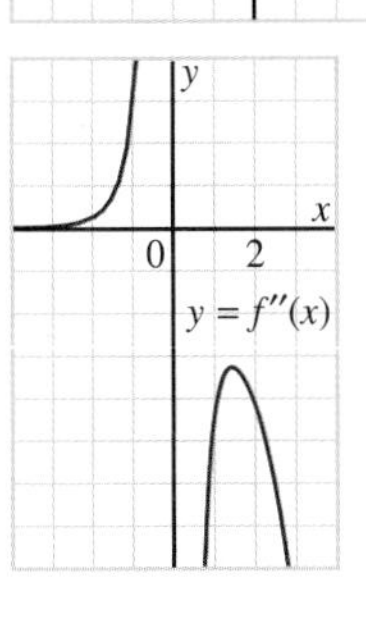

16. a) Graph 1:

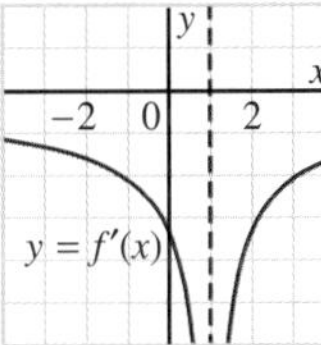

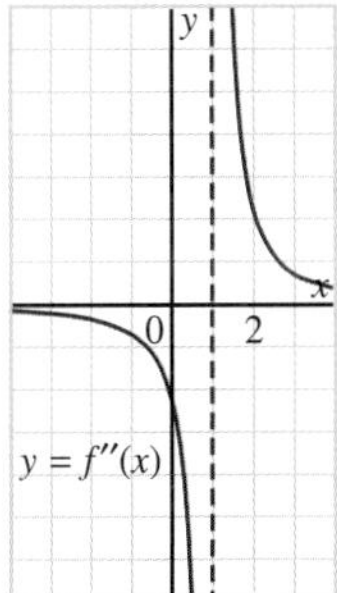

Graph 2:

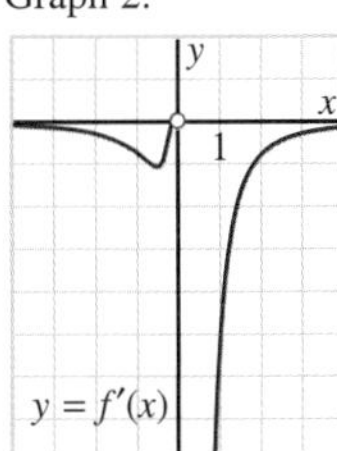

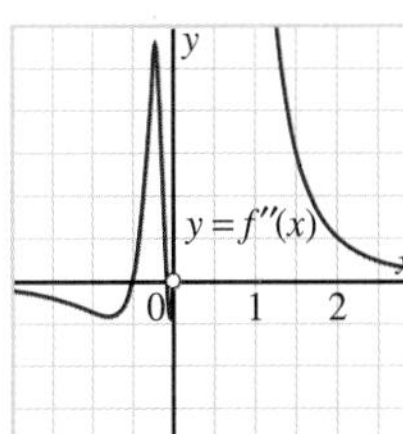

b) i) False **ii)** True **iii)** True
iv) False **v)** False

17.

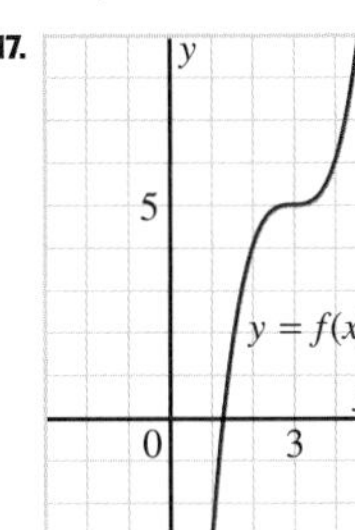

18. a) i)

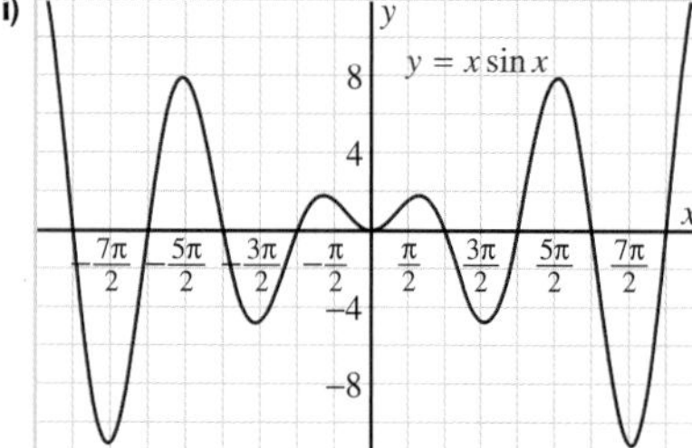

ii)

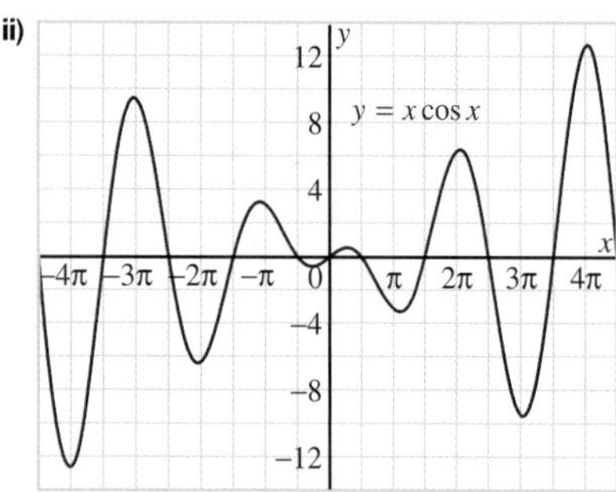

b) i) $f(x) = x \sin x$ is symmetrical about the y-axis.
ii) $f(x) = x \cos x$ has point symmetry about the origin.

c) i) $x = \pm\frac{\pi}{2}, \pm\frac{5\pi}{2}, \pm\frac{9\pi}{2}, \pm\frac{13\pi}{2}, \ldots$
ii) $x = 0, \pm 2\pi, \pm 4\pi, \pm 6\pi, \ldots$

d) i) $x = \pm\frac{3\pi}{2}, \pm\frac{7\pi}{2}, \pm\frac{11\pi}{2}, \ldots$
ii) $x = \pm\pi, \pm 3\pi, \pm 5\pi, \ldots$

19. b) i) $3 < x < 5$ **ii)** $-3 < x < -\frac{7}{3}$
iii) $-1 < x < 1$ **iv)** $-3 < x < -1$

8.2 Exercises, page 506

1. a) $A = 4\pi r^2$ **b)** $A = lw$
c) $A = 2\pi rh + 2\pi r^2$ **d)** $V = \pi r^2 h$
e) $A = \frac{bh}{2}$ **f)** $V = lwh$
g) $A = 2(lw + lh + wh)$ **h)** $V = \frac{1}{3}\pi r^2 h$

2. a) $l + w = 15$ **b)** $bh = 42$
c) $\pi r^2 h = 500$ **d)** $b^2 + 2bh = 400$
e) $b^2 h = 1200$

4. 53 679 bacteria

5. a) Volume of the box: $V = l^2 h$
b) $l^2 + 4lh = 3600$
c) $V = 900l - \frac{l^3}{4}$
d) The dimensions of the box are approximately 34.6 cm by 34.6 cm by 17.3 cm.
e) 20 785 cm^3

6. b) 14 697 cm^3

7. 12.5 and 12.5

8. a) $\sqrt[3]{2} \doteq 1.26$

9. a) 7.8 cm **b)** 8450 cm^3

10. a) 175 m by 87.5 m **b)** 15 312.5 m^2

11. 500 m by 333 m

12. Each is 16 cm.

13. b) Yes
c) For any constant length l of hypotenuse AC, right △ABC has a maximum area when AB = BC.

14. 300 m by 200 m

15. a) Use all the rope for the circular garden to maximize the area enclosed.
b) Use approximately 56 m of the rope for the square garden and 44 m of the rope for the circular garden to minimize the area enclosed.

16. b) $23 600

17. Radius is approximately 8.16 cm. Height is approximately 11.55 cm.

18. Approximately 1240 cm^3

19. Width is approximately 17.3 cm. Depth is approximately 24.5 cm.

Self-Check 8.1, 8.2, page 509

1. a) i) $D = \Re$
ii) x-intercept: –0.57; y-intercept: 1
b) i) $D = \Re$
ii) x-intercept: –0.70; y-intercept: –1
c) i) $D = \{x \mid x \in \Re, x \neq 0\}$
ii) x-intercept: 0.57; y-intercept: none

2. a), b), c) No critical values

3. a) $y' = -e^{-x}(x - 1)$
b)

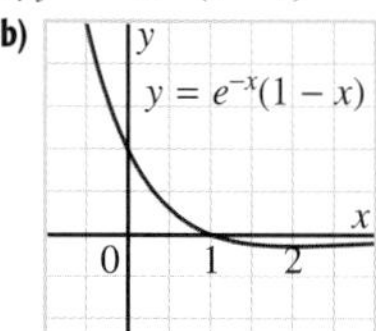

c) Maximum point: $\left(1, \frac{1}{e}\right)$; point of inflection: $\left(2, \frac{2}{e^2}\right)$
d)

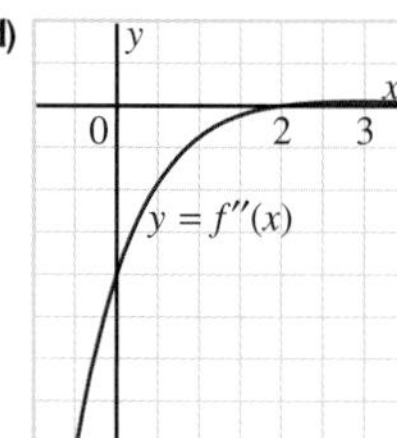

4. x-intercepts: –1, 0; y-intercept: 0; domain: $x \in \Re$; minimum point: (–0.62, –0.44); maximum point: (1.62, 0.84); increase: $-0.62 < x < 1.62$; decrease: $x < -0.62$ and $x > 1.62$; points of inflection: (0, 0), $\left(3, \frac{12}{e^3}\right)$; concave up: $x < 0$ and $x > 3$; concave down: $0 < x < 3$; horizontal asymptote: $y = 0$

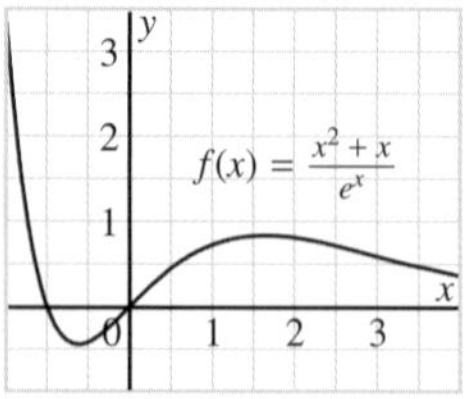

5. 4 and –6

6. a) Square: 16.8 cm; circle: 13.2 cm
b) Square: 0 cm; circle: 30 cm

7. 15 811 cm^3

8. Radius is approximately 4.15 cm and height is approximately 8.31 cm.

9. 4.24 m by 2.12 m

8.3 Exercises, page 517

1. a) $\frac{dr}{dt} = \frac{1}{8\pi r}\frac{dA}{dt}$ **b)** $\frac{dr}{dt} = \frac{1}{2\pi r}\frac{dA}{dt}$
c) $\frac{dr}{dt} = \frac{1}{4\pi r}\frac{dV}{dt}$ **d)** $\frac{dh}{dt} = \frac{9}{4\pi h^2}\frac{dV}{dt}$
e) $\frac{ds}{dt} = \frac{1}{4}\frac{dP}{dt}$ **f)** $\frac{dh}{dt} = \frac{25}{48\pi h^2}\frac{dV}{dt}$

2. a) $P = 4s$ **b)** $V = \frac{4}{3}\pi r^3$
c) $A = 6s^2$ **d)** $C = 2\pi r$ or $C = \pi d$
e) $A = \pi r^2$

3. Approximately 0.005 cm/min

4. 0.19 m/min

5. 5.6 m/min

6. a) 226.2 cm^3/s **b)** 150.8 cm^2/s

8. –96 mm^2/min

9. a) 0.7 cm/min **b)** 2.7 cm/min

10. 1206 cm^2/s

11. 45 cm^2/s

12. 65.8 cm^2/s

13. 0.05 m/s

14. a) 0.04 m/min **b)** 0.12 m/min

15. 0.16 km

16. 16 800 cm^3/min

8.4 Exercises, page 523

1. a) $\frac{dy}{dt} = \frac{-x}{y}\frac{dx}{dt}$ **b)** $\frac{dy}{dt} = \frac{z}{y}\frac{dz}{dt} - \frac{x}{y}\frac{dx}{dt}$
c) $\frac{dl}{dt} = \frac{1}{2}\frac{dP}{dt} - \frac{dw}{dt}$ **d)** $\frac{db}{dt} = \frac{2}{h}\frac{dA}{dt} - \frac{b}{h}\frac{dh}{dt}$

2. a) d **b)** a **c)** c **d)** b

3. 0.35 m/s

4. 0.14 m/s

5. 94.5 km/h

7. 17.9 m/s

8. 12.5 m/s

9. a) 17.1 km **b)** 102.6 km/h

10. 0 cm/min

11. 1.2 cm/s

12. 100.5 cm^3/s

13. 70.9 km/h

14. a) 25.7 km/h **b)** 49 min

15. 0.74 m/s

16. –74.0 mm/s

17. 21.3 km/h

18. 136.5 cm^2/min

19. a) Toward the spotlight **b)** 285.7 m/s

Self-Check 8.3, 8.4, page 526

1. Radius: 0.05 m/min; height: 0.04 m/min

2. Approximately 47 mm^2/s

3. a) 0.78 cm/min **b)** $2.\overline{3}$ cm/min

4. a) 1909 cm^3/s **b)** 545 cm^2/s

5. a) 0.19 m/s **b)** 1.49 m/s

6. 57 cm/s

7. a) 7.07 km **b)** Approximately 119 km/h

8. –0.51 cm/s

Chapter 8 Review Exercises, page 527

1. a) Local maximum value of –1
b) No points of inflection

2. a) None **b)** None
c) Vertical asymptote: $x = -2$; horizontal asymptote: $y = 0$

3. a) $D = \{x \mid x \in R, x \neq 0\}$; horizontal asymptote: $y = 0$; oblique asymptote: $y = -x$
b) $D = \{x \mid x \in R, x \neq 0\}$; vertical asymptote: $x = 0$; horizontal asymptote: $y = 0$
c) $D = \{x \mid x \in R, x \neq 0\}$; vertical asymptote: $x = 0$; horizontal asymptotes: $y = \pm 1$

4. a) Maximum point: $\left(-2, \frac{4}{e^2}\right)$; minimum point: (0, 0); points of inflection: (–3.41, 0.38) and (–0.59, 0.19)
b) $f'(x) = xe^x(x + 2)$

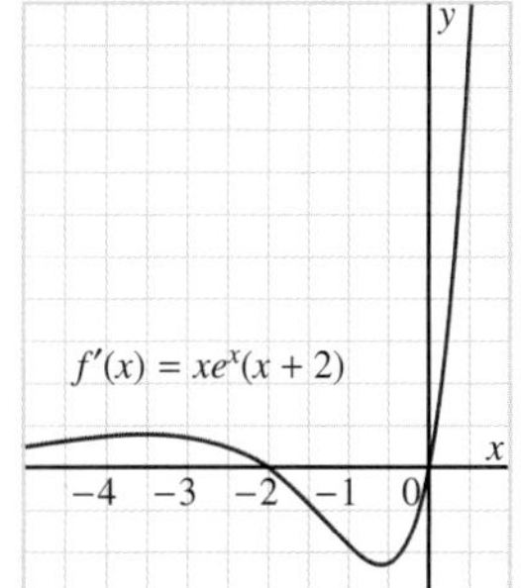

c)

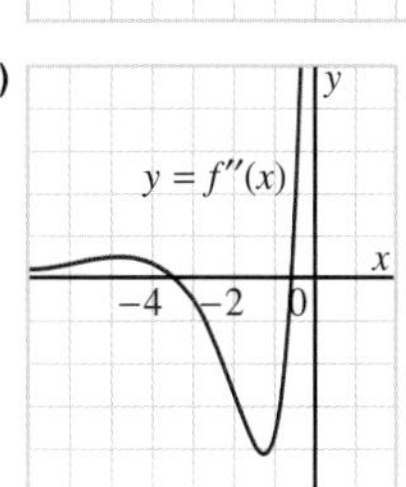

5. a) Domain: $\{x \mid x \in \text{R}, x \neq 0\}$; minimum point: $\left(2, \frac{e^2}{4}\right)$; increase: $x < 0$ and $x > 2$; decrease: $0 < x < 2$; concave up: $x \in \text{R}$ $(x \neq 0)$; vertical asymptote: $x = 0$; horizontal asymptote: $y = 0$

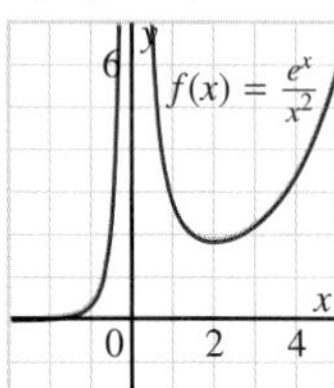

b) x-intercepts: ± 1.32; y-intercept: -2; domain: $x \in \Re$; minimum point: $(0, -2)$; increase: $x > 0$; decrease: $x < 0$; concave up: $x \in \Re$

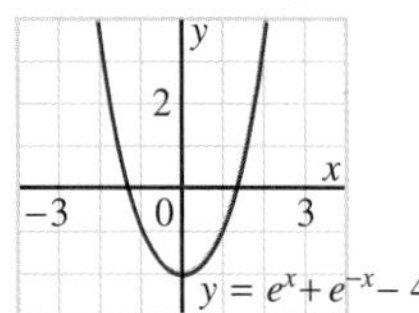

c) y-intercept: 1; domain: $x \in \Re$; minimum point: $(0, 1)$; increase: $x > 0$; decrease: $x < 0$; concave up: $x \in \Re$; oblique asymptote: $y = x$

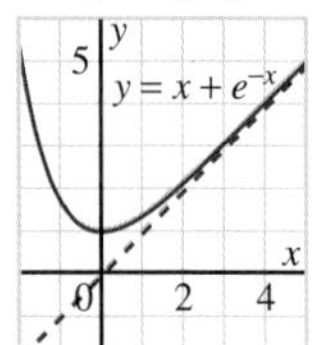

6. 28 and 28

7. Approximately 11.3 cm

8. 580 m by 362.5 m

9. 30 cm, 30 cm, 30 cm

10. a) 0.004 m/min **b)** -1 m^2/min

11. 6.8 cm/min

12. a) Approximately 1.0 cm/min **b)** Approximately 2.5 cm/min

13. 0.03 m/s

14. 0.23 m/s

15. 0.15 m/s

16. Approximately 67 km/h

17. -1.70 cm/min

Chapter 8 Self-Test, page 529

1. a) Local minimum: (3.24, 3.93); local maximum: $(-1.24, -0.12)$

b) Vertical asymptotes: $x = \pm 2$; horizontal asymptote: $y = 0$

2. y-intercept: 1; domain: $\{x \mid x \in \Re, x \neq -\frac{1}{2}\}$; minimum point: $\left(0.5, \frac{\sqrt{e}}{2}\right)$; increase: $x > 0.5$; decrease: $x < 0.5$ $(x \neq -0.5)$; concave up: $x > -0.5$; concave down: $x < -0.5$; vertical asymptote: $x = -0.5$; horizontal asymptote: $y = 0$

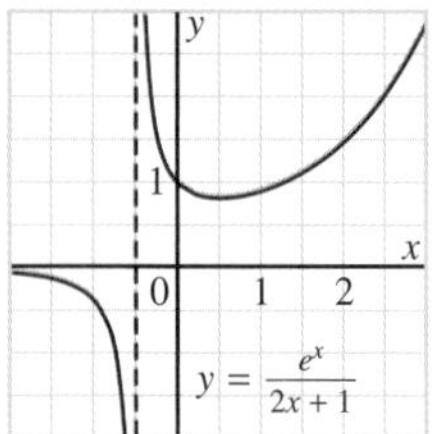

3. 10 m by 10 m and 10 m by 5 m

4. Approximate dimensions: 63.25 cm by 21.08 cm by 15.81 cm; maximum volume: 21 081.85 cm^3

6. a) Approximately 2384 cm^3/s **b)** Approximately 619 cm^2/s

7. 15.2 m/min

8. Approximately 104 cm^2/s

9. a) 0.59 cm/s **b)** 3.09 cm/s **c)** 2.94 cm/s

10. 25.38 m/min

11. 24.08 knots

12. a) 6.75 km **b)** 63.7 km/h

13. a) Graph 1: $g(x) = e^x - x$; Graph 2: $f(x) = e^x + x$

14. a) -0.015 cm/s **b)** 0.011 cm/s

15. 2.80 m

Mathematical Modelling: Optimal Consumption Time

Exercises, page 534

1. a) $t = 15 + 5(1.04)^E$ **b)** $R = \frac{\ln(0.2t - 3)}{t \ln 1.04}$

c) Approximately 40 min **d)** Approximately 20 min

2. a) Optimal cycle time is approximately 23 min.

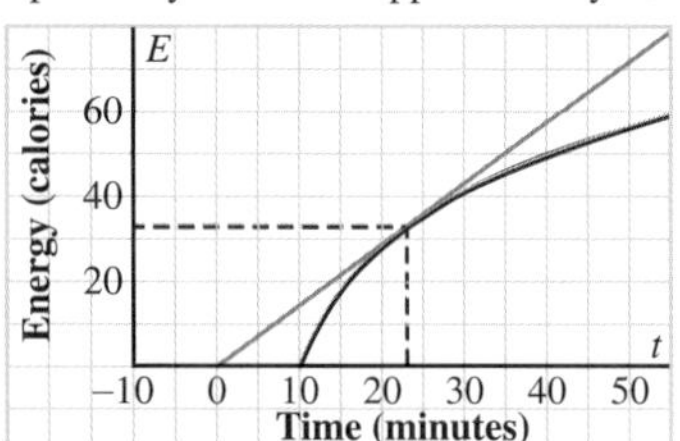

3. a) $t = 0.01E^2 + 0.09E + 10$

b) $R = \frac{-9 + \sqrt{400t - 3919}}{2t}$

c) $t \doteq 23$ min, $E \doteq 32$ calories

d) $t \doteq 22.85$ min, $E \doteq 31.63$ calories

Cumulative Review (Chapters 1–8), page 536

1. a) $f'(x) = -8x$ **b)** $f'(x) = \frac{-4}{(4x - 1)^2}$

c) $f'(x) = \frac{-1}{\sqrt{3 - 2x}}$

2. a) 3.25 **b)** 3

3. Graphs may vary for $y = f(x)$. Possible graphs follow.

a)

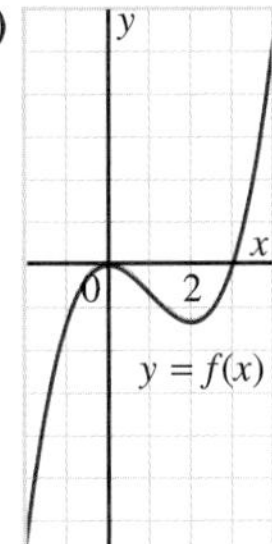

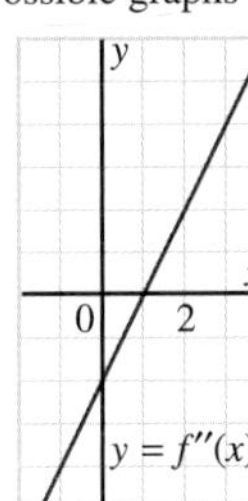

b)

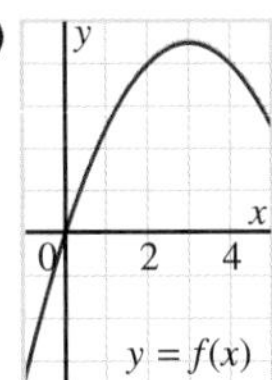

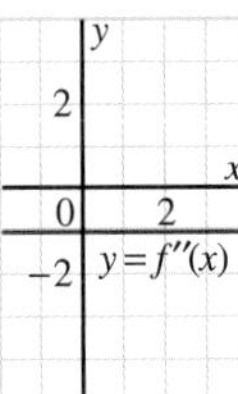

c)

y = f(x)

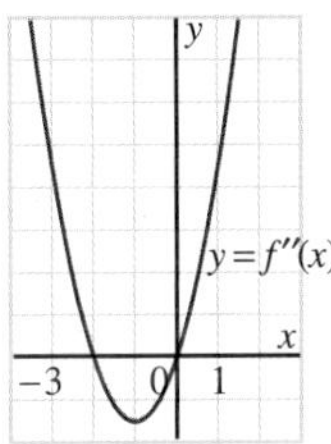

4. a) $\frac{dy}{dx} = 8x - 3$ **b)** $\frac{dy}{dx} = \frac{-1}{x^3}$

c) $\frac{dy}{dx} = \frac{-9}{(x-4)^4}$ **d)** $\frac{dy}{dx} = \frac{3}{\sqrt{6x-5}}$

e) $\frac{dy}{dx} = 2\cos 2x$ **f)** $\frac{dy}{dx} = -\sin x \cos(\cos x)$

g) $\frac{dy}{dx} = -2^{4-x} \times \ln 2$ **h)** $\frac{dy}{dx} = 2xe^{x^2-2}$

i) $\frac{dy}{dx} = \frac{2}{2x+1}$ **j)** $\frac{dy}{dx} = 0$

k) $\frac{dy}{dx} = -12\cos^2(4x+3)\sin(4x+3)$

l) $\frac{dy}{dx} = 16x - 4 + \frac{1}{x^2}$

5. a) $\frac{dy}{dx} = \frac{4x(x^2+1)^3(9x^2-1)}{\sqrt{4x^2-1}}$ **b)** $\frac{dy}{dx} = \frac{-3(8x^3+20x^2+1)}{(3x+5)^2}$

c) $\frac{dy}{dx} = \frac{-110(4-3x)^4}{(7x-2)^6}$ **d)** $\frac{dy}{dx} = \frac{-2}{1-x^2}$

e) $\frac{dy}{dx} = 3x^2e^{3x}(x+1)$ **f)** $\frac{dy}{dx} = \frac{1-2\ln 2x}{x^3}$

6. a) $\frac{dy}{dx} = \frac{y-3x}{-x-y}$ **b)** $\frac{dy}{dx} = \frac{y^2 - \sin y}{x(\cos y - 2y)}$

c) $\frac{dy}{dx} = -\frac{\cos 2x}{y \sin y^2}$

7. a) $y = 6x - 4$ **b)** $y = -\frac{5}{3}x - 4$

c) $y = 2x - e$

8. a) $x < -3$ and $0 < x < 1$ and $x > 1$

b) $x \le -2$ and $-1 \le x \le 3$

9. $k = 1$

10. a) $(0, 0)$ and $(4, 16)$

11. $y = 8x + 8$ and $y = -8x + 8$

12. a) $v(t) = 6t^2 - 18t + 12$; $a(t) = 12t - 18$

b) 1 s and 2 s **c)** 3 m/s **d)** 182 m

13. a) $f(g(x)) = 3x^2 - 22x + 40$ **b)** $f'(g(x)) = 6x - 22$

c) $f'(g(x)) = 6x - 22$

14. a) 27 **b)** −8 **c)** $\frac{3}{2}$

15. a) x-intercepts: ±3, ±2; y-intercept: 36; domain: $x \in \Re$; maximum point: (0, 36); minimum points: $(\pm\sqrt{6.5}, -6.25)$; increase: $-2.55 < x < 0$ and $x > 2.55$; decrease: $x < -2.55$ and $0 < x < 2.55$; points of inflection: (±1.47, 12.53); concave up: $|x| > 1.47$; concave down: $|x| < 1.47$

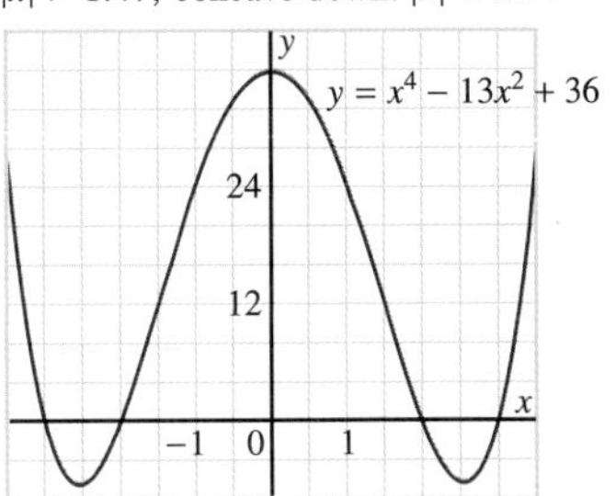

b) x-intercepts: −1, 4; y-intercept: 4; domain: $x \in \Re$; minimum point: (−1, 0); maximum point: $\left(\frac{7}{3}, 18.52\right)$; increase: $-1 < x < \frac{7}{3}$; decrease: $x < -1$ and $x > \frac{7}{3}$; point of inflection: (0.67, 9.26); concave up: $x < 0.67$; concave down: $x > 0.67$

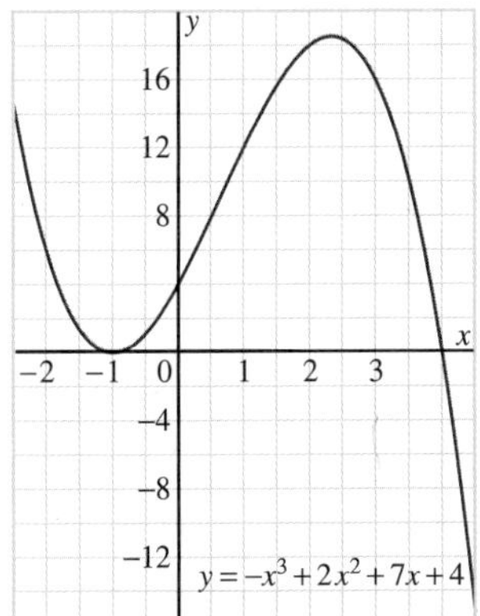

c) x-intercepts: −3, 1; domain: $\{x \mid x \in \Re, x \ne 0\}$; maximum point: $\left(3, \frac{4}{3}\right)$; increase: $0 < x < 3$; decrease: $x < 0$ and $x > 3$; point of inflection: (4.5, 1.30); concave up: $x > 4.5$; concave down: $x < 4.5$ $(x \ne 0)$; vertical asymptote: $x = 0$; horizontal asymptote: $y = 1$

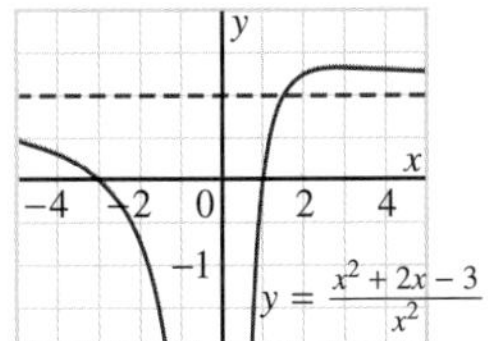

d) x-intercepts: ± 2; y-intercept: $\frac{4}{3}$; domain: $\{x \mid x \in \Re, x \neq 6\}$; maximum point: (0.3, 1.4); minimum point: (11.7, 46.6); increase: $x < 0.3$ and $x > 11.7$; decrease: $0.3 < x < 11.7$ $(x \neq 6)$; concave up: $x > 6$; concave down: $x < 6$; vertical asymptote: $x = 6$; oblique asymptote: $y = 2x + 12$

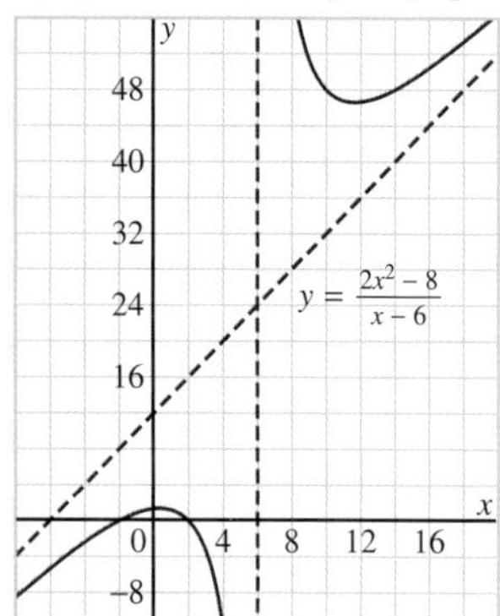

16. a) $\left(\pm\frac{1}{\sqrt{3}}, \frac{4}{9}\right)$ **b)** $(e^{-1.5}, 4.5e^{-3})$

17. Graphs may vary.

a)

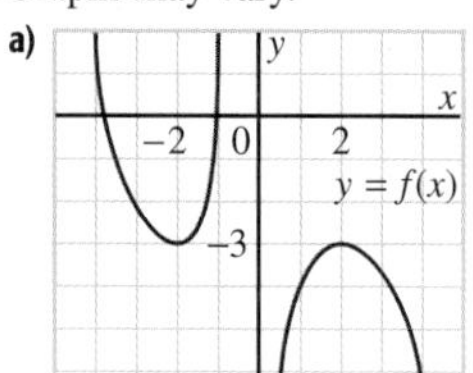

b)

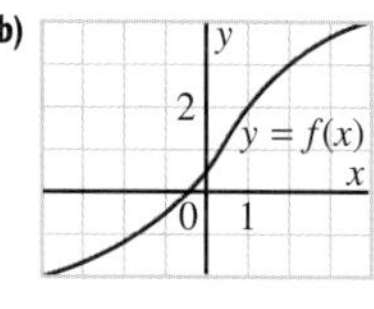

18. Equations may vary.

a) $y = \frac{-3x^2 + 9}{x^2 - 1}$ **b)** $y = \frac{2x^2 - 6x + 3}{x - 3}$

19. a) $\frac{13}{16}$ **b)** $\frac{21}{4}$ **c)** -3

d) 2 **e)** $\frac{1}{4}$ **f)** 15

20. a) $x = \frac{1}{2}$ **b)** $x = \frac{1}{8}$ **c)** $x = 4$

21. a) $x \doteq 10.32$ **b)** $x \doteq 3.28$ **c)** $x \doteq 6.82$

22. Approximately 16 years

23. a) x-intercept: 0; y-intercept: 0; domain: $x \in \Re$; minimum point: (0, 0); maximum point: $\left(\frac{1}{2}, \frac{1}{4e^2}\right)$ increase: $0 < x < \frac{1}{2}$; decrease: $x < 0$ and $x > \frac{1}{2}$; points of inflection: (0.85, 0.02), (0.15, 0.01); concave up: $x < 0.15$ and $x > 0.85$; concave down: $0.15 < x < 0.85$; horizontal asymptote: $y = 0$

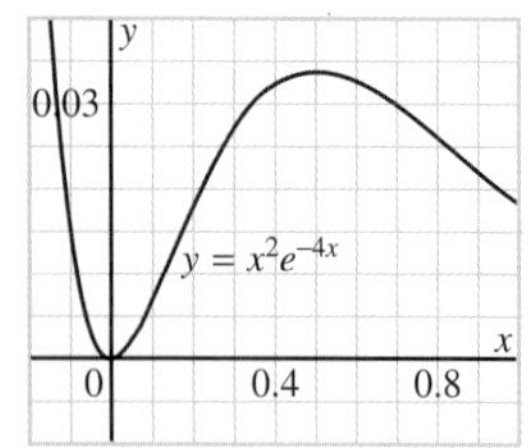

b) y-intercept: 2; domain: $x \in \Re$; minimum point: (–0.39, 1.62); increase: $x > -0.39$; decrease: $x < -0.39$; concave up: $x \in \Re$

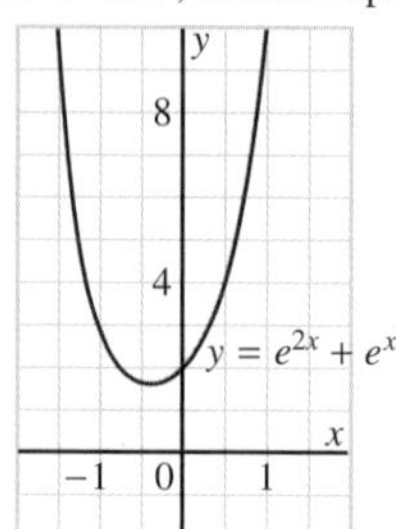

c) y-intercept: 1; domain: $x \in \Re$; maximum point: $(1, e)$; increase: $x < 1$; decrease: $x > 1$; points of inflection: (0.29, 1.64), (1.71, 1.64); concave up: $x < 0.29$ and $x > 1.71$; concave down: $0.29 < x < 1.71$; horizontal asymptote: $y = 0$

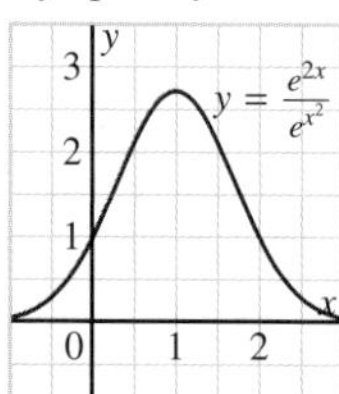

24. 2 m/s^2

25. $1469.69

26. 10 cm, 10 cm, and 10 cm

27. 30.4 cm by 52.6 cm

28. 2 m by 2 m by 0.5 m

29. a) 0.07 m/min **b)** 12 m^3/min

30. –15 cm^3/s

31. 13.5 cm^2/s

32. 0.02 mm/s

33. 3.46 cm^2/s

34. 0.0008 cm/s

35. Approximately 1452 cm^2/s

36. 6.97 m/s

Glossary

absolute maximum point: the point on the graph of a function $y = f(x)$ where the value of y is greater than all other possible values of y

absolute minimum point: the point on the graph of a function $y = f(x)$ where the value of y is less than all other possible values of y

absolute value: the distance between any real number and 0 on a number line; for example, $|-3| = 3$, $|3| = 3$

acceleration: rate of change of velocity with respect to time for an object moving in a straight line

altitude: the perpendicular distance from the base of a figure to the opposite side or vertex

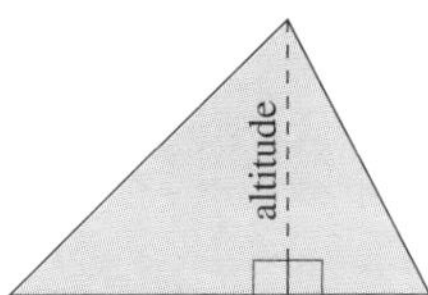

amount: the value of the principal plus interest

approximation: a number close to the exact value; the symbol $\doteq$ means "is approximately equal to"

arithmetic sequence: a sequence of numbers in which each term after the first term is calculated by adding the same number to the preceding term; for example, in the sequence 1, 4, 7, 10, …, each number is calculated by adding 3 to the previous number

arithmetic series: the indicated sum of the terms of an arithmetic sequence

asymptote to a curve: a line with the property that the distance from a point P on the curve to the line approaches zero as the distance from P to the origin increases indefinitely

axis of symmetry: a line that divides a curve into two congruent parts

base: the side of a polygon or the face of a solid from which the height is measured; the factor repeated in a power

binomial: a polynomial with two terms; for example, $3x - 8$

branches: the two distinct parts of a hyperbola

chord: a line segment whose endpoints lie on a conic

circle: the curve that results when a plane intersects a cone, parallel to the base of the cone; or the locus of a point P that moves so it is the same distance from a fixed point (the centre of the circle)

coefficient: the numerical factor of a term; for example, in the terms $3x$ and $3x^2$, the coefficient is 3

collinear points: points that lie on the same line

common denominator: a number that is a multiple of each of the given denominators; for example, 12 is a common denominator for the fractions $\frac{1}{3}, \frac{5}{4}, \frac{7}{12}$

common difference: the number obtained by subtracting any term from the next term in an arithmetic sequence

common factor: a monomial that is a factor of each of the given monomials; for example, $3x$ is a common factor of $15x$, $9x^2$, and $21xy$

common ratio: the ratio formed by dividing any term after the first one in a geometric sequence by the preceding term

complete the square: add or subtract constants to rewrite a quadratic expression $ax^2 + bx + c$ as $a(x - p)^2 + q$

complex number: any number of the form $a + bi$, where a and b are real numbers and $i^2 = -1$

composite function: a function of the form $y = f(g(x))$ or $y = f \circ g(x)$; that is, the function that is formed when the function f is applied on the function g

compound interest: see *interest*; if the interest due is added to the principal and thereafter earns interest, the interest earned is compound interest

compounding period: the time interval for which interest is calculated

compression of a function: a transformation that results in the graph of a function being compressed horizontally or vertically

concavity of a graph: the interval in which the graph opens up or opens down

cone: the surface generated when a line is rotated in space about a fixed point P on the line

congruent: figures that have the same size and shape

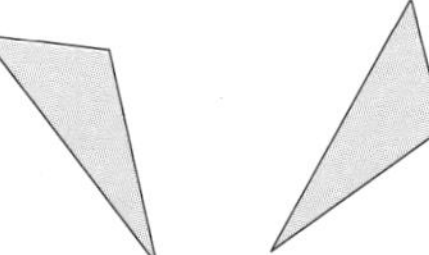

conic: one of four curves (circle, ellipse, parabola, hyperbola) that results when a plane intersects a cone

conjugate complex numbers: two numbers of the form $a + bi$ and $a - bi$, where a and b are real numbers and $i^2 = -1$

consecutive numbers: integers that come one after the other without any integers missing; for example, 34, 35, 36 are consecutive numbers, so are –2, –1, 0, and 1

constant function: $f(x) = c$, where c is a constant

constant term: a number

continuous function: a function whose graph can be drawn without lifting the pencil from the paper

corresponding angles in similar triangles: two angles, one in each triangle, that are equal

cosine of θ: the first coordinate of point P, where P is on the unit circle, centre (0, 0), and θ is the measure of the angle in standard position

critical point: the point on the graph of a function at which the tangent is horizontal; for $y = f(x)$, a point at which $f'(x) = 0$

critical value: the value of x at a critical point on the graph of a function $y = f(x)$

cube root: a number which, when raised to the power 3, results in a given number; for example, 4 is the cube root of 64, and –4 is the cube root of –64

cylinder: a solid with two parallel, congruent, circular bases

decreasing function: a function $y = f(x)$ is decreasing if y decreases as x increases

degree of a polynomial: the degree of its highest-degree term

The degree of $x^3 + 2xy + y^2$ is 3.
The degree of $3x - 2y + 5$ is 1.

degree of a term: the sum of the exponents of the variables in that term

The degree of $3x^2y^3$ is 5.
The degree of $2ab^2$ is 3.
The degree of $-4x$ is 1.
The degree of a constant, such as 7, is 0.

denominator: the term below the line in a fraction

dependent variable: the output for a function, often denoted by y

derivative: of a function $y = f(x)$ is a new function denoted $y = f'(x)$ (or y', or $\frac{dy}{dx}$) where $f'(x)$ is the instantaneous rate of change of f at x when f' exists

diagonal: a line segment that joins two vertices of a polygon, but is not a side

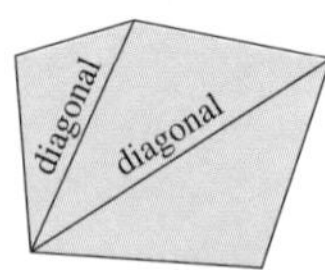

difference quotient: an estimate of the instantaneous rate of change of y with respect to x for the function $y = f(x)$ at a point $P(x_1, y_1)$, usually denoted $\frac{f(x_1 + h) - f(x_1)}{h}$, where h is a very small number

difference of squares: a polynomial that can be expressed in the form $x^2 - y^2$; the product of two monomials that are the sum and difference of the same quantities, $(x + y)(x - y)$

discontinuous function: a function whose graph cannot be drawn without lifting the pencil from the paper

discriminant: the discriminant of the quadratic equation $ax^2 + bx + c = 0$ is $b^2 - 4ac$

displacement: the distance of an object from a certain point (the origin) at a certain time, when the object is moving in a straight line

division statement: in any division problem, dividend = divisor × quotient + remainder

domain of a function: the set of x-values (or valid input numbers) represented by the graph or the equation of a function

doubling time: the time for a quantity to increase to two times its value, when the quantity is increasing exponentially

***e*:** the number such that $\lim_{h \to 0} \frac{e^h - 1}{h} = 1$; or $\lim_{n \to \infty} \left(1 + \frac{1}{n}\right)^n$; $e \doteq 2.718\,28$

ellipse: the closed curve that results when a plane intersects a cone

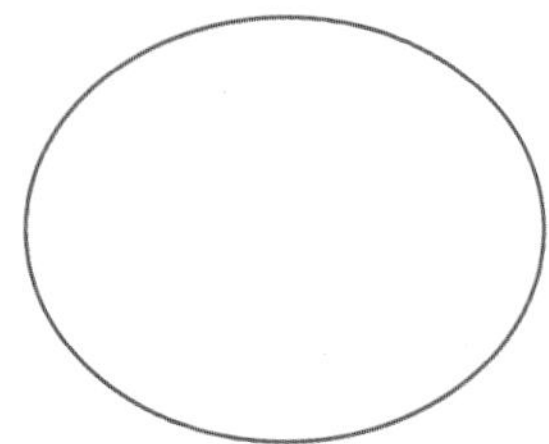

end behaviour of a function: a consideration of the value of $y = f(x)$ when x is positive and very large, and when x is negative and has a very large absolute value

equidistant: the same distance

even function: a function with the property $f(-x) = f(x)$

even number: an integer that has 2 as a factor; for example, 2, 4, –6

expansion of a function: a transformation that results in the graph of a function being expanded horizontally or vertically

explicit equation: an equation in which the dependent variable, y, is isolated and expressed in terms of the independent variable, x; for example, $y = 3x^2 + 5x + 4$

exponent: a number, shown in a smaller size and raised, that tells how many times the number before it is used as a factor; for example, 2 is the exponent in 6^2

exponential decay: a situation where an original amount is repeatedly multiplied by a constant a, where $0 < a < 1$

exponential equation: an equation where the variables occur in the exponents; for example, $2^x - 3 = 0$

exponential function: a function of the form $y = a^x$, $a > 0$, $a \neq 1$; that is, the variable is in an exponent

exponential growth: a situation where an original amount is repeatedly multiplied by a constant a, where $a > 1$

extrapolate: to estimate a value beyond known values

factor: to factor means to write as a product; to factor a given integer means to write it as a product of integers, the integers in the product are the factors of the given integer; to factor a polynomial with integer coefficients means to write it as a product of polynomials with integer coefficients

fifth root: a number which, when raised to the power 5, results in a given number; for example, 3 is the fifth root of 243 since $3^5 = 243$, and –3 is the fifth root of –243 since $(-3)^5 = -243$

first derivative: see *derivative*

formula: a rule that is expressed as an equation

fourth root: a number which, when raised to the power 4, results in a given number; for example, –2 and 2 are the fourth roots of 16

fraction: an indicated quotient of two quantities

function: a rule that gives a single output number for every valid input number

future value: see *amount*

general term of an arithmetic sequence (or series): used to determine any term by substitution; that is, t_n is the nth term in the sequence, and $t_n = a + (n - 1)d$, where a is the first term, n is the number of terms, and d is the common difference

general term of a geometric sequence (or series): used to determine any term by substitution; that is, t_n is the nth term in the sequence, and $t_n = ar^{n-1}$, where a is the first term, n is the number of terms, and r is the common ratio

geometric growth: see *exponential growth*

geometric sequence: a sequence of numbers in which each term after the first term is calculated by multiplying the preceding term by the same number; for example, in the sequence 1, 3, 9, 27, 81, …, each number is calculated by multiplying the preceding number by 3

geometric series: the indicated sum of the terms of a geometric sequence

greatest common factor: the greatest factor that two or more terms have in common; for example, $4x^2$ is the greatest common factor of $4x^3 + 16x^2 - 64x^4$

half-life: the time for a quantity to reduce to one-half of its value, when the quantity is decaying exponentially

horizontal intercept: the horizontal coordinate of the point where the graph of a function or a relation intersects the horizontal axis

hyperbola: the curve that results when a plane intersects both nappes of a cone

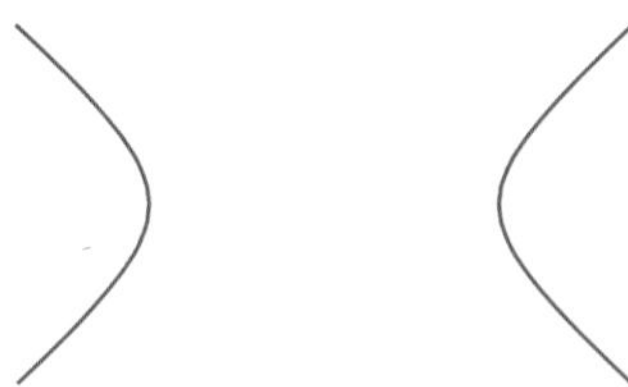

hypotenuse: the side that is opposite the right angle in a right triangle

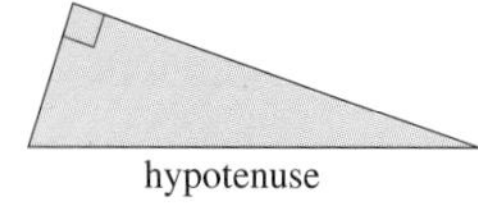

hypothesis: something that seems likely to be true

identity: an equation that is satisfied for all values of the variable for which both sides of the equation are defined

image: the figure or graph that results from a transformation

implicit equation: an equation in which the dependent variable, y, is not isolated; for example, $x^2 + y^2 = 9$

increasing function: a function $y = f(x)$ is increasing if y increases as x increases

independent variable: the input for a function; often denoted by x

inequality: a statement that one quantity is greater than (or less than) another quantity

infinite discontinuity: occurs where the graph of a function has a vertical asymptote

infinity: denoted ∞, a value greater than any computable value

integers: the set of numbers… −3, −2, −1, 0, +1, +2, +3, …

interest: money that is paid for the use of money, usually according to a predetermined percent

interpolate: to estimate a value between two known values

interval: a regular distance or space between numbers

inverse of a function: a relation whose rule is obtained from that of a function by interchanging x and y

irrational number: a number that cannot be written in the form $\frac{m}{n}$, where m and n are integers ($n \neq 0$)

isosceles triangle: a triangle with at least two equal sides

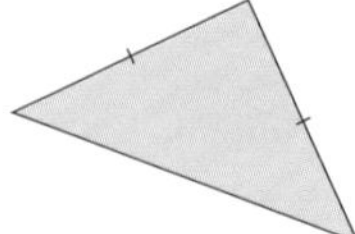

lattice point: on a coordinate grid, a point at the intersection of two grid lines

leading coefficient: the coefficient of the highest power of a polynomial expression

like radicals: radicals that have the same radical part; for example, $\sqrt{5}$, $3\sqrt{5}$, and $-7\sqrt{5}$

like terms: terms that have the same variables; for example, $4x$ and $-3x$ are like terms

limit of a function: for $y = f(x)$, the limit of $f(x)$ is the value of y when x approaches a stated value

line of best fit: a line that passes as close as possible to a set of plotted points

line segment: the part of a line between two points on the line

line symmetry: see *axis of symmetry*

linear equation: an equation in which the degree of the highest-degree term is 1; for example, $x = 6$, $y = 2x - 3$, and $4x + 2y - 5 = 0$

linear function: a function whose defining equation can be written in the form $y = mx + b$, where m and b are constants

linear growth: a situation where a constant is repeatedly added to an original amount; see *arithmetic sequence*

linear inequality: an inequality in which the degree of the highest-degree term is 1; for example, $2x - 3 > 5 - 4x$

linear system: two or more linear equations in the same variables

local maximum point: the point on the graph of a function $y = f(x)$ where the value of y is greater than the values of y in a small interval containing that point

local minimum point: the point on the graph of a function $y = f(x)$ where the value of y is less than the values of y in a small interval containing that point

logarithm: an exponent with the form $\log_a x$; $\log_a x$ is equal to a number b, where $a^b = x$, $a > 0$, $a \neq 1$

logarithmic equation: an equation that contains the logarithm of the variable; for example, $3 \log 4x + 5 = 0$

logarithmic function: has the form $y = \log_a x$ and it is the inverse of the exponential function $y = a^x$, $a > 0$, $a \neq 1$

logarithmic scale: a scale on which lengths are proportional to the logarithms of the numbers on the scale; the logarithms are usually to base 10; for example, each 1 unit increase on the scale represents a 10-fold increase in the measured quantity

median of a triangle: a line from one vertex to the midpoint of the opposite side

midpoint: the point that divides a line segment into two equal parts

mixed radical: an expression of the form $a\sqrt{x}$; for example, $2\sqrt{5}$ is a mixed radical

monomial: a polynomial with one term; for example, 14 and $5x^2$ are monomials

multiple: the product of a given number and a natural number; for example, some multiples of 8 are 8, 16, 24, …

multiplicative decay: see *exponential decay*

multiplicative growth: see *exponential growth*

multiplicative inverses: a number and its reciprocal; the product of multiplicative inverses is 1; for example, $3 \times \frac{1}{3} = 1$

nappes of a cone: the two symmetrical parts of a cone on either side of the vertex

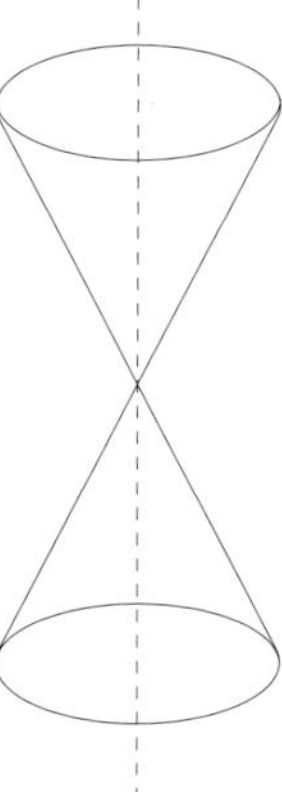

natural logarithm: a logarithm with base e, denoted $\ln x$; see *logarithm* and *e*

natural numbers: the set of numbers 1, 2, 3, 4, 5, …

negative number: a number less than 0

negative reciprocals: two numbers whose product is –1; for example, $\frac{3}{4}$ is the negative reciprocal of $-\frac{4}{3}$, and vice versa

non-linear relation: a relation that cannot be represented by a straight-line graph

numerator: the term above the line in a fraction

odd function: a function with the property $f(-x) = -f(x)$

odd number: an integer that does not have 2 as a factor; for example, 1, 3, –7

operation: a mathematical process or action such as addition, subtraction, multiplication, or division

operator notation: the derivative of $y = f(x)$ is written $\frac{d}{dx}f(x)$; in some texts, this is written as Df

opposite angles: the equal angles that are formed by two intersecting lines

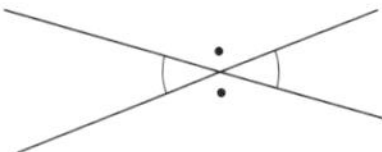

opposites: two numbers whose sum is zero; each number is the opposite of the other

optimization: determining the maximum or minimum value of a quantity

order of operations: the rules that are followed when simplifying or evaluating an expression

ordered pair: a pair of numbers, written as (x, y), that represents a point on a coordinate grid

parabola: the name given to the shape of the graph of a quadratic function

parallel lines: lines in the same plane that do not intersect

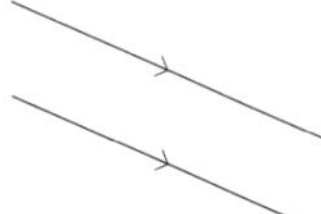

perfect square: a number that is the square of a whole number; a polynomial that is the square of another polynomial

perimeter: the distance around a closed figure

perpendicular: intersecting at right angles

pi (π): the ratio of the circumference of a circle to its diameter; $\pi \doteq 3.1416$

point discontinuity: occurs where the graph of a function has a "hole"

point of inflection: a point on a graph where the graph changes concavity; that is, changes from opens up to opens down, or vice versa

point of intersection: a point that lies on two or more figures

point slope form: the equation of a line in the form $y = m(x - p) + q$, where (p, q) are the coordinates of a point on the line and m is its slope

point symmetry: a figure that maps onto itself after a rotation of 180° about a point is said to have point symmetry

polygon: a closed figure that consists of line segments; for example, triangles and quadrilaterals

polynomial: a mathematical expression with one or more terms, in which the exponents are whole numbers and the coefficients are numbers

polynomial function: a function where $f(x)$ is a polynomial expression

positive number: a number greater than 0

power: an expression of the form a^n, where a is the base and n is the exponent; it represents a product of equal factors; for example, $4 \times 4 \times 4$ can be expressed as 4^3

power function: a function that has the form $f(x) = x^n$, where n is a constant

present value: the principal required to obtain a specific amount in the future

prime number: a natural number with exactly two factors, itself and 1; for example, 2, 5, 11

principal: money invested or loaned

product: the quantity resulting from the multiplication of two or more quantities

proportional functions: two functions of the same variable that have a constant ratio, called the constant of proportionality

Pythagorean Theorem: for any right triangle, the area of the square on the hypotenuse is equal to the sum of the areas of the squares on the other two sides

quadrant: one of the four regions into which coordinate axes divide a plane

quadratic equation: an equation of the form $ax^2 + bx + c = 0$, where $a \neq 0$; for example, $x^2 + 5x + 6 = 0$ is a quadratic equation

quadratic formula: the roots of the quadratic equation $ax^2 + bx + c = 0$ are $x = \frac{-b \pm \sqrt{b^2 - 4ac}}{2a}$, where $a \neq 0$

quadratic function: a function with defining equation $y = ax^2 + bx + c$, where a, b, and c are constants, and $a \neq 0$

quadratic inequality: an inequality of the form $ax^2 + bx + c > 0$, $ax^2 + bx + c \geq 0$; $ax^2 + bx + c < 0$, or $ax^2 + bx + c \leq 0$, where a, b, and c are constants and $a \neq 0$

quotient: the quantity resulting from the division of one quantity by another

radical: the root of a number; for example, $\sqrt{400}$, $\sqrt[3]{8}$, $\sqrt[4]{72}$, $\sqrt[5]{24}$

radical equation: an equation where the variable occurs under a radical sign

range of a function: the set of y-values (or output numbers) represented by the graph or the equation of a function

rate: a certain quantity or amount of one thing considered in relation to a unit of another thing

ratio: the quotient of two numbers or quantities

rational expression: a fraction where both numerator and denominator are polynomials

rational function: a function of the form $f(x) = \frac{p(x)}{q(x)}$, where $p(x)$ and $q(x)$ are polynomial functions, and $q(x) \neq 0$

rational number: a number that can be written in the form $\frac{m}{n}$, where m and n are integers ($n \neq 0$)

rationalize the numerator: write the numerator as a rational number, to replace the irrational number; for example, $\frac{\sqrt{2}}{6}$ is written $\frac{2}{6\sqrt{2}}$, or $\frac{1}{3\sqrt{2}}$

ray: a figure formed by a point P on a line and all the points on the line on one side of P; one part of the graph of an absolute-value function

real numbers: the set of rational numbers and the set of irrational numbers; that is, all numbers that can be expressed as decimals

reciprocal function: a rational function with numerator 1

reciprocals: two numbers whose product is 1; for example, $\frac{3}{4}$ and $\frac{4}{3}$ are reciprocals, 2 and $\frac{1}{2}$ are reciprocals

reflection of a function: a transformation that results in the graph of a function being reflected in a horizontal or vertical line

regression line: see *line of best fit*

regular polygon: a polygon that has all sides equal and all angles equal

relation: a rule that produces one or more output numbers for every valid input number

right circular cone: a cone in which a line segment from the centre of the circular base to the vertex is perpendicular to the base; see *cone*

rise: the vertical distance between two points

root of an equation: a value of the variable that satisfies the equation

rotational symmetry: a figure that maps onto itself in less than one full turn is said to have rotational symmetry; for example, a square has rotational symmetry about its centre O

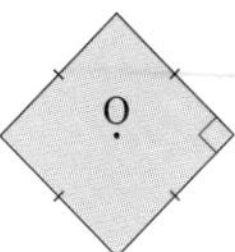

run: the horizontal distance between two points

scale: the ratio of the distance between two points on a map, model, or diagram to the distance between the actual locations; the numbers on the axes of a graph

scatter plot: a graph of data that is a set of points

scientific notation: a way to express a number as the product of a number greater than –10 and less than –1 or greater than 1 and less than 10, and a power of 10; for example, 4700 is written as 4.7×10^3

secant: a line through two points on a graph

second derivative: the instantaneous rate of change of the first derivative; for $y = f(x)$ the notation is y'', or $f''(x)$, or $\frac{d^2y}{dx^2}$

semi-annual: every six months

set notation: a description of a set of numbers that uses a typical element and lists its characteristics; often used to describe the domain or range of a function; for example, $D = \{x | x \in \Re, x \neq 1\}$

significant digits: the meaningful digits of a number representing a measurement

similar figures: figures with the same shape, but not necessarily the same size

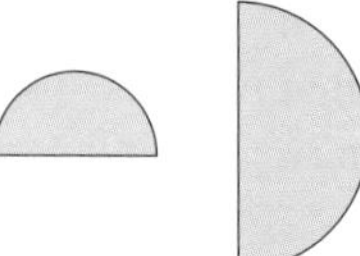

simple interest: see *interest*; interest calculated according to the formula $I = Prt$

sine of θ: the second coordinate of P, where P is on the unit circle, centre (0, 0) and θ is the measure of the angle in standard position

slope: describes the steepness of a line or line segment; the ratio of the rise of a line or line segment to its run

slope *y*-intercept form: the equation of a line in the form $y = mx + b$, where m is the slope and b is the y-intercept

square of a number: the product of a number multiplied by itself; for example, 25 is the square of 5

square root: a number which, when multiplied by itself, results in a given number; for example, 5 and –5 are the square roots of 25

standard form of the equation of a line: $Ax + By + C = 0$, where A, B, and C are integers, and $A > 0$

sum of a geometric series: the formula for the sum of the first n terms is $S_n = \frac{a(r^n - 1)}{r - 1}$, $r \neq 1$, where a is the first term, r is the common ratio, and n is the number of terms

sum of an arithmetic series: the formula for the sum of the first n terms is $S_n = \left(\frac{a + t_n}{2}\right) \times n$, where a is the first term, t_n is the nth term, and n is the number of terms; or $S_n = \frac{n}{2}[2a + (n - 1)d]$, where d is the common difference

surface area of a cone: $A = \pi r^2 + \pi rs$, where r is the base radius and s is the slant height

surface area of a cylinder: $A = 2\pi r^2 + 2\pi rh$, where r is the base radius and h is the height

surface area of a sphere: $A = 4\pi r^2$, where r is the radius

symmetrical: possessing symmetry; see *line symmetry*, *point symmetry*, and *rotational symmetry*

tangent of θ: the y-coordinate of Q, where P is on the unit circle and AQ is perpendicular to OA

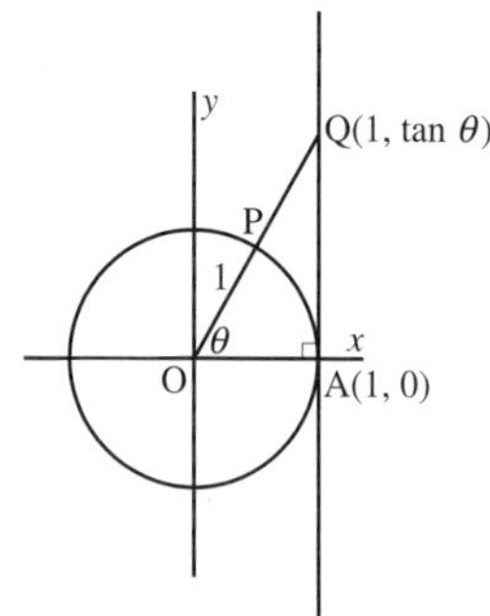

tangent to a curve: a line that touches the curve at a point and has the same slope as the curve at that point; the tangent at P is the limiting position of the secant PQ as Q approaches P

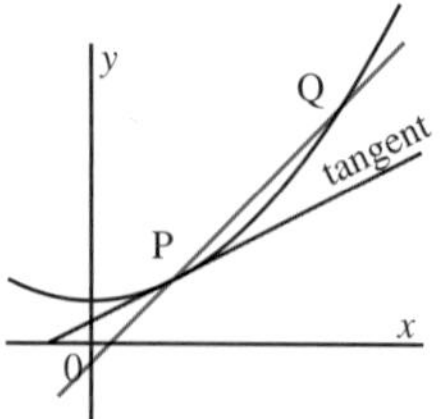

term: of a fraction is the numerator or the denominator of the fraction; when an expression is written as the sum of several quantities, each quantity is a term of the expression

three-dimensional: having length, width, and depth or height

trajectory: the curved path of an object moving through space

transformation of a function: the change in a function that results in changes to its equation and its graph; see *compression*, *expansion*, *reflection*, and *translation*

translation of a function: a transformation that results in the graph of a function being moved horizontally or vertically

trigonometric equation: an equation involving one or more trigonometric functions of a variable

trigonometric identity: an identity involving one or more trigonometric functions of a variable

trinomial: a polynomial with three terms; for example, $3x^2 + 6x + 9$

two-dimensional: having length and width, but no thickness, height, or depth

unlike radicals: radicals that have different radical parts; for example, $\sqrt{5}$ and $\sqrt{11}$

unlike terms: terms that have different variables, or the same variable but different exponents; for example, $3x$, $-4y$ and $3x^2$, $-3x$

variable: a letter or symbol representing a quantity that can vary

velocity: the rate of change of displacement with respect to time for an object moving in a straight line

vertex (plural, vertices): the corner of a figure or a solid

vertex of a parabola: the point where the axis of symmetry of a parabola intersects the parabola

vertical intercept: the vertical coordinate of the point where the graph of a function or a relation intersects the vertical axis

vertical line test: if no two points on a graph can be joined by a vertical line, then the graph represents a function

volume of a cone: $V = \frac{1}{3}\pi r^2 h$, where r is the base radius and h is the height

volume of a cylinder: $V = \pi r^2 h$, where r is the base radius and h is the height

volume of a sphere: $V = \frac{4}{3}\pi r^3$, where r is the radius

whole numbers: the set of numbers 0, 1, 2, 3, …

x-axis: the horizontal number line on a coordinate grid

x-intercept: the x-coordinate of the point where the graph of a function or a relation intersects the x-axis

y-axis: the vertical number line on a coordinate grid

y-intercept: the y-coordinate of the point where the graph of a function or a relation intersects the y-axis

zero of a function: the horizontal coordinate of the point where the graph of a function intersects the horizontal axis drawn through the origin

Index

F

G

H

PHOTO CREDITS AND ACKNOWLEDGMENTS

The publisher wishes to thank the following sources for photographs, illustrations, articles, and other materials used in this book. Care has been taken to determine and locate ownership of copyright material used in the text. We will gladly receive information enabling us to rectify any errors or omissions in credits.

PHOTOS

Cover, Gerry Johansson/Photonica and Artbase Inc.; **Inside Front Page** (centre), Gerry Johansson/Photonica and Artbase Inc.; **p. 1** (centre), Patrice Loiez/Science Photo Library/Photo Researchers, Inc.; **p. 8** (top right), Jim Cummins/Getty Images/FPG; **p. 9** (centre right), Artbase Inc.; **p. 12** (bottom right), © Ted Horowitz/The Stock Market/Firstlight.ca; **p. 70** (centre right), Artbase Inc.; **p. 71** (top), © Horrillo I. Riola/Firstlight.ca; **p. 90** (bottom right), Dave Starrett; **p. 93** (centre right), © Mark D. Maziarz/Firstlight.ca; **p. 111** (bottom right), © Tom Stewart/ The Stock Market/Firstlight.ca; **p. 113** (top right), © FoodPix/Paul Poplis; **p. 116** (bottom), COMSTOCK IMAGES/Russ Kinny; **p. 117** (top), © Lester Lefkowitz/CORBIS/Stock Market/Firstlight.ca; **p. 132** (bottom right), Dave Starrett; **p. 210** (top right), Artbase Inc.; **p. 211** (top right), Little Blue Wolf Productions/CORBIS/MAGMA; **p. 213** (bottom right), © Rommel/Masterfile; **p. 233** (top), COMSTOCK IMAGES/Russ Kinny; **p. 241** (centre right), Francesco Ruggeri/Getty Images/Image Bank; **p. 258** (centre right), Holly Harris/Getty Images/Stone; **p. 273** (top right), Dick Luria/Getty Images/FPG; **p. 283** (bottom right), Ed Gifford/Masterfile; **p. 287** (bottom right), Artbase Inc.; **p. 289** (bottom right), © Bob Anderson/ Masterfile; **p. 293** (top), © Artbase Inc; **p. 360** (centre right), © Allen Birnbach/Masterfile; **p. 363** (top), © Wei Yan/Masterfile; **p. 373** (centre right), © John Feingersh/The Stock Market/ Firstlight.ca; **p. 374** (centre right), Artbase Inc.; **p. 377** (bottom right), © Laurelidji/Firstlight.ca; **p. 388** (bottom right), © Pierre Tremblay/ Masterfile; **p. 394** (bottom right), © Reuters NewMedia Inc./CORBIS/MAGMA; **p. 406** (centre right), David Muir/Masterfile; **p. 407** (top left), AFP/CORBIS/MAGMA; **p. 416** (centre right), Artbase Inc.; **p. 431** (top), Firstlight.ca; **p. 450** (bottom right), © Jon Feingersh/The Stock Market/Firstlight.ca; **p. 453** (bottom right), Artbase Inc.; **p. 457** (bottom right), © Darwin Wiggett/Firstlight.ca; **p. 477** (top right), © Reuters NewMedia Inc./CORBIS/MAGMA; **p. 480** (top right), © David Lee/CORBIS/MAGMA; **p. 481** (top), John Gillmore/The Stock Market/Firstlight.ca; **p. 502** (top right), Artbase Inc.; **p. 517** (bottom right), Artbase Inc.; **p. 524** (bottom right), © Daryl Benson/Masterfile; **p. 525** (top right), © Guy Grenier/Masterfile; **p. 531** (bottom right), © Ron Sandford/CORBIS/Stock Market/Firstlight.ca.

ILLUSTRATIONS

Pronk&Associates Inc. **p. 64** (centre right), **p. 379** (centre right), **p. 391** (bottom left), **p. 480** (centre right), **p. 506** (bottom left), **p. 519** (bottom left), **p. 521** (top left), **p. 522** (centre left)